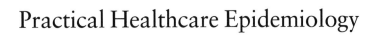

Practical Healthcare Epidemiology

Practical Healthcare Epidemiology

Third Edition

Ebbing Lautenbach, MD, MPH, MSCE
Keith F. Woeltje, MD, PhD
Preeti N. Malani, MD, MSJ

Published for the Society for Healthcare Epidemiology of America
by the University of Chicago Press

The University of Chicago Press, Chicago 60637
The University of Chicago Press, Ltd., London

14 13 12 11 10   5 4 3 2 1

ISBN: 978-0-226-47102-0
E-ISBN: 978-0-226-47104-4

The procedures and practices described in this book should be implemented in a manner consistent with the professional standards set for the circumstances that apply in each specific situation. Every effort has been made to confirm the accuracy of the information presented and to correctly relate generally accepted practices. The authors, editors, sponsoring society, and publisher cannot accept responsibility for errors or exclusions or for the outcome of the application of the material presented in this book. There is no expressed or implied warranty of this book or information imparted by it.

Care has been taken to ensure that drug selection and drug dosages are in accordance with currently accepted and/or recommended practice. Because research continues, government policy and regulations continue to change, and the effects of drug reactions and interactions vary, it is recommended that the reader review all materials and literature provided for each drug, especially those that are new or not frequently used.

Any review or mention of specific companies or products is not intended as an endorsement by the authors, editors, sponsoring society, or publisher.

Library of Congress Cataloging-in-Publication Data

The Society of Healthcare Epidemiology of America practical healthcare. — 3rd ed. / edited by Ebbing Lautenbach, Keith F. Woeltje, Preeti N. Malani.
    p. cm.
  Includes bibliographical references and index.
  ISBN 978-0-226-47102-0
  1. Epidemiology—Methodology. I. Lautenbach, Ebbing. II. Woeltje, Keith F. III. Malani, Preeti N.
  RA652.4.S63 2010
  614.4—dc22

                                        09045566

For permission to reprint material in another publication, please contact the University of Chicago Press.

Contents

Acknowledgments

We would not be able to function in healthcare epidemiology without the tireless efforts of the many infection prevention-ists with whom we work—we cannot thank them enough. We'd like to thank our bosses—Harvey Friedman and Brian Strom at the University of Pennsylvania, Victoria Fraser at Washington University, and Jeffrey Halter, Powel Kazanjian, and Carol Kauffman at the University of Michigan—for their encouragement and support. Gordon Rudy and the editorial team at the University of Chicago Press were outstanding to work with and provided inestimable support. And most especially, we'd like to offer thanks to all of our colleagues who contributed chapters to this book. We appreciate their adding even more hours to their already over-packed schedules (did someone mention H1N1?) to participate in this project.

Ebbing Lautenbach
Keith Woeltje
Preeti Malani

About the Editors

Ebbing Lautenbach, MD, MPH, MSCE, is an Associate Professor of Medicine in the Division of Infectious Diseases, Associate Professor of Epidemiology in the Department of Biostatistics and Epidemiology, and Senior Scholar in the Center for Clinical Epidemiology and Biostatistics at the University of Pennsylvania School of Medicine. He is the Associate Hospital Epidemiologist and Co-Director of Antimicrobial Management at the Hospital of the University of Pennsylvania. Dr. Lautenbach is a Fellow in the Institute on Aging, Senior Fellow at the Leonard Davis Institute for Health Economics, and a Senior Fellow at the Center for Public Health Initiatives, all at the University of Pennsylvania. Dr. Lautenbach received his MD and MPH degrees from Columbia University in New York. He completed his Internal Medicine residency and Infectious Diseases fellowship at the University of Pennsylvania, where he also received a Masters Degree in Clinical Epidemiology. He is board certified in Internal Medicine, Infectious Diseases, and Epidemiology. Dr. Lautenbach's clinical and research interests focus on the epidemiology and prevention of healthcare-acquired infections, as well as the emergence of antimicrobial resistance. He serves as an Associate Editor of the journal *Infection Control and Hospital Epidemiology*.

Keith F. Woeltje, MD, PhD, is an Associate Professor of Medicine in Infectious Diseases, as well as a member of the Institute of Clinical and Translational Sciences (ICTS) and a Scholar at the Institute for Public Health at Washington University in St. Louis. Dr. Woeltje serves as the Medical Director of Infection Prevention for BJC HealthCare. He received his MD and PhD degrees from the University of Texas Southwestern Medical Center in Dallas. He did his residency in Internal Medicine and his fellowship in Infectious Diseases at Barnes-Jewish Hospital and the Washington University School of Medicine. After a year as Medicine Chief Resident at Barnes-Jewish Hospital, Dr. Woeltje joined the faculty at the Medical College of Georgia. He served there as Hospital Epidemiologist until returning to Washington University in 2004. He is board certified in Internal Medicine and Infectious Diseases. Dr. Woeltje's research interests are in the epidemiology and prevention of healthcare-associated infections. In particular, he is interested in the application of medical informatics tools to automated surveillance.

Preeti N. Malani, MD, MSJ, is an Associate Professor of Medicine in the Divisions of Infectious Diseases and Geriatric Medicine at the University of Michigan and a Research Scientist at the Veterans Affairs Ann Arbor Healthcare System's Geriatric Research Education and Clinical Center (GRECC). Dr. Malani received her MD degree from Wayne State University School of Medicine. Prior to medical school, she completed a Masters in Journalism at Northwestern University's Medill School of Journalism. She completed her Internal Medicine residency and Infectious Diseases fellowship at the University of Michigan, where she also received a Masters Degree in Clinical Research Design and Statistical Analysis. Dr. Malani completed her fellowship training in Geriatric Medicine at the Oregon Health and Science University. She is board certified in Infectious Diseases and Geriatric Medicine. Dr. Malani's clinical and research interests include infections in older adults, surgical site infections, and infections in transplantation. She serves as an Associate Editor of the journal *Infection Control and Hospital Epidemiology*.

Contributing Authors

M. Cristina Ajenjo, MD
Department of Internal Medicine, Infectious Diseases Program
Pontificia Universidad Católica de Chile
Santiago, Chile

Anucha Apisarnthanarak, MD
Division of Infectious Diseases, Department of Internal Medicine
Thammasat University Hospital
Pratumthani, Thailand

Natasha Bagdasarian, MD, MPH
Department of Internal Medicine, Division of Infectious Diseases
University of Michigan Health System
Ann Arbor, Michigan

Judene Bartley, MS, MPH, CIC
Vice President, Epidemiology Consulting Services, Inc.
Clinical Consultant, Premier Safety Institute
Beverly Hills, Michigan

Gonzalo Bearman, MD, MPH
Associate Professor of Internal Medicine, Epidemiology, and
    Community Health
Virginia Commonwealth University School of Medicine
Associate Hospital Epidemiologist
Virginia Commonwealth University Health System
Richmond, Virginia

Werner E. Bischoff, MD, MS, PhD
Section on Infectious Diseases, Department of Internal Medicine
Department of Infection Control and Epidemiology
Wake Forest University School of Medicine
Winston-Salem, North Carolina

Henry M. Blumberg, MD
Professor of Medicine and Epidemiology, Division of Infectious
    Diseases
Emory University School of Medicine
Director, Clinical and Translational Research Training Programs
Atlanta Clinical and Translational Science Institute
Emory University
Hospital Epidemiologist, Grady Memorial Hospital
Atlanta, Georgia

John M. Boyce, MD
Clinical Professor of Medicine, Yale University School of Medicine
Chief, Infectious Diseases Section, and Hospital Epidemiologist
Hospital of Saint Raphael
New Haven, Connecticut

P. J. Brennan, MD
Professor of Medicine, Division of Infectious Diseases
Chief Medical Officer
University of Pennsylvania Health System
Philadelphia, Pennsylvania

Frederick A. Browne, MD, MBA
Chief Medical Officer
Hospital Epidemiologist
New Milford Hospital
New Milford, Connecticut

Anne M. Butler, MS
Washington University School of Medicine
St. Louis, Missouri

Romanee Chaiwarith, MD, MHS
Division of Infectious Diseases
Department of Medicine
Faculty of Medicine
Chiang Mai University
Chaing Mai, Thailand

Laura Chandler, PhD
Department of Pathology and Laboratory Medicine
Veterans Affairs Medical Center
Philadelphia, Pennsylvania

Carol E. Chenoweth, MD
Professor, Division of Infectious Diseases
Department of Internal Medicine and Department of
    Epidemiology
Hospital Epidemiologist
University of Michigan Health System
Ann Arbor, Michigan

Sandro Cinti, MD
Associate Professor, Infectious Diseases
University of Michigan Hospitals and Health Systems
Veterans Affairs Ann Arbor Health System
Ann Arbor, Michigan

Eric Cober, MD
Division of Infectious Diseases, Department of Internal Medicine
University of Michigan Health System
Ann Arbor, Michigan

Louise-Marie Dembry, MD, MS, MBA
Associate Professor of Medicine and Epidemiology
Yale University School of Medicine
Hospital Epidemiologist and Co-Director, Quality Improvement
    Support Services
Yale-New Haven Hospital
New Haven, Connecticut

Robert A. Duncan, MD, MPH
Associate Professor of Medicine
Tufts University School of Medicine
Hospital Epidemiologist, Center for Infectious Diseases
Division of Internal Medicine
Lahey Clinic
Burlington, Massachusetts

Michael Edmond, MD, MPH, MPA
Professor of Internal Medicine, Epidemiology, and
    Community Health
Virginia Commonwealth University School of Medicine
Hospital Epidemiologist
Virginia Commonwealth University Health System
Richmond, Virginia

Jeffrey Hafkin, MD
Department of Medicine, Division of Infectious Diseases
University of Pennsylvania School of Medicine
Philadelphia, Pennsylvania

Anthony D. Harris, MD, MPH
Associate Hospital Epidemiologist
University of Maryland Medical Center
Baltimore, Maryland

David K. Henderson, MD
Hospital Epidemiology Service, Clinical Center
Office of the Director, Clinical Center
National Institutes of Health
Bethesda, Maryland

Loreen A. Herwaldt, MD
Department of Internal Medicine
University of Iowa Carver College of Medicine
Professor of Epidemiology
University of Iowa College of Public Health
Hospital Epidemiologist
University of Iowa Hospitals and Clinics
Iowa City, Iowa

Stephanie Holley, RN, BSN, CIC
Program of Hospital Epidemiology
University of Iowa Hospitals and Clinics
Iowa City, Iowa

Hitoshi Honda, MD
Division of Infectious Diseases, Department of Medicine
Washington University School of Medicine
St. Louis, Missouri

William R. Jarvis, MD
Jason and Jarvis Associates
Hilton Head Island, South Carolina

Lauris C. Kaldjian, MD, PhD
Associate Professor, Department of Internal Medicine
Director, Program in Bioethics and Humanities
University of Iowa Carver College of Medicine
Iowa City, Iowa

Kenneth R. Lawrence, PharmD
Clinical Pharmacy Specialist, Department of Pharmacy
Division of Geographic Medicine and Infectious Diseases
Tufts Medical Center
Boston, Massachusetts

Darren R. Linkin, MD, MSCE
Assistant Professor, Division of Infectious Diseases, Department
    of Medicine
Assistant Professor, Department of Biostatistics and
    Epidemiology
Senior Scholar, Center for Clinical Epidemiology and Biostatistics
Center for Education and Research on Therapeutics
University of Pennsylvania School of Medicine
Philadelphia, Pennsylvania

Mark Loeb, MD, MSc
Department of Pathology and Molecular Medicine
McMaster University
Hamilton, Ontario, Canada

Tammy Lundstrom, MD, JD
Vice President, Chief Medical Officer
Providence and Providence Park Hospitals–St. John Health
    System
Associate Professor, Wayne State University School of Medicine
Adjunct Professor, Wayne State University School of Law
Southfield, Michigan

Susan MacArthur, RN, CIC, CPHQ, MPH
Director, Infection Prevention
Connecticut Children's Medical Center
Hartford, Connecticut

Lisa L. Maragakis, MD, MPH
Assistant Professor of Medicine
Division of Infectious Diseases, Department of Medicine
Johns Hopkins University School of Medicine
Department of Hospital Epidemiology and Infection Control
The Johns Hopkins Medical Institutions
Baltimore, Maryland

Joel Maslow, MD, PhD, MBA
Section of Infectious Diseases
Veterans Affairs Medical Center
Division of Infectious Diseases, Department of Medicine
University of Pennsylvania
Philadelphia, Pennsylvania

Mary Nettleman, MD, MS, MACP
Professor and Chair, Department of Medicine
Michigan State University
East Lansing, Michigan

Lindsay E. Nicolle, MD
Professor, Department of Internal Medicine, Department of
    Medical Microbiology
Health Sciences Centre
University of Manitoba
Winnipeg, Manitoba, Canada

Deborah M. Nihill, RN, MS, CIC
Interventional Epidemiology Manager
BJC Healthcare
St. Louis, Missouri

Tara N. Palmore, MD
Deputy Hospital Epidemiologist
National Institutes of Health Clinical Center
Bethesda, Maryland

Trish M. Perl, MD, MSc
Professor of Medicine, Pathology, and Epidemiology
Johns Hopkins University
Hospital Epidemiologist
The Johns Hopkins Hospital
Baltimore, Maryland

Lance R. Peterson, MD
Director of Microbiology and Infectious Diseases Research and
    Epidemiologist
NorthShore University HealthSystem
Evanston, Illinois
Clinical Professor
Pritzker School of Medicine
University of Chicago
Chicago, Illinois

Russell M. Petrak, MD
Managing Partner
Metro Infectious Disease Consultants
Chicago, Illinois

Didier Pittet, MD, MS
Professor of Medicine and Director, Infection Control Program
The University of Geneva Hospitals
Geneva, Switzerland
Honorary Professor, Division of Investigative Sciences
Hammersmith Hospital, Imperial College of Medicine, Science,
    and Technology
London, England

Jean M. Pottinger, RN, MA, CIC
Program of Hospital Epidemiology
University of Iowa Hospitals and Clinics
Iowa City, Iowa

Gina Pugliese, RN, MS
Vice President, Premier Safety Institute
Associate Faculty, University of Illinois School of Public Health
Rush University College of Nursing
Chicago, Illinois

Virginia R. Roth, MD
Director, Infection Prevention and Control Program
The Ottawa Hospital
Associate Professor of Medicine, Epidemiology, and Community
    Medicine
University of Ottawa
Ottawa, Ontario, Canada

Loie Ruhl, RN, BS, CIC
Barnes-Jewish Hospital
St. Louis, Missouri

Mark E. Rupp, MD
Professor, Infectious Diseases
Medical Director, Department of Healthcare Epidemiology
Medical Director, Clinical Trials Office
Nebraska Medical Center
Omaha, Nebraska

William A. Rutala, PhD, MPH
Director, Hospital Epidemiology, Occupational Health and Safety
    Program
University of North Carolina Health Care
Professor of Medicine, University of North Carolina School of
    Medicine
Director, Statewide Program in Infection Control and
    Epidemiology
University of North Carolina School of Medicine
Chapel Hill, North Carolina

Daniel J. Sexton, MD
Professor of Medicine, Division of Infectious Diseases
Director, Duke Infection Control Outreach Network
Director, Duke Hospital Infection Prevention Program
Duke University Medical Center
Durham, North Carolina

Emily K. Shuman, MD
Division of Infectious Diseases, Department of Internal Medicine
University of Michigan Health System
Ann Arbor, Michigan

Arjun Srinivasan, MD
Division of Healthcare Quality Promotion
Centers for Disease Control and Prevention
Atlanta, Georgia

Kerri A. Thom, MD, MS
Associate Hospital Epidemiologist
University of Maryland Medical Center
Baltimore, Maryland

William M. Valenti, MD
Clinical Associate Professor of Medicine
University of Rochester School of Medicine and Dentistry
Founding Medical Director and Staff Physician
AIDS Community Health Center
Rochester, New York

Trevor Van Schooneveld, MD
Assistant Professor, Infectious Disease
Medical Director, Antimicrobial Stewardship Program
Nebraska Medical Center
Omaha, Nebraska

Eden Wells, MD, MPH
Bureau of Epidemiology
Michigan Department of Community Health
Lansing, Michigan

David J. Weber, MD, MPH
Professor of Medicine, Pediatrics, and Epidemiology
Associate Chief of Staff
Medical Director, Hospital Epidemiology

University of North Carolina Health Care
University of North Carolina at Chapel Hill
Chapel Hill, North Carolina

Stephen G. Weber, MD, MSc
Associate Professor, Section of Infectious Diseases and Global
  Health
Chief Healthcare Epidemiologist

Medical Director of Infection Control and Clinical Quality
The University of Chicago Medical Center
Chicago, Illinois

Marc Oliver Wright, MT (ASCP), MS, CIC
Department of Infection Control
NorthShore University HealthSystem
Evanston, Illinois

Getting Started

Chapter 1 Practical Hospital Epidemiology: An Introduction

Ebbing Lautenbach, MD, MPH, MSCE,
Preeti N. Malani, MD, MSJ, and Keith F. Woeltje, MD, PhD

It is with great pleasure that we introduce the third edition of *Practical Healthcare Epidemiology*. As noted by Dr. Loreen Herwaldt in the introduction to the first edition of this textbook, in 1998, "Hospital epidemiology and infection control have become increasingly complex fields."[1] While certainly true then, it is even truer now. Today's healthcare epidemiologist faces an abundance of both challenges and opportunities. One needs to look no further than the recent emergence of influenza virus A subtype H1N1, toxigenic *Clostridium difficile*, and community-associated methicillin-resistant *Staphylococcus aureus* strains to appreciate the dynamic nature of this field. The ongoing emphasis on such issues as pandemic influenza preparedness, public reporting, and patient safety, all closely related to infection control, highlights the need for the expertise of the healthcare epidemiologist in many arenas. The need for knowledgeable, well-trained healthcare epidemiologists and infection preventionists has never been greater.

Healthcare-associated infections exact a tremendous toll in morbidity, mortality, and costs.[2] Indeed, the number of nosocomial infections per 1,000 patient-days has actually increased during the past 2 decades,[2] and there are reasons to expect this rate will continue to go up in coming years. These include the use of newer instruments and devices, an increasing number of invasive procedures, sicker patient populations, stretched resources, and new patterns of antimicrobial resistance.

The primary focus of the healthcare epidemiologist remains the prevention of healthcare-associated infections. Effective infection control efforts have been shown to dramatically reduce the incidence of nosocomial infections, with decreased morbidity, improvement in survival, and shorter duration of hospitalization.[3] Indeed, the healthcare epidemiologist must deal with all aspects of the healthcare setting to prevent patients or staff from acquiring infection. These include outbreak investigation, surveillance, development of policies, audits, teaching, advice, consultation, community links, and research. With the increasing acuity of the hospitalized patient population and the growing utilization of other healthcare facilities (eg, long-term care, outpatient care, and home care), the demands on the healthcare epidemiologist will continue to increase dramatically in the coming years.[4,5]

The knowledge and skills of the healthcare epidemiologist also lend themselves extremely well to addressing many other issues at the forefront of patient care today. Knowledge of healthcare epidemiology is useful for drug use management, quality assessment, technology assessment, product evaluation, and risk management. In particular, application of healthcare epidemiology–based practices is directly relevant to the patient safety movement. These practices include establishing clear definitions of adverse events, standardizing methods for detecting and reporting events, creating appropriate rate adjustments for case-mix differences, and instituting evidence-based intervention programs.[6,7]

We recognize that several comprehensive textbooks of healthcare epidemiology and infectious diseases exist and are excellent resources for infection control professionals. This textbook is not meant to replicate these books, but rather to complement them, as a pragmatic, easy-to-use reference emphasizing the essentials of healthcare epidemiology.

The reader may note that the title of this textbook has changed in a small but important way from the second edition.[8] We have eliminated the term "handbook" from the title. We felt that this designation suggested something that might fit into a coat pocket. Clearly this text is much more than a pocket reference. Our goal is to provide a user-friendly and straightforward introduction to the field of healthcare epidemiology, one that fosters understanding of the basics of this increasingly complex field. We anticipate that as the reader gains a greater knowledge of the essentials of healthcare epidemiology he or she will desire, and should pursue, more in-depth knowledge of specific issues. Consultation with colleagues, as well as reference to a comprehensive infection control textbook, would be essential components in expanding one's expertise in the field. But as a starting point, this overview of the fundamental aspects of healthcare epidemiology should provide a solid foundation for those entering the field of infection control. The practical nature of the textbook lends itself well to the very nature of healthcare epidemiology, a field that requires constant action (eg, surveillance and interventions). While daily decisions must be based on a thorough evaluation of the data, such decisions must also be practical in the context of the healthcare setting and surroundings of the practitioner.

This textbook is also distinguished by its focus on practical experience. While based solidly on the existing medical literature, it also offers real-world advice and suggestions from professionals who have grappled with many of the long-standing and newer issues in infection control. As with earlier editions of this textbook, we asked the authors of the chapters to write as if they were speaking to an individual who would be running an infection control program and who was just starting in this field. The authors' task was to prepare the future hospital epidemiologist for their new career by summarizing basic data from the literature and by providing essential references and resources. In addition, we asked the authors to share their own experiences of what works and what does not work in particular situations.

To help you make the most of this textbook, we will briefly summarize the overall content of the chapters and how they are organized. Even in the 6 years since publication of the second edition,[8] many changes have occurred in health care in general and in healthcare epidemiology in particular. Reflecting this, all chapters have been substantially revised. Furthermore, many new chapters have been added, to address emerging issues in this rapidly changing field.

This textbook is divided into 6 sections. The first section, Getting Started, includes several chapters that provide essential background for the beginning healthcare epidemiologist. Specific topics include ethical aspects of infection control, isolation, and disinfection and sterilization, as well as chapters on the use of the microbiology laboratory, molecular methods, and computer support for healthcare epidemiology. A new chapter on quality improvement provides a concise introduction to this important topic.

The second section, Surveillance and Prevention, includes chapters on the basics of surveillance and outbreak investigations and chapters addressing the most common and important healthcare-acquired infections, including a new chapter on urinary tract infection.

The next section, Antimicrobial-Resistant Organisms, recognizes the continually increasing threat of multidrug-resistant organisms to patients in all healthcare settings. Two chapters detail specific issues related to gram-positive and gram-negative organisms. The healthcare epidemiologist is often called on to make decisions about an institution's antimicrobial formulary; recognizing the close relationship between infection control and antimicrobial stewardship in preventing infections in hospitalized patients, an updated chapter on antimicrobial stewardship is also included in this section. A new chapter on *C. difficile* infection has been added because of the unique infection control challenges associated with this emerging infection.

The fourth section, Special Topics, includes a variety of chapters that, while exceedingly important in healthcare epidemiology, do not fit neatly into one of the other sections. Many of these chapters have been reorganized to respond to issues that have emerged only in the past few years—issues to which the healthcare epidemiologist must respond. An updated chapter on hand hygiene reviews the latest World Health Organization recommendations. In response to issues about which awareness has grown dramatically in recent years, there are also reorganized chapters on patient safety and bioterrorism preparedness. Recognizing that the scope of the "healthcare" epidemiologist is ever expanding and often includes non–acute care settings, chapters on infection control in long-term care facilities and in the outpatient setting have been updated. Because of the recognition that resources are not equal across different healthcare settings and across different geographic regions, a chapter on infection control in resource-limited settings rounds out this section.

The final section, Administrative Issues, highlights the many tasks that the healthcare epidemiologist and infection control committee must tackle on an ongoing basis to ensure that their program remains responsive to current issues, mandates, and emerging problems. There are chapters that specifically describe the development of policies and guidelines and the preparation for surveys by the Joint Commission and the Occupational Safety and Health Administration. A new chapter on government mandates offers an overview of current and future mandates and highlights the need for healthcare epidemiologists to be proactive in regard to political activism.

We believe that this textbook will provide trainees and professionals in infection control, particularly the fledgling healthcare epidemiologist, the knowledge and tools to establish and maintain a successful and effective healthcare epidemiology program. Ours is a vibrant and exciting field that presents new challenges and opportunities daily. The prospects for the healthcare epidemiologist are virtually limitless, whether they are in infection control, antimicrobial stewardship, patient safety, or beyond. We hope that this textbook provides the foundation upon which many future years of further learning, innovation, and advancement can be based.

References

1. Herwaldt LA, Decker MD. An introduction to practical hospital epidemiology. In: Herwaldt LA, Decker M, eds. *A Practical Handbook for Hospital Epidemiologists.* 1st ed. Thorofare, NJ: Slack; 1998.
2. Weinstein RA. Nosocomial infection update. *Emerg Infect Dis* 1998;4:416–420.
3. Scheckler WE, Brimhall D, Buck AS, et al. Requirements for infrastructure and essential activities of infection control and epidemiology in hospitals: a consensus panel report. Society for Healthcare Epidemiology of America. *Infect Control Hosp Epidemiol* 1998;19:114–124.
4. Jarvis WR. Infection control and changing health-care delivery systems. *Emerg Infect Dis* 2001;7:170–173.
5. Perencevich EN, Stone PW, Wright SB, et al. Raising standards while watching the bottom line: making a business case for infection control. *Infect Control Hosp Epidemiol* 2007;28:1121–1133.
6. Scheckler WE. Healthcare epidemiology is the paradigm for patient safety. *Infect Control Hosp Epidemiol* 2002; 23:47–51.
7. Gerberding JL. Hospital-onset infections: a patient safety issue. *Ann Intern Med* 2002;137:665–670.
8. Lautenbach E, Woeltje K, eds. *Practical Handbook for Healthcare Epidemiologists.* 2nd ed. Thorofare, NJ: Slack; 2004.

Chapter 2 How to Get Paid for Healthcare Epidemiology: A Practical Guide

Daniel J. Sexton, MD, and Russell M. Petrak, MD

The professional and economic opportunities in infection control and healthcare epidemiology have never been greater. Mandates from the Occupational Safety and Health Administration (OSHA) and The Joint Commission (formerly the Joint Commission for Hospital Accreditation) require that all hospitals develop and maintain programs to reduce the risk of transmission of infectious diseases to both patients and hospital employees. Public reporting of data on selected healthcare-associated infections is now the law more than 20 states. Furthermore, the risk of litigation alleging failure to provide for the safety of hospitalized patients has grown exponentially following the release of 2 landmark studies by the Institute of Medicine that each emphasize the lack of patient safety programs in the US healthcare system. We believe these legal obligations will continue to increase as the public, the press, and the plaintiff bar incrementally increase their scrutiny of the causes and prevention of adverse patient outcomes. These developments, the growing concerns of the public about their risk of getting a healthcare-associated infection, and the increasing risk of infection secondary to complex medical care have made infection control and prevention a critical component of health care. Real and theoretical threats of bioterrorism, pervasive problems with antimicrobial resistance, and a host of new and emerging infections add to the challenges facing US hospitals. Indeed, each of the challenges and problems just listed represents an opportunity for infectious diseases physicians with an interest and training in hospital epidemiology.

Yet, despite these trends, many US hospitals currently do not have a paid hospital epidemiologist. Only 23% of members of the Infectious Diseases Society of America who responded to a survey in 1998 reported that they received payment for their role in their hospital's infection control program.[1] Many American hospitals currently ask specialists in infectious diseases to donate their time and expertise related to infection prevention; some rely on other specialists, such as pathologists, to chair their infection control committee. It is ironic that most infectious diseases fellows, who become well trained in the science of infectious diseases, receive minimal, if any, training or guidance in how to get paid for practicing healthcare epidemiology. In essence, learning how to get paid is left out of most infectious diseases training program curriculums.

Our goal in this chapter is to inspire you to think of yourself as valuable and useful to the mission of the infection control and patient safety program at your local institution. We firmly believe that your expertise in meeting the challenges mentioned above deserves appropriate compensation. Although in the past many infectious diseases specialists provided hundreds or even thousands of uncompensated hours of work and shouldered significant responsibility without monetary compensation, this model is as outdated as the technique of transtracheal aspiration. In the words of an experienced hospital epidemiologist: "If you still work for free [after reading this chapter]—shame on thee!"

Healthcare Epidemiology Services as a Product: Create a Vision of Value

Healthcare-associated infections are serious, common, and important patient safety issues in health care today.[2] Healthcare epidemiologists are in the primary business of infection prevention. Infection prevention is a cornerstone of continuous quality improvement. It is critical that the concept of healthcare epidemiology services as a "product" be carefully defined and explained to hospital administrators. However, the concept of healthcare epidemiology services as a product may not be intuitive or immediately obvious to all infectious diseases specialists and many hospital administrators. Thus, the services of a healthcare epidemiologist must be explicitly defined; in addition, the processes of care, the measurement of infection rates, and the necessity to explain, monitor, and design systems to promote compliance with infection prevention measures need to be explained and emphasized. Such explanations can help hospital administrators understand the value of and rationale for providing compensation for healthcare epidemiology. Value is often not apparent unless there is a clear understanding and explanation of healthcare epidemiology services.

Because individual expertise in, interest in, and time commitments to infection control vary widely among infectious diseases specialists, and because the need for infection control also varies between individual institutions, your income for infection control services is highly dependent on how *you* define your job in infection control. This obvious but important matter should be examined carefully before you attempt to negotiate a contract for providing healthcare epidemiology services. In addition, as described below in the section The Process of Negotiating, it is sometimes necessary to modify your offer for services after you have discovered the interests, needs, and concerns of the other side involved in the negotiation process.

One of the best methods to define your product is to write a business plan. A typical business plan includes an introduction; a detailed description of the program (or product); a time line containing short-term and long-term goals; costs and required resources, such as equipment, space, and personnel; a timetable; criteria for measuring success; and a summary statement. Writing a business plan requires that you set realistic goals, assess resources (including your compensation), outline strategies to achieve your goals, and most importantly, establish criteria to measure performance and success.

Business plans are valuable to hospital administrators in several ways. First, they explain your product in terms they understand. Second, they provide a way to rationalize the cost of a healthcare epidemiology contract. Finally, business plans should outline and provide an objective way to measure performance. Most business plans need to be updated and improved over time. For example, with time your product may change and grow to include a role in antimicrobial stewardship, staff education, or patient safety issues not directly related to infectious diseases (Table 2-1).

Healthcare Epidemiology Customer: Who Is the Purchaser of Your Product?

In most instances, the purchaser of your product is a hospital administrator. Thus it is axiomatic that the crucial first step is to accurately identify the key administrator or administrators who will "purchase" your product. It is equally essential that this individual understand the scope of healthcare epidemiology, including the benefits and barriers to achieving specific goals. Until this individual understands your plan, your responsibilities, and the criteria used to assess your performance, it is unlikely that he or she will be able to accurately value your product and justify the cost of your services to his or her colleagues, to senior leadership, or to the hospital governing board.

Before undertaking the educational and promotional process just described, it is a good idea to find convincing objective answers to several obvious questions. Why should the administration buy this product? What is the benefit to the institution? How do you explain the relationship between the costs of healthcare-associated infections and the costs of an infection control program? Hospital administrators can usually calculate the direct costs of an infection control program, but it is exceedingly difficult for most hospital administrators to quantify the potential benefits. Before negotiating the costs of an infection control program, it is wise to have extensive discussions and develop logical and simple explanations about the benefits of such a program.

The task of explaining the costs of healthcare-associated infections is difficult because the current total attributable cost of such infections to society is poorly understood and difficult to measure. However, in 1985 Haley[3] estimated that the hospital-related financial burden of nosocomial infections in the United States was approximately $3.9 billion per year ($9.0 billion per year, in 2009 dollars).[4] Although there are few published studies that have rigorously assessed the costs and benefits of infection control programs, the existing literature clearly demonstrates that infection control interventions can produce enormous direct and indirect cost savings.[2,5] For example, Fraser[6] estimated that an infection control program similar to that advocated by current standards from The Joint Commission was cost-effective, compared with standard health interventions such as cancer screening. Fraser[6] estimated that the cost of infection control was $2,000–$8,000 per year of life saved, which is considerably less than the estimated cost of performing Papanicolaou smears every 2 years ($650,000 per year of

Table 2-1. A hospital epidemiologist's marketable skills

Supervises and collaborates with the infection preventionists

Chairs infection control committee

Supervises collection of surveillance data

Analyzes surveillance data and implements control strategies

Validates the surveillance system

Supervises projects to decrease endemic "common-cause" nosocomial infections

Investigates outbreaks

Analyzes antimicrobial resistance patterns and interprets their significance relative to antimicrobial utilization

Evaluates new products (eg, intravenous catheters)

Evaluates the scientific validity of policies and procedures

Interprets and implements, in a cost-effective manner, standards from regulatory agencies that are applicable to infection control, employee health, etc

Consults with medical staff, hospital staff, and administration on epidemiological matters

Analyzes trends in employee illness

Supervises vaccination programs for medical staff and hospital employees

Supervises preemployment screening of new hospital employees

Supervises the follow-up process for hospital employees who are exposed to infectious diseases or pathogens (eg, bloodborne pathogens, tuberculosis, varicella, herpes zoster, or meningococcus)

Directs and supervises AIDS-related issues (eg, hospital employee education and case reporting)

Consults with staff in the microbiology laboratory regarding the appropriate use of laboratory tests, antibiotic susceptibility reports, analysis of new rapid laboratory tests, etc

Consults with architects, engineers, and contractors regarding infection risks associated with construction projects

Consults with staff in risk management regarding actual or potential malpractice claims related to nosocomial infections

Consults with the hospital's public relations personnel regarding release of information about outbreaks, endemic nosocomial infections, quality of care, epidemiology-based report cards, etc

Serves as a liaison between hospital and medical staff in quality improvement initiatives

Consults with other staff regarding epidemiological evaluations of noninfectious disease problems (eg, falls and pressure ulcers)

Supervises or consults with the hospital's pharmacokinetic dosing service

Supervises the epidemiological initiatives in the hospital's long-term care facility and transitional care and rehabilitation facility

Directs the infection control and employee health initiatives in an expanded healthcare setting that includes the hospital or healthcare system, a home nursing program, and a home infusion-therapy program

Helps to develop an epidemiology-based quality management approach to noninfectious adverse outcomes of care

Assists in outcome management initiatives

Assists other staff as they develop clinical practice guidelines

Serves as a liaison with staff from the local public health department

Assists with epidemiology of medical error prevention and safety initiatives

Assists with policy and program evaluation of reprocessing disposable or single-use items

Assists with managing mask fit testing program

Manages smallpox or other bioterrorism vaccination programs

Assists with the development of bioterrroism and emerging infection policies and procedures

Assists with the development of alcohol-based hand hygiene program

Collects data on and assess adequacy of processes related to infection control

Designs, supervises, or oversees the scope, curriculum, and implementation of staff education related to infection control life saved), or even the cost of cholesterol reduction in high-risk persons 40 years of age or older ($32,500 per year of life saved).[6]

A single deep sternal wound surgical site infection may increase the direct cost of care by $20,000–$30,000.[7] Other studies have estimated that the economic impact of deep sternal wound surgical site infection is $14,211, after controlling for selection bias.[8] Even if the lower estimate of $14,211 per infection is used, the prevention of 4 deep surgical site infections each year at a single institution could save more than $50,000 per year.

Stone et al.[2] reviewed 55 studies published from 1990 through 2000 that contained original cost estimates for healthcare-associated infections. The mean attributable cost of all types of healthcare-associated infection was $13,973 per infection. The mean costs of specific types of infection were as follows: surgical site infection, $15,640; bloodstream infection, $38,703; pneumonia, $17,677; and methicillin-resistant *Staphylococcus aureus* (MRSA) infection, $35,367. A prospective controlled study of ventilator-associated pneumonia found the attributable mean cost to be $11,897 per infection.[9] These data further affirm the point made above: prevention of only a small fraction of serious nosocomial infections each year can justify the costs of the entire infection control program in most community and tertiary care hospitals. Until recently, Medicare and many third-party payers reimbursed hospitals for the cost of complications of care. Those days are gone. New Medicare reimbursement rules now restrict payment to costs directly related to selected healthcare-associated infections, such as catheter-associated bloodstream infection, urinary tract infection, and mediastinitis following open heart surgery.

In 1999, Kirkland et al.[10] reported that the total excess direct cost attributable to a case of surgical site infection was $5,038. In another study from the same institution, Whitehouse et al.[11] reported that the median direct excess cost of orthopedic surgical site infection was $17,708 per patient, and the total (direct and indirect) median cost per patient was even higher, $27,969 per patient. The 1-year study period reported by Whitehouse et al.[11] involved a cohort of fewer than 100 orthopedic patients with surgical site infection, yet these patients' infections cumulatively resulted in total extra costs of $867,039. Undoubtedly, these estimates underestimated the total cost of nosocomial infection, as these investigators were unable to measure the cost of outpatient services, such as outpatient intravenous antibiotic therapy, skilled nursing care, and physical therapy.

Kaye et al.[12] reported that a structured and focused infection control program for small and medium-sized community hospitals in North Carolina and South Caro-lina, which was administrated by a network employing trained physician epidemiologists and infection control nurse practitioners, resulted in dramatic reductions in the incidence of nosocomial infection over a 3-year span. During the first 3 years of affiliation in the Duke Infection Control Outreach Network, study hospitals reported that the annual rates of nosocomial bloodstream infection decreased by 23% ($P = .009$) from year 1 to year 3, annual rates of nosocomial infection or colonization with MRSA decreased by 22% ($P = .002$), and rates of ventilator-associated pneumonia decreased by 40% ($P = .001$). The rates of employee exposure to bloodborne pathogens decreased by 18% ($P = .003$).[12] These successes were attributed to the feedback on nosocomial infection rates provided to the healthcare providers, which was then linked to specific educational initiatives. This entire process resulted in education of healthcare providers about the nature of the existing problems. As a consequence, providers were secondarily motivated and stimulated to embrace effective strategies for preventing future infections.

Other safety initiatives can also produce remarkable cost savings. The Centers for Disease Control and Prevention estimates that 385,000 needle sticks and other injuries related to sharp devices and objects are sustained by hospital-based healthcare personnel every year.[13] In 1999, treating a healthcare worker to prevent disease from a needlestick injury was estimated to cost $500–$3,000 per event ($792–$4,863 per event, in 2009 dollars).[4,14]

These data can be used to estimate the current cost of nosocomial infections to an individual hospital. Using a simple spreadsheet, it is possible to estimate the hypothetical cost of various infection control strategies. Various degrees of sophistication can be applied to such cost estimates, depending upon the availability and reliability of existing data on the incidence of healthcare-associated infections, the payer mix, the rehospitalization rates, and so on. Even if such data are not available, it is usually possible to develop a basic spreadsheet that provides meaningful data that can be used to define and explain your "product" to hospital administrators. Such spreadsheets are also useful to monitor subsequent cost savings (or increases) and to develop prevention strategies and set priorities.

It is important to realize that the process of assessing the costs of healthcare-associated infections is substantially more complicated than simply assessing the aggregate costs of individual healthcare-associated infections. Numerous other cost considerations are important when assessing the total costs associated with healthcare-associated infection. The institutional costs of a poor-quality infection control program can be considerable: for example, the costs of poor or ineffective prevention measures (eg, ineffective antibiotic prophylaxis); the existing appraisal costs of ineffective, meaningless, or improperly circulated surveillance data; internal failure costs (eg, repeat sterilization of loads of instruments that fail sterilization checks) and external failure costs (eg, repeat operation because of a wound infection); all these are cumulatively substantial.[15]

One often-neglected aspect of a well-designed infection control program is the economic benefit of a reduction in the costs of poor-quality care. Although patient isolation is both necessary and effective to control potential intrafacility spread of multidrug-resistant organisms, inappropriate use of isolation may lead to decreased interaction between patients and healthcare workers, increased use of unnecessary or ineffective isolation equipment, unnecessary performance of cultures (of specimens from staff or patients previously colonized with organisms such as vancomycin-resistant enterococci), and inappropriate and expensive use of antibiotics, and the costs can be considerable in many poorly managed infection control programs.

The elements comprising the intrinsic monetary value of an effective infection control program have been reviewed by several authors.[15-17] Monetary estimates that focus exclusively on measures to reduce direct operating costs caused by infections overlook savings to be gained by reducing the number of malpractice claims, as well as lost opportunity costs resulting from failure of patients to return for further care. In addition, it is important to focus on programs, policies, and practices that eliminate waste through wise product selection, that avoid inappropriate use of expensive technology, and that address employee safety. There is intrinsic monetary value in maintaining regulatory compliance, in patient safety programs, and in programs that decrease the risk of infection with drug-resistant pathogens. Finally, it is important to link monetary arguments for effective infection control programs with an ethical rationale about the need to prevent the morbidity and mortality associated with medical errors. In other words, an effective infection control program not only makes economic sense, it is also the right ethical decision for administrators.

Perencevich et al.[17] have provided a step-by-step approach to making a business case for infection control, which can be individualized for local hospitals. Further, they critically review basic economic concepts used in healthcare decision making, such as cost-effectiveness analyses, cost-utility analyses, and cost-benefit analyses. Such analyses are beyond the scope of this chapter, but a basic familiarity with this information can be useful when discussing and explaining your business plan with knowledgeable hospital administrators.

Finally, a monograph on the direct medical costs of healthcare-associated infections is available free of charge from the Centers for Disease Control and Prevention.[18] This monograph provides reliable and recent information on the range of direct costs attributable to common healthcare-associated infections and adjusts these data to 2007 dollars using data from the Consumer Price Index for Urban Consumers.

Fundamental Skills

Knowledge about the principles and practice of infectious diseases medicine is the essential requirement for practicing hospital epidemiology, but in itself, this is not sufficient to achieve success. Specific training in infection control and epidemiology during your fellowship greatly enhances your ability to take advantage of the opportunities in healthcare epidemiology in your subsequent career. We highly recommend taking the Society for Healthcare Epidemiology (SHEA) and Centers for Disease Control and Prevention training course in healthcare epidemiology during your fellowship or prior to attempting to become a hospital epidemiologist. Such training is helpful, but the most critical requirement is an interest in infection control and healthcare epidemiology. That interest must be accompanied by a willingness to learn the fundamental principles of hospital epidemiology. Additional reading, attendance at national meetings, and regular interaction with infection control nurse practitioners and other hospital epidemiologists are necessary and vital for the professional growth and development of all hospital epidemiologists—even those with advanced specific training and years of experience in epidemiology. As in all aspects of infectious diseases medicine, life-long learning is fundamental to success, professional satisfaction, and excellence.

Although knowledge about the diagnosis, epidemiology, and prevention of nosocomial infection is a cornerstone of success in hospital epidemiology, other skills are of equal importance. The most important of these skills is the ability to communicate. Arrogance, egotism, or "holier-than-thou" attitudes may maim or even kill otherwise sincere attempts to communicate. Learn to be humble, thoughtful, and appreciative in your interactions.

Orderly rational problem-solving skills are also invaluable. Understanding the processes used for root-cause analysis and routine quality improvement activities are important and useful skills. These skills are widely used in business and medicine, and it is important that hospital epidemiologists be comfortable with the vocabulary used in the planning and implementation of these concepts.

All successful hospital epidemiologists learn to delegate responsibility and understand that the morale, professional education, growth, respect, and nurturing of their infection control team are priorities in virtually every management decision.

The skills required for success in healthcare epidemiology are similar in many ways to the skills required for success in infectious diseases practice. Skill and respect earned as a specialist in clinical infectious diseases have direct benefits and bearing on your effectiveness and reputation as a hospital epidemiologist, and vice versa. In most cases, the roles of infectious diseases specialist and hospital epidemiologist are complementary and mutually beneficial. However, it is important to clearly remember that the 2 roles are different. These different roles must sometimes be explicitly emphasized. The good of the entire institution and all patients is the paramount concern of a hospital epidemiologist, whereas the clinician is primarily concerned about the welfare of his or her individual patients.

It is vital that hospital epidemiologists understand how hospitals or healthcare systems function. Knowing lines of authority is fundamental to being effective as a hospital epidemiologist. Taking a request to the wrong person in a chain of command can doom a good idea or substantially slow the process of approval and acceptance. It is equally important to understand how decision making occurs in most hospitals. Sometimes knowing something as simple as the local hospital's budget process can be a key factor in achieving approval of an important project or endeavor.

Having an active imagination and ability to "think out of the box" may provide new opportunities or overcome long-standing opposition to good ideas. Medical training programs seldom teach residents and fellows how to understand and solve the political and personal problems that beset all professionals in virtually any professional endeavor. Effective skills in these areas are vital in hospital epidemiology. It is beyond the scope of this chapter and beyond our abilities to effectively teach you these skills in a didactic manner. However, such skills are learnable (usually from one or more wise mentors). Also self-learning is possible if there is true motivation, self-awareness, and openness to feedback.

The Process of Negotiating

The old axiom that states "In business, you do not get what you deserve, you get what you negotiate" is widely quoted because it contains more than a modicum of truth. Actually, negotiation is a process of joint problem solving. When undertaken with careful forethought and a true understanding of basic negotiation skills, the process can be rewarding and useful to both parties in the negotiation. A successful negotiation educates others about one's individual concerns, interests, and needs. Just as important, the process of negotiation allows you to understand the concerns, interests, and needs of the other side. Thereafter, these considerations logically become the basis for a final agreement.

In practice, 2 basic models of negotiation exist: a confrontational (win-lose) approach and a collaborative (win-win) approach. Although, in practice, outcomes are often neither "wins" nor "losses" but something in between, that terminology has utility in explaining 2 basic approaches to dispute resolution. The win-lose model is primarily based on taking an initial well-defined positional approach. Beginning a negotiation with a clear-cut position or a demand often seems the most efficient and desirable approach; thus, it is the usual approach taken by many people in traditional disputes concerning money and other contentious issues. In fact, positional bargaining is the only form of negotiation many people know. However, such positional bargaining is often unsuccessful, highly contentious, and stressful. Paradoxically, a win-loss (positional) approach produces angry losers who will resist vigorously, resent the final agreement, and look for ways to undermine its implementation or success.

A win-win negotiation approach is slower and more difficult; however, it is more likely to help you build long-term relationships, trust, and favorable outcomes. Such an approach is preferred in negotiating a contract for healthcare epidemiology services, even though this approach, like any negotiation strategy, sometimes fails.

Negotiation techniques, pitfalls, and styles are extensively described in a practical manner in the excellent book *Getting to Yes* by Fisher et al.[19] This book is a de facto practical primer on the fundamentals of negotiation. The authors discuss a number of key principles, of which the most important may be how to avoid positional bargaining; in other words, how to avoid talking only about what you are willing and unwilling to accept at the beginning of a negotiation. This approach is frequently a recipe for an unsuccessful negotiation (even if you happen to get what you request). Rather than bargaining over positions, Fisher et al.[19] advocate a time-tested and widely used alternative approach called "principled negotiation" or "negotiation on the merits." This method has 4 components: (1) separate the people from the problem ("go to the balcony"); (2) focus on interests, not positions; (3) generate a variety of possibilities before deciding what to do; and (4) insist that the results be based on some objective standard (to be the basis for the final agreement).

The first and most critical organizational function prior to negotiation is preparation. Many authors differ as to how to organize the multiple and critical aspects of this discussion, but most can be compartmentalized into

4 steps. The first step is to *define your needs*. A need is defined as something you must possess to be successful. This list should be extensive and objectively defined so that all issues are considered. Examples could be a need to control the intrahospital spread of MRSA, a need to minimize use of a particular antimicrobial agent, or a need to increase revenue to support a new employee or project.

The second step in preparation for negotiation is to *define your wants*. A want is something you would like to possess but don't absolutely need to be successful. An example could be your desire to hire a clinical pharmacist to act as facilitator to decrease the use of unnecessary or unwise preoperative antibiotic prophylactic therapy.

The third step in preparation for negotiation is to *delineate the costs* inherent in the contract or process you are negotiating. Costs can be broken down into the present costs, the costs of inaction, and the costs of definitive action. Present costs, defined as those that are actually being incurred, are a benchmark by which future or proposed financial performance may be measured. The costs of inaction are a projection of future costs should no action be taken. Finally, the costs of definitive action are the incremental costs that would be realized should the negotiation goal be reached. An individual's ability to delineate and articulate the financial scenario described above, markedly increases the likelihood of success. This is specifically true when your proposal is anticipated to lead to a significant costs savings.

The fourth and perhaps the most important step, as outlined by the authors of *Getting to Yes*[19] is to understand the concept of the BATNA ("Best Alternative To a Negotiated Agreement"). Whether you should or should not agree on something in a negotiation depends entirely upon the attractiveness of your BATNA. It is also helpful to understand and know the other side's BATNA. Sometimes it is helpful for the other side to understand your BATNA; yet, other times, this information is best withheld during the early phases of negotiation. A clear understanding of one's BATNA greatly helps the members of each side of a negotiation feel satisfied when their negotiation has been completed.

A final result better than one's BATNA is often a success, even if the final agreement was less than the original bargaining goal. For example, a hospital epidemiology contract for $20,000 per year may be better than one's BATNA (no income), even if the goal of the original contract negotiation was twice that amount. Similarly, the hospital administrator may leave the same negotiation realizing that his BATNA (no qualified hospital epidemiologist) was worse than his original position (no payment for hospital epidemiology oversight and management). If objective criteria were used as a basis for the final agreement, the hospital epidemiologist and the administrator have objective criteria to revisit the agreement and modify it at a later date.

Getting past "no," when a negotiation is going badly, requires moving from "positions" to "merits"; that is, emphasizing and explaining interests, options, and standards. When negotiations are stalled or otherwise unsuccessful, it is generally best to concentrate on the merits of your proposal and their proposal, rather than the positions you or the other side have expressed. In essence, it is often necessary to change the game by starting to play a new one, when the original negotiation has stalled or failed.[19]

Successful negotiators must know their goals and link them to their interests as well as understand their strengths and those of the other side. It is fundamental that a negotiator understands the needs, attitudes, interests, and positions of the person with whom they are negotiating, and whom each negotiator represents. Generally it is unwise to argue about "feelings." On the other hand, it is wise and good to choose your words carefully and to listen carefully when the other side talks and then let them know you heard them. Successful negotiators always try to align interests and negotiate for objective criteria to decide contentious issues.

When initial negotiations fail, it is often a good idea to undertake brainstorming sessions to develop proposals attractive to the other side. These proposals should address common interests and they should be explicitly presented as proposals rather than final positions. In general, it is a bad idea to make threats or to close the door to future negotiations. It is always wise to be aware of and emphasize the importance of the relationship that precedes and will follow the negotiation.

Negotiating well requires preparation and experience. However, it is not necessary to take advanced training to gain such experience. In fact, everyone uses negotiating skills in innumerable interactions with family, neighbors, employees, and an array of other individuals, such as sales personnel and even bureaucrats. When entering a negotiation for hospital epidemiology service or any professional contract, it is useful to remember these 3 maxims: "You can't control the other side, you can't control the marketplace, but you can control your preparation"[19]; "Much is lost for want of asking"[20(para.112)]; and "Let us never negotiate out of fear, but let us never fear to negotiate" (John F. Kennedy).[21(p.799)]

Functional Models of Healthcare Epidemiology Infection Control Management

There are at least 4 ways that infectious diseases specialists have modeled their functional roles as healthcare epidemiologists. The basic model is a limited role that includes functioning as chair of the hospital infection control committee, being available and willing to answer periodic questions of the infection control practitioner, and troubleshooting ad hoc infection control–related problems while providing general support for existing infection control policies. Most infectious diseases specialists who follow this basic model are local opinion leaders who do not have a formal role in the institution's hierarchy or management.

Other infectious diseases specialists may chose an expanded role in hospital epidemiology. This expanded model includes all the "basic model" functions plus a commitment to meet with the infection control practitioner(s) on a regular basis, to participate directly in the education of staff related to infection control issues, and to provide general oversight of the local infection control program. The precise goals and expectations of the infectious diseases specialist who functions in this expanded model may or may not be clearly defined. Although in some cases a time commitment is specified in the written or oral agreement for services, in other cases payment is provided on an as-needed basis. Regardless, reimbursement is usually directly linked to time commitment or time spent functioning as a hospital epidemiologist. If the goals and objectives of oversight and management of the infection control program can be clearly defined and objectively assessed, this model can have a definable value and can be considered a "product" according to the criteria discussed previously. However, common problems with this model (such as hourly compensation) are discussed below, in the section Common Dilemmas.

The advanced model includes all the functions described in the expanded model, as well as administrative and managerial oversight of the infection control activities of the local institution's healthcare system or of an expanded healthcare system of multiple hospitals. Such systems may include one or more long-term care facilities or rehabilitation units, as well as various outpatient facilities and/or affiliated hospitals. This model has significant potential to provide a valuable "product" to a healthcare system, with commensurate compensation for the healthcare epidemiologist.

A fourth model is a network or consortium model, which is an advanced model that is duplicated at multiple unaffiliated hospitals by offering a menu of services to a potentially unlimited number of institutions. A prototype of this model is the Duke Infection Control Outreach Network.[12] This model provides a wide variety of resources to affiliated community hospitals, including training and support of local infection control practitioners; assistance with outbreak investigations; education of staff and physicians; statistical expertise and analysis of local surveillance data; and detailed programs to provide feedback to local practitioners about surveillance data

on nosocomial infections, regional and national trends in nosocomial infections, and antimicrobial resistance. In addition, the Duke Infection Control Outreach Network utilizes all surveillance data collected in community hospitals, with the explicit goal of providing comparative data showing trends over time for individual hospitals. These data are also used for "benchmarking" of surveillance data on surgical site infections, bloodstream infections, exposures of hospital employees to blood-borne pathogens, and rates of device-related infections among patients hospitalized in intensive care units, so that individual hospitals affiliated with the Duke Infection Control Outreach Network can compare and contrast their data with data from hospitals of similar size and type in the region.

How to Get Paid: What Is the "Market Value" of a Hospital Epidemiologist?

Your potential income from hospital epidemiology is directly related to your imagination, interests, expertise, and negotiating skills. Your income also depends on the degree of your involvement and the amount of responsibility and accountability for the hospital epidemiology program you agree to accept. It has been our experience and impression that too many infectious diseases physicians literally give away their time and expertise in infection control and hospital epidemiology activities.

We believe that working as a healthcare epidemiologist and/or a chair of a hospital infection control committee without compensation is unwise. If you do not receive compensation for these responsibilities, it may be reasonable to resign your role as chair of the committee. Such a resignation could eliminate several meetings each year from a schedule that is already overloaded. If you feel a "citizenship" obligation to your hospital, offer to be on the education committee or another committee that does not represent the bread and butter of your specialty. If your colleagues or hospital administrators challenge the wisdom of your decision not to work for free you might ask them the following questions: "Do surgical colleagues operate gratis for the hospital?" or "Do cardiologists donate a catheterization per day to the hospital?" Finally, it may be useful to point out that hospital and healthcare systems that rely on volunteerism for their infection control programs will likely ultimately fail at or miss important opportunities to improve care.

Determining appropriate compensation for supervision and management of a hospital infection control program may seem difficult at first glance, but it is not impossible. For obvious reasons, there are no published data on the market-rate salary range for a hospital epidemiologist. As mentioned previously, the value of your work as hospital epidemiologist has to be defined in terms of your specific and unique job description, the goals of the program, your responsibility, the size of the local hospital, the type of infection control and/or patient safety problems at the specific institution, and objective criteria for performance. Networking with other epidemiologists or your SHEA liaison can provide a general idea of the range of compensation in your region or state.

To our knowledge, no one has attempted to measure the economic value of infectious diseases specialists who function as healthcare epidemiologists or hospital epidemiologists. A SHEA membership survey conducted in 1995 that had 545 respondents reported that 44% of respondents received compensation for epidemiology services, with an annual mean compensation of $52,000 and median of $40,000.[22] A second membership survey in 1998 that had 494 respondents reported that 60% of respondents received compensation for epidemiological services, with an annual mean compensation of $45,700 and median of $40,000 (SHEA staff, personal communication). Our experience with colleagues involved in hospital epidemiology suggests that annual physician income for hospital epidemiology ranges from $2,000 to more than $250,000. Many healthcare epidemiologists work as employees in academic institutions. Their salaries often include income for multiple activities, such as research and teaching. Thus, it is frequently difficult or impossible to ascertain the percentage of annual income directly attributable to epidemiology services.

Many hospitals prefer to pay healthcare epidemiologists for each hour of service. Also, many hospital administrators contend that it is easier to defend compensation for hospital epidemiology service when logs of hours spent are provided to the administration, who in turn can provide this data to the Center for Medicaid and Medicare Service auditors. We disagree with this approach. One does not "punch in and punch out" of hospital epidemiology activities like a shift or contract worker. Hospital epidemiology activities are integrated into the daily work of most infectious diseases specialists. Conversations related to infection control issues, controversies, and policies often occur in the context of patient care or in the daily routine of community hospitals. Billing by the hour, as is done by attorneys, is not feasible or practical. Also, the responsibility for infection control issues is continuous and unpredictable. Decisions and guidance about patient isolation decisions are often made at night, on weekends, or on holidays. Others in positions of leadership in the hospital, such as hospital administrators, are not paid by the hour and neither should a hospital epidemiologist. Indeed, the value of a hospital epidemiologist relates to his or her leadership, responsibility, and obligation to solve problems after hours or during emergencies.

We believe the issue of compensation must be directly related to the overall job description and the value of the total "product." We also believe establishing compensation that is linked to a bounty payment, such as the percentage of money saved by appropriate use of antibiotics or by a specific reduction in the rate of any one infection or type of infection, has the potential to undermine credibility and distort motives. However, this concept is not the same as linking compensation to the overall goals and objectives of a sound business plan. Achieving the goals for such a program is a rational basis for compensation and subsequent increases in compensation. The process of negotiation described above (in the section The Process of Negotiating) is the best way to resolve the fundamental question of the market value of a hospital epidemiologist.

Common Dilemmas

If your hospital and/or healthcare system says they can't (or won't) pay for hospital epidemiology and/or infection control services—what do you do? Obviously, the answer to this question will vary depending upon the personalities and perceptions of the local administrators and infectious diseases specialists. We can provide the following general advice.

First, it is often wise to solicit support for your proposals from key hospital staff (surgeons, pathologists, and nursing personnel), to ask for their help and support in reopening negotiations and/or reviewing the respective positions of the involved parties. Second, it is also important to understand and then remember your BATNA (Best Alternative To a Negotiated Agreement). Finally, if an agreement that provides adequate compensation for time, effort, and responsibility cannot be successfully negotiated, it may sometimes be necessary either simply to refuse all further interaction with the infection control program or to resign from the hospital staff and work at a neighboring hospital or change your status from active staff to consulting staff. It is important to remember that the preceding is only a partial list of your options, should the hospital refuse to negotiate a contract for infection control services. Indeed, you could consider some or all of the following options:

1. Maintain a relationship with the hospital's infection control practitioner and look for opportunities to revisit the original negotiation.
2. Offer to provide further assistance to the hospital if the need arises … for a price.
3. Negotiate an agreement to revisit the pay issue in a specified number of months, in exchange for continued participation in infection control and/or epidemiology activities.

4. Barter your support for existing programs (such as formulary restrictions, a prior approval program for prescription of antibiotics, or a review program for use of high-cost drugs) in exchange for the acceptance of an infection control contract.

5. Negotiate an agreement to examine cost estimates for healthcare-associated infections at your institution (with or without compensation) in return for your continued participation.

6. Negotiate an agreement to sponsor a position paper on the deficiencies and strengths of the current infection control program in return for your continued participation.

7. Negotiate an agreement for the review of the current infection control program from local, regional, or outside experts in exchange for your continued participation in the current infection control program.

8. Undertake a survey of infection control oversight processes and practices in local and regional peer institutions and provide these data to administration and to key members of the medical staff.

9. Alternately, it may be best to devise new and innovative proposals, such as the following:

    a. Offer to provide a specified number of presentations to medical and other professional staff about specific topics related to rational antimicrobial use, drug resistance, and nosocomial infections.

    b. Offer to review antimicrobial therapy for each intensive care unit patient on Monday, Wednesday, and Friday mornings and make nonbinding recommendations in the patient's chart.

    c. Offer to supervise the establishment of a database on all patients with MRSA infection or colonization to develop a "scorecard" on nosocomial transmission of MRSA at your hospital.

Numerous other options and ideas undoubtedly exist, and some of these alternatives may be uniquely applicable to an individual hospital or healthcare system. If faced with a refusal to provide reasonable compensation for healthcare epidemiology services, use your imagination and energy to devise alternatives to the rejected plan.

Hospital Epidemiology Contracts

A contract is simply a written set of promises: "I promise to do something for you and, in return, you promise to do something for me." The core components of any contract for healthcare epidemiology services should define the healthcare epidemiologist's roles and responsibilities, the lines of authority, and the compensation package. A typical contract for hospital epidemiology services often specifies access to data, computers, computer software, space, laboratory support, and secretarial support. It is also reasonable to contract for additional money for books, journals, and continuing medical education.

In general, contracts for hospital epidemiology services should delineate roles and responsibilities either generically or specifically. Some of the many specific activities suitable for inclusion in a generic contract are outlined in Table 2-1. Although the language and style of most contracts is largely controlled by the attorneys who draft and review such agreements, we have a few specific caveats. (1) You should never agree to do things that cannot be accomplished or for which you have responsibility but no authority. (2) You should not spread yourself so thin that you cannot do anything well. It is better to commit to doing a few specific things well than an array of things that are not achievable. (3) In general, it is best to have authority to supervise the people actually involved in infection control activities rather than to rely on the good will of other administrators.

Tips and Tricks of the Trade: How to Keep Your Job

We believe the following practical tips and tricks are useful:

- Pick 1 or 2 major surveillance or intervention projects each year.
- Share the knowledge of and credit for your victories with administration, medical, and hospital staff.
- Communicate, communicate, communicate (especially with administration—they pay the bill).
- Provide feedback data to doctors, nurses, and unit and floor administrators at every opportunity. Repetition in this endeavor is rarely a tactical mistake.
- Only collect data that you intend to use.
- Use your data as often as possible.
- Pick your friends and adversaries carefully (rule of thumb—keep the ratio greater than 10 : 1).
- Listen carefully to your adversaries and, when feasible and practical, collect data to refute ideas you know or suspect are wrong.
- Bargain with the other side.
- Little successes are important.
- Success begets success.
- Acknowledge and praise good behavior and good outcomes and use them as building blocks for future endeavors and relationships.
- Tell the truth politely and tactfully.
- Make all decisions based on the "60 Minutes rule" (i.e., whatever you say or do should play well with Mike Wallace interviewing you on 60 Minutes).

- Build slowly: for example, address bloodstream infection and surgical site infection surveillance first.
- Set goals and develop an annual plan.
- Document everything that is important or potentially important, including responses to memos, meetings, and your successes.
- Keep good records about what you did, when you did it, and why you did it.
- Create a monthly newsletter.
- Create an annual report for the hospital administration and other key people.
- Sponsor an annual symposium on infection prevention.
- Learn to write well and with brevity. Short memos are more effective than long memos.
- How you say something is as important, if not more important, than what you say.
- Avoid holding meetings when possible; when not possible, keep them short and focused on specific agenda items.
- Always set the meeting agenda and define the purpose at the beginning of each meeting.
- Whenever possible, make your decisions and get your votes *before* a meeting is held.
- Control the meeting agenda and watch the clock.
- Start meetings on time and finish early.
- Summarize at the end of the meeting.
- If there are tasks to be done after the meeting, do your task and follow up on the tasks assigned to others.
- Do not play the "They do not appreciate me" game.
- Do not fight with people when the outcome is not worth the fight (choose your enemies carefully).
- Learn to simplify your message (but have a message).
- Document your positions and big decisions with follow-up memos and letters.
- Do not ask for big things at big meetings unless you have notified your allies and solicited their support.
- Never say never and never say always.
- Note all cost savings.
- Establish local cost estimates using local infection control data and track the trends in these data over time.
- Assume that most people want to do the right thing. If they are not doing the right thing, assume the reason is lack of understanding rather than malice.
- Don't assume that most or all misunderstandings are someone else's fault. Examine your communication skills if a misunderstanding occurs.
- Don't assume that a refusal of a request is always a permanent decision.
- Learn the fundamentals of basic statistics or draw on the skills of a trained statistician when the need arises.

Conclusion

The opportunities available to you as a healthcare epidemiologist are often related to the limits of your imagination and energy. Your skills as an infectious diseases specialist are valuable and often directly linked with your value as a hospital epidemiologist. If you believe in your value as a healthcare epidemiologist and if you can explain this value to others and negotiate effectively, you can obtain appropriate monetary compensation and successfully fill an important and satisfying role. Finally, we advise you to remember these words: "A fair day's wages for a fair day's work, it is as just a demand as governed men ever made of government. It is the everlasting right of man" (Thomas Carlyle, 1843).[21(p.434)]

References

1. Slama TG, Sexton DJ, Ingram CW, Petrak RM, Joseph P. Findings of the 1998 Infectious Diseases Society of America membership survey. *Clin Infect Dis* 2000;31:1396–1402.
2. Stone WP, Larson E, Kawar LN. A systematic audit of economic evidence linking nosocomial infections and infectious control interventions: 1990–2000. *Am J Infect Control* 2002;30(3):145–152.
3. Haley RW. Incidence and nature of endemic and epidemic nosocomial infections. In: Bennett JV, Brachman P, eds. *Hospital Infections*. Boston: Little, Brown; 1985:359–374.
4. Historical CPI [consumer price index]. http://www.Inflation Data.comhttp://inflationdata.com/Inflation/Consumer_Price_Index/HistoricalCPI.aspx. Accessed May 9, 2009.
5. Graves N. Economics and preventing hospital-acquired infection. *Emerg Infect Dis* 2004;10(4):561–566.
6. Fraser V. The business of healthcare epidemiology: creating a vision for service excellence. *Am J Infect Control* 2002;30(2):77–85.
7. Hollenbeak CS, Murphy DM, Koenig S, Woodward RS, Dunagan WC, Fraser VS. The clinical and economic impact of deep chest surgical-site infections following coronary artery bypass graft surgery. *Chest* 2000;118:397–402.
8. Hollenbeak CS, Murphy D, Dunagan WC, Fraser VS. Nonrandom selection and the attributable cost of surgical-site infections. *Infect Control Hosp Epidemiol* 2002;23(4):177–182.
9. Warren DK, Shukla SJ, Olsen MA, et al. Outcome and attributable cost of ventilator-associated pneumonia among intensive care unit patients in a suburban medical center. *Crit Care Med* 2003;31(5):1312–1317.
10. Kirkland KB, Briggs JR, Trivette SL, Wilkinson WP, Sexton DJ. The impact of surgical-site infections in the 1990s: attributable mortality, excess length of hospitalizations, and extra costs. *Infect Control Hosp Epidemiol* 1999;20(11):725–730.
11. Whitehouse JD, Friedman ND, Kirkland KB, Richardson WI, Sexton DJ. The impact of surgical-site infections following orthopedic surgery at a community hospital and a university hospital: adverse quality of life, excess length of stay, and extra cost. *Infect Control Hosp Epidemiol* 2002;23(4):183–189.

12. Kaye KS, Engemann JE, Fulmer EM, Clark CC, Noga EM, Sexton DJ. Favorable impact of an infection control network on nosocomial infection rates in community hospital. *Infect Control Hosp Epidemiol* 2006;27:228–232.

13. Panlilo AL, Cardo DM, Campbell S, Srivastava PU, Jagger H, Orelien JG. Estimate of the annual number of percutaneous injuries in U.S. healthcare workers. In: Program and abstracts of the 4th International Conference on Nosocomial and Healthcare-Associated Infections; March 5–9, 2000; Atlanta, GA. Abstract S-T2-01.

14. United States General Accounting Office. Occupational safety: selected cost and benefit implications of needle stick prevention devices for hospitals. GAO-01–6OR (November 17, 2000).

15. Dunagan WC, Murphy DM, Hollenbeak CS, Miller SB. Making the business case for infection control: pitfalls and opportunities. *Am J Infect Control* 2002;30(2):86–92.

16. McQuillen DP, Petrak RM, Wasserman RB, Nahass RG, Scull JA, Martinelli LP. The value of infectious diseases specialists: non–patient care activities. *Clin Infect Dis* 2008; 47:1051–1063.

17. Perencevich EN, Stone PW, Wright SB, et al. Raising standards while watching the bottom line: making a business case for infection control. *Infect Control Hosp Epidemiol* 2007; 28:1121–1133.

18. Scott RD. The direct medical costs of healthcare-associated infections in U.S. hospitals and the benefits of prevention. http://www.cdc.gov/ncidod/dhqp/pdf/Scott_CostPaper.pdf. Accessed May 12, 2009.

19. Fisher R, Ury W, Patton B. *Getting to Yes*. New York: Penguin Books; 1991.

20. Bunyan J. *The Pilgrim's Progress, in the Similitude of a Dream: The Second Part*. In: Eliot CW ed. The Harvard Classics. Vol. 15, part 1. New York: PF Collier and Son; 1909–1914.

21. Kaplan J, ed. *Bartlett's Familiar Quotations*. 17th ed. Boston, New York, and London: Little, Brown and Company; 2002.

22. Society for Healthcare Epidemiology of America (SHEA). Membership survey. *SHEA Newsletter* 1996;6:5.

Chapter 3 Ethical Aspects of Infection Prevention

Loreen A. Herwaldt, MD, and Lauris C. Kaldjian, MD, PhD

Hospital epidemiologists and infection preventionists make countless decisions every day. In general, we do not make life-or-death decisions, such as whether to withdraw life support or whether to withhold possibly life-sustaining therapies. Few of our decisions require court injunctions or provide the fodder for eager journalists. We simply decide whether to isolate patients, whether to let healthcare workers continue to work, or whether to investigate clusters of infections—all very routine decisions in the life of anyone who practices infection control. These decisions are so ordinary that they could not possibly have any ethical implications. Or could they?

In fact, many of the decisions we make every day, even those we consider quite straightforward, are also ethical decisions—which is to say, they compel us to choose between competing moral values. Such choices are rarely easy, and their intrinsic difficulty is not eased by the fact that few of us have received more than cursory training in ethics. Moreover, if we attempt to train ourselves, we find that very little has been written about the ethics of our specialty, infection control.

Common Infection Prevention Decisions That Have Ethical Aspects

We may easily overlook the ethical component of our everyday decisions; thus, we may misconstrue the decision confronting us, thinking that it is without ethical consequences when, in fact, ethical principles are at stake. Take, for example, the practice of isolating a patient colonized with a drug-resistant organism. Isolating a patient constrains the patient's freedom of movement but protects the rights of other patients to be treated in an environment without unnecessary risk. Similarly, restricting healthcare workers with contagious diseases from patient care follows from epidemiologic data but also from the ethical concept of utility—which means that one should strive to maximize good outcomes and minimize harm. In such cases, we restrict the freedom of healthcare workers in order to obtain the greater benefit of protecting patients and fellow workers. Or, when stocking the hospital formulary, we consider the efficacy and cost of drugs, but we also balance the benefit of lower cost to the patient and the hospital against the risk of selecting resistant microorganisms and against physicians' freedom to prescribe any available drug.

Infection prevention personnel confront additional ethical dilemmas in many of their daily activities. For example, when managing an outbreak, infection prevention personnel must identify the source and mode of transmission of the offending pathogen and then intervene appropriately. This is simple enough, if the reservoir is a contaminated drain that is easy to replace or a nursing assistant with no political clout in the hospital. But what if the reservoir is a powerful physician with a large practice and tremendous influence with the administration? Or what if the administration thinks your recommendations are too expensive and excessive? Would you bow to the pressures and recommend interventions that you think are less than optimal, or would you risk the wrath of the physician or the administration and state your best advice regardless of the consequences?

Infection control personnel frequently must inform patients or healthcare workers that they have been exposed to an infectious disease. When the pathogen is varicella zoster virus, the problem is relatively simple. Yet infection prevention personnel must still consider ethical issues. Do you permit some susceptible employees to continue working, if they wear masks, but restrict others? Or do you restrict all susceptible healthcare workers regardless of their position or their economic status? If you are very busy at work or have plans for the evening, do you delay your response or ignore the exposure altogether? Other exposures, such as those to the hepatitis B virus, the human immunodeficiency virus (HIV), or to the prion agent that causes Creutzfeldt-Jakob disease, provoke volatile emotions and raise challenging ethical questions. For example, what do you tell employees in the pathology laboratory who were not informed that the patient might have Creutzfeldt-Jakob disease and, therefore, did not use special precautions when they processed the brain tissue? Do you recall and resterilize instruments used for the implicated brain biopsy? Do you notify patients who subsequently had surgical procedures and might have been exposed to instruments that were not sterilized in the manner recommended to kill the infectious agent?

We hope these examples enable you to see that ethical considerations abound within the practice of infection prevention. Clearly, ethics is not the esoteric discipline some misunderstand it to be. Ethics is part of our daily practice. We should not delegate ethical deliberations to others, though we will need to include professional ethicists, hospital managers, accountants, and lawyers in our discussions. We all must make maintaining our ethical integrity a routine part of our professional responsibility. This chapter is a brief introduction to the intricate intersection of ethics and infection prevention.

Taxonomy

In the introductory paragraphs, we described some routine infection prevention activities that have ethical implications, which is, in essence, a "narrative taxonomy" of ethical problems in infection control and hospital epidemiology. A taxonomy is an orderly listing or categorization of things. Most infection prevention personnel are probably familiar with taxonomy as it refers to microorganisms, but not with respect to our profession. The reason for that sense of unfamiliarity is that, to date, no one has developed a taxonomy of ethical problems in infection prevention. On the basis of our experience in infection control (L.A.H.) and ethics (L.C.K.), we have developed a taxonomy that we think will be helpful to

infection prevention personnel as they think about their own work (Table 3-1).

This taxonomy not only describes the most important ethical problems in infection control but also helps us define the individuals, groups, and organizations to which infection control personnel have specific obligations. In particular, infection control personnel have obligations to inpatients and outpatients as groups, to individual patients, to visitors as a group, to individual visitors, to healthcare workers as a group, to individual healthcare workers, to the healthcare facility for which they work, to public health entities both local and federal, to facilities to which their facility refers or transfers patients, to referring or transferring facilities, and to the public in general. The interests of these different groups are often in competition. We can use the taxonomy to help us identify the type of ethical problem that we are facing and the competing obligations that may surround that problem.

An Approach to Ethical Problems in Infection Prevention

Most discussions of medical ethics ignore the epidemiologist-population relationship and concentrate instead on the clinician-patient relationship.[1,2] Infection prevention personnel are frequently clinicians; however, we must differentiate their clinical and epidemiologic roles because their fiduciary duties do not always coincide. Medical ethics are "person-oriented," while epidemiologic ethics are "population-oriented" (Table 3-2).[3,4] Even so, the standard principles of medical ethics also apply to hospital epidemiology. These principles are as follows[5,6]:

- Autonomy (respecting the decisions of a competent patient)
- Beneficence (doing good)
- Nonmaleficence (doing no harm)
- Justice (fairness and the equitable allocation of resources)

However, the principles are applied according to the public health model,[6] which requires commitment to improving the health of populations, not only individual patients.[7] Although both medical ethics and epidemiologic ethics stress nonmaleficence and confidentiality, medical ethics emphasizes privacy, and epidemiologic ethics emphasizes investigation and reporting to protect the population. Furthermore, medical ethics stresses patient autonomy, whereas epidemiologic ethics places special priority on justice. Put more practically, medical ethics demands that the clinician treat an infected patient while maintaining the patient's confidentiality, privacy, dignity, freedom, and contact with other human beings

Table 3-1. A taxonomy of ethical problems in infection prevention

Control of the patient to limit the spread of pathogenic organisms

Isolate of patients who are colonized or infected with drug-resistant organisms
Isolate of patients who are infected with highly infectious and/or dangerous organisms

Control of healthcare workers to limit the spread of pathogenic organisms

Restrict the activities of healthcare workers who have been exposed to infectious diseases
Restrict the activities of healthcare workers who have infectious diseases

Control of medications to limit the selection and spread of antimicrobial resistance

Limit the antimicrobial agents included on the hospital formulary
Develop guidelines regarding the use of antimicrobial agents
Provide computer decision support to guide the selection of antimicrobial agents

Mandating or recommending best practice and interventions to reduce the risk of infection

Mandate or recommend treatment to eradicate carriage of resistant pathogens
Mandate implementation of isolation precautions
Mandate preemployment vaccination and/or immunity to certain pathogens
Organize and promote yearly influenza vaccination campaigns
Develop policies and procedures
Mandate postexposure testing of patients and healthcare workers
Mandate postexposure prophylactic treatment of patients and healthcare workers

Resource allocation

Establish a threshold for investigating clusters of infections
Evaluate products to assess their cost relative to their safety and efficacy
Determine whether single-use items may be reused
Guide choices regarding materials, design, number of sinks, etc., for construction projects (cost vs safety)
Limit hospital formularies to reduce costs and control antimicrobial resistance

Information disclosure

Report exposure risks to staff and patients
Report outbreaks and cases of reportable diseases to the public health department
Provide access to data on nosocomial infection rates
Identify patients colonized with resistant organisms before intra- or inter-institutional transfers
Protect the confidentiality of patients' medical records and laboratory specimens
Protect the identity of index patients in outbreaks
Protect confidentiality of patients who test positive for human immunodeficiency virus

Conflicting and competing interests

Manage outbreaks
    Staff, especially institutional leaders, may refuse to comply
    The administration may balk at the cost of investigating outbreaks
    Hospital epidemiologists who chose unpopular interventions may lose referrals and revenue
Manage exposures
    Staff, especially institutional leaders, may refuse to comply
Select the hospital formulary
    Relationships between the staff on the formulary committee and the pharmaceutical industry may compromise decisions
    Staff physicians may prefer specific antimicrobial agents not on the formulary

Individual professionalism

Act altruistically (personal convenience vs prompt intervention)
Mediate in-house disputes between administrators, clinicians, unions, and the hospital
Act when necessary, despite inadequate or conflicting data
Keep up with new developments in the field

Personal

Protect yourself from acquiring infectious diseases
Protect your family from acquiring secondary infections

(Table 3-3). In contrast, epidemiologic ethics might stress treating both infected and colonized patients to protect patients and healthcare workers. In particular cases, epidemiologic ethics might require healthcare workers to post labels on medical records and on the doors to the patients' rooms; might insist that patients stay in their rooms except when going to essential tests, in which case they must wear gowns, gloves, and masks; or might

Table 3-2. Differences in emphasis between epidemiologic ethics and medical ethics

Variable	Epidemiologic ethics	Medical ethics
Scope of concern	Populations	Individuals
Goal	Prevent infection	Treat infection
Typical principles	Nonmaleficence Justice (fairness)	Beneficence Respect for patient autonomy
Purpose of disclosure	Investigation	Diagnosis
Information handling	Confidential reporting	Confidential documentation

require healthcare workers to wear gowns, gloves and masks to avoid direct contact with patients.

By now it should be clear that ethically challenging situations are common in the practice of infection prevention and hospital epidemiology. To respond effectively to these challenges, infection prevention staff must address each problem systematically. Kaldjian et al.[51] developed an approach to ethics that is clinically oriented and helps the user state the problem clearly, collect data comprehensively, formulate an impression, and, finally, articulate a justified plan. In outline form, we present a modified version of this approach tailored to the particular demands of infection prevention (Table 3-4), and we employ this approach (in abbreviated form) as we discuss 4 core topics.

Core Ethical Topics in Infection Prevention

Staff Vaccination Programs

Vaccines were one of the public health movement's major triumphs during the 20th century, and in that very triumph are the seeds of a substantial controversy and an ethical problem. Because use of vaccines effectively decreased the incidence of many infectious diseases, the public no longer remembers how dreadful these diseases were and how many complications and deaths these infections caused. The public is now more aware of the complications of vaccines than they are of the diseases the vaccines were developed to prevent. In addition, parents of "vaccine damaged children," the natural health movement, television, radio talk shows, and the Internet have all become important "players" or "instruments" in this debate.[8,9]

The controversy about the pertussis vaccine is illustrative. In the 1940s, pertussis was the leading cause of death among children under 14 years of age. Pertussis, in fact, killed more children than measles, scarlet fever, diphtheria, polio, and meningitis combined.[10] The incidence of pertussis was already decreasing before the killed whole-cell vaccine was introduced, which was probably related to changes in social conditions, hygiene, and nutrition. However, the incidence declined significantly after the vaccine was introduced.[11]

The whole cell pertussis vaccine includes many toxic components, because it is composed of dead gram-negative bacteria and is, thus, quite reactogenic. Recipients often have significant pain, swelling, and erythema at the vaccination site, and they may develop fever, anorexia, irritability, and vomiting.[12] In addition, some children may develop inconsolable crying, excessive somnolence, short seizures, or hypotonic-hyporesponsive episodes.[12] The most severe complication of pertussis vaccination is encephalopathy, which is very rare.[12] Opponents of the vaccine also allege that the vaccine not infrequently causes serious permanent neurological damage. In some countries, such as Sweden, Japan, and the United Kingdom, the antivaccine movements gained such prominence that the countries have either stopped vaccinating children or the rate of vaccination has dropped significantly. All of these countries had outbreaks of pertussis that affected thousands of children and caused numerous deaths.[12]

The controversy over the pertussis vaccine suggests that the ethical debate over vaccines in both the public health arena and in the hospital revolves around providing the greatest good for the greatest number of people (ie, protecting them against harmful infections) and protecting the individual from harm that could be caused by a vaccination. The ethical dilemma occurs because, in general, the group gets the benefit (ie, an immunized population that is less susceptible to infection), but the individual person bears all the risk of vaccine complications.[13-17] In populations with a high rate of vaccination, a single person can elect to refuse a vaccine and may be able to avoid both the potential complications of the vaccination and the infection itself, because he or she is protected by the

Table 3-3. Differences in approach between infection prevention and medical care in the care of a patient with a transmissible microorganism

Variable	Epidemiologic approach	Medical approach
Microbial colonization	Possible treatment	Observation
Confidentiality	Qualified (eg, use of label[s] on door and/or in medical chart)	Maintained
Freedom of movement	May limit with isolation precautions	Maintained
Freedom of dress	May limit with isolation precautions	Maintained
Freedom of contact	May limit with isolation precautions	Maintained

Table 3-4. An approach to ethical problems in infection prevention

1. State the problem plainly
2. Gather and organize data
    a. Medical facts
    b. Goals and procedures of infection control
    c. Interests of patients, healthcare workers, hospital, and community
    d. Context
3. Ask: Is the problem ethical?
4. Ask: Is more information or discussion needed?
5. Determine the best course of action and support it with reference to one or more sources of ethical value
    a. Ethical principles: beneficence, nonmaleficence, respect for autonomy, justice
    b. Rights: protections that are independent of professional obligations
    c. Consequences: estimating the goodness or desirability of likely outcomes
    d. Comparable cases: reasoning by analogy from prior "clear" cases
    e. Professional guidelines: for example, APIC/CHICA-Canada professional practice standards[48]
    f. Conscientious practice: preserving epidemiologists' personal and professional integrity
6. Confirm the adequacy and coherence of the conclusion

NOTE: APIC, Association for Professionals in Infection Control and Epidemiology; CHICA-Canada, Community and Hospital Infection Control Association–Canada.

vaccinated group. However, the question arises as to whether this is fair to the persons who took the risk and were vaccinated.[13] Furthermore, if this scenario is repeated many times, the vaccination rate in the population will drop, and the nonimmune people will be at risk.

The ethical dilemma just described also occurs in healthcare facilities that mandate that healthcare workers must be immune to certain infections. For example, most healthcare facilities require that healthcare workers be immune to rubella, which means that employees must present proof that they either had the infection or that they had at least 2 rubella vaccinations. The reasons healthcare facilities have this requirement are that rubella is easily transmitted within healthcare facilities and that this virus can cause severe congenital defects if a pregnant woman becomes infected.[18,19] Thus, healthcare facilities that care for pregnant women seek to protect them by having a staff that is immune to this infection. Pregnant employees also benefit from this requirement. However, the individual healthcare provider may not benefit from receiving this vaccine, because rubella causes very mild disease in adults, and an adult vaccine recipient might develop complications. Thus, the hospital puts limits on the autonomy of its staff members and limits their freedom in order to be beneficent to pregnant patients and employees and to honor their preference to

be treated (or work) in a facility that limits the risk that their unborn child will acquire rubella.

The approach many facilities take to influenza vaccine illustrates another extreme. The influenza virus is quite contagious and can cause serious complications, hospitalization, and death among elderly people and people with significant underlying diseases. Healthcare facilities, particularly hospitals, care for many people who are at risk for complications of influenza. Thus, many hospitals offer the vaccine free of charge to employees each fall. But employees usually are not required to take the vaccine in order to work with high-risk patients.[20] Consequently, outbreaks of influenza have occurred in healthcare facilities. These outbreaks are difficult to recognize and, therefore, are underreported.[21] In this case, hospitals have elected not to mandate vaccination with a safe and effective vaccine that could prevent at least as many severe complications as does the rubella vaccine. Instead, they have elected to preserve their healthcare workers' autonomy and freedom rather than insist that vulnerable patients must be treated in an environment with the least risk of acquiring influenza.[20]

Why do hospitals chose to manage rubella one way and influenza another? To our knowledge, no one has studied this issue. However, we speculate about why this might be so. A single child who is born with congenital rubella is very dramatic, is noticed, and is considered a tragedy. The deaths of 100 elderly people who get influenza and then die of secondary bacterial pneumonia or congestive heart failure are far less dramatic because we expect "old people" to get sick and die. Similarly, a damaged child represents many impaired life-years, whereas a frail elderly person who dies represents very few life-years lost. In addition, because influenza outbreaks in healthcare facilities are rarely recognized, most hospitals probably feel that the risk to the patients is very low and, thus, do not make a big effort to encourage staff to take the vaccine. On the other hand, the hospital would face a huge lawsuit if a woman could document that she acquired rubella while receiving prenatal care in that facility. The different approaches to rubella vaccine and influenza vaccine present major ethical issues. But these issues are rarely recognized and discussed, even though employees' autonomy and freedom of choice do become prominent issues in healthcare facilities.

We believe that healthcare workers have a moral obligation to restrict their own freedom when it comes to complying with interventions such as influenza vaccine if in so doing they might help preserve their patients' health. Rea and Upshur[20] take this position in their commentary on the issue:

As Harris and Holm wrote of society in general: "There seems to be a strong prima facie obligation not to harm others by making them ill where this is avoidable." But there is a special duty of care for us as physicians not simply to avoid transmission once infected, but to avoid infection in the first place whenever reasonable. Our patients come to us specifically for help in staying or getting well. We have not just the general obligation of any member of our community, but a particular trust: *first* do no harm.[20]

The hepatitis B vaccine illustrates another approach to vaccines within the healthcare setting. The US Occupational Safety and Health Administration requires healthcare facilities to offer hepatitis B vaccine to all employees who will have contact with blood and body fluids, to protect them from acquiring this virus through an occupational exposure.[22] In this case, the individual vaccinated gets the benefit and bears the risk associated with the vaccine. In addition, employees are not required to take the vaccine. If they do not want it, they simply sign a waiver stating that they decline the vaccine, in which case they bear the risk if they are exposed to hepatitis B. The institution, thereby, fulfills its ethical and legal obligation to the employee, and the employee maintains his or her freedom to chose whether to be vaccinated.

Another question remains regarding hepatitis B vaccine, and that is whether all healthcare workers should be required to be immune to this virus to protect patients from becoming infected. Given that the risk of transmitting hepatitis B virus is very low with most healthcare-associated activities, there does not seem to be a strong ethical argument for requiring vaccination. However, more than 400 patients have acquired hepatitis B from infected healthcare workers who performed invasive procedures.[23] It is, therefore, appropriate to ask whether all healthcare workers who perform invasive procedures that could expose the patient to the healthcare workers' blood should be vaccinated against hepatitis B. Though some healthcare workers might argue that mandatory hepatitis B vaccination infringes on their right to choose, we think that mandatory vaccination for this group of healthcare workers is ethically justifiable, given the known benefits of vaccinating healthcare workers, the minimal risks associated with the vaccine, and the possible benefits to patients. Because many medical schools now require medical students to be vaccinated and the Centers for Disease Control and Prevention recommends vaccinating all infants, in the near future this question may become moot.

Isolating Patients Who Carry or Are Infected with Resistant Organisms

The incidence of colonization or infection with drug-resistant microorganisms, particularly methicillin-resistant *Staphylococcus aureus* (MRSA) and vancomycin-resistant enterococci (VRE), has increased substantially over time. One of the primary goals of infection prevention personnel is to protect patients from acquiring pathogenic organisms, including these resistant organisms, from other patients, the environment, and healthcare workers. Infection prevention personnel have several means to accomplish this goal: educating staff; implementing isolation precautions, with or without active screening programs to identify carriers (see Chapter 6, on isolation precautions); implementing hand hygiene programs; controlling the use of antimicrobial agents (see Chapter 20, on antimicrobial stewardship); and developing cleaning protocols for patients' rooms and equipment. Of these methods for controlling the spread of resistant organisms, implementing isolation precautions, with or without active screening, and controlling use of antimicrobial agents have been quite controversial and are associated with significant ethical issues. We discuss the ethical implications of isolation precautions in this section and the ethical implications of formularies in the next. We first address the arguments for and against controlling spread of MRSA and VRE and then the arguments for and against using isolation precautions to limit their spread. Subsequently we discuss the ethical issues associated with isolating patients who are colonized or infected with MRSA or VRE.

There are numerous reasons to control the spread of MRSA and VRE. Both organisms can cause serious infections.[24-27] Because MRSA and VRE are resistant to the first-line antimicrobial agents used to treat serious infections caused by *S. aureus* and enterococci, these infections may be difficult and expensive to treat. Moreover, if MRSA becomes resistant to vancomycin (ie, if the resistance gene is transferred from VRE to MRSA), infection with such strains might be virtually untreatable with currently available antimicrobial agents. Furthermore, in hospitals where the incidence of MRSA colonization and/or infection increases, there is often an associated increase in the overall incidence of nosocomial *S. aureus* infection. This occurs because MRSA infections do not replace infections caused by methicillin-susceptible *S. aureus,* rather they are added to them.[24] If MRSA and VRE are transmitted in the hospital, other organisms, such as *Clostridium difficile* and gram-negative organisms that are resistant to extended-spectrum β-lactam agents, will also be transmitted, indicating that the overall infection prevention practice in the hospital is lax and that it is not a safe environment for patients. Data from numerous institutions document the effectiveness of aggressive control measures.[25] Infection prevention personnel who take this position would also argue that, as healthcare professionals, we should first do no harm. MRSA and VRE harm many patients.

Therefore, infection prevention programs are obliged to use reasonable means to prevent selection and spread of these organisms.[25]

Other infection prevention personnel argue, to the contrary, that there are numerous reasons not to invest substantial resources and time into efforts to control MRSA and VRE.[27,28] They insist that the incidence of colonization or infection with these organisms is already so high that control measures are ineffective and waste precious resources. They would agree that aggressive measures have worked in some instances, primarily in outbreaks, but that the data on the overall incidence of MRSA and VRE colonization or infection indicate that infection control efforts have failed to stop transmission. They also argue that many colonized patients never become infected, colonization per se does not hurt these patients, and MRSA and VRE are not more virulent nor cause greater morbidity and mortality than methicillin-susceptible *S. aureus* and vancomycin-susceptible enterococci. Thus, these patients should not be subjected to treatment or to isolation. These infection prevention personnel also state that efforts to control MRSA and VRE impair patient care and, therefore, may actually cause worse patient outcomes than would have occurred if the patients were not isolated.[29-31] Finally, they would argue that eradication programs with antimicrobial agents such as mupirocin may actually increase antimicrobial resistance.[32]

There are various arguments for using isolation precautions to control the spread of MRSA and VRE.[33] (1) Contact precautions have been shown by numerous investigators to stop transmission of these organisms during outbreaks. (2) Contact precautions have also reduced transmission of MRSA and VRE in situations where they are endemic. (3) Data from several studies suggest that proximity to a patient who carries MRSA or VRE is a risk factor for acquiring these organisms.[25] (4) Common sense suggests that housing infected or colonized patients in separate rooms from patients who do not carry these organisms should reduce spread of the resistant organisms.

There are also arguments against using isolation precautions to control the spread of MRSA and VRE.[27-31] (1) MRSA and VRE are spreading despite these precautions. (2) Patients in contact isolation do not receive the same level of care as do patients with similar problems who are not in contact isolation. (3) Contact isolation may actually prevent patients from getting appropriate treatments (eg, aggressive physical rehabilitation) or from being transferred out of an acute care facility to a facility better suited to the patients' needs. (4) Contact isolation creates social isolation that impairs patients' psychological well-being.

Other infection prevention experts would argue that the real question is not *whether* to invest resources in attempts to control MRSA and VRE, but *which means* should be used to control spread. The major issue in this discussion is whether to use intensive active surveillance coupled with contact isolation precautions to control the spread of these organisms[25,34] or to enhance compliance with standard precautions and hand hygiene.[28] The crux of this debate revolves around differing interpretations of the extant data. Those who support active surveillance and use of contact isolation believe that the data strongly support this approach,[25,34] while those who support enhancing general infection prevention precautions believe either that current data suggest these measures are not effective[28] or that more data are needed before hospitals spend large amounts of money and time performing active surveillance.[35]

As suggested in the preceding paragraphs, the major ethical dilemma with respect to using contact isolation to control the spread of resistant organisms is that the health interests of patients who are not colonized or infected with a resistant organism conflict with those of the patients who are colonized or infected with one or more of these organisms. That is, the patients who are not colonized or infected expect to be treated in the safest possible environment, one that is free of organisms that could complicate or prolong their hospitalizations or could add costs to their bills. They also desire to avoid such untoward consequences or complications of hospitalization. On the other hand, patients who are colonized or infected with one of these organisms have the right to full treatment for their medical problems, which includes receiving adequate attention from staff and having access to all tests and therapies that are necessary for their care. These patients want to avoid complications of inadequate care, such as slower or impaired rehabilitation, and complications of social isolation, such as depression, anger, and noncompliance with recommendations. Each side in this debate refers to different ethical principles to support their case. Those in favor of contact precautions argue that this type of isolation protects patients from acquiring organisms that could eventually harm them and thus supports the ethical principle of nonmaleficence. The opposition argues that contact isolation violates the patients' autonomy and may violate the principle of beneficence, as well.

Both sides in this debate tell horror stories of what happened to patients when contact precautions were not used or when they were used. We are aware of these arguments and stories and, in general, believe the arguments are stronger on the side of using contact isolation to protect patients from acquiring resistant organisms. We believe that the problems caused by contact precautions can be eliminated or ameliorated significantly if healthcare workers are educated properly and are taught to be flexible in the way they apply contact precautions, and if there is an appropriate number of staff to care for patients in isolation. In addition, we believe that resistant organisms are often spread because contact precautions are breached. Thus, contact precautions have not failed; healthcare workers have failed to use contact precautions properly.

Inclusion and Restriction of Drugs by Formulary Committees

Infection prevention personnel routinely serve as members of hospital formulary committees that determine whether the pharmacy will stock particular antimicrobial agents and whether to restrict use of particular agents. Data from the literature suggest that up to 50% of antimicrobial usage in US hospitals is inappropriate. Thus, numerous investigators have attempted to identify mechanisms by which inappropriate use of antimicrobial agents can be reduced to curb costs and to reduce the rate at which drug-resistant organisms are selected by antimicrobial pressure. Limiting which antimicrobial agents are on the healthcare organization's formulary appears to be the most direct method of accomplishing these goals[36] (see Chapter 20). Though cost concerns have historically been the primary motivation for regulating the use of antimicrobial agents, the increasing prevalence of antimicrobial resistance has made the mission of formulary committees more urgent. Committees assessing antimicrobial agents now can use both cost-benefit analyses and risk-benefit analyses to help determine whether an agent should be deleted from the formulary, included but only for particular uses, or replaced with another agent in the same class.[37] Except in the case of specific outbreaks of antibiotic-resistant infections, investigators have had difficulty proving that formulary restrictions curb the overall emergence of antibiotic resistance.[38] Nevertheless, formulary committees often use evidence from the literature to guide their decisions (ie, "evidence-based formulary control"), hoping that these decisions will reduce antibiotic selection pressure and thereby reduce the incidence of infection with resistant organisms.

Ideally, formulary committees would make their decisions about antimicrobials on the basis of explicit criteria and data from clinical, epidemiological, and pharmacoeconomic studies. However, some observers have suggested that additional factors, which are often not made explicit, may influence the decision-making process, such as clinicians' anecdotal clinical experience (positive or negative); personal or institutional financial interests; relationships with pharmaceutical industry representatives developed through interactions about clinical, educational, or research issues; or the demands of managed care organizations. The influence of the pharmaceutical industry is of special concern, because studies have demonstrated that industry marketing strategies powerfully influence physicians' choices. For example, it has been demonstrated that physicians who interact with pharmaceutical representatives are more likely request that a drug be added to the formulary, even though it has little or no therapeutic advantage over agents already on the formulary.[39] Moreover, a key strategy of the pharmaceutical industry is to ensure that physicians recognize their particular brand. Companies accomplish this through a variety of means, not least of which is distributing gifts, such as note pads and pens, that display their antimicrobial agent's trade name. Janknegt and Steenhoek[40] have shown that the more ubiquitous and memorable a drug's name, the more likely it will become an agent in a clinician's so-called "evoked set": the limited set of drugs that automatically comes to mind whenever the physician considers a certain therapeutic class or clinical indication (eg, third-generation cephalosporins, or antibiotics to treat community-acquired pneumonia).

Janknegt and Steenhoek[40] have proposed that formulary committees use an evidence-based process to select drugs. These investigators created a "System of Objectified Judgement Analysis" that they believe helps formulary committees minimize the subjective biases that may hinder the committee's ability to make rational decisions. Most importantly, Janknegt and Steenhoek[40] insist that the committees *prospectively* define their selection criteria. These investigators recommend using 8 general criteria: clinical efficacy, incidence and severity of adverse effects, dosage frequency, drug interactions, acquisition cost, documentation (strength of evidence), pharmacokinetics, and pharmaceutical aspects.[40] In addition to these criteria, the investigators added group-specific criteria (eg, development of resistance, when considering antimicrobial agents) for different classes of drugs. A panel of experts in a given field scores each drug in a class and gives each criterion a relative weight. Clinicians and pharmacists can apply this evidence-based expertise within the context of their local situation. Software has been developed to accommodate clinicians who want to modify the relative weights assigned to specific criteria by the experts, because some decisions about relative weights may be controversial.[40] *Transparency* is a particular virtue of this selection process: committees must make explicit the criteria (and their relative weights) on which they base their decisions. This strategy also prevents committees from basing decisions on only a single criterion, such as cost.

Subjective interpretation cannot be entirely eliminated from the process that formulary committees use to make decisions; however, committees that implement procedures such as the System of Objectified Judgement Analysis[40] can minimize the influence of personal or financial biases. Healthcare organizations can increase the transparency of this process by requiring committee members to disclose conflicts of interests, such as financial interests in pharmaceutical companies or products, in a manner parallel to the disclosure statements required by the editors of medical journals and the planners of scientific conferences and meetings. Healthcare organizations should hold to the highest standards of integrity and, thus, should require persons who have conflicts of interest to recuse themselves from voting on inclusion of drugs or drug classes that pertain to those conflicts.[41] Similarly, healthcare organizations should consider whether the criteria for membership on the formulary committee should include the absence of such conflicts of interest.

Postexposure Testing and Prophylaxis for HIV

Healthcare workers knowingly accept the risks that they may be exposed to HIV through their work and that, if they are exposed, they may become infected with this virus. The risk of HIV infection is small but measurable. The average risk of HIV transmission is estimated to be 0.3% after percutaneous exposure to HIV-infected blood and 0.09% after mucous membrane exposure. Fortunately, exposure of intact skin to contaminated blood has not been found to be a risk for transmission. The Centers for Disease Control and Prevention recommends postexposure prophylaxis (PEP) to diminish the risk of infection, as evidenced by seroconversion. This recommendation is based on knowledge of the pathogenesis of HIV infection, the biological plausibility of the idea that infection can be prevented or ameliorated with therapy, direct and indirect evidence for the efficacy of prophylactic treatment, and a favorable overall risk-benefit ratio for treating exposed persons prophylactically.[42]

Infection prevention staff members often help guide decisions about PEP, and they must make these decisions despite the fact that the medical issues are quite complex and scientific understanding is incomplete. Gerberding[43] reviewed the medical complexity surrounding PEP. After an exposure, PEP may be instituted either empirically or after, and depending on, a thorough assessment of the risk of infection, the expected benefit of treatment, the risks of treatment, and the likelihood that the virus is susceptible to antiretroviral agents. The results of the source patient's HIV antibody testing are essential: negative test results mean that the risk of transmission can be assumed to be zero, because the interval between the onset of viremia and the detection of HIV antibody is a few days at most. The exception to this rule arises when the source patient has symptoms consistent with acute HIV infection. In contrast, the absence of detectable HIV RNA in the source patient does not mean that the risk of transmission is zero.

PEP decreases the risk of transmission. However, it is not 100% protective; there have been 21 reported cases of HIV infection that occurred even though the exposed healthcare worker received appropriate PEP. PEP is not without risks; 50% of treated persons experience adverse drug reactions, and about one-third do not adhere to the treatment protocol. In addition, PEP may adversely affect the fetus in a pregnant woman treated for HIV exposure. Moreover, there is still much about PEP that we do not understand. For example, we do not know why 99.7% of exposed persons do not become infected, and we do not know how long after exposure PEP can be started and still be effective. Furthermore, we have not identified the optimal duration of PEP or which regimens are most effective and safe.

In the midst of this medical complexity and uncertainty, infection prevention staff members also encounter ethical and legal considerations, particularly with respect to testing patients who are the source of exposure to determine whether or not they are HIV positive, or, for persons who are known to be infected, to gather additional information about their viral load and the susceptibility of the virus to antiretroviral agents. Patients who voluntarily agree to be tested should receive standard counseling, and their results must be kept confidential. However, if the exposed healthcare worker knows which patient was the source of his or her exposure, infection prevention personnel will not be able to conceal the identity of the patient whose test result is being disclosed to the exposed healthcare worker. Thus, as part of obtaining informed consent, the person who counsels the source patient must explain the limits of confidentiality.

Conflict can arise if a potential source patient refuses to be tested. This situation presents a stressful ethical and legal dilemma that requires the healthcare worker who is requesting permission to obtain the test to sensitively identify and address the causes of the patient's reluctance and still attempt to persuade the source patient to be tested. Such persuasion is justified, because the source patient's serologic status determines whether the exposed healthcare worker should take PEP, a fairly toxic combination of medications.

Healthcare workers who evaluate the source patient must follow relevant local policies and regulations. These policies can vary substantially. Some states require consent from the source patient before blood is tested for HIV antibody and make stipulations about subsequent confidentiality surrounding a positive test result.[44] Other states presume consent, holding that "the individual to whom the care provider was exposed is deemed to consent to a test to determine the presence of HIV infection in that individual and is deemed to consent to notification of the care provider of the HIV test result."[45]

If the source patient gives consent for HIV testing or if the facility has statutory permission for involuntary testing, the infection prevention or employee health personnel dealing with the exposure do not necessarily have permission to disclose the results to affected individuals or to share other relevant medical information, such as the probability that the patient's virus is resistant to antiretroviral drugs.[46] However, since the second decade of the HIV/AIDS epidemic, public and political attitudes regarding confidentiality about HIV status have generally shifted toward accepting the more traditional public health approach, which places limits on patient confidentiality to maximize benefits to those who have been exposed. This shift is evidenced by current procedures for notifying partners.[47]

Situations in which source patients refuse to be tested are clearly ethical, in that they pose problems related to ethical values, obligations, and rights. Infection prevention personnel who wish to use an organized ethical approach to these situations must carefully and sensitively assess the medical facts, the interests of all parties involved, and the particular context that surrounds the patient's refusal. To do this, infection prevention personnel may need to talk in depth with the source patient about his or her concerns regarding testing, to explain the rationale behind testing, and to help enhance trust between the patient and the healthcare team. After this process is complete, not only may the risk of transmission be better defined but also the interpersonal issues may be resolved to the point that the source patient and the healthcare team reach a consensus. In some cases, the source patient and the healthcare team may still disagree. At this point, infection prevention personnel may need to help identify a course of action that balances the interests of all involved persons.

In such discussions, different ethical principles pertain to the exposed healthcare worker and to the source patient: beneficence, nonmaleficence, and justice (ie, fairness) are relevant for the exposed healthcare worker, and autonomy is the primary principle that pertains to the source patient and his or her insistence on privacy and noninterference. Both the exposed healthcare worker and the source patient have rights in these situations. The exposed worker has the positive right to assistance that does not impose significantly on others, and the source patient has the negative right to be left alone. Working through ethical conflicts surrounding similar situations, such as prenatal HIV testing, may help infection prevention personnel reason by analogy about testing after a healthcare worker is exposed. Professional guidelines, publications by national organizations, or institutional policies, may also help infection control personnel work through these ethical issues.

By now we hope that the reader understands that HIV testing has important ethical implications and that testing and not testing a source patient each have substantial consequences for a healthcare teams' ability to make medical decisions and to protect confidential information. These situations are never easy, but infection prevention personnel who prospectively think through the potential areas of conflict will be better able to resolve the situations when they occur.

Ethical Codes

Ethical codes emphasize a profession's core values and may help guide decisions and behavior. To our knowledge, neither the Society for Healthcare Epidemiology of America nor the Association for Professionals in Infection Control and Epidemiology, the 2 societies concerned with infection control, have developed codes of ethics. Recently, the Association for Professionals in Infection Control and Epidemiology and the Community and Hospital Infection Control Association–Canada (CHICA-Canada) published a document describing "professional and practice standards" for persons practicing infection control.[48]

A well-developed and clearly stated ethical code is an essential guide, yet it is also insufficient. A code of ethics cannot identify all of the ethical dilemmas that individuals will face in the course of their practice. Nor, despite the fond hopes of professional school administrators, does the graduation recitation of such a code guarantee ethical conduct. Alone, an ethical code cannot ensure ethical behavior. It must be taught, learned, affirmed, and lived, if it is to affect our practice. As William Diehl writes: "Formal codes of ethics are hot items these days. [But one] thing is certain: any organization that requires all its employees to review and sign its ethics code each year, and then does nothing else to encourage high moral behavior, is wasting its time on the code."[49]

Any institution that acts not as it preaches wastes time and also, at least implicitly, encourages unethical behavior. Institutions reward the conduct they prize. It should be a warning to us that, at present, we are more likely to hear of inconsiderate behavior excused on the grounds of a colleague's brilliance than to hear an individual praised for making a difficult but ethically sound decision. No wonder—we were taught that intelligence and success within the system are the highest values, whereas perhaps none of our textbooks quoted Ralph Waldo Emerson's startling and humbling words: "character is higher than intellect."

As our financial and staff resources are stressed without limit and as the pressures under which we work intensify, temptation amplifies. Barbara Ley Toffler of Resources for Responsible Management states:

For many employees, being ethical is getting to be too risky—something they can't afford any more. ...The problem grows out of what I call the "move it" syndrome. ...That's when the boss tells a subordinate to "move it"—just get it done, meet the deadline, don't ask for more money, time, or people, just do it—and so it goes on down the line.

For American companies, this peril from within is as serious as outside threats from competitors. As more employers are forced to "move it," companies are increasingly vulnerable—legally, financially, and morally—to the unethical actions of decent people trying to [move it just to keep their jobs].[50]

To "move it," we may find ourselves declining to issue appropriate sanctions in an outbreak because we are loath to alienate an important doctor or lose referrals from a powerful practice group. Or, fearing management anger over bad publicity and loss of revenue, we may decide against closing a ward affected by an outbreak. Under pressure to reduce budgets, we may approve questionable practices or eliminate effective infection prevention programs. During this time of uncertainty, we may be tempted to treat influential administrators or practice groups preferentially because they control our budgets or could curtail our programs. We may be tempted to recommend a particular product because we have received grants from the company or we have purchased their stock. Or perhaps we condone altering hospital records to avoid losing accreditation. Finally, we may withhold information regarding resistant organisms so that we can transfer patients to other institutions and shorten their length of stay in our hospital.

Practical Advice

What can you as an individual hospital epidemiologist or infection preventionist do? We would recommend that you think about your job and identify the most common questions that you answer. Once you have identified the questions, you can try to identify the ethical dilemmas presented by those decisions. You can then develop a plan for dealing with the issues before you face them again, as one can usually think more clearly and dispassionately when not in the middle of a crisis. When designing such plans, you should obtain help, if necessary or prudent, from experts in medicine, law, ethics, or other appropriate disciplines.

We have described but a few of the manifold ethical challenges that confront us. Against our ambitions and our fears, we have only our values, commitments, and continual self-examination to rely on. Are we here to serve ourselves or to protect the health of patients and healthcare workers? Are we seeking to keep our jobs or are we pursuing beneficial knowledge?

As difficult as these questions may be, we must ask them or risk unethical conduct. In the quiet of our consciences, we must grade our answers candidly, guarding against our capacity to rationalize decisions that are expedient. We cannot afford to ignore the ethical aspects of infection prevention, because in neglecting ethics we risk losing sight of our profession's goal—the health of individual persons and populations.

Acknowledgement

We thank Dr. Daniel Diekema for critiquing the chapter and suggesting helpful revisions.

References

1. Jonsen AR. Do no harm. *Ann Intern Med* 1978;88:827–832.
2. Last JM. Ethical issues in public health. In: Last JM. *Public Health and Human Ecology*. East Norwalk, CT: Appleton & Lange; 1987:351–370.
3. IEA Workshop on Ethics, Health Policy and Epidemiology. Proposed ethics guidelines for epidemiologists. *American Public Health Association Newsletter* (Epidemiology Section) 1990:4–6.
4. Beauchamp TL, Childress JF. *Principles of Biomedical Ethics*. 2nd ed. New York: Oxford University Press; 1983.
5. Soskolne CL. Epidemiology: questions of science, ethics, morality, and law. *Am J Epidemiol* 1989;129:1–18.
6. Herman AA, Soskolne CL, Malcoe L, Lilienfeld DE. Guidelines on ethics for epidemiologists. *Int J Epidemiol* 1991;20: 571–572.
7. Beauchamp TL, Cook RR, Fayerweather WE, et al. Appendix: ethical guidelines for epidemiologists. *J Clin Epidemiol* 1991; 44(1 suppl):151S–169S.
8. Clements CJ, Evans G, Dittman S, Reeler AV. Vaccine safety concerns everyone. *Vaccine* 1999;17:S90–S94.
9. Freed GL, Katz SL, Clark SJ. Safety of vaccinations: Miss America, the media, and the public health. *JAMA* 1996;276: 1869–1872.
10. Gordon JE, Hood HI. Whooping cough and its epidemiological anomalies. *Am J Med Sci* 1951;222:333–361.
11. Cherry JD. The epidemiology of pertussis and pertussis immunization in the United Kingdom and the United States: a comparative study. *Curr Probl Pediatr* 1984;14:1–78.
12. Mortimer EA. Pertussis vaccine. In: Plotkin SA, Mortimer EA, eds. *Vaccine*. Philadelphia, PA: W.B. Saunders Company; 1988.
13. Diekema DS. Public health issues in pediatrics. In: Post SG, ed. *The Encyclopedia of Bioethics*. 3rd ed. Farmington Hills, MI: Thomson Gale; 2003.
14. Bazin H. The ethics of vaccine usage in society: lessons from the past. *Endeavour* 2001;25:104–108.
15. Ulmer JB, Liu MA. Ethical issues for vaccines and immunization. *Nat Rev Immunol* 2002;2:291–296.
16. Vermeersch E. Individual rights versus societal duties. *Vaccine* 1999;17:S14–S17.

17. Hodges FM, Svoboda JS, Van Howe RS. Prophylactic interventions on children: balancing human rights with public health. *J Med Ethics* 2002;28:10–16.

18. Poland GA, Nichol KL. Medical students as sources of rubella and measles outbreaks. *Arch Intern Med* 1990;150: 44–46.

19. Centers for Disease Control and Prevention (CDC). Control and prevention of rubella: evaluation and management of suspected outbreaks, rubella in pregnant women, and surveillance for congenital rubella syndrome. *MMWR Morb Mortal Wkly Rep* 2001;50:1–23.

20. Rea E, Upshur R. Semmelweis revisited: the ethics of infection prevention among health care workers. *Can Med Assoc J* 2001;164:1447–1448.

21. Evans ME, Hall KL, Berry SE. Influenza control in acute care hospitals. *Am J Infect Control* 1997;25:357–362.

22. Department of Labor, Occupational Safety and Health Administration. 29 CFR Part 1920.1030, Occupational exposure to bloodborne pathogens, final rule. *Fed Reg* December 6, 1991;56:64, 004–64, 182.

23. Sepkowitz KA. Nosocomial hepatitis and other infections transmitted by blood and blood products. In: Mandell GL, Bennett JE, Dolin R, eds. *Principles and Practice or Infectious Diseases.* 5th ed. New York: Churchill Livingstone; 2000: 3039–3052.

24. Herwaldt LA. Control of methicillin-resistant *Staphylococcus aureus* in the hospital setting. *Am J Med* 1999;106 (5A): 11S–18S.

25. Muto CA, Jernigan JA, Ostrowsky BE, et al. SHEA guideline for preventing nosocomial transmission of multidrug-resistant strains of *Staphylococcus aureus* and *Enterococcus. Infect Control Hosp Epidemiol* 2003;24:362–386.

26. Farr BM. Protecting long-term care patients from antibiotic resistant infections: ethics, cost-effectiveness, and reimbursement issues. *J Am Geriatr Soc* 2000;48:1340–1342.

27. Ostrowsky B, Steinberg JT, Farr B, Sohn AH, Sinkowitz-Cochran RL, Jarvis WR. Reality check: should we try to detect and isolate vancomycin-resistant enterococci patients. *Infect Control Hosp Epidemiol* 2001;22:116–119.

28. Teare EL, Barrett SP. Stop the ritual of tracing colonized people. *Br Med J* 1997;314:665–666.

29. Peel RK, Stolarek I, Elder AT. Is it time to stop searching for MRSA? Isolating patients with MRSA can have long term implications. *Br Med J* 1997;315:58.

30. Kirkland KB, Weinstein JM. Adverse effects of contact isolation. *Lancet* 1999;354:1177–1178.

31. Pike JH, McLean D. Ethical concerns in isolating patients with methicillin-resistant *Staphylococcus aureus* on the rehabilitation ward: a case report. *Arch Phys Med Rehabil* 2002;83: 1028–1030.

32. Vasquez JE, Walker ES, Franzus BW, Overbay BK, Reagan DR, Sarubbi FA. The epidemiology of mupirocin resistance among methicillin-resistant *Staphylococcus aureus* at a Veterans' Affairs hospital. *Infect Control Hosp Epidemiol* 2000; 21:459–464.

33. Farr BM, Jarvis WR. Would active surveillance cultures help control healthcare-related methicillin-resistant *Staphylococcus aureus* infections? *Infect Control Hosp Epidemiol* 2002;23: 65–68.

34. Calfee DP, Giannetta ET, Durbin LJ, Germanson TP, Farr BM. Control of endemic vancomycin-resistant *Enterococcus* among inpatients at a university hospital. *Clin Infect Dis* 2003;37:326–332.

35. Diekema DJ. Active surveillance cultures for control of vancomycin-resistant *Enterococcus. Clin Infect Dis* 2003;37: 1400–1402.

36. John JF, Fishman NO. Programmatic role of the infectious diseases physician in controlling antimicrobial costs in the hospital. *Clin Infect Dis* 1997;24:471–485.

37. Shlaes DM, Gerding DN, John JF, et al. Society for Healthcare Epidemiology of America and Infectious Diseases Society of America Joint Committee on the prevention of antimicrobial resistance: guidelines for the prevention of antimicrobial resistance in hospitals. *Infect Control Hosp Epidemiol* 1997;18: 275–291.

38. Kollef MH, Fraser VJ. Antibiotic resistance in the intensive care unit. *Ann Intern Med* 2001;134:298–314.

39. Wazana A. Physicians and the pharmaceutical industry: is a gift ever just a gift? *JAMA* 2000;283:373–380.

40. Janknegt R, Steenhoek A. The system of objectified judgement analysis (SOJA): a tool in rational drug selection for formulary inclusion. *Drugs* 1997;53:550–562.

41. Fijn R, van Epenhuysen LS, Peijnenburg AJM, Brouwers JRBJ, de Jong-van den Berg TW. Is there a need for critical ethical and philosophical evaluation of hospital drugs and therapeutics (D&T) committees? *Pharmacoepidemiol Drug Saf* 2002; 11:247–252.

42. Centers for Disease Control and Prevention (CDC). Updated US Public Health Service guidelines for the management of occupational exposures to HBV, HCV, and HIV and recommendations for postexposure prophylaxis. *MMWR Recomm Rep* 2001;50(RR-11):1–52.

43. Gerberding JL. Occupational exposure to HIV in health care settings. *N Engl J Med* 2003;348:826–833.

44. Catalano MT. Postexposure prophylaxis implementation issues: programmatic concerns in hospitals. *Am J Med* 1997; 102(5B):95–97.

45. Acquired immune deficiency syndrome: care provider notification. Iowa Code 2003 §141A.8. http://www.legis.state.ia.us/IACODE/2003/141A/8.html. Accessed September 11, 2009.

46. Danila RN. Recommendations for chemoprophylaxis after occupational exposure to human immunodeficiency virus: a public health agency perspective. *Am J Med* 1997;102(5B): 98–101.

47. Kaldjian LC. HIV testing and partner notification: physicians' ethical responsibilities in a persistent epidemic. *Adv Stud Med* 2003;3:413–414.

48. Horan-Murphy E, Barnard B, Chenoweth C, et al. APIC/CHICA-Canada infection control and epidemiology: professional and practice standards. *Am J Infect Control* 1999; 27:47–51.

49. Diehl WE. *The Monday connection: a spirituality of competence, affirmation, and support in the workplace.* New York: Harper Collins Publishers; 1991:85–135.

50. Toffler BL. When the signal is "move it or lose it." *NY Times* 1991;Sect F:13.

51. Kaldjian LC, Weir RF, Duffy TP. A clinician's approach to clinical ethical reasoning. *J Gen Intern Med* 2005;20:306–311

Chapter 4 Epidemiologic Methods in Infection Control

Ebbing Lautenbach, MD, MPH, MSCE

A strong working knowledge of basic epidemiologic principles and approaches is critical for the healthcare epidemiologist. The ability to accurately quantify new patterns of infectious diseases, design rigorous studies to characterize the factors associated with disease, and devise and evaluate interventions to address emerging issues are vital to effective job performance.

Epidemiology is commonly defined as the study of the distribution and determinants of disease frequency in human populations. This definition concisely encompasses the 3 main components of the discipline of epidemiology. The first, "disease frequency," involves identifying the existence of a disease and quantifying its occurrence. The second, "distribution of disease," characterizes in whom the disease is occurring, where it is occurring, and when it is occurring. Finally, "determinants of disease" focuses on formulating and testing hypotheses with regard to the possible risk factors for disease.

The value of epidemiological methods in the study of healthcare infections has been recognized for some time.[1-4] Indeed, the past 5 years has seen a renewed interest and vitality in efforts to explore previously unstudied aspects of epidemiological methods in the study of healthcare infections and antimicrobial resistance.[5-8] While this chapter is meant to provide a broad overview, the reader is also directed to numerous published textbooks which are solely dedicated to general epidemiology, infectious diseases epidemiology, and statistical analysis.[9-15]

Measures of Disease Frequency

Before setting out to identify the possible causes of a disease, one must first quantify the frequency with which the disease occurs. This is important both for measuring the scope of the problem (ie, how many people are affected by the disease) and for subsequently allowing comparison between different groups (ie, people with and people without a particular risk factor of interest). The most commonly used measures of disease frequency in epidemiology are prevalence and incidence.

Prevalence

Prevalence is defined as the proportion of people with disease at a given point in time (eg, the proportion of hospitalized patients who have a nosocomial infection). This is also sometimes referred to as the "point prevalence." It is calculated as the number of individuals with disease divided by the total number of individuals in the population observed.

$$\text{Prevalence} = \frac{\text{number of diseased individuals}}{\text{total number of individuals in the population}}$$

(A related, although infrequently used, measure is the "period prevalence," which is defined as the number of persons with disease in a given *period* of time divided by the number of persons observed during the period.) Prevalence is a proportion and as such has no units. This measure of disease frequency is dependent on both the incidence (ie, the number of new cases which develop) as well as the duration of disease (ie, how long a disease lasts once it has developed). The greater the incidence and the greater the duration of disease, the higher the prevalence. Prevalence is useful for measuring the burden of disease in a population (ie, the overall proportion of

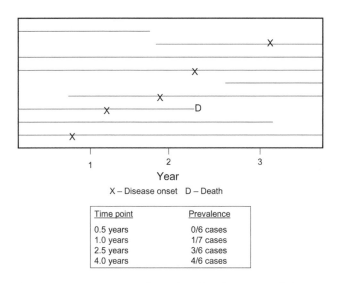

Time point	Prevalence
0.5 years	0/6 cases
1.0 years	1/7 cases
2.5 years	3/6 cases
4.0 years	4/6 cases

Figure 4-1. Measurement of prevalence in a dynamic population.

persons affected by the disease), which may in turn inform decisions regarding such issues as allocation of resources and funding of research initiatives.

All populations are dynamic; individuals are constantly entering and leaving the population. Depending on the population, the prevalence may vary depending on when it is measured (Figure 4-1). If a dynamic population is at steady state (ie, the number of individuals leaving is equal to the number of individuals entering the population), the prevalence will be constant over time.

Incidence

Incidence is defined as the number of new cases of diseases occurring in a specified period of time. Incidence may be described in several ways. Cumulative incidence is defined as the number of new cases of disease in a particular time period divided by the total number of disease-free individuals at risk of the disease at the beginning of the time period (eg, the proportion of patients who develop a nosocomial infection during hospitalization). In infectious disease epidemiology, this traditionally has been termed the "attack rate."

$$\text{Cumulative incidence} = \frac{\substack{\text{number of new cases of} \\ \text{disease between } t_0 \text{ and } t_1}}{\substack{\text{total number of disease-} \\ \text{free individuals at risk} \\ \text{of disease at } t_0}}$$

A cumulative incidence, like a prevalence, is simply a proportion and thus has no units. In order to calculate the cumulative incidence, one must have complete follow-up data on all observed individuals, such that

their final disposition with regard to having or not having the disease may be determined. Although this measure describes the total proportion of new cases occurring in a time period, it does not describe when in the time period the cases occurred (Figure 4-2).

For the cumulative incidence of nosocomial infections, the time period implied is the course of hospitalization until a first infection event or until discharge without a first infection event. However, patients do not all stay in the hospital and remain at risk for exactly the same period of time. Furthermore, most nosocomial infections are time related, and comparing the cumulative incidence of nosocomial infection among patient groups with differing lengths of stay may be very misleading. By contrast, if one is investigating infection events that have a point source and are not time related (eg, tuberculosis acquired from a contaminated bronchoscope), then the cumulative incidence is an excellent measure of incidence. Surgical site infections are also usually thought of as having a point source (ie, the operation).

Historically, nosocomial infection rates were often reported as a cumulative incidence (eg, the number of infections per 100 discharges). This definition had no unique quantitative meaning, as it did not separate first infections from multiple infections in the same patient, and allowed undefined multiple counting of individuals. The implications of a finding of 5 infections per 100 discharges would be entirely different if it represented 5 sequential infections in a single moribund patient or 5 first infections in 5 different but otherwise healthy patients, such as women with normal deliveries.

The incidence rate (or incidence density) is defined as the number of new cases of disease in a specified quantity of person-time of observation among individuals at risk

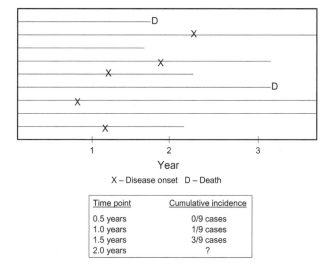

Time point	Cumulative incidence
0.5 years	0/9 cases
1.0 years	1/9 cases
1.5 years	3/9 cases
2.0 years	?

Figure 4-2. Measurement of cumulative incidence.

(eg, the number of nosocomial infections per 1,000 hospital-days).

$$\text{Incidence rate} = \frac{\begin{array}{c}\text{number of new cases of disease}\\\text{during a given time period}\end{array}}{\begin{array}{c}\text{total person-time of observation}\\\text{among individuals at risk}\end{array}}$$

The primary value of this measure can be seen when comparing nosocomial infection rates between groups that differ in their time at risk (eg, short-stay patients versus long-stay patients). When the time at risk in one group is much greater than the time at risk in another, the incidence rate, or risk per day, is the most convenient way to correct for time, and thus separate the effect of time (ie, duration of exposure) from the effect of daily risk. For convenience, in hospital epidemiology, incidence rates for nosocomial infections are usually expressed as the number of first infection events in a certain number of days at risk (eg, the number of nosocomial infections per 1,000 hospital-days,) because this usually produces a small single-digit or double-digit number.

An incidence rate is usually restricted to counting first infection events (eg, the first episode of nosocomial infection in a given patient). It is standard to consider only first events because second events are not statistically independent from first events in the same individuals (ie, patients with a first infection event are more likely to experience a second event). For example, the group of all hospitalized patients who have not yet developed a nosocomial infection would compose the population at risk. After a patient develops an infection, that patient would then be withdrawn from the analysis and would not be a part of the population still at risk for a first event. Each hospitalized patient who never develops an infection would contribute all their hospital-days (ie, the sum of days the patient is in the hospital) to the total count of days at risk for a first event. However, a patient who develops an infection would contribute only their hospital-days before the onset of the infection.

Unlike cumulative incidence, the incidence rate does not assume complete follow-up for all subjects and thus accounts for different entry and dropout rates. However, even if follow-up data are complete (and thus the cumulative incidence could be calculated), reporting the incidence rate may still be preferable. The cumulative incidence reports only the overall number of new cases occurring during the time period, regardless of whether they occur early or late in the time period. By comparison, the incidence rate, by incorporating the time at risk, accounts for a potential difference in the time to occurrence of the infection event. In considering the 2 examples in Figure 4-3, one will note that despite the fact that the cumulative incidence of disease at 4 years is the same

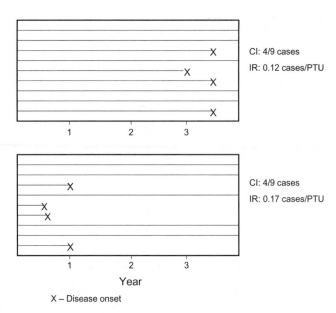

CI: 4/9 cases
IR: 0.12 cases/PTU

CI: 4/9 cases
IR: 0.17 cases/PTU

X – Disease onset

Figure 4-3. Comparison of the cumulative incidence (CI) and the incidence rate (IR). PTU, person-time unit.

for the 2 groups, subjects in the second group clearly acquire disease earlier. This information is reflected in the different incidence rates.

Since the incidence rate counts time at risk in the denominator, the implicit assumption is that all time at risk is equal (eg, the likelihood of developing a nosocomial infection in the first 5 days after hospital admission is the same as the likelihood of developing an infection during days 6–10 of hospitalization). If all time periods are not equivalent, the incidence rate may be misleading, depending on when in the course of their time at risk patients are observed for the outcome.

Study Design

One of the critical components of the field of epidemiology is identifying the determinants of disease (ie, risk factors for a particular outcome of interest). This aspect of the field focuses on formulating and testing hypotheses with regard to the possible risk factors for disease. A number of study designs are available to the hospital epidemiologist when attempting to test a hypothesis as to the causes of a disease. These study designs, in order of increasing methodological rigor, include the following types: case report, case series, ecologic study, cross-sectional study, case-control study, cohort study, and randomized controlled trial. Randomized controlled trials, case-control studies, and cohort studies are considered analytic studies, as opposed to the other study designs, which are considered descriptive studies. Analytic studies are most useful in identifying the determinants of disease. In determining the correct study design to use, the hospital epidemiologist must first carefully consider "What is the question?" Once this critical question has been clearly formulated, the optimal study design will likely also become evident. Other considerations (eg, available time, access to financial support, and/or ethical considerations) may also influence the decision as to the type of study that should be undertaken.

Case Report or Case Series

A case report is the clinical description of a single patient (eg, a single patient with a case of bloodstream infection due to vancomycin-resistant enterococcus [VRE]). A case series is simply a report of more than 1 patient with the disease of interest. One advantage of a case report or case series is its relative ease of preparation. In addition, a case report or case series may serve as a clinical or therapeutic example for other healthcare epidemiologists who may be faced with similar cases. Perhaps most importantly, a case report or case series can serve to generate hypotheses that may then be tested in future analytic studies. For example, if a case report notes that a patient had been exposed to several courses of vancomycin therapy in the month prior to the onset of VRE infection, one hypothesis might be that vancomycin use is associated with VRE infection. The primary limitation of a case report or case series is that it describes, at most, a few patients and may not be generalizable. In addition, since a case report or case series does not include a comparison group, one cannot determine which characteristics in the description of the cases are unique to the illness. While case reports are thus usually of limited interest, there are exceptions, particularly when they identify a new disease or describe the index case of an important outbreak (eg, the first report of clinical VRE infection).

Ecologic Study

In an ecologic study, one compares geographic and/or time trends of an illness with trends in risk factors (ie, a comparison of the annual amount of vancomycin used hospital-wide with the annual prevalence of VRE among enterococcal isolates from cases of nosocomial infection). Ecologic studies most often use aggregate data that are routinely collected for other purposes (eg, antimicrobial susceptibility patterns from a hospital's clinical microbiology laboratory, or antimicrobial drug dispensing data from the inpatient pharmacy). This ready availability of data provides one advantage to the ecologic study, in that such studies are often relatively quick and easy to do. Thus, such a study may provide early support for or against a hypothesis. However, one cannot distinguish between various hypotheses that might be consistent with the data. Perhaps most importantly, ecologic

studies do not incorporate patient-level data. For example, although both the annual hospital-wide use of vancomycin and the yearly prevalence of VRE among enterococcal isolates from cases of nosocomial infection might have increased significantly over a 5-year period, one cannot tell from these data whether the actual patients who were infected with VRE received vancomycin.

Cross-sectional Study

A cross-sectional study is a survey of a sample of the population in which the status of subjects with regard to the risk factor and disease is assessed at the same point in time. For example, a cross-sectional study to assess VRE infection might involve identifying all patients currently hospitalized and assessing each patient with regard to whether he or she has a VRE infection, as well as whether he or she is receiving vancomycin. One advantage of a cross-sectional study is it is relatively easy to carry out, given that all subjects are simply assessed at one point in time. Accordingly, this type of study may provide early evidence for or against a hypothesis. A major disadvantage of a cross-sectional study is that this study design does not capture information about temporal sequence (ie, it is not possible to determine which came first, the proposed risk factor or the outcome). Furthermore, a cross-sectional study does not provide information about the transition between health states (eg, development of new VRE infection or resolution of VRE infection).

Case-Control Study

In distinguishing between the various types of analytic studies (ie, case-control, cohort, and experimental) it is useful to consider the traditional 2 x 2 table (Figure 4-4). While all 3 study designs seek to investigate the potential association between a risk factor (or exposure) and an

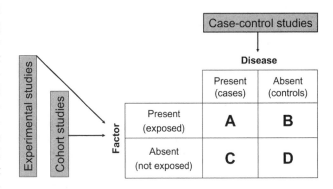

Figure 4-4. Summary of study designs used in infection control and epidemiological studies.

outcome of interest, they differ fundamentally in the way that patients are enrolled into the study. In a case-control study, patients are entered into the study based on the presence or absence of the outcome (or disease) of interest. The 2 groups (ie, the case patients with the disease and the control patients without the disease) are then compared to determine if they differ with regard to the presence of risk factors of interest. Case-control studies are always retrospective.

A case-control study design is particularly attractive when the outcome being studied is rare, because one may enroll into the study all patients with the outcome of interest. Accordingly, this study design is much more efficient and economical than the comparable cohort study, in which a group of patients with and without an exposure of interest would need to undergo follow-up for a period of time to determine who develops the outcome of interest. Even if a large cohort is available, it may be more economical to conduct a small case-control study within the cohort. Such a "nested" case-control study may produce the same information as would the larger cohort study, at a fraction of the cost. Another advantage of the case-control study is that one may study any number of risk factors for the outcome of interest. One disadvantage of a case-control study is that only one outcome may be studied. Another disadvantage of this approach is that one cannot directly calculate the incidence or relative risk from a case-control study, because the investigator fixes the number of case and control patients to be studied.

Thoughtful consideration must be taken when selecting case and control patients in a case-control study. Cases may be restricted to any group of diseased individuals. However, they must derive from a theoretical source population such that a diseased person not selected is presumed to have derived from a different source population. For example, in studying risk factors for nosocomial VRE infection, the theoretical source population could be considered to be the population of patients hospitalized at one institution. Thus, if any patient at that institution were to have a clinical isolate that represented VRE infection, they would be included as a case. However, a patient with VRE infection at a different hospital would not be included. Finally, cases must be chosen in a manner independent of the patient's status with regard to an exposure of interest.

Careful attention is also required when selecting control patients for a case-control study. Controls should be representative of the theoretical source population from which the cases were derived. Thus, if a control patient were to have developed the disease of interest, they would have been selected as a case. In the example above, control patients may be randomly selected from among all patients in the hospital not infected with VRE. In investigating the possible association between prior vancomycin use and VRE infection, these 2 groups (ie, patients with VRE infection and a random sample of all other hospitalized patients) could be compared to determine what proportion of patients in each group had experienced recent vancomycin exposure. Finally, like cases, controls must be chosen in a manner independent of the patient's status with regard to an exposure of interest and should not be selected because they have characteristics similar to those of case patients.

Cohort Study

Unlike a case-control study, in which study subjects are selected based on the presence or absence of an outcome or disease of interest, a cohort study selects subjects on the basis of the presence or absence of an exposure (or risk factor) of interest (Figure 4-4). The 2 groups (ie, the subjects with the exposure and the subjects without the exposure) are then compared to determine if they differ with regard to whether they develop the outcome of interest. The investigator may select subjects randomly or according to exposure.

A cohort study may be either prospective or retrospective. Whether a cohort study is prospective or retrospective depends on when it is conducted with regard to when the outcome of interest occurs. If patients are identified as exposed or unexposed and then follow-up is conducted forward in time to determine whether the patients develop the disease, it is a prospective cohort study. If the study is conducted after the time the outcome has already occurred, it is a retrospective cohort study. In either case, subjects are entered into the study on the basis of their exposure status with regard to a variable of interest, and these groups are then compared on the basis of the outcome of interest. For example, one might identify all patients who receive vancomycin in a hospital (ie, the exposed group) and compare them to a randomly selected group of patients who do not receive vancomycin (ie, the unexposed group). These groups could then be followed-up forward in time to determine what proportion of patients in each group develops the outcome of interest (eg, VRE infection).

One advantage of a cohort study is that one may study multiple outcomes of a single risk factor or exposure. In addition, a cohort study allows the investigator to calculate both an incidence and a relative risk in comparing the 2 groups. Potential disadvantages of a cohort study include heavy cost and time requirements, because patients must be followed-up forward in time until a sufficient number develop the outcome of interest. Depending on the course of the disease, this may be a lengthy period. In addition, if the outcome is rare, a great many subjects will need to be followed-up until the necessary number develop the disease. Finally, the longer the study duration, the more likely that subjects will be lost to follow-up, potentially biasing the results of the study. Some of these limitations are lessened in a retrospective cohort study, since outcomes have already occurred and patients do not need to be followed-up prospectively.

Randomized Controlled Trial

In clinical investigation, the closest approximation to a standard experiment in bench biology is the randomized controlled trial. In comparing the randomized controlled trial to other analytic study designs (Figure 4-4), it is very similar to the cohort study. However, in a cohort study, when patients are enrolled, they already either have or do not have the exposure of interest. In a randomized controlled trial, the investigator assigns the exposure according to some scheme, such as randomization. This study design provides the most convincing demonstration of causality, because patients in both groups should (provided randomization has worked appropriately) be equal with respect to all important variables except the one variable (exposure) manipulated by the investigator. While randomized controlled trials may provide the strongest support for or against an association of interest, they are costly studies, and there may be ethical issues which preclude conducting one. For example, in elucidating the association between vancomycin use and VRE infection, it would be unethical to randomly assign patients to receive vancomycin if they did not require the drug.

Bias and Confounding

Two common issues which arise when designing a study are the potentials for bias and confounding. Bias is the systematic error in the collection or interpretation of data. Types of bias include information bias (ie, distortion in the estimate of effect because of measurement error or misclassification of subjects with respect to one or more variables) and selection bias (ie, distortion in the estimate of effect resulting from the manner in which subjects are selected for the study). The potential for bias must be addressed at the time the study is designed, since it cannot be corrected during the analysis of the study. In randomized controlled trials, blinding is a commonly used method to minimize the potential for bias in such studies. In addition to evaluating whether bias may exist, one much also consider the likely impact of the bias on the study results. Bias may be nondifferential (ie, biasing toward the null hypothesis and making the 2 groups being compared look artificially similar), or differential (ie, biasing away from the null hypothesis and making the 2 groups being compared look artificially dissimilar).

Confounding occurs when the estimate of the effect of the exposure is distorted because it is mixed with the effect of an extraneous factor. To be a confounder, a variable must be associated with both the exposure and the outcome of interest, but it cannot be a result of the exposure. Unlike bias, a confounding variable may be controlled for in the study analysis. However, to do this, data regarding the presence or absence of the confounder must be collected during the study. Thus, it is also important to consider the potential for confounding variables in the design of the study.

Measures of Effect

Risk Versus Odds

Depending on which type of study one conducts, one will generally calculate either a relative risk (ie, in a cohort study or a randomized controlled trial) or an odds ratio (ie, in a case-control study) to characterize the strength of an association between an exposure and an outcome. Before describing these statistical measures in greater detail, it is useful to briefly compare the concepts of risk and odds. In a risk (also referred to as a probability), the numerator contains the event of interest, while the denominator contains all possible events. For example, in throwing a die, the *risk* of throwing a 3 is 1 divided by 6 (since there are 6 possible events when throwing a die). Thus, the risk of throwing a 3 is 0.167, or 16.7%. In an odds, the numerator again contains the event of interest, while the denominator contains all possible events minus the event of interest. Using again the example of throwing a die, the *odds* of throwing a 3 is 1 divided by 5 (ie, 6 minus 1). Thus, the odds of throwing a 3 is 0.2, or 20%. Since the denominator in an odds is always smaller, the value for the odds is always somewhat greater than the comparable risk.

Relative Risk

The relative risk (also called the risk ratio) is the ratio of 2 probabilities: the probability of the outcome among the exposed subjects divided by the probability of the outcome in the unexposed subjects (Figure 4-5). A relative risk can be calculated from a cohort study or a randomized controlled trial, because from these study designs one can derive population-based rates or proportions. A relative risk of 1.0 is called the value of no effect or the null value. A relative risk equal to 2.0 means the exposed subjects were twice as likely to have the outcome of interest as were the unexposed subjects. On the other hand, a relative risk of 0.5 means that the exposed subjects were half as likely to experience the outcome as

Disease

	Present (cases)	Absent (controls)
Present (exposed)	**A**	**B**
Absent (not exposed)	**C**	**D**

Factor

Relative risk

Risk of disease among exposed persons = A / (A+B)
Risk of disease among unexposed persons = C / (C+D)

Relative risk = $\dfrac{A / (A+B)}{C / (C+D)}$

Odds ratio

Odds of exposure, given disease = A / C
Odds of exposure, given no disease = B / D

Disease odds ratio = $\dfrac{A/C}{B/D}$ = $\dfrac{AD}{BC}$

Relationship between relative risk and odds ratio

When disease is rare, B >> A and D >>> C

Relative risk = $\dfrac{A / (A+B)}{C / (C+D)}$ ~ $\dfrac{AD}{BC}$ = odds ratio

Figure 4-5. Comparison of relative risk and the odds ratio, showing how the case-control formula approaches the formula for relative risk when the rare outcome criterion is met.

the unexposed subjects, indicating that the exposure had a protective effect.

Odds Ratio

As noted previously, in a case-control study, subjects are enrolled into the study on the basis of the outcome of interest. One then compares the 2 groups (ie, the subjects with the outcome and the subjects without the outcome) to determine what proportion of subjects in each group demonstrate a risk factor of interest. In this type of study, without additional information, one cannot determine how common the outcomes or the exposures are in the entire study population. Thus, unlike in a cohort study, one cannot directly calculate a relative risk. What one can calculate in a case-control study is the odds ratio. The odds ratio is defined as the odds of exposure in subjects with the outcome divided by the odds of exposure in subjects without the outcome (Figure 4-5). An odds ratio of 1.0 is called the value of no effect or the null value.

As noted above, one cannot calculate a relative risk from a case-control study because the case-control study

offers no insights into the absolute rates or proportions of disease among subjects. However, in situations in which the disease under study is rare (ie, a prevalence of less than 10% in the study population), the odds ratio derived from a case-control study closely approximates the relative risk that would have been derived from the comparable cohort study. Figure 4-5 shows how the case-control formula approaches the formula for relative risk when the rare outcome criterion is met.

Measures of Strength of Association

P Value

The most common method of measuring the strength of association in a 2 x 2 table is to use the chi-squared (χ^2) test for the comparison of 2 binomial proportions. This calculation is identical for all 2 x 2 tables, whether or not the data were derived from a cohort study or case-control study. When one has calculated the value for the χ^2 test, one can then look up the associated probability that the observed difference between binomial proportions could have arisen by chance alone. The conventional interpretation of these probabilities is that a P value of less than .05 indicates an effect at least as extreme as that observed in the study is unlikely to have occurred by chance alone, given that there is truly no relationship between the exposure and the disease. Although this is the conventional interpretation, there is nothing magical about the .05 cutoff for statistical significance. One limitation of the P value is that it reflects both the magnitude of the difference between the groups being compared as well as the sample size. Consequently, even a small difference between groups (if the sample size is large enough) may be statistically significant, even if it is not clinically important. Conversely, a larger effect that would be clinically important may not achieve statistical significance if the sample size is insufficient.

95% Confidence Interval

Given the limitations of the P value, it is generally preferable to report the 95% confidence interval (95% CI) for a given relative risk or odds ratio (depending on whether the study performed was a cohort study or case-control study, respectively). The 95% CI provides a range within which the true magnitude of the effect (ie, either the relative risk or the odds ratio) lies with a certain degree of assurance. Observing whether the 95% CI crosses 1.0 (ie, the value of null effect), provides the same information as the P value. If the 95% CI crosses 1.0, the P value will almost never be less than .05. In addition, the effect of the sample size can be ascertained from the width of the confidence interval. The narrower the confidence interval, the less variability was present in the estimate of the effect, reflecting a larger sample size. The wider the confidence interval, the greater the variability in the estimate of the effect and the smaller the sample size. When interpreting results that are not significant, the width of the confidence interval may be very helpful. A narrow confidence interval implies that there is most likely no real effect or exposure, whereas a wide interval suggests that the data are also compatible with a true effect and that the sample size was simply not adequate.

Special Issues in Healthcare Epidemiology Methods

Quasi-experimental Study Design

In addition to the study designs reviewed previously, the quasi-experimental study is a design frequently employed in healthcare epidemiology investigations.[16] This design is also frequently referred to as a "before-after" or "pre-post intervention" study.[17,18] The goal of a quasi-experimental study is to evaluate an intervention without using randomization. The most basic type of quasi-experimental study involves the collection of baseline data, the implementation of an intervention, and the collection of the same data after the intervention. For example, the baseline prevalence of VRE infection in a hospital would be calculated, an intervention to limit use of vancomycin would then be instituted, and, after some prespecified time period, the prevalence of VRE infection would again measured. Numerous variations of quasi-experimental studies exist and can include the following features: (1) use of multiple pretests (ie, collection of baseline data on more than one occasion), (2) use of repeated interventions (ie, instituting and removing the intervention multiple times in sequence), and (3) inclusion of a control group (ie, a group from whom baseline data and subsequent data are collected but for whom no intervention is implemented).

While often employed in evaluations of interventions in hospital infections, critical evaluation of the advantages and disadvantages of quasi-experimental studies has only recently been conducted.[16,19] A systematic review of 4 infectious diseases journals found that, during a 2-year period, 73 articles focusing on infection control and/or antimicrobial resistance used a quasi-experimental study design.[16] Of these 73 articles, only 12 (16%) used a control group, 3 (4%) provided justification for the use of the quasi-experimental study design, and 17 (23%) mentioned at least 1 of the potential limi-

tations of such a design.[16] Greater attention has recently been focused on increasing the quality of the design and performance of quasi-experimental studies to enhance the validity of the conclusions drawn regarding the effectiveness of interventions in the areas of infection control and antibiotic resistance.[16]

The quasi-experimental study design offers several advantages. Few study designs are available when one wishes to study the impact of an intervention. In general, a well-designed and adequately powered randomized controlled trial provides the strongest evidence for or against the efficacy of an intervention. However, there are several reasons why a randomized controlled trial may not be feasible in the study of infection control interventions. Randomizing individual patients to receive an infection control intervention is often not a reasonable approach, given the person-to-person transmission of resistant pathogens. One might consider randomizing specific units or floors within one institution to receive the intervention. However, these units are not self-contained, and patients and healthcare workers frequently move from unit to unit. Thus, any reduction noted in the number of transmissions or acquisitions of new drug-resistant infections in the intervention units is likely to also result in some reduction in the number of drug-resistant infections in nonintervention areas (ie, because of contamination). This would bias the results toward the null hypothesis (ie, the intervention had no effect). In such a situation, a well designed quasi-experimental study offers a compelling alternative approach. In addition, this study design is frequently used when it is not ethical to conduct a randomized controlled trial. Finally, when an intervention must be instituted rapidly in response to an emerging issue (eg, an outbreak), the first priority is to address and resolve the issue. In this case, it would be unethical to randomize patient groups to receive an intervention.

Potential limitations of quasi-experimental studies include regression to the mean, uncontrolled confounding, and maturation effects. Implementation of an intervention is often triggered in response to a rise in the rate of the outcome of interest above the norm.[20] The principle of regression to the mean predicts that such an elevated rate will tend to decline, even without intervention. This may serve to bias the results of a quasi-experimental study, as it may be falsely concluded that an effect is due to the intervention.[17,18] Several approaches may be employed to address this potential limitation. First, incorporating a prolonged baseline period prior to the implementation of the intervention permits an evaluation of the natural fluctuation in the rate of the outcome of interest over time and permits a more comprehensive assessment of possible regression to the mean. Second, changes in the rate of the outcome of interest may be measured at a control site (eg, another institution) during the same time period. Finally, the use of segmented regression analysis may assist in addressing possible regression to the mean, in that it will assess both the immediate change in prevalence coincident with the intervention and also the change in slope over time.[21-23]

Another potential limitation in quasi-experimental studies is uncontrolled confounding, which is most likely to occur when variables other than the intervention change over time or differ between the preintervention and postintervention periods.[17,18] This limitation can be addressed by measuring known confounders (eg, hospital census or number of admissions) and controlling for them in analyses. However, not all confounders are known or easily measured (eg, the quality of medical and nursing care). To address this, one may assess a nonequivalent dependent variable to evaluate the possibility that factors other than the intervention influenced the outcome.[16,19] A nonequivalent dependent variable should have similar potential causal and confounding variables as the primary dependent variable, except for the effect of the intervention. For example, in assessing the impact of an intervention to limit fluoroquinolone use on the prevalence of fluoroquinolone-resistant *Escherichia coli* infection, one might consider the incidence of catheter-associated bloodstream infection as a nonequivalent dependent variable. Although the prevalence of fluoroquinolone-resistant *Escherichia coli* infection and the incidence of catheter-associated bloodstream infection might both be affected by such factors as the patient census, it is unlikely that fluoroquinolone use specifically would affect the incidence of catheter-associated bloodstream infection.

Maturation effects are related to natural changes that patients experience with the passage of time.[17,18] In addition, there are cyclical trends (eg, seasonal variation) that may be a threat to the validity of attributing an observed outcome to an intervention. This potential limitation may be addressed through the approaches noted above, including assessment for a prolonged baseline period, use of control study sites, implementing interventions at different time periods at different sites, and assessing a nonequivalent dependent variable.

Control Group Selection in Studies of Antimicrobial Resistance

Many studies have focused on identifying risk factors for infection or colonization with an antimicrobial-resistant organism. The majority of these studies have had a case-control design. As noted previously, how controls are selected in case-control studies is critical in ensuring the validity of study results. Recent work has highlighted this issue of control group selection specifically for studies of antimicrobial-resistant pathogens.[5,24-27]

Two types of control groups have historically been used in studies of antimicrobial-resistant organisms.[5] The first type of control group is selected from patients who do not harbor the resistant pathogen. The second type of control group is selected from subjects infected with a susceptible strain of the organism of interest. For example, in a study of risk factors for infection with VRE in hospitalized patients, the first type of control group would be selected from among the general hospitalized patient population, whereas the second control group would be selected from among those patients infected with vancomycin-susceptible enterococci. The choice of control group should be based primarily on the clinical question being asked. Although use of this second type of control group (eg, patients infected with the susceptible form of the organism) has historically been a more common approach, it has recently been demonstrated that it may result in an overestimate of the association between antimicrobial exposure and infection with a resistant strain.[26,27] For our example of VRE infection, the postulated explanation for this finding is as follows: if the control patients are infected with vancomycin-susceptible enterococci, it is very unlikely that these patients would have recently received vancomycin (ie, the risk factor of interest), since exposure to vancomycin may have eradicated colonization with vancomycin-susceptible enterococci. Thus, the association between vancomycin use and VRE infection would be overestimated.[28] A limitation of using the first type of approach (ie, using patients without infection as controls) is that, in addition to identifying risk factors for infection with a resistant strain of the organism, this approach also identifies risk factors for infection with that organism in general (regardless of whether the strain is resistant or susceptible). Thus, there is no way to distinguish between the degree to which a risk factor is associated with being infected with the resistance phenotype and the degree to which it is associated with being infected with the organism in general.[25]

One concern with using the second type of control group (ie, selected from the group of all hospitalized patients) is the potential for misclassification bias. Specifically, subjects selected as controls who have never had a clinical culture performed may be in fact harbor unrecognized colonization with the resistant organism under study.[24] Since it is probable that patients colonized with the resistant organism would likely have had greater prior exposure to antimicrobials than did subjects not so colonized, this misclassification would likely result in a bias toward the null (ie, the case and control subjects would appear falsely similar with regard to prior antimicrobial use). Another concern with using the second type of control group (ie, identifying as controls those patients who have never had a clinical culture performed), is that differences between the case and control groups may reflect the fact that clinical cultures were performed for case patients but not for control subjects. Since procurement of samples for culture is not a random process but based on the clinical characteristics of patients, it is possible that the severity of illness or the level of antibiotic exposure may be greater among cases, regardless of the presence of infection with the antibiotic resistant organism.[5] One potential approach would be to limit eligible controls to those patients for whom at least 1 clinical culture has been performed and has not revealed the resistant organism of interest. Such a negative culture result would suggest that the patient is likely not colonized with the resistant organism. However, recent work has demonstrated that using clinical cultures to identify eligible controls leads to the selection of a control group with a higher comorbidity score and greater exposure to antibiotics, compared with a control group for whom clinical cultures were not performed.[24]

One proposed approach to addressing the difficulties in control-group selection in studies of infection with antimicrobial-resistant organisms is to use the case-case-control study design.[25,29-31] In this design, effectively 2 case-control studies are performed. In the first, cases are defined as those patients who harbor the resistant organism, and controls are defined as those patients who do not harbor the pathogen of interest. In the second, cases are instead defined as those patients harboring a susceptible strain of the pathogen of interest, and controls, as in the first approach, are defined as those patients who do not harbor the pathogen of interest.[25] These two separate studies are then carried out with risk factors from the two studies compared qualitatively. This approach allows for the comparison of risk factors identified from the 2 studies to indicate the relative contribution of the resistant infection, over and above simply having the susceptible infection. A potential limitation in this approach is the difficulty of matching for potential confounders. because of the use of only one control group.[25] Since there are 2 different case groups, variables relevant to the case group (eg, the duration of hospitalization and patient location) cannot be used for matching. In addition, the qualitative comparison of results from the 2 studies in this design leaves open the question of how much of a difference in results in meaningful.

Definitions of Antibiotic Exposure

Many studies have sought to uncover risk factors for infection or colonization with antimicrobial-resistant organisms.[8,32] Elucidating such risk factors is essential to inform interventions designed to curb the emergence of resistance. Past studies have particularly focused on antimicrobial use as a risk factor, as it can be modified in the clinical setting.[33,34] However, the approaches used to define prior antibiotic exposure vary considerably between studies.[5] Only recently have attempts been made to identify the impact of differences in these approaches on study conclusions.

A recent study by Hyle et al.[35] investigated methods used in past studies to describe the extent of prior antibiotic use (eg, presence or absence of exposure versus duration of exposure), as well as the impact of different methods on study conclusions. A systematic review of all studies investigating risk factors for harboring extended-spectrum β-lactamase–producing E. coli and Klebsiella species (ESBL-EK) was conducted. Of the 25 studies included, 18 defined prior antibiotic use as a categorical variable, 4 studies defined prior antibiotic exposure as a continuous variable, and 3 studies included both a categorical and a continuous variable to describe prior antibiotic exposure. Only 1 study provided an explicit justification for its choice of variable to describe prior antibiotic exposure. Hyle et al.[35] then reanalyzed a data set from a prior study of risk factors associated with ESBL-EK infection[36] and developed 2 separate multivariable models, one in which prior antibiotic use was described as a categorical variable (ie, exposure present or absent) and one in which antibiotic use was described as a continuous variable (ie, number of antibiotic-days of exposure). Results of the 2 multivariable models differed substantially: specifically, third-generation cephalosporin use was a risk factor for ESBL-EK infection when antibiotic use was described as a continuous variable but not when antibiotic use was described as a categorical variable.[35]

These results suggest that describing prior antibiotic use as a categorical variable may mask significant associations between prior antibiotic use and infection with a resistant organism. For example, when the categorical variable is used, a subject who received an antibiotic for only 1 day would be considered identical to a subject who received the same antibiotic for 30 days. However, the risk of infection with a resistant organism is almost certainly not the same in these 2 individuals. Describing prior antibiotic use as a continuous variable allows for a more detailed characterization of the association between length of exposure and presence of a resistant pathogen. Recent work in the medical statistics literature emphasizes that the use of cutoff values can result in misinterpretation and that dichotomizing continuous variables reduces analytic power and makes it impossible to detect nonlinear relationships.[37] Indeed, the relationship between prior antimicrobial use and infection with a resistant organism may not be linear (ie, the risk of such infection may not increase at a constant rate with increasing antimicrobial exposure). It is possible that the risk of infection with a resistant organism does not increase substantially until a certain amount of antimicrobial exposure has been attained (eg, a "lower threshold"). A more precise characterization of this "lower threshold" would serve to better inform antibiotic use strategies.

Another issue regarding defining prior antimicrobial use centers around how specific antimicrobial agents are grouped. For example, antibiotic use could be classified by agent (eg, cefazolin), class (eg, cephalosporins) or spectrum of activity (eg, activity against gram-negative organisms). Antibiotics are frequently grouped together in classes, even though individual agents within the class may differ significantly,[38] and such categorizations may mask important associations. It is unknown whether using different categorization schemes results in different conclusions regarding the association between antibiotic use and infection with a resistant organism. A recent study explored these issues, focusing on ESBL-EK infection as a model.[39] In a systematic review, 20 studies of risk factors for ESBL-EK infection that met the inclusion criteria revealed tremendous variability in how prior antibiotic use was categorized. Categorization of prior antibiotic use was defined in terms of the specific agents, drug class, and often a combination of both. No study justified its choice of categorization method. There was also marked variability across studies with regard to which specific antibiotics or antibiotic classes were assessed. As expected, a majority (16 studies) specifically investigated the use of β-lactam antibiotics as risk factors for ESBL-EK infection. A variable number of studies also examined the association between use of other antibiotics and ESBL-EK infection: aminoglycosides (9 studies), fluoroquinolones (10 studies), and trimethoprim-sulfamethoxazole (7 studies). In a reanalysis of data from a prior study of risk factors for ESBL-EK infection,[36] 2 separate multivariable models of risk factors were constructed: one with prior antibiotic use categorized by class and the other with prior antibiotic use categorized by spectrum of activity.[39] The results of these multivariable models differed substantially. Recent work has reported similar findings when focusing on risk factors for infection with carbapenem-resistant *Pseudomonas aeruginosa*.[40]

Another final important issue is how antibiotic use at a remote time is assessed. The recent systematic review of studies investigating risk factors for ESBL-EK infection (noted above)[35] found that the time window during which antibiotic use was reviewed ranged from 48 hours to 1 year prior to the onset of the drug-resistant infection. Furthermore, studies often did not explicitly state how far back in time prior antibiotic use was assessed.[35]

Conclusion

A basic understanding of epidemiologic principles and approaches is essential for the healthcare epidemiologist. The ability to compute measures of disease occurrence, to design and conduct appropriate studies to characterize the factors associated with disease, and to rigorously evaluate the results of such studies are increasingly vital functions of someone in this position. To build on the foundation provided in this chapter, the healthcare epidemiologist is encouraged to refer to more comprehensive texts and to consult with other professionals (eg, epidemiologists and biostatisticians) as needed.

References

1. Haley RW, Quade D, Freeman HE, Bennett JV. The SENIC Project. Study on the efficacy of nosocomial infection control (SENIC Project): summary of study design. *Am J Epidemiol* 1980;111:472–485.
2. Haley RW, Schaberg DR, McClish DK, et al. The accuracy of retrospective chart review in measuring nosocomial infection rates: results of validation studies in pilot hospitals. *Am J Epidemiol* 1980;111:516–533.
3. Freeman J, McGowan JE Jr. Methodologic issues in hospital epidemiology. I. Rates, case-finding, and interpretation. *Rev Infect Dis* 1981;3:658–667.
4. Freeman J, McGowan JE Jr. Methodologic issues in hospital epidemiology. II. Time and accuracy in estimation. *Rev Infect Dis* 1981;3:668–677.
5. D'Agata EM. Methodologic issues of case-control studies: a review of established and newly recognized limitations. *Infect Control Hosp Epidemiol* 2005;26:338–341.
6. Paterson DL. Looking for risk factors for the acquisition of antibiotic resistance: a 21st century approach. *Clin Infect Dis* 2002;34:1564–1567.
7. Schwaber MJ, De-Medina T, Carmeli Y. Epidemiological interpretation of antibiotic resistance studies—what are we missing? *Nat Rev Microbiol* 2004;2:979–983.
8. Harbarth S, Samore M. Antimicrobial resistance determinants and future control. *Emerg Infect Dis* 2005;11:794–801.
9. Agresti A. *Categorical Data Analysis*. Vol. 2. New York: Wiley Interscience; 2002.
10. Hennekens CH, Buring JE, Mayrent SL. *Epidemiology in Medicine*. 1st ed. Philadelphia: Lippincott Williams & Wilkins; 1987.
11. Hosmer DW, Lemeshow SL. *Applied Logistic Regression*. 2nd ed. New York: Wiley Interscience; 2000.
12. Kleinbaum DG, Kupper LL, Morgenstern H. *Epidemiologic Research: Principles and Quantitative Methods*. New York: Van Nostrand Reinhold; 1982.
13. Nelson KE, Williams CM, Graham NMH. *Infectious Disease Epidemiology: Theory and Practice*. 1st ed. New York: Aspen Publishers; 2000.
14. Rothman KJ, Greenland S. *Modern Epidemiology*. Philadelphia: Lippincott Williams & Wilkins; 1998.

15. Thomas JC, Weber DJ. *Epidemiologic Methods for the Study of Infectious Diseases.* 1st ed. Oxford: Oxford University Press; 2001.

16. Harris AD, Lautenbach E, Perencevich E. A systematic review of quasi-experimental study designs in the fields of infection control and antibiotic resistance. *Clin Infect Dis* 2005;41:77–82.

17. Shadish WR, Cook TD, Campbell DT. *Experimental and Quasi-experimental Designs for Generalized Causal Inference.* Boston: Houghton Mifflin Company; 2002.

18. Cook TD, Campbell DT. *Quasi-experimentation: Design and Analysis Issues for Field Settings.* Chicago: Rand McNally Publishing; 1979.

19. Harris AD, Bradham DD, Baumgarten M, Zuckerman IH, Fink JC, Perencevich EN. The use and interpretation of quasi-experimental studies in infectious diseases. *Clin Infect Dis* 2004;38:1586–1591.

20. Morton V, Torgerson DJ. Effect of regression to the mean on decision making in health care. *BMJ* 2003;326:1083–1084.

21. Ramsay CR, Matowe L, Grilli R, Grimshaw JM, Thomas RE. Interrupted time series designs in health technology assessment: lessons from two systematic reviews of behavior change strategies. *Int J Technol Assess Health Care* 2003;19:613–623.

22. Wagner AK, Soumerai SB, Zhang F, Ross-Degnan D. Segmented regression analysis of interrupted time series studies in medication use research. *J Clin Pharm Ther* 2002;27:299–309.

23. Matowe LK LC, Crivera C, Korth-Bradley JM. Interrupted time series analysis in clinical research. *Ann Pharmacother* 2003;37:1110–1116.

24. Harris AD, Carmeli Y, Samore MH, Kaye KS, Perencevich E. Impact of severity of illness bias and control group misclassification bias in case-control studies of antimicrobial-resistant organisms. *Infect Control Hosp Epidemiol* 2005;26:342–345.

25. Kaye KS, Harris AD, Samore M, Carmeli Y. The case-case-control study design: addressing the limitations of risk factor studies for antimicrobial resistance. *Infect Control Hosp Epidemiol* 2005;26:346–351.

26. Harris AD, Karchmer TB, Carmeli Y, Samore MH. Methodological principles of case-control studies that analyzed risk factors for antibiotic resistance: a systematic review. *Clin Infect Dis* 2001;32:1055–1061.

27. Harris AD, Samore MH, Lipsitch M, Kaye KS, Perencevich E, Carmeli Y. Control-group selection importance in studies of antimicrobial resistance: examples applied to *Pseudomonas aeruginosa,* Enterococci, and *Escherichia coli. Clin Infect Dis* 2002;34:1558–1563.

28. Carmeli Y, Samore MH, Huskins C. The association between antecedent vancomycin treatment and hospital-acquired vancomycin-resistant enterococci. *Arch Intern Med* 1999;159: 2461–2468.

29. Kaye KS, Harris AD, Gold H, Carmeli Y. Risk factors for recovery of ampicillin-sulbactam-resistant *Escherichia coli* in hospitalized patients. *Antimicrob Agents Chemother* 2000; 44:1004–1009.

30. Harris AD, Smith D, Johnson JA, Bradham DD, Roghmann MC. Risk factors for imipenem-resistant *Pseudomonas aeruginosa* among hospitalized patients. *Clin Infect Dis* 2002; 34:340–345.

31. Harris AD, Perencevich E, Roghmann MC, Morris G, Kaye KS, Johnson JA. Risk factors for piperacillin-tazobactam-resistant *Pseudomonas aeruginosa* among hospitalized patients. *Antimicrob Agents Chemother* 2002; 46:854–858.

32. Livermore DM. Can better prescribing turn the tide of resistance? *Nat Rev Microbiol* 2004;2:73–78.

33. Patterson JE. Antibiotic utilization: is there an effect on antimicrobial resistance? *Chest* 2001;119 (Suppl 2):426S-430S.

34. Safdar N, Maki DG. The commonality of risk factors for nosocomial colonization and infection with antimicrobial-resistant *Staphylococcus aureus, Enterococcus,* gram-negative bacilli, *Clostridium difficile,* and *Candida. Ann Intern Med* 2002;136:834–844.

35. Hyle EP, Bilker WB, Gasink LB, Lautenbach E. Impact of different methods for describing the extent of prior antibiotic exposure on the association between antibiotic use and antibiotic-resistant infection. *Infect Control Hosp Epidemiol* 2007;28:647–654.

36. Lautenbach E, Patel JB, Bilker WB, Edelstein PH, Fishman NO. Extended-spectrum β-lactamase-producing *Escherichia coli* and *Klebsiella pneumoniae*: risk factors for infection and impact of resistance on outcomes. *Clin Infect Dis* 2001;32: 1162–1171.

37. Royston P, Altman D, Sauerbrei W. Dichotomizing continuous predictors in multiple regression: a bad idea. *Stat Med* 2006;25:127–141.

38. Donskey CJ. The role of the intestinal tract as a reservoir and source for transmission of nosocomial pathogens. *Clin Infect Dis* 2004;39:219–226.

39. MacAdam H, Zaoutis TE, Gasink LB, Bilker WB, Lautenbach E. Investigating the association between antibiotic use and antibiotic resistance: impact of different methods of categorizing prior antibiotic use. *Int J Antimicrob Agents* 2006;28:325–332.

40. Gasink LB, Zaoutis TE, Bilker WB, Lautenbach E. The categorization of prior antibiotic use: Impact on the identification of risk factors for drug resistance in case control studies. *Am J Infect Control* 2007;35:638–642.

Chapter 5 Quality Improvement in Healthcare Epidemiology

Susan MacArthur, RN, CIC, CPHQ, MPH,
Frederick A. Browne, MD,
and Louise-Marie Dembry, MD, MS, MBA

The centers for disease control and Prevention (CDC) estimates that 5%–10% of hospitalized patients develop a healthcare-associated infection.[1] Healthcare-associated infections are an important measure of quality, and epidemiologists play a significant role by leading initiatives to prevent them. There is increasing pressure on hospital epidemiology and infection prevention programs to demonstrate compliance with evidence-based standards and to reduce the incidence of healthcare-associated infections to the "irreducible minimum." This chapter is a primer on quality improvement principles and techniques. These quality principles, when integrated with epidemiologic techniques, strengthen infection prevention programs and ensure patient safety and high-quality patient care.

The Birth of Quality Improvement

According to the American Society for Quality,[2] the birth of quality improvement in the United States resulted from the quality movement in Japan after World War II, when Japanese manufacturers were converting from producing military goods to producing civilian products for trade. Japan had a widely held reputation for poor quality products, which negatively impacted their ability to sell their products internationally. Japanese organizations sought new ways to improve the quality of their products, and these strategies represented the new "total quality" approach. Rather than relying purely on product inspection to ensure quality, Japanese manufacturers focused on improving all organizational processes through the people involved in them. This led to higher-quality products at lower prices, which made Japanese exports competitive in the world market.

Two American quality experts, W. Edwards Deming and Joseph M. Juran, were aware of Japan's progress in the quality arena and predicted that the quality of Japanese goods would overtake the quality of goods produced in the United States by the mid-1970s.[2] Japanese manufacturers began increasing their share in American markets, causing widespread economic effects in the United States. "Total quality management," which emphasized approaches that went beyond just statistics and embraced the entire organization, came out of this movement. Several other quality initiatives, such as "continuous quality management" followed. American companies were slow to adopt the principles of quality improvement,[2] and hospitals were even slower.

The Rise of Handwashing: A Quality Improvement Exercise

As epidemiologists, one of the first stories we are taught is that of Dr. Ignaz Semmelweis, a Hungarian obstetrician who in the 1840s showed that cases of puerperal fever could be prevented if doctors washed their hands in a chlorinated lime solution before examining patients.[3] His approach closely followed what we now consider modern quality improvement techniques. First, Semmelweis identified that a problem existed by examining mortality rates in 2 different obstetrical services. Next, he hypothesized that puerperal fever was contagious, and he made observations and collected data. He then implemented a practice change, or intervention (hand-washing); last, he demonstrated a significant decrease in infection (mortality) through continuous monitoring. Unfortunately, his findings were neither widely adopted nor were his recommendations implemented by the medical community, and as a result, women in 19th-century Vienna continued to die a preventable death. Even today, the existence of evidence-based practices and quality improvement techniques do not guarantee their adoption nor the successful prevention of errors.

The Birth of Hospital Epidemiology and Infection Prevention Programs

During the 1950s, epidemics of staphylococcal infection and nosocomial infections in hospitals emerged as a major public health issue. At the urging of the American Hospital Association, the CDC, and the Joint Commission (then called the Joint Commission on Accreditation of Healthcare Organizations [JCAHO]), infection control programs were instituted in thousands of hospitals across the country during the 1960s and 1970s. Each program implemented its own prevention and control strategies, with little evidence to definitively determine which interventions, if any, effectively reduced the incidence of infections and the associated costs. In 1974, the CDC initiated the 10-year Study on the Efficacy of Nosocomial Infection Control (SENIC).[4] The study showed that the incidence rate of nosocomial infections decreased and remained lower in hospitals that conducted surveillance for nosocomial infection and that used evidence-based infection prevention patient care practices. As hospital epidemiology and infection prevention developed as a discipline, quality principles were incorporated and became integral to the functioning of many programs.

The Perfect Storm

During the 1980s and early 1990s, as hospital epidemiology programs continued to evolve, they were impacted by the emergence of the human immunodeficiency virus and the implementation of universal precautions and the Occupational Safety and Health Administration's Bloodborne Pathogens Standard,[5] as well as the resurgence in tuberculosis, including nosocomial outbreaks of multidrug-resistant tuberculosis. At the same time, hospital epidemiology programs were tasked to address the increasing burden of multidrug-resistant organisms and other emerging organisms that posed a threat to hospitals. It was the Institute of Medicine's reports—in 1999, *To Err Is Human: Building a Safer Health System*[6]; in 2000, *Crossing the Quality Chasm: A New Health System for the 21st Century*[7]; and in 2003, *Transforming Health Care Quality*[8]—that targeted the prevention of hospital-acquired infections as a high priority for American hospitals. Very quickly, hospital epidemiology programs found themselves in the crosshairs of hospital administrators, quality management departments, managed-care entities, regulatory and accrediting agencies, lawmakers, and an increasingly worried public, all requesting evidence of effectiveness and increased accountability.

Although collaboration with other disciplines is nothing new to hospital epidemiology, this "perfect storm" sometimes required epidemiologists and infection preventionists to use a different language and methodology. "Multidisciplinary collaborations are essential to instigate innovative prevention research, identify new applications for old prevention strategies, maximize synergy among the broad array of professionals engaged in quality promotion efforts, minimize overlap, and conserve scarce resources."[9(p366)] Table 5-1 presents perspectives on the quality of health care from the points of view of various stakeholders.

Table 5-1. Perspectives on healthcare quality from the point of view of various stakeholders

Perspective	Infection control	Quality management	Managed care and accreditation agencies
Focus	Adverse health events	Indicators	Errors, near misses
Determinants	Risk factors	Patient mix	Root cause analysis, human factors
Monitoring	Surveillance, response	Performance measurement, improvement	Reporting, learning
Goal	Prevention	Performance improvement	System improvement
Key professionals	Healthcare epidemiologists, infection preventionists	Quality managers, accreditation officials	System engineers, healthcare purchasers and consumers

NOTE: Information is from Gerberding.[9]

The Institute of Medicine's *To Err is Human*[6] presented an overall framework to improve the delivery of health care. The concepts in this framework can be easily applied to infection prevention strategies as well:

Safe: "avoiding injuries to patients from the care that is intended to help them" (eg, ensuring compliance with hand hygiene)

Timely: "reducing waits and sometimes harmful delays for both those who receive and those who give care" (eg, administering surgical antibiotic prophylaxis 1 hour prior to the first incision)

Effective: "providing services based on scientific knowledge to all who could benefit and refraining from providing services to those not likely to benefit" (eg, use maximal sterile barrier precautions during central venous catheter insertion)

Efficient: "avoiding waste, including waste of equipment, supplies, ideas and energy" (eg, replace administration sets not used for blood, blood products, or lipids at intervals not longer than 96 hours)

Equitable: "providing care that doesn't vary in quality because of personal characteristics such as gender, ethnicity, geographic location, and socioeconomic status" (eg, implementation of standard precautions and transmission-based precautions)

Patient-centered: "providing care that is respectful of and responsive to individual patient's preferences, needs, and values ensuring that patient values guide all clinical decisions." (eg, partnering with patients and family members to "speak up" and ask questions, and educating them about what they can expect from healthcare providers and what is being done to reduce their risk of healthcare-associated infection)

The authors of *To Err is Human* postulated that "a health care system that achieves major gains in these six areas would be far better at meeting patient needs. Patients would experience care that is safer, more reliable, more responsive to their needs, more integrated, and more available, and they could count on receiving the full array of preventive, acute, and chronic services that are likely to prove beneficial. Clinicians and other health workers also would benefit through their increased satisfaction at being better able to do their jobs and thereby bring improved health, greater longevity, less pain and suffering, and increased personal productivity to those who receive their care."[6]

Public Reporting and Transparency

Healthcare consumers, regulatory and accreditation agencies, and the insurance industry, including managed-care entities, have an increasing interest in greater transparency in the reporting of healthcare-related outcomes. Infection prevention outcomes may include rates of central line–associated bloodstream infection, ventilator-associated pneumonia, surgical site infection, or hand hygiene compliance. The desire for transparency is based on the belief that these indicators can be used to evaluate those organizations that offer the highest quality of care most cost-effectively. Reported data are becoming increasingly available to the public by way of the Internet. However, questions remain about the accuracy of the reported price, process, and outcome information; the comparability of the results across different populations; and whether and how patients and others use the information in making decisions.[10]

The Concept of Value in Health Care

When we shop for a car, or anything for that matter, we want to purchase the item that gives us the best value. By value, we mean the best quality that we are able to afford. For health care, assigning value is a more difficult task but is no less important. We have learned that better care does not always mean higher-cost care, and providers are facing steadily increasing pressure to take excess cost out of the system (ie, reduce waste) while maintaining or increasing the quality of care.[11]

A Culture of Safety

The most effective quality programs provide an atmosphere that clearly places patient safety at the center of all that is done. Expectations are that processes will be in place that will enable healthcare providers to "do the right thing, the right way, the first time, every time." When an error is made, a "no name, no blame, no shame" culture encourages a focus on the improvement of processes, not a focus on individual persons. As a step toward greater transparency, some organizations invite members of the public to join patient safety committees and other initiatives.

Quality Tool Kit

As with any discipline, quality management has developed tools that are routinely used for cause analysis, data collection and analysis, evaluation and decision making, idea creation analysis, and project planning and implementation. In this section we discuss 7 basic tools of quality. A list of Web sites that contain a wealth of information relevant to quality control is given in Table 5-2.

Table 5-2.   Web sites with information relevant to quality control

Organization	URL
Centers for Disease Control and Prevention (CDC)	http://www.cdc.gov/
Institute for Healthcare Improvement (IHI)	http://www.ihi.org/ihi
Agency for Healthcare Research and Quality (AHRQ)	http://www.ahrq.gov/
Centers for Medicare and Medicaid Services (CMS)	http://www.cms.hhs.gov/
The Joint Commission (TJC)	http://www.jointcommission.org/
National Committee for Quality Assurance (NCQA)	http://www.ncqa.org/
American Society for Quality (ASQ)	http://www.asq.org/learn-about-quality/data-collection-analysis-tools/overview/overview.html
The American Society for Healthcare Risk Management (ASHRM)	http://www.ashrm.org/
The World Health Organization (WHO)	http://www.who.int
The Robert Wood Johnson Foundation (RWJF)	http://www.rwjf.org
Web sites of individual state health departments	. . .

Brainstorming

Brainstorming, affinity grouping, and multivoting are tools for generating, categorizing, and choosing among ideas from a group of people.[12] There are a number of potential benefits from using these techniques:

- Every member of the group participates.
- Many people contribute, instead of just 1 or 2 people.
- Creativity is sparked as group members listen to the ideas of others.
- A substantial list of ideas is generated, rather than just the first few things that come to mind; it categorizes ideas creatively; and it allows the group to thoughtfully choose among ideas or options.

Cause-and-Effect Diagrams

A cause-and-effect diagram, also known as an Ishikawa or "fishbone" diagram (Figure 5-1), is a graphic tool used to explore and display the possible causes of a certain effect.[13] It is used when causes group naturally under the 5 categories of materials, methods, equipment, environment, and personnel. The causes of problems at each step in the process can be shown by a process-type

cause-and-effect diagram. The benefits include the following:

- Assisting teams in understanding that there are many causes that contribute to an effect
- Providing a graphic display of the relationship of the causes to the effect and to each other
- Helping to identify areas for improvement

Control Charts

Control charts are graphs that plot data in temporal sequence, to show how a process changes over time (Figure 5-2). A control chart always has a central line for the mean value, an upper line for the upper control limit, and a lower line for the lower control limit; these lines are determined from historical data. Current data are plotted and compared with these lines. From the graph, one determines whether the process variation is consistent (in control) or is unpredictable (out of control, or affected by special causes of variation). Specific data points may signal that a process is out of control and may require action. Such signals are suggested by the following features[14(pp155–158)]:

- A single point is outside the control limits
- Two of 3 successive points are on the same side of the center line and farther than 2 standard deviations from it
- Four of 5 successive points are on the same side of the center line and farther than 1 standard deviation from it
- A run of 8 points in a row are on the same side of the center line, or 10 of 11, or 12 of 14, or 16 of 20 points
- Obvious consistent or persistent patterns suggest something unusual about the data and the measured process

Flow Charts

A flow chart shows the separate steps of a process in sequence (Figure 5-3). The following elements may be included[14(pp255–257)]:

- The sequence of actions
- Materials or services entering or leaving the process (inputs and outputs)
- Decisions that need to be made
- People who need to be involved
- The time involved at each step
- Process measurements

Flow charts are a generic tool that can be adapted for a wide variety of purposes and processes (eg, manufacturing, administrative, or service processes or a project plan)[14(pp255–257)]:

- To understand how a process is performed
- To study a process for improvement
- To communicate to others how a process is performed
- To improve communication between people involved with the same process
- To document a process

One should not be concerned about drawing the flow chart the "right way." The right way is one that helps those involved understand the process. All key persons involved with the process should be included in developing the flow chart. This includes involving them in the actual flow chart drafting sessions, interviewing them before the sessions and/or showing them the developing flow chart between work sessions, to get their feedback. The people who actually perform the process, not a "technical expert," should draw the flow chart.[14(pp255–257)]

Histograms

A histogram is a type of bar chart used to display variation in continuous data, such as time, weight, size, or temperature. A histogram assists in the recognition of patterns that may not be apparent when data are presented in a table format, or from the mean or median values of a data set. A frequency distribution shows how often each different value in a data set occurs. Although histograms resemble bar charts, there are important differences between them. A histogram should be used for the following purposes:[14(pp292–299)]

- To display numerical data
- To see the shape of the distribution of the data, especially when determining whether the output of a process is distributed approximately normally
- To see whether a process change has occurred from one time period to another
- To determine whether the outputs of 2 or more processes are different
- To communicate the distribution of data in a quick and easy format

Pareto Charts

The "Pareto Principle" refers to the concept that, in any group of things that contribute to a common effect, a relatively few contributors account for the majority of the

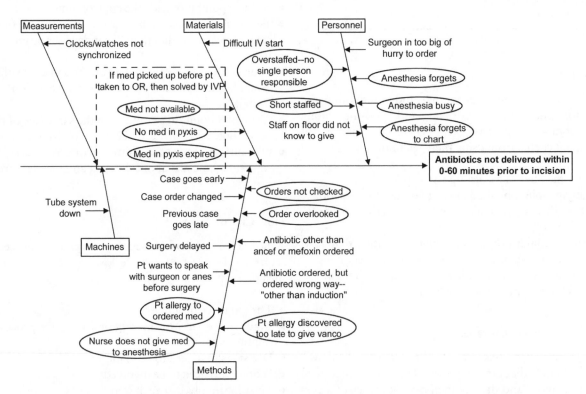

Figure 5-1. An example of a cause-and-effect diagram, also known as an Ishikawa or "fishbone" diagram, in this instance showing the causes contributing to the failure to administer antibiotic prophylaxis before surgical incision. *Ovals* designate causes not solved by intravenous prophylaxis (IVP) on induction. Ancef, cefazolin; anes, anesthesiologist; IV, intravenous line; med, medication; OR, operating room; pt, patient; vanco, vancomycin. Reproduced with permission of M. Jones, BJC HealthCare (St. Louis, Missouri).

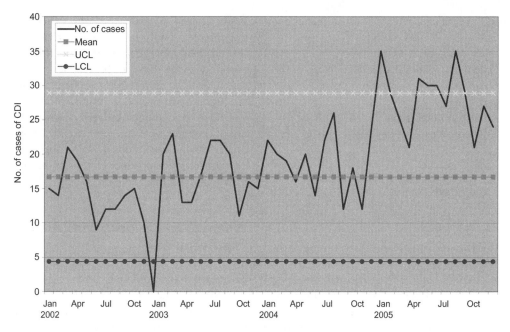

Figure 5-2. An example of a control chart, in this instance showing the number of cases of *Clostridium difficile* infection (CDI) each month during 4 years *(solid line)* with the upper control limit (UCL) and lower control limit (LCL) *(symbol lines)*.

effect. It is commonly known as the "80/20 Rule" meaning that roughly 80% of the effect comes from 20% of the causes. A Pareto diagram, which is a type of bar chart, displays the various factors that contribute to an overall effect (Figure 5-4). The factors are arranged in order according to the magnitude of their effect. This helps distinguish the factors that warrant the most attention (the "vital few"), which a team should concentrate on to achieve the greatest impact, from factors that may be useful to know about (the "useful many") but that have a relatively smaller effect.[15]

Prioritization Matrices

A priority matrix is a valuable tool to use when a complex decision needs to be made and when many variables must be considered before making the decision. Each variable is described using weighted criteria that allow the team to rate each possible decision in an objective manner.[16]

Quality Management Tools

Root Cause Analysis

A root cause analysis is conducted when a serious error occurs, to find out what happened, why it happened, and what can be done to prevent it from happening again. Root cause analysis is a *tool* for identifying

prevention strategies, and it is a *process* that is part of building a culture of safety and moving beyond a culture of blame. In root cause analysis, basic and contributing causes are discovered with the goal of preventing a recurrence.

The basic features of a root cause analysis are as follows[17]:

1. It is interdisciplinary, involving experts from the frontline services.
2. It involves those who are the most familiar with the situation.
3. It continuously digs deeper by asking "why, why, why" at each level of cause and effect.
4. It is a process that identifies changes needed to be made to systems.
5. It is a process that strives to be impartial.

The Joint Commission requires that root cause analyses be thorough and credible.[17] To be thorough, a root cause analysis must have the following features:

1. It must determine human and other factors.
2. It must determine related processes and systems.
3. It must analyze underlying cause and effect systems through a series of "Why?" questions.
4. It must identify risks and their potential contributions.
5. It must determine potential improvement in processes or systems.

To be credible, a root cause analysis must have the following features:

1. It must include participation by the leadership of the organization and those most closely involved in the processes and systems.
2. It must be internally consistent.
3. It must include consideration of the relevant literature.

Failure Modes and Effects Analysis

Failure modes and effects analysis was initially used in the 1940s by the US military and was later developed further by the aerospace and automotive industries.[14(pp236–240)] It is a step-by-step method used to identify all the possible failures in a design, a process, a product, or a service. It is different from root cause analysis in that it is used when no error has actually occurred; rather, it is undertaken to look at ways in which something might fail and to study the consequences of those potential failures. Failures are prioritized according to how serious their consequences are, how frequently they occur, and how easily they can be detected. The goal is to institute actions to eliminate or reduce failures, starting with the highest-priority ones. Failure modes and effects analysis also documents current knowledge and actions about the risks of failures for use in continuous improvement. Ideally, the analysis begins during the earliest conceptual stages of product or service design and continues throughout its life.

Where to Begin: Identifying Areas for Improvement

There are many guidelines and recommendations for the prevention of healthcare-associated infections; however, they tend to be very detailed, not user friendly, and confusing to implement. To assist acute care hospitals in

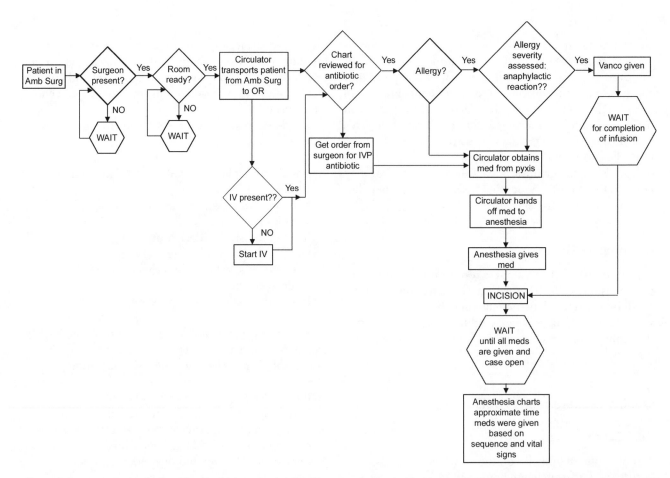

Figure 5-3.   An example of a flow chart, in this instance showing the steps associated with administration of antibiotic prophylaxis before surgical incision. Amb Surg, ambulatory surgery; IV, intravenous line; IVP, intravenous prophylaxis; med, medication; OR, operating room; vanco, vancomycin. Reproduced with permission of M. Jones, BJC HealtCare (St. Louis, Missouri).

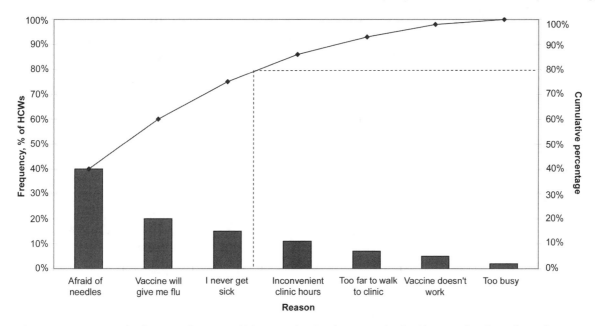

Figure 5-4. An example of a Pareto diagram, in this instance showing the reasons that healthcare workers (HCWs) most frequently give for refusing influenza vaccination (*bars* and *left axis*) and identifying the reasons that contribute 80% of the effect (*curved line* and *right axis*) and therefore warrant the most attention.

focusing and prioritizing efforts related to evidence-based practices for the prevention of healthcare-associated infections, the Society for Healthcare Epidemiology of America (SHEA) and the Infectious Diseases Society of America (IDSA) Standards and Practice Guidelines Committee appointed a task force that assembled a concise compendium of recommendations that is focused on implementation of interventions, which was published in 2009.[18] The healthcare-associated infections compendium also includes proposed performance measures for internal quality improvement efforts and recommends that accountability for implementing infection prevention practices should be assigned to specific groups and individuals in the organization.

The SHEA/IDSA compendium[18] targets 4 categories of healthcare-associated infections associated with devices and procedures (central line–associated bloodstream infection, ventilator-associated pneumonia, catheter-associated urinary tract infections, and surgical site infection) and 2 categories of healthcare-associated infection based on the infecting organism (methicillin-resistant *Staphylococcus aureus* infection and *Clostridium difficile* infection). Although the compendium is a great beginning, other topics may also be considered for process improvement projects, such as compliance with hand hygiene, immunization rates among patients and healthcare workers, planning and preparation for a surge of patients with communicable diseases, or an organizational response to infection with novel agents (eg, influenza A virus subtype H1N1, or "swine flu").

The Quality Improvement Process

Healthcare organizations operate in a milieu where the patient mix, patient risk profiles, medical care delivery systems, and medical technology are ever changing. Thus, to remain on a path of continuous improvement, organizations should always be asking themselves the following questions:

1. Why do we use this process?
2. Does it add value?
3. How can we improve outcomes?
4. What tools can be used to help make a lasting change?

Most quality improvement models have similar components, including multidisciplinary participation, planning before action, testing before wide-scale implementation, and using continuous as well as rapid measurement to provide feedback on progress to all stakeholders.

Quality improvement recognizes that both resources (inputs) and activities (processes) must be addressed together to ensure that the quality of care is high or improve it. In general, quality improvement processes have 4 key steps[19]:

Step 1. Identify the issue, and plan and prioritize.
Step 2. Collect and analyze data to further the understanding of the problem; hypothesize what changes

will resolve the problem and develop a solution strategy.

Step 3. Test and deploy: test the hypothesized solution with a small sample that becomes progressively larger.

Step 4. Report and adjust: compare results with internal and external benchmarks and make adjustments to the process to move closer to the desired goal.

One of the most challenging aspects of change is sustaining an intervention until it is "hard-wired" into the fabric and workings of the organization, such that it becomes part of the organizational routine.

Before starting any quality improvement process, it is wise to invest time in assembling and aligning a team to lead the initiative. A typical team is made up of a change sponsor or sponsors (usually a senior-level administrator), a change agent or agents (usually a mid-level administrator or manager), and "champions" (well-respected persons "in the trenches" who advocate for and model the change or intervention). Champions can be especially valuable since their credibility with colleagues and their positive expressions of support can enhance the receptivity of others to new ideas. It is important to ensure that all members of the team can clearly articulate what the change (intervention) is, why it is being made, and what success will look like. Each team member needs to understand their role and responsibilities in the change implementation process.[20]

Step 1: Identify the Issue, and Plan and Prioritize

Determine what needs to be improved then prioritize and plan the intervention and set goals. "What to improve" may involve a problem that needs a solution, an opportunity for improvement, or a process or system that needs to be changed. The change (improvement) process begins with asking the following questions:

- What is the problem?
- How do you know that it is a problem?
- How frequently does it occur, or how long has it existed?
- How will we know when it is resolved?

Once a project is identified, a plan on how to address it should be developed. The plan should define the team members, a list of objectives that meet the "SMART" criteria of the CDC (discussed in the next paragraph), the deadlines for deliverables, and the end goals. The end goal(s) should be well defined, using real numerical targets when feasible, and should be clearly stated ahead of time.

Objectives are mileposts that the project needs to achieve in order to accomplish its goals by the end of a defined period of time. Objectives are the basis for monitoring implementation of strategies and monitoring progress toward achieving the project goals. Objectives also help set targets for accountability and can be used to evaluate the project as it progresses. "SMART" is an acronym used by the CDC to summarize the 5 key qualities of an objective: it is specific, measurable, achievable, realistic, and time-phased.[21]

1. Specific
   - Objectives should provide the "who" and "what" of project activities.
   - Use only a single action verb, since objectives with more than one verb imply that more than one activity or behavior is being measured.
   - Avoid verbs that may have vague meanings to describe intended outcomes (eg, "understand" or "know"), since it may prove difficult to measure them. Instead, use verbs that document action (eg, "At the end of the session, the students will list 3 concerns...").
   - The greater the specificity, the greater the measurability.
2. Measurable
   - The focus is on how much change is expected. Objectives should quantify the amount of change expected in order to determine whether an objective has been met.
   - The objective provides a reference point from which a change in the target population can clearly be measured.
3. Achievable
   - Objectives should be attainable within a given time frame and with the available resources.
4. Realistic
   - Objectives should accurately address the scope of the problem and steps that can be implemented within a specific time frame.
   - Objectives should be directly related to the program goal.
5. Time-phased
   - Objectives should provide a time frame indicating when the objective will be measured or a time by which the objective will be met.
   - Including a time frame in the objectives helps in planning and evaluating the program.

Step 2: Collect and Analyze Data

Once an area for improvement has been identified, the second step is to collect data. Clear, well-crafted definitions for what will be measured and how it will be measured should be developed. Analyze the data before considering changes or interventions. It is important to study the details of the process and to carefully measure the end points for future "benchmarking" (defined below, in Step 4). Recruiting performance-improvement specialists can be very valuable in this step. Their experience and knowledge can make the difference between a well-designed intervention that is successful and one whose design is weak and not likely to succeed, leading to the inefficient use of resources and time and ultimately to team frustration and discouragement.

The objectives of the analysis stage can be any combination of the following:

1. Clarifying why the process or system produces the effect
2. Measuring the performance of the process or system that produces the effect
3. Formulating research questions, such as:
   A. Who is involved or affected?
   B. Where does the problem occur?
   C. When does the problem occur?
   D. What happens when the problem occurs?
   E. Why does the problem occur?

To reach these objectives, this step requires the use of existing data or collection of additional data. Data should be analyzed over an appropriate period of time to ensure the sample represents a true picture for the change initiative. The extent to which data are needed depends on the quality improvement approach chosen. Techniques may include clarifying processes through flow charts or cause-and-effect analyses, reviewing existing data, and collecting additional data if needed.

Step 3: Test and Deploy the Hypothesized Solution

This step uses the information gathered during the first 2 steps of the cycle. A hypothesis is formulated about which changes, interventions, or solutions would have the desired impact on the problem. A strategy is developed on the basis of the hypothesis generated. Developing and deploying a pilot test is often beneficial. The pilot helps identify if the intervention will likely help achieve the stated goals without expending a lot of resources. If the pilot is successful, a full-scale deployment should then be initiated.

Step 4: Report and Adjust

Once a process change has been made, a summary report should be prepared that documents the changes, procedures, findings, and performance levels for the intervention. This is an appropriate time to compare results with a "benchmark," if one is available. "Benchmarking" is the process of identifying the best practices and comparing performance relative to that reported by or observed at other hospitals or organizations. Benchmarking takes one of two forms: comparison on the basis of first-hand observations of organizations, or direct comparison with secondary published data. Using published data is the most common way to compare performance across multiple organizations simultaneously, although detailed on-site benchmarking visits to other hospitals can be very valuable and are worth considering.

Benchmarking involves 4 primary steps:

1. Select appropriate hospitals or other organizations to use for comparison.
2. Collect or observe data and processes.
3. Identify sources of differential performance.
4. Incorporate these benchmarks into performance scorecards and daily management processes.

Performance should be monitored and tracked continuously to ensure that the results achieved during the initial deployment period continue and that problems are addressed immediately to keep the change initiative on track. The feedback and adjustment process should be continued for at least 3–6 months.

Assessments of Benefit, Risk, and Cost

Quality improvement projects should include an assessment of benefit, risk, and/or cost, depending on the type of intervention being planned. Economic assessment of technologies, clinical programs and practices, and patient devices and equipment should be conducted before introducing them into clinical practice. Table 5-3 describes various types of benefit, risk, and economic assessments and when they should be utilized.[22]

Cycle of Improvement

The quality improvement process can be time consuming and can take many months, if not longer, to complete. This can lead to disengagement of team members, loss of momentum, and loss of interest in the project over time. Disadvantages of traditional quality improvement methods include the need to collect a large amount of data (baseline data) before planning the intervention; the fact that testing interventions and collecting data to document change can take a long time; the possible difficulty of determining which change contributed to improvement; and the fact that, if the intervention proves unsuccessful, then improvement in quality has been delayed. Rapid cycle improvement[23] is a methodology that attempts to address these issues and keep

Table 5-3. Various types of benefit, risk, and economic assessments and when they should be utilized

Type of assessment	Definition and comment
Cost-benefit analysis	Analysis of healthcare resource expenditures relative to possible medical benefit. This analysis may be helpful and necessary in setting priorities when choices must be made in the face of limited resources. This analysis is used in determining the degree of access to, or benefits of, the health care to be provided. Clarity of the medical decision-making process demands that cost-benefit analysis be separated and differentiated from risk-benefit analysis, as well as from determinations of efficient and cost-effective medical care during medical decision making.
Risk-benefit analysis	Weighs the potential for undesirable outcomes and side effects against the potential for positive outcomes of a treatment and is an integral part of the process of determining medical necessity in the delivery of quality medical care.
Efficient medical care	Is correlated to the timeliness of medically necessary services and supplies that are delivered at the least cost and are consistent with the applicable standard of care.
Cost-effective medical care	The selection of the least expensive medically necessary treatment from 2 or more that are equally efficacious in achieving the desired healthcare outcome.

NOTE: Information is from American Medical Association policy compendium.[22]

improvement activities moving. The goal is to pilot-test ideas rapidly using small sample sizes. Interventions are tested alongside the current process, many ideas can be tested at the same time, unsuccessful interventions are recognized sooner and abandoned, and the rapidity of the process and opportunity to show positive results quickly helps overcome the inevitable resistance to change. The Shewhart/Deming "plan-do-check-act" cycle of improvement, first used in business process improvement, can be used for rapid cycle improvement projects.[23] It follows a trial-and-error framework that is easily adapted for this purpose. It has 4 steps for problem solving, starting with a "plan" stage, in which a process needing change is identified. The "do" stage is the intervention; a certain level of imperfection is accepted as the experience itself can lead to learning and improvement. The "check" stage is used to focus on what worked well and what did not. The "act" stage includes an analysis of the difference between actual and expected results. The process is then repeated, constantly flowing in a circular manner.

Industry to Health Care

Many quality improvement strategies that have been used more recently in business and manufacturing have also been applied over the past decade in health care. They include Six Sigma, "lean production," and the "balanced scorecard." Concerns regarding barriers to translation of such methods from industry to health care have largely been overcome with experience and more widespread use in healthcare settings.

Six Sigma

Six Sigma is a methodology developed at Motorola in the 1980s and further developed by General Electric to improve processes and eliminate defects in performance by focusing on sigma (ie, standard deviations from the mean).[23,24] It aims to reduce variation, to achieve stable and predictable process results. Achieving a level of 6 sigma implies near perfection in an operational process; only 3.4 defects per 1 million opportunities, or 99.99966% accuracy. Most processes in industry function at 3 sigma (66,807 defects per 1 million opportunities). In health care, 1 patient represents 50–100 opportunities for error ("defects") during each hospital stay.

Although Six Sigma follows quality management and statistical methods, it offers a leadership approach to changing the culture of an organization.[23] Commitment from the entire organization, most importantly from the highest levels of the organization, is necessary to sustain improvements in quality. Six Sigma is a highly structured process that involves staff training for different learning levels or "belts," from yellow to green and then on to black. Teams are made up of the process owner, the staff who have been trained and have a "belt" rating, and the staff responsible for the process. Teams follow a process improvement methodology abbreviated DMAIC: define, measure, analyze, improve, and control.[23,24] It starts by identifying the problem, then measuring the problem, identifying the root cause of the problem, mitigating the root cause, and finally maintaining the gains. This methodology focuses on finding sources of variation inherent to a process and eliminating them to achieve more-consistent results.

Lean Process

"Lean process" is a quality improvement method pioneered by Toyota that focuses on improving quality by dramatically changing operational processes to become faster and more flexible and to reduce waste.[23] It does this by identifying and reducing steps that do not add value in a process. Waste is any non–value added activity that can be classified into any of the following 8 categories: transportation, motion, waiting, processing, inventory, overproduction, corrections, and defects. In health care, there are several time components of work that do not add value and that affect an organization's ability to function with speed and flexibility:

1. Process time: actual time spent performing work
2. Idle time: time when patients and staff are not performing work
3. Wait time: time spent waiting because of lines that form in parts of the facility
4. Transit time: time spent walking from one department or unit to another
5. Transition time: time interval necessary between productive work where a conversion, clean up, or changeover prepares a resource to switch from one state to another

Transition time is one of the largest components of waste in health care and the target of "lean" projects.

Balanced Scorecard

The "balanced scorecard" is a planning and management tool used to align an organization's activities to its vision and business strategy, to improve internal and external communications, and to monitor organization performance.[23,24] It translates an organization's vision and mission into actionable activities at all levels. The balanced scorecard was originated by Dr. Robert Kaplan and Dr. David Norton as a performance measurement framework that added strategic nonfinancial performance measures to traditional financial metrics to give managers and executives a more "balanced" view of organizational performance.[23] Typically, performance is looked at from 4 different perspectives or domains: financial performance, customer opinions and perspective, internal operational performance, and employee learning and growth. In clarifying an organization's key objectives, the balanced scorecard method leads to improved prioritization and alignment of quality improvement efforts. It is not unusual for infection prevention measures to be reported on an institution's balanced scorecard (eg, the incidence of catheter-related bloodstream infection and the rate of hand hygiene compliance).

Conclusion

Traditionally, hospital epidemiology departments and quality departments had little overlap, despite the fact that they both worked toward improving patient safety and the quality of care. To many epidemiologists, infection prevention has always been the premier program for quality promotion in US hospitals.[25] One significant difference today is that the activities of an infection prevention program are much more visible, and the expectation of excellent outcomes has never been higher.

To quote Julie Gerberding, former director of the CDC, "The core activities in health-care epidemiology and infection control—cluster and outbreak investigations, case-control studies to identify risk factors, surveillance and response, laboratory investigation, intervention efficacy and effectiveness studies—are tools with broad applicability to many domains of health-care quality. We can lend these tools to our colleagues in other disciplines and, in turn, benefit from their tools—root cause analysis, human factors research, hazards analysis, economic assessment—as we pursue common goals. We have a unique opportunity to experience, and, more importantly, to lead the development of consilience, the linkage of facts and fact-based theory across disciplines to create a common basis for new explanation or action, in health-care quality promotion."[9(p366)]

Acknowledgment

We thank Marilyn Jones, RN, MPH, CIC, Director, Interventional Epidemiology, BJC HealthCare (St. Louis, MO), for allowing us to reproduce Figure 5-1 and Figure 5-3, which she made.

References

1. Weinstein RA. Nosocomial infection update. *Emerg Infect Dis* 1998;4:416–420.
2. American Society for Quality (ASQ). The History of Quality—Beyond Total Quality http://www.asq.org/. Accessed October 21, 2009.
3. About the Cover. Ignaz Philipp Semmelweis (1818–65). *Emerg Infect Dis* 2001;7(2). http://www.cdc.gov/ncidod/EID/vol7no2/cover.htm. Accessed October 21, 2009.
4. The SENIC Project: study on the efficacy of nosocomial infection control (SENIC Project). Summary of study designs. *Am J Epidemiol* 1980;111(5):473.
5. Occupational Safety and Health Administration (OSHA). Bloodborne pathogens standard. http://www.osha.gov/SLTC/bloodbornepathogens/index.html. Accessed October 21, 2009.
6. Kohn LT, Corrigan JM, Donaldson MS, eds. *To Err Is Human: Building a Safer Health System*. Washington, DC: National Academy Press, Institute of Medicine; 2000.

7. *Crossing the Quality Chasm: A New Health System for the 21st Century*. Washington, DC: National Academy Press, Institute of Medicine; 2001.

8. Adams K, Corrigan JM, eds.; Committee on Identifying Priority Areas for Quality Improvement. *Priority Areas for National Action: Transforming Health Care Quality*. Washington, DC: National Academy Press, Institute of Medicine; 2003.

9. Gerberding J. Health-care quality promotion through infection prevention: beyond 2000. *Emerg Infect Dis* 2001; 7(2):363–366.

10. Colmers JM. Public reporting and transparency. The Commonwealth Fund; 2007. http://www.commonwealthfund.org/Publications/Fund-Reports/2007/Feb/Public-Reporting-and-Transparency.aspx. Accessed October 21, 2009.

11. Martin LA, Neumann CW, Mountford J, Bisognano M, Nolan TW. Increasing efficiency and enhancing value in health care: ways to achieve savings in operating costs per year. Institute for Healthcare Improvement, Innovation Series White Paper. Cambridge, MA: Institute for Healthcare Improvement; 2009.

12. Institute for Healthcare Improvement (ISI). Idea Generation Tools: Brainstorming, Affinity Grouping, and Multivoting (ISI Tool). Cambridge, MA: ISI. http://www.ihi.org/IHI/Topics/Improvement/ImprovementMethods/Tools/Brainstorming+Affinity+Grouping+Multivoting.htm. Accessed October 21, 2009.

13. Institute for Healthcare Improvement (ISI). Idea Generation Tools: Cause and Effect Diagrams (ISI Tool). Cambridge, MA: ISI. http://www.ihi.org/IHI/Topics/Improvement/ImprovementMethods/Tools/Cause+and+Effect+Diagram.htm. Accessed October 21, 2009.

14. Tague NR. *The Quality Toolbox*. 2nd ed. ASQ Quality Press; 2004.

15. Institute for Healthcare Improvement (ISI). Process Analysis Tools: Cause Pareto Diagram. Cambridge, MA: ISI. http://www.ihi.org/NR/rdonlyres/6A94E0FF-80FB-4382-094 D6-AA7E8733D746/649/ParetoDiagram1.pdf. Accessed October 21, 2009.

16. Brassard M, Ritter D. *The Memory Jogger II*. Salem, NH: Goal/QPC; 1994.

17. United States Department of Veteran Affairs, National Center for Patient Safety. Root cause analysis (RCA). http://www.va.gov/NCPS/rca.html. Accessed October 21, 2009.

18. Yokoe DS, Mermel LA, Anderson DJ, et al. Executive summary: a compendium of strategies to prevent healthcare-associated infections in acute care hospitals. *Infect Control Hosp Epidemiol* 2008;29(suppl 1):S12–S21.

19. Langabeer JR. *Health Care Operations Management: A Quantitative Approach to Business and Logistics*. Sudbury, MA: Jones and Bartlett; 2008.

20. Kornacki MJ, Silversin J. *Leading Physicians through Change: How to Achieve and Sustain Results*. Tampa, FL: American College of Physician Executives; 2000.

21. Centers for Disease Control and Prevention. Writing smart objectives. http://www.cdc.gov/HealthyYouth/evaluation/resources.htm. Accessed October 21, 2009.

22. American Medical Association Council on Long Range Planning and Development in Cooperation with the Council on Constitution and Bylaws, and the Council on Ethical and Judicial Affairs. Policy compendium of the American Medical Association. Chicago: American Medical Association, 1999: E-2.03, E-2.095.

23. Vonderheide-Liem DN, Pate B. *Applying Quality Methodologies to Improve Healthcare: Six Sigma, Lean Thinking, Balanced Scorecard, and More*. Marblehead, MA: HCPro; 2004.

24. Ruiz U, Simon J. Selecting successful health system managing approaches. In: Mayhall CG, ed. *Hospital Epidemiology and Infection Control*. 3rd ed. Philadelphia: Lippincott Williams & Wilkins; 2004:177–183.

25. Wenzel RP, Pfaller MA. Infection control: the premier quality assessment program in United States hospitals. *Am J Med* 1991;91:27S–31S.

Chapter 6 Isolation

Michael Edmond, MD, MPH, MPA, and
Gonzalo M. L. Bearman, MD, MPH

Patients infected or colonized with certain microorganisms must be placed in isolation while in a healthcare facility, to prevent nosocomial transmission of these pathogens. Isolation systems enable healthcare workers to more readily identify patients who need to be isolated and to institute the appropriate precautions. This chapter presents an overview of isolation precautions, emphasizing the recommendations of the isolation guidelines from the Centers for Disease Control and Prevention (CDC).[1] Resources listed at the end of this chapter should be consulted on specific issues related to isolation.

The goal of isolation is to prevent transmission of microorganisms from infected or colonized patients to other patients, hospital visitors, and healthcare workers. Use of personal protective equipment (eg, masks, eye wear, gloves, and gowns) and specific room requirements help to accomplish this goal.

The importance of appropriate isolation cannot be overstated. The medical literature is replete with examples of nosocomial outbreaks of influenza, tuberculosis, varicella, severe acute respiratory syndrome (SARS), and even hepatitis A infection, all of which could have been prevented if isolation practices had been optimal. Isolation efforts incur costs, but the direct and indirect costs of nosocomial outbreaks are often substantial. Hand hygiene and appropriate isolation remain the cornerstones of infection control and are assuming greater importance as the prevalence of multidrug-resistant organisms increases.

The ideal isolation system is described in Table 6-1. While no system meets all of these characteristics, infection control personnel should consider these ideals when designing and implementing a system.

In assessing appropriate infection control for any infectious disease, one needs to know the mode of transmission (via the air, droplets, contact, blood and body fluids, or a combination of any of these). Moreover, it is necessary to also know the times of onset and termination of infectivity, since the patient may be infectious before and after the symptomatic period. The emergence of SARS demonstrated the difficulties of implementing appropriate infection control measures when the mode of transmission and infective period are unclear.

The CDC has led the effort to formalize guidelines for isolation. They published their first guidelines in 1970. Subsequently, the CDC has modified and streamlined these guidelines several times to address emerging problems in infectious diseases, such as multidrug-resistant *Mycobacterium tuberculosis* infection, pandemic influenza, and vancomycin-resistant enterococcal infection, as well as to incorporate an increased understanding about the mechanisms of transmission for some diseases.

The CDC and the Hospital Infection Control Practices Advisory Committee (HICPAC) issued a guideline in 1996 for a new system of isolation. This system replaced the previous category-specific and disease-specific systems and integrated universal precautions and body substance isolation. It remains the basis for typical practices in US hospitals. The guideline was updated in 2007.[1] Still, individual healthcare institutions may find it necessary to modify the basic guidelines.

Table 6-1. Characteristics of the ideal isolation system

Utilizes current understanding of the mechanisms of transmission of infectious pathogens
Requires isolation precautions for all patients with infectious diseases that may be nosocomially transmitted (ie, eliminates transmission of infection in the hospital)
Avoids isolation of patients who do not require it ("over-isolation")
Is easily understood by all members of the healthcare team
Is easily implemented
Encourages compliance
Is environmentally friendly (avoids unnecessary use of disposable products)
Is inexpensive
Interferes minimally with patient care
Poses no detrimental impact on patient safety
Minimizes patient discomfort

This chapter presents an overview of isolation so that infection control personnel can implement an appropriate system. Infection control personnel should also consult the detailed guidelines for implementing isolation precautions that are referenced at the end of this chapter.

Current CDC Guidelines

The CDC and HICPAC developed a system for isolation that has 2 levels of precautions: standard precautions, which apply to all patients, and transmission-based precautions, employed for patients with documented or suspected colonization or infection with certain microorganisms.[1]

Standard Precautions

Standard precautions apply to blood; to all body fluids, secretions, and excretions, except sweat, whether or not they are visibly bloody; to nonintact skin; and to mucous membranes. The intent of standard precautions is primarily to protect the healthcare worker from pathogens transmitted via blood and body fluids. Requirements for standard precautions are outlined in Table 6-2.

Transmission-Based Precautions

Whereas standard precautions apply to all patients, transmission-based precautions apply to selected patients on the basis of either a clinical syndrome or a suspected or confirmed specific diagnosis.[1] Transmission-based precautions are divided into 3 categories that reflect the major modes of transmission of infectious agents in the healthcare setting: airborne transmission precautions (hereafter, "airborne precautions"), droplet

Table 6-2. Requirements for standard precautions

Hand hygiene

After touching blood, body fluids, secretions, excretions, and/or contaminated items or inanimate objects in the immediate vicinity of the patient
Immediately after removing gloves
Before and after patient contact

Gloves

For touching blood, body fluids, secretions, excretions, and/or contaminated items
For touching mucous membranes and/or nonintact skin

Mask, eye protection, face shield

To protect mucous membranes of the eyes, nose, and mouth during procedures and patient-care activities likely to generate splashes or sprays of blood, body fluids, secretions, and/or excretions

Gown

To protect skin and prevent soiling of clothing during procedures and patient-care activities likely to generate splashes or sprays of blood, body fluids, secretions, and/or excretions

Patient-care equipment

Soiled patient-care equipment should be handled in a manner to prevent skin and mucous membrane exposures, contamination of clothing, and transfer of microorganisms to other patients and environments
Reusable equipment must be cleaned and reprocessed before being used in the care of another patient

Environmental control

Requires procedures for routine care, cleaning, and disinfection of patient furniture and the environment

Linen

Soiled linen should be handled in a manner to prevent skin and mucous membrane exposures, contamination of clothing, and transfer of microorganisms to other patients and environments

Sharp devices

Avoid recapping used needles
Avoid removing used needles from disposable syringes by hand
Avoid bending, breaking, or manipulating used needles by hand
Place used sharp devices in puncture-resistant containers

Patient resuscitation

Use mouthpieces, resuscitation bags, or other ventilation devices to avoid mouth-to-mouth resuscitation

Patient placement

Patients who contaminate the environment or cannot maintain appropriate hygiene should be placed in a private room

NOTE: Modified from Siegel et al.[1]

precautions, and contact precautions (Table 6-3 and Table 6-4). Some diseases require more than 1 isolation category.

Airborne Precautions

Airborne precautions prevent diseases transmitted by aerosols containing droplet nuclei or contaminated dust particles.[1] Droplet nuclei are less than 5 μm in size and may remain suspended in the air, allowing them to migrate for long periods of time. Aerosol transmission of pathogens may be obligate, preferential, or opportunistic.[2] M. tuberculosis is probably the only pathogen that is transmitted exclusively via fine-particle aerosol and is thus an example of obligate aerosol transmission. Pathogens that are preferentially but not exclusively transmitted via aerosols include rubeola (measles) virus and varicella virus. Opportunistic pathogens typically are transmitted by other routes but under special circumstances may be transmitted by the airborne route; examples include smallpox virus, SARS-associated coronavirus, influenza virus, and noroviruses.

Patients with suspected or confirmed tuberculosis (pulmonary or laryngeal), measles, varicella, or disseminated zoster should be placed under airborne precautions. In addition, empirical use of airborne precautions should be strongly considered for human immunodeficiency virus–infected patients with cough, fever, and unexplained pulmonary infiltrates in any location until tuberculosis can be ruled out. Appropriate isolation requires an airborne infection isolation room (AIIR): a private room with negative air-pressure and at least 6 air exchanges per hour (preferably 12). Air from the room should be exhausted directly to the outside or through a high-efficiency particulate air (HEPA) filter. The door to the room must be kept closed at all times.

If the patient must be transported from the isolation room to another area of the hospital, the patient should put on a standard surgical mask before leaving the isolation room. All persons entering the room should wear respirators. In the United States, the Occupational Safety and Health Administration (OSHA) requires respirator to meet the following 4 performance criteria[3]:

1. It filters 1-μm particles with an efficiency of at least 95%.
2. It fits different facial sizes and characteristics.
3. It can be fit tested to obtain a leakage rate of less than 10%.
4. It can be checked for fit each time the healthcare worker puts on the mask.

There are numerous products available that are certified by the National Institute for Occupational Safety and Health as meeting the N-95 standard (ie, filters 95% of airborne particles). A not uncommon problem is the issue of respiratory protection for men with beards, since they cannot achieve an adequate fit with an N-95 mask. For these individuals, a powered air-purifying respirator can be used.

Patients with suspected or confirmed tuberculosis should be instructed to cover their mouth and nose with a tissue when coughing or sneezing. Those with suspected tuberculosis should remain in isolation until tuberculosis can be ruled out. Patients with confirmed tuberculosis who are receiving effective antituberculous treatment can be moved out of the negative air-pressure rooms when they are improving clinically and when 3

Table 6-3. Summary of transmission-based precautions

Variable	Airborne precautions	Droplet precautions	Contact precautions
Room	Negative air-pressure, single-patient room required with air exhausted to outside or through HEPA filters; door must be closed	Single-patient room preferred; door may remain open	Single-patient room preferred; door may remain open; use disposable non-critical patient-care equipment or dedicate equipment to a single patient
Masks	N-95 or portable respirator for those entering room; place surgical mask on patient if transport out of room is required	Surgical or isolation mask for those entering room; place surgical or isolation mask on patient if transport out of room is required	NA
Gowns	NA	NA	When entering room
Gloves	NA	NA	When entering room

NOTE: Modified from Siegel et al.[1] HEPA, high-efficiency particulate air filter; NA, not applicable.

Table 6-4. Isolation precautions required for various diseases and pathogens

Airborne precautions

Measles
Monkeypox[a]
Tuberculosis, pulmonary or laryngeal; draining lesion[a]
SARS[a]
Smallpox[a]
Varicella[a]
Zoster, disseminated or in an immunocompromised patient until
    dissemination ruled out[a]

Droplet precautions

Adenovirus pneumonia[a]
Diphtheria, pharyngeal
Haemophilus influenzae meningitis, epiglottitis; pneumonia (in an
    infant or child)
Influenza
Meningococcal infections
Mumps
Mycoplasma pneumonia
Parvovirus B19 infection
Pertussis
Plague, pneumonic
Rhinovirus infection[a]
Rubella
SARS[a]
Group A streptococcal pneumonia; serious invasive disease; major
    skin, wound, or burn infection[a]; pharyngitis, scarlet fever (in an
    infant or young child)
Viral hemorrhagic fever[a]

Contact precautions

Adenovirus conjunctivitis
Adenovirus pneumonia[a]
Burkholderia cepacia pneumonia in cystic fibrosis

Clostridium difficile diarrhea
Conjunctivitis, acute viral
Decubitus ulcer, infected and drainage not contained
Diarrhea, infectious (in a diapered or incontinent patient)
Diphtheria, cutaneous
Enterovirus infection (in an infant or young child)
Furunculosis (in an infant or young child)
Hepatitis A, hepatitis E (in a diapered or incontinent patient)
HSV infection, neonatal or disseminated or severe primary
    mucocutaneous
Human metapneumovirus infection
Impetigo
Lice
Infection or colonization with MDR bacteria (eg, MRSA, VRE,
    VISA, VRSA, ESBL producers, KPC producers, drug-resistant
    Streptococcus pneumoniae)
Monkeypox[a]
Parainfluenza infection (in an infant or child)
Rhinovirus infection[a]
Rotavirus infection
RSV infection (in an infant, child, or immunocompromised
    patient)
Rubella, congenital
SARS[a]
Scabies
Smallpox[a]
Staphylococcus aureus major skin, wound or burn infection
Group A streptococcal major skin, burn or wound infection[a]
Tuberculous draining lesion
Vaccinia, fetal, generalized, progressive, or eczema vaccinatum
Varicella[a]
Viral hemorrhagic fever[a]
Zoster, disseminated or in an immunocompromised patient

NOTE: Modified from Siegel et al.[1] ESBL, extended-spectrum β-lactamases; HSV, herpes simplex virus; KPC, *Klebsiella pneumoniae* cephalosporinase; MDR, multidrug-resistant; MRSA, methicillin-resistant *S. aureus*; RSV, respiratory syncytial virus; SARS, severe acute respiratory syndrome; VISA, vancomycin-intermediate *S. aureus*; VRE, vancomycin-resistant enterococci; VRSA, vancomycin-resistant *S. aureus*.
a Condition requires 2 types of precautions.

consecutive sputum smears of samples collected at least 8 hours apart have no detectable acid-fast bacilli. Patients with multidrug-resistant tuberculosis may need to be isolated for the duration of their hospital stay.

If the patient has suspected or confirmed measles, varicella, or disseminated zoster, nonimmune individuals should not enter the room. If a nonimmune healthcare worker must enter the room, he or she should wear a respirator (as described above). For immune healthcare workers, there are no clear guidelines. Some facilities require respirators for all healthcare workers entering any AIIR, for the sake of consistency. Other facilities do not require respirators for immune healthcare workers to enter the room of a patient with measles or varicella.

Droplet Precautions

Droplet precautions prevent the transmission of microorganisms by particles larger than 5 μm.[1] These droplets are produced when the patient talks, coughs, or sneezes. Droplets also may be produced during some medical procedures. Some illnesses that require droplet precautions include bacterial diseases, such as invasive *Haemophilus influenzae* type B infections, meningococcal infections, multidrug-resistant pneumococcal disease, pharyngeal diphtheria, *Mycoplasma* pneumonia, and pertussis. Some viral diseases, including seasonal influenza, mumps, rubella, and parvovirus infection, also require these precautions.

Droplet precautions require patients to be placed in a private room or cohorted with another patient who is infected with the same organism. The door to the room may remain open. Those entering the room should wear standard surgical or isolation masks. When transported out of the isolation room, the patient should wear a mask, but a mask is not required for those transporting the patient.

Contact Precautions

Contact precautions prevent transmission of epidemiologically important organisms from an infected or colonized patient through direct contact (touching the patient) or indirect contact (touching surfaces or objects in the patient's environment).[1] Contact precautions require patients to be placed in a private room or cohorted with another patient who is infected with the same organism. Healthcare workers should wear a gown and gloves when entering the room. They should change the gloves while caring for the patient if they touch materials that have high concentrations of microorganisms. While still in the isolation room, healthcare workers should remove their gown and gloves, taking care to not contaminate clothing or skin, and perform hand hygiene with a medicated handwashing agent; they must take care not to contaminate their hands before leaving the room. Noncritical patient care items (eg, stethoscopes and bedside commodes) that are used for the patients who are in contact isolation should not be used for other patients. If such items must be shared, they should be cleaned and disinfected before reuse. Patients should leave isolation rooms infrequently.

Contact isolation is indicated for patients infected or colonized with multidrug-resistant bacteria (eg, methicillin-resistant *Staphylococcus aureus* or vancomycin-resistant enterococci). It is also indicated for patients with *Clostridium difficile* or rotavirus enteritis and for diapered or incontinent patients who are infected or colonized with other agents transmitted by the oral-fecal route (eg, *Escherichia coli* O157:H7, *Shigella* species, rotavirus, or hepatitis A virus). Infants and young children with respiratory syncytial virus, parainfluenza, or enteroviral infection also require contact isolation, as do patients with severe herpes simplex virus infection (ie, neonatal, disseminated, or severe primary mucocutaneous disease), impetigo, scabies, or pediculosis. Patients with varicella or disseminated zoster infection require both contact and airborne precautions.

Instituting Empirical Isolation Precautions

Frequently, patients are admitted to the hospital without a definitive diagnosis. However, they may have an infectious process that may place other patients and healthcare workers at risk. Therefore, patients with certain clinical syndromes should be placed in isolation while a definitive diagnosis is pending. Table 6-5 delineates appropriate empirical isolation precautions for various clinical syndromes on the basis of the potential mechanisms of transmission.

Discontinuing Isolation Precautions

The discontinuation of isolation precautions is pathogen specific and based on the duration of infectivity. For some types of infection (eg, acute bacterial infection), the duration of infectivity is highly impacted by the initiation of effective antimicrobial therapy. For other infections (eg, viral infections) or colonization with multidrug-resistant pathogens, the impact of therapy on the duration of infectivity is less. The CDC isolation guideline[1] should be consulted for pathogen-specific recommendations on the duration of isolation precautions.

Table 6-5. Appropriate empirical isolation precautions for various clinical syndromes

Airborne precautions

Vesicular rash[a]

Maculopapular rash with cough, coryza, and fever

Cough, fever, and upper lobe pulmonary infiltrate

Cough, fever, any pulmonary infiltrate in a patient infected with HIV (or at high risk for HIV infection)

Cough, fever, any pulmonary infiltrate, and recent travel to countries with outbreaks of SARS or avian influenza[a]

Droplet precautions

Meningitis

Petechial or ecchymotic rash with fever

Paroxysmal or severe persistent cough during periods of pertussis activity

Contact precautions

Acute diarrhea with likely infectious etiology in incontinent or diapered patient

Vesicular rash[a]

Respiratory infections in infants and young children

History of infection or colonization with MDR organisms

Skin, wound, or urinary tract infection in a patient with a recent hospital or nursing home stay in a facility where MDR organisms are prevalent

Abscess or draining wound that cannot be covered

Cough, fever, any pulmonary infiltrate, and recent travel to countries with outbreaks of SARS or avian influenza[a]

NOTE: Modified from Siegel et al.[1] HIV, human immunodeficiency virus; MDR, multidrug-resistant; SARS, severe acute respiratory syndrome.
a Condition requires 2 types of precautions.

References

1. Siegel JD, Rhinehart E, Jackson M, Chiarello L, the Healthcare Infection Control Practices Advisory Committee (HICPAC). 2007 Guideline for isolation precautions: preventing transmission of infectious agents in healthcare settings, June 2007. http://www.cdc.gov/ncidod/dhqp/gl_isolation.html. Accessed May 25, 2009.
2. Roy CJ, Milton DK. Airborne transmission of communicable infection—the elusive pathway. *N Engl J Med* 2004;350: 1710–1712.
3. Jensen PA, Lambert LA, Iademarco MF, Ridzon R. Guidelines for preventing the transmission of *Mycobacterium tuberculosis* in health-care settings, 2005. *MMWR Morb Mortal Wkly Rep* 2005;54(RR-17):1–141.

Suggested Reading

Centers for Disease Control and Prevention (CDC). Guidelines for environmental infection control in health-care facilities: recommendations of CDC and the Healthcare Infection Control Practices Advisory Committee (HICPAC). *MMWR Morb Mortal Wkly Rep* 2003;52(RR-10):1–42. http://www.cdc.gov/ncidod/dhqp/pdf/guidelines/Enviro_guide_03.pdf. Accessed May 25, 2009.

Centers for Disease Control and Prevention. Guideline for hand hygiene in health-care settings: recommendations of the Healthcare Infection Control Practices Advisory Committee and the HICPAC/SHEA/APIC/IDSA Hand Hygiene Task Force. *MMWR Morb Mortal Wkly Rep* 2002;51(RR-16):1–44. http://www.cdc.gov/mmwr/PDF/rr/rr5116.pdf. Accessed May 25, 2009.

Centers for Disease Control and Prevention. Bioterrorism Web Site. http://www.bt.cdc.gov/bioterrorism/. Accessed May 25, 2009.

Centers for Disease Control and Prevention. Severe Acute Respiratory Syndrome Web site. http://www.cdc.gov/ncidod/sars/ic.htm. Accessed May 25, 2009.

Jarvis WR, ed. *Bennett & Brachman's Hospital Infections*. 5th ed. Philadelphia, PA: Lippincott Williams & Wilkins; 2007.

Mayhall CG, ed. *Hospital Epidemiology and Infection Control*. 3rd ed. Philadelphia, PA: Lippincott Williams & Wilkins; 2004.

Pickering LK, ed. *2009 Red Book: Report of the Committee on Infectious Diseases*. 28th ed. Elk Grove Village, IL: American Academy of Pediatrics; 2009.

Siegel JD, Rhinehart E, Jackson M, Chiarello L, the Healthcare Infection Control Practices Advisory Committee. Management of Multidrug-Resistant Organisms in Healthcare Settings, 2006. http://www.cdc.gov/ncidod/dhqp/pdf/ar/MDROGuideline2006.pdf. Accessed May 25, 2009.

Wenzel RP, ed. *Prevention and Control of Nosocomial Infections*. 4th ed. Philadelphia, PA: Lippincott Williams & Wilkins; 2003.

Chapter 7 Disinfection and Sterilization in Healthcare Facilities

William A. Rutala, PhD, MPH, and David J. Weber, MD, MPH

In the United States in 1996, approximately 46,500,000 surgical procedures and an even larger number of invasive medical procedures were performed.[1] For example, there are about 5 million gastrointestinal endoscopies performed per year.[1] Each of these procedures involves contact by a medical device or surgical instrument with a patient's sterile tissue or mucous membranes. A major risk of all such procedures is the introduction of pathogenic microbes, which can lead to infection. Failure to properly disinfect or sterilize equipment may lead to transmission via contaminated medical and surgical devices (eg, bronchoscopes contaminated with *Mycobaterium tuberculosis*).

Achieving disinfection and sterilization through the use of disinfectants and sterilization practices is essential for ensuring that medical and surgical instruments do not transmit infectious pathogens to patients. Since it is not necessary to sterilize all patient-care items, healthcare policies must identify whether cleaning, disinfection, or sterilization is indicated, based primarily on each item's intended use.

Multiple studies in many countries have documented lack of compliance with established guidelines for disinfection and sterilization.[2,3] Failure to comply with scientifically based guidelines has led to numerous outbreaks.[3-7] In this chapter, which is an updated and modified version of other chapters,[8,9] a pragmatic approach to the judicious selection and proper use of disinfection and sterilization processes is presented, based on well-designed studies assessing the efficacy (through laboratory investigation) and effectiveness (through clinical studies) of disinfection and sterilization procedures.

A Rational Approach to Disinfection and Sterilization

Over 40 years ago, Earle H. Spaulding[10] devised a rational approach to disinfection and sterilization of patient-care items or equipment. This classification scheme is so clear and logical that it has been retained, refined, and successfully used by infection preventionists and others planning methods for disinfection or sterilization.[8,11-13] Spaulding believed that the nature of disinfection could be understood more readily if instruments and items for patient care were divided into 3 categories based on the degree of risk of infection involved in the use of the items: critical, semicritical, and noncritical.[10] This terminology is employed by the Centers for Disease Control and Prevention (CDC) guidelines for environmental infection control and for disinfection and sterilization.[14,11]

Critical Items

Critical items are so called because of the high risk of infection if such an item is contaminated with any microorganism, including bacterial spores. Thus, it is critical that objects that enter sterile tissue or the vascular system be sterile, because any microbial contamination could result in disease transmission. This category includes surgical instruments, cardiac and urinary catheters, implants, and ultrasound probes used in sterile body cavities. The items in this category should be purchased in a sterile state or should be sterilized by steam sterilization, if possible. If heat sensitive, the object may be treated with ethylene oxide (ETO), hydrogen peroxide gas plasma, ozone, vaporized hydrogen peroxide, or liquid chemical sterilants, if other methods are unsuitable. Table 7-1 and Table 7-2 list several germicides categorized as chemical sterilants, and Table 7-3 summarizes the advantages and disadvantages of various sterilization processes. Chemical sterilants include the following agents: formulations of glutaraldehyde at a concentration of 2.4% or higher; 3.4% glutaraldehyde with 26% isopropanol; ortho-phthalaldehyde at a concentration of 0.55% or higher; 1.12% glutaraldehyde with 1.93% phenol/phenate; 7.5% hydrogen peroxide; 7.35% hydrogen peroxide with 0.23% peracetic acid; 0.2% peracetic acid; 1.0% hydrogen peroxide with 0.08% peracetic acid; 2.0% hydrogen peroxide; hypochlorous acid or hypochlorite with 650–675 ppm active free chlorine; and 8.3% hydrogen peroxide with 7.0% peracetic acid. The indicated exposure times range from 3 to 12 hours, except for 0.2% peracetic acid, which has an exposure time of 12 minutes at 50°C–56°C.[15] Liquid chemical sterilants can be relied upon to produce sterility only if cleaning to eliminate organic and inorganic material precedes treatment and proper guidelines on concentration, contact time, temperature, and pH are met. Another limitation to sterilization of devices with liquid chemical sterilants is that the devices cannot be wrapped during processing in a liquid chemical sterilant; thus it is impossible to maintain sterility after processing and during storage. Furthermore, after exposure to the liquid chemical sterilant, devices may require rinsing with water that generally is not sterile. Therefore, because of the inherent limitations of using liquid chemical sterilants in a nonautomated reprocessor, their use should be restricted to reprocessing critical devices that are heat sensitive and incompatible with other sterilization methods.

Semicritical Items

Semicritical items are those that come in contact with mucous membranes or nonintact skin. Included in this category are respiratory therapy and anesthesia equipment, most endoscopes (both gastrointestinal and respiratory), endocavitary probes (eg, rectal or vaginal probes), laryngoscope blades, esophageal manometry probes, anorectal manometry catheters, and diaphragm fitting rings. These medical devices should be free of all microorganisms (ie, mycobacteria, fungi, viruses, and bacteria), although small numbers of bacterial spores may be present. Intact mucous membranes, such as those

of the lungs or the gastrointestinal tract, generally are resistant to infection by common bacterial spores but are susceptible to infection with other organisms, such as bacteria, mycobacteria, and viruses. Semicritical items minimally require high-level disinfection with chemical disinfectants. Glutaraldehyde (both with other disinfectants [alcohol or phenol/phenate] and without), hydrogen peroxide, ortho-phthalaldehyde, peracetic acid with hydrogen peroxide, and chlorine have been cleared for use as disinfectants by the US Food and Drug Administration (FDA), and they are dependable high-level disinfectants, provided the factors influencing germicidal procedures are met (Table 7-1 and Table 7-2).[15] The exposure time for most high-level disinfectants varies from 8 to 45 minutes at 20°C–25°C. Outbreaks occur when ineffective disinfectants—including iodophor, alcohol, and over-diluted glutaraldehyde[5]—are used for high-level disinfection. When a disinfectant is selected for use with certain patient-care items, its chemical compatibility after extended use with the items to be disinfected also must be considered. For example, compatibility testing by Olympus America of 7.5% hydrogen peroxide with endoscopes found cosmetic and functional changes (Olympus, October 15, 1999, personal communication). Similarly, Olympus does not endorse the use of the hydrogen peroxide with peracetic acid products because of cosmetic and functional damage (Olympus America, April 15, 1998, and September 13, 2000, personal communication).

Semicritical items that will have contact with the mucous membranes of the respiratory tract or gastrointestinal tract should be rinsed with sterile water, filtered water, or tap water, followed by an alcohol rinse.[11,16] Use of an alcohol rinse and forced-air drying markedly reduces the likelihood of contamination of the instrument (eg, an endoscope), most likely by removing the wet environment favorable to bacterial growth.[17] After being rinsed, items should be dried and stored in a manner that protects them from damage or contamination. There is no recommendation to use sterile or filtered water rather than tap water for rinsing semicritical equipment that will have contact with the mucous membranes of the rectum (eg, rectal probes, or anoscope) or vagina (eg, vaginal probes).[11]

Noncritical Items

Noncritical items are those that come in contact with intact skin but not mucous membranes. Intact skin acts as an effective barrier to most microorganisms; therefore, the sterility of items coming in contact with intact skin is not critical. Examples of noncritical items are bedpans, blood pressure cuffs, crutches, bed rails, linens, bedside

Table 7-1. Methods for disinfection and sterilization of patient-care items and environmental surfaces

Process, level of microbial inactivation, method	Examples of methods and processing times	Healthcare application(s) and examples
Sterilization: destroys all microorganisms, including bacterial spores		
High temperature	Steam ($\sim$40 min), dry heat (1–6 h, depending on temperature)	Heat-tolerant critical patient-care items (surgical instruments) and semicritical items
Low temperature	Ethylene oxide gas ($\sim$15 h), H_2O_2 gas plasma (28–52 min), ozone ($\sim$4 h), H_2O_2 vapor (55 min)	Heat-sensitive critical and semicritical patient-care items
Liquid immersion	Chemical sterilants[a]: >2% glut ($\sim$10 h); 1.12% glut and 1.93% phenol (12 h); 7.35% H_2O_2 and 0.23% PAA (3 h); 8.3% H_2O_2 with 7.0% PAA (5 h); 7.5% H_2O_2 (6 h); 1.0% H_2O_2 and 0.08% PAA (8 h); $\geq$0.2% PAA (12 min at 50°C–56°C)	Heat-sensitive critical and semicritical patient-care items that can be immersed
High-level disinfection: destroys all microorganisms except high numbers of bacterial spores		
Heat, automated	Pasteurization (30 min at 65°C–77°C)	Heat-sensitive semicritical items (eg, respiratory therapy equipment)
Liquid immersion	Chemical sterilants / HLDs[a]: >2% glut (20–45 min); 0.55% OPA (12 min); 1.12% glut and 1.93% phenol (20 min); 7.35% H_2O_2 and 0.23% PAA (15 min); 7.5% H_2O_2 (30 min); 1.0% H_2O_2 and 0.08% PAA (25 min); 650–675 ppm chlorine (10 min); 2.0% H_2O_2 (8 min); 3.4% glut with 26% isopropanol (10 min)	Heat-sensitive semicritical items (eg, gastrointestinal endoscopes, bronchoscopes, and endocavitary probes)
Intermediate-level disinfection: destroys vegetative bacteria, mycobacteria, most viruses, and most fungi, but not bacterial spores		
Liquid contact	EPA-registered hospital disinfectant with a label claim regarding tuberculocidal activity (eg, chlorine-based products, phenolics, or accelerated H_2O_2 with exposure times of at least 1 min)	Noncritical patient-care items (eg, blood pressure cuffs) or surfaces with visible blood
Low-level disinfection: destroys vegetative bacteria and some fungi and viruses but not mycobacteria or spores		
Liquid contact	EPA-registered hospital disinfectant with no tuberculocidal claim, such as chlorine-based products, phenolics, accelerated H_2O_2, quaternary ammonium compounds (for at least 1 min), or 70%–90% alcohol	Noncritical patient care items (eg, blood pressure cuffs) or surfaces (eg, bedside table) with no visible blood

NOTE: Modified from Rutala and Weber,[8] Rutala et al.,[11] and Kohn et al.[154] EPA, Environmental Protection Agency; FDA, Food and Drug Administration; Glut, glutaraldehyde; H_2O_2, hydrogen peroxide; HLD, high-level disinfectant; OPA, ortho-phthalaldehyde; PAA, peracetic acid.
a Consult the FDA-cleared package insert for information about the cleared contact time and temperature; also, see the text of this chapter for discussion of why one product is used at a reduced exposure time (2% glutaraldehyde for 20 min at 20°C). Increasing the temperature with an automated endoscope reprocessor will reduce the contact time (5 min at 25°C, compared with, eg, 12 min at 20°C with OPA). The exposure temperature for some high-level disinfectants varies from 20°C to 25°C; higher temperatures are noted. Tubing must be completely filled for high-level disinfection and liquid chemical sterilization. Material compatibility should be investigated if appropriate (eg, use of H_2O_2 and H_2O_2 with PAA will cause functional damage to endoscopes).

tables, patient furniture, and floors. In contrast to critical and some semicritical items, most noncritical reusable items may be decontaminated where they are used and do not need to be transported to a central processing area. There is virtually no documented risk of transmitting infectious agents to patients by means of noncritical items,[18] when they are used as noncritical items and do not contact nonintact skin and/or mucous membranes.

Table 7-2. Summary of the advantages and disadvantages of using chemical agents as sterilants or as high-level disinfectants

Sterilization method	Advantages	Disadvantages
Peracetic acid and hydrogen peroxide	• No activation required • Odor or irritation not significant	• Material compatibility concerns (with, lead, brass, copper, and zinc), both cosmetic and functional • Limited clinical experience • Potential for eye and skin damage
Glutaraldehyde	• Numerous published studies on its use • Relatively inexpensive • Excellent material compatibility	• Respiratory irritation from glutaraldehyde vapor • Pungent and irritating odor • Relatively slow mycobactericidal activity • Coagulates blood and fixes tissue to surfaces • Allergic contact dermatitis
Hydrogen peroxide	• No activation required • May enhance removal of organic matter and organisms • No disposal issues • No odor or irritation issues • Does not coagulate blood or fix tissues to surfaces • Inactivates *Cryptosporidium* species • Published studies on its use	• Material compatibility concerns (with brass, zinc, copper, and nickel silver plating), both cosmetic and functional • Serious eye damage with contact
Ortho-phthalaldehyde	• Fast-acting, high-level disinfectant • No activation required • Odor not significant • Excellent materials compatibility claimed • Does not coagulate blood or fix tissues to surfaces claimed	• Stains protein gray (eg, skin, mucous membranes, clothing, and environmental surfaces) • Limited clinical experience • More expensive than glutaraldehyde • Eye irritation with contact • Slow sporicidal activity • Exposure may result in hypersensitivity
Peracetic acid	• Rapid sterilization cycle time (30–45 min) • Low temperature (50°C–55°C) liquid immersion sterilization • Environmental friendly by-products (acetic acid, oxygen, and water) • Fully automated • Single-use system eliminates need for concentration testing • Standardized cycle • May enhance removal of organic material and endotoxin • No adverse health effects to operators under normal operating conditions • Compatible with many materials and instruments • Does not coagulate blood or fix tissues to surfaces • Sterilant flows through scope, facilitating removal of salt, protein, and microbes • Rapidly sporicidal • Provides procedure standardization (constant dilution, perfusion of channel, temperatures, and exposure)	• Potential material incompatibility (eg, aluminum anodized coating becomes dull) • Used for immersible instruments only • Biological indicator may not be suitable for routine monitoring • One scope or a small number of instruments can be processed in a cycle • More expensive than high-level disinfection (because of endoscope repairs, operating costs, and purchase costs) • Serious eye and skin damage (concentrated solution) with contact • Point-of-use system, no sterile storage

NOTE: Modified from Rutala and Weber.[155] All products listed here are effective in the presence of some organic soil, are relatively easy to use, and have a broad spectrum of antimicrobial activity (against bacteria, fungi, viruses, bacterial spores, and mycobacteria). These characteristics are documented in the literature; contact the manufacturer of the instrument and sterilant for additional information. The Food and Drug Administration has cleared all products listed in this table as chemical sterilants, except ortho-phthalaldehyde, which it has cleared as a high-level disinfectant.

However, these items (eg, bedside tables and bed rails) could potentially contribute to secondary transmission by contaminating hands of healthcare workers or by contact with medical equipment that will subsequently come in contact with patients.[19] Table 7-1 lists several low-level disinfectants that may be used for noncritical items; note that the exposure times listed are at least 1 minute.

Current Issues in Disinfection and Sterilization

Reprocessing of Endoscopes

Physicians use endoscopes to diagnose and treat numerous medical disorders. Although endoscopes are a valuable diagnostic and therapeutic tool in modern medicine and the incidence of infection associated with use has been reported as very low (about 1 case per 1.8 million procedures[20]), more outbreaks of healthcare-associated infection have been linked to contaminated endoscopes than to any other medical device.[3-5] To prevent the spread of healthcare-associated infections, all heat-sensitive endoscopes (eg, gastrointestinal endoscopes, bronchoscopes, and nasopharyngoscopes) must be properly cleaned and, at a minimum, subjected to high-level disinfection after each use. High-level disinfection can be expected to destroy all microorganisms, although, when high numbers of bacterial spores are present, a few spores may survive.

Recommendations for the cleaning and disinfection of endoscopic equipment have been published and should be strictly followed.[11,16] Unfortunately, audits have shown that personnel do not adhere to guidelines on reprocessing,[21-23] and outbreaks of infection continue to occur.[24,25] To ensure that reprocessing personnel are properly trained, there should be initial and annual competency testing for each individual who is involved in reprocessing endoscopic instruments.[11,16]

In general, endoscope disinfection or sterilization with a liquid chemical sterilant or high-level disinfectant involves 5 steps after leak testing:

1. *Clean:* mechanically clean internal and external surfaces; brush internal channels and flush each internal channel with water and an enzymatic cleaner or detergent.
2. *Disinfect:* immerse endoscope in high-level disinfectant (or chemical sterilant) and perfuse disinfectant (which eliminates air pockets and ensures contact of the germicide with the internal channels) into all accessible channels, such as the suction/biopsy channel and the air/water channel; ensure exposure for the time recommended for the specific product.
3. *Rinse* the endoscope and all channels with sterile water, filtered water (commonly used with automated endoscope reprocessors), or tap water.
4. *Dry:* rinse the insertion tube and inner channels with alcohol and dry it with forced air after disinfection and before storage.
5. *Store* the endoscope in a way that prevents recontamination and promotes drying (eg, hung vertically).

Unfortunately, there is suboptimal compliance with the recommendations for reprocessing endoscopes. In addition, there are instances where the scientific literature and recommendations from professional organizations regarding the use of disinfectants and sterilants may differ from the manufacturer's label claim. One example is the contact time used to achieve high-level disinfection with 2% glutaraldehyde. On the basis of the requirements of the FDA (which regulates liquid sterilants and high-level disinfectants used on critical and semicritical medical devices), manufacturers test the efficacy of their germicide formulations under worst-case conditions (ie, using the minimum recommended concentration of the active ingredient) and in the presence of organic soil (typically 5% serum). The soil is used to represent the organic load to which the device is exposed during actual use and that would remain on the device in the absence of cleaning. These stringent test conditions are designed to provide a margin of safety by ensuring that the contact conditions for the germicide provide complete elimination of the test bacteria (eg, 10^5–10^6 M. *tuberculosis* organisms in organic soil and dried on an endoscope) inoculated into the areas most difficult for the disinfectant to penetrate and in the absence of cleaning. However, the scientific data demonstrate that M. *tuberculosis* levels can be reduced by at least 8 $\log_{10}$ with cleaning (4 $\log_{10}$) followed by chemical disinfection for 20 minutes at 20°C (4–6 $\log_{10}$).[11,15,16,26,27] Because of these data, at least 14 professional organizations worldwide that have endorsed an endoscope reprocessing guideline recommend contact with 2% glutaraldehyde for 20 minutes at 20°C (or less than 20 minutes, outside the United States) to achieve high-level disinfection that differs from that specified by the manufacturer's label.[16,28-30] It is important to emphasize that the FDA tests do not include cleaning, a critical component of the disinfection process. When cleaning has been included in the test methodology, use of 2% glutaraldehyde for 20 minutes has been demonstrated to be effective in eliminating all vegetative bacteria.

Inactivation of the Creutzfeldt-Jakob Disease Prion

Creutzfeldt-Jakob disease (CJD) is a degenerative neurologic disorder of humans that has an annual incidence in the United States of approximately 1 case per 1 million population.[31-33] CJD is caused by a proteinaceous infectious agent or prion. CJD is related to other human transmissible spongiform encephalopathies that include kuru (incidence, 0; now eradicated), Gerstmann-Sträussler-Scheinker syndrome (incidence, 1 case per 40 million population), and fatal familial insomnia syndrome (incidence, less than 1 case per 40 million population). Prion diseases do not elicit an immune response, result in a noninflammatory pathologic process confined to the central nervous

Table 7-3. Summary of advantages and disadvantages of commonly used sterilization technologies

Method	Advantages	Disadvantages
Steam	• Nontoxic to patient, staff, and the environment • Cycle easy to control and monitor • Rapidly microbicidal • Least affected by organic and/or inorganic soils, among sterilization processes listed • Rapid cycle time • Penetrates medical packing and device lumens	• Deleterious for heat-sensitive instruments • Microsurgical instruments are damaged by repeated exposure • May leave instruments wet, causing them to rust • Potential for burns
Hydrogen peroxide gas plasma	• Safe for the environment • Leaves no toxic residuals • Cycle time is 28–55 minutes and no aeration necessary • Used for heat- and moisture-sensitive items since process temperature <50°C • Simple to operate, install (requires 208 V outlet), and monitor • Compatible with most medical devices • Only requires electrical outlet	• Cellulose (paper), linens, and liquids cannot be processed • Sterilization chamber is small: approximately 0.1–0.21 m^3 (3.5–7.3 ft^3) • Endoscope or medical device restrictions based on lumen internal diameter and length (see manufacturer's recommendations) • Requires synthetic packaging (polypropylene wraps and polyolefin pouches) and special container tray • Hydrogen peroxide may be toxic at levels greater than 1 ppm TWA
100% ETO	• Penetrates packaging materials and device lumens • Single-dose cartridge and negative-pressure chamber minimizes the potential for gas leakage and ETO exposure • Simple to operate and monitor • Compatible with most medical materials	• Requires aeration time to remove ETO residue • Sterilization chamber is small: 0.11–0.25 m^3 (4–8.8 ft^3) • ETO is toxic, a carcinogen, and flammable • ETO emission is regulated by states, but a catalytic cell removes 99.9% of ETO and converts it to CO_2 and H_2O • ETO cartridges require storage in a flammable-liquid storage cabinet • Lengthy cycle and aeration time
ETO mixtures: 8.6% ETO and 91.4% HCFC; 10% ETO and 90% HCFC; 8.5% ETO and 91.5% CO_2	• Penetrates medical packaging and many plastics • Compatible with most medical materials • Cycle easy to control and monitor	• Some states (eg, CA, NY, and MI) require ETO emission reduction of 90%–99.9% • CFC (inert gas that eliminates explosion hazard) banned in 1995 • Potential hazards to staff and patients • Lengthy cycle and aeration time • ETO is toxic, a carcinogen, and flammable
Peracetic acid	• Rapid cycle time (30–45 min) • Low-temperature (50°C–55°C) liquid immersion sterilization • Environmentally friendly by-products • Sterilant flows through endoscope, which facilitates salt, protein, and microbe removal	• Point-of-use system, requires no sterile storage • Biological indicator may not be suitable for routine monitoring • Used for immersible instruments only • Some material incompatibility (eg, anodized aluminum coating becomes dull) • One scope or a small number of instruments processed in a cycle • Potential for serious eye and skin damage (concentrated solution) with contact • Must use connector between system and scope to ensure infusion of sterilant to all channels
Ozone	• Used for moisture-sensitive and heat-sensitive items	• Sterilization chamber is small: 0.11 m^3 (4 ft^3)

(continued)

Table 7-3. (continued)

Method	Advantages	Disadvantages
	• Ozone generated from oxygen and water (oxidizing) • No aeration needed because there are no toxic by-products • FDA-cleared for metal and plastic surgical instruments, including some instruments with lumens	• Uncertain of material compatibility, penetrability, and organic material resistance because of limited use and limited data • Limited microbicidal efficacy data
Hydrogen peroxide vapor	• Safe for the environment and healthcare workers; leaves no toxic residuals • Fast: cycle time is 55 min, and no aeration is necessary • Used for heat-sensitive and moisture-sensitive items (metal and nonmetal devices)	• Sterilization chamber is small: 0.14 m³ (4.8 ft³) • Medical devices restrictions based on lumen internal diameter and length (eg, 1 mm diameter and 125 mm length; see manufacturer's recommendations) • Not used for liquid, linens, powders, or any cellulose materials • Requires synthetic packaging (polypropylene) • Limited use and limited comparative microbicidal efficacy data

NOTE: Modified from Rutala and Weber.[156] CFC, chlorofluorocarbon; ETO, ethylene oxide; HCFC, hydrochlorofluorocarbon; TWA, time-weighted average concentration.

system, have an incubation period of years, and usually are fatal within 1 year after diagnosis.

CJD occurs as both a sporadic disease (approximately 85% of cases) and as familial or inherited disease (approximately 15% of cases). Fewer than 1% of CJD episodes have resulted from healthcare-associated transmission; the majority result from use of contaminated tissues or grafts.[34] Iatrogenic CJD has been described in humans in 3 circumstances: in patients for whom contaminated medical equipment was used during intracranial placement of contaminated electroencephalography electrodes (2 cases in Switzerland) or neurosurgical procedures (4 suspected cases; 3 cases in United Kingdom and 1 case in France); in patients who received hormone therapy with cadaveric human growth hormones or gonadotropin (more than 190 cases [26 in the United States]; since 1985, human growth hormone has been manufactured by recombinant DNA technology and this risk has been eliminated); and in patients who received an implant of contaminated grafts from humans (cornea, 2 cases; and dura mater, more than 190 cases [3 cases in the United States for which the risk factor was use of the Lyodura brand graft processed before 1987]).[34-37] All known instances of iatrogenic CJD have resulted from exposure to infectious brain, pituitary, or eye tissue. Tissue infectivity studies in experimental animals have determined the infectiousness of different body tissues.[37,38]

The prions that cause CJD and other transmissible spongiform encephalopathies exhibit an unusual resistance to conventional chemical and physical decontamination methods. In order for a surgical instrument to act as a vehicle of prion transmission, it must come into contact with infective tissue (eg, brain) during surgery on the infected patient, it must maintain any adhered infectivity after being decontaminated and sterilized, and it must contact the receptive tissue in the recipient.[39] The disinfection and sterilization recommendations for prevention of CJD transmission in this chapter should break this chain of events. They are based on the belief that infection prevention measures should be predicated on the following factors: epidemiologic evidence that links specific body tissues or fluids to transmission of CJD; infectivity assay results that demonstrate body tissues or fluids are contaminated with infectious prions; data on cleaning with detergents, microbiological indicators, and proteins[27,40-42]; data on inactivation of prions; data on the risk of disease transmission associated with the use of the instrument or device; and a review of other recommendations[35,43-45] (L.M. Sehulster, written communication, 2000). Other CJD recommendations have been based primarily on inactivation studies.[13,46,47]

The 3 parameters integrated into disinfection and sterilization processing for prion-contaminated medical instruments are as follows: the risk that the patient has a prion disease, the comparative infectivity of different body tissues, and the intended use of the medical device[43-45] (L.M. Sehulster, written communication, 2000). Patients at high risk include the following: patients with known prion disease; patients with rapidly progressive dementia consistent with possible prion disease; patients with a familial history of CJD, Gerstmann-Sträussler-Scheinker syndrome, or fatal familial insomnia syndrome; patients known to carry a mutation in the *PrP* gene, which is involved in familial transmissible spongiform encephalopathies; patients with a history of dura mater transplantation; patients with electroencephalography findings or laboratory evidence suggestive of a transmissible spongiform encephalopathy (eg, markers of neuronal injury, such as 14-3-3 protein); and patients with a known history of cadaver-derived pituitary hormone injection. Tissues at high risk of carrying prions include brain tissue, spinal cord tissue, pituitary tissue, and posterior eye tissue (that involves the retina or optic nerve); all other tissues are considered to have low or no risk. Critical devices are any that enter sterile tissue or the vascular system (eg, surgical instruments). Semicritical devices are those that contact nonintact skin or mucous membranes (eg, gastrointestinal endoscopes). The Association of peri-Operative Registered Nurses and the Association for the Advancement of Medical Instrumentation recommended practices for reprocessing surgical instruments exposed to CJD prions are consistent with the following recommendations.[43,48,49]

Recommendations for disinfection and sterilization of prion-contaminated medical devices are as follows. Instruments should be kept wet (eg, immersed in water or a prionicidal detergent) or damp after use and until they are decontaminated. They should be decontaminated (eg, in an automated washer-disinfector) as soon as possible after use. Dried films of tissue are more resistant to prion inactivation by steam sterilization than are tissues that are kept moist.[40-42,50-60] This may relate to the rapid heating that occurs in the film of dried material, compared with the bulk of the sample, and the rapid fixation or dehydration of the prion protein in the dried film.[61] It appears that prions in the dried portions of brain macerates are less efficiently inactivated than undisturbed tissue. In addition, certain disinfectants (eg, glutaraldehyde, formaldehyde, and ethanol) can fix or dehydrate the protein and make it more difficult to inactivate.[62-65] A procedure that uses formalin–formic acid has been recommended for inactivation of prion infectivity in tissue samples obtained from patients with CJD.[66]

The high resistance of prions to standard sterilization methods warrants special procedures in the reprocessing of surgical instruments. Special prion reprocessing is necessary when reprocessing critical or semicritical medical devices that have had contact with high-risk tissues from high-risk patients. After the device has been cleaned, it should be sterilized by either autoclaving (ie, steam sterilization) or using a combination of sodium hydroxide and autoclaving,[35] using 1 of the following 4 options:

Option 1. Autoclave at 134°C for 18 minutes in a prevacuum sterilizer.[51,52,55,56,63,67-71]

Option 2. Autoclave at 132°C for 1 hour in a gravity displacement sterilizer.[64,72-75]

Option 3. Immerse in 1 N NaOH (1 N NaOH is a solution of 40 g NaOH in 1 liter of water) for 1 hour; remove and rinse in water, then transfer to an open pan and autoclave (121°C gravity displacement or 134°C porous or prevacuum sterilizer) for 1 hour.[55,67,73,74,76]

Option 4. Immerse instruments in 1 N NaOH for 1 hour and heat in a gravity displacement sterilizer at 121°C for 30 minutes.[77]

It is essential with any sterilization process, and especially one for a prion-contaminated device, that the instrument is fully accessible to the sterilant (eg, steam).[56] Prion-contaminated medical devices that are impossible to clean or to expose fully to steam and other sterilants should be discarded. Flash sterilization should not be used for reprocessing. Always discard single-use devices. To minimize environmental contamination, noncritical environmental surfaces should be covered with plastic-backed paper and, if contaminated with high-risk tissues, the paper should be properly discarded. There are no antimicrobial products registered by the Environmental Protection Agency (EPA) specifically for inactivation of prions on environmental surfaces, and there are no sterilization processes cleared by the FDA for sterilization of reusable surgical instruments. However, the EPA has issued quarantine exemptions to several states to permit the temporary use of a phenolic preparation (containing 6.4% ortho-benzyl-para-chlorophenol, 3.0% para-tertiary-amylphenol, 0.5% ortho-phenyl phenol, 4.9% hexylene glycol, 12.6% gycoloic acid, and 8% isopropanol) for inactivation of prions on hard, nonporous surfaces in laboratories that handle contaminated or potentially contaminated animal tissues and waste.[78] If no EPA-registered or exempted products are available, then noncritical environmental surfaces (eg, laboratory surfaces) contaminated with high-risk tissues (eg, brain tissue) should be cleaned and then spot-decontaminated with a 1:5 to 1:10 dilution of hypochlorite solutions, ideally, for a contact time of at least 15 minutes.[68,72,75,79-82]

No recommendation can be made regarding the use of special prion reprocessing for reprocessing critical or semicritical devices contaminated with low-risk tissues from high-risk patients. Although low-risk tissue has been found to transmit CJD at a low frequency, this has only been demonstrated when low-risk tissue is inoculated into the brain of a susceptible animal. However, in humans, medical instruments contaminated with low-risk tissue would be unlikely to transmit infection after standard cleaning and sterilization, since the instruments would not be used in the central nervous system. Environmental surfaces contaminated with low-risk tissues require only standard disinfection.[43-45] Since noncritical surfaces are not involved in disease transmission, the normal exposure time (1 minute or longer) is recommended.[11]

To minimize the possibility of using neurosurgical instruments that potentially were contaminated during procedures performed on patients in whom CJD was later diagnosed, healthcare facilities should consider using the sterilization guidelines outlined above for neurosurgical instruments used during brain biopsy performed on patients in whom a specific lesion has not been demonstrated (eg, by magnetic resonance imaging or computerized tomography scans). Alternatively, disposable neurosurgical instruments could be used for such patients[43] or instruments could be quarantined until the pathology of the brain biopsy specimen is reviewed and the diagnosis of CJD is excluded. If disposable instruments are used, they should be of the same quality as reusable devices. Some countries (eg, France and Switzerland) have implemented enhanced sterilization rules to prevent transmission of CJD by means of surgical instruments, requiring steam sterilization at 134°C for 18 minutes for all surgical instruments. Other countries (eg, the United Kingdom) discard all surgical instruments used on high-risk tissues from patients known to have CJD.[60]

When strictly followed, these recommendations should eliminate the risk of transmitting infection by means of prion-contaminated medical and surgical instruments.

Clostridium difficile Infection: Role of the Environment and Transmission Prevention

C. difficile is an enteric bacterial pathogen that causes an infection that results in a broad spectrum of disease, ranging from mild diarrhea to life-threatening pseudomembraneous colitis. Although *C. difficile*–associated diarrhea (CDAD) has been frequently encountered in hospitals and long-term care facilities for many years, the rates in the United States have tripled from 2000 to 2005, and morbidity and mortality have also increased.[83-86] This trend has been associated with the emergence of a new, highly virulent strain of *C. difficile* that produces greater quantities of toxins A and B and a separate binary toxin. To effectively manage this disease and keep informed of its changing epidemiology, optimal strategies in CDAD surveillance, diagnosis, treatment, antibiotic control, and infection prevention are warranted.[87] This chapter only considers the role of the environment in transmission and the infection prevention strategies that prevent transmission (see Chapter 19, on prevention of *C. difficile* disease).

The 2 major reservoirs of *C. difficile* in healthcare settings are humans who are colonized or infected, and inanimate objects. Patients with symptomatic intestinal infection are thought to be a major reservoir. There are 3 mechanisms of transfer of *C. difficile* in the healthcare setting: (1) direct transfer of *C. difficile* from a colonized or infected patient to the environment (eg, a rectal thermometer, commode, or over-the-bed table), followed by contact with this environmental surface by another patient and subsequent inoculation into the mouth (Figure 7-1, section A); (2) direct transfer from a colonized or infected patient to a healthcare worker by direct contact and transfer on the healthcare worker's hands to a noncolonized or noninfected patient (Figure 7-1, section B); and (3) indirect transfer by a healthcare worker (or any other person) having contact with the contaminated environment and then transferring the organism to a noncolonized or noninfected patient (Figure 7-1, section C).[88] These modes of transmission can be prevented by infection prevention strategies such as contact precautions, hand hygiene with soap and water, and removal or inactivation of *C. difficile* spores from the inanimate environment (environmental surfaces or patient-care equipment).

Several factors facilitate the environmental route of transmission of *C. difficile*. First, the organism contaminates the environment of patients colonized or infected with *C. difficile*. Second, the *C. difficile* spore can survive in the hospital environment for up to 5 months,[89] whereas the vegetative bacteria die, because of desiccation, within 15 minutes in room air.[90] Vegetative *C. difficile* can remain viable on moist surfaces for up to 3 hours in room air. These data suggest that moist surfaces in hospitals (eg, toilets, sinks, and moist dressings) may provide a suitable environment for vegetative *C. difficile* to persist for several hours.[90] The spore is also more resistant to the effect of the gastric acids in the stomach.[91] Thus, the spore is the bacterial form that is more likely important in disease transmission and that must be inactivated and/or removed by surface disinfection. Accordingly, a claim based only on the vegetative bacteria would likely be potentially misleading and be incompletely effective in preventing disease transmission. Thus, the recent EPA letter preventing claims based on the inactivation of vegetative bacteria is both soundly based in science and judicious public health policy (F. Sanders, EPA, written communication, September 2008). Third, since spores are relatively resistant to inactivation by low-level disinfectants, a higher level of disinfection is needed to prevent environmental spread. At present, a sodium hypochlorite solution is the only product registered by the EPA for killing of *C. difficile* spores.

Transmission of *C. difficile* to a patient by means of transient hand carriage on healthcare workers' hands is thought to be the most likely mode of transmission (Figure 7-1, sections B and C). Transient hand carriage can occur through patient or environmental contact. Fifty-nine percent of 35 healthcare workers had cultures of specimens from their hands that were positive for

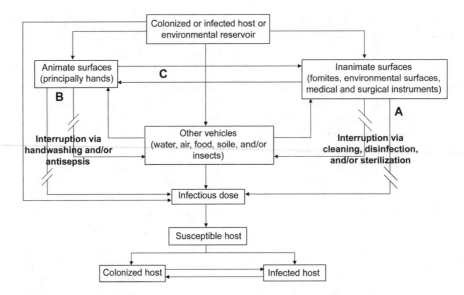

Figure 7-1. Mechanisms of transmission of *Clostridium difficile* by way of animate and inanimate surfaces in the healthcare setting. *Section A:* Direct transfer of *C. difficile* from a colonized or infected patient to the environment (eg, to a rectal thermometer, commode, or over-the-bed table) and subsequent contact with that object by another patient and inoculation into the mouth. *Section B:* Direct transfer from a colonized or infected patient to a healthcare worker by way of contact and transfer on the hands to a noncolonized or noninfected patient. *Section C:* Indirect transfer by a healthcare worker who contacts the contaminated environment and transfers the organism to a noncolonized or noninfected patient.

C. difficile after direct contact with a culture-positive patient.[92] *C. difficile* can be found at multiple body sites on patients with CDAD, including the groin, chest, abdomen, forearm, and hands, and can be transferred to the care provider's hands.[93]

Role of the Environment
While healthcare workers' hands are the most likely vehicle of transmission, the hands may become contaminated by either patient contact or contact with a contaminated environment, or both. *C. difficile* contamination has been found in the rooms of patients who are colonized or infected with *C. difficile,* and the spores can persist on hard surfaces for months.[89] For example, *C. difficile* contamination has been found on 49% of sites in rooms occupied by patients with *C. difficile* infection and on 29% of sites in rooms occupied by asymptomatic carriers.[92] Contamination of the environment and patient-care equipment occurs through fecal shedding or by means of the contaminated hands of the patient or healthcare workers.[91] There are several observations that demonstrate that contaminated environmental surfaces are important in the acquisition of *C. difficile,* to include that the incidence of CDAD is significantly associated with the proportion of culture-positive environmental sites, and that there is epidemiological evidence that the use of sodium hypochlorite for

environmental cleaning may significantly reduce the incidence of CDAD. Data also demonstrate that the proportion of personnel hand-specimen cultures positive for *C. difficile* was strongly correlated with the density of environmental contamination.[94] For example, the proportion of positive hand-specimen cultures was 0% when the proportion of environmental sites contaminated was 0%–25%, it was 8% when the environmental contamination rate was 25%–50%, and it was 36% when the environmental contamination rate was greater than 50%.[95] Additionally, the use of an effective antimicrobial (ie, sodium hypochlorite) significantly decreased environmental contamination rates in rooms of patients with *C. difficile* infection. For example, Eckstein et al.[96] observed 9 (90%) of 10 rooms of patients with CDAD had 1 or more *C. difficile*–positive cultures prior to cleaning with a 1 : 10 dilution of bleach, compared with 2 (20%) of the rooms after cleaning. Kaatz et al.[97] recovered *C. difficile* from 31% of environmental samples in their study. After the ward was disinfected with unbuffered hypochlorite (with 500 ppm available chlorine), the surface contamination decreased to 21% of the initial levels, and the CDAD outbreak ended. Phosphate buffered hypochlorite (with 1,600 ppm available chlorine) was even more effective in reducing environmental levels of *C. difficile,* with a 98% reduction in the level surface contamination.

Prevention Strategies

With CDAD rates increasing, clearly there is a need for more-effective infection prevention strategies. Strategies to prevent patient ingestion of spores consist of traditional infection prevention strategies that target the environment, hand hygiene, and barrier precautions, such as contact precautions.[85] Two strategies have been shown to be effective at interrupting disease transmission during CDAD clusters or epidemic periods: effective room decontamination by surface disinfection with sodium hypochlorite to minimize environmental contamination, and the use of effective barrier precautions (especially gloves) during patient contact to prevent transmission.[85]

Studies have shown that admission to a room previously occupied by a patient with methicillin-resistant *Staphylococcus aureus* (MRSA),[98] vancomycin-resistant enterococci (VRE),[99] or *C. difficile*[100] significantly increases the odds of acquiring the drug-resistant organisms. These studies demonstrate the importance of effective room disinfection or elimination of the pathogen from the environment. Pathogen survival on environmental surfaces or patient care equipment may be attributable to use of ineffective cleaning products (ie, disinfectants that do not kill the pathogen) or poor practices (ie, failure to wipe all surfaces or use of poor technique that does not remove the pathogen).[101]

Since *C. difficile* is shed in the feces, any surface or device that becomes contaminated by feces or hands can serve as a reservoir for its spores. The percentage of environmental surface samples contaminated with *C. difficile* in patients' rooms can vary from approximately 10% to greater than 50%.[89,94-97,102-109] The *C. difficile* spore load on environmental surfaces in healthcare facilities is low. To our knowledge, there are 7 studies that have assessed the microbial load of *C. difficile* on environmental surfaces, and most usually found less than 10 colonies of *C. difficile* in culture of a sample from a surface found to be contaminated.[89,95-97,103,107,109] Two studies reported more than 100 colonies; one reported a range of "1 to >200" colonies, and one study that sampled several sites with a sponge found 1,300 colonies. The heaviest contamination is found on floors, but other sites frequently found to be contaminated are windowsills, commodes, toilets, call buttons, scales, blood pressure cuffs, toys, bathtubs, tables, light switches, phones, door handles, mops, electronic thermometers, and feeding tube equipment. These spores will remain in the environment for months unless physically removed or inactivated by disinfectants. Most low-level disinfectants used in healthcare facilities (eg, alcohol, quaternary ammonium compounds, and phenolics) are not effective against *C. difficile* spores, although higher-level disinfectants do kill the spores (eg, glutaraldehyde and chlorine at a concentration of 5,000 ppm) (unpublished data, W.A.R., December 2008).[110]

The importance of environmental contamination in disease transmission is emphasized by the epidemiological findings that disinfection with sodium hypochlorite (ie, bleach) has been shown to be effective in reducing environmental contamination in patient rooms and in reducing CDAD rates in hospital units where the rate of CDAD is high (defined as more than 3 cases per 1,000 patient days). In an intervention study, the incidence of CDAD for bone marrow transplant patients decreased significantly, from 8.6 to 3.3 cases per 1,000 patient days, after the preparation used for environmental disinfection in the rooms of patients with CDAD was switched from a quaternary ammonium compound to a 1:10 dilution of concentrated hypochlorite solution.[104] When the protocol was reversed and use of a quaternary ammonium compound was reintroduced to those units, CDAD rates returned to the high baseline rate of 8.1 cases per 1,000 patient-days. No reduction in CDAD rates was seen among neurosurgical ICU and medicine patients, for whom baseline rates were 3.0 and 1.3 cases per 1,000 patient-days, respectively. Three other studies have also provided epidemiological data that demonstrate the effectiveness of chlorine disinfection in reducing the concentration of *C. difficile* spores on environmental surfaces and in reducing the incidence of *C. difficile* infection.[94,97,111] For this reason, the CDC recommends use of a 1:10 dilution of concentrated bleach during outbreaks of CDAD.[11] One application of bleach, covering surfaces to allow sufficient wetness for 1 minute of contact time, is recommended. A dilution of bleach with water normally takes 1–3 minutes to dry. For control of sporadic CDAD cases when there is no epidemic or recognized cross-transmission of *C. difficile*, hospitals can use their regular EPA-registered disinfectant for disinfection of patient rooms. Recently, room disinfection with vaporized hydrogen peroxide has also been found to reduce *C. difficile* incidence rates.[107]

In summary, environmental interventions are an important part of a comprehensive strategy for preventing transmission of *C. difficile* in the healthcare setting. The use of chlorine during hyperendemic and epidemic periods has been shown to reduce environmental contamination with *C. difficile* and to reduce the incidence of *C. difficile* infection. Interventions, such as use of chlorine, aimed at optimizing environmental disinfection are an important component of our infection prevention strategies.

New Technology and Issues

Human Papilloma Virus

Emerging pathogens are of growing concern to the general public and infection preventionists. Human papilloma virus is an extremely common sexually acquired pathogen and is considered the cause of cervical cancer. While there are limited data regarding the inactivation of human papilloma virus by disinfectants, because in vitro replication of complete virions has only been achieved recently, a pseudotype human papilloma virus 16 and a bovine papilloma virus were used in an infectivity assay to evaluate potential methods of disinfecting surfaces contaminated with human papilloma virus.[112] In that study, the bovine papilloma virus demonstrated substantial sensitivity to 70% ethanol, and all infectivity was eliminated for pseudotype human papilloma virus 16 virions.

Class 6 Chemical Indicator

The new CDC guideline on disinfection and sterilization[14] does not discuss the Class 6 indicator, because the guideline was prepared before the Class 6 indicator was offered in the United States. Class 6–emulating indicators are cycle-specific indicator strips. For example, the Class 6–emulating indicators for steam sterilization use a thermochromic ink that is claimed to visibly react to the critical parameters of steam sterilization (time, temperature, and the presence of saturated steam) and to provide an integrated response to steam sterilization operating cycles at 132°C for 3 minutes or longer. The indicator gives a visible confirmation of the achievement of sterilization conditions at the location within the sterilization chamber. The CDC guideline does say that biological indicator spores, intended for the type of sterilizer, should be used to monitor the effectiveness of sterilizers at least weekly. Thus, Class 6–emulating indicators are not a substitute for a biological indicator. No professional organization has recommended the use of Class 6–emulating indicators as a substitute for biological indicators, and there are no data that demonstrate that a Class 6 indicator mimics a biological indicator at suboptimal sterilization times and/or temperatures.

Contact Time for Disinfectant on Noncritical Surfaces

In order to get EPA clearance of the CDC guideline, it was necessary to insert 2 sentences: "By law, all applicable label instructions on EPA-registered products must be followed. If the user selects exposure conditions that differ from those on the EPA-registered product label, the user assumes liability from any injuries resulting from off-label use and is potentially subject to enforcement action under the Federal Insecticide, Fungicide, and Rodenticide Act (FIFRA)."[14(p54)] There are several points that should be made about this apparent disconnect between label instructions and what studies demonstrate. (1) Multiple scientific studies have demonstrated the efficacy of hospital disinfectants against pathogens causing healthcare-associated infections, with a contact time of at least 1 minute.[13,19,113-128] (2) The only way an institution can achieve a contact time of 10 minutes is to reapply the surface disinfectant 5–6 times to the surface, since the typical drying time for a water-based disinfectant is 1.5–2 minutes, and currently healthcare facilities (eg, University of North Caroline Health Care) are achieving surface disinfection of noncritical patient care items and environmental surfaces with a single application of a disinfectant and requiring a drying time of more than 1 minute. (3) As important as disinfectant contact time is the proper application of the disinfectant to the surface or equipment to ensure that all contaminated surfaces and noncritical patient care equipment are wiped; current studies demonstrate that only approximately 50% of high-risk objects are cleaned at terminal cleaning. (4) There are no data that demonstrate improved infection prevention with a 10-minute contact time, compared with a 1-minute contact time. (5) Finally, we are not aware of an enforcement action by the EPA against health care facilities for "off-label" use of a surface disinfectant.

Thus, we believe the guideline allows us to continue our use of low-level disinfectants for noncritical environmental surfaces and patient care equipment with a 1-minute contact time.[11] Additionally, all healthcare facilities should reemphasize the thoroughness of cleaning to ensure that all contaminated surfaces are wiped.

Room Decontamination Units: Hydrogen Peroxide Vapor and UV Light

Surface disinfection of noncritical surfaces and equipment is normally performed by manually applying a liquid disinfectant to the surface with a cloth, wipe, or mop. Recent studies have identified significant opportunities in hospitals to improve the cleaning of frequently touched objects in the patient's immediate environment.[101,129,130] For example, one study found that, of 20,646 standardized environmental surfaces (14 types of objects), only 9,910 (48%) were cleaned at terminal room cleaning.[130] Epidemiologic studies have shown that patients admitted to rooms previously occupied by individuals infected or colonized with MRSA,[98] VRE,[99] or C. difficile[100] are at significant risk of acquiring these organisms from contaminated environmental surfaces. These data have led to the development of room decontamination units that avoid the problems associated with the thoroughness of terminal cleaning activities in patient rooms.

Hydrogen peroxide vapor has been used increasingly for the decontamination of biological safety cabinets and rooms in healthcare facilities.[107,131-139] Investigators have found that use of hydrogen peroxide vapor is highly effective for eradicating various pathogens (eg, MRSA, M. tuberculosis, Serratia species, C. difficile spores, and Clostridium botulinum spores) from rooms, furniture, and equipment. This room decontamination method has not only been found to be effective in eradicating pathogens from contaminated surfaces but has also been found to significantly reduced the incidence of *C. difficile* infection.[107] A summary of the advantages and disadvantages of using hydrogen peroxide vapor for room decontamination is shown in Table 7-4.

Use of UV-C light units has also been proposed for room decontamination. One unit (Tru-D; Lumalier) uses an array of UV sensors that determines and targets shadowed areas and delivers a measured dose of UV radiation that destroys microorganisms. This unit is fully automated, activated by a hand-held remote controller, and does not require the room ventilation to be modified. It uses UV-C radiation (with a wavelength in the range of 254 nm) to decontaminate surfaces. It measures UV radiation reflected from the walls, ceiling, floors, and other treated areas and calculates the operation time to deliver the programmed lethal dose for pathogens.[140] After the dose of UV light is delivered, the device turns off and an audible alarm notifies the operator. In preliminary studies, it has reduced colony counts of MRSA, VRE, and

---

Table 7-4. Advantages and disadvantages of using hydrogen peroxide vapor (HPV) for room decontamination

Advantages

Efficacious (ie, has reliable biocidal activity) against a wide range of pathogens

Surfaces and equipment are decontaminated

Decreases incidence of disease (eg, *Clostridium difficile* infection)

Does not leave residue and does not give rise to health and safety concerns (is converted by aeration units into oxygen and water)

Uniform distribution in the room by means of an automated dispersal system

Useful for disinfecting complex equipment and furniture

Disadvantages

Contribution of the environment to disease transmission is approximately 5%

Only used at terminal disinfection (not during daily cleaning)

Environment is rapidly recontaminated

All patients must be removed from the area

Decontamination takes approximately 3–5 h (bed turnover time is 72 min)

HVAC system must be disabled to prevent unwanted dilution of HPV during the exposure; room must be sealed with tape

Costs

Does not remove dust and stains, which are important to patients and visitors

Sensitive parameters: gas concentration, 280 ppm; temperature, 26°C–28°C; relative humidity, 48%–57%

With long-term use, exposure to microcondensation may damage sensitive electronics

---

NOTE: HVAC, heating, ventilation, and air-conditioning.

Acinetobacter species by approximately 3.5 $\log_{10}$ in approximately 15 minutes. Sixty minutes is needed to achieve a 2.7 $\log_{10}$ reduction of *C. difficile* spores (W.A.R., D.J.W., and M. F. Gergen, unpublished data, 2009).

Low-Temperature Hydrogen Peroxide Vapor Sterilization

A new low-temperature sterilization system (V-Pro; Steris) uses vaporized hydrogen peroxide to sterilize reusable metal and nonmetal devices used in healthcare facilities. The system is compatible with a wide range of medical instruments and materials (eg, polypropylene, brass, and polyethylene). The by-products are nontoxic: only water vapor and oxygen are produced. The system is not intended to process liquids, linens, powders, or any cellulose materials. The system can sterilize instruments with diffusion-restricted spaces (eg, scissors) and medical devices with a single stainless steel lumen on the basis of the lumen internal diameter and length (eg, an inside diameter of 1 mm or larger and a length of 125 mm or shorter). Thus, gastrointestinal endoscopes and bronchoscopes cannot be sterilized in this system at the current time. Although this system has not been comparatively evaluated against other sterilization processes, vaporized hydrogen peroxide has been shown to be effective in killing spores, viruses, mycobacteria, fungi, and bacteria.[141]

Automated Endoscope Reprocessors

The potential for transmission of infection during endoscopy remains a concern for healthcare workers and patients.[142] Automated endoscope reprocessors offer many potential advantages (eg, they automate and standardize reprocessing steps, reduce personnel exposure to chemicals, and filter tap water) but some disadvantages (eg, they do not eliminate the need for cleaning, disinfection failure has been linked to poor design, and the devices do not monitor the concentration of high-level disinfectant). Two new automated endoscope reprocessors should offer benefits over older models. One integrates cleaning and has achieved an FDA-cleared cleaning claim (Evo-Tech; Advanced Sterilization Products). The user must do the "bedside" cleaning (ie, wipe external surfaces and flush each lumen with a detergent solution) and then must place the scope directly (within 1 hour) into the machine. This eliminates the labor-intensive manual cleaning. The device also automatically detects leaks, flushes alcohol through the channels prior to cycle completion to promote drying, and integrates minimum effective concentration monitoring. In addition, it has a printer that provides complete monitoring of critical cycle parameters, including the minimum effective concentration of the high-level disinfectant (ortho-phthalaldehyde), the disinfection time, and channel blockage, as well as temperature, pressure, and time, to ensure compliance throughout the process. The manufacturer's data show that the amount of residual organic material in the internal channels and external insertion tube surfaces was below the limit of 8.5 μg/cm^2.

The other automated endoscope reprocessor (Reliance; Steris) requires a minimal number of connections to the endoscope channels and uses a control boot (housing apparatus that creates pressure differentials to ensure connector-less fluid flow through all channels that are accessible through the endoscope's control handle channel ports). Data demonstrate that the soil and microbial removal effected by the washing phase of the device was equivalent to that achieved by optimal manual cleaning.[143] For example, there was reduction of more than 99% in protein and hemoglobin levels, and both methods reduced the level of residual organic material to below 6.4 μg/cm^2.

Accelerated Hydrogen Peroxide

Accelerated hydrogen peroxide is a newer disinfectant that contains very low levels of anionic and nonionic surfactants, which act with hydrogen peroxide to produce microbicidal activity. These ingredients are considered safe for humans and benign for the environment. It is prepared and marketed in several concentrations from 0.5% to 7%. The lower concentrations (0.5%: eg, Oxivir; JohnsonDiversey) are designed for the disinfection of hard surfaces, while the higher concentrations are recommended for use as high-level disinfectants. A 0.5% concentration of accelerated hydrogen peroxide demonstrated bactericidal and virucidal activity in 1 minute and mycobactericidal and fungicidal activity in 5 minutes.[144] It is more costly than other low-level disinfectants, such as quaternary ammonium compounds.

A high-level disinfectant based on accelerated hydrogen peroxide (Resert; Steris) is also available for disinfection of heat-sensitive semicritical medical devices, including manual and automatic reprocessing of flexible endoscopes. This product has demonstrated sporicidal activity, with a reduction in viability titer of 6 log$_{10}$ in 6 hours at 20°C; it has also demonstrated mycobactericidal, fungicidal, and virucidal activity with a contact time of 8 minutes. It is reported to be a relatively mild solution for users and is considered to be compatible with flexible endoscopes. It is slightly irritating to skin and mildly irritating to eyes, according to accepted standard test methods (the same as 3% topical hydrogen peroxide).[145,146] A 7% concentration of accelerated hydrogen peroxide can be reused for several days and retain its broad-spectrum antimicrobial activity.[146,147] This product has no special shipping or venting requirements.

Assessing Risk to Patients from Disinfection and Sterilization Failures

Disinfection and sterilization are critical components of infection control. Unfortunately, breaches of disinfection and sterilization guidelines are not uncommon. A 14-step algorithm has been constructed to aid infection preventionists in the evaluation of potential disinfection and sterilization failures.[148] The need to notify patients because of improper reprocessing of semicritical medical devices (eg, endoscopes) and critical medical instruments has occurred regularly.[148] In this chapter we provide a method for assessing patient risk for adverse events, especially infection; the algorithm given in Table 7-5 can guide an institution in managing potential disinfection and sterilization failures.

Emerging Pathogens, Antibiotic-Resistant Bacteria, and Bioterrorism Agents

Emerging pathogens are of growing concern to the general public and infection preventionists. Relevant pathogens include *Cryptosporidium parvum, Helicobacter pylori, Escherichia coli* O157:H7, human immunodeficiency virus, hepatitis C virus, rotavirus, multidrug-resistant *M. tuberculosis,* human papilloma virus, and nontuberculous mycobacteria (eg, *Mycobacterium chelonae*). Similarly, recent publications have highlighted the concern about the potential for biological terrorism.[149] The CDC has categorized several pathogens as "high priority" because they can be easily disseminated or transmitted from person to person, they cause high mortality, and infections are likely to cause public panic

---

Table 7-5.   Algorithm for an exposure investigation after a failure to follow disinfection and sterilization principles

1. Confirm the disinfection or sterilization reprocessing failure
2. Embargo any improperly disinfected or sterilized items
3. Do not use the questionable disinfection or sterilization unit (eg, a sterilizer or automated endoscope reprocessor)
4. Inform key stakeholders
5. Conduct a complete and thorough evaluation of the cause of the disinfection and/or sterilization failure
6. Prepare a line listing of potentially exposed patients
7. Assess whether the disinfection or sterilization failure increases patients' risk for infection
8. Inform expanded list of stakeholders of the reprocessing issue
9. Develop a hypothesis for the disinfection or sterilization failure and initiate corrective action
10. Develop a method to assess potential adverse patient events
11. Consider notification of state and federal authorities
12. Consider patient notification
13. Develop a long-term follow-up plan
14. Generate an after-action report

and social disruption.[150] They include *Bacillus anthracis* (which causes anthrax), *Yersinia pestis* (which causes plague), variola major virus (which causes smallpox), *Francisella tularensis* (which causes tularemia), the filoviruses Ebola virus and Marburg virus (which cause Ebola hemorrhagic fever and Marburg hemorrhagic fever, respectively); and the arenaviruses Lassa virus and Junin virus (which cause Lassa fever and Argentine hemorrhagic fever, respectively), and related viruses.[150]

The susceptibility of each of these pathogens to chemical disinfectants and/or sterilants has been studied, and all of these pathogens (or surrogate microbes such as feline-calicivirus for Norwalk virus, vaccinia for variola virus,[151] and *Bacillus atrophaeus* [formerly *Bacillus subtilis*] for *B. anthracis*) are susceptible to currently available chemical disinfectants and/or sterilants.[152] Standard sterilization and disinfection procedures for patient-care equipment (as recommended in this chapter) are adequate to sterilize or disinfect instruments or devices contaminated with blood or other body fluids from persons infected with bloodborne pathogens, emerging pathogens, and bioterrorism agents, with the exception of prions (see the section in this chapter on inactivation of prions, above). No changes in procedures for cleaning, disinfecting, or sterilizing need to be made.[11,26]

In addition, there are no data to show that antibiotic-resistant bacteria (MRSA, VRE, or multidrug-resistant *M. tuberculosis*) are less sensitive to the liquid chemical germicides than antibiotic-sensitive bacteria, with the contact conditions and germicide concentrations currently used.[11,26,128,153]

Conclusion

When properly used, disinfection and sterilization can ensure the safe use of invasive and noninvasive medical devices. The method of disinfection and sterilization depends on the intended use of the medical device: critical items (those that contact sterile tissue) must be sterilized prior to use; semicritical items (those that contact mucous membranes or nonintact skin) must undergo high-level disinfection; and noncritical items (those that contact intact skin) should undergo low-level disinfection. Cleaning should always precede high-level disinfection and sterilization. Current disinfection and sterilization guidelines must be strictly followed.

References

1. Centers for Disease Control and Prevention. Ambulatory and inpatient procedures in the United States, 1996. Series 13, No. 139. Atlanta, GA; 1998. http://www.cdc.gov/nchs/pressroom/99facts/ambinpat.htm. Accessed July 15, 2009.

2. McCarthy GM, Koval JJ, John MA, MacDonald JK. Infection control practices across Canada: do dentists follow the recommendations? *J Can Dent Assoc* 1999;65:506–511.

3. Spach DH, Silverstein FE, Stamm WE. Transmission of infection by gastrointestinal endoscopy and bronchoscopy. *Ann Intern Med* 1993;118:117–128.

4. Weber DJ, Rutala WA. Lessons from outbreaks associated with bronchoscopy. *Infect Control Hosp Epidemiol* 2001;22:403–408.

5. Weber DJ, Rutala WA, DiMarino AJ Jr. The prevention of infection following gastrointestinal endoscopy: the importance of prophylaxis and reprocessing. In: DiMarino AJ Jr, Benjamin SB, eds. *Gastrointestinal Diseases: An Endoscopic Approach*. Thorofare, NJ: Slack; 2002:87–106.

6. Meyers H, Brown-Elliott BA, Moore D, et al. An outbreak of *Mycobacterium chelonae* infection following liposuction. *Clin Infect Dis* 2002;34:1500–1507.

7. Lowry PW, Jarvis WR, Oberle AD, et al. *Mycobacterium chelonae* causing otitis media in an ear-nose-and-throat practice. *N Engl J Med* 1988;319:978–982.

8. Rutala WA, Weber DJ. Disinfection and sterilization in health care facilities: what clinicians need to know. *Clin Infect Dis* 2004;39:702–709.

9. Rutala WA, Weber DJ. Disinfection and sterilization: what's new. *APIC Text of Infection Control and Epidemiology*. Washington, DC: Association of Professionals in Infection Control; 2009.

10. Spaulding EH. Chemical disinfection of medical and surgical materials. In: Lawrence C, Block SS, eds. *Disinfection, Sterilization, and Preservation*. Philadelphia: Lea & Febiger, 1968:517–531.

11. Rutala WA, Weber DJ, Healthcare Infection Control Practices Advisory Committee. Guideline for disinfection and sterilization in healthcare facilities, 2008. http://www.cdc.gov/ncidod/dhqp/. Accessed October 23, 2009.

12. Simmons BP. CDC guidelines for the prevention and control of nosocomial infections. Guideline for hospital environmental control. *Am J Infect Control* 1983;11:97–120.

13. Rutala WA; 1994, 1995, and 1996 APIC Guidelines Committee. APIC guideline for selection and use of disinfectants. Association for Professionals in Infection Control and Epidemiology. *Am J Infect Control* 1996;24:313–342.

14. Sehulster L, Chinn RYW, Healthcare Infection Control Practices Advisory Committee. Guidelines for environmental infection control in health-care facilities: recommendations of CDC and the Healthcare Infection Control Practices Advisory Committee (HICPAC). *MMWR Morb Mortal Wkly Rep* 2003;52(RR-10):1–42.

15. Food and Drug Administration. FDA-cleared sterilant and high-level disinfectants with general claims for processing reusable medical and dental devices—March 2009. http://www.fda.gov/cdrh/ode/germlab.html. Accessed July 17, 2009.

16. Nelson DB, Jarvis WR, Rutala WA, et al. Multi-society guideline for reprocessing flexible gastrointestinal endoscopes. *Infect Control Hosp Epidemiol* 2003;24:532–537.

17. Gerding DN, Peterson LR, Vennes JA. Cleaning and disinfection of fiberoptic endoscopes: evaluation of glutaraldehyde exposure time and forced-air drying. *Gastroenterology* 1982;83:613–618.

18. Weber DJ, Rutala WA. Environmental issues and nosocomial infections. In: Wenzel RP, ed. *Prevention and Control of*

Nosocomial Infections. Baltimore: Williams and Wilkins; 1997:491–514.

19. Weber DJ, Rutala WA. Role of environmental contamination in the transmission of vancomycin-resistant enterococci. *Infect Control Hosp Epidemiol* 1997;18:306–309.

20. Schembre DB. Infectious complications associated with gastrointestinal endoscopy. *Gastrointest Endosc Clin N Am* 2000;10:215–232.

21. Jackson FW, Ball MD. Correction of deficiencies in flexible fiberoptic sigmoidoscope cleaning and disinfection technique in family practice and internal medicine offices. *Arch Fam Med* 1997;6:578–582.

22. Orsi GB, Filocamo A, Di Stefano L, Tittobello A. Italian National Survey of Digestive Endoscopy Disinfection Procedures. *Endoscopy* 1997;29:732–738.

23. Honeybourne D, Neumann CS. An audit of bronchoscopy practice in the United Kingdom: a survey of adherence to national guidelines. *Thorax* 1997;52:709–713.

24. Srinivasan A, Wolfenden LL, Song X, et al. An outbreak of *Pseudomonas aeruginosa* infections associated with flexible bronchoscopes. *N Engl J Med* 2003;348:221–227.

25. Cetse JC, Vanhems P. Outbreak of infection associated with bronchoscopes. *N Engl J Med* 2003;348:2039–2040.

26. Rutala WA, Weber DJ. Selection and use of disinfectants in healthcare. In: Mayhall CG, ed. *Hospital Epidemiology and Infection Control*. Philadelphia: Lippincott Williams & Wilkins; 2004:1473–1522.

27. Rutala WA, Weber DJ. FDA labeling requirements for disinfection of endoscopes: a counterpoint. *Infect Control Hosp Epidemiol* 1995;16:231–235.

28. Kruse A, Rey JF. Guidelines on cleaning and disinfection in GI endoscopy. Update 1999. The European Society of Gastrointestinal Endoscopy. *Endoscopy* 2000;32:77–80.

29. British Society of Gastroenterology. Cleaning and disinfection of equipment for gastrointestinal endoscopy. Report of a working party of the British Society of Gastroenterology Endoscope Committee. *Gut* 1998;42:585–593.

30. British Thoracic Society. British Thoracic Society guidelines on diagnostic flexible bronchoscopy. *Thorax* 2001;56:1–21.

31. Centers for Disease Control and Prevention. Surveillance for Creutzfeldt-Jakob disease—United States. *MMWR Morb Mortal Wkly Rep* 1996;45:665–668.

32. Johnson RT, Gibbs CJ Jr. Creutzfeldt-Jakob disease and related transmissible spongiform encephalopathies. *N Engl J Med* 1998;339:1994–2004.

33. Collins SJ, Lawson VA, Masters CL. Transmissible spongiform encephalopathies. *Lancet* 2004;363:51–61.

34. Brown P, Brandel J-P, Preece M, Sato T. Iatrogenic Creutzfeldt-Jakob disease: the waning of an era. *Neurology* 2006;67:389–393.

35. World Health Organization. WHO infection control guidelines for transmissible spongiform encephalopathies. http://www.who.int/csr/resources/publications/bse/en. Accessed July 17, 2009.

36. Brown P, Preece M, Brandel JP, et al. Iatrogenic Creutzfeldt-Jakob disease at the millennium. *Neurology* 2000;55:1075–1081.

37. Brown P. Environmental causes of human spongiform encephalopathy. In: Baker H, Ridley RM, eds. *Methods in Molecular Medicine: Prion Diseases*. Totowa, NJ: Humana Press, 1996:139–154.

38. Brown P, Gibbs CJ, Rodgers-Johnson P, et al. Human spongiform encephalopathy: the National Institutes of Health series of 300 cases of experimentally transmitted disease. *Ann Neurol* 1994;35:513–529.

39. Rabano A, de Pedro-Cuesta J, Molbak K, et al. Tissue classification for the epidemiological assessment of surgical transmission of sporadic Creutzfeldt-Jakob disease: a proposal on hypothetical risk levels. *BMC Public Health* 2005;5:9.

40. Jacobs P. Cleaning: principles, methods and benefits. In: Rutala WA, ed. *Disinfection, Sterilization, and Antisepsis in Healthcare*. Champlain, NY: Polyscience Publications; 1998:165–181.

41. Merritt K, Hitchins VM, Brown SA. Safety and cleaning of medical materials and devices. *J Biomed Mater Res* 2000;53:131–136.

42. Alfa MJ, Jackson M. A new hydrogen peroxide-based medical-device detergent with germicidal properties: comparison with enzymatic cleaners. *Am J Infect Control* 2001;29:168–177.

43. Rutala WA, Weber DJ. Creutzfeldt-Jakob disease: recommendations for disinfection and sterilization. *Clin Infect Dis* 2001;32:1348–1356.

44. Favero MS. Current issues in hospital hygiene and sterilization technology. *J Infect Control (Asia Pacific Edition)* 1998;1:8–10.

45. Favero MS, Bond WW. Chemical disinfection of medical and surgical materials. In: Block SS, ed. *Disinfection, Sterilization, and Preservation*. Philadelphia: Lippincott Williams & Wilkins; 2001:881–917.

46. Steelman VM. Activity of sterilization processes and disinfectants against prions (Creutzfeldt-Jakob disease agent). In: Rutala WA, ed. *Disinfection, Sterilization, and Antisepsis in Healthcare*. Champlain, NY: Polyscience Publications; 1998:255–271.

47. Committee on Health Care Issues. American Neurological Association. Precautions in handling tissues, fluids, and other contaminated materials from patients with documented or suspected Creutzfeldt-Jakob disease. *Ann Neurol* 1986;19:75–77.

48. Association of Operating Room Nurses (AORN). Recommended practices for cleaning and care of surgical instruments and powered equipment. In: AORN, ed. *Perioperative Standards and Recommended Practices, 2009 edition*. Denver, CO: AORN; 2009:626–630.

49. Association for the Advancement of Medical Instrumentation (AAMI). Comprehensive guide to steam sterilization and sterility assurance in health care facilities. ANSI/AAMI ST79:2006. http://www.aami.org/standards/. Accessed July 17, 2009.

50. Lipscomb IP, Pinchin H, Collin R, Kevil CW. Effect of drying time, ambient temperature and pre-soaks on prion-infected tissue contamination levels on surgical stainless steel: concerns over prolonged transportation of instruments from theatre to central sterile service departments. *J Hosp Infect* 2007;65:72–77.

51. Fichet G, Comoy C, Duval C, et al. Novel methods for disinfection of prion-contaminated medical devices. *Lancet* 2004;364:521–526.

52. Fichet G, Comoy E, Dehen C, et al. Investigations of a prion infectivity assay to evaluate methods of decontamination. *J Microbiol Methods* 2007;70:511–518.

53. Baier M, Schwarz A, Mielke M. Activity of an alkaline 'cleaner' in the inactivation of the scrapie agent. *J Hosp Infect* 2004;57:80–84.

54. Yoshioka M, Murayama Y, Miwa T, et al. Assessment of prion inactivation by combined use of *Bacillus*-derived protease and SDS. *Biosci Biotechnol Biochem* 2007;71: 2565–2568.

55. Yan ZX, Stitz L, Heeg P, Roth K, Mauz P-S. Low-temperature inactivation of prion-protein on surgical steel surfaces with hydrogen peroxide gas plasma sterilization. *Zentr Steril* 2008;16:26–34.

56. Jackson GS, McKintosh E, Flechsig E, et al. An enzyme-detergent method for effective prion decontamination of surgical steel. *J Gen Virol* 2005;86:869–878.

57. Lawson VA, Stewart JD, Masters CL. Enzymatic detergent treatment protocol that reduces protease-resistant prion load and infectivity from surgical-steel monofilaments contaminated with a human-derived prion strain. *J Gen Virol* 2007; 88:2905–2914.

58. McDonnell G. Prion disease transmission: can we apply standard precautions to prevent or reduce risks? *Healthcare Challenges* 2008;18:298–304.

59. United Kingdom Department of Health. New technologies working group report on prion inactivating agents. London: 2008:1–30. http://www.dh.gov.uk/en/Publicationsandstatistics/index.htm. Accessed July 17, 2009.

60. Fichet G, Harrison J, McDonnell G. Reduction of risk of prion transmission on surgical devices with effective cleaning processes. *Zentr Steril* 2007;15:418–437.

61. Taylor DM. Inactivation of transmissible degenerative encephalopathy agents: a review. *Veterinary J* 2000;159: 10–17.

62. Kampf G, Blob R, Martiny H. Surface fixation of dried blood by glutaraldehyde and peracetic acid. *J Hosp Infect* 2004; 57:139–143.

63. Taylor DM, McConnell I. Autoclaving does not decontaminate formol-fixed scrapie tissues. *Lancet* 1988; 1(8600): 1463–1464.

64. Brown P, Liberski PP, Wolff A, et al. Resistance of scrapie infectivity to steam autoclaving after formaldehyde fixation and limited survival after ashing at 360°C: practical and theoretical implications. *J Infect Dis* 1990;161: 467–472.

65. Fernie K, Steele PJ, Taylor DM, Somerville RA. Comparative studies on the thermostability of five strains of transmissible-spongiform-encephalopathy agent. *Biotechnol Appl Biochem* 2007;47:175–183.

66. Brown P, Wolff A, Gajdusek DC. A simple and effective method for inactivating virus infectivity in formalin-fixed tissue samples from patients with Creutzfeldt-Jakob disease. *Neurology* 1990;40:887–890.

67. Yan Z, Stitz L, Heeg P, Pfaff E, Roth K. Infectivity of prion protein bound to stainless steel wires: a model for testing decontamination procedures for transmissible spongiform encephalopathies. *Infect Control Hosp Epidemiol* 2004;25: 280–283.

68. Taylor DM, Fraser H, McConnell I, et al. Decontamination studies with the agents of bovine spongiform encephalopathy and scrapie. *Arch Virol* 1994;139:313–326.

69. Vadrot C, Darbord J-C. Quantitative evaluation of prion inactivation comparing steam sterilization and chemical steri-lants: proposed method for test standardization. *J Hosp Infect* 2006;64:143–148.

70. Taylor DM. Inactivation of prions by physical and chemical means. *J Hosp Infect* 1999;43 (suppl):S69–S76.

71. Peretz D, Supattapone S, Giles K, et al. Inactivation of prions by acidic sodium dodecyl sulfate. *J Virol* 2006;80: 322–331.

72. Kimberlin RH, Walker CA, Millson GC, et al. Disinfection studies with two strains of mouse-passaged scrapie agent. Guidelines for Creutzfeldt-Jakob and related agents. *J Neurol Sci* 1983;59:355–369.

73. Taguchi F, Tamai Y, Uchida A, et al. Proposal for a procedure for complete inactivation of the Creutzfeldt-Jakob disease agent. *Arch Virol* 1991;119:297–301.

74. Ernst DR, Race RE. Comparative analysis of scrapie agent inactivation methods. *J Virol Methods* 1993;41:193–201.

75. Brown P, Rohwer RG, Gajdusek DC. Newer data on the inactivation of scrapie virus or Creutzfeldt-Jakob disease virus in brain tissue. *J Infect Dis* 1986;153:1145–1148.

76. Taylor DM, Fernie K, McConnell I. Inactivation of the 22A strain of scrapie agent by autoclaving in sodium hydroxide. *Vet Microbiol* 1997;58:87–91.

77. Brown SA, Merritt K. Use of containment pans and lids for autoclaving caustic solutions. *Am J Infect Control* 2003;31: 257–260.

78. Race RE, Raymond GJ. Inactivation of transmissible spongiform encephalopathy (prion) agents by Environ LpH. *J Virol* 2004;78:2164–2165.

79. Brown P, Rohwer RG, Green EM, Gajdusek DC. Effect of chemicals, heat, and histopathologic processing on high-infectivity hamster-adapted scrapie virus. *J Infect Dis* 1982; 145:683–687.

80. Brown P, Gibbs CJ Jr, Amyx HL, et al. Chemical disinfection of Creutzfeldt-Jakob disease virus. *N Engl J Med* 1982; 306:1279–1282.

81. Lemmer K, Mielke M, Pauli G, Beekes M. Decontamination of surgical instruments from prion proteins: in vitro studies on the detachment, destabilization and degradation of PrP^{sc} bound to steel surfaces. *J Gen Virol* 2004;85:3805–3816.

82. Clinical and Laboratory Standards Institute (CLSA). Protection of laboratory workers from occupationally acquired infections: approved guideline. CLSA M29-A3. 2005.

83. Warny M, Pepin J, Fang A, et al. Toxin production by an emerging strain of *Clostridium difficile* associated with outbreaks of severe disease in North America and Europe. *Lancet* 2005;366:1079–1084.

84. McDonald LC, Killgore GE, Thompson A, et al. An epidemic, toxin gene-variant strain of *Clostridium difficile*. *N Engl J Med* 2005;353:2433–2441.

85. Gerding DN, Muto CA, Owens RC Jr. Measures to control and prevent *Clostridium difficile* infection. *Clin Infect Dis* 2008;46(Suppl 1):S43–S49.

86. Kazakova SV, Ware K, Baughman B, et al. A hospital outbreak of diarrhea due to an emerging epidemic strain of *Clostridium difficile*. *Arch Intern Med* 2006;166:2518–2524.

87. Bartlett JG. Narrative review: the new epidemic of *Clostridium difficile*-associated enteric disease. *Ann Intern Med* 2006;145:758–764.

88. Sattar SA, Springthorpe VS. Transmission of viral infections through animate and inanimate surfaces and infection control through chemical disinfection. In: Hurst CJ, ed. *Modeling*

Disease Transmission and Its Prevention by Disinfection. Cambridge, UK: Cambridge University Press, 1996:224–257.

89. Kim KH, Fekety R, Batts DH, et al. Isolation of *Clostridium difficile* from the environment and contacts of patients with antibiotic-associated colitis. *J Infect Dis* 1981;143:42–50.

90. Jump RLP, Pultz MJ, Donskey CJ. Vegetative *Clostridium difficile* survives in room air on moist surfaces and in gastric contents with reduced acidity: a potential mechanism to explain the association between proton pump inhibitors and *C. difficile*-associated diarrhea. *Antimicrob Agents Chemother* 2007;51:2883–2887.

91. Association for Professionals in Infection Control and Epidemiology (APIC). Guide to the elimination of *Clostridium difficile* in healthcare settings. Washington DC; APIC; 2008:1–66.

92. McFarland LV, Mulligan ME, Kwok RY, Stamm WE. Nosocomial acquisition of *Clostridium difficile* infection. *N Engl J Med* 1989;320:204–210.

93. Bobulsky GS, Al-Nassir WN, Riggs MM, Sethi AK, Donskey CJ. *Clostridium difficile* skin contamination in patients with *C. difficile*-associated disease. *Clin Infect Dis* 2008;46:447–450.

94. Wilcox MH, Fawley WN, Wigglesworth N, Parnell P, Verity P, Freeman J. Comparison of the effect of detergent versus hypochlorite cleaning on environmental contamination and incidence of *Clostridium difficile* infection. *J Hosp Infect* 2003;54:109–114.

95. Samore MH, Venkataraman L, DeGirolami PC, Arbeit RD, Karchmer AW. Clinical and molecular epidemiology of sporadic and clustered cases of nosocomial *Clostridium difficile* diarrhea. *Am J Med* 1996;100:32–40.

96. Eckstein BC, Adams DA, Eckstein EC, et al. Reduction of *Clostridium difficile* and vancomycin-resistant *Enterococcus* contamination of environmental surfaces after an intervention to improve cleaning methods. *BMC Infect Dis* 2007;7:61.

97. Kaatz GW, Gitlin SD, Schaberg DR, et al. Acquisition of *Clostridium difficile* from the hospital environment. *Am J Epidemiol* 1988;127:1289–1294.

98. Huang SS, Datta R, Platt R. Risk of acquiring antibiotic-resistant bacteria from prior room occupants. *Arch Intern Med* 2006;166:1945–1951.

99. Drees M, Snydman DR, Schmid CH, et al. Prior environmental contamination increases the risk of acquisition of vancomycin-resistant enterococci. *Clin Infect Dis* 2008;46:678–685.

100. Shaughnessy M, Micielli R, Depestel D, et al. Evaluation of hospital room assignment and acquisition of *Clostridium difficile* associated diarrhea (CDAD). In: Program and abstracts of the 48th Annual Interscience Conference on Antimicrobial Agents and Chemotherapy and the Infections Disease Society of America; Washington, DC; October 25–28, 2008. Abstract K-4194.

101. Carling PC, Parry MF, Von Beheren SM, Healthcare Environmental Hygiene Study Group. Identifying opportunities to enhance environmental cleaning in 23 acute care hospitals. *Infect Control Hosp Epidemiol* 2008;29:1–7.

102. Cohen SH, Tang YJ, Rahmani D, Silva J Jr. Persistence of an endemic (toxigenic) isolate of *Clostridium difficile* in the environment of a general medicine ward. *Clin Infect Dis* 2000;30:952–954.

103. Fekety R, Kim K-H, Brown D, Batts DH, Cudmore M, Silva J Jr. Epidemiology of antibiotic-associated colitis: isolation of *Clostridium difficile* from the hospital environment. *Am J Med* 1981;70:906–908.

104. Mayfield JL, Leet T, Miller J, Mundy LM. Environmental control to reduce transmission of *Clostridium difficile*. *Clin Infect Dis* 2000;31:995–1000.

105. Cohen SH, Tang YJ, Muenzer J, Gumerlock PH, Silva J Jr. Isolation of various genotypes of *Clostridium difficile* from patients and the environment in an oncology ward. *Clin Infect Dis* 1997;24:889–893.

106. Dubberke ER, Reske KA, Noble-Wang J, et al. Prevalence of *Clostridium difficile* environmental contamination and strain variability in multiple health care facilities. *Am J Infect Control* 2007;35:315–318.

107. Boyce JM, Havill NL, Otter JA, et al. Impact of hydrogen peroxide vapor room decontamination on *Clostridium difficile* environmental contamination and transmission in a healthcare setting. *Infect Control Hosp Epidemiol* 2008;29:723–729.

108. Verity P, Wilcox MH, Fawley W, Parnell P. Prospective evaluation of environmental contamination by *Clostridium difficile* in isolation side rooms. *J Hosp Infect* 2001;49:204–209.

109. Mulligan ME, George WL, Rolfe RD, Finegold SM. Epidemiological aspects of *Clostridium difficile* induced diarrhea and colitis. *Am J Clin Nutr* 1980;33:2533–2538.

110. Perez J, Springthorpe S, Sattar SA. Activity of selected oxidizing microbicides against spores of *Clostridium difficile*: relevance to environmental control. *Am J Infect Control* 2005;33:320–325.

111. McMullen KM, Zack J, Coopersmith CM, Kollef M, Dubberke E, Warren DK. Use of hypochlorite solution to decrease rates of *Clostridium difficile*-associated diarrhea. *Infect Control Hosp Epidemiol* 2007;28:205–207.

112. Roden RBS, Lowy DR, Schiller JT. Papillomavirus is resistant to desiccation. *J Infect Dis* 1997;176:1076–1079.

113. Rutala WA, Weber DJ. Surface disinfection: should we do it? *J Hosp Infect* 2001;48(Suppl A):S64–S68.

114. Dharan S, Mourouga P, Copin P, Bessmer G, Tschanz B, Pittet D. Routine disinfection of patients' environmental surfaces: myth or reality? *J Hosp Infect* 1999;42:113–117.

115. Ward RL, Bernstein DI, Knowlton DR, et al. Prevention of surface-to-human transmission of rotaviruses by treatment with disinfectant spray. *J Clin Microbiol* 1991;29:1991–1996.

116. Gwaltney JM Jr, Hendley JO. Transmission of experimental rhinovirus infection by contaminated surfaces. *Am J Epidemiol* 1982;116:828–833.

117. Sattar SA, Jacobsen H, Springthorpe VS, Cusack TM, Rubino JR. Chemical disinfection to interrupt transfer of rhinovirus type 14 from environmental surfaces to hands. *Appl Environ Microbiol* 1993;59:1579–1585.

118. Rutala WA, Barbee SL, Aguiar NC, Sobsey MD, Weber DJ. Antimicrobial activity of home disinfectants and natural products against potential human pathogens. *Infect Control Hosp Epidemiol* 2000;21:33–38.

119. Silverman J, Vazquez JA, Sobel JD, Zervos MJ. Comparative in vitro activity of antiseptics and disinfectants versus clinical isolates of *Candida* species. *Infect Control Hosp Epidemiol* 1999;20:676–684.

120. Best M, Sattar SA, Springthorpe VS, Kennedy ME. Efficacies of selected disinfectants against *Mycobacterium tuberculosis*. *J Clin Microbiol* 1990;28:2234–2239.

121. Best M, Kennedy ME, Coates F. Efficacy of a variety of disinfectants against *Listeria* spp. *Appl Environ Microbiol* 1990; 56:377–380.

122. Best M, Springthorpe VS, Sattar SA. Feasibility of a combined carrier test for disinfectants: studies with a mixture of five types of microorganisms. *Am J Infect Control* 1994;22: 152–162.

123. Springthorpe VS, Grenier JL, Lloyd-Evans N, Sattar SA. Chemical disinfection of human rotaviruses: efficacy of commercially-available products in suspension tests. *J Hyg (Lond)* 1986;97:139–161.

124. Akamatsu T, Tabata K, Hironga M, Kawakami H, Uyeda M. Transmission of *Helicobacter pylori* infection via flexible fiber-optic endoscopy. *Am J Infect Control* 1996;24:396–401.

125. Resnick L, Veren K, Salahuddin SZ, Tondreau S, Markham PD. Stability and inactivation of HTLV-III/LAV under clinical and laboratory environments. *JAMA* 1986;255: 1887–1891.

126. Weber DJ, Barbee SL, Sobsey MD, Rutala WA. The effect of blood on the antiviral activity of sodium hypochlorite, a phenolic, and a quaternary ammonium compound. *Infect Control Hosp Epidemiol* 1999;20:821–827.

127. Rice EW, Clark RM, Johnson CH. Chlorine inactivation of *Escherichia coli* O157:H7. *Emerg Infect Dis* 1999;5: 461–463.

128. Anderson RL, Carr JH, Bond WW, Favero MS. Susceptibility of vancomycin-resistant enterococci to environmental disinfectants. *Infect Control Hosp Epidemiol* 1997;18:195–199.

129. Carling PC, Briggs JL, Perkins J, Highlander D. Improved cleaning of patient rooms using a new targeting method. *Clin Infect Dis* 2006;42:385–388.

130. Carling PC, Parry MF, Rupp ME, et al. Improving cleaning of the environment surrounding patients in 36 acute care hospitals. *Infect Control Hosp Epidemiol* 2008;29: 1035–1041.

131. French GL, Otter JA, Shannon KP, Adams NMT, Watling D, Parks MJ. Tackling contamination of the hospital environment by methicillin-resistant *Staphylococcus aureus* (MRSA): a comparison between conventional terminal cleaning and hydrogen peroxide vapour decontamination. *J Hosp Infect* 2004;57:31–37.

132. Bartels MD, Kristofferson K, Slotsbjerg T, Rohde SM, Lunfgren B, Westh H. Environmental methicillin-resistant *Staphylococcus aureus* (MRSA) disinfection using dry-mist-generated hydrogen peroxide. *J Hosp Infect* 2008;70:35–41.

133. Hall L, Otter JA, Chewins J, Wengenack NL. Use of hydrogen peroxide vapor for deactivation of *Mycobacterium tuberculosis* in a biological safety cabinet and a room. *J Clin Microbiol* 2007;45:810–815.

134. Hardy KJ, Gossain S, Henderson N, et al. Rapid recontamination with MRSA of the environment of an intensive care unit after decontamination with hydrogen peroxide vapour. *J Hosp Infect* 2007;66:360–368.

135. Johnston MD, Lawson S, Otter JA. Evaluation of hydrogen peroxide vapour as a method for the decontamination of surfaces contaminated with *Clostridium botulinum* spores. *J Microbiol Methods* 2005;60:403–411.

136. Heckert RA, Best M, Jordan LT, Dulac GC, Eddington DL, Sterritt WG. Efficacy of vaporized hydrogen peroxide against exotic animal viruses. *Appl Environ Microbiol* 1997;63: 3916–3918.

137. Klapes NA, Vesley D. Vapor-phase hydrogen peroxide as a surface decontaminant and sterilant. *Appl Environ Microbiol* 1990;56:503–506.

138. Bates CJ, Pearse R. Use of hydrogen peroxide vapour for environmental control during a *Serratia* outbreak in a neonatal intensive care unit. *J Hosp Infect* 2005;61: 364–366.

139. Shapey S, Machin K, Levi K, Boswell TC. Activity of a dry mist hydrogen peroxide system against environmental *Clostridium difficile* contamination in elderly care wards. *J Hosp Infect* 2008;70:136–141.

140. Owens MU, Deal DR, Shoemaker MO, Knudson GB, Meszaros JE, Deal JL. High-dose ultraviolet C light inactivates spores of *Bacillus subtilis* var. *niger* and *Bacillus anthracis* Sterne on non-reflective surfaces. *Appl Biosafety: J Am Biological Safety Assoc* 2005:1–6.

141. Steris. VHP MD series low temperature sterilizers for medical devices. Technical data monograph SD771. Steris, 2008. http://www.steris.com/documents.cfm?id=SD771. Accessed July 23, 2009.

142. American Society for Gastrointestinal Endoscopy. Infection control during GI endoscopy. *Gastrointest Endosc* 2008;67: 781–790.

143. Alfa MJ, Olson N, Degagne P. Automated washing with the Reliance endoscope processing system and its equivalence to optimal manual cleaning. *Am J Infect Control* 2006;34: 561–570.

144. Omidbakhsh N, Sattar SA. Broad-spectrum microbicidal activity, toxicologic assessment, and materials compatibility of a new generation of accelerated hydrogen peroxide-based environmental surface disinfectant. *Am J Infect Control* 2006;34:251–257.

145. Omidbakhsh N. A new peroxide-based flexible endoscope-compatible high-level disinfectant. *Am J Infect Control* 2006;34:571–577.

146. Howie R, Alfa MJ, Coombs K. Survival of enveloped and non-enveloped viruses on surfaces compared with other micro-organisms and impact of suboptimal disinfectant exposure. *J Hosp Infect* 2008;69: 368–376.

147. Sattar SA, Adegbunrin O, Ramirez J. Combined application of simulated reuse and quantitative carrier test to assess high-level disinfection: experiments with an accelerated hydrogen peroxide-based formulation. *Am J Infect Control* 2002; 30:449–457.

148. Rutala WA, Weber DJ. How to assess disease transmission when there is a failure to follow recommended disinfection and sterilization principles. *Infect Control Hosp Epidemiol* 2007;28:519–524.

149. Henderson DA. The looming threat of bioterrorism. *Science* 1999;283:1279–1282.

150. Centers for Disease Control and Prevention (CDC). Biological and chemical terrorism: strategic plan for preparedness and response. Recommendations of the CDC Strategic Planning Workgroup. *MMWR Recomm Rep* 2000;49(RR-4):1–14.

151. Klein M, DeForest A. The inactivation of viruses by germicides. *Chem Specialists Manuf Assoc Proc* 1963;49:116–118.

152. Rutala WA, Weber DJ. Infection control: the role of disinfection and sterilization. *J Hosp Infect* 1999;43:S43–S55.

153. Rutala WA, Stiegel MM, Sarubbi FA, Weber DJ. Susceptibility of antibiotic-susceptible and antibiotic-resistant hospital

bacteria to disinfectants. *Infect Control Hosp Epidemiol* 1997;18:417–421.

154. Kohn WG, Collins AS, Cleveland JL, Harte JA, Eklund KJ, Malvitz DM. Guidelines for infection control in dental health-care settings—2003. *MMWR Recomm Rep* 2003; 52(RR-17):1–67.

155. Rutala WA, Weber DJ. Disinfection of endoscopes: review of new chemical sterilants used for high-level disinfection. *Infect Control Hosp Epidemiol* 1999;20:69–76.

156. Rutala WA, Weber DJ. Clinical effectiveness of low-temperature sterilization technologies. *Infect Control Hosp Epidemiol* 1998;19:798–804.

Chapter 8 Twenty-First Century Microbiology Laboratory Support for Healthcare-Associated Infection Control and Prevention

Lance R. Peterson, MD, and
Marc Oliver Wright, MT (ASCP), MS, CIC

The clinical microbiology laboratory continues to grow in its pivotal role for the support of infection control and prevention activities that improve the safety of patients during their journey through our healthcare organizations. Much has changed, even in the first decade of this new century. Molecular diagnostic tests have shortened the time for detection of important pathogens and facilitated widespread surveillance, when necessary.[1,2] Electronic monitoring systems facilitate surveillance from the laboratory that can detect organization-wide (hospital and outpatient) occurrences of infectious disease.[3] The electronic medical record not only permits access to the medical record whenever it is needed but also can be used for comprehensive data collection; for example, for automated collection of data on use of invasive medical devices, which are often associated with healthcare-associated infectious diseases.[4] The optimal approach for managing healthcare-associated infections is to gather the broadest possible surveillance and infection-risk data,[5] as originally recommended by the Centers for Disease Control and Prevention. The principal hindrance to achieving this goal has been that attempting universal surveillance manually is very resource intensive, prone to error, and produces large amounts of data that, in the absence of analysis, are rendered meaningless.[6] However, as an increasing number of analytical tools become available, the twenty-first century is demonstrating that use of laboratory data is the most cost-effective, comprehensive surveillance approach for defining and tracking the overall risk of healthcare-associated infection.[3]

Two consensus reports published in 1998 and 1999 set the current standard requirement for microbiology laboratory services as part of a comprehensive infection control program that monitors both acute care and out-of-hospital healthcare services.[7,8] The final reports indicated that necessary contributions from the laboratory should include all aspects of surveillance (eg, providing for systematic observance and measurement of cases of disease) encompassing all the diverse areas of care, since this is the critical data management activity for any infection control plan; the necessary aspects of this included molecular typing of microbial pathogens.[7] More than 10 years after that report, we now recognize that even more is required. Although a comprehensive report on resources available to infection control programs on the World Wide Web was recently published,[9] no current assessment of specific laboratory support needs has been compiled. On the basis of our experience and past publications relating to this topic, present and future needs for laboratory-based infection control programs will require reliable surveillance and accurate identification of both common microbes and emerging pathogens; recognition of emerging antimicrobial resistance; participation in active surveillance for epidemiologically significant organisms, including preparation of specialized growth media or other novel tools that can rapidly detect pathogens or resistance mechanisms of interest; support for outbreak iznvestigations; and molecular typing of subspecies.[10] This evolving role dictates a strong collaboration between the hospital epidemiologist and the clinical microbiologist and microbiology laboratory, with a consequent positive impact on both the infection control program and the diagnostic laboratory.[11]

In their 2009 standard, the Joint Commission states that "The hospital provides laboratory resources when

Table 8-1. Necessary functions for clinical laboratory support of infection control and prevention

Participate in developing organization-wide and laboratory-specific infection control plans

Ensure qualified personnel are available to perform infection control activities

Detect and track rates of and trends in infection for the organization; this includes periodic reporting of an antibiogram for the organization

Report results in a way that permits surveillance of nosocomial infections; this may include streamlining efforts for automated surveillance activities

Detect critical pathogens, notify appropriate infection control personnel, and help implement precautions for reducing in-hospital and healthcare-associated disease transmission

Provide resources for outbreak investigations, when needed (surveillance testing, molecular typing, species identification, expanded susceptibility testing, and other needs that arise)

Provide resources for ongoing surveillance testing, when needed

Serve as an institutional expert with regards to appropriate culturing methods and interpretation of significant laboratory results

Assist with implementation of interventions

Have an effective system for reporting within the organization and to public health officials

Assist in monitoring outcomes to determine if program goals are met; provide help to modify interventions if necessary to meet goals

Participate in the infection control committee, seminars, personal interactions, and newsletters

Coordinate the infection control activities of the pathology department and report in departmental meetings

NOTE: Modified from Scheckler et al.[7] with additional data from the Joint Commission,[13] and Pearson et al.[14]

needed to support the infection prevention and control program."[12(p81)] A summary of the needed contributions from clinical laboratory operations, based on prior standards, publications, and our own experience is shown in Table 8-1. These contributions range from developing a formal, department-specific infection control plan for the pathology department; to providing traditional microbial identification and susceptibility testing; to incorporating sophisticated molecular typing and data analysis systems for detection, tracking, and monitoring of healthcare-associated infections for the entire organization.[14] Each of these elements is addressed in the following sections of this chapter.

Development of an Infection Control Plan

The pathology and laboratory medicine department should have a written, department-specific infection control and prevention plan for preventing the transmission of infections that is consistent with the plan for the entire organization. The plan should identify the likely hazards faced by those working in the department and should describe risk reduction strategies for employee and patient safety. Within the departmental plan, there should be specific named individual(s) who are responsible for developing, implementing, and updating the plan. Systems for reporting of appropriate information, both to the key personnel in the organization and to local, state, and federal public health officials as legally required, need to be included. The departmental plan should also delineate activities for patients, workers, and trainees that are required to reduce the transmission of infections, facilitate hand hygiene, address screening for pathogen exposure and assess immunity to such potential exposures, minimize the risk from handling of medical devices, outline the management of infected personnel who may pose a risk to others, and describe the availability and use of personal protective equipment. The proper handling, maintenance, disinfection, and storage of equipment used in the department should be described. Importantly, the plan should include a description of how staff and trainees will receive information about infection prevention and management. All policies, procedures, and activities in the departmental plan should be based upon accepted guidelines and practices that have been demonstrated to be successful. Included in this plan should be a mechanism for regularly reporting the status of infection control activities that impact the department.

Personnel

Qualified personnel who are knowledgeable, either through experience or training, in infection prevention and control and in data analysis are required for the laboratory program. Experience can be gained from the recognized need for the participation of microbiologists in the organizational infection control program.[11,15] Recognized training and certification, either in medical microbiology or in infection control (preferably both, for those involved in the infection control program), can be obtained from several resources. Postdoctoral training with specialty certification in medical microbiology is available from the American Board of Pathology[16] and the American Academy for Medical Microbiology.[17] Specialty courses in infection control can be obtained through the Society for Healthcare Epidemiology of America[18] and the Association for Professionals in Infection Control and Epidemiology,[19] with certification available through the Certification Board of Infection Control and Epidemiology.[20] Healthcare organizations are encouraged to provide incentives for such training through developing a salary structure that rewards additional education, experience, and certification in infection control and medical (clinical) microbiology. The extent of experience and/or training needed to support the laboratory's infection control plan will be dictated by the size of the healthcare organization and the complexity of the patients being cared for within the system.

Pathogen Detection and Data Collection From Laboratory Resources

A cornerstone of managing the spread of infections is surveillance that can detect outbreaks and unfavorable disease trends. Since all material that is obtained to be cultured for potentially infected persons is sent to the clinical microbiology laboratory for processing, this information resource is critical to an effective infection prevention program. The basic requirement for the microbiology laboratory's participation in the infection control and prevention program is the accurate microbial identification and antimicrobial-susceptibility testing of medically relevant pathogens.[21,22] For today's laboratory, this has been simplified by the availability of many semiautomated instruments that reliably perform these functions. Also, there are reliable reference texts that provide direction to the clinical laboratory in specimen processing, pathogen detection, and antimicrobial agent susceptibility determination. However, one must be aware of the limitations of instruments and test methods,[22] which are placing ever-growing challenges on the laboratory as new pathogens emerge, antimicrobial resistance disseminates, and new awareness grows as to the deficiencies of well-accepted testing approaches. During this past decade we have all become aware of the importance of accurate species identification of nonfermentative gram-negative bacilli and drug-resistant streptococci (eg, enterococci) and detection of methicillin resistance in staphylococci, as these organisms become more prevalent.[23] Additional challenges are accurate detection of vancomycin-resistant staphylococci, along with accurate delineation of multidrug-resistant Enterobacteriaceae, such as extended-spectrum β-lactamase (ESBL)–producing *Escherichia coli* and *Klebsiella* species, AmpC enzyme–containing Enterobacteriaceae with inducible expression of cephalosporinases, and the new emerging pathogen, carbapenemase-producing *Klebsiella pneumoniae*. It is these very organisms that can pose difficulties for both identification[24] and susceptibility testing[25-28] by automated systems and usually require that manual confirmation methods be available in clinical laboratories.[29]

In addition to these formidable challenges, many organizations have increasing difficulty with control of multidrug-resistant *Acinetobacter baumannii* in the hospital,[30] and organisms producing CTX-M ESBLs (ie, ESBLs carrying the CTX-M enzyme) that are often

seen in community settings.[31] *Clostridium difficile* is a reemerging pathogen that is gaining worldwide recognition. Accurate test results are not the sole responsibility of the laboratory; with regard to *C. difficile* infection, one newly recognized problem is caused by the practice of ordering repeat tests for a given patient when the result of the initial evaluation for *C. difficile* stool toxin is negative.[32] Not only does this practice lead to false-positive test results and potential misdiagnosis for the patient, but it can also falsely elevate the rate of healthcare-associated *C. difficile* infection by as much as 25%.[33] In the calculation of the rate of healthcare-associated infections, it is important not only that the appropriate denominator (reflecting the population at risk) is chosen but also that laboratory testing and reporting is performed in a way that minimizes the number of false-positive results included in the numerator of the rate calculation.

A crucial role of the laboratory is to be an "early warning center" for infection control problems arising in the organization.[21] Thus, once the laboratory generates diagnostic data, the next step is to organize them into useful information that facilitates detection of outbreaks and tracking of infection trends. Some of the semiautomated identification and susceptibility testing systems have available surveillance programs to aid in infection control activities, but these have not been well studied and are not widely used. They also require input of any information gathered that is not generated by the identification and susceptibility testing performed, such as admission-discharge-transfer data. Use of all the data in the microbiology laboratory information system is the ideal, as it has the potential to allow monitoring of changes in antimicrobial susceptibility rates, detecting infection trends, and alerting the infection control program to the appearance of new pathogens. A pilot study analyzing these data using simple algorithms reported a capacity to detect clonal outbreaks of infection with only a small fraction (2%) of the personnel resources needed by traditional surveillance,[34] which demonstrates the potential of this approach.

Surveillance and Data Analysis

Perhaps the most important phase in an epidemiologic investigation is the initial step of realizing that a problem exists and beginning to gather preliminary data. Surveillance should be used to identify potential problems, to prospectively monitor trends, and to assess the quality of care in a critical area of patient safety—preventing healthcare-associated infectious diseases. To do this optimally requires high-quality laboratory data that are timely, accessible, and easily searched so that relevant information can be obtained.[35] Historically, microbiology laboratory data were reviewed daily in paper form by infection preventionists. However, the identification of important patterns relies on establishing associations between results that may be separated in time by weeks or months, or in separate nursing units or even hospitals, in a large healthcare organization. Other factors that must be taken into consideration are details about the isolate, such as antimicrobial susceptibility phenotype, source of the culture specimen, or the admitting physician and/or practice specialty; all must be put into context with the historical data for each facility. Occasionally, the laboratory personnel may recognize that a problem exists and notify infection control personnel. However, planned use of the data generated by the microbiology laboratory information system will be the most efficient and effective approach to ensure the broadest monitoring for potential infection control issues. Complicating this approach is the fact that, over the last few decades, the volume of electronic data managed by laboratory staff has increased exponentially. In the 1980s and 1990s, the challenge faced by laboratorians was one of work flow, data entry, access, and physical storage. Today, the challenge is to unlock the tremendous potential of the vast electronic data sets clinical practice creates and to provide added value to our patients and their healthcare facilities.

To effectively use the microbiology data for infection control and prevention management, an effective computerized approach is needed (see Chapter 10). Currently, there are 2 data-management concepts that have been developed for application to the microbiology laboratory information system: hypothesis-based knowledge discovery (HBKD) and data mining.[3,36,37,38,39] The key factor differentiating these 2 concepts is that the hypothesis-based knowledge discovery requires that the analyst first ask a question or design a query. The analyst is then presented with the results of the question asked. This type of automated or semiautomated surveillance works well for low-dimensional data over limited periods of time and even can be very sophisticated if a high number of the "right" questions are posed. However, the process remains user dependent, with trillions of potential surveillance patterns of interest. Often the key to identifying an important trend is the identification of complex, subtle, or unexpected relationships that emerge over time. Data-mining systems incorporate mathematical and statistical techniques to generate complex algorithms for searching immense quantities of data; they may incorporate historical data time frames and data fields, and they do not require that the user specify criteria in order for important patterns to be recognized. When properly designed, automated surveillance systems will add patient location tracking information to the microbiology laboratory data so that an elegant computer analysis can place patients carrying pathogens of importance together geographically and temporally in ways that are not possible with any type of manual surveillance approach. A challenge for these types of automated systems is to achieve an acceptable level of sensitivity and specificity, such that significant trends are able to be detected without inundating the recipient with output that is statistically significant yet clinically meaningless.

The ability to use computer-generated information for determination of infection rates to compare current practice against past performance or national standards may also be desirable, and systems are becoming available that show promise for reliable healthcare-associated disease detection, as well as a basis for comparison within or between organizations.[36,37] Use of large databases containing this type of objectively collected data may also provide reliable estimates of healthcare-associated infection costs and may assist in decisions regarding which interventions are both medically beneficial and economically justified.[40]

When automated surveillance systems are designed or developed for healthcare organizations, the clinical laboratory plays a pivotal role in both the implementation and effective utilization of said systems. The involvement of the laboratory often begins in the preplanning stages, in which the data from the laboratory information system are described and perhaps modified for compatibility with automated systems. An effective implementation strategy also includes review of work processes in the clinical laboratory and may suggest modifications, particularly in the reporting stage, to maximize efficiencies of the system. After implementation, the laboratory serves a key role in aiding in the interpretation of potential disease patterns. This is particularly important in ruling out potential outbreaks and communicating work flow process changes with the infection control department or primary users and/or owners of the automated surveillance system.[41] For example, if the laboratory modifies its antimicrobial susceptibility testing panel by replacing antibiotic A with antibiotic B, an automated surveillance system may detect a pseudo-outbreak of infection with organisms resistant to drug B (ie, may detect a rate greater than the baseline of zero), or hypothesis-based knowledge discovery queries may fail to detect an outbreak if the query is based on a relationship between the organism of interest and antibiotic A (which is no longer included in the testing panel). These misleading results can be avoided by communicating such changes in advance (allowing the users and/or owners of the automated system to modify algorithms or queries) or, less optimally, by noting such changes when there is false-positive detection of an outbreak.

For laboratories that serve a large number of community physician offices, the use of technology that effectively screens the microbiology laboratory information system database also can detect infection pattern trends among the patients cared for in the community offices and can alert the laboratory's healthcare organization to the emergence of infections (eg, infections due to influenza virus, *Salmonella* species, or rotavirus) that are presenting in the outpatient population.[42] This not only serves to aid in meeting requirements for mandatory reporting to public health agencies but also forewarns the healthcare organization of the potential for an influx of patients requiring transmission precautions. Identification of postdischarge infections that are confirmed by culture and detected only in the community physician's office can aid in acute care surveillance activities, such as postdischarge surgical-site infection surveillance. Similarly, clinical laboratories that serve residents of long-term care facilities and have sophisticated monitoring of their information-system data can play an important role in disease surveillance, outbreak detection, and antimicrobial resistance monitoring for this major part of the healthcare system, whose infection control and prevention needs are only beginning to be recognized.[43,44]

One aspect of surveillance that is very useful for programmatic planning is use of a point prevalence survey as a risk assessment tool. This can range from a chart review for all patients in a particular hospital or nursing unit,[45] to performance of surveillance cultures to determine the colonization burden in the patient population.[1] In using this latter tool, the infection control staff determines if there appears to be an increasing problem with one or more pathogens, such as methicillin-resistant *Staphylococcus aureus* (MRSA), vancomycin-resistant enterococci (VRE), or ESBL-producing Enterobacteriaceae. The laboratory then prepares or purchases selective media capable of detecting the target microbes, and the infection control staff collect appropriate specimens for surveillance cultures. Since this is a survey of microbial burden in the population, there is usually no need for rapid testing and reporting, which reduces the laboratory's resource utilization, while at the same time it provides infection control personnel with very useful planning information regarding the need for future action. Recognizing change in the rate of colonization over time, particularly if that change is correlated with the rate of clinical disease, assists with planning for a potential intervention; if the rate of colonization with a specific microbe remains acceptably low, there may be no need for any change in practice.

Outbreak Investigation

The clinical microbiology laboratory needs to have the capacity (either internally or through a reference labora-

tory) for effective outbreak investigation. A component of this that may not be immediately apparent is the requirement for ongoing services relating to outbreak monitoring once an intervention is initiated by the infection control department and there is a need to determine the results after the intervention. In addition to the information needed for effective surveillance described earlier, the laboratory should have the capacity to perform subspecies identification (often referred to as "DNA fingerprinting") that can aid the infection preventionists in the outbreak investigation.[46] Subspecies identification to determine microbial clonality can rapidly focus the infection preventionists' attention on the biology of an outbreak during the initial investigation. For example, transmission of microbes from patient to patient is likely if strains are identical or closely related. In this event, enhancement and/or retraining in the appropriate infection control practices are needed. Examples of this are typically endemic and epidemic infection with MRSA, *C. difficile,* or VRE.[2,47] This scenario is less likely if strains are unrelated at the subspecies level. The causes of an outbreak of infection with unrelated strains is often found to be a change in the prescribing of antimicrobials, such as is seen when there is an increasing number of infections due to ESBL-producing bacteria or there are breaks in standard nursing practice techniques, such as handling of ventilator tubing, that can lead to an increase in the number of cases of pneumonia due to various gram-negative pathogens in ventilated patients.[46,48-50] A focus on these factors often can rapidly resolve the outbreak.

There are several methods for the laboratory determination of subspecies relatedness or DNA fingerprinting[51] (these are discussed in detail in Chapter 9). The important issue to consider when choosing a system is to inquire whether it has been validated against clinical epidemiologic information. To date, the most information regarding clinical validation and pathogen discrimination continues to support the utility of pulsed-field gel electrophoresis and genomic restriction endonuclease analysis. Both typing systems have been validated using organisms from documented outbreak scenarios,[52,53] and pulsed-field gel electrophoresis is available as a commercial product. An automated ribotyping system is also commercially available, but the clinical epidemiologic utility of this system has not been consistently confirmed.[54,55] The various subspecies typing options have been recently reviewed for MRSA, and pulsed-field gel electrophoresis remains the most discriminatory method,[56] a finding likely true for many other pathogens of concern. The decade of the 1990s saw the introduction of this technology as a powerful tool for outbreak investigation, and the accessibility of subspecies identification is now widespread: 81% of laboratories had access to this service by 2003.[57,58] In order for a laboratory to adequately support the infection control program in this aspect of outbreak investigation, all potentially relevant microbial isolates need to be saved for a period of time so that they can be recovered from storage in case molecular typing is needed once a potential outbreak is recognized. In our experience, most as-yet-unidentified potential outbreaks picked up by our data mining system are detected within a month. Therefore, we have found it sufficient to store any bacterial isolate that had undergone susceptibility testing for 2 months. This storage period has facilitated retrospective DNA fingerprint analysis whenever needed, and has not unnecessarily burdened the laboratory personnel resources or storage (space) capacity. Storage for 2 months can be accomplished for most organisms by using primary purity-plate cultures held at room temperature or refrigerator temperature, after which the organisms and media can be safely discarded (in accordance with the established local and federal regulations for disposal of biohazardous material).

Another important role for the clinical laboratory during outbreak investigation and infection control management is supporting active surveillance to detect carriers of the pathogen(s) of concern. This capacity has increased recently: 51.7% of clinical laboratories provided this service in 1993, and nearly 95% did in 2003.[58] The most frequently encountered organisms for which performance of active surveillance cultures is recommended are MRSA and VRE.[47,59-64] However, during an outbreak, special media and recovery techniques may be required to detect a wide range of bacteria: for example, Enterobacteriaceae,[61,64] *Pseudomonas aeruginosa*,[65] *A. baumannii*,[66] carbapenemase-producing *K. pneumoniae*,[67,68] and *C. difficile*.[69] Here, the laboratory's role is to prepare special screening media for rapid detection of the target organisms and to report the findings to the infection preventionists.

A new, rapidly developing technology is expanding the role of surveillance and the participation of the clinical laboratory in effective infection control interventions: use of polymerase chain reaction (PCR) analysis in surveillance activities has great potential, as it can quickly detect specific emerging pathogens, genetic determinants of resistance, and unique virulence factors that are markers for the presence of pathogenic microbes of importance to the management of healthcare-associated infections. The technology of real-time PCR has been applied to surveillance for MRSA.[1,2,70,71] This test can be completed in as little as 2 hours from the time of patient sampling; initial reports found that the PCR test performed directly on an anterior nares swab specimen is nearly as sensitive as culture for detecting *S. aureus*.[72,73] Technology such as this has the potential to revolutionize active surveillance screening (followed by isolation or

decolonization of patients who have important pathogens detected), to limit spread of nosocomially acquired bacteria in ways not previously attempted.[1,2] PCR has been suggested in the past as a means to reliably detect *C. difficile* by targeting its pathogenic toxin genes.[74] Commercial and laboratory-derived real-time PCR tests for *C. difficile* infection have the potential to rapidly detect this microbe and may offer the opportunity for both improved management and more accurate determination of the epidemiology of this healthcare-associated infection, which has worldwide importance.[31,75-77] Using PCR for VRE surveillance remains problematic because of the high number of results that are false-positive for the vanB genotype.[7]

When very rapid testing and reporting of an emerging infection from the clinical laboratory is possible, the impact on organization-wide infection control measures can be dramatic. We experienced the benefit of such rapid testing capabilities during the outbreak of infection with the novel influenza A virus subtype H1N1, or "swine flu," in the spring of 2009. Our laboratory had serendipitously developed a quantitative PCR assay during the earlier influenza season that was able to detect changing influenza virus patterns, including this novel influenza virus, within 24 hours after obtaining a sample from the patient.[79] With daily updates of disease trends (Figure 8-1), we were able to move appropriately to an effective infection prevention strategy. It included 24-hour per day screening for syndromic disease of all persons entering our facilities at monitored hospital entrances, followed by limited monitoring of employees during personnel shift changes. Subsequently we used reminder signs (with no monitoring) for appropriate influenza containment practices, and finally we withdrew high-alert infection control practices for potential pandemic influenza. All this occurred within a 3-week period.[79] This further allowed rapid reporting of specific results for infection with influenza A (H1N1) to clinicians, who could modify treatment regimens as appropriate. Lastly, infection preventionists were able to report cases of infection with the novel influenza virus to local public health authorities in a timely manner with a high degree of specificity, so as to not overwhelm those agencies with a combination of cases of influenza A (H1N1) infection and cases of seasonal influenza, pending confirmation.

Additional resource needs may be required from the clinical microbiology laboratory when an outbreak investigation or control intervention is under way. These can include personnel and supplies, when the volume of tests to be performed increases several-fold. Other resources rapidly needed may vary from relatively simple tasks, such as performance of additional susceptibility testing, to more complex tasks, such as detection of airborne or environmental pathogens. Whenever such

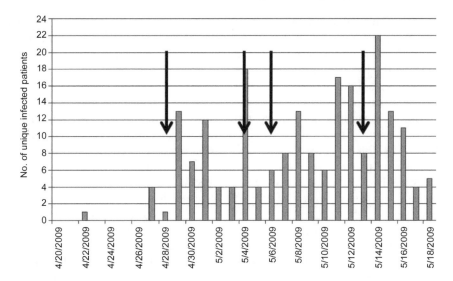

Figure 8-1. Tracking of infections due to the novel influenza A virus subtype H1N1 on the basis of daily laboratory reports during the spring of 2009. The arrows indicate when major decisions were made regarding the influenza prevention program by the infection control department for the healthcare organization.

additional tasks are considered, the protocols should be developed jointly between the clinical laboratory, infection control staff, and hospital administration. It is important to not overutilize the resources of the clinical laboratory; this is best accomplished through careful planning of what is needed and how the goals are to be met before the surveillance project begins. A good example of this is in the area of environmental sampling. Such testing is not generally recommended in the absence of a potential outbreak, since some microorganisms are always recovered. However, it can be invaluable when the source of a clonal outbreak is thought to be an environmental reservoir. An example is searching for carbapenemase-producing *K. pneumoniae* or multidrug-resistant *A. baumannii* when clonal dissemination of this pathogen occurs in an intensive care unit, in order to develop a successful intervention program.[67,68,80] When such sampling is considered, it is imperative to carefully define ahead of time what pathogen(s) are being sought and then to confine the search to detect only those organisms defined before the specimens for surveillance culture were obtained. When such advanced planning is not done, many resources are wasted, both by the laboratory, through recovery and identification of a wide array of bacterial and fungal species, and by the infection preventionists, through trying to determine the meaning and impact of the detected microbes.

Interventions

Interventions to limit the spread of healthcare-associated infections and antimicrobial resistance are among the most critical activities of the infection control and prevention program and are carried out by the infection preventionists. Interventions are most successful if there is a thorough understanding of the microbiology of the problem being confronted and knowledge of what has been useful for past management of similar occurrences. Because modern health care is increasingly complex and the nature of any given problem is often unclear, frequent meetings between the key infection control stakeholders—typically the infection control, clinical microbiology, and infectious diseases departments—facilitate interpretation of information on trends in the microbes detected, the course of the outbreak, and antibiotic resistance.[45,47,81]

During an intervention, decisions regarding the type of testing performed by the laboratory can have a profound impact on the success or failure of the intervention. This is well demonstrated with regard to active surveillance and isolation of patients as a means of reducing the incidence of hospital-acquired MRSA colonization and infection. In a healthcare organization that has a high prevalence of MRSA disease (eg, 13.9 clinical MRSA isolates per 1,000 discharges or 4.6 cases of MRSA bacteremia per 1,000 patients), an action such as improvement in compliance with hand hygiene (without specific surveillance for asymptomatic carriers) or performance of routine surveillance cultures to detect colonized persons can have a major impact, and the role of the laboratory may be minimal.[82,83] Contrasting with this are organizations that have a modest disease prevalence (eg, 2.1 clinical MRSA isolates per 1,000 discharges, or 0.5 cases of MRSA bacteremia per 1,000 patients, or 1–4 MRSA infections per 1,000 patient-days), in which case the choice of laboratory testing may have a more critical role.[2,84,85] It has been found that widespread testing for MRSA colonization using a rapid method with a high sensitivity that captures more than 85% of potential MRSA isolation-days has a major impact on the reduction of disease.[2] However, testing fewer patients[2] or using less sensitive laboratory test methods[84,85] that only facilitate isolation of MRSA-positive patients for 33%–70% of the potential in-hospital days can result in programs that have no impact on reduction of MRSA disease, when implemented in a setting with this lower burden of MRSA disease. Thus, in this example, the 3 components that determine the percentage of isolation-days captured are (1) the breadth of the population tested, (2) the sensitivity of the laboratory test used, and (3) the time from when the sample is obtained from the patient until the test result for the target organism is reported. Because the latter 2 components involve laboratory testing, it is easy to see how the right (or wrong) decision can have a dramatic impact.

When an intervention is designed and ready for implementation, a monitoring system, usually based on the same surveillance methods that initially detected the problem, and a monitoring plan must be in place so that the outcome of the intervention is tracked and the effectiveness of the effort is measured.

Reporting Systems

The clinical microbiology laboratory has many requirements for effectively transmitting the information it generates and collects. As mentioned earlier, a key element is to ensure that the methods used to detect potential infection with healthcare-associated pathogens are highly accurate.[32] In addition to the direct patient-care reporting that is its traditional role, the clinical microbiology laboratory needs to effectively report relevant information to the infection preventionists responsible for infection management and prevention within the organization. Beyond this, the entire US communicable disease surveillance system depends on reporting by clinical and public health laboratories.[86] Both the Centers for Disease Control and Prevention and local public health departments set regulations for reportable diseases, and the clinical microbiology laboratory must work closely with the infection preventionists of their organization to make sure the reporting system is well organized and ensures that all required reports are efficiently transmitted, while avoiding duplication of effort that can waste personnel resources. The reporting requirements are the same for healthcare organizations, regardless of whether their microbiologic testing is done on site or at a

referral laboratory, and this must be taken into account when the system for implementing infection control notification is designed.

A novel extension of the electronic medical record pertains to involving laboratory professionals in directly altering patient care as a result of detection of organisms that require an intervention. Beginning in 2007, our organization introduced the practice of having technologists in the microbiology laboratory enter test results into the patient's electronic medical record if a specific, predetermined pathogen (eg, MRSA or VRE) is detected in a clinical or surveillance culture; the technologists add this information to the patient's problem list and order the appropriate isolation cart to be delivered to the patient's room. This action causes an alert banner to appear in the electronic medical record (so healthcare workers visualize the problem pathogen when they log into the specific medical record), generates a "resistant pathogen list" for infection control, and places the order for the isolation cart to be delivered to the patient's room, ready for use. All this occurs with no phone calls and results in virtually instantaneous communication.

Monitoring Outcome

Tracking the medical and economic benefits of a comprehensive infection management and prevention program is perhaps one of the most difficult tasks faced by the program, but it is critical if adequate long-term resources are to be provided for all necessary aspects of this important patient safety effort.[87] Continued monitoring of process measures and/or patient outcomes with appropriate statistical analyses will allow the healthcare organization to monitor the success of its infection prevention and control program. Ongoing work in health economics research suggests that enhanced surveillance also reduces morbidity, mortality, and cost.[1,37,88] The Deficit Reduction Act of 2008[89] suspended additional reimbursement for "reasonably preventable" events, including some healthcare-associated infections. Accordingly, many healthcare organizations are more willing to commit resources aimed at preventing these occurrences, but with the expectation of a return on this investment. If a formal economic analysis (eg, cost-effectiveness analysis) is not feasible, a less stringent, business-case analysis with appropriate follow-up may be sufficient. More-detailed descriptions of performing economic or business-case analyses with regards to infection control programs are available.[90,91] Finally, as computer systems for monitoring the microbiology laboratory information system data become more sophisticated, they will enable the laboratory to provide various types of outcome analysis information as a routine function, and further document the medical and economic benefits of the infection control program.

Communicating Infection Prevention

The need for ongoing, continuing education is a well-recognized requirement for an effective infection control and prevention program.[7] The laboratory contributes to this through formal participation in the organization's infection control committee[11] and through active involvement in the departmental infection control plan.[13] In a less formal manner, laboratory personnel can help in the ongoing education of healthcare personnel through daily intrapersonal interactions while reporting patient laboratory results; by presenting healthcare infection information at conferences and seminars that inform practitioners of key events, such as the start and end of the influenza season (this knowledge impacts when to use antiviral agents empirically) or a switch in influenza strain types from influenza A to B or strain H1 to H3 (this information affects which antivirals to prescribe); and by highlighting the activities of the organization's infection control and prevention activities in newsletters throughout the year. Within the microbiology laboratory, there is a need for ongoing annual competency testing for personnel (eg, *Mycobacterium tuberculosis* biological safety, bioterrorism preparedness, and *C. difficile* infection testing practices). Importantly, the laboratory leadership must be aware of the potential for laboratory-acquired infections[92] and ensure the safety of workers by fostering safe practices and providing appropriate personal protective equipment, education regarding effective hand hygiene, and effective vaccines. These practices need to be updated whenever new issues relating to laboratory safety are recognized.[92] Such ongoing communication of the many aspects of the infection control program is necessary and beneficial, and is an expected contribution of the clinical microbiology laboratory for improving patient care.

The Coordinated Program

As an aspect of high quality medical care, infection control and prevention has an importance that cannot be underestimated. It requires continued effort not only in the performance of a myriad of managerial and technical activities but also in the maintenance of effective interactions between all the personnel and departments involved in the organizations' program. Two critical elements of this interactive management plan are the infection preventionists and the clinical microbiology profes-

sionals, who must work closely together and collaborate for the benefit of patients.[93] The collaboration takes place in formal committee meetings, as well as informal daily discussions of infection control–related activities.

Conclusion

The twenty-first century is beginning with increased opportunities for the diagnostic clinical microbiology laboratory to impact the effectiveness of the healthcare organization's infection control and prevention program. Traditional requirements for accurate and timely microbial identification and susceptibility testing remain key pillars in the support of infection preventionists. However, emerging technologies now offer even more opportunities for lowering the risk of a patient developing a healthcare-associated infection, if fully implemented by the diagnostic laboratory. Molecular testing technology is readily available that can quickly determine microbial clonality and focus the direction of an epidemiologic investigation so that the proper intervention can be applied more quickly. Available molecular tests also allow for real-time active surveillance to effectively limit the spread of important healthcare pathogens more quickly than ever possible before. Novel information systems with improving capabilities offer the capacity to track trends and detect outbreaks in various settings, from outbreaks in communities that cover a wide geographic area to transmission of a healthcare-associated pathogen between 2 or 3 persons on a single nursing unit. Once even more widely applied, these tests can alert us to critical events that indicate newly emerging antimicrobial resistance and perhaps the emergence of a novel, previously unrecognized infectious disease. The future is very bright for an ever-strengthening partnership between the infection control and clinical microbiology professionals working as a team to lower the number of healthcare infections, to enhance patient safety, and to improve the quality of care.

References

1. Peterson LR, Hacek DM, Robicsek A. 5 Million Lives Campaign. Case study: an MRSA intervention at Evanston Northwestern Healthcare. *Jt Comm J Qual Patient Saf* 2007;33:732–738.
2. Robicsek A, Beaumont JL, Paule SM, et al. Universal surveillance for methicillin-resistant *Staphylococcus aureus* in 3 affiliated hospitals. *Ann Intern Med* 2008;148:409–418.
3. Brossette SE, Hacek DM, Gavin PJ, et al. A laboratory-based, hospital-wide, electronic marker for nosocomial infection: the future of infection control surveillance? *Am J Clin Pathol* 2006;125:34–39.

4. Wright MO, Fisher A, John M, Reynolds K, Peterson LR, Robicsek A. The electronic medical record as a tool for infection surveillance: successful automation of device-days. *Am J Infect Control* 2009;37(5):364–370.

5. Emori TG, Culver DH, Horan TC, et al; National nosocomial infections surveillance system. NNIS: description of surveillance methods. *Am J Infect Control* 1991;19:19–35.

6. Haley RW, Aber RC, Bennett JV. Surveillance of nosocomial infections. In: Bennett JV, Brachman PS, eds. *Hospital Infections.* New York: Little, Brown & Company; 1986:51–71.

7. Scheckler WE, Brimhall D, Buck AS, et al. Requirements for infrastructure and essential activities of infection control and epidemiology in hospitals: a consensus panel report. *Infect Control Hosp Epidemiol* 1998;19:114–124.

8. Friedman C, Barnette M, Buck AS, et al. Requirements for infrastructure and essential activities of infection control and epidemiology in out-of-hospital settings: a consensus panel report. Association for Professionals in Infection Control and Epidemiology and Society for Healthcare Epidemiology of America. *Infect Control Hosp Epidemiol* 1999;20:695–705.

9. Johnson LE, Reyes K, Zervos MJ. Resources for infection prevention and control on the World Wide Web. *Clin Infect Dis* 2009;48:1585–1595.

10. Pfaller MA, Cormican MG. Microbiology: the role of the clinical laboratory. In: Wenzel RP, ed. *Prevention and Control of Nosocomial Infections.* 3rd ed. Baltimore, MD: Williams & Wilkins; 1997:95–118.

11. Peterson LR, Hamilton JD, Baron EJ, et al. Role of clinical microbiology laboratories in the management and control of infectious diseases and the delivery of healthcare. *Clin Infect Dis* 2001;32:605–610.

12. The Joint Commission. Standard IC.01.02.01, EP 1. In: The Joint Commission. *2009 Hospital Accreditation Standards.* 2009. http://www.jointcommission.org/standards/. Accessed May 8, 2009.

13. The Joint Commission. Proposed Standards, Prevention and Control of Infection: Laboratory. http://www.jcaho.org/accredited+organizations/accredited+organizations+.htm. Accessed July 30, 2003.

14. Pearson A, Becker L, Almaraz J. Departmental role and scope in infection control: use of a template that meets Joint Commission requirements. *Am J Infect Control* 1996;24:53–56.

15. Reller LB, Weinstein MP. Policy statement on consolidation of clinical microbiology laboratories. *Clin Infect Dis* 2001; 32:604.

16. Accreditation Council for Graduate Medical Education. Program Requirements for Residency Education in Medical Microbiology. http://www.acgme.org/. Accessed May 12, 2009.

17. The American College of Microbiology. Microbiology Training Standards. http://www.asm.org/Academy/index.asp?bid=2253. Accessed May 12, 2009.

18. The Society for Healthcare Epidemiology of America (SHEA). SHEA/CDC Training Course in Healthcare Epidemiology. http://www.shea-online.org/about/shea_courses.cfm. Accessed May 12, 2009.

19. Association for Professionals in Infection Control and Epidemiology. Infection Control Education. http://www.apic.org/AM/Template.cfm?Section=Education. Accessed May 12, 2009.

20. Certification Board of Infection Control and Epidemiology Web page. http://www.cbic.org/. Accessed July 7, 2009.

21. Weinstein RA, Mallison GF. The role of the microbiology laboratory in surveillance and control of nosocomial infections. *Am J Clin Pathol* 1978;69:130–136.

22. Diekema DJ, Pfaller MA. Infection Control Epidemiology and Clinical Microbiology. In: Murray PR, Baron EJ, Jorgensen JH, Landry ML, Pfaller MA, eds. *Manual of Clinical Microbiology.* 9th ed. Washington, DC: American Society for Microbiology Press; 2007:118–128.

23. Zhanel GG, Decorby M, Nichol KA, et al. Characterization of methicillin-resistant *Staphylococcus aureus*, vancomycin-resistant enterococci and extended-spectrum beta-lactamase-producing *Escherichia coli* in intensive care units in Canada: results of the Canadian National Intensive Care Unit (CAN-ICU) study (2005–2006). *Can J Infect Dis Med Microbiol* 2008;19:243–249.

24. Gavin PJ, Warren JR, Obias AA, Collins SM, Peterson LR. Evaluation of the Vitek 2 system for rapid identification of clinical isolates of gram-negative bacilli and members of the family Streptococcaceae. *Eur J Clin Microbiol Infect Dis* 2002;21:869–874.

25. Ender PT, Durning SJ, Woelk WK, et al. Pseudo-outbreak of methicillin-resistant *Staphylococcus aureus*. *Mayo Clinic Proc* 1999;74:855–859.

26. Tsakris A, Pantazi A, Pournaras S, Maniatis A, Polyzou A, Sofianou D. Pseudo-outbreak of imipenem-resistant *Acinetobacter baumannii* resulting from false susceptibility testing by a rapid automated system. *J Clin Microbiol* 2000;38: 3505–3507.

27. Steward CD, Mohammed JM, Swenson JM, et al. Antimicrobial susceptibility testing of carbapenems: multicenter validity testing and accuracy levels of five antimicrobial test methods for detecting resistance in Enterobacteriaceae and *Pseudomonas aeruginosa* isolates. *J Clin Microbiol* 2003; 41:351–358.

28. Pasteran F, Mendez T, Guerriero L, Rapoport M, Corso A. Sensitive screening tests for suspected class A carbapenemase production in species of Enterobacteriaceae. *J Clin Microbiol* 2009;47(6):1631–1639.

29. Sunenshine RH, Wright MO, Maragakis LL, et al. Multidrug-resistant *Acinetobacter* infection mortality rate and length of hospitalization. *Emerg Infect Dis* 2007;13:97–103.

30. Rossolini GM, D'Andrea MM, Mugnaioli C. The spread of CTX-M-type extended-spectrum beta-lactamases. *Clin Microbiol Infect* 2008;14(Suppl 1):33–41.

31. Peterson LR, Robicsek A. Does my patient have *Clostridium difficile* infection? *Ann Intern Med* 2009;151(3):176–179.

32. Hacek DM, Robicsek A, Thomson RB Jr, Peterson LR. A myth of *Clostridium difficile* infection (CDI) diagnosis: the illusion of repeat EIA testing reliability. In: Program and abstracts of the 49th Annual Meeting of the Interscience Conference on Antimicrobial Agents and Chemotherapy; September 12–15, 2009; San Francisco, CA. Abstract K-1910.

33. Swenson JM, Hindler J, Peterson LR. Special phenotypic methods for detecting antibacterial resistance. In: Murray PR, Baron EJ, Jorgensen JH, Pfaller MA, Tenover FC, Yolken RH, eds. *Manual of Clinical Microbiology.* 7th ed. Washington, DC: American Society for Microbiology Press; 1999: 1563–1577.

34. Hacek DM, Cordel RL, Noskin GA, Peterson LR. Computer-assisted surveillance for detecting clonal outbreaks of nosocomial infection. *J Clin Microbiol* 2004;42:1170–1175.

35. Emori TG, Gaynes RP. An overview of nosocomial infections, including the role of the microbiology laboratory. *Clin Microbiol Rev* 1993;6:428–442.

36. Wright MO, Perencevich EN, Novak C, Hebden JN, Standiford HC, Harris AD. Preliminary assessment of an automated surveillance system for infection control. *Infect Control Hosp Epidemiol* 2004;25:325–332.

37. Brossette SE, Hymel PA Jr. Data mining and infection control. *Clin Lab Med* 2008;28:119–126.

38. Obenshain MK. Application of data mining techniques to healthcare data. *Infect Control Hosp Epidemiol* 2004;25(8): 690–695.

39. Wright MO. Automated surveillance and infection control: toward a better tomorrow. *Am J Infect Control* 2008; 36(Suppl 3):S1–S6.

40. Kilgore ML, Ghosh K, Beavers CM, Wong DY, Hymel PA Jr, Brossette SE. The costs of nosocomial infections. *Med Care* 2008;46:101–104.

41. Gotham IJ, Smith PF, Birkhead GS, Daisson MC. Policy issues in developing information systems for public health surveillance of communicable diseases. In: O'Carroll PW, Yasnoff WA, Ward ME, Ripp LH, Martin EL, eds. *Public Health Informatics and Information Systems*. New York: Springer; 2003:550–556.

42. Peterson LR, Hacek DM, Rolland D, Brossette SE. Detection of a community infection outbreak with virtual surveillance. *Lancet* 2003;362:1587–1588.

43. Simor AE. The role of the laboratory in infection prevention and control programs in long-term–care facilities for the elderly. *Infect Control Hosp Epidemiol* 2001;22(7):459–463.

44. Robicsek A, Vosburg J, Beaumont J, Hacek DM, Peterson LR. Universal surveillance as a tool for studying the community epidemiology of methicillin-resistant *Staphylococcus aureus* (MRSA). In: Program and abstracts of the 44th annual meeting of the Infectious Diseases Society of America; Toronto, Canada; October 12–15, 2006. Abstract 392.

45. Scheckler WE. Surveillance, foundation for the future: a historical overview and evolution of methodologies. *Am J Infect Control* 1997;25:106–111.

46. Peterson LR, Noskin GA. New technology for detecting multidrug-resistant pathogens in the clinical microbiology laboratory. *Emerg Infect Dis* 2001;7:306–311.

47. Price CS, Paule S, Noskin GA, Peterson LR. 2003. Active surveillance reduces the incidence of vancomycin-resistant enterococcal bacteremia. *Clin Infect Dis* 2003;37:921–928.

48. Hacek DM, Suriano T, Noskin GA, Kruszynski J, Reisberg B, Peterson LR. Medical and economic benefit of a comprehensive infection control program that includes routine determination of microbial clonality. *Am J Clin Pathol* 1999; 111:647–654.

49. Gardam MA, Burrows LL, Kus JV, et al. Is surveillance for multidrug-resistant Enterobacteriaceae an effective infection control strategy in the absence of an outbreak? *J Infect Dis* 2002;186:1754–1760.

50. Palmer LB. Ventilator-associated infection. *Curr Opin Pulm Med* 2009;15:230–235.

51. Soll DR, Pujol C, Lockhart SB. Laboratory procedures for the epidemiological analysis of microorganisms. In: Murray PR, Baron EJ, Jorgensen JH, Landry ML, Pfaller MA, eds. *Manual of Clinical Microbiology*. 9th ed. Washington, DC: American Society for Microbiology Press; 2007:129–151.

52. Clabots CR, Johnson S, Bettin KM, et al. Development of a rapid and efficient restriction endonuclease analysis (REA) typing system for *Clostridium difficile* and correlation with other typing systems. *J Clin Microbiol* 1993;31:1870–1875.

53. Savor C, Pfaller MA, Kruszynski JA, Hollis RJ, Noskin GA, Peterson LR. Genomic methods for differentiating strains of *Enterococcus faecium*: an assessment using clinical epidemiologic data. *J Clin Microbiol* 1998;36:3327–3331.

54. Price CS, Huynh H, Paule S, et al. Comparison of an automated ribotyping system to restriction endonuclease analysis and pulsed-field gel electrophoresis for differentiating vancomycin-resistant *Enterococcus faecium*. *J Clin Microbiol* 2002;40:1858–1861.

55. Fontana J, Stout A, Bolstorff B, Timperi R. Automated ribotyping and pulsed-field gel electrophoresis for rapid identification of multidrug-resistant *Salmonella* serotype Newport. *Emerg Infect Dis* 2003;9:496–499.

56. Struelens MJ, Hawkey PM, French GL, Witte W, Tacconelli E. Laboratory tools and strategies for methicillin-resistant *Staphylococcus aureus* screening, surveillance and typing: state of the art and unmet needs. *Clin Microbiol Infect* 2009;15: 112–119.

57. McGowan JE Jr. New laboratory techniques for hospital infection control. *Am J Med* 1991;91(3B):245S–251S.

58. Peterson L, Miller JM, e-mail to Microbiology Directors' Forum, ClinMicroNet internet discussion group, February, 2003.

59. Muto CA, Jernigan JA, Ostrowsky BE, et al. SHEA guideline for preventing nosocomial transmission of multidrug-resistant strains of *Staphylococcus aureus* and *Enterococcus*. *Infect Control Hosp Epidemiol* 2003;24:362–386.

60. Noskin GA, Bednarz P, Suriano T, Reiner S, Peterson LR. Persistent contamination of fabric-covered furniture by vancomycin-resistant enterococci: implications for upholstery selection in hospitals. *Am J Infect Control* 2000;28:311–313.

61. Collins SM, Hacek DM, Degan LA, Wright MO, Noskin GA, Peterson LR. Contamination of the clinical microbiology laboratory with vancomycin-resistant enterococci and multidrug-resistant Enterobacteriaceae: implications for hospital and laboratory workers. *J Clin Microbiol* 2001;39: 3772–3774.

62. Trick WE, Temple RS, Chen D, Wright MO, Solomon SL, Peterson LR. Patient colonization and environmental contamination by vancomycin-resistant enterococci in a rehabilitation facility. *Arch Phys Med Rehabil* 2002;83:899–902.

63. Singh K, Gavin PJ, Vescio T, et al. Microbiologic surveillance using nasal cultures alone is sufficient for detection of methicillin-resistant *Staphylococcus aureus* isolates in neonates. *J Clin Microbiol* 2003;41:2755–2757.

64. Hacek DM, Trick WE, Collins SM, Noskin GA, Peterson LR. Comparison of the RODAC imprint method with selective enrichment broth for the recovery of vancomycin-resistant enterococci and drug resistant Enterobacteriaceae from environmental surfaces. *J Clin Microbiol* 2000;38:4646–4648.

65. Sorin M, Segal-Maurer S, Mariano N, Urban C, Combest A, Rahal JJ. Nosocomial transmission of imipenem-resistant *Pseudomonas aeruginosa* following bronchoscopy associated with improper connection to the Steris System 1 processor. *Infect Control Hosp Epidemiol* 2001;22:409–413.

66. Urban C, Segal-Maurer S, Rahal JJ. Considerations in control and treatment of nosocomial infections due to

multidrug-resistant *Acinetobacter baumannii*. *Clin Infect Dis* 2003;36:1268–1274.

67. Kochar S, Sheard T, Sharma R, et al. Success of an infection control program to reduce the spread of carbapenem-resistant *Klebsiella pneumoniae*. *Infect Control Hosp Epidemiol* 2009;30:447–452.

68. Centers for Disease Control and Prevention (CDC). Guidance for control of infections with carbapenem-resistant or carbapenemase-producing Enterobacteriaceae in acute care facilities. *MMWR Morb Mortal Wkly Rep* 2009;58(10):256–260.

69. Clabots CE, Bettin KM, Peterson, LR, Gerding DN. Evaluation of cycloserine-cefoxitin-fructose agar and cycloserine-cefoxitin-fructose broth for recovery of *Clostridium difficile* from environmental sites. *J Clin Microbiol* 1991;29:2633–2635.

70. Paule SM, Hacek DM, Kufner B, et al. Performance of the BD GeneOhm methicillin-resistant *Staphylococcus aureus* test before and during high-volume clinical use. *J Clin Microbiol* 2007;45:2993–2998.

71. Wolk DM, Picton E, Johnson D, et al. Multicenter evaluation of the Cepheid Xpert methicillin-resistant *Staphylococcus aureus* (MRSA) test as a rapid screening method for detection of MRSA in nares. *J Clin Microbiol* 2009;47:758–764.

72. Shrestha NK, Shermock KM, Gordon SM, et al. Predictive value and cost-effectiveness analysis of a rapid polymerase chain reaction for preoperative detection of nasal carriage of *Staphylococcus aureus*. *Infect Control Hosp Epidemiol* 2003;24:327–333.

73. Paule SM, Pasquariello AC, Hacek DM, et al. Direct detection of *Staphylococcus aureus* from adult and neonate nasal swab specimens using real-time polymerase chain reaction. *J Mol Diagn* 2004;6:191–196.

74. Tang YJ, Gumerlock PH, Weiss JB, Silva J Jr. Specific detection of *Clostridium difficile* toxin A gene sequences in clinical isolates. *Mol Cell Probes* 1994;8:463–467.

75. Peterson LR, Manson RU, Paule SM, et al. Detection of toxigenic *Clostridium difficile* in stool samples by real-time polymerase chain reaction for the diagnosis of C. *difficile*–associated diarrhea. *Clin Infect Dis* 2007;45:1152–1160.

76. Sloan LM, Duresko BJ, Gustafson DR, Rosenblatt JE. Comparison of real-time PCR for detection of the *tcd*C gene with four toxin immunoassays and culture in diagnosis of *Clostridium difficile* infection. *J Clin Microbiol* 2008;46:1996–2001.

77. Stamper PD, Alcabasa R, Aird D, et al. Comparison of a commercial real-time PCR assay for tcdB detection to a cell culture cytotoxicity assay and toxigenic culture for direct detection of toxin-producing *Clostridium difficile* in clinical samples. *J Clin Microbiol* 2009;47:373–378.

78. Paule SM, Trick WE, Tenover FC, et al. Comparison of polymerase chain reaction to culture for surveillance detection of vancomycin-resistant enterococci. *J Clin Microbiol* 2003;41:4805–4807.

79. Kaul KL, Nowak J, Peterson LR. Swine flu: testing the lab response to an epidemic—now and a glimpse of possibilities. ASCP Hot Topic. American Society for Clinical Pathology, Chicago, IL. May 27, 2009. http://www.ascp.org/MainMenu/AboutASCP/Newsroom/Swine-Flu-Links.aspx. Accessed July 8, 2009.

80. Carlo La Forgia C, Franke J, Hacek DM, Thomson RB Jr, Robicsek A, Peterson LR. Management of a multidrug-resistant (MDR) *Acinetobacter baumannii* outbreak in an intensive care unit using novel environmental disinfection: a 38-month report. *Am J Infect Control* 2009 (in press).

81. Kolmos HJ. Role of the clinical microbiology laboratory in infection control—a Danish perspective. *J Hosp Infect* 2001;48(Suppl A):S50–S54.

82. Grayson ML, Jarvie LJ, Martin R, et al.; Hand Hygiene Study Group and Hand Hygiene Statewide Roll-out Group, Victorian Quality Council. Significant reductions in methicillin-resistant *Staphylococcus aureus* bacteraemia and clinical isolates associated with a multisite, hand hygiene culture–change program and subsequent successful statewide roll-out. *Med J Australia* 2008;188:633–640.

83. Huang SS, Yokoe DS, Hinrichsen VL, et al. Impact of routine intensive care unit surveillance cultures and resultant barrier precautions on hospital-wide methicillin-resistant *Staphylococcus aureus* bacteremia. *Clin Infect Dis* 2006;43:971–978.

84. Harbarth S, Masuet-Aumatell C, Schrenzel J, et al. Evaluation of rapid screening and pre-emptive contact isolation for detecting and controlling methicillin-resistant *Staphylococcus aureus* in critical care: an interventional cohort study. *Crit Care* 2006;10(1):R25.

85. Jeyaratnam D, Whitty CJ, Phillips K, et al. Impact of rapid screening tests on acquisition of methicillin resistant *Staphylococcus aureus*: cluster randomized crossover trial. *BMJ* 2008;336(7650):927–930.

86. Skeels MR. Laboratories and disease surveillance. *Mil Med* 2000;165(7 suppl 2):16–19.

87. Roberts RR, Scott RD 2nd, Cordell R, et al. The use of economic modeling to determine the hospital costs associated with nosocomial infections. *Clin Infect Dis* 2003;36:1424–1432.

88. Lee TA, Hacek DM, Stroupe KT, Collins SM, Peterson LR. Three surveillance strategies for vancomycin-resistant enterococci in hospitalized patients: detection of colonization efficiency and a cost-effectiveness model. *Infect Control Hosp Epidemiol* 2005;26:39–46.

89. Center for Medicare and Medicaid Services. Deficit Reduction Act. http://www.cms.hhs.gov. Accessed May 27, 2009.

90. Wright MO, Perencevich EN. Cost-effectiveness of infection control programs. In Zervos M, Simjee S, Chen A, Foley S, eds. *Techniques for the Study of Hospital Acquired Infection*. Hoboken, NJ: J Wiley and Sons; 2009.

91. Perencevich EN, Stone PW, Wright SB, Carmeli Y, Fisman DN, Cosgrove SE; Society for Healthcare Epidemiology of America. Raising the standards while watching the bottom line: making the business case for infection control. *Infect Control Hosp Epidemiol* 2007;28:1121–1133.

92. Centers for Disease Control and Prevention. Laboratory-acquired meningococcal disease—United States, 2000. *MMWR Morb Mortal Wkly Rep* 2002;51:141–144.

93. Pfaller MA, Herwaldt LA. The clinical microbiology laboratory and infection control: emerging pathogens, antimicrobial resistance, and new technology. *Clin Infect Dis* 1997;25:858–870.

Chapter 9 Molecular Typing Systems

Jeffrey Hafkin, MD, Laura Chandler, PhD, and
Joel Maslow, MD, PhD, MBA

Infectious disease clinicians, hospital epidemiologists, and infection prevention specialists frequently encounter outbreak scenarios in which patients are likely to be infected with genetically similar organisms. Although many of these suspected outbreaks are likely due to a common source (eg, because they are limited to a specific time and place), the source of other outbreaks is not so obvious. For these latter cases, it is helpful to be able to determine with some certainty whether groups of organisms are genetically the same or similar (ie, represent acquisition from a common source or person-to-person transmission).

Molecular typing methods have become powerful tools for epidemiologic investigations of healthcare-associated infections.[1-4] When used properly as part of an epidemiologic investigation, strain typing facilitates the job of the healthcare epidemiologist. Practitioners in this field must be familiar with the available methods and must understand their limitations, their costs, and the applicability of the various techniques in order to choose the most appropriate method for a given purpose. Results of molecular typing should not be interpreted alone; strain typing data should be included as part of a complete, well-designed epidemiologic investigation. Although molecular studies can be used to prove or disprove epidemiologic hypotheses, if they are not used appropriately, incorrect conclusions may be drawn and laboratory resources wasted. The goal of strain typing is to provide laboratory evidence that epidemiologically related isolates implicated in an outbreak are also genetically related and therefore represent the same strain, presumably transmitted by a common source of exposure or from patient to patient.

General Principles of Epidemiologic Typing

Three assumptions are generally made in using molecular typing for the purpose of epidemiologic studies of microorganisms: (1) the isolates responsible for an outbreak are the recent progeny of a common ancestor, (2) these isolates have the same genotype, and (3) epidemiologically unrelated strains have different genotypes.[5] The concept of "clonality" is a fundamental principle in strain typing: microbial isolates recovered from different sources are considered to be clones if they have "so many identical phenotypic and genetic traits that the most likely explanation for this identity is a common origin."[6] Therefore, the determination of clonality cannot be made with absolute certainty, but instead represents a statistical likelihood of identity among a group of independently isolated organisms.[6] The better the discriminatory power of the technique used to identify the isolate, the more likely that clonality can be accurately determined. However, because of the inherent instability of phenotypic traits in all organisms, especially in the presence of environmental selective pressure, clonality must be considered in a relative rather than an absolute sense.[6] Because of natural genetic drift, it is important to consider the length of time over which isolates have been collected when you are interpreting results. In general, isolates from a potential outbreak spanning 1–3 months are appropriate for strain typing analyses.[3,5] The rate of genetic drift will determine the likelihood that clonally derived progeny maintain similarity, when analysis is done on isolates obtained over prolonged periods.

Wide genetic variability within a bacterial species usually allows for the differentiation of unrelated strains by a number of different techniques.[3] However, because the most virulent bacterial strains often are clonally restricted (ie, they represent only a small fraction of the number of clones within a species[7]) the bacteria that are most likely to cause infection may have little genetic diversity, making them difficult to differentiate. An example of this uniformity is seen in methicillin-resistant *Staphylococcus aureus* (MRSA) isolates, which are derived from a small number of clones.[8]

Molecular typing systems have added tremendously to the clinical, epidemiologic, and laboratory investigation of *Mycobacterium tuberculosis* infection. In addition to aiding in the epidemiologic assessment of outbreaks—whether in an institution or in the community—typing also can provide useful information for laboratories diagnosing *M. tuberculosis* infection.[9] For example, molecular typing can be used to detect test results that are false-positive for *M. tuberculosis* because of cross-contamination, and hence it can reduce the potential for needlessly exposing patients to toxic treatment. For public health programs, molecular typing provides important information to aid in control of *M. tuberculosis* infection in the community. Through its ability to rapidly detect clustering of transmitted cases, molecular typing greatly enhances contact investigations. Typing is useful in assessing the burden of *M. tuberculosis* infection in the community that represents recent transmission, as distinct from endogenous reactivation. Prudent use of molecular typing can therefore help to channel resources into programs that are most likely to have a significant public health impact.

Epidemiologic Analysis: Outbreak Investigation Versus Surveillance

There are 2 major epidemiologic reasons for doing strain typing: to conduct an outbreak investigation, and to perform long-term prospective surveillance to detect and monitor emerging or reemerging pathogens. Outbreak investigations typically involve the analysis of a limited number of isolates collected over a relatively short period of time.[10] Results of these comparative typing systems are usually applicable only to a particular local problem. In contrast, monitoring the prevalence and geographic spread of epidemic and endemic clones in a population over longer periods of time requires the development of databases and

requires the use of typing systems that are standardized, highly reproducible, and uniform in nomenclature.[10]

General Concepts for Strain Typing

The most useful typing methods fulfill 3 criteria: they have *typeability*, which is the ability to give an unambiguous result (type) for each isolate; they have *reproducibility*, which is the capacity to give the same result upon repeat testing; and they have *discriminatory power*, which is the ability to differentiate among epidemiologically unrelated strains.[11,12] Choosing a typing method depends not only on the strengths of the particular method in each of these areas but also on the organism in question. Both the *ease of performance*, which is the technical complexity of the technique and the costs of reagents and equipment, and the *ease of interpretation* are also important criteria for choosing a typing system.[3,5,13]

Although phenotypic methods have been successfully used for epidemiologic purposes, they have largely been replaced by molecular methods. Phenotypic methods for typing include biotyping, serotyping, bacteriophage susceptibility testing, antimicrobial susceptibility testing, and bacteriocin production testing (Table 9-1). Biotyping and antibiotic susceptibility testing are routinely automated as part of the routine workup of clinical specimens. The remaining techniques have ceased to be routinely available and are now relegated to certain research laboratories. Since many phenotypic traits are particularly susceptible to environmental pressure, phenotypic methods are limited in their discriminatory power and reproducibility. For example, antibiotic resistance traits, often encoded on plasmids, are strongly affected by the ubiquitous presence of antibiotics in the hospital environment.

Genotypic techniques involve direct DNA analysis of chromosomal or extrachromosomal genetic elements (Table 9-1). The introduction of DNA-based methods in molecular typing has vastly improved the ability of investigators to distinguish among different strains within a species. The techniques offer improved typeability, reproducibility, and discriminatory power, compared with phenotypic methods. Genotypic methods are also less affected by natural variation in a population, although changes such as insertion of DNA into or deletion of DNA from the chromosome, the gain or loss of extrachromosomal DNA, or random mutations may introduce difficulties in interpretation.[1]

Choosing Controls

In all typing systems, well-characterized control strains should be analyzed together with the clinical isolates in

Table 9-1. Phenotypic and genotypic typing methods used for epidemiologic purposes

Phenotypic methods

  Biotyping
  Antimicrobial susceptibility testing
  Serotyping
  Bacteriophage typing
  Bacteriocin typing
  Multilocus enzyme electrophoresis (MLEE)

Genotypic methods

  Plasmid fingerprinting, with or without restriction enzyme
    analysis
  Chromosomal restriction enzyme analysis (REA) with conven-
    tional electrophoresis
  Genome restriction fragment–length polymorphism analysis
    (RFLP)
  Insertion sequence (IS) probe fingerprinting
  Ribotyping
  Pulsed-field gel electrophoresis
  Polymerase chain reaction (PCR) analysis
  Arbitrarily primed PCR (AP-PCR)
  Repetitive chromosomal element PCR (Rep-PCR)
  PCR-RFLP
  Spacer oligonucleotide typing ("spoligotyping")
  Mycobacterial interspersed repeat unit (MIRU)–based
    genotyping

question. This ensures that all steps in the protocol are working, and that the data are both reproducible from testing run to testing run within the laboratory and consistent with the data obtained by other investigators for the same strain.[3] The use of reference strains provides for consistency between analyses performed at different times. In the investigation of a potential outbreak of nosocomial infection, epidemiologically unrelated strains should be included to ensure that the procedure is able to differentiate between epidemic and endemic strains. Control strains should be geographically and temporally similar to the strains of interest. This is particularly important for strains with limited genetic variability, such as MRSA. In such situations, it may appear that groups of patients are infected with the same strain; however, because MRSA tends to be clonally restricted, unrelated strains may not be easily differentiated by typing.

The use of appropriate controls is critical for polymerase chain reaction (PCR)–based methods as well. Amplification of a highly conserved region or gene may give the appearance that unrelated organisms are related. Alternatively, amplification of hypervariable targets may yield differences that are due to rapid genetic drift (such as happens with some viral antigens) and obscure genetic relatedness between strains. Thus, the choice of control strains or regions should be considered carefully.

Genotyping Methods

Plasmid Restriction Enzyme Analysis

Plasmids (extrachromosomal DNA elements) are present in most bacteria. In plasmid typing, plasmid DNA is extracted and subjected to restriction-enzyme digestion. The digests are electrophoresed, and the numbers and sizes of the fragments are compared. Plasmid restriction enzyme analysis (REA) is the simplest of the DNA-based methods and is relatively inexpensive. Limitations of plasmid REA include the following: (1) plasmids can be spread to multiple species ("plasmid outbreak"), (2) plasmids may be lost spontaneously, and (3) plasmids may be altered by mobile genetic elements, such as transposons. The strong selective pressure exerted by the hospital environment for organisms to carry antibiotic-resistance or virulence factors can also drive these changes. Plasmid REA is generally most effective in analyzing whether the presence of antibiotic-resistance elements carried by genetically distinct strains results from the dissemination of resistance plasmids.[14]

Chromosomal REA

In this technique, restriction enzymes cleave bacterial chromosomal DNA at particular sequences that may be repeated multiple times in the genome, generating numerous fragments that can be separated using agarose gel electrophoresis. Different strains of the same bacterial species have different REA profiles because of variations in DNA sequences that alter the frequency and distribution of restriction sites. The advantage of this technique is that all isolates are typeable by REA. The disadvantage, however, is the difficulty in interpreting complex profiles that may contain hundreds of unresolved and overlapping bands. REA patterns may also be confounded by the presence of contaminating plasmids. Currently, the primary use of chromosomal REA is as an alternative technique for evaluating *Clostridium difficile* infection.[15]

Southern Blot Analysis of Chromosomal DNA

In this method, the restriction fragments separated by conventional agarose gel electrophoresis, described above, are transferred onto a nitrocellulose or nylon membrane, and a chemically or radioactively labeled fragment of DNA or RNA is used as a probe to detect fragment(s) that contain sequences (loci) homologous to the probe. Variations in the number and size of these fragments are called "restriction fragment–length polymorphisms" (RFLPs). While this technique is generally too laborious and time consuming for typing most hospital-associated pathogens, it is useful for typing

M. tuberculosis isolates using the DNA insertion element IS6110, a mobile sequence present in most strains of *M. tuberculosis*.[15,16] IS6110 is a 1,365–base pair (bp) insertion sequence unique to the *M. tuberculosis* complex and related to the IS3 family of enterobacterial insertion sequence elements. Insertion sequences are small, mobile, genetic elements (often less than 2.5 kilobases in length) widely distributed in most bacterial genomes. Unlike transposons, which encode phenotypic markers, insertion sequence elements carry only the genetic material related to their transposition and regulation. Southern blot analysis of chromosomal DNA involves extraction of DNA from *M. tuberculosis* isolates derived from culture, followed by digestion using *Pvu*II restriction endonucleases, agarose gel electrophoresis, and Southern blot analysis using IS6110-based probes.

IS6110-based genotyping remains the gold standard technique for fingerprinting *M. tuberculosis*. While it has a significant ability to discriminate between strains containing more than 6 copies of IS6110, it has limited resolution in analyzing clinical strains with 6 or fewer copies of IS6110. Additional limitations of this technique include its inability to distinguish among members of the *M. tuberculosis* complex, the fact that it is labor intensive (eg, it requires subculturing and DNA isolation), the long turnaround time (30–40 days), and finally, the tediousness of interlaboratory comparison of RFLP patterns.

In addition, RFLP analysis of *mec* (the gene encoding methicillin resistance) and Tn*554* (which carries the gene encoding erythromycin resistance) has been used for typing *S. aureus*.[8]

Ribotyping is a variation of this method that uses a ribosomal RNA probe for detecting DNA sequences that encode ribosomal RNA loci.[17] All bacterial isolates can be typed using this method, because they all have at least 1 chromosomal ribosomal RNA operon and the sequences are highly conserved. While ribotyping has only moderate discriminatory power in relation to the number of ribosomal operons, its usefulness as a tool to group phylogenetically related organisms is excellent.[11]

Pulsed-Field Gel Electrophoresis

Pulsed-field gel electrophoresis (PFGE) is similar to chromosomal REA but uses restriction enzymes that have fewer recognition sites in the bacterial chromosomal DNA, yielding a smaller number of restriction fragments (10–30). The fragments are resolved using angulated alternating current pulses, allowing for a higher level of band resolution than does conventional electrophoresis. PFGE is highly discriminatory and reproducible and can be used for many bacterial species.[3] A notable exception is that certain strains of *C. difficile* are not typeable using PFGE because of spontaneous degradation of chromosomal DNA during the protocol.[19]

Major limitations of PFGE include the need for technical expertise and the initial costs of the equipment, although many laboratories are skilled in this technique. While commercial products exist for DNA preparation, these are not universally used (including in our laboratory at the Veterans Affairs Medical Center in Philadelphia, PA). On the other hand, interpretation of restriction-fragment profiles produced by PFGE is more straightforward than interpretation of the profiles produced by chromosomal REA, and guidelines with criteria for strain typing with PFGE have been published.[3]

PCR-Based Methods

PCR is a powerful molecular method that has been adapted for use as a typing tool for epidemiologic investigations.[13,18] It amplifies minute quantities of microbial DNA present in a sample, which allows for detection and identification of microbes and can generate a "molecular fingerprint" for each organism based on its unique sequences of DNA (or RNA). Compared with other methods, PCR offers the advantage of speed (results are obtained in minutes to hours) and is unrivaled in sensitivity. PCR can also be extremely specific, depending on the choice of primers. Specific PCR assays use primers that are complementary to unique DNA sequences in a particular microbe's genome, while consensus PCR assays are designed to amplify conserved regions of microbial DNA in order to detect a wider variety of organisms. Unique sequences of DNA within these conserved regions may then be used to identify specific microbes by sequencing or restriction enzyme analysis. For example, the bacterial 16S ribosomal RNA gene is a useful gene for both broad-range and organism-specific PCR. The high sensitivity of PCR can lead to its greatest limitation, in that false-positive results can occur because of amplification of contaminating DNA present in a sample. Therefore, PCR should be performed only by personnel trained in the use of the technique in a laboratory specifically set up for the procedure, and negative controls must always be used. Numerous automated, closed systems have been developed and are commercially available, which eliminate or limit many of the technical problems associated with DNA purification and cross-contamination. Several modifications of PCR useful for epidemiologic purposes are outlined below.

Arbitrarily Primed PCR Analysis

In arbitrarily primed PCR (AP-PCR), random primers are used to amplify the DNA. Because random primers are short and hybridize to many areas on the genome, multiple fragments are produced. The DNA product is electrophoresed, and the resulting band pattern is used to determine whether isolates are related or not related.[1]

AP-PCR has several advantages: it is widely applicable to many bacterial species and is rapid and fairly easy to perform, compared with other typing systems. A disadvantage of AP-PCR is the difficulty of standardization; small changes in reaction conditions may give variability in banding patterns, resulting in low reproducibility (patterns may differ from experiment to experiment and between different laboratories).[18] Therefore, the most reliable results are obtained from testing a set of isolates in a single experiment. In addition, standardized guidelines have not yet been developed for interpreting data obtained by AP-PCR.

Amplified Fragment–Length Polymorphism Analysis
In amplified fragment–length polymorphism (AFLP) analysis, chromosomal DNA is digested with a restriction endonuclease, generating a large number of fragments. Adapter sequences are ligated to the ends of the fragments, and PCR primers designed to anneal to the adapter sequences are used to amplify the fragments in a PCR reaction. PCR products are analyzed by gel electrophoresis. Because of the amplification, AFLP is a highly sensitive method. It also has the advantages of being highly reproducible and relatively easy to interpret.[18]

Repetitive PCR Analysis
Repetitive PCR analysis (Rep-PCR) is based on the analysis of repetitive sequence elements (noncoding). These sequences (generally 30–500 bp) are dispersed throughout the genome of eubacteria.[20] The repetitive elements are amplified by PCR, generating multiple amplicons of various sizes. The amplicons are analyzed by gel electrophoresis; the banding patterns generated are highly reproducible and specific. Rep-PCR has been adapted to automated microfluidic systems that are commercially available.

Rep-PCR is relatively rapid and easy to perform. The availability of kits and the automated format of this procedure can facilitate the implementation of strain typing in laboratories that do not have the personnel to perform the more labor-intensive techniques, such as PFGE. Rep-PCR has been used for several hospital-associated pathogens, including *S. aureus*, *Enterococcus*, and gram-negative organisms. One of the limitations of Rep-PCR is that its discriminatory power is low for the species that exhibit clonality, such as *S. aureus*.

Spacer Oligonucleotide Typing
Spacer oligonucleotide typing ("spoligotyping") is the most commonly used PCR-based technique for subspeciating *M. tuberculosis* strains.[21] Spoligotyping involves PCR amplification with labeled primers of the direct repeat locus of the *M. tuberculosis* chromosome. These labeled PCR products are used to probe a membrane containing covalently bound oligonucleotides corresponding to each of the 43 variable-spacer sequences in *M. tuberculosis* strains H37Rv and in *Mycobacterium bovis* (bacille Calmette-Guérin strain P3). Compared with IS6110-based genotyping, spoligotyping is much quicker, is less labor intensive, requires much less DNA, and can be performed on clinical specimens. Furthermore, results are highly reproducible and can be computerized and displayed in a binary code. Hence, they are more amenable to intralaboratory comparison. Spoligotyping does not discriminate among *M. tuberculosis* strains as well as IS6110-based RFLP analysis does, however. Therefore, IS6110-based genotyping is often used to add additional information, if strain patterns are identical by spoligotyping and clonality is questioned.

Mycobacterial Interspersed Repetitive Unit Typing
A new typing technique known as mycobacterial interspersed repetitive unit (MIRU) analysis has proven useful for genotyping of *M. tuberculosis*. The *M. tuberculosis* genome contains 41 loci with direct tandem repeats of 50–70 bp that are known as mycobacterial interspersed repeat units, or "MIRUs." The number of repeats per locus varies between strains. MIRU typing measures variability in 12 of the 41 MIRUs by employing PCR amplification followed by gel electrophoresis. Since there are 2–8 alleles present at each of the 12 loci, this allows for 1.6×10^6 possible combinations. Advantages of MIRU typing include the fact that it can be automated and thus used to evaluate a large number of strains, and results can be digitized and catalogued in a computer database. Furthermore, MIRU typing is less labor intensive than IS6110-based genotyping is and can be applied to *M. tuberculosis* isolates from culture without DNA purification. As with spoligotyping, MIRU typing is less discriminatory than is IS6110-based genotyping.

Multilocus Sequence Typing
Multilocus sequence typing (MLST) is based on the sequencing of multiple housekeeping genes and is the genetic equivalent of multilocus enzyme electrophoresis.[22] The targeted housekeeping genes are under stabilizing selection; thus, genetic variation is accumulated relatively slowly. The sequences of these genes are determined by routine DNA sequencing methods, after which they are analyzed with publicly available databases. MLST has been used for strain typing for epidemiologic studies and offers the advantage of a standardized approach. If a laboratory has DNA sequencing capabilities, data collection for MLST analysis is straightforward and does not require specialized protocols. Data management is relatively easy, as the databases may be accessed via the Internet and they are curated. Large

amounts of data may be collected, used, and then archived for future use. MLST typing schemes and publicly available databases are available for many organisms.

Guidelines for Selection of a Typing Method

The choice of a strain typing system or method will depend on the capability and expertise of the laboratory that serves the hospital. To perform strain typing in-house, the laboratory must have expertise in molecular methods and have the capability to develop and maintain a database of strain typing patterns. In this case, the choice of a system will depend on the cost and the performance characteristics of the method (discriminatory power, reproducibility, and the availability of the relevant database[s]).[12]

Smaller hospitals that do not have on-site laboratories that can perform strain typing may choose to use the services of a commercial laboratory or a public health laboratory, which may assist with epidemiologic typing, depending on the organism and the question. Preferred strain typing techniques for particular microorganisms are presented in Table 9-2.

Practical Uses of Strain Typing: Outbreak Investigation

Molecular epidemiologic techniques are employed to determine whether organisms in a cluster originate from a common source. The following case histories are illustrative of how molecular epidemiologic techniques can aid in an outbreak investigation.

Example 1. A healthcare worker has active tuberculosis diagnosed. Surveillance of exposed patients and staff shows no evidence of purified protein derivative (PPD) test conversion or active disease. Two months after diagnosis of the index case, a coworker of the index patient receives a diagnosis of tuberculous lymphadenitis. For both clinical and legal reasons, it is important to determine whether isolates from the 2 cases belong to the same strain or not. In this instance, molecular analysis of the 2 infecting strains by RFLP analysis of IS6110 revealed that the strains were be distinct, therefore excluding the possibility of transmission between hospital staff.[23]

Example 2. Over a time span of 3 weeks, 5 patients receiving care at a dialysis unit developed bloodstream infection with MRSA. Antibiotic susceptibility and biotyping patterns were identical for all isolates, and a

Table 9-2. Preferred strain typing techniques for healthcare-associated and community-acquired pathogens

Species	Reference method	Alternative method(s)
Staphylococcus aureus	PFGE	Arbitrarily primed PCR, plasmid fingerprinting[a]
Coagulase-negative staphylococci	PFGE	Plasmid fingerprinting[a]
Streptococcus pneumoniae	PFGE	Serotyping
Enterococci	PFGE	None
Escherichia coli,[b] *Citrobacter* species, *Proteus* species, *Providencia* species	PFGE	Arbitrarily primed PCR, Rep-PCR
Klebsiella species, *Enterobacter* species, *Serratia* species	PFGE	Plasmid fingerprinting[a]
Salmonella species, *Shigella* species	Serotyping	PFGE
Pseudomonas aeruginosa	PFGE	None
Burkholderia species, *Stenotrophomonas maltophilia,* *Acinetobacter* species	PFGE	None
Clostridium difficile	Arbitrarily primed PCR	REA, PFGE[c]
Mycobacterium tuberculosis	IS6110 RFLP	Spoligotyping, MIRU typing
Mycobacteria other than *M. tuberculosis*	PFGE	None

NOTE: IS6110 RFLP, restriction fragment–length polymorphism typing using IS6110; MIRU, mycobacterial interspersed repeat unit; PCR, polymerase chain reaction; PFGE, pulsed-field gel electrophoresis; REA, restriction endonuclease analysis of chromosomal DNA using conventional electrophoresis; Rep-PCR, repetitive PCR.
a With or without restriction analysis.
b *E. coli* 0157:H7 must be identified by serotyping.
c Many strains of *C. difficile* are not typeable by PFGE because of DNA degradation.

common source of infection was suspected. If the isolates have identical patterns and differ from control strains of *S. aureus,* then the dialysis equipment and/or staff may be implicated as the source of infection. If, however, the isolates belong to different strains, then it is likely that the infections derive from colonizing flora, and infection control practices should be reviewed with the staff. In this case, PFGE demonstrated that the 5 bloodstream isolates were unique, again ruling out a common source of infection.

Example 3. Molecular epidemiologic techniques have been employed to characterize sources and confirm laboratory contamination, as well as to identify otherwise unsuspected clusters of infection. A study of IS6110 patterns of *M. tuberculosis* isolates from patients in San Francisco found previously unsuspected clusters of isolates with identical genotypes. In this study, subsequent epidemiologic follow-up demonstrated epidemiologic associations for most clusters; in contrast, cross-contamination within the laboratory was demonstrated as the cause for other clusters.[23]

References

1. Singh A, Goering RV, Simjee S, Foley SL, Zervos MJ. Application of molecular techniques to the study of hospital infection. *Clin Microbiol Rev* 2006;19(3):512–530.

2. Fey PD, Rupp ME. Molecular epidemiology in the public health and hospital environments. *Clin Lab Med* 2003;23(4): 885–901.

3. Tenover FC, Arbeit RD, Goering RV. How to select and interpret molecular strain typing methods for epidemiological studies of bacterial infections: a review for healthcare epidemiologists. Molecular Typing Working Group of the Society for Healthcare Epidemiology of America. *Infect Control Hosp Epidemiol* 1997;18(6):426–439.

4. Pfaller MA. Molecular epidemiology in the care of patients. *Arch Pathol Lab Med* 1999;123(11):1007–1010.

5. Tenover FC, Arbeit RD, Goering RV, et al. Interpreting chromosomal DNA restriction patterns produced by pulsed-field gel electrophoresis: criteria for bacterial strain typing. *J Clin Microbiol* 1995;33(9):2233–2239.

6. Orskov F, Orskov I. From the National Institutes of Health: summary of a workshop on the clone concept in the epidemiology, taxonomy, and evolution of the Enterobacteriaceae and other bacteria. *J Infect Dis* 1983;148(2):346–357.

7. Musser JM. Molecular population genetic analysis of emerged bacterial pathogens: selected insights. *Emerg Infect Dis* 1996; 2(1):1–17.

8. Kreiswirth B, Kornblum J, Arbeit RD, et al. Evidence for a clonal origin of methicillin resistance in *Staphylococcus aureus*. *Science* 1993;259(5092):227–230.

9. Mathema B, Kurepina NE, Bifani PJ, Kreiswirth BN. Molecular epidemiology of tuberculosis: current insights. *Clin Microbiol Rev* 2006;19(4):658–685.

10. Struelens MJ, De Gheldre Y, Deplano A. Comparative and library epidemiological typing systems: outbreak investigations versus surveillance systems. *Infect Control Hosp Epidemiol* 1998;19(8):565–569.

11. Maslow JN, Mulligan ME, Arbeit RD. Molecular epidemiology: application of contemporary techniques to the typing of microorganisms. *Clin Infect Dis* 1993;17(2):153–162.

12. Weber S, Pfaller MA, Herwaldt LA. Role of molecular epidemiology in infection control. *Infect Dis Clin North Am* 1997;11(2):257–278.

13. van Belkum A, Tassios PT, Dijkshoorn L, et al. Guidelines for the validation and application of typing methods for use in bacterial epidemiology. *Clin Microbiol Infect* 2007;13(Suppl 3): 1–46.

14. Lautenbach E, Patel JB, Bilker WB, Edelstein PH, Fishman NO. Extended-spectrum β-lactamase–producing *Escherichia coli* and *Klebsiella pneumoniae*: risk factors for infection and impact of resistance on outcomes. *Clin Infect Dis* 2001; 32(8):1162–1171.

15. Maslow J, Mulligan ME. Epidemiologic typing systems. *Infect Control Hosp Epidemiol* 1996;17(9):595–604.

16. van Embden JD, Cave MD, Crawford JT, et al. Strain identification of *Mycobacterium tuberculosis* by DNA fingerprinting: recommendations for a standardized methodology. *J Clin Microbiol* 1993;31(2):406–409.

17. Stull TL, LiPuma JJ, Edlind TD. A broad-spectrum probe for molecular epidemiology of bacteria: ribosomal RNA. *J Infect Dis* 1988;157(2):280–286.

18. van Belkum A. DNA fingerprinting of medically important microorganisms by use of PCR. *Clin Microbiol Rev* 1994; 7(2):174–184.

19. Kato H, Kato N, Watanabe K, et al. Application of typing by pulsed-field gel electrophoresis to the study of *Clostridium difficile* in a neonatal intensive care unit. *J Clin Microbiol* 1994;32(9):2067–70.

20. Versalovic J, Koeuth T, Lupski JR. Distribution of repetitive DNA sequences in eubacteria and application to fingerprinting of bacterial genomes. *Nucleic Acids Res* 1991;19(24): 6823–6831.

21. Vitol I, Driscoll J, Kreiswirth B, Kurepina N, Bennett KP. Identifying *Mycobacterium tuberculosis* complex strain families using spoligotypes. *Infect Genet Evol* 2006;6(6):491–504.

22. Maiden MC. Multilocus sequence typing of bacteria. *Annu Rev Microbiol* 2006;60:561–588.

23. Small PM, McClenny NB, Singh SP, Schoolnik GK, Tompkins LS, Mickelsen PA. Molecular strain typing of *Mycobacterium tuberculosis* to confirm cross-contamination in the mycobacteriology laboratory and modification of procedures to minimize occurrence of false-positive cultures. *J Clin Microbiol* 1993;31(7):1677–1682.

Chapter 10 Informatics for Infection Prevention

Keith F. Woeltje, MD, PhD, and Anne M. Butler, MS

Computers have become nearly ubiquitous in modern offices, including hospitals. Unlike even a few years ago, it would now be rare to find an infection preventionist or healthcare epidemiologist who did not have a computer on their desk. However, the use of computers for healthcare epidemiology is moving beyond the desktop computer. With the increasing amount of patient information that is electronically available, the computer is no longer just a tool for analyzing and reporting on data, but it is increasingly a tool for data acquisition and automation, even to the point of fully automated surveillance for healthcare-associated infections (HAIs). Automated surveillance is advantageous because it can reduce the time spent by healthcare epidemiologists on routine surveillance and reallocate it to prevention efforts. This chapter will review the use of computers in assisting the healthcare epidemiologist.

Informatics in Practice

Because of the increasing availability of electronic medical data, numerous hospitals have implemented automated surveillance systems for HAIs. Data for electronic surveillance is obtainable from a wide variety of sources. Microbiology laboratory systems can provide information on culture results, and other laboratory systems may have additional information that is useful, such as white blood cell counts with differential counts. Electronic nursing records or electronic medical records may record pertinent signs, such as temperature and blood pressure. Administrative systems, such as the admissions, discharge,

transfer (ADT) system, often provide useful information regarding patient location and demographic characteristics. Pharmacy records can be used to determine whether patients received antibiotics. Billing data often provide information on comorbidities and surgical procedures.

An ideal system would be for fully automated electronic surveillance for all HAIs. In theory, such a system could be built by evaluating patients' discharge diagnoses, since these are nearly universally available electronically. Unfortunately there are often significant delays before the diagnosis codes are finalized, making timely surveillance difficult. More significantly, discharge codes have proven unreliable for HAI surveillance in practice.[1-4] Reasons for this include differences between clinical and surveillance definitions for infections, as well as the fact that, from a billing perspective, only a limited number of codes can be applied, and HAI indicators may not be included for a given patient.

Many electronics surveillance systems for HAI are centered around microbiologic data. While for some HAIs (eg, central-line–associated bloodstream infection [CLABSI]) this may be very reasonable, for others (eg, ventilator-associated pneumonia), culture results really are not part of the surveillance definition.[5] Even positive blood culture results may represent a secondary rather than primary infection; this ambiguity can be partially dealt with by evaluating cultures of specimens from other body sites. There is also the issue of contamination, although software rules have been shown to deal with this issue reasonably well.[6,7]

Although there may be concerns about automated surveillance, it should be remembered that routine manual surveillance has its own limitations—because definitions have a subjective component, there can be considerable variation between raters in the assessment of whether a patient has an HAI.[8-11] In one validation study by the Centers for Disease Control and Prevention (CDC) for the old National Nosocomial Infections Surveillance system, investigators found twice as many infections as did hospital personnel.[12] Despite very careful chart review, even the investigators in phase I of this study failed to identify infections that the hospitals had detected; 37% of such patients were determined to have an HAI on further review.

One way to utilize electronic systems to overcome these issues is to use them for "augmented surveillance," rather than fully automated surveillance: the electronic systems help infection preventionists perform manual surveillance by making them more efficient. Many "homegrown" systems (ie, developed independently by a healthcare facility) and commercial systems can provide alerts on positive culture results, thereby aiding case-finding. At Barnes-Jewish Hospital (St. Louis, Missouri), specialized reports from its GermWatcher system[13] are used to evaluate for clusters of infection due to specified organisms on hospital units. Beyond these simple systems, more-sophisticated rule sets can be implemented that assist the infection preventionist by reducing the number of patients that need to be evaluated. At Barnes-Jewish Hospital, we have begun to implement a system that presents data on candidate patients to the infection preventionist, who then applies traditional National Healthcare Safety Network definitions. During validation trials, by implementing screening rules, we were able to eliminate 68% of positive culture results from having to be evaluated by the infection preventionist because they were highly unlikely to represent a true CLABSI (negative predictive value, 99.3%).[14] We are evaluating similar systems to improve surveillance for ventilator-associated pneumonia and surgical site infection by reducing the number of low-likelihood patients that need to be evaluated the infection preventionist, while at the same time reducing the chance of overlooking patients with true HAI. At Brigham and Women's Hospital (Boston, Massachusetts), a computer-assisted system to perform surveillance for ventilator-associated pneumonia has been evaluated.[15]

Because manual surveillance for HAIs is so labor intensive, in many areas of the hospital there is effectively no surveillance for HAIs. In such a circumstance, a fully automated surveillance system might be very valuable. Several authors have described surveillance systems for detection of nosocomial bacteremia in general.[16-18] But most hospitals would prefer to focus on CLABSI, because this infection may be more amenable to interventions to reduce the incidence. Trick et al.[8] reported on an automated surveillance system that performed comparably to manual surveillance; although not identical, the results of using the automated system trended well over time. It was not fully automated, in that data on central venous catheter usage had to be manually acquired. Nevertheless, such a system could help focus interventions on areas where rates were most problematic and could be used to track the impact of such interventions. Other investigators have reported on automated systems for detection of CLABSI as well.[19,20] CLABSI is particularly amenable to automated surveillance, so undoubtedly much more will be published on this topic in the near future.

Efforts have been made to detect other HAIs as well. For surgical site infection, a combination of antibiotic exposure data and diagnosis codes may provide a reasonable measure.[21-25] At the very least, this can be used to augment surveillance, in order to minimize missed cases while reducing the overall number of patients whose cases have to be reviewed by the infection preventionist.[26]

Resources Necessary for Implementation

Computers

With the incredible advances made in computer processor technology and data storage in the past decades, even computers that are several years old can readily handle the routine needs of the healthcare epidemiologist or infection preventionist. Only the most intensive statistical software continues to need the very latest central processing units (CPUs) for optimal performance. When a computer does need to be replaced, inexpensive lower-end machines will usually be perfectly adequate for routine use. Decisions on hardware specifications and operating systems are generally made by the hospital information technology department.

An entire computer does not need to be replaced in order to improve its utility. Increasing the amount of random-access memory (RAM) can often improve the performance of a computer significantly. File sizes seem to get larger and larger, and over time we accumulate more of them. Fortunately, the price of hard drives has dropped considerably. Adding a new hard drive may extend the useful life of a computer, if lack of storage space has become a significant issue. Inexpensive external hard drives are also a critical investment, along with software to back up essential data (unless this is done automatically on the hospital's internal network). As the prices of liquid crystal displays (LCDs) have decreased, adding a second screen (or simply getting a single large monitor) can dramatically improve efficiency by reducing the amount of scrolling the user needs to do,[27] potentially quickly repaying the cost of the monitor.

Minimum software for healthcare epidemiology, as for all business users, includes word processing and presentation software. The Microsoft Office suite of products (Microsoft) is a de facto standard and is available for both the Microsoft Windows operating system and the Macintosh OS X operating system (Apple). Alternatives do exist, including the freely available OpenOffice.org suite (OpenOffice.org).[28] Many of these packages can read and write Microsoft Office files, allowing for documents to be shared between people who use different programs.

A connection to the Internet is essential. E-mail allows collaboration both within the healthcare facility and with colleagues outside it. Access to the World Wide Web is equally important. Many facilities block access to the Web, fearing inappropriate use by employees. However the abundance of useful information from the CDC, the World Health Organization (WHO), professional organizations such as the Society for Healthcare Epidemiologists of America (SHEA) and the Association for Professionals in Infection Control and Epidemiology (APIC), as well as county and state health departments, makes Web access an absolutely essential resource, without which a healthcare epidemiologist or infection preventionist is severely hampered. A request for an exception to an institution's Web blocking policy should be made. Security may be improved by avoiding use of the Internet Explorer browser (Microsoft) and instead using a more secure alternative, such as the freely available Firefox browser.[29,30,31]

Data

The technical details of establishing an informatics database or "data warehouse" are best left to specialists trained in informatics and information technology. But in order to convey your needs and understand what their limitations are, some understanding of databases and informatics concepts is useful. An overview text, such as *Biomedical Informatics: Computer Applications in Health Care and Biomedicine* by Shortliffe and Cimino,[32] provides a good background, as does participation in an American Medical Informatics Association (AMIA) 10 x 10 program.[33] A conceptual understanding of relational databases is also very useful; Hernandez's *Database Design for Mere Mortals*[34] is a nice introduction that is not focused on any particular database system.

Data Needs

At the most basic level, electronic access to data on positive microbiologic culture results is the foundation for an electronic surveillance system. At BJC HealthCare, we have recently begun to also save data on negative culture results in our informatics database. Although we do not yet incorporate negative culture results into any surveillance algorithms, you might imagine how having these data could help distinguish true infection from contamination (eg, by identifying common skin contaminants). The fact that samples for culture were sent to the laboratory (even if the culture results were negative) may also help distinguish community-onset infection from healthcare facility–onset infection. For example, if a specimen for *Clostridium difficile* culture was sent on the day of admission, but the result was negative, and then a second specimen was sent on hospital day 3 that had a positive culture result, this would suggest that the patient had been admitted with symptoms suggestive of *C. difficile* infection, and that the infection was not hospital acquired, as might seem to be the case if only the positive culture result was available. Having quantitative culture results, where available, rather than simply an indication that result was positive or negative, is also potentially useful.

Other laboratory data might be helpful, such as the white blood cell count with a differential count. Patients' vital signs, such as temperature and blood pressure, may be used to assist in determining whether an infection is present.

Data Sources

Electronic data for surveillance is increasingly available from a wide variety of sources, including microbiology laboratory, pharmacy, and billing databases. However, all of the information necessary for electronic surveillance may not be available in electronic format. For instance, some information needed to stratify patients by level of risk (eg, ASA score, wound class, and/or duration of surgery) will have to come from another source, such as a surgery documentation system. This is also true of denominator data (eg, the number of central-line days and ventilator-days). The information can be made available by providing for manual entry of these data into the system used for surveillance. But such provisions should be seen as a temporary measure, and plans should be formed to provide some mechanism for obtaining the information electronically. In particular, being able to determine whether a given patient had an indwelling device (eg, a central venous catheter) present on a given day is also critical for optimal rule-based determination of a whether there is a device infection (ie, numerator data).

It should be noted that data on the number of device-days obtained by some electronic systems may differ from those obtained manually. Traditionally, the presence of a device in patients on a hospital unit is determined at the same time every day. So devices removed before the counting is done for that day are not included in the count, nor are devices inserted after the counting.

Electronic systems may only determine that a patient did or did not have a device in place at some point during a given day. This may give slightly higher counts of device-days (and consequently lower infection rates) than the traditional manual counting method does. The National Healthcare Safety Network allows for use of electronically obtained counts of device-days, as long as they do not differ from manual counts by more than 5%.[5]

Hospitals are increasingly implementing use of electronic medical records, and institutions that have systems in place may adopt second- or third-generation systems as their needs evolve. It is important for infection preventionists and healthcare epidemiologists (who will directly use the information), as well as end users such as intensive care unit managers and medical directors (who will want the results from electronic surveillance), to be involved as electronic medical record systems are being developed. This is to ensure that relevant information, such as data on device usage, is captured in a usable format. Ideally, essential information is captured in a discrete data field, and not just as words typed into the electronic record. Entries in such "plain text" data fields are typically not standardized, and are difficult to use to obtain consistent data. Likewise, as institutions aggregate selected information into a clinical "data warehouse," it is essential that the key users be involved so that the necessary data are kept. While electronic medical records are potentially very useful, their utility is defined by how much of the information needed by infection prevention personnel is actually present.

Data Storage and Nomenclatures
Once data have been gathered from hospital systems, they must be stored. Although the technical details are best left to information technology professionals, healthcare epidemiologists who hope to use the data must work with these professionals to ensure that the data can be useful.

A wide variety of standardized nomenclatures exist for medical purposes: for example, those in SNOMED,[35] and the *International Classification of Diseases, Ninth Revision*,[36] and other resources. For the most part it does not matter what standard is chosen to be used at a facility, as long as it is used consistently. However, if data are being aggregated from multiple facilities, there may need to be translation from one nomenclature to another, which is a potential source of error or ambiguity. Another issue is how do deal with a newly discovered organism before it is recognized in the official nomenclature.

Some design choices may affect usability, and should be made in cooperation with the information technology staff. For example, a database system may link organism genus and species designations with antimicrobial sensitivity data. With such a system, it would be possible to determine whether an isolate of *S. aureus* was methicillin resistant or methicillin sensitive, but it would require extra steps. For flagging and alerting purposes, some facilities choose to consider methicillin-resistant *S. aureus* (MRSA) as a distinct organism. While this facilitates some uses, it complicates matters if you want to query the system for data on all *S. aureus* isolates. If you want to know about all staphylococcal isolates, that adds additional complexity. None of these issues are insurmountable, but give careful consideration to the database design and possibly include the linking of multiple concepts with an organism designation (eg, "MRSA," "Staphylococcus," "aureus," and "gram-positive coccus") to facilitate later analysis.

Data Validation and Reliability
As electronic sources of data are being arranged, careful validation must be done to ensure that the data are complete. Some data may change after initial release; in particular, microbiology laboratory testing may yield intermediate results (eg, "gram-positive cocci in clusters") before the final result. Decisions will have to be made about whether to wait and only process final result data or to provide a mechanism to accommodate intermediate result data; also recognize that sometimes even "final" results are later changed.

Once systems are in place, they may fail: for example, reports may no longer be generated, formatting may change, or the computer network may be unavailable for a period of time. Provisions should be made for detecting when this occurs and for replacing any missing data. Other changes may not result in missing data but instead result in changed data. For example, the microbiology laboratory may begin speciating isolates of coagulase-negative staphylococci, or an organism's name may change, so that suddenly new organism names are appearing that the system is not set up to process. Again, processes will need to be in place to determine when this occurs so that matters can be rectified.

Software for Healthcare Epidemiology

General Statistical Software
Although counts and some basic statistics can be done using spreadsheet software, even modestly advanced statistics (see Chapter 4 on epidemiologic methods) will require actual statistical software. Some basic programs frequently come bundled with introductory statistical texts. These programs will often handle basic needs (eg, the χ^2 test and simple analysis of variance).

Epidemiologists who require more advanced statistics may need to use one of the many advanced statistical packages that are available, such as SPSS (SPSS Inc), SAS (SAS Institute), or STATA (Stata Corporation). These programs are extremely versatile, but they require significant time to learn. In addition, they can be quite expensive. A free, open-source alternative is the program R,[37] which is available for the Windows, Macintosh OS X, and Linux operating systems. A variety of resources for learning to use the software are available online (eg, the Web page of the teaching course "Applied Epi Using R"[38]).

Epi Info

Epi Info is a computer program designed by the CDC and is available as a free download from the CDC Web site.[39] Epi Info is only available for Microsoft Windows (older versions ran on the Microsoft MS-DOS operating system). It was originally designed as a program for field epidemiologists, but has become versatile enough for myriad uses. It sits somewhere between a general purpose statistical package and a specialized infection control package in its scope.

Epi Info consists of a number of modules. One, *Nut-Stat,* is designed for nutritional anthropometry, and is of limited use to most healthcare epidemiologists. During program installation you can choose which modules to install, so you can elect not to install this component. The other modules are *MakeView, Enter, Analysis,* and *EpiMap.* A number of small utility programs are also included, including a program to quickly calculate 2 x 2 contingency-table statistics and a program to add password protection to files that might contain sensitive information.

The *MakeView* module allows the user to create data-entry screens. These screens are then used by another module, *Enter,* as we will see. The screens can be laid out to simplify data entry by being made attractive and by grouping of similar information. During the design, constraints can be placed on the various data entry fields so that only appropriate answers can be entered (eg, only "Y" or "N" would be allowed for a yes-or-no question—the person entering the data could be prevented from entering a different letter or a number). The designer can also specify whether a particular field can be left blank. Epi Info was built on the basis of the Microsoft Access database program (the user need not have Access installed to use Epi Info). Because the records are stored in a true relational database, fairly sophisticated data collection can be managed. Once the data entry screens are designed, the *Enter* program module can be used to enter data into Epi Info. Data entry can be done at various times. It is even possible to allow different persons to enter data on their own computers and then consolidate the data later. Data can also be updated or changed once it has been entered.

The *Analysis* module provides statistical analysis of the data. In addition to analyzing data collected using the *MakeView* and *Enter* modules, *Analysis* can also read files created by some other programs, including Microsoft Excel and older versions of Epi Info. *Analysis* can perform basic statistical analysis, such as calculation of frequencies, the χ^2 test (and the Fisher exact test), and analysis of variance. Advanced statistical analyses include logistic regression and Kaplan-Meier survival curves. Within *Analysis* the user can create new variables and can recode variables to simplify analysis. For example, if you were investigating an outbreak of surgical site infection and had a list of surgeons for each patient, you could create a new variable "DrSmith" and then code it as either "Y" or "N" according to whether a given patient had Dr. Smith as a surgeon. This would allow you to create a 2 x 2 contingency table. (How to use Epi Info for outbreak investigations is also discussed in Chapter 12.) This facility can greatly simplify analysis, and allows the user not to have to guess when designing the data entry screens at all of the possible analyses that will eventually be wanted.

The *EpiMap* module allows the user to link data in Epi Info files to geographic maps. Maps for all US states and many countries are available from the Epi Info Web site.[39] The site also has links to other resources for finding other existing maps. Commercial tools are available for creating maps in the appropriate format. It would be possible, for example, to create a map of a hospital and use color-coding to display rates of MRSA colonization and infection.

Other General Software

A variety of other programs are available that are useful to healthcare epidemiologists. Some of these are smaller programs that are more suitable for approximate, "quick-and-dirty" calculations than are full-fledged statistics packages. Epi Info comes with a program called StatCalc that allows for quick entry of 2 x 2 tables and for calculations of the sample size needed to reach a designated statistical power. A similar (and more versatile) program is EpiCalc[40] (created by Mark Myatt; the Web site also has links to other useful small programs). Additional software resources can be found online; useful links include Tomás Aragón's Medical Epidemiology Web site,[41] the "Free Statistical Software" page on the StatPages.org Web site,[42] and Girish Singh's listing of statistical software freely available on the Internet.[43] One unique Web site, Open Source Epidemiologic Statistics for Public Health, actually makes epidemiologic calculators available online.[44]

Specialized Software for Infection Prevention

"Homegrown" systems. When computer systems were first introduced into hospitals, early visionaries saw how they could assist in surveillance for HAIs.[45-47] Since

commercial software for this purpose did not exist, of necessity hospitals and universities developed their own systems. An early pioneer was LDS Hospital, in Salt Lake City, Utah, who developed the HELP system and Computerized Infectious Disease Monitor (CIDM).[47,48] The CIDM system used microbiology laboratory data, admission date data, and rules regarding whether a detected organism was likely a true pathogen to determine whether a positive culture result represented a nosocomial infection or not. In the early 1990s Barnes Hospital (now Barnes-Jewish Hospital) implemented a system called GermWatcher.[6,49,50,13] It was different in architecture from the LDS CIDM but similar in function, providing reports of positive blood culture results that likely represented true nosocomial pathogens. Another such surveillance system is MONI, developed at the Allgemeines Krankenhaus der Stadt Wien (Vienna General Hospital).[51]

Such "homegrown" systems have the advantage that they can be highly customized to the needs of the institution, and by their very nature they integrate well into the existing data infrastructure. However, they require significant local expertise and significant investments of time and money by the parent institution for development and maintenance. And, in many cases, the developers of these systems need to "reinvent the wheel" as they implement systems to deal with common issues. Of course, all of the software need not be written from scratch; large portions of the systems may be built up from free or commercial components of varying sizes and complexity. Brigham and Women's Hospital is investigating the integration of 2 freely available software programs, WHONet and SatScan, to apply rigorous statistical analysis to detection of clusters of infections.[52]

Commercial systems. A growing number of commercial applications for infection prevention surveillance are available. Some of these systems started out with a focus on desktop personal computers, whereas others were designed from the beginning to be run on corporate computer systems. Some commercial systems send encrypted hospital data to the software vendor for remote processing. Almost all of the commercial software systems can import microbiology laboratory data and other hospital data to facilitate reporting, and many allow the user to develop rules to apply to the information, to assist in automating surveillance. Many of these systems are quite sophisticated, although published scientific literature about them is somewhat sparse.[53,54]

In evaluating these systems, hospitals should determine what features they need and whether a vendor can provide these features. Speaking to current users of the software can be very helpful to determine how they are using the system and learn about any issues with implementation; vendors are usually very willing to put potential clients in touch with happy users. It can also be valuable to talk with those who are unhappy with their system, or those who quit using a particular system. These users are usually found by word of mouth or by "networking" with colleagues at professional society meetings. It may well be that the issues those users had have been fixed subsequently or do not apply to your situation. In any case, having a comprehensive view of the pros and cons of any system is helpful. Another consideration is that some of the commercial systems are components of larger software systems for hospitals that may have other patient-safety issues, such as pharmacy errors. So the hospital should evaluate its overall needs and not just those of infection prevention. Obviously, the hospital information technology department needs to be involved from the beginning in making decisions regarding these systems.

Conclusions

Electronic surveillance for HAI is clearly coming. It will likely be a mix of fully automated surveillance and augmented surveillance. In fact, it seems likely that solely manual surveillance will be replaced completely. Clearly, some degree of manual surveillance will always be necessary. In addition, new terms may be needed to distinguish between rates calculated from data obtained in a fully automated manner and data obtained in an augmented manner. Furthermore, as HAI rates decline, continual refinement of electronic surveillance methods will be needed, to combat worsening positive predictive values. Overall, electronic surveillance should extend the oversight of infection preventionist and hospital epidemiologists to more areas and more procedures, while taking less time, thus allowing more time for active interventions.

References

1. Stevenson KB, Khan Y, Dickman J, et al. Administrative coding data, compared with CDC/NHSN criteria, are poor indicators of health care-associated infections. *Am J Infect Control* 2008;36:155–164.
2. Sherman ER, Heydon KH, St John KH, et al. Administrative data fail to accurately identify cases of healthcare-associated infection. *Infect Control Hosp Epidemiol* 2006;27:332–337.
3. Moro ML, Morsillo F. Can hospital discharge diagnoses be used for surveillance of surgical-site infections? *J Hosp Infect* 2004;56:239–241.
4. Madsen KM, Schonheyder HC, Kristensen B, Nielsen GL, Sorensen HT. Can hospital discharge diagnosis be used for surveillance of bacteremia? A data quality study of a Danish

hospital discharge registry. *Infect Control Hosp Epidemiol* 1998;19:175–180.

5. Centers for Disease Control and Prevention. The National Healthcare Safety Network (NHSN) Manual; Patient Safety Component Protocol. March 2009. http://www.cdc.gov/nhsn/psc.html. Accessed July 12, 2009.

6. Kahn MG, Steib SA, Dunagan WC, Fraser VJ. Monitoring expert system performance using continuous user feedback. *J Am Med Inform Assoc* 1996;3:216–223.

7. Yokoe DS, Anderson J, Chambers R, et al. Simplified surveillance for nosocomial bloodstream infections. *Infect Control Hosp Epidemiol* 1998;19:657–660.

8. Trick WE, Zagorski BM, Tokars JI, et al. Computer algorithms to detect bloodstream infections. *Emerg Infect Dis* 2004;10:1612–1620.

9. Wenzel RP, Osterman CA, Hunting KJ. Hospital-acquired infections. II. Infection rates by site, service and common procedures in a university hospital. *Am J Epidemiol* 1976;104:645–651.

10. Haley RW, Schaberg DR, McClish DK, et al. The accuracy of retrospective chart review in measuring nosocomial infection rates: results of validation studies in pilot hospitals. *Am J Epidemiol* 1980;111: 516–533.

11. Burke JP. Infection control—a problem for patient safety. *N Engl J Med* 2003;348:651–656.

12. Emori TG, Edwards JR, Culver DH, et al. Accuracy of reporting nosocomial infections in intensive-care-unit patients to the National Nosocomial Infections Surveillance System: a pilot study. *Infect Control Hosp Epidemiol* 1998;19:308–316.

13. GermWatcher. Open Clinical Web site. http://www.openclinical.com/aisp_germwatcher.html. Accessed September 23, 2009.

14. Woeltje KF, Butler AM, Goris AJ, et al. Automated surveillance for central line–associated bloodstream infection in intensive care units. *Infect Control Hosp Epidemiol* 2008; 29:842–846.

15. Klompas M, Kleinman K, Platt R. Development of an algorithm for surveillance of ventilator-associated pneumonia with electronic data and comparison of algorithm results with clinician diagnoses. *Infect Control Hosp Epidemiol* 2008;29:31–37.

16. Leth RA, Moller JK. Surveillance of hospital-acquired infections based on electronic hospital registries. *J Hosp Infect* 2006;62:71–79.

17. Wang SJ, Kuperman GJ, Ohno-Machado, Onderdonk A, Sandige H, Bates DW. Using electronic data to predict the probability of true bacteremia from positive blood cultures. *Proc AMIA Symp* 2000;893–897.

18. Graham PL III, San Gabriel P, Lutwick S, Haas J, Saiman L. Validation of a multicenter computer-based surveillance system for hospital-acquired bloodstream infections in neonatal intensive care departments. *Am J Infect Control* 2004; 32:232–234.

19. Bellini C, Petignat C, Francioli P, et al. Comparison of automated strategies for surveillance of nosocomial bacteremia. *Infect Control Hosp Epidemiol* 2007;28:1030–1035.

20. Woeltje KF, Goris AJ, Butler AM, et al. Automated surveillance for catheter-associated bloodstream infections outside of the intensive care unit. In: Program and abstracts of the 17th Annual Scientific Meeting of the Society for Healthcare Epidemiology of America; Baltimore, MD; April 2007.

21. Platt R, Yokoe DS, Sands KE, CDC Eastern Massachusetts Prevention Epicenter Investigators. Automated methods for surveillance of surgical site infections. *Emerg Infect Dis* 2001;7:212–216.

22. Sands K, Vineyard G, Platt R. Surgical site infections occurring after hospital discharge. *J Infect Dis* 1996;173:963–970.

23. Sands K, Vineyard G, Livingston J. Efficient identification of postdischarge surgical site infections using automated medical records. *J Infect Dis* 1999;179:434–441.

24. Sands KE, Yokoe DS, Hooper DC, et al. Detection of postoperative surgical-site infections: comparison of health plan-based surveillance with hospital-based programs. *Infect Control Hosp Epidemiol* 2003;24:741–743.

25. Yokoe DS, Noskin GA, Cunningham SM, et al. Enhanced identification of postoperative infections among inpatients. *Emerg Infect Dis* 2004;10:1924–1930.

26. Bolon MK, Hooper D, Stevenson KB, et al.; Centers for Disease Control and Prevention Epicenters Program. Improved surveillance for surgical site infections after orthopedic implantation procedures: extending applications for automated data. *Clin Infect Dis* 2009;48:1223–1229.

27. Thompson C. Meet the life hackers. *New York Times Magazine*. October 16, 2005. http://www.nytimes.com/2005/10/16/magazine/16guru.html. Accessed August 15, 2009.

28. OpenOffice.org Web site. http://www.openoffice.org/. Accessed September 23, 2009.

29. Mozilla Firefox Web site. http://www.mozilla.com/en-US/firefox/personal.html. Accessed September 23, 2009.

30. Krebs B. Security fix. *The Washington Post*. January 29, 2009. http://voices.washingtonpost.com/securityfix/2009/01/blog fight_the_truth_about_ie_v.html. Accessed August 15, 2009.

31. BBC. Serious security flaw found in IE. *BBC News*. December 16, 2008. http://news.bbc.co.uk/2/hi/technology/7784908.stm. Accessed August 15, 2009.

32. Shortliffe EH, Cimino JJ, eds. *Biomedical Informatics: Computer Applications in Health Care and Biomedicine*. 3rd Ed. New York: Springer; 2006.

33. AMIA 10 x 10. American Medical Informatics Association (AMIA) Web site. http://www.amia.org/10x10. Accessed September 23, 2009.

34. Hernandez MJ. *Database Design for Mere Mortals*. 2nd ed. Boston: Addison-Wesley Professional; 2003.

35. SNOMED (Systematized Nomenclature of Medicine—Clinical Terms) Web page. http://www.nlm.nih.gov/research/umls/Snomed/snomed_main.html. Accessed September 23, 2009.

36. *International Classification of Diseases, Ninth Revision*. http://www.cdc.gov/nchs/icd.htm. Accessed September 23, 2009.

37. The R Project for Statistical Computing. http://www.r-project.org/. Accessed September 23, 2009.

38. Applied Epi Using R. Medical Epidemiology Web site. http://sites.google.com/site/medepi/epir. Accessed September 23, 2009.

39. Epi Info. Centers for Disease Control and Prevention Web site. http://www.cdc.gov/epiinfo/. http://www.cdc.gov/epiinfo/. Accessed September 23, 2009.

40. EpiCalc. Brixton Health Web site. http://www.brixtonhealth.com/index.html. Accessed September 23, 2009.

41. Software recs. Medical Epidemiology Web site. http://sites.google.com/site/medepi/recs. Accessed September 23, 2009.

42. Free Statistical Software. StatPages.org Web site. http://stat pages.org/javasta2.html. Accessed September 23, 2009.

43. Singh G. Freely available statistical software on Internet. *Internet Journal of Medical Simulation* 2009;2(2). http://www. ispub.com/journal/the_internet_journal_of_medical_simula tion/volume_2_number_2_62/article_printable/freely_availa ble_statistical_software_on_internet.html. Accessed September 23, 2009.

44. Dean AG, Sullivan KM, Soe MM. OpenEpi: open source epidemiologic statistics for public health, version 2.3, updated May 20, 2009. http://www.openepi.com/Menu/OpenEpi Menu.htm. Accessed September 23, 2009.

45. Schifman RB, Palmer RA. Surveillance of nosocomial infections by computer analysis of positive culture rates. *J Clin Microbiol* 1985;21:493–495.

46. Mertens R, Ceusters W. Quality assurance, infection surveillance, and hospital information systems: avoiding the Bermuda Triangle. *Infect Control Hosp Epidemiol* 1994;15:203–209.

47. Evans RS, Larsen RA, Burke JP, et al. Computer surveillance of hospital-acquired infections and antibiotic use. *JAMA* 1986;256:1007–1011.

48. Evans RS, Gardner RM, Bush AR, et al. Development of a computerized infectious disease monitor (CIDM). *Comput Biomed Res* 1985;18:103–113.

49. Kahn MG, Steib SA, Fraser VJ, Dunagan WC. An expert system for culture-based infection control surveillance. In: Proceedings of the Annual Symposium on Computer Applications in Medical Care. 1993:171–175.

50. Kahn MG, Bailey TC, Steib SA, Fraser VJ, Dunagan WC. Statistical process control methods for expert system performance monitoring. *J Am Med Inform Assoc* 1996;3:258–269.

51. Chizzali-Bonfadin C, Adlassnig K-P, Koller W. An intelligent database and monitoring system for surveillance of nosocomial infections. In: Greens RA, Peterson HE, Protti DJ, eds. *Proceedings of the 8th World Congress on medical informatics.* MEDINFO 95 Edmonton, Canada: Healthcare Computing and Communications; 1995.

52. Huang SS, Yokoe DS, Stelling J, et al. Automated cluster detection in hospitals. In: Program and abstracts of the 46th annual meeting of the Infectious Diseases Society of America; Washington, DC; October 25–28, 2008.

53. Brossette SE, Sprague AP, Hardin JM, Waites KB, Jones WT, Moser SA. Association rules and data mining in hospital infection control and public health surveillance. *J Am Med Inform Assoc* 1998;5:373–381.

54. Wright MO, Perencevich EN, Novak C, Hebden JN, Standiford HC, Harris AD. Preliminary assessment of an automated surveillance system for infection control. *Infect Control Hosp Epidemiol* 2004;25:325–332.

Surveillance and Prevention

Chapter 11 Surveillance: An Overview

Trish M. Perl, MD, MSc, and Romanee Chaiwarith, MD, MHS

A good surveillance system does not guarantee that you make the right decisions but it reduces the chance of making the wrong ones.
—Alexander Langmuir[1]

The cornerstone of any clinical outcomes program, including one responsible for healthcare epidemiology and infection control, is surveillance for complications of medical care. This chapter reviews basic principles of surveillance for healthcare-associated infections (HAIs) and colonization or infection with epidemiologically important organisms. In the current era of emphasis on patient safety, this activity has been elevated from the dark corners of hospitals into the public, regulatory, and legislative realms. In this chapter we describe the importance of these activities, including the components of a surveillance system, methods for surveillance, methods for finding healthcare-associated events of interest, and data sources. We describe methods used to stratify patients by their risk of developing an HAI or acquiring an epidemiologically significant organism. We also discuss the importance of calculating rates of infection and/or colonization or other outcome rates in a standardized fashion to ensure that appropriate comparisons can be made. Finally, we review the importance of using computers and information technology, which are becoming integral to efficient and effective surveillance. Although we focus on surveillance as it applies to HAIs and colonization or infection with epidemiologically significant organisms, the methods we discuss in this chapter are

well established in healthcare settings and can also be applied to noninfectious complications of health care. We refer to surveillance systems that collect information about HAIs, colonization or infection with epidemiologically significant organisms, and the related processes of care. For simplicity, we use the term "HAI" to encompass these 3 areas of interest to the healthcare epidemiologist. We encourage healthcare epidemiology and infection control teams to use this information as they design surveillance systems or integrate informatics systems into their daily activities so that they meet the goals of their individual institution's program. The information in this chapter should be supplemented with the training and resources needed to provide healthcare institutions with accurate collection and analysis of these complex data.

Background

In 1847, Ignaz Semmelweis reported on an unusual difference in mortality rates among mothers delivering babies at the Allgemeines Krankenhaus der Stadt Wien, in Austria (the Vienna Lying-In Hospital). Semmelweis had made the general observation that mothers delivering in the hospital's First Obstetrical Clinic were more likely to develop puerperal fever than were those delivering in the Second Obstetrical Clinic. This difference was so pronounced that it was common knowledge on the streets, and women admitted to the hospital would plead for admission to the Second Clinic, believing they would die if they gave birth in the First Clinic. Troubled by this seemingly inexplicable disparity, Semmelweis documented the mortality rates from puerperal fever in the 2 clinics and began to collect data on the differences in patients and practices. He examined numerous variables without identifying a possible cause, until he noted that women in the First Clinic were treated by interns, who began the morning examining cadavers. Midwives, who were not involved in cadaver dissections, treated women in the Second Clinic. Hypothesizing that some element transferred from cadavers to the interns' hands and then to the pregnant women was responsible for the puerperal fever, Semmelweis instituted the practice of having all interns wash their hands with a chlorinated lime solution after dissection and prior to examining patients. With this single intervention, rates of puerperal fever in the First Clinic declined dramatically, until they generally equaled those in the Second Clinic.

Semmelweis's observation of the difference in rates of death due to puerperal fever was a basic form of surveillance. Surveillance can be described as the process of identifying rates of complications and ultimately intervening to reduce those rates. Surveillance is a dynamic process for collecting, concatenating, analyzing, and disseminating data concerning specific healthcare events that occur in a specific population.[2] Findings from surveillance are commonly linked to or should result in actions or decisions, often at the level of hospital policy. As the cornerstone of hospital epidemiology and infection control programs, surveillance provides data which are used to determine baseline rates of HAIs, colonization or infection with epidemiologically important organisms, or other adverse events, and to detect changes in previously measured rates or distributions of these events. The changes that have been found may lead to investigations of each case, including the determination of whether such events or significantly increased rates were clustered in time and/or space, the generation of hypotheses about risk factors, the institution of prevention and control measures, and, ultimately, the determination of whether the interventions instituted to positively affect these changes were effective. Appropriately performed surveillance requires defined events and systematic case-finding, and it commonly uses stratification of risk to identify trends and to evaluate the impact of interventions over time. Surveillance data should also be used to determine the risk factors for the outcome of interest, to monitor compliance with established hospital policies and practices, to evaluate changes in practice, and to identify topics for further study. Importantly, such processes are critical to ensure that appropriate data on rates are generated for interhospital comparisons.

Historically, in the United States, it was accepted by the Centers for Disease Control and Prevention (CDC), accrediting agencies, and hospital administrators that surveillance for nosocomial infections (now called healthcare-associated infections) was an important element of an infection prevention and control program. In 1974, the CDC initiated the Study on the Efficacy of Nosocomial Infection Control (SENIC) to determine the magnitude of the problem with HAIs, to evaluate the extent to which hospitals had adopted surveillance and control programs, and to examine whether infection prevention and control programs reduced rates of surgical site infection (SSI), ventilator-associated pneumonia (VAP), urinary tract infection (UTI), and bloodstream infection (BSI).[3] The SENIC investigators found that different combinations of infection control practices helped reduce the incidence of each of these types of infections.[3] However, surveillance was the only component found essential for reducing all 4 types of infection. Of note, this study examined the rates of the 4 most common infections, but data suggest that the incidences of other types of HAIs are reduced with comprehensive surveillance activities.[3-5] In addition, the SENIC project concluded that effective programs included surveillance for HAIs, adequate numbers of infection control practitioners or infection preventionists, feedback of data to healthcare providers, and a trained hospital epidemiologist (a physician trained in epidemiologic methods and infection control and prevention strategies).

Since the SENIC[3] was published, surveillance for HAIs and infection or colonization with epidemiologically important organisms, including those resistant to antimicrobial agents, has taken on even greater importance in facilitating the prevention of transmission among an increasingly ill population of patients. In addition, although much of the experience with surveillance for healthcare-associated events has taken place in North America, European and other international groups have recently developed large surveillance programs for HAIs and antimicrobial-resistant organisms that have supported the fundamental findings of the SENIC project and that have enhanced our understanding of the expanding roles of surveillance.[6-10] Such sophisticated surveillance programs that involve large numbers of hospitals have proved extremely effective in identifying new trends in the spread of antimicrobial-resistant organisms and in measuring the impact of interventions.

Several other new trends in surveillance should be mentioned. "Syndromic surveillance" uses health-related data to track events of potential public health significance, such as infection with an agent of bioterrorism and newly emerging diseases (eg, severe acute respiratory syndrome).[11] Data that can be captured in these systems include admission diagnoses, emergency department chief complaints, prescriptions written or filled, and test utilization patterns. Such data, once concatenated, are processed using sophisticated algorithms to identify syndromes (eg, respiratory syndromes, febrile influenza-like illness, or gastroenteritis) that mimic significant infectious and noninfectious diseases. The utility of this type of surveillance is that it detects outbreaks early and tracks disease trends and patterns. Hence, its potential application in healthcare epidemiology and infection control programs is being investigated, and it is being used increasingly.[12] Such data would be routinely transmitted to the public health authorities, as an "early warning" system for a bioterrorism event or to alert clinicians and, over time, assess the effectiveness of interventions. Currently, there are few data on how effective these efforts are or what components are required to make such surveillance efficient, valid, and effective.

Because of the impact and importance of HAIs and multidrug-resistant organisms, transparency about frequency and trends is being called for increasingly in the United States and other developed nations. France, the United Kingdom, and many states in the United States have passed legislation requiring healthcare facilities to report HAIs, detection of multidrug-resistant organisms, and/or process-of-care measures to public health authorities or to agencies that can publicly display the data, once they are reviewed and verified. The intended goal is to release these data to the public and encourage improvements in the quality of health care and patient safety.[13-15] With the looming requirement to publicly report HAIs, infection and colonization with epidemiologically significant organisms, and proxies of these 2 measures, the importance of using standard definitions, identifying cases systematically, and choosing the appropriate population at risk to form the denominator in rate calculations is paramount.[16]

What Does Surveillance Entail?

Surveillance requires that relevant information be collected systematically. This includes the careful collection and validation of both numerator and denominator data. The purpose of and time frame for data collection should be specific. Data need to be analyzed and displayed to enhance interpretation and facilitate any necessary interventions. This chapter focuses on surveillance for HAIs, infection or colonization with epidemiologically important organisms, and relevant processes of care. However, these principles can be applied to noninfectious adverse outcomes of medical care, such as falls and medication errors. This process and its epidemiologic aspects are important to maintain, given the increasing pressure to publicly report these data and to reimburse healthcare providers for their ability to provide safe care. Additionally, although we recognize that an increasing number of patients receive medical care and surgical procedures in the outpatient setting, this chapter primarily describes surveillance in the hospital. The principles set out here, however, can be used in any healthcare setting.

Why Conduct Surveillance?

Surveillance is conducted for a myriad of reasons. Some are more important than others. Conducting surveillance or establishing a surveillance program allows an infection prevention and control program to achieve multiple goals:

- To establish the baseline rate
- To detect clustering in time and space of infections or healthcare-related events (ie, outbreaks)
- To convince clinicians and administrators that there is a potential problem (that may require additional resources)
- To generate hypotheses concerning risk factors
- To identify a source of cases with which to test hypotheses concerning risk factors
- To assess the impact of prevention and control measures (ie, interventions)
- To guide treatment (eg, the choice of antimicrobial agents) and/or prevention strategies (eg, administration of vaccine or chemoprophylaxis)
- To reinforce practices and procedures
- To satisfy patient-care standards, guidelines, and/or regulatory requirements
- To defend law suits
- To conduct research
- To reduce the incidence of HAIs
- To make comparisons within and between hospitals or healthcare systems

What Is Needed to Plan for and Conduct Surveillance for HAIs and Multidrug-Resistant Organisms?

Many infection prevention and control programs establish surveillance systems because of recommendations from the CDC or another federal agency, legislative directives, regulatory agency requirements, or other external pressure, such as competition from hospitals in the community that have already established programs. Hospitals that have established programs under such circumstances may not have established their own goals and priorities prior to undertaking surveillance. Consequently, data collection becomes an end unto itself.

Unfortunately, in these instances, the surveillance data have little influence on the infection rates because clarity as to their purpose and practical application is lacking. On the other hand, healthcare epidemiology and infection control programs with defined objectives and goals and administrative support can be extremely important and effective and can use data to motivate clinicians and enhance quality improvement efforts.

There are a number of requirements for a successful surveillance program (Table 11-1). First among them is a set of clear and specific primary objectives. When developing a new surveillance system or revising an existing system, the staff must first define the priorities of the infection control program. In this way, both the type of surveillance they should conduct and the types of data they should collect are determined. After the hospital epidemiology and infection control staff have analyzed preliminary data from their own institution (ie, data obtained either through the previous surveillance system or through a hospital-wide prevalence survey), they can custom design a surveillance system specific for their own facility. Federal and state (or provincial) regulatory issues, national guidelines, and local patient care standards that may dictate special surveillance needs must be considered. For example, some states require environmental surveillance for *Legionella* species, while others require reporting of HAI rates.

When developing a surveillance program, hospital epidemiology and infection control personnel should consider characteristics of the institution, including the size, the hospital type (eg, private, university, or federal, and teaching or nonteaching), the patient populations served, the procedures and treatments offered, and the proportions of inpatient care and outpatient care provided at that facility. These staff also should consider the resources available to the infection control program, including the budget, the number of personnel and their level of training and experience, and the computer hardware and software that can be used by the staff. Hospital epidemiology and infection control staff should design a surveillance system that accomplishes the stated objectives with the most efficient use of resources, keeping in mind that additional resources must remain available to appropriately utilize the surveillance data gathered (eg, for analysis, reporting, developing interventions, and monitoring efficacy). Because of the trend toward increasing transparency and public reporting of these data, verification of numerators and denominators will become increasingly important to administrators, which will require additional resources.

A healthcare epidemiology and infection control program must determine which events to study and the data sources available in their hospital when they choose the data sources and case-finding methods they will use. As they design their surveillance system, the staff should consider the advantages and disadvantages of different surveillance methods (Table 11-2) and the sensitivity of different case-finding methods (Table 11-3). Definitions must be standard and applied in the same fashion. In general, in the United States, definitions promoted by the CDC are used.[17] These definitions, although not perfect, have been used for years, and their utility is well understood by the healthcare epidemiology community. They are discussed in more detail below (in the discussion of how to target outcomes and populations). Measures of occurrence should be determined and calculated using appropriate numerators and denominators. The infection preventionist should collect basic information on all patients with HAIs (Table 11-4). For some infections, one may want to collect additional data (eg, on central venous catheter–associated BSIs) or collect information during certain time periods (eg, when conducting a study to evaluate the prevalence of certain infections and to identify risk factors for those infections).

Infection prevention and control programs that have bountiful resources may want to continue doing hospital-wide surveillance so they can detect HAIs in all patient populations. However, programs that find themselves in this enviable position should develop innovative methods for conducting hospital-wide surveillance and not just use the traditional labor-intensive method of total chart review. Most infection control programs, however, have severely limited budgets. Therefore, the staff must decide how to use these precious resources to their greatest possible advantage. We believe that most infection control programs should not conduct hospital-wide surveillance. Instead, they should limit their surveillance to specific infections, pathogens, or patient populations.

Ideally, infection prevention and control staff should focus on infections or transmission of organisms that can

Table 11-1. Needs and resources for systems of surveillance for healthcare-associated infections, infection or colonization with epidemiologically important organisms, and associated processes of care

Statement of primary objective(s) and goal(s)
Standardized application of case definitions
Measurement(s) of occurrence
Numerator and denominator data
Access to data
Standardized collection method
Human and financial resources (ie, trained personnel to collect data, define events, and manage data; computer hardware and software)
Data (information) to stratify by risk
Mechanism to report results broadly and efficiently
Leadership support

Table 11-2. Advantages and disadvantages of 4 basic surveillance strategies for healthcare-associated infections and infection or colonization with epidemiologically significant organisms

Strategy	Advantages	Disadvantages
Hospital-wide surveillance		
Incidence	• Provides data on infections due to all organisms, on all infection sites, and on all units • Identifies clusters • Establishes baseline infection rates • Allows outbreaks to be recognized early • Identifies risk factors	• Expensive and labor intensive • Large amounts of data are collected, and there is little time to analyze it • No defined prevention objectives; it is difficult to develop interventions • Not all infections are preventable
Prevalence	• Inexpensive • Uses time efficiently; can be done periodically	• Overestimates infection rates and does not capture data on important differences • Has limited value in small institutions
Targeted surveillance		
Site specific	• Flexible and can be combined with other strategies • Can include a postdischarge component • Identifies risk factors • Easily adapted to interventions	• No defined prevention strategies or objectives • May miss clusters • Denominator data may be inadequate • No baseline rates in other units
Unit specific	• Focuses on patients at greater risk • Requires fewer personnel • Simplifies surveillance effort	• May miss clusters • Denominator data may be inadequate
Rotating	• Less expensive • Less time consuming and labor intensive	• May miss clusters, or underestimate or overestimate rates • Includes all hospital areas • Risk stratification may be difficult • Baseline infection rates may be unreliable
Outbreak	• Valuable when used with other strategies • Does not provide baseline infection rate	• Thresholds are institution-specific types of surveillance • Baseline infection rates not available
Limited periodic	• Decreases possibility of missing a significant problem • Liberates infection preventionist for other activities, including interventions • Increases the efficiency of surveillance	• May miss clusters • Baseline infection rates may be unreliable
Objective or priority based	• Can be adaptable to institutions with special populations and resources • Focuses on specific issues at the institution • Identifies risk factors • Easily adapted to interventions • Can include a postdischarge component	• Baseline infection rates not available • May miss clusters or outbreaks
Syndromic	• Can identify events that may be missed by traditional surveillance systems • May facilitate early recognition of an outbreak or event • Uses data from administrative systems • Automated	• Nonspecific and not sensitive • Not validated in most types of settings • Validation is very time consuming • Less resource intensive • Baseline infection rates not available

be prevented, occur frequently, cause serious morbidity, increase mortality, are costly to treat, or are caused by organisms resistant to multiple antimicrobial agents. For example, because infections associated with medical devices are preventable, consider surveying those UTIs associated with indwelling catheters or those infections caused by antimicrobial-resistant organisms. Another strategy is to limit surveillance to device-related infec-tions in intensive care units or limit it to step-down units where such devices are commonly used. HAIs caused by *Legionella* species or *Aspergillus* species occur infre-quently. However, these infections cause substantial mortality, and environmental controls can prevent most cases. Therefore, the infection control staff may want to use microbiology laboratory data (and other sources of data if necessary) to identify all of these cases

Table 11-3. Sensitivities and time demands of various case-finding methods used for healthcare-associated infection surveillance

Data source reviewed and/or method of review	Sensitivity, range, %	Estimated time required for surveillance[a]
Review of physician self-report forms	14–34	3 h/w
Review of reports of fever	47–56[b]	8 h/w
Review of reports of antibiotic use	48–81[b]	13.8 h/w
Review of reports of fever and antibiotic use	70[b]	13.4 h/w
Review of microbiology laboratory reports	33–84[b]	23.2 h/w
Review of criterion standard test results ("gold standard")	94–100	35.7–45 h/w
Selective review of medical charts using "kardex clues"	82–94[b]	35.7 h/w
Medical chart review		
Prospective	76–94	53.6 h/w
Retrospective (University of Virginia)	79	35.7 h/w
Retrospective (SENIC)	74–96	Not specified
Infection control sentinel sheet survey[c]	73–87	1 min per chart
Ward liaison surveillance	62	17.6 h/w
Laboratory-based ward liaison surveillance	76–89	32.0 h/w
Risk factor–based surveillance	50–89	32.4 h/w
Selective surveillance based on review of physician reports	74	Not specified
Automated computer algorithm review of antibiotic exposure and *ICD-9* codes (for surgical site infection only)	88–91	Not specified
Electronic surveillance		
For bloodstream infection	48–100	Not specified
For-surgical site infection	8–97	Not specified
For ventilator-associated pneumonia	50–100	Not specified

NOTE: H/W, hours per week; *ICD-9, International Classification of Diseases, Ninth Revision;* PICC, peripherally inserted central catheter; SENIC, Study on the Efficacy of Nosocomial Infection Control.
a Time required for an infection preventionist to perform surveillance in a 500-bed acute care hospital.
b The criterion standard ("gold standard") for healthcare-associated infection surveillance was the report of a trained physician who examined each patient, each medical record, and all "kardexes" (nursing notes) and who verified microbiologic data.
c For intensive care units or unit-based surveillance.

and set up an intervention designed to lower the rates. Vancomycin-resistant enterococci (VRE), methicillin-resistant *Staphylococcus aureus* (MRSA), *Clostridium difficile,* multidrug-resistant *Pseudomonas aeruginosa* or *Acinetobacter* species, and other bacteria resistant to multiple antimicrobials can spread rapidly within hospitals. Colonization or infection caused by these organisms can be very costly to treat or may not be treatable. Therefore, microbiology laboratory data may be used to perform surveillance for patients who are infected or colonized with these organisms. Infection control personnel may choose to study infections that are relatively minor but occur frequently, because these infections will increase the total cost to the healthcare system substantially. For example, saphenous vein harvest site infections are less severe than sternal wound infections after coronary artery bypass graft procedures. However, at least two-thirds of SSIs that occur after coronary artery bypass graft are harvest site infections, with an attribut-

able cost of nearly $7,000 per infection.[18] Because harvest site infections occur much more frequently than do serious sternal wound infections, the total cost to a healthcare system of the former approximates that of the latter. Furthermore, harvest site infections may be caused by problems with surgical technique and thus could be prevented if surgical technique was improved. Therefore, infection control personnel might want to develop a surveillance system that is able to detect harvest site infections instead of focusing on the more serious but less common sternal wound infections. So the challenge is to determine how a program can get the most from the resources that it has to use—that is, should the surveillance focus on serious infections that are rare or more common infections that are less serious?

As medical care moves from the hospital to outpatient settings and alternative care settings, healthcare practitioners will be challenged with how to identify HAIs that develop in the ambulatory-care setting.[19,20] Unless healthcare epidemiology and infection control teams expand their boundaries, they will underestimate the frequency of infections associated with medical care. At present, however, some infection prevention and control programs that monitor patients who develop SSI after ambulatory operative procedures remain in a quandary about how to best find them.[20] In addition, infection prevention and control staff might consider using surveillance to identify patients who acquire infections associated with outpatient treatments, such as dialysis, chemotherapy, and intravenous therapy (eg, antimicrobial or antiviral therapy, or parenteral nutrition) or associated with programs such as hospital-based home care. These latter areas of treatment remain of interest not only because increasingly-ill patients are receiving care in these areas but also because recently there have been incidents of use of contaminated products or poor infection control practices at home that have put patients at risk.

How Do You Assess the Surveillance System and Develop Priorities?

At least annually, a healthcare epidemiology and infection control program should evaluate its surveillance system to determine if it provided meaningful and actionable data. The staff should ask themselves a series of questions:

- Did the surveillance system measure meaningful outcomes or proxies of best practice? Are these outcomes relevant to the hospital population and infectious diseases that are prevalent or emerging in the community?
- Did the surveillance system detect clusters or outbreaks?
- How good was the system at identifying the events of interest (ie, the sensitivity, specificity, and/or positive and negative predictive values)?
- How representative was the system, if it was not 100% sensitive? Could the findings be generalized to other institutions?
- Were patient-care practices changed on the basis of the surveillance data?
- Were the data used to develop and implement interventions to decrease the endemic rate of infection?
- Were surveillance findings distributed to the administrative and clinical staff?
- Were the data used to encourage interventions or to assess their efficacy?
- Were the data used to ensure that rates of infection or colonization did not increase when procedures were changed, new products were introduced, etc.?
- How easy was it to collect data? What burden of data collection (ie, the time required to collect valid data,

given the importance of the outcome) is acceptable to the infection preventionists? What is the burden for other groups within the institution?
- Were the data made available in a timely fashion?
- How flexible was the surveillance system?

If no one, including the infection prevention and control staff, uses the data to alter practice, one must conclude that the current system is ineffective. At this point, it is more fruitful to abandon the surveillance system and devise a new strategic plan with surveillance goals and objectives in mind. This strategic plan should clearly identify infection prevention and control objectives, outline the surveillance data needed to address those objectives, and include specific actions that use the collected data to achieve those goals. The goals should focus on infections that are truly preventable and cause harm ("low-hanging fruit") or those for which surveillance is required because of legislative mandates and should be targeted with an established time line and goals. If appropriate, one could develop interventions on the basis of currently available data. The staff could then plan how they will use the revised surveillance system to monitor the efficacy of the proposed interventions.

To determine the surveillance needs, strategic thinking is necessary. The infection prevention and control team should meet off-site for a day and discuss issues and priorities and should use documents to accomplish the goals. As part of this, each institution should assess their "risk." This process should occur annually and allows the hospital epidemiology team to think critically about where problems and vulnerabilities lie, what regulatory requirements they must fulfill, and which areas have the greatest chance of being improved (Figure 11-1). The institution's risk can include the frequency of healthcare-related events of interest and the risk of patient harm and/or its severity. Also included in the risk assessment is an evaluation of the institution's response to a situation. For example, the infection prevention and control team may consider a problem's importance to be lower if the hospital leadership and hospital units are engaged in solving the problem than if they deny a problem exists.

Validation of data ensures the credibility of the infection prevention team. Periodically, the team must verify that the surveillance system is actually capturing the data the staff believes they are receiving. Hospital departments that provide data for the surveillance program may change their procedures and thereby cause what appears to be a change in rates. Surveillance systems that use data from information systems are particularly vulnerable to this. This issue is becoming an increasingly important problem for systems in which both numerator and denominator data are generated by computer from administrative data systems. If the departments that provide data

Table 11-4. Information to collect about patients who have a healthcare-associated infection, infection or colonization with an epidemiologically significant organism, and/or a process-of-care or disease syndrome

General information for all sites of infection and infections with epidemiologically significant organisms

Patient name
Patient identification number
Patient age
Patient sex
Nursing unit
Service
Admission date
Date of infection onset and/or date specimen for culture was obtained
Site of infection
Organism(s) isolated
Antimicrobial susceptibility pattern of isolates

Additional information

Presence of a risk factor (eg, CVC, urinary catheter, or ventilation)
Date of exposure or risk factor[a]
Primary diagnosis[a]
Comorbidities present, if any
Medications received (antibiotics, steroids, and/or chemotherapeutic agents)[a]
Exposure or risk factor (eg, immunosuppression, instrumentation, and/or procedure[s])[a]
Antibiotic therapy and duration of antibiotic therapy
Comments

General information for infection with resistant or epidemiologically important organisms

Date culture specimen was obtained
Site from which culture specimen was obtained
Current and/or previous roommates
Previous rooms during current hospitalization
Previous hospitalization(s) in this hospital
Intensive care unit stay

Additional information

Underlying disease(s) or condition(s)
History of antimicrobial use
Previous hospitalization(s) in another hospital
Previous stay in a long-term care facility
Previous vaccinations or history of infectious disease(s)[a]
Results of molecular typing of isolate(s)

General information for syndromic surveillance

Patient name
Patient identification number
Patient age
Service date
Chief complaint (with or without *ICD-9* code[s])
Laboratory and radiologic tests ordered and their results and findings
Prescriptions made
Other

General information for infection site–specific surveillance and some process measures

Surgical site infection
   Surgical procedure
   Surgery date
   Surgeons (attending and resident)
   Patient's ASA score
   Wound classification
   Time of incision
   Time procedure finished
   Antibiotic(s) administered perioperatively
   Timing of perioperative antibiotic administration, any redosing needed, and dose[a]
   Intraoperative oxygenation and/or glucose administered; patient temperature
   Other members of the surgical team[a]
   Operating room number[a]

(continued)

Table 11-4. (continued)

Bloodstream infection
    Intravascular catheters present (yes or no)
    Type of intravascular catheter (central placement, PICC vs peripheral)
    Location of intravascular catheter insertion site
    Number of days catheter in place
    Person(s) who inserted catheter[a]
    Any secondary source of infection
    Checklist completed (yes or no)
Lower respiratory infection or ventilator-associated pneumonia
    Endotracheal intubation (yes or no)
    Ventilation (yes or no)
    Number of ventilator days
    Date patient intubated
    Checklist completed (yes or no)
Urinary tract infection
    Urethral catheter present (yes or no)
    External catheter present (yes or no)
    Number of days catheter in place
    Person(s) who inserted catheter[a]
    Other urinary tract instrumentation[a]
    Checklist completed (yes or no)
General information for processes of care
    Process was performed (yes or no)
    There was an opportunity to perform the process (yes or no)
    Aspects of or steps in process being surveyed

NOTE: ASA, American Society of Anesthesiologists; CVC, central venous catheter; ICD-9, International Classification of Diseases, Ninth Revision; PICC, peripherally inserted central catheter.
a Information to be collected under particular circumstances, such as during an outbreak.

notify infection prevention and control staff about procedural changes, system upgrades, and coding changes, the infection preventionists can revalidate data and modify surveillance appropriately. However, departments often change important procedures without informing staff in other departments. These changes could affect relevant rates of infection or colonization substantially. Other changes do not affect the infection rates directly but alter the surveillance system's ability to obtain the necessary data. Because changes in procedures instituted by other departments can be invisible and can affect rates of infection or colonization substantially, infection control personnel would be wise to investigate dramatic changes in the rates or other important results before assuming that there is an outbreak or that an intervention has been very successful (Table 11-5).

For example, you calculate the proportion of S. aureus isolates that are resistant to methicillin and find that it has dropped precipitously from 34% to 0%. However, you suspect that the decrease is not real. Most of the MRSA isolates were recovered from surgical wounds, so you check to see whether the surgeons had changed their management of infected wounds. You discover that the surgeons are now treating SSI empirically without first sending a wound specimen to be cultured. On further

investigation, you find that the laboratory has changed the criteria for culturing wound specimens: laboratory personnel no longer plate the specimen if the Gram stain does not show any white blood cells. The 2 unrelated changes factitiously reduced the proportion of S. aureus isolates that are resistant to methicillin.

Another example: the overall infection rate in another hospital suddenly decreases (see Table 11-5). Ever skeptical of numbers that seem too good to be true, the hospital epidemiology and infection control team searches for a possible factitious cause for this rapid decline. They eventually discover that the fiscal department had changed the bed-count procedure so that an admission was counted each time a patient was transferred to another unit. Thus, the denominator was inflated, and the resulting infection rate appeared low.

How Should the Outcomes of Interest and Targeted Populations Be Determined?

To begin collecting surveillance data, the healthcare epidemiology and infection control staff first must identify the outcomes and the population they will study. An

HEIC Risk Assessment

Review date: _____ Reviewed by: _____

Hazard identification	Probability					Outcome severity						Assessment score	Needed[a]				Achieved				Preparedness score
	High	Med	Low	None	x	Very High	High Disrup	Mod Disrup	Low Disrup	None	=	=	High	Med	Low	x	High	Med	Low	=	=
Score:	3	2	1	0		4	3	2	1	0			3	2	1		1	2	3		
Healthcare-associated infections (examples)																					
Surgical site infection**																					
CABG																					
Laminectomy or fusion																					
Craniotomy																					
Cesarean section																					
Colon surgery																					
Outpatient surgery																					
Hysterectomy																					
Other																					
MDRO infection																					
MRSA																					
Clostridium difficile																					
VRE																					
MDR GNRs																					
Central line–associated BSI																					
VAP																					
Foley catheter–associated UTI																					
Legionella infection																					
Tuberculosis																					
Rotavirus infection																					
RSV infection																					
Influenza																					
Other infection control issues																					
Hand hygiene noncompliance																					
CDC MDRO GDL (2006)																					
Surveillance of team member illness																					
Isolation policy noncompliance																					
Construction-related issues																					
Staff influenza vaccination rates**																					
Staff sharps injuries																					
Pandemic influenza																					
Outbreaks																					
Influx or surge of patients																					
Catheter laboratory																					
Respiratory protection																					
OR GDL or standard noncompliance																					
Surgical site scrub																					
Surgical attire																					
Antibiotic prophylaxis																					
Flash sterilization																					
Traffic control																					
Room cleaning																					

Figure 11-1. A, Healthcare epidemiology and infection control (HEIC) risk assessment form (Infection Control Plan Risk assessment template 2 8 08). (Continued)

On the basis of the risk assessment, Johns Hopkins has identified those items with a Preparedness Score of 6 or greater as priority focus areas for HEIC. They are prioritized below in descending order.

Priority	Risk
1	
2	
3	
4	
5	
6	
7	
8	
9	
10	
11	
12	
13	
14	
15	
16	
17	
18	
19	
20	
21	
22	
23	
24	
25	

Figure 11-1. (Continued) B, Form listing the HEIC priorities identified with part A of the form. *Level of preparedness needed: determine the level of preparedness needed on the basis of the assessment score, utilizing the following guidelines: score ≤2, rating is "low"; score 3–5, rating is "medium"; score ≥6, rating is "high." Note that all healthcare-associated infections are scored at a minimum of 2 for "Level of preparedness needed." BSI, bloodstream infection; CABG, coronary artery bypass graft; CDC, Centers for Disease Control and Prevention; disrup, disruption; GDL, guideline; GNR, gram-negative rod; MDR, multidrug-resistant; MDRO, multidrug-resistant organisms; MRSA, methicillin-resistant *Staphylococcus aureus*; OR, operating room; RSV, respiratory syncytial virus; UTI, urinary tract infection; VAP, ventilator-associated pneumonia; VRE, vancomycin-resistant enterococci.

Table 11-5. Examples of practices that affect observed rates of infection

Change in practice	Apparent effect on infection rate
Locus of treatment is shifted from the hospital to the outpatient setting	Decrease in the overall infection rate, because surveillance rarely is performed in the outpatient setting
Patients are discharged earlier, and the length of stay in the hospital decreases	Decrease in the overall infection rate
Patients are discharged earlier after operative procedures, and the length of stay in the hospital decreases	Decrease in the rate of surgical site infection, because surveillance is rarely performed in the outpatient setting
Low-risk operative procedures are performed in a separate ambulatory surgery facility rather than in the hospital	Increase in the rate of surgical site infection, because patients who have a higher risk of infection have operative procedures performed in the hospital
Patients residing on a boarding unit are not counted toward the hospital's denominator (ie, the number of patients hospitalized)	Increased infection rate, if surveillance is performed on these admitted patients, because the denominator appears to decrease
The accounting department changes procedure and counts an admission each time a patient is transferred to another unit	Decreased overall infection rate, because the denominator for a different unit (ie, the number of patients occupying beds) is inflated artificially
The accounting department changes from charging one general cost for a CVC insertion to charging a cost for each time a CVC is accessed	Decreased rate of central line–associated infection, because the denominator (ie, the number of CVCs inserted) appears to increase
The business office assigns each surgical procedure to the admitting physician, regardless of that physician's specialty, rather than to the surgeon who performs the procedure	Inaccurate surgeon-specific infection rates, because some surgical site infections will be assigned to the wrong physician
Physicians treat patients empirically for possible infection	Decreased infection rates, if case-finding relies solely on microbiology laboratory reports, without performance of culture
Microbiology laboratory changes screening criteria for processing specimens	Decreased rates of infection, if case-finding relies solely on microbiology laboratory reports
Definitions of infection are used inconsistently and/or written definitions are absent	Inaccurate infection rates
ICD-9 codes are used to identify patients with healthcare-associated infections	Inaccurate infection rates
A new computer system is used to count catheter-days	Inaccurate infection rates, because the denominator (ie, the number of catheter-days) is overestimated or underestimated

NOTE: CVC, central venous catheter; ICD-9, *International Classification of Diseases, Ninth Revision.*

outcome of interest should be identified on the basis of the impact of the event on patients (ie, the associated morbidity and/or mortality), its frequency, its impact on the institution (financial and other resource burden), and any regulatory requirements. For example, healthcare facilities with only oncology, pediatric, ophthalmology, or trauma patients may prioritize outcomes of interest differently than general hospitals.

Next, the staff should develop written definitions that are precise, concise, and nonambiguous. One of the greatest challenges is to choose definitions to help differentiate patients who are colonized from those who are infected (Table 11-6). The CDC developed definitions for HAIs that were introduced in 1988 and are used widely and accepted as a standard.[21] The CDC and the National Health Safety Network (NHSN; formerly the

National Nosocomial Infection Survey [NNIS]) recently updated the previously proposed definitions for HAIs.[22] More recently, with the interest in multidrug-resistant organisms, such as MRSA, VRE, and *C. difficile*, more-refined definitions have been proposed to stratify according to the place of acquisition, as community acquired, healthcare associated, or healthcare acquired.[23-26] For some events, such as SSI in a transplant recipient or BSI in a hematopoietic transplant recipient, the CDC/NHSN definitions[22] need to be studied and may need to be modified. In other cases, hospitals may need to develop their own definitions. Others can use the definitions as they are written or slightly modified. Importantly, CDC/NHSN definitions should be used if infection control personnel wish to compare their institution's infection rates to those published by the NHSN.[21] It is also of note that,

Table 11-6. Examples of definitions of healthcare-associated infection used for surveillance

Site of infection	Criteria for infection	Source of data	Comments	Issues	Reporting refinements
Blood	Positive culture result	Laboratory tests	Must rule out contaminant(s)	Infection with CoNS may require different definition	State if infection is primary or secondary to an infection at another site
Urine	105 cfu of bacteria per mL; WBC in urine	Laboratory tests	Lower bacterial counts may be accepted if associated with compatible symptoms and pyuria. A finding of *Candida* species and mixed organisms frequently represents contamination	May miss many infections	Note if there is current or prior bladder catheterization
Postoperative wound or surgical site	Pus at the incision site	Laboratory tests, clinical findings	Cellulitis is classified separately; depth of SSI is classified as organ/space, deep, or incisional	May miss many infections	Note if it is a stitch abscess, in contrast to a more involved infection
Other wounds	Presence of pus	Laboratory tests, clinical findings	Includes decubitus site and tracheotomy site	May require tissue biopsy	
Burn	≥106 organisms per gram of biopsied tissue; alternatively, new inflammation or new pus not present on admission	Laboratory tests, clinical findings	Success of skin grafts is reported to be greater if they are placed over burn sites with bacterial counts of ≤105 organisms per gram of tissue	Commonly requires laboratories with specialized capacity	Note the antibiogram; organisms isolated are frequently resistant to multiple antibiotics
Pulmonary system	New sputum production with new inflammation or new pus not present on admission	Clinical and radiographic findings	Clinical picture must be compatible; other entities (eg, atelectasis and pulmonary emboli or infarction) must be ruled out	No "gold standard" diagnosis for infection with *Candida*, enterococci, and CoNS, which generally represent contaminants	Note if pneumonia is associated with assisted ventilation; rule out colonization; note the antibiogram since pulmonary infection is a frequent source of drug-resistant organisms
Gastrointestinal tract	Culture that grows a pathogen, or unexplained diarrhea with a duration of ≥2 days	Laboratory tests, clinical findings	Pathogens are defined as: *Salmonella* spp, *Shigella* spp, pathogenic *Escherichia coli*, and viruses (rotavirus); also *Clostridium difficile* toxin	Infection can have shorter incubation periods with some pathogens (generally foodborne)	List any antibiotics patient is receiving
Skin and/or IV catheter site	Pus at site	Clinical findings	Includes permanent CVCs and artificial catheters; catheter-tip culture may be helpful; diagnosis of tunnel infections may be based on different clinical parameters	Neutropenic patients rarely have clinical signs of infection; criteria for such patients are not well studied	Note site of IV catheter, duration of placement, and the type of catheter; obtaining bacterial colony counts on the catheter can be helpful
Miscellaneous[a]	Clinical picture	Laboratory tests, clinical findings	Need to perform appropriate diagnostic tests	Viral infections can have longer incubation periods	Note if findings are associated temporarily with any hospital procedure or receipt of any blood products or drugs
Nasal and/or perirectal sites	Culture that grows a pathogen such as MRSA or VRE	Laboratory tests	Findings are usually the result of surveillance program(s); need to perform appropriate diagnostic tests	Patients rarely have signs of infection	Culture findings usually denote colonization

NOTE: Information in the table is drawn from Wenzel et al.[50] and Centers for Disease Control and Prevention and National Healthcare Safety Network recommendations. CFU, colony-forming units; CoNS, coagulase-negative staphylococci; CVC, central venous catheter; IV, intravenous; MRSA, methicillin-resistant *Staphylococcus aureus*; spp, species; SSI, surgical site infection; VRE, vancomycin-resistant enterococci; WBC, white blood cells.
a For example, hepatitis, upper respiratory infection, or peritonitis.

although these definitions are currently being used for reimbursement in some settings, they have not been validated for this purpose.

The NHSN, supported by the CDC, maintains a database that accepts surveillance data on HAIs, infection and colonization with multidrug-resistant organisms, and infections in healthcare workers that are collected in a systematic fashion from hospitals and other healthcare facilities.[27] Having a large repository of data that have been collected in a similar fashion allows for comparisons of some rates between institutions ("benchmarking"), so that infection prevention and control programs can determine whether their rates are similar to those at other institutions and can evaluate the success of infection prevention and control interventions. Some larger institutions, or those with large volumes, can use control charts or 95% confidence intervals to compare rates.[28] The advantage of using internal rates for comparison is that the patient population is likely similarly heterogeneous with respect to underlying illnesses. The major disadvantage is that one cannot discern whether the rates (eg, rates of HAI or infection with multidrug-resistant organisms) are higher or lower than they should be, in comparison with other institutions, because of the differences in definitions.

To collect meaningful data, infection prevention and control personnel not only must use clear definitions but also must apply the definitions consistently. Training of healthcare epidemiology and infection control personnel enhances their ability to identify infections appropriately and consistently, as has been shown by Cardo and colleagues.[29,30] These investigators demonstrated that, with training, the sensitivity and specificity of surveillance for SSI increased from approximately 84% to greater than 93%. This training is critical to ensure that appropriate definitions, sources of information, and case-finding strategies are used. Ehrenkranz et al.[31] described the severe consequences that may occur if collected data do not reflect objective, written definitions that are applied systematically. An infection control team recorded SSI rates of 3%–11% for a particular surgeon over a 3-year period. The infection control committee considered these rates excessive and repeatedly investigated this surgeon. After the surgeon decided to stop performing operations, the hospital administrator asked a consultant to review the findings. During the investigation, the consultant found that infection control staff did not use a specific definition of SSI. The consultant concluded that a surveillance error, not poor surgical technique, accounted for the surgeon's reported high infection rates.

Because of the changing paradigm of health care, infections that have been labeled "nosocomial" in the past are now called "healthcare associated." The term "healthcare-associated infection" describes an infection that is not present or incubating at the time the patient is admitted to the healthcare institution. Thus, an infection is not considered healthcare associated if it represents a complication or extension of an infectious process present at admission. In general, infections with onset that occurs more than 48 hours after admission and within 7–30 days after hospital discharge are defined as healthcare associated. The time frame is modified for infections that have incubation periods of less than 48–72 hours (eg, gastroenteritis caused by norovirus) or more than 10 days (eg, hepatitis A or hepatitis C). An SSI is considered healthcare associated if the onset of infection occurs within 30 days after the operative procedure, or within 1 year after the procedure if a device or foreign material is implanted.

Infections should be considered healthcare associated if they are related to procedures, treatments, or other events that occur immediately after the patient is admitted to the hospital or that are performed in an alternative healthcare setting, such as a surgical center. For example, BSI associated with a central venous catheter, pneumonia associated with mechanical ventilation, and UTI associated with urethral catheterization should be considered healthcare associated even if the onset of infection occurs within the first 48 hours after hospitalization. Additionally, within the term "HAI," infection control professionals can include infections or pathogen transmission resulting from a surgical or medical procedure that did not require hospitalization. For example, if a patient has a surgical procedure in the outpatient surgical suite and develops an SSI, it is considered healthcare associated. Or, for another example, because bone marrow transplantation, chemotherapy, and other procedures are now being performed in the outpatient setting, high-risk patients treated in such alternative settings develop BSIs; these infections are almost always related to indwelling intravenous catheters. On the other hand, if a patient admitted with shortness of breath and fever has a chest radiograph that reveals consolidation in the left lower lobe of the lung on day 4 after admission and has a positive result of a test for *Legionella* antigen on day 6 after admission, the infection is not healthcare associated, because it was incubating when the patient was admitted.

Because of these situations, infection prevention and control professionals will need to refine and validate definitions in order to capture infectious complications in new healthcare delivery settings. Fortunately, Lessler and colleagues[32] have proposed a systematic method to determine whether or not an infection was likely acquired in the hospital or in the community. Using the incubation period and the incidence rate ratio of infection, one can obtain a quantitative result that allows one to mathematically estimate the likelihood that an infection was acquired in a certain setting. One can select disease-specific cutoff values to distinguish community-acquired from hospital-acquired infections that perform well for important illnesses. For example, patients who develop influenza symptoms in the first 1.5 days of their hospital stay are classified as having acquired infection in the community. If patients develop symptoms later in their hospital stay, then the likelihood that the infection was acquired in the hospital is 87%. Methods of this type will improve the application of HAI definitions.

Collecting Data

One of the first mantras of surveillance is that its purpose is to determine the burden of disease, to identify trends and potential problems, and to establish the epidemiologic features of the illness or event of interest. Hence, only the information needed to adequately analyze and interpret the data should be collected. If these data suggest a potential problem, the healthcare epidemiology and infection control team can design a more comprehensive study. Data can be collected prospectively, retrospectively, or using both strategies. In prospective or concurrent surveillance, data are collected at the time the event occurs or shortly thereafter. Concurrent surveillance requires the infection prevention and control staff to review the medical record or electronic databases, to assess the patient(s) affected, and to discuss the event with caregivers at the time of the infection. Because the data are obtained close to the time the event occurs, additional information not normally a part of the medical record may be available, such as ward logbooks and nursing reports. This becomes less of a barrier as records become available electronically. The advantage of this form of data collection is that clusters of the event of interest can be identified as they occur. Importantly, providing feedback of information about any adverse event as it is occurring helps healthcare workers appreciate the significance of the event; it becomes "real" and less theoretical and helps them identify other potential prevention strategies. If data are collected after the patient is discharged or retrospectively, clusters or potential outbreaks are not identified as promptly. In some cases, the "distance" of the event in time sometimes impedes interest and interventions. Nonetheless, both methods have similar sensitivities, but retrospective surveillance depends on the completeness, accuracy, and quality of the medical records[33] (Table 11-3). Commonly, programs mix and match data collection strategies to ensure the most complete data collection.

To identify cases, highly sensitive methods for case-finding are preferred, so that important cases are not missed. Commonly, infection prevention and control practitioners employ several case-finding methods simultaneously. The practice of using information collected simultaneously from different sources has increased with the shift to early patient discharge and to provision of care in the outpatient setting. However, with this method, one must identify strategies to increase the specificity of the surveillance process and thus reduce the time wasted collecting irrelevant data. If one uses currently available computer systems that can identify patients who may have an HAI or may have acquired an epidemiologically important organism, the time spent in reviewing charts can be reduced; thus, computer-based surveillance strategies are rapidly emerging as important adjuncts to surveillance.[19,20,34,35] As computer hardware and software become more sophisticated and as computer-based decision algorithms are developed and tested, these can be integrated into surveillance systems to identify HAIs.[19,20,34,35] For example, Yokoe and colleagues[34] showed that the rate of agreement between a surveillance method that uses microbiology laboratory data and the CDC's NNIS review method was 91%. The method that uses microbiology laboratory data requires approximately 20 minutes less time per isolate than does routine surveillance, which is a substantial reduction in the burden of this surveillance on infection control staff. This group has done further work with other institutions in the use of electronic methods to identify SSI after multiple procedures, including coronary arterial bypass grafting and cesarean delivery.[36] They have validated the use of key data elements (eg, the number of antibiotic-days and *International Classification of Diseases, Ninth Revision* codes) to identify SSIs in several institutions.

When an infection prevention and control program begins a new surveillance project, the staff should periodically look for flaws in the data, the collection tool, the data sources, and the surveillance process. In this manner, they can identify problems or errors and correct them before reaching the end of the study. Infection control staff can determine the surveillance project's sensitivity and specificity by examining a random subset of medical records for a defined time period and comparing the number of events identified by this review with the number identified by the usual surveillance system. In addition, changes in the surveillance system—such as identifying a new data source or a new item, modifying definitions, or changing the personnel who collect the data—can impact the integrity of the surveillance progress. Such changes require validation to ensure that the data are high quality. Validation needs to be done for numerators and, commonly, denominators. The latter is most important for device-associated infections, now that these data are commonly reported publicly in many countries. Validation is also used to ensure that practitioners apply the definitions systematically and uniformly.[37]

Managing Data

Managing surveillance data and organizing them in a meaningful fashion are necessary to identify patterns and trends. One of the first organizational processes is for the "cases" identified to be to catalogued systematically on a flow sheet or in a computer spread sheet. This is called a line-listing and will include data pertinent to the problem being examined (see Chapter 12, on outbreak investigations). For example, a line listing may include, in a single row, the patient name, hospital or medical record number, the admission data, type or site of infection, date of infection onset, organism(s), and surgical procedure, if relevant. Pads of columnar accounting paper may suffice for a written database in small hospitals, but relational databases on personal computers are generally used for larger hospitals. Many of the infection control software vendors provide this feature in the software. Once the data are in a database, infection control personnel can easily plot numbers or rates over time so they can identify possible trends. Miniprograms or "macros" for commonly available relational databases can be developed to present data to staff and facilitate work flow.

Recently, several exciting computer software programs have been developed that can integrate information from various hospital computer systems (by means of health level 7 [HL7] messages) and can provide data that are concatenated and presented in epidemiologic terms. These electronic infection control software programs facilitate obtaining the information needed for surveillance and can allow electronic labeling of patients with HAIs. Queries can automatically examine specified hospital units or geographic areas, looking for important time-and-space relationships or clusters of cases. The power of these programs is just beginning to be understood and utilized in infection control.

Analyzing Data

If the infection control and prevention team do not analyze their data, they have wasted the time, money, and effort they spent collecting and recording the data. The purpose of surveillance is not merely to count and record infections but to identify problems quickly and to intervene so that the risk of infection is reduced. The time factor inherent in this process requires that the data be analyzed promptly. The frequency of data analysis is based on the nature and frequency of the healthcare event of interest and the purpose of surveillance. The goal is to strike a balance between analyzing the data frequently enough to detect clusters promptly and collecting data for a long enough period of time to ensure that variations in rates are real. Computer programs are facil-

itating collection of data for a long period, since standard analyses can be generated automatically. In addition, the infection control and prevention team must also ensure that an adequate sample of cases has been reviewed, so that the data are meaningful. For example, if 3 patients had a procedure performed by a particular surgeon in 1 month, and 1 of the patients develops an SSI, it is difficult to interpret the infection rate because there are so few cases. In general, data should be collected for 50 procedures or processes for analysis to be useful.

Infection prevention and control personnel commonly report only the number of events that occur in a specified time period (ie, the numerator). While there is merit to providing units with immediate feedback during an intervention or outbreak, to compare data over time one must calculate the incidence, or the proportion of patients being studied who have a new instance of the event of interest. This calculation requires both the number of events studied in the defined population (the numerator) and the number of patients at risk during the same time period (the denominator). The following example illustrates why the healthcare epidemiology and infection control team must select the appropriate denominator and why having a denominator is important. Suppose 10 patients in a hospital develop MRSA BSI during 1 month. If the hospital discharged 1,000 patients that month, the incidence rate of new MRSA BSI acquired in the hospital would be 1%. However, if all 10 patients were hospitalized on the medical service, which discharged 600 patients that month, the incidence rate of patients with MRSA BSI on the medical service would be 1.7%. If 8 of the 10 patients developed MRSA BSI while in the medical intensive care unit, which discharged 90 patients that month, the incidence rate in the medical intensive care unit would be 9%. If, however, the number of admissions in the medical intensive care unit drops to 45 patients for the month, the incidence of MRSA BSI is 18%. Thus, one can assess the true incidence of an event in a defined population only if one uses a denominator that accurately represents the patients who are at risk of experiencing the event.

Similarly, if one evaluates only summary reports of microbiology laboratory data, important trends in specific units may be missed. For example, the summary reports may obscure the fact that 90% of the *Pseudomonas aeruginosa* isolates recovered from patients in the medical intensive care unit are resistant to certain antibiotics. Stratton et al.[38,39] demonstrated that yearly summaries showed little variation in antimicrobial susceptibility patterns within the whole hospital. Focused microbiologic surveillance on specific units, in contrast, demonstrated that the predominant pathogens and their antimicrobial susceptibility patterns differed among specialty units and between those units and the entire hospital. Similarly, Srinivasan and colleagues[40] would have missed a doubling of *P. aeruginosa* infection among patients undergoing bronchoalveolar lavage had they calculated the overall hospital rate and not procedure-specific rates. In both instances, the safety and subsequent care of patients is altered by the prompt identification of changes in the microbiology.

Another important variable to determine in surveillance is the endemic (baseline) rate of all types of healthcare-associated events of interest, including HAIs. One can discern this rate after data have been collected for a sufficiently long period of time. Subsequently, it is easier to determine whether the current rates are substantially different from the baseline rate. In addition to calculating overall rates for the population of patients in the institution, the infection prevention and control staff can analyze the data further by calculating attack rates for specific nursing units, services, and/or procedures. These rates enable the staff to identify significant changes and important trends within subgroups of patients that might be missed if the entire population were analyzed as a whole. When comparing data, either within an institution or with data from another institution, one must use comparable surveillance methods, definitions, and time frames. One can then use statistical tests of differences to determine whether the rates have changed significantly over time.

Finally, the infection prevention and control team must interpret the data. If the incidence of a particular event increases substantially, they should analyze the data thoroughly to determine if a problem really exists. The analysis should include assessing whether the increase is statistically significant. However, even if the increase is not statistically significant, it may be clinically significant and warrant initiation of control measures. Furthermore, the team should assess whether the incidence of an event is acceptable. For example, even if the rate of SSI is stable, the incidence may be higher than that reported by comparable institutions or may be higher than it would be if the process of care was improved. A study by Classen et al.[41] demonstrated that examining the process of care can allow the rate of SSI to be decreased significantly. In their study, the SSI rate among patients who received antibiotic prophylaxis within 2 hours before the start of surgery was significantly lower than it was among patients who received antibiotic prophylaxis either early (2–24 hours before the start of surgery) or after the operation was completed (ie, more than 3 hours after first incision but less than 24 hours after the end of surgery). Therefore, an infection control team that wants to decrease SSI rates in their hospital might want to review the time at which antimicrobial prophylaxis for surgery is given. Another example is shown in the study by Berenholtz et al.,[42] who systematically implemented evidence-based interventions, pro-vided feedback of the data, and decreased the rate of catheter-associated BSI by 80%. Such strategies have been implemented in multiple institutions, with similar improvements.[43,44]

Communicating Results

Finally, the infection prevention and control staff must communicate the data to the persons who need the information, such as the clinical staff, and who have the power to authorize changes. To ensure that regulatory requirements are met and that results are communicated to the organization's leadership, infection prevention and control practitioners should regularly report to the institution's healthcare epidemiology and infection control committee and also to the quality or performance improvement committee. Data to be communicated include appropriate rates and counts, which should be shared with key persons on individual nursing units, in each clinical service, in the nursing administration, and in the hospital administration. For example, as part of an intervention to reduce catheter-related BSI, we report BSI data weekly to intensive care unit personnel.[42] A banner is posted in the intensive care unit that lists the number of weeks since the last infection. In addition, infection control personnel may need to report their data to the education service, the intensive care unit committee, the safety committee, or an external agency, such as the local health authority. Commonly, committee reports should include simple but well-labeled graphic displays of trends over time. Changes in rates over months, quarter-years, or years may be important to display. Comparison ("benchmarking") with national data is helpful to promote ongoing process-improvement activities. When reporting data, in any circumstance, epidemiology personnel must maintain the confidentiality of data related to both patients and employees.

Simple reports that the target audience can understand in a few seconds (the amount of time usually given to a report at a busy committee meeting) are most effective. Graphical or tabular displays of the data can present important trends in pictures and help clinicians and administrative personnel grasp key points quickly (Figure 11-1). The use of 95% confidence intervals, to help in understanding the significance of a variation in rates, or comparison with rates such as those published by the NHSN, can be helpful in comparisons against other groups. Benchmarking rates against rates from an external organization has proved useful. While most North American institutions compare their rates with those published by the NHSN, other groups are using similar methods in other areas of the world, including in the developing world.[45-49] Moreover, individuals who are unfamiliar with issues in infection prevention and control may not understand the importance of a problem if they are presented only with the data. This is particularly true if the number of cases is small or the etiologic agent has not been discussed in the popular press. Thus, the infection prevention and control team should include their assessment and conclusions in the report, so that they can persuade clinicians or hospital administrators that corrective action is necessary to reduce the number of cases.

Surveillance for HAIs

Data Sources

Many different sources provide information about patients with infections (see Table 11-7). In addition, infection prevention and control personnel can obtain data from databases maintained by other departments, such as the departments of medical records, pharmacy, respiratory therapy, admissions, risk control, and financial management. However, it must be remembered that these databases are generally designed for billing or administrative purposes, not for collecting data on infections or multidrug-resistant organisms. Therefore, one must determine whether those databases include data needed for surveillance and whether the data are com-

---

Table 11-7. General sources of data for surveillance

Patient data
    Clinical ward rounds that include questioning of nurses and other healthcare providers and review of temperature curves, symptoms, and other data
    Electronic Medical Record: review notes, vital signs, antimicrobial use, surgical procedure data, pharmacy orders, and Radiology Department and Laboratory reports
    Hospital employees
    Laboratory, Radiology Department, and Pathology Department reports
    Pharmacy Department
    Admissions Department
    Emergency Department and emergency transfer personnel
    Operating room
    Outpatient clinics, including surgical centers
    Medical Records Department
    Employee or Occupational Health Department
    Incident reports
    Postdischarge clinic visits
    Local, state, and/or provincial public health officials
    National public health sources
Laboratory data
    Microbiologic, virologic, and serologic test reports
    Morbidity and Mortality Weekly Report
    National Healthcare Safety Network and other national databases
    Antimicrobial susceptibility patterns

---

plete and accurate. This situation is changing, because, increasingly, vendors are developing products that collect data for epidemiologic purposes; nonetheless, the sources of data must be validated to ensure that accurate information is generated. For instance, an infection prevention and control professional who conducts surveillance and uses the daily surgery schedule to obtain the number and classification of operative procedures, instead of using the list of completed operative procedures, will not calculate accurate rates, because if surgeons add, cancel, and/or change operative procedures during the day, the denominator for calculating SSI rates will be inaccurate.

Surveillance Methods

Each infection prevention and control team must determine which of many surveillance methods is best suited to their hospital. To help infection prevention and control personnel choose the most appropriate approach, we describe 4 basic surveillance methods; we have summarized their advantages and disadvantages in Table 11-2.

Hospital-wide Traditional Surveillance
Hospital-wide surveillance, the most comprehensive method, requires the infection prevention and control practitioner to prospectively and continuously survey all care areas to identify patients who have acquired infections or epidemiologically significant organisms during hospitalization.[50,51] The infection preventionist gathers information from daily microbiology laboratory reports and from the medical records of patients who have fever or cultures growing organisms and of patients who are receiving antibiotics or are under isolation precautions. The infection preventionist also garners important information by talking frequently (daily if possible) with nursing staff and by seeing patients occasionally. In addition, the infection preventionist periodically reviews all autopsy reports and employee health records. The infection prevention and control team regularly calculates the overall hospital rate of HAI and infection with multidrug-resistant organisms and also calculates infection rates according to type or site of infection, nursing unit, physician service, pathogen, and operative procedure. This could be done monthly, quarterly, or semiannually, depending on the hospital size and number of HAIs or infections with multidrug-resistant organisms.

Traditional hospital-wide surveillance is comprehensive. However, this system is very costly, and it identifies many infections that cannot be prevented. Consequently, many infection prevention and control programs have developed other surveillance methods that require fewer resources. With access to increasingly sophisticated electronic patient records, it is possible to capture many clinical details that facilitate and enhance surveillance for HAIs and infections with multidrug-resistant organisms. With such improved clinical informatics systems, the time commitment required for surveillance will need to be reevaluated.

Prevalence Survey

A prevalence survey can be hospital-wide or can focus on a specific area of the hospital. In a prevalence survey, the infection control professional counts the number of active infections or cases of infection with epidemiologically significant organisms during a specified time period.[52] The total number of active infections is defined as all infections present during the time of the survey, including those that are newly diagnosed and those being treated when the survey begins. This total number is divided by the number of patients present and at risk of the event of interest (eg, present in the hospital for more than 48 hours) during the survey.

Because new *and* existing infections are counted, the rates obtained from prevalence surveys are usually higher than incidence rates, which consider only new cases within a given time period. Prevalence surveys can focus on particular populations, such as patients with central venous catheters or patients receiving antimicrobials.[53] Prevalence studies also are useful for monitoring the number of patients colonized or infected with epidemiologically important organisms such as *C. difficile*, VRE, or MRSA.

Infection prevention and control programs also can use prevalence studies to assess the risk factors for infection with multidrug-resistant organisms in a particular population. To determine why patients in this population are developing infections, the epidemiology staff could collect additional data about potential risk factors from all patients surveyed. Because prevalence studies assess all patients in the target population, regardless of whether they have an infection, infection control personnel can compare the prevalence of infection among patients who have the potential risk factor with the prevalence among patients who do not have the potential risk factor.

Targeted Surveillance

There are several approaches to targeted surveillance. Many infection prevention and control programs focus their efforts on selected areas of the hospital, such as critical-care units, or selected services, such as the cardiothoracic surgery service. Other programs focus surveillance on specific populations, such as patients at high risk of acquiring infection (eg, patients undergoing transplant or pediatric patients), patients undergoing specific medical interventions (eg, hemodialysis patients), or

patients with specific types or sites of infection (eg, BSI, catheter-associated infection, or SSI). Some infection prevention and control teams target surveillance on infections associated with specific devices, such as VAP. By limiting the scope of surveillance, infection control personnel can collect data on entire patient populations. Thus, they can accurately assess the incidence of infection in the surveyed populations.

Some infection prevention and control programs use data from the microbiology laboratory to limit the scope of surveillance. For example, the epidemiology team may focus either on specific microorganisms, such as *Legionella* species or *C. difficile*, or on organisms with particular antimicrobial susceptibility patterns, such as VRE or MRSA. This type of surveillance allows programs to focus on patients at increased risk, on areas of the hospital with an elevated rate of infection, or on patient populations identified by regulatory agencies. Importantly, however, because surveillance activities are limited, the surveillance program can develop interventions to prevent the development of infection or transmission of multidrug-resistant organisms and can assess the impact of these interventions.

Periodic Surveillance

There are several ways to conduct periodic surveillance. In one method, the infection control program conducts hospital-wide surveillance only during specified time intervals, such as 1 month each calendar quarter. Infection control programs that use this method frequently conduct targeted surveillance during the alternate periods. In another method, the infection control program conducts surveillance on one or a few units for a specified time period and then shifts to another unit or units. By rotating surveillance from unit to unit, the infection control team is able to survey the entire hospital during the year.

Outbreak Thresholds

Some investigators have conducted surveillance to assess baseline infection rates at their institution and, on the basis of their data, developed threshold rates to identify outbreaks. Subsequently, they stopped conducting routine surveillance and evaluated problems only when the number of isolates of a particular species or the number of cultures positive for a pathogen exceeded those outbreak thresholds.[54,55] For example, McGuckin et al.[55] used a threshold of the 80th percentile above the baseline for each bacterial species from a particular nursing ward for a specified time period. Similarly, Schifman et al.[54] established a threshold of double the baseline positive culture rate. Wright et al.[56] used a computer-based program and set a threshold of 3 sigma (ie, standard deviations from the mean).

Objective- or Priority-Based Surveillance

Objective- or priority-based surveillance is a form of targeted surveillance that is primarily used in hospitals with special populations and resources, such as children's hospitals, ophthalmologic hospitals, orthopedic institutions, or other specialty institutions. These facilities can focus their interest on problems or events that are specific to their types of institution. For example, many children's hospitals have active programs of surveillance for respiratory virus infections, especially respiratory syncytial virus infection, that can be particularly devastating in children.[5]

Case-Finding Methods

Infection prevention and control personnel should collect data only for infections and/or multidrug-resistant organisms that were acquired in their facility or as a consequence of procedures performed or treatments administered in their hospital or clinics. For example, a patient may become infected with *C. difficile* while in Hospital A and then be transferred to Hospital B while still infected. Hospital B personnel should not include this infection in their HAI rate, even though it was acquired in a hospital. If such infections are included, it will cause the extent of the problem to be overestimated and the efficacy of the prevention and control programs to be underestimated. Although not included in HAI rates, all patients who on admission carry or are infected with particular organisms of interest, such as *C. difficile*, VRE, MRSA, and respiratory syncytial virus, should be identified. Such data allow infection control personnel to estimate the entire population of patients affected by these organisms. By determining the proportion of patients who acquire the organism in the hospital, infection control personnel can evaluate the efficacy of their infection control efforts. However, to improve patient care, notifying an institution about the infection is helpful and, some argue, should be required by regulations.

Investigators have described various methods used to identify patients with HAIs. We review some of these case-finding methods in this section; their sensitivities and the time they require are summarized in Table 11-3. Of note, institutional resources should be considered, to best determine how to choose the most sensitive case-finding method.

Total Chart Review

In total chart review, the infection preventionist reviews nurses' and physicians' notes, medication and treatment records, and radiologic and laboratory reports for each patient 1 or 2 times per week.[50,51] In addition, many infection preventionists review notes from the specialties of respiratory therapy, physical and occupational therapy, dietetics, and any other specialty service caring for the patient. Because the infection preventionist requires 10–30 minutes to review each medical record, this method is time-consuming and costly. As mentioned above, this type of review likely requires less time if it is possible to use improved electronic patient records that can display all the information needed for surveillance on a few computer screens. Clinical information systems typically provide demographic and administrative information, data from the microbiology laboratory and the radiology and pharmacy departments, and, in some instances, physician, physical therapy, respiratory therapy, and nursing notes and information about use of intravascular lines. Although many institutions have moved away from using total chart review, new algorithms that use data captured from electronic medical records may change that approach in the next few years.[35,36,57]

Review of Laboratory Reports

Clinical laboratory reports often are the primary source of data for identifying infections, particularly if the infection preventionist reviews virologic and serologic testing reports in addition to bacteriologic test results.[50,58,59] The infection preventionist may find some HAIs directly from reports of cultures that yield pathogens. For example, a patient who has a blood culture that grows *S. aureus* likely has an HAI if the blood sample was obtained 10 days after admission. A laboratory report might prompt the infection preventionist to review the patient's medical record. While reviewing the medical record, the infection preventionist might identify an HAI for which a culture was not performed. For example, a patient might have a blood culture from which *Klebsiella pneumoniae* was isolated. In the medical record, the infection preventionist might learn that chest radiographs revealed a new pulmonary infiltrate and that Gram stain of a sputum sample revealed many white blood cells and gram-negative rods. This information might lead the infection preventionist to conclude that the patient had pneumonia and secondary bacteremia caused by *K. pneumoniae*. Alternatively, a urine culture result might prompt the infection preventionist to review a patient's medical record; while reviewing the record, the infection preventionist might discover healthcare-associated pneumonia caused by another organism (Table 11-8).

Most laboratories maintain logbooks or notebooks that infection preventionists can scan quickly to obtain preliminary results. In addition, the laboratory staff often will notify infection control personnel about positive test results. This is especially true if the epidemiology personnel visit the laboratory frequently and develop a rapport with the laboratory staff. In fact, the laboratory may note outbreaks that were not identified either by standard surveillance techniques or by outbreak

Table 11-8. Clinical clues to help identify patients at high risk who have a healthcare-associated infection

Hospital unit and infection or infection site	Patients and/or clinical finding to assess
Medical unit	
Urinary tract	Patients with fever
	Patients with urinary tract catheters
Surgical unit	
Surgical site, lower respiratory tract, urinary tract	Patients with fever
	Patients with productive cough
	Patients with draining or erythematous wounds
	Patients who returned to surgery
	Patients who had blood cultured
	Patients with abnormal intravenous catheter sites
	Patients who received prolonged antibiotic therapy
	Patients who were readmitted
Pediatric unit	
Gastrointestinal tract, respiratory tract	Patients with fever
	Patients with diarrhea
	Patients with episodes of apnea or bradycardia
	Patients with lethargy or irritability
	Patients with new respiratory symptoms
Hematology/oncology unit	
Bloodstream, permanent catheter site, fungal or viral infection	Patients with fever or new fever
	Patients with new radiographic abnormalities or respiratory symptoms
	Patients receiving antibiotic therapy
	Patients who had blood cultured
	Patients with erythematous intravenous catheter sites
Intensive care unit	
Pneumonia, intravenous catheter site, urinary tract	Patients with fever
	Patients with diarrhea
	Patients receiving antibiotic therapy
	Patients who had blood cultured
	Patients with erythematous intravenous catheter sites or catheters that were changed
Obstetrics/gynecology unit	
Surgical site, lower respiratory tract	Patients with fever
	Patients with foul-smelling wound drainage
	Patients with erythematous wounds
	Patients receiving new antibiotic therapy
	Patients with productive cough
Extended or long-term care unit	
Urinary tract, lower respiratory tract	Patients with new fever
	Patients with new mental status changes
	Patients who recently started antibiotic therapy
	Patients with productive cough or new respiratory symptoms
	Patients with foul-smelling urine
	Patients with new diarrhea

detection algorithms using newer electronic databases.[60-62] Nonetheless, laboratory reports have some substantial limitations, and the infection prevention and control program should not use them as the sole source of data for identifying patients with HAIs. Laboratory reports do not capture infections for which the clinician does not order that a culture be performed, but instead the clinician treats a patient with clinical evidence of infection empirically. This commonly occurs with SSI or UTI. In addition, cultures of specimens from some sites of infection may yield negative results. This is particularly true if the patient is receiving antimicrobial therapy or if the organism is fastidious or does not grow on routinely used culture media. Consequently, the sensitivity of laboratory records is directly affected by the number of infections for which culture is performed and by the culture methods used by the laboratory.[63] The sensitivity of using microbiology laboratory reports for case-finding is 33%–84%. Of note, as patients are discharged from the hospital earlier, patients with signs and symptoms of infection may be less likely to have culture performed.[63]

Kardex Screening

Nursing "kardexes" or notes are cards or documents that contain patient information, which can provide "clues" to the infection preventionist. These documents can be surveyed 1 or 2 times each week to determine whether a patient is receiving antibiotics, intravenous fluids, or parenteral nutrition and whether a patients has an indwelling urinary catheter, special orders for wound-dressing changes, or orders for isolation precautions.[50] If the infection preventionist identifies one of these "kardex clues" or other information that suggests the patient is at risk of having an HAI, the infection preventionist reviews the patient's record. If the kardex record is complete and current, the method is highly sensitive and enables the infection preventionist to spend less time reviewing charts.

Clinical Ward Rounds

Infection preventionists who regularly visit clinical wards can gain excellent information about patients, infections, and other adverse events, because much important information is not included in the patients' records.[64,65] This method allows the infection preventionist to be highly visible in patient-care areas, to observe infection control practices directly, and to talk with the healthcare workers caring for patients. In this manner, the infection preventionist not only can collect data on patients with HAIs but also can assess compliance with isolation precautions, can answer questions on infection control issues, and can conduct informal educational sessions. One variation of this method is to use trained personnel who function as liaisons to the hospital epidemiology and infection control group. They can notify the infection preventionist of patients with potential nosocomial infections.

Computer Alerts, Personal Device Assistants, and Computer-Based Automated Surveillance

Many organizations have "home grown" computer programs or purchased software in place to facilitate surveillance or facilitate identification of patients colonized or infected with epidemiologically important organisms ("flags"). These tools are important to enhance the infection preventionist's efficiency and to protect patients. Although individual successes are reported in the literature, infection preventionists need to deter-mine the utility of these tools to them and their practices. Some groups have developed "data marts" that concatenate data from a hospital's transactional computer servers. These clinical data "warehouses" are created by importing patient-level data from the information systems, and they can be used to monitor antimicrobial resistance, to measure antimicrobial use, to detect HAIs, to measure the cost of infections, and to detect antimicrobial prescribing errors.[66] Recently, several investigators have used small hand-held computers to facilitate surveillance.[67] Typically, these devices include administrative data about patients of interest. The infection preventionist adds other details. These data then are downloaded into relational databases that allow the infection preventionist to manipulate and analyze the data. Additionally, several companies have developed cutting-edge computer software that allows microbiologic data to be supplemented with other clinically relevant data, such as pharmacy or radiology department reports and data on device use.[68] These data can be used to enhance and automate surveillance for HAIs; they are presented on a single screen and can be viewed easily by the infection preventionist, and the data are collected in relational databases. These easy-to-use programs allow practitioners to "query" the system and to display data graphically. Some of these programs have data mining capabilities. The impact of one of these programs has been measured, and it has been shown to be more efficient at identifying outbreaks than routine surveillance is.[56] Eleven of the 18 alerts identified by the software were determined to be potential outbreaks, which corresponded to a positive predictive value of 0.61; routine surveillance identified 5 of these 11 alerts during this time period.[56] These systems have various degrees of sophistication, and some can include data from electronic patient records. Such advances are likely to decrease the time required to perform surveillance, which will allow the infection preventionist to spend time with interventions and other activities.

Other investigators have mined large databases to look for novel ways of identifying HAIs.[36,57] As computerized medical records and administrative databases have become more common, there has been substantial interest in using such systems to conduct surveillance for HAIs. The goals are to increase the sensitivity of surveillance, to decrease the need for chart review, and to reduce costs. Most of the work has been done with electronic systems that support large healthcare systems providing integrated health care. This facilitates identifying infections treated in the outpatient setting or alternative settings. One common strategy is to use automated claims data to identify patients on the basis of infection-related *International Classification of Diseases, Ninth*

Revision codes and/or records of antibiotic use, readmission, emergency department visits, or the use of other medical tests during a specified period of time. The overall sensitivity, specificity, and predictive values of these novel surveillance systems need to be assessed.[19,34,35,69] Of note, Sands et al.[19] examined the sensitivity of using the health plan administrative database in this manner and found that the overall sensitivity was 72%, compared with 49% for the hospital-based surveillance system. Importantly, the sensitivity for detection of SSIs after discharge was 99%, far superior to that of routine surveillance. Trick and colleagues[57] compared 5 manual and computer-assisted methods of surveillance for hospital-acquired, central venous catheter–associated BSI. Rates of infection were 1.0–12.5 infections per 1,000 patient-days by investigator review and 1.4–10.2 infections per 1,000 patient-days by computer algorithm ($P = .004$). A recent review summarizes the performance of various selected algorithms to identify cases of catheter-associated bloodstream infection, surgical site infection, and ventilator-associated pneumonia. While these algorithms are exciting and the use of techniques such as natural language processing show promise, their validity and generalizability across disparate information technology systems are not known.[70]

Postdischarge Surveillance
As patients are discharged from hospitals earlier, the infection prevention and control team will have increasing difficulty detecting HAIs. One way of obtaining the data is to perform surveillance after patients are discharged. Infection prevention and control teams who do not conduct postdischarge surveillance will report spuriously low HAI rates, because traditional hospital-based surveillance methods identify only events that occur while the patient is in the hospital or institution. In fact, studies have documented that postdischarge surveillance identifies 13%–70% more SSIs than do methods that survey only inpatients.[19,69,71]

Most investigators who have studied methods for postdischarge surveillance have not evaluated all discharged patients but have focused on specific populations, such as postoperative patients, postpartum women, or neonates.[19,69,72] Investigators have assessed various methods for identifying HAIs after such patients are discharged, including directly assessing patients, reviewing records of visits to clinics or emergency departments, and contacting physicians or patients by mail or telephone. Although all these methods identify patients who develop infections after discharge, the methods are time consuming and can lack sensitivity. None of these methods have been accepted widely. Sands et al.[19,69] used administrative billing databases from an integrated healthcare system to study the best methods to identify SSIs, 84% of which develop after

discharge. Unfortunately, most infection prevention and control programs do not have access to such resources.

Which Case-Finding Method Is Best?

Each case-finding method has some merit, but each also has limitations. There is no agreement on which case-finding method is best. Some infection prevention and control personnel consider total chart review to be the "gold standard" (criterion standard) for identifying HAIs. However, in 2 studies that compared total chart review with combinations of 2 or more case-finding methods, it identified only 74%–94% of the infections identified by the combined methods.[33,50] Investigators were unable to identify all HAIs by reviewing only the medical record, for 4 reasons:

1. Records did not document all data required to determine whether the patients met the criteria for having specific infections.
2. Laboratory or radiology department reports were missing.
3. Records were not available.
4. The reviewer could not examine the patient.

Consequently, it is clear that total chart review is no more sensitive than other case-finding methods or combinations of methods. Of note, these issues will change as more and more electronic medical records become available.

Nettleman and Nelson[73] conducted surveillance to identify adverse events among patients hospitalized on general medical wards. They used numerous data sources and found that no single source identified all adverse occurrences. In fact, the number of adverse events the investigators identified in each category was dependent on which data source they used. Certain data sources efficiently identified specific adverse occurrences. For example, the investigators identified 77% of medication-related errors by reviewing the medication administration record, but they detected only 10% of these events by reviewing the physicians' progress notes. Conversely, they identified 100% of procedure-related adverse occurrences by reviewing the physicians' progress notes, but they did not detect any of these events by reviewing the medication administration record.[73]

Therefore, the infection control team must select case-finding methods that best identify the HAIs they choose to study. For example, the infection preventionist could identify most BSIs by reviewing microbiology laboratory reports but would find very few BSIs by observing patients directly. On the other hand, the infection preventionist might identify most SSIs by observing surgical wounds directly but not by reviewing microbiology laboratory reports.

NHSN

In 1970, the CDC enrolled a sample of hospitals, which voluntarily agreed to collect data on nosocomial infections, into the NNIS system. Restructured into the NHSN in 2005, it currently has almost 3,000 participating hospitals, and the NNIS/NHSN is the only source of national data on HAIs in the United States.[74] The participating hospitals range in size from 80 to 1,200 beds and include state, federal, profit, and not-for-profit institutions. The NNIS/NHSN program has several goals:

- To estimate the incidence of HAIs and infections with multidrug-resistant organisms
- To identify changes in the pathogens causing HAIs, the frequency of HAIs of specific types and at specific sites, the predominant risk factors, and the antimicrobial susceptibility patterns
- To provide data on HAIs with which hospitals can compare their data, including the distribution of HAIs by major types and sites, device-associated infection rates by type of intensive care unit, and SSI rates by operative procedure
- To develop strategies that infection prevention and control personnel can use for surveillance and assessment of HAIs

Initially, all hospitals participating in NNIS conducted prospective, traditional hospital-wide surveillance. The CDC investigators subsequently identified methodological problems that made comparisons of data among hospitals unreliable. Thus, in 1986, the NNIS program created 3 surveillance components in addition to hospital-wide surveillance: the adult intensive care unit or the pediatric intensive care unit, the high-risk nursery, and the population of all surgical patients or patients who underwent specific procedures. The revised system has some advantages. First, the newer components require infection prevention and control personnel to collect data on exposure to devices or to specific operative procedures, which allows the CDC to adjust HAI rates in the surveyed units for exposure to these devices and procedures. Hospitals that participate in NNIS/NHSN regularly receive reports that compare their adjusted data with the aggregate adjusted data. Infection prevention and control personnel in hospitals not participating in NNIS/NHSN can compare their adjusted rates with the adjusted rates published by the CDC. Also, each hospital can choose the surveillance component in which it will participate. Thus, each infection prevention and control program can both

design a surveillance program that meets the needs of their institution and participate in NNIS/NHSN.

In 2005, NNIS was restructured into the NHSN, which retained some features of NNIS but expanded others. National open enrollment for hospitals and outpatient hemodialysis centers occurred in 2007. NHSN is a secure, Internet-based surveillance system that integrates patient and healthcare personnel safety surveillance systems. In addition to fulfilling the NNIS goals, the NHSN provides surveillance data on both outcome measures and process measures known to be associated with prevention of HAI, provides facilities with risk-adjusted data that can be used for interfacility comparisons, and provides local quality improvement activities that can assist facilities in developing surveillance and analysis methods that permit timely recognition of patient and healthcare personnel safety problems and prompt intervention with appropriate measures. The NHSN also has the capacity to allow healthcare facilities to share data in a timely manner with public health agencies, as well as with other facilities.

Switzerland, Germany, Spain, the Netherlands, and France have developed remarkable, mostly country-wide surveillance systems for HAIs that have elements of the NNIS/NHSN system.[10,52,75-77]

HAI Rates

Measures of frequency of HAIs have a myriad of names but can be categorized as incidence measures and prevalence measures. The merits of these measures and some of the controversies about them are described in this section. Incidence is the number of new events divided by the number of patients at risk during a defined period of interest. Prevalence is number of events (new and old) that are present during a defined period of interest. Prevalence is usually ascertained by surveys. Commonly, the prevalence is obtained at a given point in time. It is calculated as the number of active current infections divided by the number of patients at risk or studied. To measure incidence, the most common measures include the crude cumulative incidence (the number of infections per 100 admissions or discharges), the crude incidence density or the adjusted infection rate (the number of infections per 1,000 patient-days), the specific cumulative or incidence density (according to unit, procedure, or provider), and the adjusted cumulative or incidence density (adjusted for intrinsic host factors, such as age). Finally, standardized infection ratios are used to compare SSI frequency[77]; these are calculated as the ratio of the observed to the expected rate of infection (the expected rate is derived from data from a reference facility or reference source) for each risk category of operations. More and more healthcare epidemiologists recommend that the denominator for such rates should be patient-days at risk, which is calculated as the total number of days of hospitalization for all patients at risk.

Overall Hospital Infection Rates

Infection prevention and control programs that conduct hospital-wide surveillance sometimes track the overall infection rate for their facility. This rate is calculated by dividing the number of HAIs identified in a given month by the number of patients admitted or discharged during the same month. The overall HAI rate has several inherent disadvantages:

- It treats all infections as though they are of equal importance. Furthermore, changes in rates of uncommon but epidemiologically important infections (eg, bacteremia) might be hidden in the larger volume of common but less important infections (eg, UTIs).
- It does not distinguish between patients who had a single infection and those who had numerous infections.
- It may not be accurate and may underestimate the true rate, because the infection preventionist often cannot identify all HAIs.
- It does not account for patients who are at increased risk for becoming infected because of underlying diseases or exposure to procedures and medical devices; therefore, it tends to obscure important trends in intensive care units or among high-risk patients.
- It does not adjust for length of stay.
- It is not adjusted for risk, and therefore it cannot be compared with rates from other hospitals.

In short, the accuracy and usefulness of the overall HAI rate is limited. Therefore, we recommend that infection prevention and control personnel stop calculating their overall infection rate and begin calculating adjusted infection rates.

Site-Specific Infection Rates

Site-specific infection rates (ie, rates of infection stratified by type or body site of infection) are a more appropriate measure, because they represent a more homogeneous group of infections. Examples of site-specific infections include BSI, catheter-associated infection, UTI, and VAP. These rates are calculated using the number of infections as the numerator and dividing it by an appropriate denominator, usually the number of device-days (eg, catheter-days or ventilator-days). The most common measure used is the specific cumulative incidence or incidence density ratio (see Chapter 4, on epidemiologic methods).

Adjusting Rates

Hospital epidemiology and infection control personnel calculate infection rates so that they can identify problems and assess the effectiveness of their interventions. In addition, they follow rates over time to identify significant increases above baseline rates and to assess the efficacy of their program. In addition, to determine whether they actually have a problem, hospital epidemiology personnel often compare their rates with those of other institutions. However, comparisons within a single hospital over time may not be valid, because the patient population or patient care may have changed substantially. Further, comparisons between hospitals may not be valid, because healthcare facilities are not standardized.[78] Patients in different hospitals have different underlying diseases and different severities of illness. In addition, patients who have the same disease and the same severity of illness but who are in different hospitals could undergo different diagnostic and therapeutic interventions and stay in the hospital for different lengths of time. Each hospital has its own unique environment, patient-care practices, and healthcare providers. Infection control programs vary substantially in the intensity of surveillance, the methods used for surveillance, the definitions of infections used, and the methods used for calculating infection rates. Consequently, infection control personnel must use adjusted rates if they want to assess their rates over time or to compare their rates with those in other hospitals. In the following paragraphs, we discuss several methods for adjusting rates.

Adjusting for Length of Stay

Infection rates more accurately reflect the risk of infection when they are adjusted for length of stay. Infection prevention and control staff attempt to control for the length of stay by calculating the number of HAIs per patient-day. This method uses the total number of HAIs in a month as the numerator and the total number of patient-days in that month (ie, the sum of the number of days that each patient was on the unit during the month) as the denominator. For example, an obstetrics ward admits many patients who stay in the hospital for a very brief time and whose risk of infection is low, but a rehabilitation ward admits a few patients who stay for long periods of time and whose risk of infection is high. If the number of patients admitted were used as the denominator, the infection rate for the obstetrics ward probably would underestimate the risk of infection, whereas the rate for the rehabilitation ward most likely would overestimate the risk of infection. By using the number of patient-days as the denominator, infection control staff control for the effect of length of stay on the infection rate. However, this method does not control for the effect of other risk factors, such as use of invasive devices or the severity of the patient's underlying illness.

Adjusting for Exposure to Devices

Device-associated infection rates control for the duration of exposure to an invasive device, which is one of the major risk factors for these infections. Therefore, device-associated rates can be compared more reliably over time and between institutions than can overall infection rates. To calculate this rate, the infection control team first specifies the type of device (eg, mechanical ventilators) and the population (eg, patients in the medical intensive care unit) to be studied. Next, the team identifies the cases of device-associated infection (eg, VAP) that occur in the selected population during a specified time period. The number of infections is the numerator. To obtain the denominator, the team sums the number of patients exposed to the device during each day of the specified period. For example, if the team surveyed the medical intensive care unit for 7 days and found that the number of patients who underwent ventilation on each of those days was 4, 3, 5, 5, 4, 6, and 4, then the number of ventilator-days would be 31. If the team identified 3 cases of VAP during the week, the VAP rate would be 3 divided by 31, or 0.097 cases per ventilator-day; this can be expressed as 97 cases per 1,000 ventilator-days.

Adjusting for Severity of Illness

One would expect that a 28-year-old man who does not have underlying medical illnesses and who is undergoing an elective herniorrhaphy would have a lower risk of acquiring an SSI than would a 65-year-old man who has chronic lung disease treated with steroids, diabetes mellitus, and heart disease and who is undergoing an emergency exploratory laparotomy. Several investigators have developed scores to determine a patient's severity of illness. These scores range from simple, subjective scales based on clinical judgment to scores obtained from commercial computer programs that use objective clinical data to assess the severity of a patient's underlying illnesses. Unfortunately, the currently available severity-of-illness scores cannot identify patients who are at high risk of developing an HAI. Thus, most infection control programs do not use these scores to adjust HAI rates for risk.

The risk index with which infection control personnel are most familiar is the surgical-wound classification. This classification system classifies procedures into 4 categories: clean, clean-contaminated, contaminated, and dirty.[79] The incidence of infection increases as the wound classification changes from clean to dirty. However, this system does not account for each patient's intrinsic susceptibility to infection. Consequently, its ability to predict which patients are at highest risk of SSI is limited.

Investigators in the SENIC project and the NNIS/NHSN have developed risk indices that include variables assessing the patients' intrinsic risk of infection.[80,81] Culver et al.[81] used the NNIS risk index to stratify the risk of SSI and to standardize SSI rates. However, other investigators tested the validity of the NNIS risk index and found that it did not predict which patients were at highest risk of developing an SSI after cardiothoracic surgery.[82] Hence, before risk indices can be used to identify patients at high risk or to adjust rates, they must be validated, and the population for which they are most predictive must be defined.

Surveillance for Process Measures

Surveillance for outcome measures or infection rates has long been used in infection control programs. However, there are several limitations to outcome-based surveillance. First, the preventable fraction of HAIs is not known, therefore making it difficult to evaluate if the infection-prevention measures are adequate in a given unit. Second, infection rates do not provide the information on what breaches in infection control measures contribute to the problem and can become the focus of prevention efforts.[83]

Surveillance for process measures may fill some of the gap. Warren et al.[44] developed a checklist tool as one of the strategies to decrease catheter-related BSI in the intensive care unit. The checklist was used to ensure adherence to infection-control practices and was 1 component of the intervention to reduce the infection rate.[43,84] This checklist was subsequently adopted by Pronovost and colleagues[84] in a statewide effort to prevent BSI. Other examples of surveillance based on process measures are the vaccination rate among healthcare personnel, the rate of compliance with recommended hand hygiene, the rate of adherence to administration of surgical antibiotic prophylaxis within 1 hour before the first incision, the rate of compliance with an implemented VAP "bundle" (ie, a combined group of infection-control measures), the rate of appropriate indwelling urinary catheter use, and the device utilization ratio (eg, a central-line utilization ratio).[85,86]

In contrast to outcome-based surveillance, process-measure surveillance provides performance targets; for example, adherence to infection control procedures for catheter insertion for every single patient. Deviations in adherence are easy to recognize, and infection control measures can be implemented early, maybe before an increase in the infection rate occurs.[83]

Process-measure surveillance data can be used as performance indicators for adherence to infection control guidelines and can be further evaluated for effects on the outcome or infection rate. Surveillance of process measures is now one of the essential elements of surveillance proposed by the NHSN.[87]

Surveillance in Developing Countries

The rationale for a basic surveillance system for HAIs and infection or colonization with multidrug-resistant organisms in developing countries is not different from the rationale for a surveillance system in developed countries. However, with limited resources, one has to concentrate efforts on the most achievable goals ("low-hanging fruit") and focus on specific areas of the hospital and specific procedures that have high rates of HAI and infection with multidrug-resistant organisms.[88] Benchmarking of rates with those in other developing countries has been demonstrated by Rosenthal et al.,[46-48] who use the NNIS/NHSN methods in less developed areas of the world extremely successfully.

Use of Surveillance Data to Meet Regulatory Requirements

Public reporting of outcome and process measures to state or national authorities is intended to enable consumers to make more informed choices for safer care. By promoting competition, a public reporting system may influence healthcare facilities to undertake efforts to improve the quality of care and may result in optimal patient outcomes. However, unintended consequences, such as the intention to avoid admission of sicker patients, may occur. The Healthcare Infection Control Practices Advisory Committee (HICPAC) found inconclusive evidence for the effectiveness of public reporting systems in improving healthcare performance.[89] Therefore, HICPAC has not recommended for or against mandatory public reporting of HAIs. HICPAC, however, proposed guidance on public reporting of HAIs in 2005, highlighting the essential elements for public reporting systems, identifying appropriate measures of healthcare performance for both process and outcome measures, and identifying patient populations to be monitored, as well as making recommendations on case-finding methods, data validation, resource and infrastructure requirements, HAI rates and risk adjustment, and production of useful reports and feedback.[14]

Illinois was the first state to enact mandatory reporting of HAIs (in 2002), and up to the present, only several states have followed suit. At the moment, not all outcome and process measures are required to be reported publicly. The National Quality Forum has made recommendations on outcome measures and process measures for public reporting of HAIs. Outcome measures include the rates of central line–associated BSI, SSI, and VAP among patients in intensive care units; the rates of catheter-associated UTI, late sepsis, and meningitis in neonates; and late sepsis or meningitis in neonates with very low birth weight. Process measures include the processes for prevention of central line–associated BSI, SSI, and VAP.[90]

The Healthcare-Associated Infection Working Group of the Joint Public Policy Committee of the Society for Healthcare Epidemiology of America also provide a tool kit with guidance on the components necessary for a meaningful reporting system.[91] They recommend identification or creation of an agency at the state level responsible for collecting and analyzing data prior to public disclosure, identification of well-trained personnel to collect data, use of strategies to prevent unintended consequences of public reporting, and use of outcome and process-measure surveillance.

These documents are dynamic and still need to be modified to represent the real situation with respect to HAIs in healthcare settings. Research is still needed to assess the impact of public reporting systems on HAIs.

Moving Forward: Electronic, Automatic, and Computer-Based Surveillance

An essential part of healthcare surveillance in the 21st century will be the integration of increasingly important and rapidly developing surveillance technologies. The Institute of Medicine, in their 1991 report (reiterated in their 1999 report), endorsed a computer-based patient record system as an essential technology and recommended widespread utilization within a decade, a target that has long since passed.[92] As healthcare epidemiology and infection control comes under more pressure from the public and from legislative, administrative, and regulatory forces, the focus of the infection preventionist and the hospital epidemiologist will likely shift away from pure prevention efforts. This will translate into a need to be more efficient and to spend less time on surveillance. Computer-based surveillance systems can save some of the time spent doing routine surveillance, can facilitate multicenter comparisons and the exchange of information between facility sites, can notify infection preventionists of potential outbreaks and clusters of infection before they would be picked up by manual surveillance, and can reduce the occurrence of the errors common in manual methods of surveillance.

At the time of the first SCENIC investigation in 1976, most of the infection preventionists surveyed spent 50% or more of their time performing surveillance.[93] A survey conducted by the CDC in 2000 found that an infection preventionist spends 35%–40% of work time performing surveillance.[94] One early study found that the use of electronic systems reduced the time spent on surveillance by 65%,[95] and more recent studies have found similar results.[96] The development of standardized, instantly recoverable, and easily shared infection control data not only will facilitate communication and comparison of infection rates between multiple facility sites and institutions but also will allow infection control programs to easily meet the demands of complying with regulatory mandates. The ability to immediately access electronic records of, for example, infection rates or trends in resistance, for an entire facility greatly eases the process of auditing, accrediting, and regulation performed by bodies such as the Joint Commission.[97] An effective electronic surveillance program can be integrated with infection control–related goals, such as managing antibiotic usage, tracking adverse drug events, and identifying emerging drug-resistant organisms; such integrated systems have already been shown to be cost-effective.[98] Both commercial and independently developed computer-based surveillance systems have the potential to increase efficiency and to reduce economic costs at multiple levels of the healthcare system. There are few peer-reviewed studies that explicitly investigate the cost-effectiveness of computer-based surveillance, but at least 2 studies have shown automated surveillance to be significantly cost-effective.[99] Several individual case studies have found that implementing an electronic surveillance system reduced both infection rates and associated economic costs.[100] Although more studies aimed specifically at analyzing cost-effectiveness are needed, the potential for computer-based surveillance to reduce expenses (both worker hours and infection costs) will be a vital part of its approval and implementation in any healthcare facility.

The use of electronic surveillance is not without its drawbacks and limitations. The introduction of any new system in a hospital is susceptible to a sharp learning curve, but this is especially the case with the introduction of new and complex electronic surveillance systems. Although there is some evidence that investing in an electronic system is cost-effective, the obvious challenge of funding the purchase of an expensive new system must be overcome, which may be difficult given the paucity of cost-benefit data.[52] The investment in implementing a new system is not only financial; although Wisniewski et al.[66] estimated that using an independently-designed electronic system saved 1,750 worker hours for each 1,500 charts reviewed, they estimated an investment of 4,000 worker-hours to develop such a system. While manual chart review can be error prone, electronic surveillance often has low thresholds and may identify false-positive outbreaks or clusters of infection. Perhaps the most important issue is to recognize that while electronic methods of surveillance may reduce the need for manual review, it is not a substitute for critical thinking or further analysis. Computer-based surveillance systems should be seen as a valuable tool by the infection preventionist to obtain, analyze, and communicate relevant information, without becoming overwhelmed by the information or overly reliant on a solely electronic system.

Conclusion

We believe that the surveillance systems of the 21st century must be extremely flexible so that they can be adapted to meet the needs of rapidly changing healthcare systems and the emerging dominance of information technology. We also think that effective healthcare epidemiology and infection control teams will not use a one-size-fits-all approach to surveillance but will mix and match different case-finding methods and surveillance methods to create a surveillance system that meets the needs not only of their entire healthcare system but also of the individual components (eg, the intensive care units and the ambulatory-surgery center). That said, as legislators pass laws requiring hospitals to report infection rates, the importance of using standard definitions and standard approaches for identifying infections and for calculating rates is paramount. In addition, we think infection control personnel must use computers and must develop electronic surveillance systems that use the computerized databases already present in their institution (eg, databases from the laboratory, surgical services, and financial management). They need to advocate for inclusion of newer information technology into their program so that they can enhance its efficiency in surveillance activities and increase interventions to prevent adverse events. Furthermore, the infection prevention and control team should collaborate with personnel from the information systems department to develop algorithms to identify patients with possible HAIs and to determine the thresholds used to identify likely outbreaks. As hospitals develop their electronic patient-record systems, computer-based surveillance should become easier. However, we would encourage infection control personnel not to wait until their hospital has electronic patient records but to use whatever resources are available currently to streamline surveillance. Finally, computerized models that predict which patients are at highest risk of infection would allow infection control personnel to improve the efficiency of surveillance and to stratify infection rates more appropriately. However, these models are not currently available. We would encourage the infection control community to make development of such models a priority.

Acknowledgments

We thank John Frederick and Abigail Carlson for their critical review of this chapter.

References

1. Langmuir AD. The surveillance of communicable diseases of national importance. *N Engl J Med* 1963;268:182–192.
2. Perl TM. Surveillance, reporting and the use of computers. In: Wenzel RP, ed. *Prevention and Control of Nosocomial Infections*. Vol 1. Baltimore, MD: Williams and Wilkins; 1997: 127–162.
3. Haley RW, Quade D, Freeman HE, Bennett JV, Committee TCSP. The SENIC project: Study on the Efficacy of Nosocomial Infection Control (SENIC Project): summary of study design. *Am J Epidemiol* 1980;111:472–485.
4. Haley RW. The "hospital epidemiologist" in U.S. hospitals, 1976–1977: a description of the head of the infection surveillance and control program. Report from the SENIC project. *Infect Control* 1980;1(1):21–32.
5. Karanfil LV, Conlon M, Lykens K, et al. Reducing the rate of nosocomially transmitted respiratory syncytial virus. *Am J Infect Control* 1999;27(2):91–96.
6. Carlet J, Astagneau P, Brun-Buisson C, et al. French national program for prevention of healthcare-associated infections and antimicrobial resistance, 1992–2008: positive trends, but perseverance needed. *Infect Control Hosp Epidemiol* 2009;30(8):737–745.
7. Gastmeier P, Kampf G, Wischnewski N, et al. Prevalence of nosocomial infections in representative German hospitals. *J Hosp Infect* 1998;38(1):37–49.
8. Harbarth S, Ruef C, Francioli P, Widmer A, Pittet D. Nosocomial infections in Swiss university hospitals: a multi-centre survey and review of the published experience. Swiss-Noso Network. *Schweiz Med Wochenschr* 1999;129(42): 1521–1528.
9. Richet H, Wiesel M, Le Gallou F, Andre-Richet B, Espaze E. Methicillin-resistant *Staphylococcus aureus* control in hospitals: the French experience. Association des Pays de la Loire pour l'Eviction des Infections Nosocomiales. *Infect Control Hosp Epidemiol* 1996;17(8):509–511.
10. Prevalence of nosocomial infections in France: results of the nationwide survey in 1996. The French Prevalence Survey Study Group. *J Hosp Infect* 2000;46(3):186–193.
11. Buehler JW, Berkelman RL, Hartley DM, Peters CJ. Syndromic surveillance and bioterrorism-related epidemics. *Emerg Infect Dis* 2003;9(10):1197–1204.
12. Bravata DM, McDonald KM, Smith WM, et al. Systematic review: surveillance systems for early detection of bioterrorism-related diseases. *Ann Intern Med* 2004;140(11):910–922.
13. McKibben L, Fowler G, Horan T, Brennan PJ. Ensuring rational public reporting systems for health care-associated infections: systematic literature review and evaluation recommendations. *Am J Infect Control* 2006;34(3):142–149.
14. McKibben L, Horan T, Tokars JI, et al. Guidance on public reporting of healthcare-associated infections: recommendations of the Healthcare Infection Control Practices Advisory Committee. *Am J Infect Control* 2005;33(4):217–226.
15. Fraser V, Murphy D, Brennan PJ, Frain J, Arias KM, Perl TM. Politically incorrect: legislation must not mandate specific healthcare epidemiology and infection prevention and control practices. *Infect Control Hosp Epidemiol* 2007;28(5): 594–595.
16. McKibben L, Horan TC, Tokars JI, et al. Guidance on public reporting of healthcare-associated infections: recommendations of the Healthcare Infection Control Practices Advisory Committee. *Infect Control Hosp Epidemiol* 2005;26(6): 580–587.
17. Garner JS. Guideline for isolation precautions in hospitals. The Hospital Infection Control Practices Advisory Committee. *Infect Control Hosp Epidemiol* 1996;17(1):53–80.
18. Morales EM, Herwaldt LA, Nettleman M, Larson C, Sanford L, Perl T. *Staphylococcus aureus* carriage and saphenous vein harvest site infection (HSI) following coronary artery bypass surgery (CABG). In: Program and abstracts of the 34th Annual Interscience Conference on Antimicrobial Agents and Chemotherapy (ICAAC); Orlando, FL; October 1994. Abstract.
19. Sands KE, Yokoe DS, Hooper DC, et al. Detection of postoperative surgical-site infections: comparison of health plan-based surveillance with hospital-based programs. *Infect Control Hosp Epidemiol* 2003;24(10):741–743.
20. Sands K, Vineyard G, Livingston J, Christiansen C, Platt R. Efficient identification of postdischarge surgical site infections: use of automated pharmacy dispensing information, administrative data, and medical record information. *J Infect Dis* 1999;179(2):434–441.
21. Garner JS, Jarvis WR, Emori TG, Horan TC, Hughes JM. CDC definitions for nosocomial infections, 1988. *Am J Infect Control* 1988;16(3):128–140.
22. Horan TC, Andrus M, Dudeck MA. CDC/NHSN surveillance definition of health care-associated infection and criteria for specific types of infections in the acute care setting. *Am J Infect Control* 2008;36(5):309–332.
23. Cohen AL, Calfee D, Fridkin SK, et al. Recommendations for metrics for multidrug-resistant organisms in healthcare settings: SHEA/HICPAC position paper. *Infect Control Hosp Epidemiol* 2008;29(10):901–913.
24. McDonald LC, Coignard B, Dubberke E, Song X, Horan T, Kutty PK. Recommendations for surveillance of *Clostridium difficile*–associated disease. *Infect Control Hosp Epidemiol* 2007;28(2):140–145.
25. Dubberke ER, Gerding DN, Classen D, et al. Strategies to prevent *Clostridium difficile* infections in acute care hospitals. *Infect Control Hosp Epidemiol* 2008;29(Suppl 1):S81–S92.
26. Calfee DP, Salgado CD, Classen D, et al. Strategies to prevent transmission of methicillin-resistant *Staphylococcus aureus* in acute care hospitals. *Infect Control Hosp Epidemiol* 2008; 29(Suppl 1):S62–S80.
27. National Health Safety Network (NHSN) Web site. http//: www.cdc.gov/ncidod/dhqp/nhsn.html. Accessed October 26, 2009.
28. Morrison AJJ, Kaiser DL, Wenzel RP. A measurement of the efficacy of nosocomial infection control using the 95 percent confidence interval for infection rates. *Am J Epidemiol* 1987;126:292–297.
29. Cardo DM, Falk PS, Mayhall CG. Validation of surgical wound classification in the operating room. *Infect Control Hosp Epidemiol* 1993;14(5):255–259.

30. Cardo DM, Falk PS, Mayhall CG. Validation of surgical wound surveillance. *Infect Control Hosp Epidemiol* 1993; 14(4):211–215.

31. Ehrenkranz NJ, Richter EI, Phillips PM, Shultz JM. An apparent excess of operative site infections: analyses to evaluate false-positive diagnosis. *Infect Control Hosp Epidemiol* 1995;16:712–716.

32. Lessler J, Brookmeyer R, Perl TM. An evaluation of classification rules based on date of symptom onset to identify healthcare associated infections. *Am J Epidemiol* 2007;166(10): 1220–1229.

33. Haley RW, Schaberg DR, McClish DK, et al. The accuracy of retrospective chart review in measuring nosocomial infection rates. *Am J Epidemiol* 1980;111:516–533.

34. Yokoe DS, Anderson J, Chambers R, et al. Simplified surveillance for nosocomial bloodstream infections. *Infect Control Hosp Epidemiol* 1998;19(9):657–660.

35. Platt R, Yokoe DS, Sands KE. Automated methods for surveillance of surgical site infections. *Emerg Infect Dis* 2001;7(2): 212–216.

36. Yokoe DS, Noskin GA, Cunnigham SM, et al. Enhanced identification of postoperative infections among inpatients. *Emerg Infect Dis* 2004;10(11):1924–1930.

37. Broderick A, Mori M, Nettleman MD, Streed SA, Wenzel RP. Nosocomial infections: validation of surveillance and computer modeling to identify patients at risk. *Am J Epidemiol* 1990;131(4):734–742.

38. Stratton CW, Ratner H, Johnston PE, Schaffner W. Focused microbiologic surveillance by specific hospital unit as a sensitive means of defining antimicrobial resistance problems. *Diagn Microbiol Infect Dis* 1992;15:11S–18S.

39. Stratton CW, Ratner H, Johnston PE, Schaffner W. Focused microbiologic surveillance by specific hospital unit: practical application and clinical utility. *Clin Ther* 1993;15(Suppl A): 12–20.

40. Srinivasan A, Wolfenden LL, Song X, et al. An outbreak of *Pseudomonas aeruginosa* infections associated with flexible bronchoscopes. *N Engl J Med* 2003;348(3):221–227.

41. Classen DC, Evans RS, Pestotnik SL, Horn SD, Menlove RL, Burke JP. The timing of prophylactic administration of antibiotics and the risk of surgical-wound infection. *N Engl J Med* 1992;326(5):281–286.

42. Berenholtz SM, Pronovost PJ, Lipsett PA, et al. Eliminating catheter-related bloodstream infections in the intensive care unit. *Crit Care Med* 2004;32(10):2014–2020.

43. Pronovost P, Needham D, Berenholtz S, et al. An intervention to decrease catheter-related bloodstream infections in the ICU. *N Engl J Med* 2006;355(26):2725–2732.

44. Warren DK, Cosgrove SE, Diekema DJ, et al. A multicenter intervention to prevent catheter-associated bloodstream infections. *Infect Control Hosp Epidemiol* 2006;27(7):662–669.

45. Edwards JR, Peterson KD, Andrus ML, Dudeck MA, Pollock DA, Horan TC. National Healthcare Safety Network (NHSN) report, data summary for 2006 through 2007, issued November 2008. *Am J Infect Control* 2008;36(9):609–626.

46. Rosenthal VD, Maki DG, Graves N. The International Nosocomial Infection Control Consortium (INICC): goals and objectives, description of surveillance methods, and operational activities. *Am J Infect Control* 2008;36(9):e1–e12.

47. Rosenthal VD, Maki DG, Mehta A, et al. International Nosocomial Infection Control Consortium report, data summary for 2002–2007, issued January 2008. *Am J Infect Control* 2008;36(9):627–637.

48. Rosenthal VD, Maki DG, Salomao R, et al. Device-associated nosocomial infections in 55 intensive care units of 8 developing countries. *Ann Intern Med* 2006;145(8):582–591.

49. Rosenthal VD, Maki DG. Prospective study of the impact of open and closed infusion systems on rates of central venous catheter-associated bacteremia. *Am J Infect Control* 2004;32(3):135–141.

50. Wenzel RP, Osterman CA, Hunting KJ, Gwaltney JM. Hospital acquired infections: I. Surveillance in a university hospital. *Am J Epidemiol* 1976;103(3):251–260.

51. Haley RW, Culver DH, White JW, et al. The efficacy of infection surveillance and control programs in preventing nosocomial infections in US hospitals. *Am J Epidemiol* 1985; 121:182–205.

52. Pittet D, Harbarth S, Ruef C, et al. Prevalence and risk factors for nosocomial infections in four university hospitals in Switzerland. *Infect Control Hosp Epidemiol* 1999;20(1): 37–42.

53. Climo M, Diekema D, Warren DK, et al. Prevalence of the use of central venous access devices within and outside of the intensive care unit: results of a survey among hospitals in the prevention epicenter program of the Centers for Disease Control and Prevention. *Infect Control Hosp Epidemiol* 2003;24(12):942–945.

54. Schifman RB, Palmer RA. Surveillance of nosocomial infections by computer analysis of positive culture rates. *J Clin Microbiol* 1985;21:493–495.

55. McGuckin MB, Abrutyn E. A surveillance method for early detection of nosocomial outbreaks. *APIC* 1979; 7:18–21.

56. Wright MO, Perencevich EN, Novak C, Hebden JN, Standiford HC, Harris AD. Preliminary assessment of an automated surveillance system for infection control. *Infect Control Hosp Epidemiol* 2004;25(4):325–332.

57. Trick WE, Zagorski BM, Tokars JI, et al. Computer algorithms to detect bloodstream infections. *Emerg Infect Dis* 2004;10(9):1612–1620.

58. Gross PA, Beaugard A, Van Antwerpen C. Surveillance for nosocomial infections: can the sources of data be reduced? *Infect Control* 1980;1:233–236.

59. Glenister HM. How do we collect data for surveillance of wound infection? *J Hosp Infect* 1993;24(4):283–289.

60. Maragakis LL, Winkler A, Tucker MG, et al. Outbreak of multidrug-resistant *Serratia marcescens* infection in a neonatal intensive care unit. *Infect Control Hosp Epidemiol* 2008; 29(5):418–423.

61. Maragakis LL, Chaiwarith R, Srinivasan A, et al. *Sphingomonas paucimobilis* bloodstream infections associated with contaminated intravenous fentanyl. *Emerg Infect Dis* 2009;15(1):12–18.

62. Maragakis LL, Cosgrove SE, Song X, et al. An outbreak of multidrug-resistant *Acinetobacter baumannii* associated with pulsatile lavage wound treatment. *JAMA* 2004;292(24): 3006–3011.

63. Manian FA, Meyer L. Comprehensive surveillance of surgical wound infections in outpatient and inpatient surgery. *Infect Control Hosp Epidemiol* 1990;11(10):515–520.

64. Glenister H, Tayor L, Bartlett C, Cooke M, Sedgwick J, Leigh D. An assessment of selective surveillance methods for

detecting hospital-acquired infection. *Am J Med* 1991;91 (Suppl 3B):121S–124S.

65. Ford-Jones EL, Mindorff CM, Pollock E, et al. Evaluation of a new method of detection of nosocomial infection in the pediatric intensive care unit: the infection control sentinel sheet system. *Infect Control Hosp Epidemiol* 1989;10:515–520.

66. Wisniewski MF, Kieszkowski P, Zagorski BM, Trick WE, Sommers M, Weinstein RA. Development of a clinical data warehouse for hospital infection control. *J Am Med Inform Assoc* 2003;10(5):454–462.

67. Farley JE, Srinivasan A, Richards A, Song X, McEachen J, Perl TM. Handheld computer surveillance: shoe-leather epidemiology in the "palm" of your hand. *Am J Infect Control* 2005; 33(8):444–449.

68. Leal J, Laupland KB. Validity of electronic surveillance systems: a systematic review. *J Hosp Infect* 2008;69(3): 220–229.

69. Sands K, Vineyard G, Platt R. Surgical site infections occurring after hospital discharge. *J Infect Dis* 1996;173(4):963–970.

70. Klompas M, Yokoe DS. Automated surveillance of healthcare associated infections. *Clin Infect Dis* 2009;48:1268–1275.

71. Holtz TH, Wenzel RP. Postdischarge surveillance for nosocomial wound infection: a brief review and commentary. *Am J Infect Control* 1992;20(4):206–213.

72. Yokoe DS, Christiansen CL, Johnson R, et al. Epidemiology of and surveillance for postpartum infections. *Emerg Infect Dis* 2001;7(5):837–841.

73. Nettleman MD, Nelson AP. Adverse occurrences during hospitalization on a general medicine service. *Clin Perform Qual Health Care* 1994;2(2):67–72.

74. Emori TG, Culver DH, Horan TC, et al. National Nosocomial Infections Surveillance System (NNIS): description of surveillance methods. *Am J Infect Control* 1991;19:19–35.

75. Gastmeier P, Geffers C, Sohr D, Dettenkofer M, Daschner F, Ruden H. Five years working with the German nosocomial infection surveillance system (Krankenhaus Infektions Surveillance System). *Am J Infect Control* 2003;31(5):316–321.

76. Gastmeier P, Weigt O, Sohr D, Ruden H. Comparison of hospital-acquired infection rates in paediatric burn patients. *J Hosp Infect* 2002;52(3):161–165.

77. Jodra VM, Rodela AR, Martinez EM, Fresnena NL. Standardized infection ratios for three general surgery procedures: a comparison between Spanish hospitals and U.S. centers participating in the National Nosocomial Infections Surveillance System. *Infect Control Hosp Epidemiol* 2003;24(10): 744–748.

78. Nosocomial infection rates for interhospital comparison: limitations and possible solutions. A Report from the National Nosocomial Infections Surveillance (NNIS) System. *Infect Control Hosp Epidemiol* 1991;12(10):609–621.

79. Garner J. CDC guidelines for the prevention and control of nosocomial infection. Guideline for prevention of surgical wound infection surveillance. *Am J Infect Control* 1986; 14:71–82.

80. Haley RW, Culver DH, Morgan WM, White JW, Emori TG, Hooton TM. Identifying patients at high risk of surgical wound infection: a simple multivariate index of patient susceptibility and wound contamination. *Am J Epidemiol* 1985;121(2):206–215.

81. Culver DH, Horan TC, Gaynes RP, et al. Surgical wound infection rates by wound class, operative procedure, and patient risk index. National Nosocomial Infections Surveillance System. *Am J Med* 1991;91(3B):152S–157S.

82. Roy MC, Herwaldt LA, Embrey R, Kuhns K, Wenzel RP, Perl TM. Does the Centers for Disease Control's NNIS system risk index stratify patients undergoing cardiothoracic operations by their risk of surgical-site infection? *Infect Control Hosp Epidemiol* 2000;21(3):186–190.

83. Tokars JI, Richards C, Andrus M, et al. The changing face of surveillance for health care-associated infections. *Clin Infect Dis* 2004;39(9):1347–1352.

84. Pronovost P. Interventions to decrease catheter-related bloodstream infections in the ICU: the Keystone Intensive Care Unit Project. *Am J Infect Control* 2008;36(10): S171–S175.

85. Marschall J, Mermel LA, Classen D, et al. Strategies to prevent central line-associated bloodstream infections in acute care hospitals. *Infect Control Hosp Epidemiol* 2008;29(Suppl 1): S22–S30.

86. Lee TB, Montgomery OG, Marx J, Olmsted RN, Scheckler WE. Recommended practices for surveillance. Association for Professionals in Infection Control and Epidemiology (APIC). *Am J Infect Control* 2007;35(7):427–440.

87. Outline for Healthcare-Associated Infections Surveillance. 2006. http://www.cdc.gov/ncidod/dhqp/nhsn.html. Accessed March 15, 2009.

88. Williams JD, Sharma KB, Atukorala SD. *Guidelines on prevention and control of hospital-associated infections.* New Dehli: World Health Organization; 2002.

89. Wong ES, Rupp ME, Mermel L, et al. Public disclosure of healthcare-associated infections: the role of the Society for Healthcare Epidemiology of America. *Infect Control Hosp Epidemiol* 2005;26(2):210–212.

90. Yokoe DS, Classen D. Improving patient safety through infection control: a new healthcare imperative. *Infect Control Hosp Epidemiol* 2008;29(Suppl 1):S3–S11.

91. Society for Healthcare Epidemiology of America; Healthcare-Associated Infection Working Group of the Joint Public Policy Committee. Essentials of public reporting of healthcare-associated infections: a tool kit. 2007. http://www.shea-online.org/ Assets/files/Essentials_of_Public_Reporting_Tool_Kit.pdf. Accessed October 26, 2009.

92. Institute of Medicine. *The Computer-Based Patient Record: An Essential Technology for Health Care.* Washington, DC: National Academy Press; 1991.

93. Haley RW, Shachtman RH. The emergence of infection surveillance and control programs in US hospitals: an assessment, 1976. *Am J Epidemiol* 1980;111(5):574–591.

94. Nguyen GT, Proctor SE, Sinkowitz-Cochran RL, Garrett DO, Jarvis WR. Status of infection surveillance and control programs in the United States, 1992–1996. Association for Professionals in Infection Control and Epidemiology, Inc. *Am J Infect Control* 2000;28(6):392–400.

95. Evans RS, Larsen RA, Burke JP, et al. Computer surveillance of hospital-acquired infections and antibiotic use. *JAMA* 1986;256(8):1007–1011.

96. Chalfine A, Cauet D, Lin WC, et al. Highly sensitive and efficient computer-assisted system for routine surveillance for surgical site infection. *Infect Control Hosp Epidemiol* 2006;27(8):794–801.

97. Association for Professionals in Infection Control (APIC). The Importance of Surveillance Technologies in the Prevention of

Healthcare-Associated Infections (HAIs). APIC Position Paper. Washington, DC: APIC; 2009. http://www.apic.org/Content/NavigationMenu/PracticeGuidance/PositionStatements/Position_Statements.htm. Accessed October 28, 2009.

98. Dellit TH, Owens RC, McGowan JE Jr, et al. Infectious Diseases Society of America and the Society for Healthcare Epidemiology of America guidelines for developing an institutional program to enhance antimicrobial stewardship. *Clin Infect Dis* 2007;44(2):159–177.

99. Furuno JP, Schweizer ML, McGregor JC, Perencevich EN. Economics of infection control surveillance technology: cost-effective or just cost? *Am J Infect Control* 2008; 36(3 Suppl):S12–S17.

100. Hess W, Finck W. Real-time infection protection. Using real-time surveillance data, payers and providers are averting infection, saving lives and reaping benefits. *Healthc Inform* 2007;24(8):63–64.

Chapter 12 Outbreak Investigations

Arjun Srinivasan, MD, and William R. Jarvis, MD

Epidemics or outbreaks of healthcare-associated infections (HAIs) or nosocomial infections are defined as hospital-acquired or healthcare facility–acquired infections among patients or staff that represent an increase in incidence over expected background rates. Epidemic-associated infections often are clustered temporally or geographically, suggesting that the infections are from a common source (eg, contaminated equipment or devices), are secondary to person-to-person transmission, or are associated with specific procedures. Epidemics of HAIs occur infrequently, but in some settings, such as intensive care units (ICUs), they account for a substantial percentage of HAIs and are associated with increased morbidity, mortality, and cost.[1] It has been estimated that approximately 5% of HAIs occur as part of epidemics.[2]

Recent developments in healthcare delivery have had competing influences on the detection and reporting of HAI outbreaks. Some factors have enhanced the ability of healthcare facilities to recognize HAI outbreaks. These include the increasing use of electronic data systems to track HAIs and the increasing availability of laboratory assays that can establish the relatedness of patient isolates through molecular or other genetic techniques.[3-5] Even pathogens that frequently cause endemic infection may be identified as likely to be transmitted as part of an outbreak if their relatedness is established on the basis of currently available molecular typing methods. However, other factors, including increasing regulation related to privacy of patient information, mandatory public reporting, hospital "report cards," pressure to achieve "zero HAIs," and the vulnerability of healthcare facilities to press scrutiny or legal liability may have resulted in a decline in the reporting of nosocomial epidemics.[6] Declining hospitalization rates, shorter durations of hospitalization, and the increasing shift toward outpatient medical care may reduce the likelihood of nosocomial epidemic transmission but may also reduce the recognition of epidemics.[7(p1250)]

The investigation of epidemics of HAI continues to play a critical role in the identification of new agents, reservoirs, and modes of transmission; in helping inform strategies for infection prevention and control; and in ensuring the safety of medical products and devices. Our understanding of the modes of transmission of many important pathogens, including methicillin-resistant *Staphylococcus aureus* (MRSA), extended spectrum β-lactamase (ESBL)–producing organisms, multidrug-resistant (MDR) *Mycobacterium tuberculosis,* carbapenem-resistant *Acinetobacter* or *Klebsiella* species, hepatitis C virus, and severe acute respiratory syndrome (SARS)–associated coronavirus, has been greatly advanced by outbreak investigations. Likewise, our understanding of the risks for occupational transmission of human immunodeficiency virus (HIV) is largely derived from investigation of epidemic transmission in healthcare facilities.[8,11-13] More recently, outbreaks of *Clostridium difficile* infection have led to the recognition of a new epidemic strain of that organism.[14] Outbreak investigations have also led to or enhanced the recognition of new routes of disease transmission in healthcare settings, such as via allograft tissues,[15] organ transplantation,[16] and compounded medical products,[17] and have led to important advances in patient safety. Finally, outbreak investigations continue to identify and to result in the recall of contaminated medical products and devices[18] and drugs.[19] Accordingly, the recognition and effective investigation of epidemic transmission in healthcare settings is among the most important activities of healthcare epidemiology and infection control.

Recognizing Outbreaks

Hospitals and healthcare systems need reliable, sensitive surveillance systems that allow infection control personnel to detect increased HAI rates in a defined time period and geographic area that suggest epidemic transmission. Although recognition of outbreaks often is believed to be among the major functions of surveillance, most outbreaks are recognized not as a part of routine surveillance but because of the occurrence of sentinel events. In most of those situations, recognition of a potential problem has come from the identification of an uncommon or newly recognized pathogen: for example, SARS-associated coronavirus, group A *Streptococcus,* or *Rhodococcus bronchialis.*[11,20,21] In other instances, an increased incidence of HAIs, especially infections caused by unusual organisms (eg, nontuberculous *Mycobacteria* species), infections caused by unusual strains of more common organisms (eg, carbapenem-resistant *Klebsiella pneumoniae*), or infections in hosts in whom they are seldom diagnosed, may indicate epidemic transmission.[22-38] Even if the pathogens or anatomic sites of infection are not unusual, an increased proportion of patients with infections in specific hospital units may indicate epidemic transmission. This is true particularly of units that care for highly vulnerable patients, such as ICUs, units for human immunodeficiency virus (HIV)–infected persons, and units for patients undergoing organ transplantation or hemodialysis. Similarly, information from ongoing surveillance for HAIs can detect temporal increases in HAI rates over well-established baseline rates, which often are the result of epidemic spread. Finally, anecdotal reports from alert clinicians or laboratorians about possible increases in HAI rates may indicate an outbreak and merit further investigation.

Investigating Outbreaks

The First Steps

In general, an outbreak investigations can be divided into 2 major sections, the initial investigation and the follow-up investigation, each of which has multiple components, as summarized in Table 12-1. Though investigations are divided into these steps for the purposes of teaching and explanation, outbreaks generally do not unfold in a linear or orderly manner. Accordingly, it is

---

Table 12-1. Steps of an outbreak investigation

Preliminary investigation and descriptive study

Review existing information
Determine the nature, location, and severity of the disease problem
Verify the diagnoses
Create a case definition
Find and ascertain cases
Request that the laboratory save isolates from affected patients and any suspected sources or vehicles
Graph an epidemic curve
Summarize case patient data in a line listing
Establish the existence of an outbreak
Institute or assess the adequacy of emergency control measures

Comparative study and definitive investigations

Review records of existing case patients
Develop hypotheses
Conduct comparative studies (case-control or cohort) to test hypotheses
Conduct microbiologic or other laboratory studies and surveys
Conduct observational studies, including interviews and questionnaire surveys
Conduct experiments to confirm the mode of transmission

Acting on results

Communicate the results of the investigation to the administration and the departments involved (as well as any necessary regulatory bodies), along with a plan for definitive control measures
Implement definitive control measures
Maintain surveillance for a sufficient time to ensure that control measures are effective

---

possible, if not likely, that many steps might have to occur simultaneously and be repeated multiple times in the course of the investigation. Likewise, not all of the steps might be applicable in all settings.

A variety of computer programs can provide invaluable assistance in investigating outbreaks (see Chapter 10 on medical informatics). Epi Info is one such software package that allows public health and other staff to collect and analyze data generated in their epidemiologic investigation.[39] It is available for the Microsoft Windows operating system. The program, which can be downloaded at no cost from the Centers for Disease Control and Prevention Web site,[40] enables the investigator to determine quickly, using summary data, whether an increase in the incidence of an HAI or a difference in risk between 2 groups of persons is statistically significant (Figure 12-1); to develop questionnaires; to enter data for descriptive or comparative studies; and to calculate statistical measures: odds ratios or relative risks, 95% confidence intervals, χ^2 test values, χ^2 test for trend values, and *P* values (on the use of statistical tests, see Chapter 4 on epidemiologic methods). The program

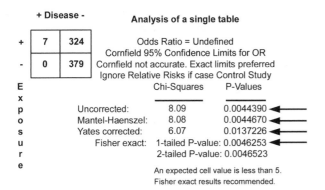

Figure 12-1. Example of analysis with Epi Info to determine whether there was a statistically significant increase in the number of *Rhodococcus bronchialis* infections during the epidemic period.[7] Using the "Statcalc" option in Epi Info "Utilities," a comparison was made between the numbers of such infections that occurred after open heart surgery procedures performed during the epidemic period (*n* = 331) and the number that occurred during the 12 months preceding the epidemic period (0 of 379 patients). The difference between the rates was significant (*P* = .0046, by the Fisher exact 2-tailed test). Data from Richet et al.[21] and Dean et al.[39]

contains training modules on how to use Epi Info to investigate epidemics. A training module for the investigation of the epidemic of *R. bronchialis* surgical site infection that is used as an example in this chapter is available on the Epi Info main screen (in the Help file; it is titled "Rhodo Tutorial").

Reviewing Existing Information and Confirming the Presence of an Outbreak

At the first suspicion of an HAI outbreak, the microbiology laboratory should be asked to save all isolates that might be part of the outbreak, in case these are needed later for further studies. It is critical to make this request quickly and early in the possible outbreak, since isolates are frequently discarded in busy clinical microbiology laboratories after results are reported to the clinician who ordered testing. In the absence of detailed information, the initial request about which isolates to save may be broad, but can be refined as more is known.

Once the infection control team suspects an outbreak, they should take steps to confirm whether it is actually occurring. Both the possibility of an outbreak of true infections and the possibility of pseudo-infections should be evaluated. Infection control personnel should review some or all medical records from putative case patients to help determine whether it is a true outbreak associated with clinical illness or a pseudo-outbreak in which there is a rise in the number of positive culture results not associated with true infection. Confirming the presence of an outbreak or pseudo-outbreak is not always straightforward, especially when dealing with pathogens that

are common causes of HAIs (eg, MRSA or *C. difficile*). When available, historical surveillance data is useful. Quick reviews of microbiologic records to assess the approximate number of cases over time also can be helpful in determining whether a reported increase in cases of HAI might represent an outbreak. If there is uncertainty, it may be necessary to embark on an investigation.

Any outbreak investigation should begin with a review of published literature. There are many thousands of published articles on HAI outbreaks available through PubMed,[41] and the information they provide will help inform investigation efforts. Another useful resource is the Worldwide Database for Nosocomial Outbreaks;[42] this is a free database that contains summary information on more than 2,000 HAI outbreaks, including information on potential sources of the HAIs and control measures that were implemented. Certain procedures, vehicles, and technical errors repeatedly are associated with outbreaks in healthcare facilities; some of these are summarized in Table 12-2. Infection control personnel will be able to investigate outbreaks more efficiently if they are aware of these associations. For example, contamination of environmental surfaces along with lapses in infection control precautions (eg, hand hygiene and contact precautions) have been implicated repeatedly in outbreaks of infection with MRSA, vancomycin-resistant enterococci (VRE), *C. difficile*, and *Acinetobacter* species. Improper use of single-dose and multiple-dose medication vials and glucometers and failure to follow other safe-injection practices have resulted in outbreaks of viral hepatitis and infection with bacterial pathogens. Likewise, improperly reprocessed hemodialyzers, inadequately processed water used to admix dialysate, and inadequacy of waste handling options have caused epidemics in hemodialysis units.

Many states require that HAI outbreaks be reported to public health officials, and some infections also must be reported to the Federal government (usually through the state health department). Infection control staff should check their local and state laws for reporting requirements. However, even in the absence of reporting requirements, it often is helpful to speak with local and/or state public health department personnel when an HAI outbreak is suspected. Not only can public health personnel provide suggestions on the structure of the investigation, but alerting them to outbreaks is critical in identifying problems that might be widespread, such as the distribution of a contaminated product or device. In addition, infection control personnel should report suspected intrinsic contamination of sterile products, fatal blood transfusion reactions, infections associated with blood or tissue products, and infections associated with defective devices to both the Centers for Disease Control and Prevention's Division of

Table 12-2. Identified causes of various types of nosocomial outbreaks, Centers for Disease Control and Prevention, 1990–2008

Outbreak	Cause	Reference(s)
Group A streptococcal surgical site infection	Dissemination by a colonized healthcare worker	Mastro et al.[20]
Tuberculosis, including with drug-resistant strains	Exposure of immunosuppressed patients to infectious tuberculosis patients in absence of adequate source or engineering controls	Jarvis[8]
SARS in healthcare workers	Hospitalization of persons with symptoms of SARS before enhanced respiratory precautions for patients and contacts were known to be necessary and implemented	Booth et al.[11]
Bloodstream infection with gram-negative organisms in patients prescribed narcotics	Use of injected narcotics by substance-dependent healthcare workers, with contamination	Ostrowsky et al.,[23] Maki et al.[24]
Bloodstream infections in patients receiving medications from single-dose multiple-dose vials	Multiple use of single-use containers with preservative-free solutions; sharing of multiple-dose containers among patients	Grohskopf et al.[25]
Pyrogenic reactions in newborns prescribed single-dose gentamicin	High pyrogen levels in injectable gentamicin	Buchholz et al.[26]
Multistate outbreaks due to a single organism	Intrinsic contamination of medications, particularly those prepared by compounding pharmacies	CDC,[17] Jhung et al.,[18] Labarca et al.[29]
Outbreaks of postoperative infections	Extrinsic contamination of equipment and medications, particularly lipid-based medications, that support rapid bacterial growth	Richet et al.,[21] Freitas et al.,[22] Wenger et al.[32]
Bacteremia and pyrogenic reactions in hemodialysis patients	Dialysis machine waste handling options; poor water quality; reprocessing of dialyzers without prior hand antisepsis and glove change	Wang et al.,[30] Jochimsen et al.[31]
Bloodborne pathogen nosocomial infections and other bloodstream infections	Multiple-dose vials and syringes used for more than one patient, and failure to follow safe injection practices; also improper use of glucose monitoring equipment	El Sayed et al.,[12] Krause et al.,[13] Thompson et al.[43]
Postoperative infections after allograft and organ transplantation	Contamination of allografts and unrecognized infections in donors	Kainer et al.,[15] Iwamoto et al.[16]

NOTE: SARS, severe acute respiratory syndrome.

Healthcare Quality Promotion (1-800-893-0485) and the Food and Drug Administration's MedWatch Program (1-800-FDA-1088).

Development of the Case Definition and Case Finding Efforts

The initial review provides the basis for a preliminary case definition. A case definition states who ("person") had the symptoms or findings, delineates a finite time period ("time") during which the symptoms began or were recognized, and specifies a location ("place") associated with the onset of symptoms. A case definition may be based on clinical, laboratory, radiologic, pathologic or other data, and the definition may change, becoming broader or more specific, as additional information

becomes available. How broad or narrow to make the case definition often depends on the frequency with which the organism or condition is encountered. Requiring microbiologic criteria for inclusion as a case will often increase the specificity of the case definition but may reduce the sensitivity. The decision to include microbiologic criteria is often driven by the pathogen involved and the type of infection. For example, investigations of outbreaks of influenza in adult care settings often do not require microbiologic confirmation of cases, given the limited sensitivity of diagnostic tests for influenza. The following statement is an example of a case definition: "A case of *R. bronchialis* surgical site infection (SSI) is defined as any Hospital A patient with a surgical site culture positive for *R. bronchialis* who had undergone open heart surgery between May 1 and December 31, 1993."[21]

Once a preliminary case definition is established, infection control personnel should develop a methodology for finding additional cases. If the case definition includes a laboratory result, laboratory records are a logical place to start and can facilitate rapid identification of possible cases. If the outbreak involves an HAI or adverse event or a multidrug-resistant pathogen for which the facility is performing surveillance, infection control and surveillance records can be useful for case-finding. Records from the radiology, pathology, or pharmacy departments might also be useful, if the infection has typical radiologic or pathologic findings or antimicrobial treatments. Finally, discussions with healthcare personnel in affected areas also can be helpful in identifying possible cases, particularly in outbreaks where the case definition is primarily clinical (ie, not based on the pathogen).

Another issue to consider in outbreaks for many healthcare-associated pathogens, is whether there might be additional patients who are only colonized with the outbreak pathogen. In those circumstances, examining only clinical culture results will underestimate the number of cases and could compromise control efforts, if the colonized patients continue to serve as a reservoir for transmission. Accordingly, surveillance culturing might be needed to identify additional cases. However, the benefits of performing surveillance cultures must be weighed against the resources required. One option that can be useful is the performance of a single round of surveillance cultures, sometimes referred to as a "point prevalence survey," which can help assess the scope of the problem and can help determine whether ongoing performance of surveillance cultures will be needed. Often such an approach uses a "concentric circle" method: that is, perform cultures for all patients in one area of the facility; if some culture results are positive, move to the next area; if all results are negative, no further surveillance cultures are necessary.

Creation of the Epidemic Curve and Line Listing

After identifying suspected case patients, infection control personnel can chart an epidemic curve. The shape of the epidemic curve may suggest the possible source and mode of transmission of the etiologic agent(s) (Figure 12-2). However, there are important caveats to interpreting epidemic curves in HAI outbreaks. First, patients often become colonized with organisms well before they develop clinical infection, and some patients will not develop infection at all. Hence, the "incubation period" suggested by the epidemic curve is often misleading, for many HAI pathogens. Second, exposures in healthcare settings are often ongoing, and organisms may be transmitted both from patient to patient and

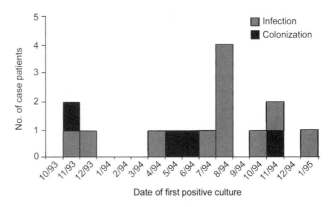

Figure 12-2. Example of an epidemic curve for a nosocomial outbreak, showing the distribution of case patients according to the date of the first culture positive for *Malassezia pachydermatis* during the period from October 1993 through January 1995. Reproduced with permission from Chang et al.[38]

from a common contaminated source. Hence, the shape of the curve in a point-source outbreak of an HAI might look very different from that seen in a point-source outbreak of a foodborne disease. Such merging of modes of transmission is more likely to occur the longer the outbreak has been ongoing.

Information on the number of cases also can be used to calculate rates of infection during the pre-outbreak and outbreak periods, also known as "attack rates" (Figure 12-1). Comparison of rates during the pre-outbreak and outbreak periods can determine if the increase is statistically significant; however, it is important to note that even increases that are not statistically significant (ie, those representing endemic infection) may have clinical implications and hence may require further investigation.

Once cases have been identified, infection control personnel should review the medical records to produce a line listing. This enumerates all affected patients and displays characteristics of the patients that may be important to the investigation; it is critical for characterizing the outbreak, assessing its extent, and generating hypotheses (Table 12-3). The line listing, arguably, is the single most important tool in any outbreak investigation and hence merits considerable early discussion and effort. In general, information that can be helpful on a line listing can include the following: details on each patient's signs or symptoms (if there is the possibility that it is a pseudo-outbreak), demographic data, underlying diseases, medications received, procedures undergone, location information, history of contact with healthcare personnel, and host factors that might have predisposed the patient to the adverse event under investigation. While the line listing is a powerful tool in guiding an investigation, developing it is a resource intensive

Table 12-3. Line listing of characteristics of case patients in the *Rhodococcus bronchialis* surgical site infection investigation, using the Epi Info 2002 command "List."

Line	Age	Sex	Smoking	CABG	Dr A	Dr B	CN A	CN B
1	66	M	Yes	Yes	Yes	Yes	Yes	No
2	60	M	Yes	Yes	Yes	No	Yes	No
3	53	M	Yes	Yes	Yes	Yes	Yes	Yes
4	56	M	Yes	Yes	Yes	Yes	Yes	Yes
5	65	M	No	Yes	Yes	No	Yes	No
6	57	M	No	Yes	Yes	No	Yes	No
7	55	M	No	Yes	Yes	No	Yes	No

NOTE: This line listing shows that all case patients were men between the ages of 53 and 66 years, all had received a coronary artery bypass graft (CABG), and in all cases, Dr. A and circulating nurse (CN) A participated in the procedure. Data from Richet et al.[21] and Dean et al.[39]

activity, because it involves a review of a variety of different sources of information, which might include medical records, patient location information (ie, admission, discharge, and transfer data), and staff interviews. Hence, it is critical to carefully weigh the benefits of any information to be included on the line listing against the resources required to obtain it. One option is to create an initial simple line listing with some very basic information on patient demographic characteristics, underlying diseases, and potential exposures (such as invasive procedures and hospital locations). This type of limited line listing can be useful in helping focus subsequent investigative efforts, if many of the patients do have a common exposure. However, it is important to remember that such a preliminary line listing can sometimes be misleading, because not every case patient might have been exposed to the common source, and some exposures might only be associated with cases and not with the actual source of the outbreak. As with any part of an outbreak investigation, it is important to continue to reassess the information on the line listing in the context of all of the other information being gathered.

In the *R. bronchialis* outbreak we are using as an example, the line listing was helpful in generating hypotheses that could be tested quickly in a comparative epidemiologic study.[21] All patients that met the case definition were men, and surgeon A and circulating nurse A were involved in all of their surgical procedures, which were all coronary artery bypass graft procedures. Among the hypotheses tested was the possibility that procedures to clip or shave chest hair (a procedure performed preferentially on men) could be implicated. Of course, the pos-

sibilities that surgeon A or circulating nurse A were involved also were selected to be tested.

Infection Control Observations

In most outbreak investigations, it is the observations of practices that ultimately identify the cause. The line listing is critical in helping guide both the type and location of observations that will need to be done. Because impressions and recollections change rapidly during times of stress, infection control personnel should interview staff and review procedures immediately after they recognize a potential epidemic. For infectious disease outbreaks, the type of pathogen and infection being investigated also can be important factors in determining the types of observations and reviews that should be performed. For example, investigations of outbreaks of *Aspergillus* infection should generally include careful review and observations of construction activities in or near patient areas. Likewise, clusters of surgical site infection will require a careful review of preoperative, intraoperative, and postoperative care. Initial observations should generally be free-form (ie, performed without a detailed observation form) and should focus on practice patterns and workflow that deviate from good infection control practices and facility or unit policies. Infection control staff also should review written protocols, interview supervisory staff, and identify procedural changes implemented before, during, and/or after the epidemic. Investigators should observe procedures that are implicated and question personnel who perform these techniques directly. Do not believe what anyone tells you; see it yourself. You may gather valuable information by observing different personnel on the same and different shifts as they perform procedures. In the *R. bronchialis* epidemic, a particularly striking finding was that the hands of circulating nurse A would become wet during the clotting-time procedure.[21] Semistructured interviews that pose similar questions to all staff members may be effective in identifying procedures that are being undertaken in different ways by specific staff members. More-detailed and focused observation tools can be developed, if needed, but should be informed by the free-form observations. In addition to specific practices, observations also should review compliance with general infection control practices, such as hand hygiene and isolation precautions. In addition to helping delineate the potential causes of outbreaks, these observations also can provide useful teachable moments in infection prevention.

Potential sources or vehicles for the outbreak may be identified during the observations and interviews. Whenever possible, these should be promptly removed from clinical areas to prevent healthcare workers from using them and possibly causing additional cases or from discarding them and preventing confirmation that they are a source of the outbreak.

Immediate Actions

Because the primary goal of any outbreak investigation is to terminate the outbreak, it not only acceptable but also important to implement a variety of infection control measures throughout the course of the investigation. Some immediate recommendations often can be made on the basis of the line listing and/or initial observations and interviews. For example, a strong association with a particular type of procedure or observations of infection control breaches during the procedure might lead to immediate alterations in the manner or facility location in which the procedure is performed, or even to a temporary cessation of the procedure. It is always appropriate to reinforce education on compliance with general infection control recommendations during any outbreak. In addition to making these initial recommendations, it is vital to develop a plan to ensure compliance with them.

One of the most difficult immediate decisions facing infection control staff will be a potential decision to temporarily close a ward to new admissions or halt a specific procedure. There are no laws or regulations dictating when wards should be closed or procedures halted; however, some general principles can be applied. Staff must consider the severity of the illness, the size of the outbreak, and the rate at which new cases are occurring. They must also weigh carefully the benefit of closing a ward or halting a procedure against the risk of decreased access to care. In general, a healthcare facility should close wards or halt procedures when an outbreak is causing severe disease and continues despite implementation of initial infection control precautions. Before deciding to close the ward or discontinue specific procedures, the infection control staff should determine what criteria must be met before the ward can reopen or suspended activities can be resumed.

Conducting an Epidemiologic Study

Reviewing the Line Listing

Before conducting a comprehensive epidemiologic study, infection control personnel should review the line listing (Table 12-3), the epidemic curve (Figure 12-2), and information obtained from interviews and observations, because these tools may suggest the cause of the outbreak. This is true particularly if the problem is an acute, self-limited, one-time incident, such as recognized or suspected contamination. In many instances, sources are

identified and outbreaks terminated without the performance of an analytic study. Infection control staff should consider several factors, including what resources are available, before they decide to conduct a comparative study.

However, there are some situations in which analytic studies can be especially useful and should be considered. First, analytic studies can often help guide further investigations and suggest new avenues for exploration in situations where the source of an outbreak remains unclear and control measures have been ineffective. Second, they might be useful in convincing clinicians that the proposed source or mechanism suggested by chart review and observations is indeed correct. This can be particularly helpful when environmental cultures do not or cannot confirm the source and when the proposed intervention(s) to address the source are resource intensive. Finally, analytic studies are powerful teaching tools and might be undertaken to further both the medical knowledge about HAI outbreaks and the educational experience of trainees in healthcare epidemiology, infection control, and/or public health.

Comparative Studies

The first step in any analytic study is the gathering of relevant data. The line listing is an important place to begin this process. Infection control personnel usually must then review the patients' medical records to determine if there are additional exposures that might be important. In some instances, investigators may learn important information from short, open-ended interviews with patients. However, hospitalized patients usually do not know or recall details essential to the investigation and may be too ill to answer questions. In general, the medical and laboratory records and other documents provide the most information (Table 12-4). Infection control personnel should design a standardized form on which to collect demographic data, information about exposures, and other data regarding the study subjects. The form ensures that personnel will collect data uniformly and that they will not need to review records repeatedly to find missed items.

In HAI outbreaks, the case-control study and, less commonly, the cohort study are the most frequently employed analytic tools in investigating the statistical association between various risk factors and case status. The relative advantages and disadvantages of these study designs are described in detail in Chapter 4 (on epidemiologic methods). Case-control studies are typically faster and easier to conduct than cohort studies.[44,45]

Each variable that is evaluated as a possible risk factor will increase the time and effort required for the analysis. Furthermore, each additional variable increases the

Table 12-4. Sources of information for an outbreak investigation.

Log books

Operating or delivery room
Emergency department
Nursing unit
Intensive care unit (census or admission and discharge log books)
Procedure room

Microbiologic record
Employee health records
Infection control surveillance data

Patient medical records

Clinical notes
Operative notes
Pathology reports
Microbiology and other laboratory reports
Radiology procedure notes and records

Pharmacy records
Hospital billing records
Central-service records
Purchasing records

likelihood that a characteristic entirely unrelated to the outbreak will appear to be a risk factor (ie, because it is statistically significant by chance alone). To avoid these pitfalls, infection control personnel should include few, if any, characteristics that are not plausible risk factors. Obviously, a narrow interpretation of biologic plausibility may inappropriately restrict the investigation to only previously suspected or confirmed risk factors. Such inappropriate restriction may be avoided by ensuring during the early phase of the studies (particularly in the design of the line listing) that a careful review of cases casts a very wide net for hypothesis generation but that the data collection and analysis for the comparative study are more limited. Risk factors to examine in an analytic study should be those that have been reported in the past and/or have clear biologic plausibility. It is advisable to avoid including characteristics that would be "interesting" but are unlikely, according to the literature, the line listing, and expert advice, to be related to the outbreak. Likewise, given that the primary goal of an outbreak investigation is to terminate the outbreak, it is often preferable to focus on risk factors that might be modifiable.

In the *R. bronchialis* epidemic that is our example, it was noted in the line listing that smoking was reported by only 4 case patients (Table 12-3). The likelihood that smoking was associated with a statistically significant increased risk of being a case patient was thus low, as was the possibility that it significantly decreased the risk of surgical site infection (ie, was protective). The biologic plausibility of a protective role for cigarette smoking also

was extremely doubtful. Therefore, it was appropriate to not pursue this variable.

In addition to deciding which risk factors to analyze, infection control personnel should consider carefully whether to perform a matched study. It should be remembered that variables on which case patients and control patients are matched cannot be evaluated as potential risk factors.[44] Furthermore, matching may make control and case patients so similar that the investigators would miss all but the most obvious risk factors. The analysis of a study that is designed as a matched study has to be conducted in a matched fashion.[44] Given that HAI outbreaks generally involve a small sample size, a matched analysis may reduce the likelihood of finding statistically significant associations. However, matching is useful if there are numerous characteristics that are associated with the epidemic condition but are not the real cause(s)—that is, that are confounders. These confounders can obscure the real causative factors. Matching also may be useful in selected investigations where a considerable amount of risk is experienced by a specific subpopulation.

Many investigators choose not to perform a matched analysis but to control for confounding variables either by stratifying the analysis by possible confounders or by using multivariate analysis. Another alternative to matching is to restrict analysis to the subpopulation at highest risk, if patients in this risk category alone appeared to be among the case patients (for example, infants with birth weight of more than 1,300 g).[38]

Because HAI outbreaks generally involve a small number of cases, the causative factor may not achieve statistical significance. Hence, infection control personnel should explore any factor with an odds ratio or relative risk that suggests an association, even if the difference between case and control patients only approaches, but does not achieve, statistical significance.

Performance of Environmental and Personnel Cultures

Environmental and/or personnel cultures can be powerful tools in confirming the source of HAI outbreaks, particularly if environmental isolates can be compared with case-patient isolates using molecular typing. However, more often than not, culture results are negative, for a variety of reasons that raise important and challenging questions. Was it because the item or person sampled for culture is actually not the source of the outbreak, or because the implicated organism was present before but not at the time the sample for culture was obtained? Or perhaps because the wrong part of the item was sampled? Or because the technique used was not sensitive enough to detect the contamination? A variety of other factors also may contribute to culture results being negative. For example, only a small proportion of the individual units of an intrinsically contaminated commercially prepared medication may be contaminated. Similarly, individuals who disseminate the epidemic strain may shed the organism only intermittently (eg, when their hands are wet[21]), may be colonized intermittently (eg, if they have a viral respiratory infection), or may be colonized in an unusual site (eg, the rectum or scalp[20] but not the hands). Even when positive, environmental and personnel culture results can be difficult to interpret, because organisms that cause HAI outbreaks (eg, gram-negative waterborne organisms; fungi, including *Aspergillus* species; and gram-positive cocci, including *S. aureus*) frequently can be isolated from nonsterile environmental sources or from healthcare workers.

There also are important methodological challenges in both obtaining and processing environmental samples. For example, the swabs used in many facilities to sample surfaces for culture can only be used on small surface areas (generally an area approximately 2.5 x 2.5 cm, or 1 square inch). Also, some environmental pathogens, particularly waterborne organisms, have adapted to survive in very low-nutrient conditions and require special media to grow in the microbiology laboratory. Finally, the yield of surface cultures may be limited by residual disinfectants, which must be neutralized before the sample is processed. Infection control personnel can save time and resources if they consult with a microbiologist before they collect specimens or consume a limited supply of the suspected vehicle.

Given the challenges in interpreting the results of environmental and personnel cultures, sampling should be directed at sources for which there is epidemiologic data from the line listing, observations, or analytic studies linking them to the outbreak. Performing cultures at random will increase the cost of an investigation substantially and may fail to identify the source or implicate the wrong source. Observations also can be critical in identifying the appropriate times and sites to sample for culture. In our example outbreak of *R. bronchialis* infection, circulating nurse A's hands were culture-positive for the outbreak strain of *R. bronchialis* only when they were wet; culture of specimens from healthcare workers' hands obtained without the epidemiologic and observational study data would probably have missed her link to the epidemic[21] (Table 12-5 and Table 12-6). Cultures of specimens from nonsterile areas (eg, floors and walls) that do not have plausible connections to the outbreak waste valuable resources and frequently yield uninterpretable data, because there are no standard limits or acceptable levels of contamination of surfaces, air, and nonsterile equipment in healthcare facilities.

The isolation of the outbreak organism in a potential vehicle or disseminator that is implicated in the epidemiologic study and not in other sources or staff members is a very compelling finding, suggesting that an implicated source is, in fact, causing the outbreak. In the investigation of the *R. bronchialis* sternal wound infection outbreak, the epidemic strain was isolated from the hands of circulating nurse A only when they were wet, and it was found during observational studies using bile esculin that her technique for warming blood specimens in the water bath was associated with contamination of sterile surgical instruments with microdroplets of water; these findings strongly supported the hypothesis that she was implicated in the outbreak.[21] On the other hand, infection control personnel should not abandon their hypothesis if cultures do not yield the outbreak organism from a source or reservoir that was implicated strongly in the epidemiologic study or by observations. Indeed, this is more often the case, though investigations in which this occurs frequently do not get published in the literature.

Table 12-5. Example analysis from the case-control study data for the epidemic of *Rhodococcus bronchialis* surgical site infection: comparison of the duration of the procedures demonstrates that it did not differ significantly between case and control patients, suggesting that airborne transmission of suspended microorganisms did not play a major role in the outbreak

Patient group	n	Total	Mean	Variance	SD	P
		Duration of procedures, minutes				
Cases	7	1785.00	255.00	4214.67	64.92	.84
Controls	28	7381.00	263.61	9886.10	99.43	

NOTE: Data from Richet et al.[21] and Dean et al.[39] SD, standard deviation.

Table 12-6. Example analysis from the case-control study data for the epidemic of *Rhodococcus bronchialis* surgical site infection: a 2 x 2 table demonstrates that case patients were significantly more likely to have had circulating nurse A (CN-A) involved in the procedure (*P* <.001, by Fisher exact test)

CN-A involved	Case patient		
	Yes	No	Total
Yes	7	6	13
No	0	22	22
Total	7	28	35

NOTE: Data are no. of patients. Data from Richet et al.[21] and Dean et al.[39]

If environmental or personnel cultures do yield the outbreak organism, infection control personnel may want to confirm that outbreak-associated isolates are genetically related to each other and to those recovered from the implicated source. Molecular typing is available and widely used for evaluating the relatedness of strains for a wide variety of organisms, including fungi, viruses, and bacteria (see Chapter 9 on the use of molecular typing).[3-5,8,12,13,20-25,28,34] An example of such an analysis using pulsed-field gel electrophoresis is shown in Figure 12-3.

Further Investigations to Confirm Hypotheses

In some instances, additional laboratory and other investigations may be useful in confirming the source of an outbreak. For example, sampling the air around a staff member may determine which circumstances and activities are associated with disseminating the pathogen.[20]

In outbreaks of infection due to an airborne pathogen, such as *M. tuberculosis* or *Aspergillus niger*, infection control personnel may need to use chemical smoke tubes to evaluate the direction of air flow.[8-10]

In other instances, novel investigative methods can be useful in identifying previously unrecognized mechanisms of transmission. For example, in the investigation of the *R. bronchialis* epidemic, the ingenious use of bile esculin in the water-bath showed how microdroplets of water from circulating nurse A's wet hands contaminated the sterile equipment; these droplets, invisible to the naked eye, fluoresced brightly in the presence of ultraviolet light.[21]

Implementing Control Efforts

Infection control personnel should focus their interventions on the immediate cause of an outbreak and should

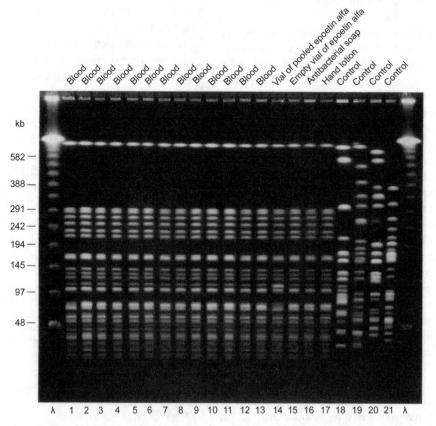

Figure 12-3. Example of analysis of the relatedness of isolates by means of pulsed-field gel electrophoresis (PFGE). The PFGE gel shows banding patterns for *Serratia liquefaciens* isolates from an outbreak of bloodstream infection in which *S. liquefaciens* infection was epidemiologically linked with receipt of epoetin alfa in a hemodialysis center. Band patterns are identical for isolates from case patients, from full and empty vials of pooled epoetin alfa, from antibacterial soap, and from hand lotion in the hemodialysis center, but band patterns for isolates from other centers in the state that were not associated with the outbreak were different from each other and from the outbreak strain. Reproduced with permission from Grohskopf et al.[25]

institute the simplest measures that will correct the problem. The more focused the control measures, the more feasible their implementation and the more likely healthcare workers will be to adhere to the measures. Infection control personnel should emphasize the specific measures required to stop the outbreak and also encourage staff to comply with routine procedures. Sometimes, during an investigation, infection control personnel will identify deficiencies other than those directly associated with the transmission of the epidemic strain and may be tempted to use the epidemic as an opportunity to revamp the entire infection control program. This must be done with caution, as it can distract staff from the implementation of essential programs to control the current outbreak. However, outbreak investigations can provide important insights into weaknesses in a facility's infection control policies and procedures that should be addressed following the termination of the outbreak. Healthcare workers and administrators often are more receptive to changes in practice at the time of an outbreak than they are in non-outbreak circumstances.

Infection control personnel should develop a plan and a time line for implementing the control measures. After implementing the control measures, they should continue to work closely with the staff in the affected area to ensure that they understand and efficiently implement the recommendations and that they continue to comply with the recommendations over time. Obviously, infection control personnel must determine whether the measures are effective, and, ideally, should assess whether the decline in the number of cases of the epidemic condition was associated with implementation of one or more measures. Often, staff can prove that the interventions were effective by demonstrating that no new cases occurred after the control measures were implemented, even though patients continued to be at risk. Once an outbreak is controlled, it is important to review the control measures that were implemented to determine which measures, if any, might not be necessary any longer.

Other Considerations in Outbreak Investigations

General Considerations in Outbreak Investigations

Outbreaks are a considerable source of stress for healthcare providers, administrators, and patients. It is natural for staff to be defensive and wary of investigations that may be seen as attempts to blame them for an outbreak. Hence, it is vital that healthcare personnel understand that the investigation is a collaboration with the health-care epidemiology personnel and not an attempt to affix blame. Techniques that are seen clearly as fact-finding and that guarantee confidentiality obviously are far more effective than interviews that appear to target specific tasks or staff members. Particularly effective strategies include questionnaire surveys of all staff in which open-ended questions are administered face-to-face by a neutral party. It is vital that all activities related to the investigation be conducted in a neutral and supportive manner, respecting the right to privacy of staff members and patients. A healthcare facility culture that focuses on systemic changes to protect patients and healthcare facility staff is clearly preferable to a culture that appears to focus primarily on blame (see Chapter 30 on administering an infection prevention program). Infection control staff must strive to create this culture of safety in the context of the outbreak investigation and emphasize that the investigation is designed to improve the healthcare systems that might be the cause. The infection prevention and control department should be the strongest advocates for patient safety.

Medico-Legal Concerns

Staff in healthcare facilities often are concerned about the protection of the privacy of living case patients, their families, and healthcare staff, particularly since the passage of the Health Insurance Portability and Accountability Act (HIPAA). It is important to note that HIPAA does allow facilities to share protected health information with public health officials for the purposes of public health investigations. Likewise, facility personnel often are concerned about the legal implications of sharing information on outbreaks with health departments, providers, and the public. There is no doubt that outbreaks in healthcare facilities have resulted and will continue to result in litigation. However, facilities that are proactive in reporting and investigating outbreaks and that openly share information often find not only that they benefit from the assistance that they receive but also that they are in a better position with respect to lawsuits that might arise. Nonetheless, lawsuits are a reality of health care; infection control staff should be aware of this and take steps to be prepared for legal action, should it arise. A critical step is to alert the facility's risk-management staff when an outbreak is suspected. Risk managers can be critical partners in working both to prevent lawsuits from being filed and to advise staff on how to prepare for potential legal inquiries. Another action that infection control staff should take is to keep a detailed outline of the events that occurred and the actions that were taken in the investigation. Lawsuits are often filed long after outbreaks have occurred, and having a detailed time line can be critical in reconstructing past events.

Communications

Given the stress engendered by HAI outbreaks, communication with various groups of people is vital during investigations. Patients, healthcare workers, and health department officials will all want regular updates on the status of the outbreak and the investigation efforts. Likewise, media may become interested if they learn of the outbreak. To optimize communications during an outbreak, healthcare facilities should identify a spokesperson who will update appropriate internal and external constituencies regularly. The spokesperson should present enough data to assure these constituencies that personnel are investigating the problem thoroughly and carefully. However, the spokesperson should not divulge prematurely the potential hypotheses being tested or offer premature reassurances or predictions that may under-estimate or over-estimate the threat posed by the suspected epidemic. Importantly, spokespersons should not conjecture about the cause of the outbreak but rather report the cause only after it has been identified by the investigative team. The credibility and effectiveness of spokespersons are ensured by frank explanations of the process that is being undertaken to explore the event and by admission that the source of the epidemic is, at the time, unclear. Conversely, confidence and credibility are likely to be irretrievably lost if there are premature assertions that the cause of the epidemic has been identified or that the outbreak is under control.

Conclusion

Most HAIs are endemic, and epidemics are relatively infrequent. However, when they occur, outbreaks can cause substantial morbidity and mortality and can increase the cost of medical care significantly. Infection control personnel who identify outbreaks quickly and investigate them thoroughly and systematically using optimal epidemiologic and laboratory methods can improve medical care, advance medical knowledge, and ensure patient safety.

References

1. Wenzel RP, Thompson RL, Landry SM, et al. Hospital-acquired infections in intensive care unit patients: an overview with emphasis on epidemics. *Infect Control* 1983;4:371–375.
2. Doebbeling BN. Epidemics: identification and management. In: Wenzel RP, ed. *Prevention and Control of Nosocomial Infections*. 2nd ed. Baltimore, MD: Williams & Wilkins; 1992:177–206.
3. Jarvis WR. Usefulness of molecular epidemiology for outbreak investigations. *Infect Control Hosp Epidemiol* 1995;15(7): 500-503.
4. Grundmann H, Schneider C, Daschner FD. Fluorescence-based DNA fingerprinting elucidates nosocomial transmission of phenotypically variable *Pseudomonas aeruginosa* in intensive care units. *Eur J Clin Microbiol Infect Dis* 1995;14: 1057–1062.
5. Llovo J, Mateo E, Munoz A, Urquijo M, On SL, Fernandez-Astorga A. Molecular typing of *Campylobacter jejuni* isolates involved in a neonatal outbreak indicates nosocomial transmission. *J Clin Microbiol* 2003;41(8):3926–3928.
6. Vermund SH, Fawal H. Emerging infectious diseases and professional integrity: thoughts for the new millennium. *Am J Infect Control* 1999;27:497–499.
7. Popovic JR, Hall MJ. 1999 National Hospital Discharge Survey: Advance Data. 2001. Publication 319.
8. Jarvis WR. Nosocomial transmission of multidrug-resistant *Mycobacterium tuberculosis*. *Am J Infect Control* 1995; 23:146–151.
9. Maloney SA, Pearson ML, Gordon MT, Del Castillo R, Boyle JF, Jarvis WR. Efficacy of control measures in preventing nosocomial transmission of multidrug-resistant tuberculosis to patients and health care workers. *Ann Intern Med* 1995; 122:90–95.
10. Manangan LP, Bennett CL, Tablan N, et al. Nosocomial tuberculosis prevention measures among two groups of US hospitals, 1992 to 1996. *Chest* 2000;117:380–384.
11. Booth CM, Matukas LM, Tomlinson GA, et al. Clinical features and short-term outcomes of 144 patients with SARS in the greater Toronto area. *JAMA* 2003;289:2801–2809.
12. El Sayed NM, Gomatos PJ, Beck-Sague CM, et al. Epidemic transmission of human immunodeficiency virus in renal dialysis centers in Egypt. *J Infect Dis* 2000;181:91–97.
13. Krause G, Trepka MJ, Whisenhunt RS, et al. Nosocomial transmission of hepatitis C virus associated with the use of multidose saline vials. *Infect Control Hosp Epidemiol* 2003; 24:122–127.
14. McDonald LC, Killgore GE, Thompson A, et al. An epidemic, toxin gene-variant strain of *Clostridium difficile*. *NEJM* 2005;353(23):2433–2441.
15. Kainer MA, Linden JV, Whaley DN. *Clostridium* infections associated with musculoskeletal-tissue allografts. *NEJM* 2004;350(25):2564–2571.
16. Iwamoto M, Jernigan DB, Guasch A. Transmission of West Nile virus from an organ donor to four transplant recipients. *NEJM* 2003;348(22):2196–2203.
17. Centers for Disease Control and Prevention. *Exophiala* infection from contaminated injectable steroids prepared by a compounding pharmacy—United States, July–November 2002. *MMWR Morb Mortal Wkly Rep* 2002;51(49): 1109–1112.
18. Jhung MA, Sunenshine RH, Noble-Wang J. A national outbreak of *Ralstonia mannitolilytica* associated with use of a contaminated oxygen-delivery device among pediatric patients. *Pediatrics* 2007;119(6):1061–1068.
19. Blossom DB, Kallen AJ, Patel PR. Outbreak of adverse reactions associated with contaminated heparin. *NEJM* 2008; 359(25):2674–2684.
20. Mastro TD, Farley TA, Elliott JA, et al. An outbreak of surgical-wound infections due to group A *Streptococcus* carried on the scalp. *NEJM* 1990;323:968–972.
21. Richet HM, Craven PC, Brown JM, et al. A cluster of *Rhodococcus (Gordona) bronchialis* sternal-wound infections after coronary-artery bypass surgery. *NEJM* 1991;324:104–109.

22. Freitas D, Alvarenga L, Sampaio J, et al. An outbreak of *Mycobacterium chelonae* infection after LASIK. *Ophthalmology* 2003;110(2):276–285.

23. Ostrowsky BE, Whitener C, Bredenberg HK, et al. *Serratia marcescens* bacteremia traced to an infused narcotic. *NEJM* 2002;346(20):1529–1537.

24. Maki DG, Klein BS, McCormick RD, et al. Nosocomial *Pseudomonas pickettii* bacteremias traced to narcotic tampering: a case for selective drug screening of health care personnel. *JAMA* 1991;265:981–6.

25. Grohskopf LA, Roth VR, Feikin DR, et al. *Serratia liquefaciens* bloodstream infections from contamination of epoetin alfa at a hemodialysis center. *NEJM* 2001;344:1491–1497.

26. Buchholz U, Richards C, Murthy R, et al. Pyrogenic reactions associated with single daily dosing of intravenous gentamicin. *Infect Control Hosp Epidemiol* 2000;21(12):771–774.

27. Duffy R, Tomashek K, Spangenberg M, et al. Multistate outbreak of hemolysis in hemodialysis patients traced to faulty blood tubing sets. *Kidney Int* 2000;57:1668–1674.

28. Wang SA, Tokars JI, Bianchine PJ, et al. *Enterobacter cloacae* bloodstream infections traced to contaminated human albumin. *Clin Infect Dis* 2000;30:35–40.

29. Labarca JA, Trick WE, Peterson CL. A multistate nosocomial outbreak of *Ralstonia pickettii* colonization associated with an intrinsically contaminated respiratory care solution. *Clin Infect Dis* 1999;29(5):1281–1286.

30. Wang SA, Levine RB, Carson LA, et al. An outbreak of gram-negative bacteremia in hemodialysis patients traced to hemodialysis machine waste drain ports. *Infect Control Hosp Epidemiol* 1999;20:746–751.

31. Jochimsen EM, Frenette C, Delorme M, et al. A cluster of bloodstream infections and pyrogenic reactions among hemodialysis patients traced to dialysis machine waste-handling option units. *Am J Nephrol* 1998;18:485–9.

32. Wenger PN, Brown JM, McNeil M, Jarvis WR. *Nocardia farcinica* sternotomy site infections in patients following open heart surgery. *J Infect Dis* 1998;178:1539–1543.

33. Mangram AJ, Archibald LK, Hupert M, et al. Outbreak of sterile peritonitis among continuous cycling peritoneal dialysis patients. *Kidney Int* 1998;54:1367–1371.

34. Welbel SF, McNeil MM, Kuykendall RJ, et al. *Candida parapsilosis* bloodstream infections in neonatal intensive care unit patients: epidemiologic and laboratory confirmation of a common source outbreak. *Pediatr Infect Dis J* 1996;15:998–1002.

35. Saiman L, Ludington E, Dawson JD, et al. The National Epidemiology of Mycoses Study Group. Risk factors for *Candida* species colonization of neonatal intensive care unit patients. *Pediatr Infect Dis J* 2001;20:1119–1124.

36. Dent A, Toltzis P. Descriptive and molecular epidemiology of gram-negative bacilli infections in the neonatal intensive care unit. *Curr Opin Infect Dis* 2003;16:279–283.

37. Schuchat A, Zywicki SS, Dinsmoor MJ, et al. Risk factors and opportunities for prevention of early-onset neonatal sepsis: a multicenter case-control study. *Pediatrics* 2000;105(1 Pt 1):21–26.

38. Chang HJ, Miller HL, Watkins N, et al. An epidemic of *Malassezia pachydermatis* in an intensive care nursery associated with colonization of health care workers' pet dogs. *NEJM* 1998;338:706–711.

39. Dean AG, Arner TG, Sangam S, et al. Epi Info 2000, a database and statistics program for public health professionals for use on Windows 95, 98, NT, and 2000 computers. Atlanta, GA: Centers for Disease Control and Prevention; 2000.

40. Epi Info. Centers for Disease Control and Prevention Web site. http://www.cdc.gov/epiinfo/. Accessed July 10, 2009.

41. PubMed Central Web site. http://www.ncbi.nlm.nih.gov/sites/entrez?db=pmc. Accessed July 10, 2009.

42. Worldwide Database for Nosocomial Outbreaks Web site. http://outbreak-database.charite.de/Outbreak/Home.aspx. Accessed July 10, 2009.

43. Thompson ND, Perz JF, Moorman AC, Holmberg SD. Nonhospital health care-associated hepatitis B and C virus transmission: United States, 1998–2008. *Ann Intern Med* 2009;150(1):33–39.

44. Rotham KJ. Types of epidemiologic study. In: Rotham KJ, ed. *Modern Epidemiology*. Boston: Little, Brown and Co; 1986:51–76.

45. Schlesselman JJ. *Case Control Studies*. New York: Oxford University Press; 1982.

Chapter 13 Urinary Tract Infection

Eric Cober, MD, Emily K.Shuman, MD, and
Carol E. Chenoweth, MD

Urinary tract infections (UTIs) account for up to 40% of healthcare-associated infections. The majority of these infections develop in patients with indwelling urinary catheters. Urinary catheters disrupt the normal host immune mechanisms and allow for the formation of biofilm, thereby affecting the frequency of bacterial colonization and the etiologic organisms found in catheter-associated UTI (CAUTI). These factors have important implications for treatment and prevention of UTI in the catheterized patient.

Pathogenesis

The human urinary tract has multiple natural defense mechanisms that prevent attachment of potential pathogens to the uroepithelium, including the length of the urethra, micturition, and urine flow.[1,2] The urinary tract mucosa has antibacterial properties and secretes inhibitors of bacterial adhesion (ie, Tamm-Horsfall proteins and bladder mucopolysaccharides) that prevent attachment of pathogens.[2] Urine osmolality and pH and organic acids inhibit growth of most microorganisms. The use of a urinary catheter interferes with these normal defenses and allows colonization and attachment of microorganisms.

The vast majority of organisms associated with CAUTI enter the bladder by ascending the urethra from the perineum. Rarely, organisms such as *Staphylococcus aureus* cause upper tract infection through hematogenous spread. In the presence of a catheter, organisms ascend into the bladder in one of two ways. First, organisms may enter through extraluminal migration in the mucous film surrounding the external aspect of the catheter. Organisms entering by this route are primarily endogenous organisms, originating from the rectum and colonizing the patient's perineum.[3] Approximately 70% of episodes of bacteriuria among catheterized women are believed to involve an extraluminal route.[2] The second route of entry into the bladder is through intraluminal reflux or migration, which occurs when organisms gain access to the internal lumen of the catheter through failure of a closed drainage system.[2,3] Most of these organisms are exogenous and result from cross-transmission of organisms introduced from the hands of healthcare personnel.[2,3]

Tambyah and colleagues[3] performed a prospective study to determine the probable route by which microorganisms gained access to the catheterized bladder. Serial paired quantitative cultures of the specimen port and the collection bag were performed. Of 173 catheter-related UTIs, 115 (66%) were thought to be acquired through extraluminal migration of organisms ascending from the perineum along the external surface of the catheter. A smaller proportion of infections (34%) were acquired from intraluminal contamination of the collection system.

While most UTIs due to Enterobacteriaceae are thought to originate from an endogenous source, microorganisms causing nosocomial UTI may be transmitted from one patient to another in an institution. An estimated 15% of episodes of healthcare-associated bacteriuria occur in clusters, often involving highly antimicrobial-resistant organisms.[2,4] Most hospital-based outbreaks have been associated with lack of proper hand hygiene by healthcare personnel. Despite these occasional clusters, most cases of nosocomial UTI are associated with a patient's own endogenous organisms.

The formation of biofilm on the inner and outer surfaces of urinary catheters has important implications for prevention and treatment of CAUTIs. Adhesion of microorganisms to catheter materials is dependent on the hydrophobic nature of organisms and the catheter surface. Once microorganisms attach and multiply, this sheet of organisms secretes an extracellular matrix of glycocalyces, imbedding the microorganisms. Organisms in the biofilm grow slower than planktonic bacteria growing within the urine itself, and yet microorganisms located within biofilm may ascend the inner surface of the catheter in 1–3 days. Some organisms in the biofilm, such as *Proteus* species, *Pseudomonas aeruginosa*, *Klebsiella pneumoniae*, and *Providencia* species, have the ability to hydrolyze urea in the urine to free ammonia. The resulting increase in pH allows precipitation of minerals, such as hydroxyapatite or struvite, which then deposit in the catheter biofilm, causing mineral encrustations along the catheter. Encrustations are a feature of biofilm uniquely associated with urinary catheters.[2]

Epidemiology

Descriptive Epidemiology

CAUTIs account for approximately 40% of all hospital-acquired infections, but UTIs make up a smaller proportion of nosocomial infections in intensive care unit (ICU) patients; CAUTIs account for 15%–21% of nosocomial infections in pediatric ICU patients, and 23% of nosocomial infections in adult ICU patients in the United States.[5]

The incidence of UTI varies by ICU type; rates of CAUTI reported through the National Healthcare Safety Network (NHSN) in 2006 and 2007 ranged from 3.1 infections per 1,000 catheter-days in medical-surgical ICUs to 7.7 infections per 1,000 catheter-days in burn ICUs. CAUTI is reported at an incidence of 5.0 infections per 1,000 catheter-days in pediatric ICUs but is infrequently identified in neonatal ICUs.[2,6-9] Rates of CAUTI in general care wards are equivalent to or higher than those in the ICU, ranging from 4.7 infections per 1,000 catheter-days in adult step-down units to 16.8 infections per 1,000 catheter-days in rehabilitation units.[9]

Microbial Etiology

Enterobacteriaceae are the pathogens most commonly associated with catheter-related UTI hospital-wide (Table 13-1). Other significant pathogens more common

Table 13-1. Ranking of selected pathogens associated with catheter-associated urinary tract infections (CAUTIs) reported to the National Healthcare Safety Network, January 2006 through October 2007

Pathogen	Percentage of CAUTIs
Escherichia coli	21.4
Candida species	21.0
Enterococci species	15.5
Pseudomonas aeruginosa	10.0
Klebsiella pneumoniae	7.7
Enterobacter species	4.1
Coagulase-negative *Staphylococcus* species	2.5
Coagulase-positive *Staphylococcus* species	2.2
Acinetobacter baumannii	1.2
Klebsiella oxytoca	0.9

NOTE: Data are from Edwards et al.[9]

in the ICU include *Candida* species, enterococci, and *Pseudomonas aeruginosa*.[2,9] Most infections (80%) associated with short-term indwelling urinary catheters are due to a single species of organism. Conversely, infections associated with long-term indwelling catheters are polymicrobial in 77%–95% of cases, and 10% have more than 5 species of organisms present. This pathogen distribution has not changed significantly from previous reports between 1986 and 1999.[9] In addition, the antimicrobial resistance profiles found in 2006 and 2007 differed only slightly from those in historical reports from 1986 to 2003.[9]

Risk Factors

Most studies on CAUTI have focused on bacteriuria, a precursor of symptomatic infection. The most important, consistent risk factor for healthcare-associated bacteriuria is the duration of catheterization; up to 97% of UTIs in ICUs are associated with an indwelling urinary catheter. Bacteriuria occurs quickly and frequently in catheterized patients, with an average daily risk of 3% to 10% per day. Of patients with a catheter in place for 2–10 days, 26% will develop bacteriuria. Nearly all patients catheterized for a month will have bacteriuria, making this duration the dividing line between short-term and long-term catheterization.[2]

Females have a higher risk of bacteriuria than males (relative risk [RR], 1.7–3.7).[5] Systemic antimicrobial therapy has a protective effect against bacteriuria

(RR, 2.0–3.9).[5] Nonadherence to catheter care recommendations has also been associated with increased risk of bacteriuria. Other risk factors identified in one or more studies include the following: rapidly fatal underlying illness, age of more than 50 years, nonsurgical disease, hospitalization on an orthopedic or urological service, catheter insertion after day 6 of hospitalization, catheter inserted outside the operating room, diabetes mellitus, and a serum creatinine level greater than 2 mg/dL at the time of catheterization. Heavy periurethral colonization with bacteria has also been associated with increased risk of bacteriuria.

Risk factors for UTI-associated bacteremia are less clearly defined than are those for catheter-associated bacteremia, because catheter-associated bacteremia occurs in fewer than 4% of infections. Krieger and colleagues[10] followed 1,233 patients with hospital-acquired UTI and found secondary bloodstream infections in 32 patients (2.6%). Risk factors for bloodstream infection from a urinary source included infection due to *Serratia marcescens*, compared with infection due to other organisms (RR, 3.5), and male sex (RR, 2.0). No other factors were found to significantly predispose to bacteremia in patients with UTIs.

Diagnosis and Surveillance

The terms "bacteriuria" and "urinary tract infection" are often used interchangeably in the published literature pertaining to healthcare-associated UTI and CAUTI. The distinction between them is important clinically, because asymptomatic catheter-associated bacteriuria is rarely associated with adverse outcomes and generally does not require treatment with antimicrobials.[11] Most studies of CAUTI use bacteriuria as the primary outcome. In general, bacteriuria in a catheterized patient is defined as growth in culture of 10^2 colony forming units (cfu) or more of a predominant pathogen per milliliter of urine specimen collected aseptically from a sampling port.[4]

The NHSN has developed surveillance definitions for healthcare-associated UTI,[12] which allow for standardization and interfacility comparison of infection rates. The definitions distinguish between symptomatic UTI and asymptomatic bacteriuria. For adult patients, there are 2 possible definitions for symptomatic UTI. According to the first definition, a patient must have at least 1 sign or symptom (temperature higher than 38°C, urinary urgency, urinary frequency, dysuria, or suprapubic tenderness) and a urine culture positive for a pathogen or pathogens (defined as yielding 10^5 cfu/mL or more, with no more than 2 species of microorganisms detected). The second definition states that a patient must have at least

2 signs and symptoms along with at least 1 laboratory finding that is not quite as compelling as a positive urine culture (positive result of a urine dipstick test for leukocyte esterase or nitrite, pyuria with detection of at least 3 white blood cells per high-power field, positive Gram stain of urine, 2 urine cultures yielding 10^2 cfu/mL or more of the same pathogen, 1 urine culture with a single pathogen detected at a concentration of 10^5 cfu/mL or less for a patient being treated with antimicrobials, or physician diagnosis and/or treatment of a UTI). Asymptomatic bacteriuria in a patient who has had a urinary catheter in the previous 7 days is defined as a urine culture positive for a pathogen in the absence of signs or symptoms. In a patient who has not had a catheter within the previous 7 days, 2 positive urine culture results are required. Note that this definition does not allow for classification of asymptomatic bacteriuria with a pathogen detected at a concentration of less than 10^5 cfu/mL, although this may occur in patients who have been treated with antimicrobials.

Clinical diagnosis of CAUTI is quite difficult. Pyuria is not a reliable indicator of UTI in a patient with catheterization. Musher and colleagues[13] found that most catheterized patients with bacteriuria had pyuria, but 30% of patients with pyuria did not have bacteriuria. Diagnosis of UTI in patients with long-term urinary catheters is particularly difficult, as bacteriuria is invariably present. Systemic symptoms of infection may be the only indications of UTI, especially in patients who have spinal cord injuries.[14]

Surveillance for CAUTI has not been a priority for most hospitals in the past, mostly because of lack of resources required to perform full hospital surveillance.[15] This is likely to change, since CAUTI has been chosen by the Centers for Medicare and Medicaid Services as one of the hospital-acquired complications for which costs will no longer be reimbursed.[16] Surveillance can be performed facility-wide, or it can be targeted at hospital locations where rates of catheter utilization are highest. Cases of symptomatic CAUTI are identified using the NHSN definitions. Data collection forms utilizing these standardized criteria are available from the NHSN. The incidence of symptomatic CAUTI per 1,000 urinary catheter–days is the most widely accepted process measure used in surveillance and is endorsed by the NHSN,[17] jointly by the Infectious Diseases Society of America–Society for Healthcare Epidemiology of America,[18] and by the Association for Professionals in Infection Control and Epidemiology.[19] Other measures that may be relevant include the incidence of asymptomatic bacteriuria, the percentage of patients with indwelling catheters, the percentage of catheterization with accepted indications, and the duration of catheter use.

Table 13-2. Key strategies for prevention of catheter-associated urinary tract infection

Avoid use of indwelling urinary catheters
    Place only for appropriate indications
    Follow institutional protocols for placement, including perioperatively
    Use alternatives to indwelling catheterization (intermittent catheterization, condom catheter, or portable bladder ultrasound scanner)
Remove indwelling catheters early
    Use nurse-based interventions
    Use electronic reminders
Use proper techniques for insertion and maintenance of catheters
    Adhere to sterile insertion practices
    Use a closed drainage system
    Avoid routine bladder irrigation
Consider use of antimicrobial-coated catheters in some patients

Prevention of CAUTIs

General Strategies for Prevention

Compliance with hand hygiene before and after patient care is recommended for prevention of all healthcare-associated infections, including UTIs.[20] The urinary tracts of hospitalized patients and patients in long-term care facilities represent a significant reservoir for multidrug-resistant organisms; hospital transmission of these organisms has been reported. Use of contact precautions, with use of gowns and gloves, is currently recommended as part of a multifaceted strategy to prevent transmission of multidrug-resistant organisms.[21] Use of indwelling devices, such as urinary catheters, has been shown to increase the risk of colonization with multidrug-resistant organisms. These devices should only be used when absolutely necessary and should be removed as soon as possible. In addition, UTI has been found to be an important cause of antibiotic use in hospitalized patients.[22] Repeated antimicrobial treatment for infections related to long-term catheterization is an important risk factor for colonization with multidrug-resistant organisms, yet some of this use of antimicrobials may be inappropriate. Reduction in use of broad-spectrum antimicrobials is an important strategy to prevent development of antimicrobial resistance associated with urinary catheters. Antimicrobial stewardship programs are another strategy for reducing overall use of antimicrobials.[23]

Specific Strategies for Prevention

Multiple guidelines have been developed to outline strategies for the prevention of CAUTI.[18,24] Nevertheless, a recent nationwide survey of hospitals revealed that more than one-half of hospitals did not have a system for moni-

toring which patients had urinary catheters, three-quarters did not monitor duration of catheterization, and nearly one-third did not conduct any surveillance for UTI.[15] Key strategies for prevention of CAUTI are summarized in Table 13-2.

Limitation of Use and Early Removal of Urinary Catheters

Urinary catheters account for the majority of nosocomial UTIs (approximately 80%), and the most effective strategy for prevention is avoidance of catheterization.[25] The incidence of placement for an inappropriate indication has been documented to be 21%–50% of catheterized patients, in the recent literature.[26-28] Physician documentation of the indications for urinary catheters and their presence has been reported to be present in less than 50% of cases.[29] Physicians are frequently unaware of the presence of urinary catheters in their patients, and this unawareness has been correlated with inappropriate catheter use.[30]

Indwelling urinary catheterization should be limited to certain indications (Table 13-3). Catheters should not be inserted for convenience or for incontinence in the absence of another compelling indication. Each institution should develop written guidelines and explicit criteria for indwelling urinary catheterization based on these widely accepted indications, although modifications based on local needs may be appropriate.[18,31] Regular education of nursing staff regarding proper indications and supporting rationale should be undertaken. If appropriate criteria for catheter placement are not met, nursing staff should be encouraged to discuss alternatives with the ordering physician. Physician orders should be required prior to any catheter insertion, and institutions should implement a system for documenting placement of catheters.[18]

Table 13-3. Appropriate indications for the placement of a urinary catheter

Accurate monitoring of urine output in a critically ill patient
Acute anatomical or functional urinary retention or obstruction
Perioperative use for selected surgical procedures
    For surgical procedures of anticipated long duration
    For urologic procedures
    For procedures in patients with urinary incontinence
    For procedures requiring intraoperative urinary monitoring or expected large volume of intravenous infusions
Urinary incontinence in patients with open perineal or sacral wounds
Improved comfort for end-of-life care, if desired

NOTE: Recommendations are from Gould et al.[24] and Saint and Lipsky.[49]

Interventions for limiting urinary catheter use should be targeted at hospital locations where initial placement often occurs, such as emergency departments and operating rooms.[28] Active surveillance that uses the number of cases of symptomatic CAUTI per 1,000 urinary catheter-days or another outcome measure should be performed before, during, and after targeted interventions, to determine whether the intervention is a success.

A number of nurse-driven interventions have demonstrated promising effectiveness in reducing the duration of catheterization. A nurse-based reminder to physicians to remove unnecessary urinary catheters in an adult ICU in a Taiwanese hospital resulted in a reduction in the incidence of CAUTI (from 11.5 to 8.3 cases per 1,000 catheter-days).[32] Nurse-initiated reminders to physicians of the presence of urinary catheters also decreases the number of catheter-days.[33,34] Such interventions are easy to implement and may consist of either a written notice or a verbal contact with the physician regarding the presence of a urinary catheter and alternative options.

The advent of electronic medical records and computerized physician order entry systems allow targeted interventions both to reduce the number of catheters placed and to reduce the duration of catheterization. Cornia and colleagues[35] found that use of a computerized reminder reduced the duration of catheterization by 3 days. In some settings, an infection prevention specialist may have the capability of working with the information technology department to integrate catheter protocols into electronic physician order entry sets.

Perioperative Management of Urinary Retention

Specific protocols for the management of postoperative urinary retention may be beneficial. Although only a limited number of prospective studies have addressed optimal postoperative bladder management strategies, indwelling urinary catheterization following surgery has become ubiquitous in some centers. In one large cohort study, the authors demonstrated that 85% of patients admitted for major surgical procedures had perioperative indwelling catheters, and the half of these patients with duration of catheterization greater than 2 days were significantly more likely to develop UTI and less likely to be discharged to home.[36] Older surgical patients are particularly at risk for prolonged catheterization; in another study, 23% of surgical patients older than 65 years of age were discharged to skilled nursing facilities with an indwelling catheter in place and were substantially more likely to be rehospitalized or die within 30 days.[37]

In a large prospective clinical trial involving orthopedic patients, incorporation of a multifaceted protocol for perioperative catheter management resulted in a two-thirds reduction in the incidence of UTI.[38] The intervention protocol consisted of limiting catheterization to patients who underwent surgery with a duration of more than 5 hours or total hip and knee replacement, if the patient met one of several conditions. Urinary catheters were removed on postoperative day 1 after total knee arthroplasty and on postoperative day 2 after total hip arthroplasty.[38] Although these procedures were effective at this particular hospital, each institution should develop protocols written by a local, multidisciplinary group.

Alternatives to Indwelling Urinary Catheters

Intermittent urinary catheterization may reduce the risk of bacteriuria and UTI, compared with indwelling urinary catheterization. Patients with neurogenic bladder and long-term urinary catheters may particularly benefit from intermittent catheterization.[24] One recent meta-analysis demonstrated reduced risk of asymptomatic and symptomatic bacteriuria in postoperative patients following hip or knee surgery with intermittent catheterization, compared with indwelling catheterization (RR, 2.90 [95% confidence interval {CI}, 1.44–5.84]), but included only 2 studies with a total of 194 patients.[39] Several studies of intermittent catheterization in postoperative patients have demonstrated increased risk of urinary retention and bladder distention.[39-41] Incorporating use of a portable bladder ultrasound scanner with intermittent catheterization may attenuate this risk and reduce the total number of catheterizations performed.[42,43]

External catheters, or condom catheters, should be considered as an alternative to indwelling catheters in appropriately selected male patients without urinary retention or bladder outlet obstruction. A recent randomized trial demonstrated a decrease in the composite outcomes of bacteriuria, symptomatic UTI, and death in patients with condom catheters, compared with patients with indwelling catheters, although the benefit was limited to those men without dementia.[44] Condom catheters may also be more comfortable and less painful than indwelling catheters.[44,45]

Proper Techniques for Insertion and Maintenance of Urinary Catheters

Once a decision has been made to proceed with catheterization, proper catheter insertion and maintenance are essential for prevention of CAUTI. Urinary catheters should be inserted using sterile equipment and aseptic technique by a trained healthcare practitioner.[24,46] Cleaning of the meatal area should be undertaken prior to catheter insertion, but there is currently no consensus regarding the use of sterile water, compared with use of an antiseptic preparation.[24,46] A recent randomized study comparing sterile water with 0.1% chlorhexidine for cleaning of the meatal area prior to insertion demonstrated no difference in the development of bacteriuria.[47] Ongoing catheter maintenance with daily meatal cleaning using an antiseptic has also not shown clear benefit, and it may actually increase rates of bacteriuria, compared with routine care with soap and water.[48,49] A single-use packet of sterile lubricant jelly should be used for insertion, to reduce urethral trauma, but it does not need to possess antiseptic properties.[24] Urinary catheters should not be routinely exchanged, except for mechanical reasons, because any reduction in the rate of bacteriuria with routine changing is generally only transient.[46,50]

Use of closed urinary catheter systems with sealed urinary catheter-tubing junctions reduce the risk of CAUTI. Breaches of the closed system should be avoided, and urine should be sampled only from a port after cleaning with an antiseptic solution or from the drainage bag using sterile technique if a large sample is required.[18,46] Breach of the closed system to routinely instill antimicrobial agents is associated with increased rates of infection, and irrigation of the bladder with antimicrobial agents potentially causes the organisms colonizing the catheter biofilm to flow into the bladder.[51] Amphotericin B bladder irrigation, delivered either through intermittent instillation into a usual catheter system or through a 3-way indwelling bladder catheter, is not routinely recommended for asymptomatic candiduria in most patients, according to a recent evidence-based national guideline.[52,53] Although amphotericin B bladder irrigation will often result in initial clearance of candiduria, there is a high rate of relapse after discontinuation of irrigation, and the clinical outcomes of morbidity and mortality do not appear to be improved with treatment.[54,55]

Anti-infective Catheters

Use of antiseptic and antibiotic-impregnated urinary catheters may have an impact on the rates of catheter-associated bacteriuria and are a possible adjunctive infection control measure. Antiseptic catheters currently available are coated with silver alloy. Earlier catheters coated with silver oxide lacked efficacy, compared with silver alloy–coated catheters, and are no longer available.[56] Other antibiotic-impregnated catheters have utilized various types of antibiotics, including nitrofurazone, minocycline, and rifampin.

In a large meta-analysis, use of silver alloy–coated catheters significantly reduced the incidence of asymptomatic bacteriuria (RR, 0.54 [95% CI, 0.43–0.67]) among adult patients catheterized for less than 7 days, compared with use of latex catheters. Among the patients catheterized for more than 7 days, a reduction in asymptomatic bacteriuria was less pronounced (RR, 0.64 [95% CI, 0.51–0.80]). In the same meta-analysis, antibiotic-impregnated catheters were compared with standard catheters and were found to decrease the rate of asymptomatic bacteriuria (RR, 0.52 [95% CI, 0.34–0.78]) for a duration of catheterization less than 7 days but demonstrated no benefit for a duration of catheterization of more than 7 days.[57] Another meta-analysis demonstrated similar reductions in asymptomatic bacteriuria in patients with short-term catheterization.[58] There are few trials assessing antiseptic- and antibiotic-coated catheters in patients with long-term urinary catheterization, and no conclusions can be drawn regarding such patients.[59]

Use of anti-infective urinary catheters appears to be one option to reduce the incidence of bacteriuria in patients with short-term urinary catheterization (for less than 7 days), but the effect on the more important outcomes of urinary catheter–associated bacteremia and symptomatic CAUTI are not clear from the current literature. There is no consensus that anti-infective urinary catheters need be used routinely to prevent CAUTI.[24] A recent national survey of infection control officers found that most hospitals that utilized anti-infective catheters based their decisions on hospital-specific pilot studies.[15]

Summary

CAUTIs account for approximately 40% of all healthcare-associated infections. Despite studies showing the benefit of interventions for prevention of CAUTI, adoption of these practices has not occurred in many healthcare facilities in the United States. Because urinary catheters account for the majority of healthcare-associated UTIs, the most important interventions are directed at avoiding placement of urinary catheters and promoting early removal when appropriate. Alternatives to use of indwelling catheters, such as intermittent catheterization and use of condom catheters, should be considered. If indwelling catheterization is appropriate, use of proper aseptic practices for catheter insertion and maintenance and use of closed urinary-catheter collection systems are essential for prevention of CAUTI. The use of anti-infective catheters may also be considered in some circumstances in which the benefit outweighs the cost. Attention to the prevention of CAUTI will likely increase as the Centers for Medicare and Medicaid Services and other third-party payers no longer reimburse for hospital-acquired CAUTI.

References

1. Warren JW. Catheter-associated urinary tract infections. *Int J Antimicrob Agents* 2001;17(4):299–303.
2. Saint S, Chenoweth CE. Biofilms and catheter-associated urinary tract infections. *Infect Dis Clin North Am* 2003; 17: 411–432.
3. Tambyah PA, Halvorson KT, Maki DG. A prospective study of pathogenesis of catheter-associated urinary tract infections. *Mayo Clin Proc* 1999; 74:131–136.
4. Maki DG, Tambyah PA. Engineering out the risk of infection with urinary catheters. *Emerging Infect Dis* 2001;7(2):1–6.
5. Chenoweth CE, Saint S. Urinary tract infections. In: Jarvis WR, ed. *Bennett & Brachman's Hospital Infections*. 5th ed. Philadelphia: Lippincott Williams & Wilkins, Wolters Kluwer; 2007:507–516.
6. Gaynes RP, Edwards JR, Jarvis WR, Culver DH, Tolson JS, Martone WJ. Nosocomial infections among neonates in high-risk nurseries in the United States. *Pediatrics* 1996;98(3): 357–361.
7. Langley JM, Hanakowski M, LeBlanc JC. Unique epidemiology of nosocomial urinary tract infection in children. *Am J Infect Control* 2001; 29:94–98.
8. National Nosocomial Infectious Surveillance (NNIS) System. National Nosocomial Infectious Surveillance (NNIS) system report, data summary from January 1992 through June 2004, issued October 2004. *Am J Infect Control* 2004; 32:470–485.
9. Edwards JR, Stat M, Peterson KD, et al. National Healthcare Safety Network (NHSN) report, data summary for 2006 through 2007, issued November 2008. *Am J Infect Control* 2008; 36:609–626.
10. Krieger JN, Kaiser DL, Wenzel RP. Nosocomial urinary tract infections: secular trends, treatment and economics in a university hospital. *J Urol* 1983;130(1):102–106.
11. Tambyah PA, Maki DG. Catheter-associated urinary tract infection is rarely symptomatic: a prospective study of 1,497 catheterized patients. *Arch Intern Med* 2000;160(5):678–682.
12. Horan TC, Andrus M, Dudeck MA. CDC/NHSN surveillance definition of health care-associated infection and criteria for specific types of infections in the acute care setting. *Am J Infect Control* 2008; 36:309–332.
13. Musher DM, Thorsteinsson SB, Airola VW II. Quantitative urinalysis: diagnosing urinary tract infection in men. *JAMA* 1976; 236:2069–2072.
14. Biering-Sorenson F, Bagi P, Hoiby N. Urinary tract infections in patients with spinal cord lesions: treatment and prevention. *Drugs* 2001;61(9):1275–1287.
15. Saint S, Kowalski CP, Forman J, et al. A multicenter qualitative study on preventing hospital-acquired urinary tract infection in US hospitals. *Infect Control Hosp Epidemiol* 2008; 29(4):333–341.
16. Saint S, Meddings JA, Calfee D, Kowalski CP, Krein SL. Catheter-associated urinary tract infection and the Medicare rule changes. *Ann Intern Med* 2009;150(12):877–884.
17. Centers for Disease Control and Prevention National Healthcare Safety Network. Catheter-Associated Urinary Tract Infection (CAUTI) Event. http://www.cdc.gov/nhsn/library.html. Accessed August 30, 2009.
18. Lo E, Nicolle L, Classen D, et al. Strategies to prevent catheter-associated urinary tract infections in acute care hospitals. *Infect Control Hosp Epidemiol* 2008;29(Suppl 1):S41–S50.
19. Greene L, Marx J, Oriola S. Guide to the elimination of catheter-associated urinary tract infections (CAUTIs). Washington, DC: Association for Professionals in Infection Control and Epidemiology; 2008. http://www.apic.org/Content/Navigation Menu/PracticeGuidance/APICEliminationGuides/CAUTI_ Guide1.htm. Accessed August 30, 2009.
20. Boyce JM, Pittet D. Guideline for hand hygiene in health-care settings: recommendations of the Healthcare Infection Control Practices Advisory Committee and the HICPAC/SHEA/APIC/ IDSA Hand Hygiene Task Force. Society for Healthcare Epidemiology of America / Association for Professionals in Infection Control / Infectious Diseases Society of America. *MMWR Recomm Rep* 2002;51(RR-16):1–45.
21. Siegel JD, Rhinehart E, Jackson M, Chiarello L. 2007 Guideline for isolation precautions: preventing transmission of infectious agents in health care settings. *Am J Infect Control* 2007;35(10 Suppl 2):S65–S164.
22. Gandhi T, Flanders SA, Markovitz E, Saint S, Kaul DR. Importance of urinary tract infection to antibiotic use among hospitalized patients. *Infect Control Hosp Epidemiol* 2009;30(2):193–195.
23. Dellit TH, Owens RC, McGowan JE Jr, et al. Infectious Diseases Society of America and the Society for Healthcare Epidemiology of America guidelines for developing an institutional program to enhance antimicrobial stewardship. *Clin Infect Dis* 2007;44(2):159–177.
24. Gould CV, Umscheid CA, Agarwal RK, Kuntz G, Pegues DA, HICPAC. Centers for Disease Control and Prevention. Guideline for prevention of catheter-associated urinary tract infections (draft). 2008. http://wwwn.cdc.gov/ PUBLICCOMMENTS/comments/guidelines-for-prevention-of-catheter-associated-urinary-tract-infections-2008.aspx. Accessed August 30, 2009.
25. Nicolle LE. The prevention of hospital-acquired urinary tract infection. *Clin Infect Dis* 2008;46(2):251–253.
26. Gardam MA, Amihod B, Orenstein P, Consolacion N, Miller MA. Overutilization of indwelling urinary catheters and the development of nosocomial urinary tract infections. *Clin Perform Qual Health Care* 1998;6(3):99–102.
27. Jain P, Parada JP, David A, Smith LG. Overuse of the indwelling urinary tract catheter in hospitalized medical patients. *Arch Intern Med* 1995;155(13):1425–1429.
28. Munasinghe RL, Yazdani H, Siddique M, Hafeez W. Appropriateness of use of indwelling urinary catheters in patients admitted to the medical service. *Infect Control Hosp Epidemiol* 2001;22(10):647–649.
29. Conybeare A, Pathak S, Imam I. The quality of hospital records of urethral catheterisation. *Ann R Coll Surg Engl* 2002;84(2):109–110.
30. Saint S, Wiese J, Amory JK, et al. Are physicians aware of which of their patients have indwelling catheters? *Am J Med* 2000; 109:476–480.
31. The Institute for Healthcare Improvement (IHI). Getting Started Kit: Preventing Catheter-Associated Urinary Tract Infections. IHI: 2009. http://www.ihi.org/IHI/Programs/ ImprovementMap/PreventCatheterAssociatedUrinary TractInfections.htm. Accessed August 30, 2009.
32. Huang WC, Wann SR, Lin SL, et al. Catheter-associated urinary tract infections in intensive care units can be reduced by prompting physicians to remove unnecessary catheters. *Infect Control Hosp Epidemiol* 2004;25(11):974–978.

33. Fakih MG, Dueweke C, Meisner S, et al. Effect of nurse-led multidisciplinary rounds on reducing the unnecessary use of urinary catheterization in hospitalized patients. *Infect Control Hosp Epidemiol* 2008;29(9):815–819.

34. Saint S, Kaufman SR, Thompson M, Rogers MA, Chenoweth CE. A reminder reduces urinary catheterization in hospitalized patients. *Jt Comm J Qual Patient Saf* 2005;31(8):455–462.

35. Cornia PB, Amory JK, Fraser S, Saint S, Lipsky BA. Computer-based order entry decreases duration of indwelling urinary catheterization in hospitalized patients. *Am J Med* 2003;114(5):404–407.

36. Wald HL, Ma A, Bratzler DW, Kramer AM. Indwelling urinary catheter use in the postoperative period: analysis of the national surgical infection prevention project data. *Arch Surg* 2008;143(6):551–557.

37. Wald HL, Epstein AM, Radcliff TA, Kramer AM. Extended use of urinary catheters in older surgical patients: a patient safety problem? *Infect Control Hosp Epidemiol* 2008;29(2):116–124.

38. Stephan F, Sax H, Wachsmuth M, Hoffmeyer P, Clergue F, Pittet D. Reduction of urinary tract infection and antibiotic use after surgery: a controlled, prospective, before-after intervention study. *Clin Infect Dis* 2006;42(11):1544–1551.

39. Niel-Weise BS, van den Broek PJ. Urinary catheter policies for short-term bladder drainage in adults. *Cochrane Database Syst Rev* 2005(3):CD004203.

40. Michelson JD, Lotke PA, Steinberg ME. Urinary-bladder management after total joint-replacement surgery. *N Engl J Med* 1988;319(6):321–326.

41. Oishi CS, Williams VJ, Hanson PB, Schneider JE, Colwell CW Jr, Walker RH. Perioperative bladder management after primary total hip arthroplasty. *J Arthroplasty* 1995;10(6):732–736.

42. Moore DA, Edwards K. Using a portable bladder scan to reduce the incidence of nosocomial urinary tract infections. *Medsurg Nurs* 1997;6(1):39–43.

43. Stevens E. Bladder ultrasound: avoiding unnecessary catheterizations. *Medsurg Nurs* 2005;14(4):249–253.

44. Saint S, Kaufman SR, Rogers MA, Baker PD, Ossenkop K, Lipsky BA. Condom versus indwelling urinary catheters: a randomized trial. *J Am Geriatr Soc* 2006;54(7):1055–1061.

45. Saint S, Lipsky BA, Baker PD, McDonald LL, Ossenkop K. Urinary catheters: what type do men and their nurses prefer? *J Am Geriatr Soc* 1999;47(12):1453–1457.

46. Guidelines for preventing infections associated with the insertion and management of short term indwelling urethral catheters in acute care. *J Hosp Infect* 2001;47(Suppl):S39–S46.

47. Webster J, Hood RH, Burridge CA, Doidge ML, Phillips KM, George N. Water or antiseptic for periurethral cleaning before urinary catheterization: a randomized controlled trial. *Am J Infect Control* 2001;29(6):389–394.

48. Burke JP, Garibaldi RA, Britt MR, Jacobson JA, Conti M, Alling DW. Prevention of catheter-associated urinary tract infections. Efficacy of daily meatal care regimens. *Am J Med* 1981;70(3):655–658.

49. Saint S, Lipsky BA. Preventing catheter-related bacteriuria: should we? Can we? How? *Arch Intern Med* 1999;159(8):800–808.

50. Tenney JH, Warren JW. Bacteriuria in women with long-term catheters: paired comparison of indwelling and replacement catheters. *J Infect Dis* 1988;157(1):199–202.

51. Warren JW, Platt R, Thomas RJ, Rosner B, Kass EH. Antibiotic irrigation and catheter-associated urinary-tract infections. *N Engl J Med* 1978;299(11):570–573.

52. Drew RH, Arthur RR, Perfect JR. Is it time to abandon the use of amphotericin B bladder irrigation? *Clin Infect Dis* 2005;40(10):1465–1470.

53. Pappas PG, Kauffman CA, Andes D, et al. Clinical practice guidelines for the management of candidiasis: 2009 update by the Infectious Diseases Society of America. *Clin Infect Dis* 2009;48(5):503–535.

54. Kauffman CA, Vazquez JA, Sobel JD, et al. Prospective multicenter surveillance study of funguria in hospitalized patients. The National Institute for Allergy and Infectious Diseases (NIAID) Mycoses Study Group. *Clin Infect Dis* 2000;30(1):14–18.

55. Simpson C, Blitz S, Shafran SD. The effect of current management on morbidity and mortality in hospitalised adults with funguria. *J Infect* 2004;49(3):248–252.

56. Saint S, Elmore J, Sullivan S, Emerson S, Koepsell T. The efficacy of silver alloy-coated urinary catheters in preventing urinary tract infection: a meta-analysis. *Am J Med* 1998;105:236–241.

57. Schumm K, Lam T. Types of urethral catheters for management of short-term voiding problems in hospitalised adults. *Cochrane Database Syst Rev* 2008(2):CD004013.

58. Johnson JR, Kuskowski MA, Wilt TJ. Systematic review: antimicrobial urinary catheters to prevent catheter-associated urinary tract infection in hospitalized patients. *Ann Intern Med* 2006;144(2):116–126.

59. Jahn P, Preuss M, Kernig A, Langer G, Seifert-Huehmer A. Types of indwelling urinary catheters for long-term bladder drainage in adults. *Cochrane Database Syst Rev* 2007(3):CD004997.

Chapter 14 Preventing Hospital-Acquired Pneumonia

Emily K. Shuman, MD, and Carol E. Chenoweth, MD

Hospital-acquired pneumonia (HAP) is the second most common healthcare-associated infection in the United States. HAP is not a universally reportable illness, but available data suggest that it occurs at a rate of 5–10 episodes per 1,000 hospital admissions.[1] The attributable mortality of HAP has been reported as anywhere from 5% to more than 30%.[2] Excess cost has been estimated at $12,000–$40,000 per case, and hospitalization is typically prolonged by an additional 7–9 days.[3] In addition to substantial mortality and cost, HAP is associated with excess use of antimicrobials.

Ventilator-associated pneumonia (VAP), a subset of HAP, is the second most common healthcare-associated infection in the intensive care unit (ICU). Previous studies estimated that VAP occurred in 10%–20% of patients undergoing mechanical ventilation. However, more recent publications report rates typically ranging from 1–4 episodes per 1,000 ventilator-days, but exceeding 10 episodes per 1,000 ventilator-days in some neonatal and surgical ICUs.[4] The incidence of pneumonia is increased by as much as 6-fold to 20-fold among patients undergoing mechanical ventilation.[2] The risk of developing pneumonia is highest during the first few days after intubation and is estimated at 3% per day for the first 5 days, 2% per day from day 5 to day 10, and 1% per day thereafter.[5] This is in part because of the increased risk of pneumonia conferred by the process of intubation itself. In addition to the outcomes listed above for HAP, VAP is associated with prolonged duration of mechanical ventilation.[4]

Definitions

HAP is defined as pneumonia with onset 48 hours or more after admission to the hospital and that was not incubating at the time of admission.[2] VAP refers to pneumonia that develops in a patient who was intubated and ventilated at the time of or within 48 hours prior to onset. "Healthcare-associated pneumonia (HCAP)" and "nursing home–associated pneumonia (NHAP)" are relatively recently defined terms, and these entities occur along a continuum with HAP and VAP. Healthcare-associated pneumonia includes cases in patients who were admitted to an acute care hospital for 2 or more days within the previous 90 days, who received intravenous antimicrobials, chemotherapy, or wound care within the previous 30 days, or who receive hemodialysis. Nursing home–associated pneumonia refers to cases in patients residing in a nursing home or long-term care facility. Because VAP has been studied more extensively than HAP, this chapter primarily addresses the epidemiology, diagnosis, and prevention of VAP. Prevention of healthcare-associated pneumonia and nursing home–associated pneumonia will not be discussed, but the principles are similar.

Pathogenesis

HAP is caused by invasion of the normally sterile lung parenchyma by microorganisms. It should be noted that this chapter focuses on pneumonia caused by bacterial organisms; however, fungal and viral organisms can cause HAP, especially in severely immunocompromised patients. The typical route of bacterial infection is by microaspiration of oropharyngeal or gastric contents. This allows for entry of bacterial pathogens into the lower respiratory tract, where infection can occur if the host's mechanical and immunological defenses are overwhelmed. Infection is more likely to occur in the presence of virulent or multidrug-resistant organisms (MDROs).

Epidemiology

Etiology

Most of what we know about the microbiology of HAP comes from studies specifically examining VAP in the ICU setting.[6] The bacteria responsible for HAP have traditionally been divided into those that cause pneumonia early after hospitalization (0–4 days) and those that cause pneumonia later after hospitalization (5 days or more). Cases of early pneumonia are typically due to community-acquired bacteria, such as *Streptococcus pneumoniae, Haemophilus influenzae, Escherichia coli,* or other drug-susceptible enteric gram-negative bacilli, or to methicillin-susceptible *Staphylococcus aureus.* Pneumonia that develops later during the hospital course is more likely to be due to MDROs, such as nonenteric gram-negative bacilli (*Pseudomonas aeruginosa, Acinetobacter* species, and *Stenotrophomonas maltophilia*) or methicillin-resistant *S. aureus* (MRSA).

Many cases of HAP are likely polymicrobial, especially in patients with adult respiratory distress syndrome.[6] In general, it is not thought that anaerobes play a significant role in HAP, even in patients with witnessed aspiration.[7] Infection with *Legionella pneumophila* typically only occurs when a hospital's water supply is contaminated or when there is ongoing construction.[1] More recent studies using molecular rather than culture-based methods for identification of organisms involved in VAP suggest that the range of bacteria associated with VAP is much more complex than we currently recognize.[8]

Risk factors for infection with specific organisms are described in the literature,[1] and clinical risk factors may be better at predicting potential causative organisms than is the time of onset of pneumonia. Patients with a history of prior hospitalization or broad-spectrum antimicrobial use are more likely to become infected with MDROs, regardless of when they develop pneumonia during their hospitalization.[9,10] It should be noted that the incidence of infection with MDROs differs among institutions and among different units in the same institution. Therefore, institution-specific epidemiologic data regarding HAP and VAP should be collected.

Risk Factors

Again, most studies examining risk factors for HAP have focused on VAP. Risk factors for VAP include conditions that lead to colonization of the respiratory and upper gastrointestinal tracts with bacteria, in particular with virulent organisms, as well as conditions that lead to increased risk of aspiration and impaired host defenses.[1,11] Risk factors can be further divided into those that are modifiable and those that are not (Table 14-1).

Use of endotracheal tubes is a modifiable risk factor that leads to bacterial colonization of the respiratory tract and impaired host defenses.[1] Use of nasogastric tubes and enteral feeding tubes predispose patients to colonization of the upper gastrointestinal tract and to development of sinusitis, which can also increase the risk of VAP. Stress ulcer prophylaxis with acid-suppressive

Table 14-1. Risk factors for ventilator-associated pneumonia

Category of risk factor	Modifiable factors	Nonmodifiable factors
Increased aspiration and/or bacterial colonization of respiratory and upper gastrointestinal tracts	• Endotracheal tubes • Nasogastric tubes • Enteral feeding tubes • Supine positioning • Sedation • Contaminated respiratory equipment • Gastric alkalinization	• Impaired mental status • Witnessed aspiration • Sinusitis
Acquisition of virulent organisms	• Prolonged use of antimicrobials • Transmission from healthcare workers because of inadequate hand hygiene	• Comorbid illness • Frequent hospitalizations • Prolonged hospitalization
Impaired host defenses	• Endotracheal tubes • Malnutrition	• Extremes of age • Immunosuppression

medications also contributes to colonization of the upper gastrointestinal tract.[1] Use of nasogastric and enteral feeding tubes and use of acid-suppressive medications for stress ulcer prophylaxis are considered to be modifiable risk factors. Sedation and supine positioning increase the risk of aspiration.[11] Use of sedation in mechanically ventilated patients is a potentially modifiable risk factor, although many patients have underlying conditions that lead to impaired consciousness, which is typically not modifiable. Patient positioning is, however, generally a modifiable risk factor.

As noted above, use of broad-spectrum antimicrobials is a modifiable risk factor for acquisition of MDROs. The presence of indwelling devices, such as central catheters and urinary catheters, is another modifiable risk factor. Prior hospitalization and comorbid conditions, such as diabetes and chronic kidney disease, are nonmodifiable risk factors that predispose patients to acquisition of MDROs. Other conditions that contribute to impaired host defenses, such as extremes of age or underlying immunosuppression, are not modifiable.[11]

Transmission

As noted previously, early cases of HAP are often due to endogenous community-acquired organisms in those patients without significant previous healthcare exposure. Patients with previous healthcare exposure are often already colonized with hospital-acquired pathogens. Sources of pathogens in the hospital include medical devices (eg, ventilator circuits) and the general hospital environment (eg, air and water). However, the most important source of pathogens is the hands of healthcare personnel, because of inadequate hand hygiene.[6]

Diagnosis and Surveillance

Diagnostic and surveillance definitions typically pertain specifically to VAP. Accurate diagnosis of VAP is very difficult and has generally been divided into clinical methods and invasive diagnostic methods. Clinical diagnosis of VAP (on the basis of fever, purulent sputum, leukocytosis, and infiltrate(s) visible on a chest x-ray) tends to be overly sensitive and not very specific. The clinical pulmonary infection score was developed to make clinical diagnosis of VAP more quantitative.[12] This score is calculated using a point system for 6 clinical variables (temperature, white blood cell count, tracheal secretions, oxygenation, chest x-ray abnormalities, and semiquantitative endotracheal aspirate findings); a score of more than 6 is considered highly suggestive of VAP. Use of this score has correlated well with use of bronchoscopy for diagnosis.

Invasive diagnostic methods typically use bronchoscopy to obtain lower respiratory tract specimens for quantitative culture, possibly by use of a protected specimen brush to limit contamination or by routine bronchoalveolar lavage. A patient is considered to have VAP if a culture grows more than 10^3 organisms from a specimen obtained by bronchoscopy with a protected specimen brush, or grows more than 10^4 organisms from a specimen obtained by bronchoscopy with bronchoalveolar lavage. Invasive diagnosis is considered to be more specific than clinical diagnosis.[6] However, false-negative results can occur for patients already receiving antimicrobial therapy. The most recent guidelines of the American Thoracic Society and the Infectious Diseases Society of America recommend that lower respiratory tract culture specimens be obtained (either by tracheal aspiration or by bronchoscopy) from all patients with clinically suspected VAP, when feasible.[1] However, invasive diagnostic testing is not indicated for routine surveillance or to document adequate treatment of VAP. It has not been definitively shown that the method of diagnosis of VAP affects clinical outcomes.[13] Use of invasive diagnostic methods is also recommended for research studies on VAP.

For hospital epidemiologists and infection preventionists, use of the National Healthcare Safety Network definitions for VAP is recommended for surveillance.[4,14] These definitions include 3 sets of criteria: clinical, radiographic, and microbiological. On the basis of these criteria, patients are categorized as meeting 1 of 3 different definitions of pneumonia: clinically defined pneumonia, pneumonia with specific laboratory findings, or pneumonia in immunocompromised patients. Flow diagrams to help determine if patients meet the criteria for pneumonia are available from the National Healthcare Safety Network. It should be noted that the surveillance definition for VAP often correlates poorly with diagnosis of VAP by clinicians. In general, surveillance for VAP must be performed actively, because reliance on discharge diagnosis codes lacks sensitivity and specificity.[15] The outcome measure of interest for infection prevention and control programs is the VAP rate, or incidence density, which is typically measured as episodes per 1,000 ventilator-days.

Treatment

The guiding principle of treatment of HAP is that early administration of appropriate antimicrobial therapy reduces mortality. Delaying treatment with antimicrobials for more than 24 hours after patients meet criteria for VAP results in increased mortality.[16] There are multiple studies demonstrating that adequate empirical antimicrobial therapy (ie, antimicrobials directed against organisms later isolated from culture) results in a lower mortality than does inadequate empirical therapy.[17,18] In addition, inadequate empirical antimicrobial therapy has been identified as a strong independent predictor of mortality in multivariate regression analyses.[19,20]

The American Thoracic Society and the Infectious Diseases Society of America guidelines for treatment of HAP recommend empirical antimicrobial therapy based on the time of onset of HAP after admission to the hospital.[1] For patients with early HAP (onset 0–4 days after admission), treatment directed against community-acquired organisms is reasonable. However, exceptions should be made for patients with risk factors for colonization with MDROs, such as previous hospitalization. These patients, as well as patients with late HAP (onset 5 days or more after admission), should receive treatment with antimicrobials that have activity against nonenteric gram-negative bacilli and MRSA. The current guidelines recommend the use of a broad-spectrum β-lactam, along with a fluoroquinolone or aminoglycoside, for empirical treatment that has coverage for gram-negative organisms, as well as vancomycin or linezolid for empirical treatment that has coverage for MRSA. Of note, evidence supporting the use of dual empirical therapy for infection with gram-negative bacilli is lacking. In general, empirical therapy should also take into account local epidemiologic data and antimicrobial resistance patterns.

The need for early appropriate antimicrobial therapy must be balanced with the need to limit the use of broad-spectrum antimicrobials. Use of these drugs leads to unintended consequences, such as spread of antimicrobial resistance and development of other healthcare-associated infections, such as *Clostridium difficile* infection. One strategy to limit the inappropriate use of broad-spectrum antimicrobials is to use narrower-spectrum empirical therapy for low-risk patients, as described above. Another strategy is to limit the duration of antimicrobial use. In the past, VAP has typically been treated for 14–21 days. However, a recent study demonstrated that there was no difference in mortality between patients treated with an 8-day course and patients treated with a 15-day course for VAP.[21] Patients infected with nonenteric gram-negative bacilli, such as *P. aeruginosa*, did have a higher risk of relapse with the shorter course. Therefore, it is now recommended that patients with VAP, with the exception of those who have non-enteric gram-negative bacilli isolated from respiratory-tract cultures, receive a shorter course of antimicrobial therapy. A final strategy for limiting the use of broad-spectrum antimicrobials is "de-escalation" of therapy, or narrowing of its spectrum, once a causative organism has been isolated from a respiratory-tract culture.[22]

Strategies for Prevention

Multiple evidence-based guidelines for prevention of HAP now exist.[2,4,23] Preventive strategies can be divided into general strategies that focus on reducing the burden of all healthcare-associated infections and targeted strategies that focus specifically on VAP.

General Strategies

General strategies for preventing HAP focus on limiting the transmission and emergence of MDROs in the hospital setting.

Preventing Transmission of Hospital-Acquired Pathogens
Adherence to performance of hand hygiene before and after patient care is recommended for prevention of healthcare-associated infections and is included in the guidelines for prevention of HAP.[24] Use of contact precautions with gowns and gloves is currently recommended by the Centers for Disease Control and Prevention as part of a multifaceted strategy for prevention of transmission of MDROs.[25] In addition, the role of the hospital environment in transmission of MDROs is now becoming increasingly clear, and interventions to improve environmental cleaning have been shown to reduce transmission of MDROs.[26]

Preventing Development of Antimicrobial Resistance
The presence of an indwelling device, such as a central catheter or urinary catheter, increases the risk of colonization with MDROs. Therefore, these devices should be used only when absolutely necessary and should be removed as soon as possible. Reduction in use of broad-spectrum antimicrobials is the most important strategy to prevent development of antimicrobial resistance. As discussed above, de-escalation of therapy and use of shorter courses of therapy are 2 approaches to achieving this goal. Antimicrobial stewardship programs are another strategy for reducing use of antimicrobials.[27] Use of antimicrobial cycling and dual therapy to prevent emergence of resistance are currently not recommended because of insufficient evidence.

Targeted Strategies

Targeted strategies focus on the mechanisms by which VAP occurs: endotracheal intubation and mechanical ventilation, aspiration, bacterial colonization of the respiratory and upper gastrointestinal tracts, and contamination of respiratory equipment (Table 14-2). Often,

Table 14-2. Targeted preventive strategies for ventilator-associated pneumonia

Targeted risk factor	Recommended	Not routinely recommended
Endotracheal intubation and mechanical ventilation	• Use of noninvasive ventilation in carefully selected patients • Minimizing the duration of mechanical ventilation with use of weaning protocols	...
Aspiration	• Semirecumbent patient positioning • Use of cuffed endotracheal tubes • Suctioning of subglottic secretions • Postpyloric feeding to avoid overdistention of stomach	• Use of oscillating beds to enhance clearance of secretions
Bacterial colonization of respiratory and upper gastrointestinal tracts	• Orotracheal intubation (over nasotracheal intubation) to reduce risk of sinusitis • Oral care with antiseptic solution • Use of acid-suppressive agents for stress ulcer prophylaxis only in high-risk patients	• Use of selective oropharyngeal or digestive decontamination • Use of antiseptic-coated endotracheal tubes
Contamination of respiratory equipment	• Removal of condensate from ventilator circuits • Proper cleaning, disinfection, and sterilization of reusable respiratory equipment	• Frequent ventilator circuit changes

many of these strategies are used together as a "bundle" approach for prevention of VAP.[28]

Noninvasive Ventilation and Minimizing the Duration of Mechanical Ventilation

Endotracheal intubation is a major risk factor for development of VAP in patients who require mechanical support because of respiratory failure. The use of noninvasive ventilation or positive-pressure mask ventilation has been shown to decrease the need for endotracheal intubation and mechanical ventilation for patients with respiratory failure due to exacerbations of chronic obstructive pulmonary disease or congestive heart failure.[29,30] Furthermore, use of noninvasive ventilation has been shown to improve the survival rate, to reduce rates of all healthcare-associated infections, including VAP, and to reduce antimicrobial use.[31,32] Therefore, it is now recommended to use noninvasive ventilation whenever possible, in carefully selected patients.

If endotracheal intubation and mechanical ventilation are necessary, it is recommended that the duration of mechanical ventilation be minimized as much as possible. Patients who are undergoing mechanical ventilation should receive daily assessments of their readiness to be weaned from ventilation, with use of weaning protocols.[2,4] This typically involves interruption of sedation to determine if a patient is ready to be weaned from the ventilator. While this approach has been shown to decrease the duration of mechanical ventilation,[33] it has not definitively been shown to reduce the incidence of VAP.

Reducing the Risk of Aspiration

To reduce the risk of aspiration, it is recommended that patients undergoing mechanical ventilation be maintained in a semirecumbent position (30°–45° elevation of the head of the bed).[2,4] This should be done unless there are specific contraindications, such as underlying conditions or procedures that require patients to remain in a supine position. It has been shown that elevation of the head of the bed to 45° reduces the risk of aspiration of gastric contents.[34] In addition, there are several studies demonstrating the effectiveness of semirecumbent positioning in decreasing the incidence of VAP, including a randomized, controlled trial.[29,35] However, maintaining patients in a semirecumbent position can be difficult to achieve,[36] and in one study this strategy was not associated with a decrease in tracheal colonization or VAP.[37] Adherence to maintaining semirecumbent positioning of mechanically ventilated patients can be facilitated by use of beds that provide continuous monitoring of the angle of incline.

For patients who are intubated, use of a cuffed endotracheal tube is recommended to reduce the risk of aspiration,[2,4,38] and cuff pressure should be maintained at a minimum of 20 cm H_2O. Suctioning of subglottic secretions is also recommended.[39] Although not addressed in any of the guidelines for prevention of VAP, early tracheotomy (within 48 hours after intubation) has been shown to reduce the risk of VAP, compared with delayed tracheotomy (more than 14 days after intubation).[40] Tracheotomy prevents aspiration of pooled oral secretions and often allows for more rapid weaning from the ventilator, as well as decreased use of sedative medications. Early tracheotomy may be a useful strategy for prevention of VAP in patients for whom prolonged mechanical ventilation (for more than 7 days) is anticipated.

Additional strategies to prevent aspiration in intubated patients include avoidance of unplanned extubation and reintubation and use of postpyloric feeding to avoid overdistention of the stomach.[41,42] Of note, postpyloric feeding has been shown to prevent microaspiration but does not appear to reduce the risk of VAP, when compared with gastric feeding. However, use of postpyloric feeding whenever possible is still recommended in the most recent guidelines for prevention of VAP.[2,4] Use of oscillating beds to enhance clearance of secretions is not routinely recommended.

Reducing Colonization of the Respiratory and Upper Gastrointestinal Tracts

The role of colonization of the respiratory and upper gastrointestinal tracts in the development of VAP continues to be debated, and there are some clear recommendations regarding this issue, as well as some topics that are not yet resolved.

For prevention of VAP, orotracheal intubation is recommended over nasotracheal intubation.[2,4] Nasotracheal intubation predisposes patients to sinusitis, which increases the risk of VAP.[43]

Patients undergoing mechanical ventilation should also receive regular oral care with an antiseptic solution, although there are no specific recommendations in the guidelines regarding which solution to use or how frequently to provide oral care.[2,4] A study of nursing home patients demonstrated that tooth brushing after each meal decreased the risk of developing pneumonia.[44] Another study performed in an ICU showed that oral care 3 times daily with swabs soaked in povidone-iodine and routine tooth brushing reduced the rate of colonization with pathogenic bacteria in the oral cavity and led to a significant decrease in the incidence of VAP.[45] Until recently, use of chlorhexidine for oral care had been studied only in patients undergoing cardiac surgery,[46] but a recent randomized controlled trial of oral decontamination with 2% chlorhexidine solution for ventilated patients demonstrated a reduced rate of oropharyngeal colonization with gram-negative bacilli and a significant decrease in the incidence of VAP.[47] The only adverse effect of using chlorhexidine noted in the study was irritation of the oral mucosa, which occurred in approximately 10% of patients.[47] It should be noted that other studies examining use of chlorhexidine solutions for oral care utilized 0.12% chlorhexidine solutions.

Acid-suppressive agents, such as histamine 2 (H2) blockers and proton pump inhibitors, that are given to hospitalized patients for stress ulcer prophylaxis can lead to colonization of the upper gastrointestinal tract with pathogenic bacteria. Sucralfate, an agent that is also used for stress ulcer prophylaxis, is not thought to lead to bacterial colonization of the upper gastrointestinal tract since it has no effect on gastric pH. Seven meta-analyses have addressed the risk of VAP in critically ill patients receiving stress ulcer prophylaxis.[48] Four of these studies demonstrated a reduction in the risk of VAP, and 3 demonstrated nonsignificant trends toward reduction, with use of sucralfate, compared with use of histamine 2 blockers. However, a large clinical trial showed that there was no difference in the risk of VAP and that patients receiving sucralfate for stress ulcer prophylaxis were more likely to have gastrointestinal bleeding events.[49] At this point, it is not clear if preferential use of sucralfate over acid-suppressive agents for stress ulcer prophylaxis is useful for prevention of VAP. However, it is recommended that acid-suppressive therapy be reserved for those patients who are at highest risk for developing stress ulcers.[2,4] Another approach that has been studied to lower gastric pH in order to prevent bacterial colonization is use of acidified tube feeding. However, this strategy is not recommended, since it has not been shown to reduce the risk of VAP but is associated with an increased risk of acidemia and gastrointestinal bleeding.[50]

Another issue that remains unresolved at this point is use of selective oropharyngeal decontamination or selective digestive tract decontamination for prevention of VAP. Selective oropharyngeal decontamination refers to the application of topical nonabsorbable antimicrobials (usually polymixin, an aminoglycoside, and amphotericin) to the oropharynx. Selective digestive tract decontamination typically includes administration of similar topical antimicrobials to both the oropharynx and the stomach (by means of a nasogastric tube), as well as administration of a short course of intravenous antimicrobials. Both methods reduce colonization with commensal bacteria, as well as enteric gram-negative bacilli that may be acquired in the hospital. A meta-analysis demonstrated that both methods reduce the risk of respiratory infections among critically ill patients and that selective digestive tract decontamination reduces overall mortality.[51] A recent large randomized controlled trial demonstrated that both methods reduced mortality, compared with standard care.[52] However, these methods have largely been studied in settings where there is a low burden of MDRO infection, so these findings may not be applicable to settings where VAP is often caused by MDROs. For this reason, and because of the reasons outlined above for limiting the use of antimicrobials as much as possible, the use of selective oropharyngeal decontamination, selective digestive tract decontamination, or systemic antimicrobial prophylaxis for prevention of VAP is currently not routinely recommended.[2,4]

Biofilm formation on the endotracheal tube is now believed to play a possible role in the development of VAP. Several studies using experimental models have demonstrated that antiseptic-coated endotracheal tubes reduce the incidence of bacterial colonization of endotracheal tubes, ventilatory circuits, and the lungs.[53,54] A recent randomized controlled trial compared use of silver-coated endotracheal tubes with standard endotracheal tubes.[55] Patients for whom the silver-coated endotracheal tubes were used had a lower risk of VAP and delayed onset of VAP, but there were no differences in other outcomes, such as duration of intubation, length of stay, or mortality. There are currently no definitive recommendations regarding use of antiseptic-coated endotracheal tubes in the published guidelines.[2,4] In the future, use of antiseptic-coated endotracheal tubes may be considered in settings where the incidence of VAP has been reduced, but not eliminated, using other preventive measures.

Minimizing Contamination of Respiratory Equipment

There has been considerable interest in the role of contaminated respiratory equipment, particularly mechanical ventilatory circuits, in the development of VAP. The tubing and condensation within ventilatory circuits are frequently colonized with bacteria, especially in those areas that are closest to the patient.[56] The bacteria isolated from ventilatory circuits have been shown to correlate closely with the bacteria isolated from patients' sputum. It is currently recommended that condensate be removed from ventilatory circuits, with the circuit remaining closed during removal.[2,4,57] However, it has not been shown that frequent ventilator circuit changes result in a lower risk of VAP.[58] Therefore, ventilator circuits should be changed only when visibly contaminated or not functioning properly. Guidelines regarding cleaning, disinfection, and sterilization of reusable respiratory equipment are available.[4]

Implementation of Preventive Strategies and Reporting

In order for preventive strategies to be implemented, healthcare personnel who care for patients at risk for VAP must be educated about the epidemiology, risk factors, and patient outcomes associated with VAP. Close involvement of physician and nursing leadership in implementation and ongoing use of preventive strategies is critical. For prevention of VAP, protocols for maintaining respiratory equipment and for promotion of noninvasive ventilation should be put in place. For other preventive measures, such as semirecumbent positioning and oral care, grouping interventions into bundles may be useful to promote adherence.[28] Often a checklist with each component of the bundle is used: items are checked off daily as they are completed, usually by nursing staff. In addition to promoting adherence, checklists also provide process measures, since the percentage of time each component of the bundle is implemented can be easily calculated.

Internal feedback regarding both process and outcome measures should be provided to hospital administration and appropriate healthcare personnel. This helps to establish which interventions are effective in reducing VAP rates and where further work is needed. Given the difficulties associated with diagnosing VAP, external reporting with the purpose of comparing VAP rates between institutions is not currently recommended. However, some states now have made it mandatory to report healthcare-associated infection rates. Local requirements for reporting are typically available from local and state health departments.

Conclusion

HAP occurs frequently and is associated with significant cost, morbidity, and mortality. Most of the research regarding prevention of HAP has focused on VAP. The following targeted strategies are recommended for prevention of VAP: use of noninvasive ventilation whenever possible; semirecumbent positioning, use of a cuffed endotracheal tube, and avoidance of overdistention of the stomach to prevent aspiration; and oral care with an antiseptic solution. Questions remain about many preventive strategies, including early tracheotomy, avoidance of acid-suppressive medications for stress ulcer prophylaxis, use of selective oropharyngeal decontamination or selective digestive tract decontamination, and use of antiseptic-coated endotracheal tubes. Further research may help to answer some of these questions in the future, but for now, definitive recommendations regarding use of these strategies are not available. In addition, data regarding prevention of pneumonia in non-ICU patients who are not mechanically ventilated are currently lacking, and further studies regarding this population are needed. Putting preventive strategies into practice requires education and teamwork. Feedback regarding process and outcome measures is essential to maintain progress in reducing the incidence of HAP and all healthcare-associated infections.

References

1. American Thoracic Society, Infectious Diseases Society of America. Guidelines for the management of adults with hospital-acquired, ventilator-associated, and healthcare-associated pneumonia. *Am J Respir Crit Care Med* 2005; 171(4):388–416.

2. Tablan OC, Anderson LJ, Besser R, Bridges C, Hajjeh R. Healthcare Infection Control Practices Advisory Committee, Centers for Disease Control and Prevention. Guidelines for preventing health-care–associated pneumonia, 2003: recommendations of the CDC and the Healthcare Infection Control Practices Advisory Committee. *MMWR Recomm Rep* 2004;53(RR-3):1–36.

3. Warren DK, Shukla SJ, Olsen MA, et al. Outcome and attributable cost of ventilator-associated pneumonia among intensive care unit patients in a suburban medical center. *Crit Care Med* 2003;31(5):1312–1317.

4. Coffin SE, Klompas M, Classen, et al. Strategies to prevent ventilator-associated pneumonia in acute care hospitals. *Infect Control Hosp Epidemiol* 2008;29(Suppl 1):S31–S40.

5. Cook DJ, Walter SD, Cook RJ, et al. Incidence of and risk factors for ventilator-associated pneumonia in critically ill patients. *Ann Intern Med* 1998;129(6):433–440.

6. Chastre J, Fagon JY. Ventilator-associated pneumonia. *Am J Respir Crit Care Med* 2002;165(7):867–903.

7. Marik PE, Careau P. The role of anaerobes in patients with ventilator-associated pneumonia and aspiration pneumonia: a prospective study. *Chest* 1999;115(1):178–183.

8. Bahrani-Meugeot FK, Paster BJ, Coleman S, et al. Molecular analysis of oral and respiratory bacterial species associated with ventilator-associated pneumonia. *J Clin Microbiol* 2007; 45(5):1588–1593.

9. Ibrahim EH, Ward S, Sherman G, Kollef MH. A comparative analysis of patients with early-onset vs. late-onset nosocomial pneumonia. *Chest* 2000;117(5):1434–1442.

10. Trouillet JL, Chastre J, Vuagnat A, et al. Ventilator-associated pneumonia caused by potentially drug-resistant bacteria. *Am J Respir Crit Care Med* 1998;157(2):531–539.

11. Kollef MH. Ventilator-associated pneumonia: a multivariate analysis. *JAMA* 1993;270(16):1965–1970.

12. Pugin J, Auckenthaler R, Mili N, Janssens JP, Lew PD, Suter PM. Diagnosis of ventilator-associated pneumonia by bacteriologic analysis of bronchoscopic and nonbronchoscopic "blind" bronchoalveolar lavage fluid. *Am Rev Respir Dis* 1991;143(5 Pt 1):1121–1129.

13. Ruiz M, Torres A, Ewig S, et al. Noninvasive versus invasive microbial investigation in ventilator-associated pneumonia: evaluation of outcome. *Am J Respir Crit Care Med* 2000; 162(1):119–125.

14. National Healthcare Safety Network Web page. http://www. cdc.gov/nhsn. Accessed June 25, 2009.

15. Sherman ER, Heydon KH, St John KH, et al. Administrative data fail to accurately identify cases of healthcare-associated infection. *Infect Control Hosp Epidemiol* 2006;27(4): 332–337.

16. Irequi M, Ward S, Sherman G, Fraser VJ, Kollef MH. Clinical importance of delays in the initiation of appropriate antibiotic treatment for ventilator-associated pneumonia. *Chest* 2002; 122(1):262–268.

17. Luna CM, Vujavich P, Niederman MS, et al. Impact of BAL data on the therapy and outcome of ventilator-associated pneumonia. *Chest* 1997;111(3):676–685.

18. Rello J, Gallego M, Mariscal D, Sonora R, Valles J. The value of routine microbial investigation in ventilator-associated pneumonia. *Am J Resp Crit Care Med* 1997; 156(1):196–200.

19. Celis R, Torres A, Gatell JM, Almela M, Rodriguez-Roisin R, Agusti-Vidal A. Nosocomial pneumonia: a multivariate analysis of risk and prognosis. *Chest* 1988;93(2):318–324.

20. Torres A, Aznar R, Gatell JM, et al. Incidence, risk, and prognosis factors of nosocomial pneumonia in mechanically ventilated patients. *Am Rev Respir Dis* 1990;142(3):523–528.

21. Chastre J, Wolff M, Fagon JY, et al. Comparison of 8 vs 15 days of antibiotic therapy for ventilator-associated pneumonia: a randomized trial. *JAMA* 2003;290(19):2588–2598.

22. Rello J, Vidaur L, Sandiumenge A, et al. De-escalation therapy in ventilator-associated pneumonia. *Crit Care Med* 2004; 32(11):2183–2190.

23. Dodek P, Keenan S, Cook D, et al. Evidence-based clinical practice guideline for the prevention of ventilator-associated pneumonia. *Ann Intern Med* 2004;141(4):305–313.

24. Pittet D, Benedetta A, Boyce J, for the World Health Organization World Alliance for Patient Safety First Global Patient Safety Challenge Core Group of Experts. The World Health Organization guidelines on hand hygiene in health care and their consensus recommendations. *Infect Control Hosp Epidemiol* 2009;30(7):611–622.

25. Siegel JD, Rhinehart E, Jackson M, Chiarello M, and the Healthcare Infection Control Practices Advisory Committee. 2007 Guideline for isolation precautions: preventing transmission of infectious agents in healthcare settings, June 2007. Atlanta, GA: Centers for Diseases Control and Prevention; 2007. http://www.cdc.gov/ncidod/dhqp/gl_isolation.html. Accessed September 24, 2009.

26. Boyce JM. Environmental contamination makes an important contribution to hospital infection. *J Hosp Infect* 2006; 65(Suppl 2):50–54.

27. Dellit TH, Owens RC, McGowan JE, et al. Infectious Diseases Society of America and the Society for Healthcare Epidemiology of America guidelines for developing an institutional program to enhance antimicrobial stewardship. *Clin Infect Dis* 2007;44(2):159–177.

28. Resar R, Pronovost P, Haraden C, Simmonds T, Rainey T, Nolan T. Using a bundle approach to improve ventilator care processes and reduce ventilator-associated pneumonia. *Jt Comm J Qual Patient Saf* 2005;31(5):243–248.

29. Ram F, Lightowler J, Wedzicha J. Non-invasive positive pressure ventilation for treatment of respiratory failure due to exacerbations of chronic obstructive pulmonary disease. Cochrane Review. In: *The Cochrane Library* 2003;3. Oxford: Update Software. CD0044104.

30. Park M, Sangean MC, Volpe MS, et al. Randomized, prospective trial of oxygen, continuous positive airway pressure, and bilevel positive airway pressure by face mask in acute cardiogenic pulmonary edema. *Crit Care Med* 2004;32(12): 2546–2548.

31. Girou E, Schortgen F, Delcalux C, et al. Association of non-invasive ventilation with nosocomial infections and survival in critically ill patients. *JAMA* 2000;284(18):2361–2367.

32. Girou E, Brun-Buisson C, Taille S, Lemaire F, Brochard L. Secular trends in nosocomial infections and mortality associated with noninvasive ventilation in patients with exacerbation of COPD and pulmonary edema. *JAMA* 2003;290(22):2985–2991.

33. Kress JP, Pohlman AS, O'Connor MF, Hall JB. Daily interruption of sedative infusions in critically ill patients undergoing mechanical ventilation. *N Engl J Med* 2000;342(20):1471–1477.

34. Torres A, Serra-Batlles J, Ros E, et al. Pulmonary aspiration of gastric contents in patients receiving mechanical ventilation: the effect of body position. *Ann Intern Med* 1992;116(7):540–543.

35. Drakulovic MB, Torres A, Bauer TT, Nicolas JM, Nogue S, Ferrer MN. Supine body position as a risk factor for mechanically ventilated patients: a randomised trial. *Lancet* 1999;354(9193):1851–1858.

36. van Nieuwenhoven CA, Vandenbroucke-Grauls C, van Tiel FH, et al. Feasibility and effects of semirecumbent position to prevent ventilator-associated pneumonia: a randomized study. *Crit Care Med* 2006;34(2):396–402.

37. Girou E, Buu-Hoi A, Stephan F, et al. Airway colonisation in long-term mechanically ventilated patients: effect of semirecumbent position and continuous subglottic suctioning. *Intensive Care Med* 2004;30(2):225–233.

38. Rello J, Sonora R, Jubert P, Artigas A, Rue M, Valles J. Pneumonia in intubated patients: role of respiratory airway care. *Am J Respir Crit Care Med* 1996;154(1):111–115.

39. Dezfulian C, Shojania K, Collard HR, Kim HM, Matthay MA, Saint S. Subglottic secretion drainage for preventing ventilator-associated pneumonia: a meta-analysis. *Am J Med* 2005;118(1):11–18.

40. Rumbak MJ, Newton M, Truncale T, Schwartz SW, Adams, JW, Hazard PB. A prospective, randomized, study comparing early percutaneous dilational tracheotomy to prolonged translaryngeal intubation (delayed tracheotomy) in critically ill medical patients. *Crit Care Med* 2004;32(8):1689–1694.

41. Torres A, Gatell JM, Aznar E, et al. Re-intubation increases the risk of nosocomial pneumonia in patients needing mechanical ventilation. *Am J Respir Crit Care Med* 1995;152(1):137–141.

42. Heyland DK, Drover JW, MacDonald S, Novak F, Lam M. Effect of postpyloric feeding on gastroesophageal regurgitation and pulmonary microaspiration: results of a randomized controlled trial. *Crit Care Med* 2001;29(8):1495–1501.

43. Holzapfel L, Chastang C, Demingeon G, Bohe J, Piralla B, Coupry A. A randomized study assessing the systematic search for maxillary sinusitis in nasotracheally mechanically ventilated patients: influence of nosocomial maxillary sinusitis on the occurrence of ventilator-associated pneumonia. *Am J Respir Crit Care Med* 1999;159(3):695–701.

44. Yoneyama T, Yoshida M, Ohrui T, et al. Oral care reduces pneumonia in older patients in nursing homes. *J Am Geriatr Soc* 2002;50(3):430–433.

45. Mori H, Hirasawa H, Oda S, Shiga H, Matsuda K, Nakamura M. Oral care reduces incidence of ventilator-associated pneumonia in ICU populations. *Intensive Care Med* 2006;32(2):230–236.

46. Houston S, Hougland P, Anderson JJ, LaRocco M, Kennedy V, Gentry LO. Effectiveness of 0.12% chlorhexidine gluconate oral rinse in reducing prevalence of nosocomial pneumonia in patients undergoing heart surgery. *Am J Crit Care* 2002;11(6):567–570.

47. Tantipong H, Morkchareonpong C, Jaiyindee S, Thamlikitkul V. Randomized controlled trial and meta-analysis of oral decontamination with 2% chlorhexidine solution for the prevention of ventilator-associated pneumonia. *Infect Control Hosp Epidemiol* 2008;29(2):131–136.

48. Collard HR, Saint S, Matthay MA. Prevention of ventilator-associated pneumonia: an evidence-based systematic review. *Ann Intern Med* 2003;138(6):494–501.

49. Cook D, Guyatt G, Marshall J, et al. A comparison of sucralfate and ranitidine for the prevention of upper gastrointestinal bleeding in patients requiring mechanical ventilation. Canadian Critical Care Trials Group. *N Engl J Med* 1994;330(6):377–381.

50. Heyland DK, Cook DJ, Schoenfeld PS, Frietag A, Varon J, Wood G. The effect of acidified enteral feeds on gastric colonization in critically ill patients: results of a multicenter randomized trial. Canadian Critical Care Trials Group. *Crit Care Med* 1999;27(11):2399–2406.

51. Liberati A, D'Amico R, Pifferi, Torri V, Brazzi L. Antibiotic prophylaxis to reduce respiratory tract infections and mortality in adults receiving intensive care. *Cochrane Database Syst Rev* 2004;(1):CD000022.

52. de Smet AMGA, Kluytmans JAJW, Cooper BS, et al. Decontamination of the digestive tract and oropharynx in ICU patients. *N Engl J Med* 2009;360(1):20–31.

53. Pacheco-Fowler V, Gaonkar T, Wyer PC, Modak S. Antiseptic impregnated endotracheal tubes for the prevention of bacterial colonization. *J Hosp Infect* 2004;57(2):170–174.

54. Berra L, De Marchi L, Yu ZX, Laquerriere P, Baccarelli A, Kolobow T. Endotracheal tubes coated with antiseptics decrease bacterial colonization of the ventilator circuits, lungs, and endotracheal tube. *Anesthesiology* 2004;100(6):1446–1456.

55. Kollef MH, Afessa B, Anzeuto A, et al. Silver-coated endotracheal tubes and incidence of ventilator-associated pneumonia: the NASCENT randomized trial. *JAMA* 2008;300(7):805–813.

56. Craven DE, Goularte TA, Make BJ. Contaminated condensate in mechanical ventilator circuits: a risk factor for nosocomial pneumonia? *Am Rev Respir Dis* 1984;129(4):625–628.

57. Kollef MH. Prevention of hospital-associated pneumonia and ventilator-associated pneumonia. *Crit Care Med* 2004;32(6):1396–1405.

58. Stamm AM. Ventilator-associated pneumonia and frequency of circuit changes. *Am J Infect Control* 1998;26(1):71–73.

Chapter 15 Basics of Surgical Site Infection Surveillance and Prevention

Lisa L. Maragakis, MD, MPH, and Trish M. Perl, MD, MSc

During the early years of modern surgery, many patients died from "wound sepsis." Despite scientific advances and improved practices, infection at the surgical site continues to cause substantial morbidity, mortality, and increased cost of health care. Surgical site infection (SSI) is the second most common type of healthcare-associated infection in the United States, accounting for approximately 20% of all healthcare-associated infections.[1] Approximately 500,000 SSIs occur annually, accounting for 3.7 million excess hospital days and more than $1.6 billion in excess hospital costs.[2] Patients who develop an SSI have a risk of death between 2- and 11-fold higher than that of patients without an SSI.[3-5] These numbers likely underestimate the magnitude of the problem, yet they highlight the tremendous human and financial costs of SSIs and the importance of SSI surveillance and prevention.

The Centers for Disease Control and Prevention (CDC) conducted the Study of the Efficacy of Nosocomial Infection Control (SENIC) in order to investigate and document the cost-effectiveness of infection prevention and control activities, including surveillance.[6] These and many subsequent investigators concluded that infection surveillance programs that report SSI rates to surgeons can decrease overall SSI rates by at least 32%–50%.[6-8] However, several important questions about surveillance for SSIs remain:

- Which surveillance methods are best?
- Which surgical patients should we survey?
- How should we conduct postdischarge surveillance?
- How can we effectively use technology and automated systems for surveillance?

- How can surveillance activities be used to develop interventions to prevent SSI?

The purpose of this chapter is to explore some of these questions and to summarize the basic steps required to implement a hospital- or healthcare system–based SSI surveillance and prevention program.

SSI Surveillance: Objectives and Definitions

Objectives

The main objective of SSI surveillance is to prevent SSIs, thereby reducing morbidity and improving patient care. To achieve this goal, infection prevention and control personnel must first determine the baseline SSI rate to assess the magnitude of the problem for each type of surgical procedure in an institution. By monitoring SSI rates over time, epidemiology staff can identify clusters of infections and discover overall trends. SSI surveillance facilitates comparisons, both within an institution and with other similar institutions. The objective of such comparisons is timely recognition of SSI rates that are above the baseline rate for the institution or above national benchmark rates, such as those published by the CDC National Healthcare Safety Network (NHSN; formerly the National Nosocomial Infection Surveillance [NNIS] system). Analysis of trends and internal and external comparisons allow facilities to identify potential patient safety problems and to promptly implement appropriate corrective measures. Infection prevention and control personnel should provide surveillance data to surgeons and other members of the surgical team as education and as an intervention to prevent SSIs. Finally, the SSI surveillance program is used to assess the impact of interventions designed and implemented for SSI prevention (eg, antimicrobial prophylaxis protocols or education to reinforce use of aseptic technique).

Definitions

Before implementing an SSI surveillance system, infection prevention and control personnel must agree on a precise definition for SSI. Ideally, in each hospital, the definition should remain unchanged so that infection prevention and surgical staff can compare data over time and evaluate interventions implemented to reduce SSI rates. The definition must be simple to use, accepted by nurses and surgeons, and applied consistently. Standard SSI definitions are available from the CDC;[9] they are shown in Figure 15-1. Use of a standard surveillance definition allows for comparison of SSI rates across institutions and between surgeons. It is important to note that a surveillance definition is not always the same as a clinical definition of wound infection, since these definitions are used for different purposes.

Ehrenkranz et al.[10] powerfully demonstrated why we must use objective criteria to identify SSIs, and why we must apply such definitions consistently. The infection preventionist at a 200-bed general community hospital reported that a neurosurgeon's SSI rate was

excessive, and the infection control committee repeatedly investigated the surgeon's practice. When the surgeon proposed to terminate his practice, the hospital administrator asked consultants to perform an independent investigation. Ehrenkranz et al.[10] determined that the infection preventionist had incorrectly categorized noninfected patients as being infected. The consultants recommended that the infection control committee establish clear standards for what constituted a confirmed SSI and that they be applied routinely and consistently. Over the next 2 years, none of the surgeon's patients was reported to have had an SSI.

According to the CDC NHSN definition, an SSI is categorized as either incisional SSI or organ-space SSI. Incisional SSI is classified further as either superficial (involving only the skin and subcutaneous tissue) or deep (involving deep soft tissues of the incision) (Figure 15-1). The definition of superficial-incisional SSI requires that at least 1 of the following 4 events occurs within 30 days after the surgical procedure:

1. Purulent drainage from the superficial incision.
2. Isolation of organisms from culture of an aseptically obtained sample of fluid or tissue from the superficial incision.
3. At least 1 of the following signs or symptoms of infection: pain or tenderness, localized swelling, redness, or heat, and deliberate opening of the superficial incision by the surgeon, unless incision is culture negative for pathogens.
4. Diagnosis of a superficial incisional SSI by the surgeon or attending physician.

Deep-incisional SSI and organ-space SSI are defined similarly.[9] When an implant is placed, infection that occurs within 1 year after the procedure is considered to be an SSI.

Surveillance Methods

Healthcare institutions use many different surveillance methods to identify healthcare-associated infections. We review several common surveillance methods and discuss whether each method can be used effectively to survey for SSIs.

100% Chart Review and Wound Examination

After conducting prospective SSI surveillance for 10 years, Olson and Lee[8] concluded that daily hospital medical-chart review and examination of postoperative wounds is probably the most sensitive and rigorous way

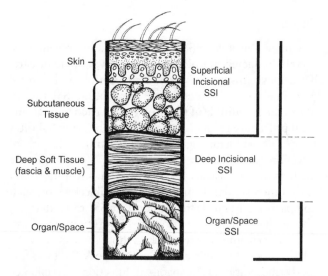

Figure 15-1. Schematic of surgical site infection (SSI) anatomy and appropriate classification. This figure depicts the cross-sectional anatomy of a surgical incision upon which is superimposed the most recent classification for SSI and the definition of an infection at each site. Reproduced with permission from Horan et al.[9]

to perform SSI surveillance. Although these methods are tedious and time-consuming, some experts still consider them to be the gold standard and recommend that SSI surveillance include daily examination of postoperative wounds. To facilitate this method, infection prevention and control personnel or surgeons can train staff nurses who see the wounds during routine care to recognize signs of infection and report all clinically suspicious wounds to the infection preventionist. The infection preventionist then examines all such wounds and determines which meet the criteria for infection.[11] Clearly, this comprehensive approach is not practical or feasible in large hospitals with many surgical patients, especially if most SSIs occur after discharge from the initial hospitalization.

100% Chart Review

Cardo et al.[12] compared surveillance performed by an infection preventionist who reviewed surgical patients' medical records and discussed each patient's progress with nurses and physicians with surveillance performed by a hospital epidemiologist who reviewed the patients' medical records and examined their wounds. The infection preventionist identified 84% of SSIs noted by the hospital epidemiologist. The authors concluded that accurate data on SSIs can be collected by persons who do not examine the operative wounds directly.[12] Haley et al.[13] corroborated this finding. However, the quality of the information gleaned from medical records depends on their completeness and on the reviewer's experience. Furthermore, if the infection prevention and control team has limited resources for surveillance, review of the medical records of all surgical patients will not be possible. Such programs must either focus on specific surgical subpopulations or use other less time-consuming surveillance methods.

Targeted SSI Surveillance: 100% Chart Review for Selected Procedures

Some infection prevention and control programs aim to survey all surgical patients by chart review, even healthy young patients undergoing minor procedures. However, such comprehensive surveillance for all surgical procedures consumes so much time that most programs must perform some form of targeted surveillance.

In most hospitals, approximately 70% of operative procedures are categorized as clean, which have a low microbial contamination burden and therefore a relatively low SSI risk (Table 15-1).[14] Some hospitals target SSI surveillance to only these clean operative procedures, on the assumption that SSIs are rarely preventable in the other wound classification categories with higher levels

Table 15-1. Wound classifications and their criteria, reflecting the extent of microbial contamination

Clean

　Elective, not emergent procedure
　No traumatic injury
　Primary closure
　No acute inflammation or break in surgical technique
　No entry into the respiratory, gastrointestinal, biliary, or genitourinary tracts
　Urgent or emergent procedure that is otherwise clean

Clean-contaminated

　Elective opening of respiratory, gastrointestinal, biliary, or genitourinary tracts with minimal spillage (eg, appendectomy)
　No infected urine or bile encountered

Contaminated

　Minor break in surgical technique during a clean procedure
　Nonpurulent inflammation noted at the time of surgery
　Gross spillage from gastrointestinal tract
　Infected urine or bile encountered
　Major break in surgical technique
　Penetrating trauma <4 hours old

Dirty

　Chronic open wounds to be grafted or covered
　Purulent inflammation noted at the time of surgery
　Preoperative perforation of respiratory, gastrointestinal, biliary, or genitourinary tracts
　Penetrating trauma >4 hours old

NOTE: Definitions adapted from Berard et al.[14]

of microbial contamination. However, the SENIC project demonstrated that SSI surveillance of contaminated or dirty procedures reduced SSI rates as effectively as did SSI surveillance of clean or clean-contaminated procedures.[15] Therefore, we advocate SSI surveillance that includes operative procedures from all wound classification categories, as resources and priorities allow.

Another potential strategy that can be combined with targeting clean procedures, or used independently, is to target surveillance to specific operative procedures that are performed frequently at an institution, since high-volume procedures pose SSI risk to a greater number of patients. Alternatively, surveillance can be targeted to those procedures for which SSIs cause substantially higher morbidity and mortality (eg, SSI after craniotomy or coronary artery bypass procedures generally has higher morbidity and mortality than does SSI after hernia repair). A third strategy is to target surveillance to operative procedures that have been identified as having high infection rates at a given institution. Depending on the resources available for surveillance, combinations of these targeting strategies may be employed, and the list of surveillance activities should be periodically reassessed to determine appropriate priorities. By targeting surveillance in this manner, infection prevention and control personnel can utilize available resources to focus SSI surveillance on high-volume or high-risk procedures, so that they can intervene immediately if the SSI rates are found to be high or to increase substantially.

SSI surveillance should not be targeted to specific hospital units, because this will likely underestimate the SSI rates and may miss problems or clusters of SSIs that occur in units not under surveillance. Hospitals that limit SSI surveillance to the surgical intensive care unit will miss many infections, because the average SSI occurs 7–12 days after surgery and most patients leave the surgical intensive care unit within a few days after surgery.

Targeted SSI Surveillance: 100% Chart Review of Patients at High Risk

Another way to perform targeted surveillance is to stratify patients according to their risk of developing an SSI and then survey a select group of high-risk patients. The ideal risk index is a simple additive scale that can be calculated at the end of surgery to predict the patients who are at high risk of SSI. Ideally, the risk index should be validated prospectively on specific services in individual hospitals to document that it predicts a patient's risk accurately. Risk indices are also used for risk-adjustment of SSI surveillance data, which is done so that infections are stratified and reported according to the relative risk for different patient populations.

Used alone, variables such as wound classification[14] have limited ability to stratify patients, because infection rates vary substantially within each group.[16] Therefore, investigators have developed risk indices that include multiple variables, to better predict which patients are at highest risk of developing SSI. In 1985, Haley et al.[17] published the SENIC risk index, which includes 4 factors: whether the operation was abdominal, whether the duration of surgery was more than 2 hours, the wound classification, and the number of discharge diagnoses. The SENIC risk index predicts the risk of SSI twice as well as the traditional wound classification, but one of the index components, the number of discharge diagnoses, must be obtained retrospectively. Therefore, Culver et al.[18] adapted the SENIC risk index and assessed the underlying severity of illness with the American Society of Anesthesiologists score, rather than with the number of discharge diagnoses. This risk index, known as the "NNIS risk index" (Table 15-2), also includes a component that accounts for the expected variability in the duration of the operative procedure: instead of using a constant cutoff point of 2 hours for length of surgery (as used by the SENIC

Table 15-2. The National Nosocomial Infections Surveillance (NNIS) System risk index

Risk factor	Score
American Society of Anesthesiologists preoperative assessment score of 3, 4, or 5	1
Operative procedure lasting longer than time T[a]	1
Operative procedure classified as either contaminated or dirty by the traditional wound classification system	1

NOTE: To calculate the total score, sum the scores for the factors present. The total score ranges from 0 to 3.
a The time T is the 75th percentile of the duration for each type of procedure and so depends on the procedure performed.

index), the NNIS index uses "time T," which is the 75th percentile of the duration for each type of operative procedure. The NNIS risk index is a simple additive scale with a score that ranges from 0 to 3, and the 3 index variables are usually available in the anesthesia record at the end of a surgical procedure.

The NNIS index has rarely been validated in populations other than in the NNIS participating hospitals, nor has it been compared with other risk indices. Some investigators report that the NNIS risk index does not stratify patients accurately by their SSI risk, and we and our colleagues found that it had low sensitivity (24%) and positive predictive value (43%) for identifying SSI after cardiothoracic operative procedures.[19]

The NNIS risk index and its predecessors may predict SSI poorly, because procedures and patients are too diverse; therefore, procedure-specific risk indices may be helpful to address this problem.[20-22] For example, Nichols et al.[20] published a risk index that accurately predicted postoperative septic complications in a subset of patients who underwent operations after penetrating abdominal trauma. Additionally, they showed that the risk factors included in the index could identify high-risk patients who benefited from prolonged antibiotic therapy (for 5 days) and delayed wound closure, and low-risk patients who did well with short-term antibiotic therapy (for 2 days) and primary wound closure.[20] Thus, this risk index not only stratifies patients by their risk of infection but also helps predict which patients will benefit from specific preventive strategies.

Selective Chart Review

Wenzel et al.[23] studied the sensitivity of reviewing selected medical records, compared with reviewing all medical records, to detect healthcare-associated infections.

The infection preventionist scanned the medical record for signs of possible healthcare-associated infection, such as fever or antibiotic use. If one or more signs of concern were noted, the infection preventionist evaluated the patient's medical record more thoroughly. Wenzel et al.[23] found that the infection preventionist, using this type of selective review, correctly identified 82%–94% of the healthcare-associated infections identified when all medical records were reviewed, and selective review saved many hours of the infection preventionist's time.[23] The success of the selective method depends upon the completeness and accuracy of the data that are screened to detect fever and utilization of antibiotics. Although the selective review method may be a good source of information about certain healthcare-associated infections, screening for fever and antibiotic utilization is probably not sufficiently sensitive to detect patients with SSI. Unfortunately, more useful indicators, such as the frequency with which wound dressings are changed and descriptions of discharge from the wound, are not readily available without a full review of each medical record. These limitations may be overcome with the increasing use of electronic medical records, which could allow algorithms and/or techniques such as natural language processing to be used to identify patients at risk of developing an SSI.

Antimicrobial utilization can be a useful indicator of healthcare-associated infection, but it has low sensitivity when used as the sole indicator.[23,24] There are several reasons why infection prevention and control programs should not rely solely upon pharmacy data when conducting SSI surveillance. Some patients may continue to receive antibiotic prophylaxis postoperatively, and some patients receive antimicrobial agents for infections that were present preoperatively (eg, peritonitis caused by a ruptured appendix). On the other hand, a patient whose infected wound is drained but who does not receive antimicrobials would not be identified by this surveillance method. This being said, Yokoe et al.[25,26] have championed the use of electronic records for surveillance of SSI, in which one of the most powerful predictive findings is use of antibiotics for more than the "expected" number of days.

Similarly, microbiologic data are an extremely useful component of SSI surveillance, but these data should not be used as the sole source for case-finding. Surveillance for healthcare-associated infections that relies only on microbiologic data has a sensitivity of only 33%–65%.[23,24] Only two-thirds of inpatients' surgical wounds and even fewer outpatients' wounds are sampled for culture, despite clinical evidence of SSI.[27] Frequently, surgeons do not obtain samples for culture because they treat SSI with operative drainage, and some feel it is not necessary to know the etiologic agent or its susceptibility

to antibiotics. It is important to note that a wound culture that yields an organism does not necessarily indicate that the patient has an SSI. Instead, the organism may be colonizing the wound. This is particularly relevant for burns or skin that is left to heal by secondary intention. The opposite is also true; a culture that does not yield organisms does not eliminate the possibility that the patient has an SSI. For example, wound cellulitis, deep wound abscess, or SSI due to an organism that is not detected by routine culture methods could result in a negative superficial-wound culture result.

Postdischarge Surveillance

The proportion of surgeries performed in the outpatient setting continues to increase, and the postoperative length of stay after inpatient surgery is decreasing. Therefore, the majority of SSIs occurs in the outpatient setting. Two studies found that 45%–72% of SSIs were detected after discharge from the hospital.[28,29] Sands et al.[30] found that 84% of 132 SSIs among 5,572 nonobstetric operative procedures in adult patients became apparent after the patients were discharged. According to one study, postdischarge SSI results in more outpatient visits, readmissions, emergency department visits, and use of home health services, compared with in-hospital SSI, and also results in vastly increased costs ($5,155 for the 8 weeks after discharge, vs $1,773 for in-hospital SSI).[31]

The cost and time required to perform postdischarge surveillance may discourage many infection prevention and control programs from instituting such systems. However, these programs must acknowledge that inpatient surveillance alone will vastly underestimate the actual incidence of SSI.[28,29,32,33] Therefore, strategies that identify SSIs after hospital discharge are necessary and will likely become increasingly available with the use of integrated electronic medical records.

The census approach—surveying each patient or physician for a defined time period—has been used in several studies to measure healthcare-associated infection rates during the postdischarge period.[29,35] Telephone surveys and questionnaires sent to patients or physicians also have been used.[30,35,36] Seaman and Lammers[35] found that patients, despite using verbal or printed instructions, were unable to recognize infections. They concluded that "reliance on printed instructions, telephone interviews, or any other means of patient self evaluation may not allow early recognition of infection"[35(p218)] and therefore should not be used for clinical investigations of wound healing. Similarly, Sands et al.[30] found that questionnaires sent to patients and to surgeons had sensitivities of 28% and 15%, respectively.

In a later study of postdischarge surveillance, Sands et al.[37] identified a method that appears more reliable.

They used automated pharmacy dispensing information, administrative data, and electronic records to identify postdischarge SSI. In this method, specific codes for diagnoses, tests, and treatments were evaluated for their ability to predict postdischarge SSI. They found that an automated system of surveillance of hospital discharge diagnosis codes and pharmacy dispensing data had a sensitivity of 77% and specificity of 94%, far better than use of questionnaires sent to patients and surgeons.[37]

Delgado et al.[29] found that most postdischarge SSIs occurred following clean surgery (eg, hernia, breast, or vascular surgery). A significant number of postdischarge SSIs also occurred after clean-contaminated surgery of the biliary tract, in their study. This may give some guidance to infection control personnel when they are choosing procedures to target for postdischarge surveillance. Delgado et al.[29] also found that risk factors for in-hospital SSI are not determinants of postdischarge SSI, with the exception of body mass index.

We suggest that infection prevention programs establish links with home healthcare agencies or other agencies that provide care for outpatients to develop mechanisms by which SSIs can be identified. Furthermore, they must not conclude prematurely that outpatient surgeries (ie, "same-day" surgeries) and endoscopic surgeries are not complicated by SSI.

The severity of illness among surgical patients is increasing, operative procedures are becoming more sophisticated, the average length of stay after surgery is decreasing, and more patients have operative procedures in the outpatient setting. While these dramatic shifts in patient populations, operative techniques, and healthcare delivery increase the need for outpatient surveillance, they also increase the complexity of performing accurate surveillance.

Electronic Data Surveillance

Computerization of medical records and technological advances promise to improve the quality of data and make surveillance more automated. Infection prevention programs are trying to find new automated methods by which they can identify patients with SSIs, particularly those whose signs and symptoms occur after discharge. Electronic data that can enhance SSI surveillance are available from a variety of sources, including data on antimicrobial prescribing, hospital readmissions, repeat surgery, and microbiologic culture results.[25,38-40] Sands et al.[37] found that individual components of an automated screening system could identify SSIs. They determined (1) that use of coded diagnoses, tests, and treatments in the medical record had a sensitivity of 74%; (2) that specific codes and combinations of codes identified a subset of 2% of all procedures, among which

74% of SSIs occurred; and (3) that use of hospital discharge diagnosis codes and pharmacy dispensing data had sensitivity of 77% and specificity of 94%.

Platt et al.[26,41] reported that automated claims and pharmacy data from several health insurance plans can be combined to allow routine monitoring for indicators of postoperative infection. Bouam et al.[42] demonstrated how cooperation between infection control and medical informatics personnel can produce an automated system of surveillance. They compared computerized prospective surveillance of the electronic medical record with standard prospective review of laboratory data and medical charts by an infection preventionist. The automated method required much less time and performed well, with a sensitivity and a specificity of 91%.[42] Chalfine et al.[40] found that using a combination of microbiologic culture results and surgeon questionnaires was much less time consuming and had a sensitivity of 84.3% and specificity of 99.9%, compared with conventional chart-review surveillance, for patients who underwent gastrointestinal surgery. Yokoe et al.[25] found that use of electronic data on antimicrobial drug exposure and diagnosis codes enhanced conventional SSI surveillance and substantially decreased the number of medical charts that infection preventionists had to review to determine patients' SSI status following coronary artery bypass surgery, cesarean delivery procedures, and breast surgery.

In a 2003 editorial, J. P. Burke[43] suggested that computers can do more than simply automate traditional surveillance methods. Computerized event monitoring can be used to trigger epidemiologic investigations, and data mining can uncover small outbreaks and trends that might otherwise be missed. Time saved by automating surveillance liberates infection control staff to interpret the new data, answer new questions, and design new interventions to prevent SSIs.[43] Several academic groups and entrepreneurial companies have developed novel systems that access administrative and clinical data to facilitate SSI surveillance. Their utility and validity have yet to be tested.

Summary of Surveillance Methods

When selecting a surveillance method, infection preventionists must consider their objectives and the various sources from which the necessary data may be obtained. Although wound examination and medical chart review for all surgical patients is considered the gold standard, most institutions will need to set priorities and perform some type of targeted SSI surveillance. Surveillance can be targeted to focus on high-volume procedures, high-risk procedures, high-risk patients, or procedures of particular interest at a given institution. Data sources available at one institution may not be available at others. Diagnostic procedures, such as computerized tomography or magnetic resonance imaging, may help to identify deep or organ-space infections; the operating room logbook may allow epidemiology staff to identify patients who return to the operating room for wound drainage and debridement. Review of the pharmacy records or microbiology reports is not recommended as the sole method for identifying patients who have an SSI. Electronic data from computerized databases can help infection preventionists perform surveillance efficiently and are particularly helpful for postdischarge surveillance. The availability of resources will determine, in part, which surveillance methods are most useful in individual healthcare facilities. No surveillance method is perfect, and any chosen strategy must be validated and periodically reassessed.

Collection, Tabulation, Analysis, and Reporting of Data

Data Sources

Data collected during SSI surveillance can be classified into 3 categories: host factors, surgical and environmental factors, and microbial factors (Figure 15-2). Host factors are conditions that reflect the patient's intrinsic susceptibility to infection. These conditions usually are present when the patient is admitted to the hospital. Some of these factors increase the risk of SSI after many different operative procedures (eg, remote infection, greater age, or greater preoperative length of stay),[44,45] while others may increase the risk only after specific operative procedures.[46] Surgical and environmental factors can increase the probability of bacterial contamination

at the time of the surgical procedure and lead to SSI. For example, a contaminated wound, a long procedure, and poor surgical technique are risk factors for SSI.[44] Microbial factors, such as the virulence of the organism or the ability of the organism to adhere to sutures, may alter the risk of SSI, but few studies have addressed these issues systematically. To conduct routine surveillance for SSI, infection control personnel rarely need to know whether the patient carries specific organisms. However, during an outbreak or when trying to answer specific research questions, they may need to obtain specimens from specific body sites for preoperative surveillance cultures.

The amount of data that infection preventionists should collect depends on the purpose of the surveillance program and on the specific issues identified at a given institution. In general, the basic data that should be collected include the following: the patient's identification, the date of admission, the date of surgery, the type of procedure, the wound classification[14] (ie, clean, clean-contaminated, contaminated, or dirty), the surgeon's identification code, the date that the SSI was diagnosed, and the type of infection (ie, superficial, deep, or organ-space). Other useful data are the American Society of Anesthesiologists score, the procedure's duration and urgency, the organism identified, and the type, timing, and duration of perioperative antibiotic administration. In addition, many institutions collect data on whether or not razors are used, the type of skin preparation used, and other data that are collected for quality improvement efforts. Data on other specific risk factors and infection prevention process measures may also be collected, especially to determine the cause(s) of elevated SSI rates and to monitor the progress of prevention efforts.

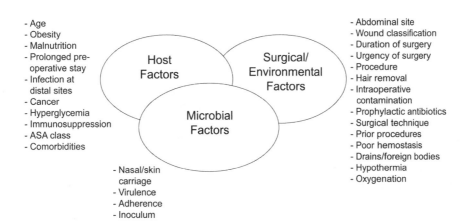

Figure 15-2. Risk factors for surgical site infection (SSI). This figure indicates that the patient's risk of developing an SSI varies with numerous host factors, surgical and environmental factors, and microbial factors. A complex interaction of these factors determines whether the patient will acquire an SSI. ASA, American Society of Anesthesiologists.

Because surgical patients are readily identifiable, the denominator to calculate the incidence—the number of patients who undergo a surgical procedure—is easier to obtain than it is for other healthcare associated infections, such as pneumonia, urinary tract infection, or bacteremia, for which the number of ventilator-days, urinary catheter–days, or intravenous catheter–days must be determined retrospectively. All surgical patients can be included in a registry at the end of the operative procedure. Indeed, many hospitals maintain operating room logbooks or have computer databases for financial management that allow infection prevention and control staff to identify the appropriate patients who have undergone surgical procedures to be included in the denominator. Hospitals that have a separate database for surgical services may collect demographic data and patient, operative, and environmental data that are useful to the infection control program. In hospitals that do not collect these data, infection prevention and control personnel should seriously consider what information they need before they spend many hours collecting data. Several investigators have found that they collected data on variables that ultimately did not help them determine which patients were at highest risk of developing an infection or help them facilitate prevention efforts.[19]

Equations

Once data are collected and entered into a computerized database, an SSI incidence can be tabulated. The following formulas are used to calculate basic SSI rates. To calculate the incidence of the outcome (eg, SSI) in a clearly delineated population for a given time (eg, 1 month, 1 quarter, or 1 year), divide the numerator (the number of patients with SSI following the procedure of interest during the specified time period) by the denominator (the total number of patients who had the surgical procedure of interest during the specified time period), and then multiply the result by 100 to obtain a percentage. Some examples follow:

1. The *service-specific incidence* is calculated as the number of patients with SSI after a procedure for a specific service (eg, neurosurgery) divided by the number of patients who had such a procedure, times 100.
2. The *surgeon-specific incidence* is calculated as the number of patients with SSI following an operation by a particular surgeon divided by the number of patients operated on by that surgeon, times 100.
3. The *procedure-specific incidence* is calculated as the number of SSIs occurring after a specific type of procedure (eg, cholecystectomy) divided by the number of procedures of that type performed (eg, cholecystectomies), times 100.

4. The *risk-specific incidence* is calculated as the number of SSIs in patients with an NNIS risk index score of 2 divided by the total number of procedures in patients with an NNIS risk index score of 2, times 100.

Once infection prevention staff identify the population to be surveyed, the proper denominator is determined by searching the operating room logbook or the hospital's computerized database. All of the patients in the defined population are followed throughout the time frame designated by the definition of SSI (ie, 30 days after the operation, or 1 year after the operation if an implant was inserted). For NHSN surveillance, the denominator needs to include the procedures with the *International Classification of Diseases, Ninth Edition (ICD-9)* procedure codes specified by the CDC protocol. Each SSI is included in the numerator. If the denominator is too broad (eg, includes all surgical patients in a large hospital), the group becomes very heterogeneous, and the calculated infection rate probably will be falsely low. Consequently, infection prevention and control personnel may not identify clusters of SSI or other problems. The surgeon-specific incidence and procedure-specific incidence more closely reflect the true SSI incidence. In general, data from at least 50–100 procedures should be included before calculating either an overall incidence or a surgeon-specific incidence.

Reporting Recommendations

The infection control program should stratify SSI rates by type of procedure or by a specific risk index, to allow comparisons among surgeons or among hospitals. The traditional wound classification system[14] has served this purpose for a quarter of a century, but it has some limitations, as mentioned earlier. The CDC NNIS risk index is used widely today, but it does not perform well in certain circumstances.[19] Despite these limitations, it is important to stratify the patients' risk by one of the available indices, particularly if the patients have numerous comorbidities or if the operative procedures performed in the hospital are quite complex.

The infection prevention and control program should report SSI rates to surgeons, operating room staff, and hospital administration. Increased transparency provides these employees with improved perspectives about the magnitude of the institutional problem. However, for surgeon-specific SSI rates, confidentiality is of utmost importance, and codes should be used instead of names if these rates are calculated and reported. Reporting SSI rates to practicing surgeons has been shown to reduce rates by means of the Hawthorne Effect—the effect of having one's performance observed.[6,8,10] Infection prevention and control personnel should periodically present the SSI data graphically and meet with surgical personnel to discuss rates, clusters, and specific cases. These discussions improve communication and cooperation between the infection prevention and control team and the surgical team. Infection preventionists may learn ways to make the data more useful to the surgical team and may also use these feedback sessions as a means to reinforce the importance of preventive measures to reduce the SSI risk.

The infection prevention and control program should report SSI rates, costs, lengths of stay associated with SSIs, and the effects of preventive measures to the hospital's administrators. Several investigators have demonstrated that SSIs are the primary independent determinant of hospital costs and length of stay after operative procedures.[47,48] Moreover, Olson and Lee[8] demonstrated a $3 million cost savings in a 10-year wound surveillance program. If an infection prevention and control program can demonstrate that surveillance and interventions reduce SSI rates, lengths of stay, and costs, then hospital administrators will likely be more willing to provide the program with resources.

Finally, infection prevention and control personnel should periodically review their data to determine whether they should change their priorities or focus their energy on specific problems. For example, if a program analyzes its data and finds, after a few months, that the sensitivity of the case-finding method used for post-discharge surveillance is very low (eg, only 2% of SSIs are identified by this method), the infection prevention and control staff may change their case-finding method, instead of spending time and energy for a year or more before realizing that the method was not effective.

Implementation of SSI Surveillance

Below we offer a few suggestions that the new hospital epidemiologist or infection preventionist might find useful when implementing or improving an SSI surveillance and prevention program. The infection preventionist is a very important component of any surveillance program and should have personality traits that facilitate a good working relationship with surgical personnel. Infection prevention and control personnel need training, feedback on performance, and time to attend conferences on healthcare-associated infections. The hospital epidemiologist and infection preventionists should be active members of their institution's infection control committee. Finally, in programs that use direct wound examination to identify SSIs, a surgeon should train infection control staff so that they can evaluate subtle nuances

of a wound's appearance. In addition, it is important to validate the SSI surveillance methodologies.

Most importantly, the infection prevention and control program must communicate and work closely with the surgical personnel so that they take the responsibility for SSI prevention. To achieve this goal, infection control staff should do the following:

- Review the standard definitions of SSI with the surgical personnel so they understand and accept the criteria.
- Validate case-finding strategies and definition application among infection preventionists to ensure that surveillance is standard within the institution.
- Review identified cases with surgical personnel.
- Ask for input from the surgical staff about the format of SSI reports and whether the surgeon-specific incidence or the procedure-specific incidence is preferred.
- Meet with surgeons and surgical nurses on a regular basis to build trust and to discuss issues such as SSI risk stratification, surveillance, and prevention methods.
- Visit the operating room routinely to identify potential problems and to develop rapport and mutual respect.
- Encourage surgical personnel to join the infection control committee.
- Join the surgical professional societies and attend annual meetings in order to understand the surgical perspective on SSIs.
- Discuss protocols and goals for studies of SSIs with surgical staff and encourage them to participate.
- Analyze SSI cases and examine risk factors that may be modifiable.
- Attempt to develop creative strategies for identifying and preventing SSIs as the medical environment changes.
- Publish results of studies in surgical journals.

In addition, the infection control program needs adequate resources to conduct effective surveillance. Clerical support, computerized databases, data analysts, and medical records personnel contribute significantly to the SSI surveillance program.

Interventions to Prevent SSI

Once data are collected, tabulated, and analyzed, the infection control program can develop and implement interventions to prevent SSI and reduce the SSI rate. These interventions will be dictated largely by the problems that have been identified through surveillance and data analysis. Infection prevention staff examine factors in the preoperative, intraoperative, and postoperative periods for possible factors and interventions to prevent SSI.[49]

It is important to become familiar with the guidelines, recommendations, requirements, and strategies published by major authorities and agencies regarding SSI surveillance and prevention. The Hospital Infection Control Practices Advisory Committee (HICPAC) published guidelines for SSI prevention in 1999.[49] The Surgical Care Improvement Project grew out of the Surgical Infection Prevention Collaborative, which was initially created in 2002 by the Centers for Medicare and Medicaid Services.[50] These surgical care improvement projects identify evidence-based process measures (eg, antimicrobial prophylaxis and proper hair removal) that can be monitored and improved to prevent SSI.[50-52] The Centers for Medicare and Medicaid Services requires hospitals to report compliance with Surgical Care Improvement Project quality measures as part of the acute care inpatient prospective payment system.[53] The World Health Organization (WHO) developed a surgery safety checklist, toolkit, and implementation guide as part of their Safe Surgery Saves Lives program. Implementation of the WHO checklist resulted in significantly decreased rates of complications and decreased rates of death among surgical patients in 8 diverse hospitals worldwide.[54] The Institute for Healthcare Improvement incorporates both the Surgical Care Improvement Project prevention measures and the WHO Surgical Safety Checklist into its "5 Million Lives" campaign to improve the quality of health care for hospitalized patients.[55] The Society for Healthcare Epidemiology of America and the Infectious Diseases Society of America published a "Compendium of Strategies to Prevent Healthcare-Associated Infections" in 2008 that includes a practical, concise implementation guide regarding SSI prevention strategies.[38,56,57]

Preoperative Intervention

Preoperative preparation of the patient is one area for intervention. Some variables cannot be modified (eg, age and sex), but others can be modified (eg, glucose control and hair removal). Several interventions, such as minimizing the duration of preoperative hospital stay and eradicating remote infections, have been shown to reduce SSI rates.[44] Infection prevention control personnel need to ensure that best practices are followed for proper hair removal and antiseptic skin preparation. The preoperative stay should be as short as possible. Sometimes it is prudent to send the patient home and readmit them for surgery. Patients who are colonized with *Staphylococcus aureus* in the anterior nares are at increased risk for the development of SSI, as well as other healthcare-associated infections.[58-61] A randomized

double-blind clinical trial demonstrated that preoperative treatment with mupirocin decolonized the nares of patients and decreased healthcare-associated infections with *S. aureus* among patients who were colonized.[61] Treated patients also had a decrease in *S. aureus* SSI rates that was not statistically significant. Further study is needed to determine if this intervention can benefit a targeted subset of surgical patients.

The surgical team should perform appropriate antiseptic scrubbing and avoid having long or artificial nails and wearing jewelry. Infection prevention and control personnel need to ensure that policies are in place (and followed) to restrict patient care by surgical staff with transmissible infectious diseases. If an epidemiologic investigation suggests that the source of the outbreak might be healthcare workers who carry the organism, infection control personnel should obtain appropriate samples for culture from the implicated individuals (eg, nares and skin lesion specimens to detect *S. aureus*; and specimens from the throat, skin lesions, and if necessary, the vagina and rectum to detect *Streptococcus pyogenes*).

Finally, antimicrobial prophylaxis is crucial for preventing SSI and must be administered according to published guidelines for each type of procedure. To be effective, the appropriate prophylactic antibiotic must be given during the appropriate time interval and for the appropriate duration. Studies have shown that there is much room for improvement in the administration of antimicrobial prophylaxis.[62,63] Several groups of investigators have successfully utilized automated alerts and decision support as an intervention to improve administration and redosing of antimicrobial prophylaxis.[64-66]

Intraoperative Intervention

Implementation of best practices during the intraoperative period can reduce SSI rates. Proper ventilation should be maintained in all operating rooms, including positive air pressure, appropriate air exchanges, and appropriate air filters. Traffic through the operating room must be kept to a minimum. Infection prevention and control personnel need to review data regarding sterilization of surgical instruments to ensure that procedures conform to guidelines. Use of flash sterilization should be limited, and it should never be used solely for convenience. Infection prevention and control personnel can work with surgical staff to ensure that proper surgical attire and drapes are used. Observation in the operating room and collaboration with surgeons, surgical nurses, and surgical staff are important to maintaining a high standard of asepsis and good surgical technique. Studies have demonstrated that intraoperative hypothermia (even mild decreases in core body temperature) approximately triples the risk of SSI and leads to many other adverse events.[67-69] Therefore, normothermia should be maintained during surgery. The main methods of warming include passive insulation, fluid warming, and active warming of the patient with a forced-air system. Hyperglycemia increases mortality in critically ill patients and increases the risk of many adverse outcomes, including SSI.[70-72] Latham et al.[70] showed that postoperative hyperglycemia and previously undiagnosed diabetes approximately doubles the risk of SSI. Interventions include screening for diabetes and hyperglycemia, as well as intensive insulin therapy. The risk of SSI is also directly related to tissue oxygenation.[73] Administration of supplemental oxygen (80% vs 30% inspired oxygen) during surgery and the immediate postoperative period has been shown to reduce the incidence of SSI by approximately half.[68,74]

Postoperative Intervention

Because most contamination happens during the operation through contact or airborne transmission, events that occur during the postoperative period (eg, improper dressing changes or isolation techniques) are less likely to contribute to SSI, though this has not been well studied. If epidemiological data indicate that postoperative care may be associated with increased SSI rates, the infection prevention and control staff may need to investigate practices used during this period.

In summary, interventions can be designed on the basis of problems identified during surveillance and close observation of current practices, knowledge of published guidelines for infection control, and close collaboration with surgeons and surgical staff. Performing surveillance and providing the results to surgeons can itself lower the incidence of SSI.

Conclusion

SSI surveillance is an important component of any infection prevention and control program, and it is a special form of continuous quality assurance in which the ultimate benefactors of control efforts are the patients. Therefore, the infection prevention and control program should define clear objectives, utilize precise definitions, and meticulously implement a surveillance system and appropriate prevention measures. Although methods of case-finding are hard to choose, the infection control team should focus on patients or procedures at high risk of infection, if their resources are limited. Collecting data and calculating rates are useless if epidemiology and surgical staff do not use the data to prevent SSIs. To succeed

in such an effort, infection prevention and control personnel must collaborate closely with surgical teams and utilize available guidelines and recommendations to implement, monitor, and improve compliance with SSI prevention measures.

As healthcare delivery shifts to the outpatient setting, numerous aspects of SSI surveillance must change, because many factors that influence the risk of SSI also will change. Surveillance methods that worked well in the past and were supported by well-designed studies may no longer be efficacious. We need creative research to determine how we should develop and apply risk indices, to identify which methods we should use for postdischarge surveillance, and to design and test new SSI prevention strategies. Indeed, exciting opportunities are open to those willing to accept the challenges.

References

1. Burke JP. Patient safety: infection control—a problem for patient safety. *N Engl J Med* 2003;348(7):651–656.
2. Martone WJ, Nichols RL. Recognition, prevention, surveillance, and management of surgical site infections: introduction to the problem and symposium overview. *Clin Infect Dis* 2001;33(Suppl 2):S67–S68.
3. Engemann JJ, Carmeli Y, Cosgrove SE, et al. Adverse clinical and economic outcomes attributable to methicillin resistance among patients with *Staphylococcus aureus* surgical site infection. *Clin Infect Dis* 2003;36(5):592–598.
4. Kirkland KB, Briggs JP, Trivette SL, Wilkinson WE, Sexton DJ. The impact of surgical-site infections in the 1990s: attributable mortality, excess length of hospitalization, and extra costs. *Infect Control Hosp Epidemiol* 1999;20(11):725–730.
5. McGarry SA, Engemann JJ, Schmader K, Sexton DJ, Kaye KS. Surgical-site infection due to *Staphylococcus aureus* among elderly patients: mortality, duration of hospitalization, and cost. *Infect Control Hosp Epidemiol* 2004;25(6):461–467.
6. Haley RW, Culver DH, White JW, et al. The efficacy of infection surveillance and control programs in preventing nosocomial infections in US hospitals. *Am J Epidemiol* 1985;121(2):182–205.
7. Cruse PJ, Foord R. The epidemiology of wound infection: a 10-year prospective study of 62,939 wounds. *Surg Clin North Am* 1980;60(1):27–40.
8. Olson MM, Lee JT Jr. Continuous, 10-year wound infection surveillance: results, advantages, and unanswered questions. *Arch Surg* 1990;125(6):794–803.
9. Horan TC, Gaynes RP, Martone WJ, Jarvis WR, Emori TG. CDC definitions of nosocomial surgical site infections, 1992: a modification of CDC definitions of surgical wound infections. *Infect Control Hosp Epidemiol* 1992;13(10):606–608.
10. Ehrenkranz NJ, Richter EI, Phillips PM, Shultz JM. An apparent excess of operative site infections: analyses to evaluate false-positive diagnoses. *Infect Control Hosp Epidemiol* 1995;16(12):712–716.
11. Lee JT. Wound infection surveillance. *Infect Dis Clin North Am* 1992;6(3):643–656.

12. Cardo DM, Falk PS, Mayhall CG. Validation of surgical wound surveillance. *Infect Control Hosp Epidemiol* 1993; 14(4):211–215.

13. Haley RW, Schaberg DR, McClish DK, et al. The accuracy of retrospective chart review in measuring nosocomial infection rates. Results of validation studies in pilot hospitals. *Am J Epidemiol* 1980;111(5):516–533.

14. Berard F, Gandon J. Postoperative wound infections: the influence of ultraviolet irradiation of the operating room and of various other factors. *Ann Surg* 1964;160(Suppl 2):1–192.

15. Haley RW. Surveillance by objective: a new priority-directed approach to the control of nosocomial infections. The National Foundation for Infectious Diseases lecture. *Am J Infect Control* 1985;13(2):78–89.

16. Ferraz EM, Bacelar TS, Aguiar JL, Ferraz AA, Pagnossin G, Batista JE. Wound infection rates in clean surgery: a potentially misleading risk classification. *Infect Control Hosp Epidemiol* 1992;13(8):457–462.

17. Haley RW, Culver DH, Morgan WM, White JW, Emori TG, Hooton TM. Identifying patients at high risk of surgical wound infection. A simple multivariate index of patient susceptibility and wound contamination. *Am J Epidemiol* 1985;121(2):206–215.

18. Culver DH, Horan TC, Gaynes RP, et al. Surgical wound infection rates by wound class, operative procedure, and patient risk index. National Nosocomial Infections Surveillance System. *Am J Med* 1991;91(3B):152S–157S.

19. Roy MC, Herwaldt LA, Embrey R, Kuhns K, Wenzel RP, Perl TM. Does the Centers for Disease Control's NNIS system risk index stratify patients undergoing cardiothoracic operations by their risk of surgical-site infection? *Infect Control Hosp Epidemiol* 2000;21(3):186–190.

20. Nichols RL, Smith JW, Robertson GD, et al. Prospective alterations in therapy for penetrating abdominal trauma. *Arch Surg* 1993;128(1):55–63.

21. Nichols RL, Smith JW, Klein DB, et al. Risk of infection after penetrating abdominal trauma. *N Engl J Med* 1984;311(17): 1065–1070.

22. Richet HM, Chidiac C, Prat A, et al. Analysis of risk factors for surgical wound infections following vascular surgery. *Am J Med* 1991;91(3B):170S–172S.

23. Wenzel RP, Osterman CA, Hunting KJ, Gwaltney JM Jr. Hospital-acquired infections. I. Surveillance in a university hospital. *Am J Epidemiol* 1976;103(3):251–260.

24. Perl TM. Surveillance, reporting and the use of computers. In: Wenzel RP, editor. *Prevention and Control of Nosocomial Infections*. 2nd ed. Baltimore: Williams & Wilkins; 1993: 139–176.

25. Yokoe DS, Noskin GA, Cunningham SM, et al. Enhanced identification of postoperative infections among inpatients. *Emerg Infect Dis* 2004;10(11):1924–1930.

26. Platt R, Kleinman K, Thompson K, et al. Using automated health plan data to assess infection risk from coronary artery bypass surgery. *Emerg Infect Dis* 2002;8(12):1433–1441.

27. Manian FA, Meyer L. Comprehensive surveillance of surgical wound infections in outpatient and inpatient surgery. *Infect Control Hosp Epidemiol* 1990;11(10):515–520.

28. Avato JL, Lai KK. Impact of postdischarge surveillance on surgical-site infection rates for coronary artery bypass procedures. *Infect Control Hosp Epidemiol* 2002;23(7):364–367.

29. Delgado-Rodriguez M, Gomez-Ortega A, Sillero-Arenas M, Llorca J. Epidemiology of surgical-site infections diagnosed after hospital discharge: a prospective cohort study. *Infect Control Hosp Epidemiol* 2001;22(1):24–30.

30. Sands K, Vineyard G, Platt R. Surgical site infections occurring after hospital discharge. *J Infect Dis* 1996;173(4):963–970.

31. Perencevich EN, Sands KE, Cosgrove SE, Guadagnoli E, Meara E, Platt R. Health and economic impact of surgical site infections diagnosed after hospital discharge. *Emerg Infect Dis* 2003;9(2):196–203.

32. Noy D, Creedy D. Postdischarge surveillance of surgical site infections: a multi-method approach to data collection. *Am J Infect Control* 2002;30(7):417–424.

33. Kent P, McDonald M, Harris O, Mason T, Spelman D. Post-discharge surgical wound infection surveillance in a provincial hospital: follow-up rates, validity of data and review of the literature. *ANZ J Surg* 2001;71(10):583–589.

34. Burns SJ, Dippe SE. Postoperative wound infections detected during hospitalization and after discharge in a community hospital. *Am J Infect Control* 1982;10(2):60–65.

35. Seaman M, Lammers R. Inability of patients to self-diagnose wound infections. *J Emerg Med* 1991;9(4):215–219.

36. Rosendorf LL, Octavio J, Estes JP. Effect of methods of postdischarge wound infection surveillance on reported infection rates. *Am J Infect Control* 1983;11(6):226–229.

37. Sands K, Vineyard G, Livingston J, Christiansen C, Platt R. Efficient identification of postdischarge surgical site infections: use of automated pharmacy dispensing information, administrative data, and medical record information. *J Infect Dis* 1999;179(2):434–441.

38. Anderson DJ, Kaye KS, Classen D, et al. Strategies to prevent surgical site infections in acute care hospitals. *Infect Control Hosp Epidemiol* 2008;29(Suppl 1):S51–S61.

39. Miner AL, Sands KE, Yokoe DS, et al. Enhanced identification of postoperative infections among outpatients. *Emerg Infect Dis* 2004;10(11):1931–1937.

40. Chalfine A, Cauet D, Lin WC, et al. Highly sensitive and efficient computer-assisted system for routine surveillance for surgical site infection. *Infect Control Hosp Epidemiol* 2006;27(8):794–801.

41. Hirschhorn LR, Currier JS, Platt R. Electronic surveillance of antibiotic exposure and coded discharge diagnoses as indicators of postoperative infection and other quality assurance measures. *Infect Control Hosp Epidemiol* 1993; 14(1):21–28.

42. Bouam S, Girou E, Brun-Buisson C, Karadimas H, Lepage E. An intranet-based automated system for the surveillance of nosocomial infections: prospective validation compared with physicians' self-reports. *Infect Control Hosp Epidemiol* 2003;24(1):51–55.

43. Burke JP. Surveillance, reporting, automation, and interventional epidemiology. *Infect Control Hosp Epidemiol* 2003; 24(1):10–12.

44. Mayhall CG. Surgical infections including burns. In: Wenzel RP, ed. *Prevention and Control of Nosocomial Infections*. 2nd ed. Baltimore: Williams & Wilkins; 1993:614–664.

45. Kernodle DS, Kaiser AB. Postoperative infections and antimicrobial prophylaxis. In: Mandell GL, Bennett JE, Dolin R, eds. *Principles and Practice of Infectious Diseases*. 4th ed. New York: Churchill Livingstone; 1995:2742–2756.

46. Nagachinta T, Stephens M, Reitz B, Polk BF. Risk factors for surgical-wound infection following cardiac surgery. *J Infect Dis* 1987;156(6):967–973.

47. Weintraub WS, Jones EL, Craver J, Guyton R, Cohen C. Determinants of prolonged length of hospital stay after coronary bypass surgery. *Circulation* 1989;80(2):276–284.

48. Taylor GJ, Mikell FL, Moses HW, et al. Determinants of hospital charges for coronary artery bypass surgery: the economic consequences of postoperative complications. *Am J Cardiol* 1990;65(5):309–313.

49. Mangram AJ, Horan TC, Pearson ML, Silver LC, Jarvis WR. Guideline for prevention of surgical site infection, 1999. Centers for Disease Control and Prevention (CDC) Hospital Infection Control Practices Advisory Committee. *Am J Infect Control* 1999;27(2):97–132.

50. Bratzler DW, Hunt DR. The surgical infection prevention and surgical care improvement projects: national initiatives to improve outcomes for patients having surgery. *Clin Infect Dis* 2006;43(3):322–330.

51. Bratzler DW, Houck PM, Richards C, et al. Use of antimicrobial prophylaxis for major surgery: baseline results from the National Surgical Infection Prevention Project. *Arch Surg* 2005;140(2):174–182.

52. Dellinger EP, Hausmann SM, Bratzler DW, et al. Hospitals collaborate to decrease surgical site infections. *Am J Surg* 2005;190(1):9–15.

53. Centers for Medicare & Medicaid Services Medicare Program. Hospital inpatient prospective payment systems and fiscal year 2009 rates: final fiscal year 2009 wage indices and payment rates including implementation of Section 124 of the Medicare Improvement for Patients and Providers Act of 2008; notice. *Federal Register* 2008;73(193):57888–58017.

54. Haynes AB, Weiser TG, Berry WR, et al. A surgical safety checklist to reduce morbidity and mortality in a global population. *N Engl J Med* 2009;360(5):491–499.

55. Institute for Healthcare Improvement. http://www.ihi.org. Accessed April 22, 2009.

56. Yokoe DS, Mermel LA, Anderson DJ, et al. A compendium of strategies to prevent healthcare-associated infections in acute care hospitals. *Infect Control Hosp Epidemiol* 2008;29(Suppl 1):S12–S21.

57. Yokoe DS, Classen D. Improving patient safety through infection control: a new healthcare imperative. *Infect Control Hosp Epidemiol* 2008;29(Suppl 1):S3–S11.

58. Perl TM, Golub JE. New approaches to reduce *Staphylococcus aureus* nosocomial infection rates: treating *S. aureus* nasal carriage. *Ann Pharmacother* 1998;32(1):S7–S16.

59. Wenzel RP, Perl TM. The significance of nasal carriage of *Staphylococcus aureus* and the incidence of postoperative wound infection. *J Hosp Infect* 1995;31(1):13–24.

60. Kluytmans J, van Belkum A, Verbrugh H. Nasal carriage of *Staphylococcus aureus*: epidemiology, underlying mecha-

nisms, and associated risks. *Clin Microbiol Rev* 1997;10(3):505–520.

61. Perl TM, Cullen JJ, Wenzel RP, et al. Intranasal mupirocin to prevent postoperative *Staphylococcus aureus* infections. *N Engl J Med* 2002;346(24):1871–1877.

62. Silver A, Eichorn A, Kral J, et al. Timeliness and use of antibiotic prophylaxis in selected inpatient surgical procedures. The Antibiotic Prophylaxis Study Group. *Am J Surg* 1996;171(6):548–552.

63. Classen DC, Evans RS, Pestotnik SL, Horn SD, Menlove RL, Burke JP. The timing of prophylactic administration of antibiotics and the risk of surgical-wound infection. *N Engl J Med* 1992;326(5):281–286.

64. Zanetti G, Flanagan HL Jr, Cohn LH, Giardina R, Platt R. Improvement of intraoperative antibiotic prophylaxis in prolonged cardiac surgery by automated alerts in the operating room. *Infect Control Hosp Epidemiol* 2003;24(1):13–16.

65. Webb AL, Flagg RL, Fink AS. Reducing surgical site infections through a multidisciplinary computerized process for preoperative prophylactic antibiotic administration. *Am J Surg* 2006;192(5):663–668.

66. Kanter G, Connelly NR, Fitzgerald J. A system and process redesign to improve perioperative antibiotic administration. *Anesth Analg* 2006;103(6):1517–1521.

67. Kurz A, Sessler DI, Lenhardt R. Perioperative normothermia to reduce the incidence of surgical-wound infection and shorten hospitalization. Study of Wound Infection and Temperature Group. *N Engl J Med* 1996;334(19):1209–1215.

68. Sessler DI, Akca O. Nonpharmacological prevention of surgical wound infections. *Clin Infect Dis* 2002;35(11):1397–1404.

69. Melling AC, Ali B, Scott EM, Leaper DJ. Effects of preoperative warming on the incidence of wound infection after clean surgery: a randomised controlled trial. *Lancet* 2001;58(9285):876–880.

70. Latham R, Lancaster AD, Covington JF, Pirolo JS, Thomas CS. The association of diabetes and glucose control with surgical-site infections among cardiothoracic surgery patients. *Infect Control Hosp Epidemiol* 2001;22(10):607–612.

71. Van den BG, Wouters P, Weekers F, et al. Intensive insulin therapy in the critically ill patients. *N Engl J Med* 2001;345(19):1359–1367.

72. Dellinger EP. Preventing surgical-site infections: the importance of timing and glucose control. *Infect Control Hosp Epidemiol* 2001;22(10):604–606.

73. Hopf HW, Hunt TK, West JM, et al. Wound tissue oxygen tension predicts the risk of wound infection in surgical patients. *Arch Surg* 1997;132(9):997–1004.

74. Greif R, Akca O, Horn EP, Kurz A, Sessler DI. Supplemental perioperative oxygen to reduce the incidence of surgical-wound infection. Outcomes Research Group. *N Engl J Med* 2000;342(3):161–167.

Chapter 16 Surveillance and Prevention of Infections Associated With Vascular Catheters

Werner E. Bischoff, MD, PhD

Intravascular catheter–related infections (IV-CRIs) are one of the predominant causes of morbidity and mortality in the United States.[1-3] It is estimated that more than 150 million intravascular devices are purchased each year, including more than 7 million central venous catheters (CVCs) alone.[4-6] These devices, and in particular CVCs, are the source of most primary bloodstream infections (BSIs)[5,6]; they account for approximately 250,000–500,000 hospital-acquired BSIs in the United States annually.[4,7-11] These infections not only impose a significant economic burden, with a marginal cost estimate of more than $36,000 per episode, but are also associated with a high mortality rate of 10% among hospitalized patients.[2,6,8,10,12-16]

Given these numbers, preventive measures are paramount for the control of and reduction in the incidence of IV-CRIs in the hospital setting. An accurate and practical surveillance system is essential for identifying the true scope of the problem and therefore allowing for implementation of effective strategies to reduce the IV-CRI rate. Surveillance can provide the basis for an effective infection control program through the routine and orderly collection of data, subsequent analysis, and timely reporting. This chapter provides healthcare epidemiologists and infection preventionists with the knowledge necessary to design and implement an effective surveillance program for IV-CRIs, to choose and execute the appropriate preventive measures, and to ensure the commitment of hospital personnel and administration in the continuing effort to reduce IV-CRI rates.

Arguments for Surveillance and Prevention of IV-CRIs

Surveillance and prevention are ongoing processes, requiring a considerable amount of time and personnel resources. Therefore, convincing the administration of the importance of these measures is an essential component to guarantee adequate resources and a successful implementation of such a program.

Evidence of the efficacy of surveillance and control programs has been best demonstrated in the Study of the Efficacy of Nosocomial Infection Control (SENIC).[17] Hospitals with a high-intensity control program, regardless of the level of surveillance activities, showed a moderate 15% reduction in nosocomial bacteremia rates. However, hospitals with a similar-intensity control program but with at least a medium level of surveillance showed a 35% reduction in nosocomial bacteremia rates. In another study, Curran et al.,[18] utilizing surveillance with feedback, found a significant reduction of 8.5%–5.3% in the number of infections associated with peripheral vascular catheters ($P < .001$). Similar results were also reported for CVCs.[19] The introduction of ongoing surveillance by an infection preventionist in a 42-bed intensive care unit (ICU) achieved a significant reduction from 7.18 infections per 1,000 catheter-related exposure-days to 4.29 infections per 1,000 catheter-related exposure-days ($P = .0045$).[20] Lastly, the National Nosocomial Infections Surveillance System of the Centers for Disease Control and Prevention (CDC) provides an indirect indication of the efficacy of a surveillance system by documenting a substantial decrease in the BSI rate in different types of ICUs during the period 1990–1999, ranging from a decrease of 31% in surgical ICUs to a decrease of 44% in medical ICUs.[21]

Another essential point is the cost-effectiveness of a surveillance and control program. Estimates of the additional costs related to nosocomial BSI range from $4,000 to $56,000 per episode.[8,13,22-24] Slater[25] calculated savings of $108,000 during the 9 months after introduction of a vascular-catheter nurse in a surgical ICU, which was accounted for by prevention of an estimated 18 BSIs. In a neonatal ICU, the cost estimate for one nosocomial BSI was approximately $22,980, with accommodation costs accounting for 70% of the total additional charges, while physician fees, pharmaceuticals, and other ancillary services accounted for less than 30%.[26] Critically ill patients with BSI are hospitalized for an average of 6.5–22 days longer than patients without BSI.[8,13,22,24] This implies that the additional length of stay caused by IV-CRI contributes substantially to the overall costs and can also be used as an indicator for cost-benefit analyses.

The substantial human suffering caused by IV-CRI and the costs attributed to them have led to a national action plan by the US Department of Health and Human Services to reduce the incidence of IV-CRI.[2] In addition, the Joint Commission for Hospital Accreditation publishes annual National Patient Safety Goals, which include surveillance and control measures addressing IV-CRI that should be implemented in set time frames.[27] These commitments from official key players and the associated potential ramifications for hospitals can also be used to reinforce the usefulness of an IV-CRI surveillance and control program.

Goals and Attributes of a Surveillance Program

There are 5 goals that a successful IV-CRI surveillance program should meet:[28]

1. To define the endemic ("background") incidence of IV-CRI
2. To identify increases in the incidence of IV-CRI above the endemic level
3. To identify specific risks for IV-CRI in patients undergoing routine hospital care or procedures
4. To inform hospital personnel of the risks of the care or procedures they provide to patients
5. To evaluate the utility and efficiency of control measures

To accomplish these 5 goals, there are a number of attributes for public health surveillance systems published by the CDC that can be easily transferred to surveillance for IV-CRI.[29] These universal attributes should be kept in mind when planning and implementing a surveillance program.

Simplicity refers to the use of understandable and easily manageable methods for data collection, analysis, and dissemination.

Flexibility describes the potential of a surveillance program to respond quickly and appropriately to changing information needs or operating conditions. Simpler systems can allow for fast reactions to these changes.

Data quality refers to the completeness and validity of data recorded in the surveillance system. Counting the number of missing responses is an easy tool for assessing the completeness of a database. Evaluation of validity requires additional studies, such as retrospective chart review of sampled data.

Acceptability refers to the willingness of persons and organizations to participate in the surveillance program. This is affected by the user-friendliness of the surveillance program and by the acceptance of the data from persons inside and outside the sponsoring institution.

Sensitivity is another attribute that refers to the percentage of true cases being reported by the surveillance program and also the ability to detect outbreaks as changes in the number of cases over time. Testing the validity of a program regarding sensitivity requires additional studies.

Positive predictive value is calculated as the proportion of reported case patients who actually have the health-related event under surveillance. The positive predictive value focuses on the sensitivity and specificity of the case definition by confirming cases reported through the surveillance system. Low values suggest that noncases are being investigated, which results in unnecessary interventions and use of resources.

Representativeness describes how accurately the data reflect the actual occurrence of events in a population over time. To achieve generalizability or high external validity, the population should be well defined, and results should be comparable to those from other surveillance systems.

Timeliness of the individual steps in a surveillance system is of paramount importance, to ensure effective responses to changing conditions. The transparency and simplicity of the surveillance program structure support the success of the program.

Stability refers to the ability of the surveillance program to collect, analyze, and report data without gaps.

Surveillance Options

Three major surveillance strategies can be distinguished: hospital-wide surveillance, objective-directed surveillance, and targeted surveillance. The most extensive effort is hospital-wide surveillance that covers the entire institution for an extended period of time.[30-32] Objective-directed surveillance is addressed to an identified problem, such as a high incidence of IV-CRI in a specific ICU, and allocates surveillance and intervention resources to address this problem.[33] Unfortunately, no long-term data are collected, since surveillance efforts are redirected once the specific objective is met. Targeted surveillance combines elements of the 2 previous approaches; it focuses long-term surveillance efforts on a type of infection, such as BSI, or on a location, such as ICUs, without having a specific problem objective.[34-36] This allows collection of data in areas of high risk, with a longitudinal surveillance component. Since financial resources are limited in most institutions, the focus has shifted toward targeted surveillance. Published data demonstrate that IV-CRIs occur most often in ICUs and oncology wards.[35,36] Therefore, it is highly recommended that targeted surveillance efforts be directed to these high-risk areas.

Selection of Surveillance Targets

There is a wide (and growing) selection of different intravascular devices, ranging from peripheral venous catheters to intra-aortic balloon pumps. CVCs are the most commonly used vascular devices in the modern ICU and, as such, offer themselves as the primary surveillance target.[3] This is also reflected in the National Healthcare Safety Network (NHSN) program and the associated benchmarks for central line–associated BSI (CLABSI).[1] However, surveillance should be extended to other vascular devices, if they are frequently used in an institution and pose a potential risk for the patients.[37]

Outpatient Surveillance

The classic approach limits surveillance to the hospital stay of patients. Admission and discharge dates mark the time points of observation activities. However, the length of hospital stay has decreased in recent years, and in many cases hospitalization is completely replaced by outpatient care.[38-41] The first reduces the probability of detecting an event, whereas the second requires an entirely different approach to access data. The extension of surveillance efforts after discharge of patients appears to be very difficult, showing a wide range of outcomes, depending on the methods utilized and the institution involved.[42-44] However, targeted postdischarge surveillance for catheter-related BSI has been suggested as a rational alternative.[45]

Outpatients are also a challenging population to monitor, since their contact with a health care provider is limited and variable. A successful example for outpatient surveillance is the Dialysis Surveillance Network initiated by the CDC in 1999.[46,47]

Individual facilities should take feasibility and cost-effectiveness into account when considering whether outpatient surveillance should be initiated and which components would provide the most valuable and accurate data.

Case Definition

A surveillance program is based on a clear definition of an event, in this instance a case of IV-CRI. Unfortunately, there is still no universal agreement as to which findings correctly indicate this particular event. The clinical signs of catheter-related BSI are unreliable, with low sensitivity and specificity values.[48-52] Fever, erythema, or purulence around the catheter insertion site can be indicators of an infection, but they are not conclusive indicators. In comparison, the microbiologic diagnosis of catheter-related infections is more precise overall. However, there are numerous definitions, based on processing and culturing of suspected catheters and other materials, that make selection of one method versus another quite difficult. One can make up one's own definition of an IV-CRI, but to ensure comparability with other studies it is best to rely on existing standards.

Definitions of CLABSI for Surveillance

The following surveillance definitions are listed in the NHSN device-associated infection module for CLABSI.[53]

Primary BSIs are classified according to the criteria used, as either laboratory-confirmed BSI or clinical sepsis. Clinical sepsis may be used to report only primary BSI in neonates (30 days old or less) and infants (1 year old or less). BSIs that are central line-associated (ie, a central line or umbilical catheter was in place at the time of or within 48 hours before the onset of the event) should be reported. *Note:* There is no minimum period of time that the central line must be in place in order for the BSI to be considered central line–associated.

Laboratory-confirmed BSI must meet 1 of the following 3 criteria:

1. The patient has a recognized pathogen cultured from 1 or more blood samples;
   and the organism cultured from blood is not related to an infection at another site.
2. The patient has at least 1 of the following 3 signs or symptoms: fever (temperature greater than 38°C), chills, or hypotension;
   and has signs and symptoms and positive laboratory results that are not related to an infection at another site
   and has a common skin contaminant (ie, diphtheroids [*Corynebacterium* species], *Bacillus* species

[not *Bacillus anthracis*], *Propionibacterium* species, coagulase-negative staphyolococci [including *Staphylococcus epidermidis*], viridans group strepococci, *Aerococcus* species, or *Micrococcus* species) recovered from cultures of 2 or more blood specimens drawn on separate occasions.

3. The patient is 1 year of age or less and has at least 1 of the following 4 signs or symptoms: fever (core temperature greater than 38°C), hypothermia (core temperature lower than 36°C), apnea, or bradycardia; *and* has signs and symptoms and positive laboratory test results that are not related to an infection at another site *and* has a common skin contaminant (ie, diphtheroids [*Corynebacterium* species], *Bacillus* species [not *B. anthracis*], *Propionibacterium* species, coagulase-negative staphylococci [including *S. epidermidis*], viridans group strepococci, *Aerococcus* species, or *Micrococcus* species) recovered from cultures of 2 or more blood specimens drawn on separate occasions.

Clinical sepsis must meet the following criterion:

1. The patient is 1 year of age or less and has at least 1 of the following clinical signs or symptoms with no other recognized cause: fever (core temperature greater than 38°C), hypothermia (core temperature lower than 36°C), apnea, or bradycardia; *and* blood culture was not done or no organisms were detected in blood culture; *and* the patient has no apparent infection at another site; *and* the patient has had treatment for sepsis initiated by the physician.

Other Definitions of CLABSI

The definitions of CLABSI given above are used for infection control surveillance. However, there are other definitions that have been created to cover different aspects of catheter-related infections; these are listed below. These definitions can be applied if the surveillance efforts need to be broadened.

Catheter colonization is considered present if there is significant growth of more than 1 microorganism in a culture of the catheter tip, a subcutaneous segment of the catheter, or the catheter hub.[37]

Phlebitis is considered present if there is an induration or erythema, warmth, and pain or tenderness along the track of a catheterized or recently catheterized vein.

Exit site infection is diagnosed on the basis of either microbiological or clinical findings, as follows:

1. Microbiological criterion: exudate from the catheter exit site that yields a microorganism, with or without concomitant BSI.
2. Clinical criteria: presence of erythema, induration, and/or tenderness within 2 cm of the catheter exit site, which may be associated with other signs and symptoms of infection, such as fever or purulent drainage emerging from the exit site, with or without concomitant BSI.

Tunnel infection is considered present if there is tenderness, erythema, and/or an induration more than 2 cm from the catheter exit site along the subcutaneous track of a tunnel catheter (eg, a Hickman or Broviac catheter), with or without concomitant BSI.

Pocket infection is considered present if a totally implanted intravascular device shows purulent fluid that might or might not be associated with spontaneous rupture and drainage or necrosis of the overlying skin, with or without concomitant BSI.

Infusate-related BSI is considered present if there is concordant growth of the same organism recovered from culture(s) of infusate and blood specimens (preferably drawn percutaneously) with no other identifiable source of infection.

Catheter-related BSI, including bacteremia or fungemia, is considered present if a patient has an intravascular device and there are microorganisms recovered from more than 1 culture of blood obtained from the peripheral vein, there are no clinical manifestations of infection (eg, fever, chills, and/or hypotension), and there is no apparent source for BSI (with the exception of the catheter). One of the following 2 findings should be present:

1. A positive result of semiquantitative catheter culture (more than 15 colony forming units [cfu] per catheter segment) or quantitative catheter culture (more than 10^2 cfu per catheter segment) and isolation of the same organism (species) from a catheter segment and from a peripheral blood sample
2. Simultaneous quantitative culture of blood from the catheter and culture of peripheral blood with colony counts (expressed in colony-forming units per milliliter) in a ratio of more than 3 : 1 (catheter vs peripheral blood) and a differential time to culture positivity (growth in a culture of a blood sample obtained through a catheter hub is detected by an automated blood culture system at least 2 hours earlier than growth in a culture of a simultaneously drawn peripheral blood sample of equal volume)

Note that this definition differs from the definition of CLABSI used for infection-control surveillance activities.

Utilizing a uniform case definition in a surveillance program promotes the acceptability of the results internally and externally and the generalizability of the outcomes. Deviations from accepted standards should be carefully considered given the potential problems in future interpretation.

Collection of Samples for Culture and of Clinical Data

Information regarding infection is generally of at least 2 types, clinical and microbiologic. Clinical data are collected routinely by infection control personnel, and microbiology data are provided by the hospital laboratory. It is paramount to coordinate the surveillance efforts of both parties in order to ensure a timely and efficient collection process. The microbiology laboratory identifies the occurrence of an IV-CRI by culturing specimens from patients, such as catheter tips or blood, and by identifying pathogens and their antimicrobial-resistance patterns. As clinical signs are mostly inconclusive, the laboratory results play a critical role in the diagnosis of catheter-related BSI. Therefore, a strong collaboration between the infection control department and the microbiology laboratory should be established before initiating surveillance. It is recommended that well-established guidelines be adopted, such as those laid out by the NHSN for CLABSI.[53]

Specimen Collection Considerations

Ideally, blood specimens for culture should be obtained from 2–4 blood draws from separate venipuncture sites (eg, the right and the left antecubital veins), not through a vascular catheter. These blood draws should be performed simultaneously or within a short period of time (ie, a few hours).

However, these recommendations are often considered impractical, particularly in an ICU, if one wants to preserve vascular access. Unfortunately, using blood samples drawn through a catheter yields a high percentage of false-positive culture results, triggering unnecessary and potentially harmful treatments.[54] At this time no clear procedural recommendation can be given to ensure the optimal sensitivity and specificity of blood culture for the diagnosis of IV-CRI.

Specimen Processing Considerations

The details of specimen processing in the laboratory should also be discussed by infection control personnel and the microbiology laboratory. Standardized laboratory techniques are provided by the National Committee for Clinical Laboratory Standards (NCCLS).[55] Use of these standards allows comparison of one's laboratory results with those of other investigators. One important caveat regarding the microbiology laboratory is the interval between sampling and reporting of results. The surveillance program should allow for at least 48 hours before conclusively determining whether an IV-CRI is present or absent in a patient.

Data Collection Considerations

Trained infection preventionists are preferred for collection of clinical data.[56] This applies in particular to IV-CRI, since the complexity of event definitions and the combination of laboratory findings and clinical signs demand a high standard of knowledge and experience.

Data collection methods range from simple spreadsheets to hand-held computer devices. Selection criteria should focus on the feasibility and acceptability of the method for the users and the available resources. At minimum, the data collected should include demographic information (the patient's name, hospital identification number, age or date of birth, and sex), admission information (date of admission to the hospital and/or unit and the dates of catheter-related procedures), and clinical and laboratory information (criteria for infection and pathogens isolated). In addition, a multitude of other kinds of information can be gathered, such as data on underlying diseases, severity of symptoms, or antimicrobial utilization.

In recent years, several vendors have offered electronic "data mining" systems for medical data. These systems can greatly simplify the data collection process for surveillance purposes and, when combined with targeted logical rules, can help infection control staff in determining events. However, before implementation of such a system, the user should make sure that the information gathered from hospital databases is accurate and complete and that the rules for identifying possible events are based on accepted standards and are transparent to all parties involved. Frequent validation of the electronic data and comparison of event detection with the findings of manual surveillance are recommended; the latter is of particular importance for data collection and analysis systems located outside the original institution.

Data Validation

A central part of the management process is validation of the collected data. Validity, or the lack of systematic error, has an internal component and an external component.[57] Applied to the surveillance of IV-CRI, internal validity focuses on the accurate measurement of these infections. Missed events or misidentified events can jeopardize the entire surveillance effort. Selection bias, confounding bias, and information bias are the 3 major threats to internal validity. Information bias, in particular, is a critical element, since all further evaluation of potential associations is based on the accuracy of the information obtained by surveillance. Therefore, routine testing of the internal validity should be implemented from the beginning.

The completeness of the data collection can be easily addressed by listing the data missing from the surveillance forms. Clusters of blank spots can pinpoint problems in the ascertainment of information and provide clues about how to improve the overall data collection process. The accuracy of the data can be determined by comparison with a concurrent surveillance activity. During these spot-checks, a dedicated staff member, such as the hospital epidemiologist or an experienced infection preventionist, repeats the data collection and compares the outcomes of both efforts. The individual routinely assigned to surveillance should be unaware of the time and location of these spot-checks, to avoid any bias. Deviations in the 2 outcomes can identify problems in the case definition or misconceptions of the observer.

External validity addresses the generalizability of the inferences derived from the surveillance efforts to people or patient populations outside that particular surveillance group. This does not imply that the more diverse patient groups better the scientific value of the surveillance data. Since it is often not feasible, and also unnecessary, to include everyone in a surveillance program, one has to make choices regarding which subgroup of the population is most likely to represent a valid sample for the specific scientific problem in question. For IV-CRI, this sample selection can be based on risk assessments, leading, for example, to the selection of patients who are in critical care or the oncology ward or receiving hemodialysis. External validity requires the sound definition of the sample group and its environment. This definition should include demographic data for the patients but also hospital demographic data, including the number of beds, the variety of services, the ratio of healthcare workers to patients, and the infection control measures in place. The collection of this information allows for the comparison of the characteristics of the patient group studied with the characteristics of other populations.

Analysis, Presentation, and Feedback of Data

The initial analysis of IV-CRI surveillance data is mostly descriptive. Trends in infection rates, outbreak detection, or comparisons with other data sets are best visualized in the form of graphics accompanied by more-detailed tables. This section is intended to provide a template for the initial analysis and presentation of data on IV-CRI rates.

The most commonly used formula to calculate IV-CRI rates is as follows:[1]

$$\frac{\text{no. of IV-CRIs (for a specific site)}}{\text{no. of intravascular catheter device-days}} \times 1{,}000 = \frac{\text{no. of IV-CRIs}}{\text{per 1{,}000 device-days.}}$$

Placing the number of device-days in the denominator reflects the actual exposure to the devices. It is also useful to specify the location in which the surveillance took place (eg, the number of IV-CRIs in the surgical ICU). Both measures help to stratify one's own data for comparison with data from other units or hospitals.

There are multiple periods that can be used for reporting: counts per week, per month, per quarter, or per year. However, one has to be aware that the shorter the interval, the less accurate and statistically valid the results may become, particularly for small data sets. On the other hand, long intervals tend to obscure more minute changes, making the detection of unusual clusters of infection difficult. For analyzing a data set, it is recommended that different intervals be applied, to observe overall trends and to detect self-limiting outbreaks of IV-CRI.

Once the rates of IV-CRI have been determined, the data from the microbiology laboratory should be more closely inspected. Species identification and resistance patterns can indicate particular trends or clusters, as well as the endemic occurrence of infection with specific pathogens, including bacteria and fungi. Additional methods, such as molecular typing, can be utilized to identify clonal similarities among pathogen strains but are costly and time-consuming.

IV-CRI rates can also be compared internally (eg, with those from other units) or externally (with those from other facilities). The majority of reported and accepted comparison data have been generated by the NHSN program.[1] As mentioned previously, it is crucial to match the underlying surveillance techniques and adjust for potentially confounding factors, to achieve a meaningful comparison.

The last step is the feedback and dissemination of the findings to the appropriate personnel. To effect change—in this instance, to reduce IV-CRI rates—individuals directly and indirectly involved in patient care must be aware of the data generated by the surveillance effort. Without the understanding of the data and the accountability of these individuals, surveillance efforts, regardless of how good they are, may be in vain. Therefore, the report of surveillance results, even if seemingly unimpressive, should be firmly integrated into the surveillance program.

Prevention of Infections Associated With Vascular Catheters

Several guidelines for the prevention of catheter-related BSIs have been published recently.[2,37,58] The guidelines range from broad statements to very detailed instructions. For example, the action plan of the US Department of Health and Human Services is divided into broad recommendations for aseptic insertion of vascular catheters and for appropriate maintenance. These are summarized below.[2]

Recommendations for aseptic insertion of vascular catheters are as follows:

1. Maintain aseptic technique during insertion and care of intravascular catheters.
2. Use aseptic technique—including the use of a cap, a mask, a sterile gown, sterile gloves, and a large sterile drape—for the insertion of a CVC, including for peripherally inserted CVCs and guidewire exchange.
3. Apply an appropriate antiseptic to the insertion site on the skin before catheter insertion and during dressing changes.
4. Although a 2% chlorhexidine-based preparation is preferred for skin antisepsis, tincture of iodine, an iodophor, or 70% alcohol can be used.
5. Select the catheter, insertion technique, and insertion site with the lowest risk for complications (infectious and noninfectious) for the anticipated type and duration of intravenous therapy.
6. Use a subclavian site (rather than a jugular or a femoral site) in adult patients to minimize infection risk for nontunneled CVC placement.
7. Weigh the risk and benefits of placing a device at a recommended site to reduce infectious complications against the risk for mechanical complications (eg, pneumothorax, subclavian artery puncture, subclavian vein laceration, subclavian vein stenosis, hemothorax, thrombosis, air embolism, and catheter misplacement).

Recommendations for appropriate maintenance of vascular catheters are as follows:

1. Use either sterile gauze or a sterile, transparent, semipermeable dressing to cover the catheter insertion site.
2. Promptly remove any intravascular catheter that is no longer essential.
3. Replace the catheter-site dressing when it becomes damp, loosened, or soiled or when inspection of the site is necessary.

All the recommendations above fall into rating category 1A: strongly recommended for implementation and strongly supported by well-designed experimental, clinical, or epidemiological studies. These recommendations were accompanied by national 5-year prevention targets, including the outcome measure of the number of CLA-BSIs per 1,000 device-days, for ICUs and other locations, as well as occurrence of rates below the current 25th percentile for NHSN data, by location type. In addition, the recommendations used the process metric of 100% compliance with the recommended CVC "bundle" of measures for nonemergency insertions.

In addition to these currently well-supported recommendations, there are numerous other interventions that still require additional evidence before gaining full acceptance. For example, a catheter checklist can be used to ensure compliance with infection control protocols at the time of insertion. It also provides accountability for the insertion process, which can reinforce adherence.[59] Bathing the patient (if older than 2 months) in a chlorhexidine preparation on a daily basis may reduce the level of skin colonization with potential pathogens and therefore reduce the risk of IV-CRI.[58] The formation of a dedicated intravenous therapy team has been shown to reduce the incidence of IV-CRI in a few studies but, again, needs further evaluation.[58]

These examples of ongoing research demonstrate the wide variety of interventions one can consider for one's own institution. Again, the feasibility of implementation in the daily routine and the cost-efficacy ratio have to be considered before a specific measure is chosen.

Conclusion

For most institutions, the primary goals for the implementation of an IV-CRI surveillance and control program are to determine the incidence of IV-CRI, to monitor the trends over time, to compare the rate with findings from other surveillance systems (such as the NHSN), and to address potential shortcomings in preventive practices. A surveillance program should start with a small, well-defined patient group, such as the patients in an ICU. Once the techniques and methods are mastered, the surveillance efforts can be extended to include all patient groups identified as being at risk. After a surveillance program has been established, the data collected should guide the selection of preventive measures, if a reduction in the incidence of IV-CRI is required. Surveillance can serve as the tool for initial detection of problem areas and as the tool for subsequent monitoring of the impact of preventive measures. The relationship between surveillance and infection control interventions should be seen as a dynamic exchange that involves adaptation to new surveillance data and preventive measures.

The directions that a surveillance and prevention program can take are numerous; they include detection of trends or outbreaks in infection rates and pathogen distribution; comparison of new techniques or devices, such as catheters with antimicrobial coatings; and testing the impact of behavioral changes among healthcare professionals.[1,21,60-62]

Surveillance for IV-CRI and active reduction of its incidence is a time-consuming, personnel-intensive undertaking. However, without the information gathered and the preventive measures taken by dedicated infection control practitioners, there would be no scientific evidence regarding the endemicity of IV-CRI, the occurrence of outbreaks, or the usefulness of new methods or techniques to reduce the incidence of IV-CRI.

References

1. National Healthcare Safety Network (NHSN) Report, data summary for 2006 through 2007, issued November 2008. *Am J Infect Control* 2008;36:609–626.

2. US Department of Health and Human Services (DHHS). Action Plan to Prevent Healthcare-Associated Infections. DHHS; 2009. http://www.hhs.gov/ophs/initiatives/hai/infection.html. Accessed July 27, 2009.

3. Maki DG, Kluger DM, Crnich CJ. The risk of bloodstream infection in adults with different intravascular devices: a systematic review of 200 published prospective studies. *Mayo Clin Proc* 2006;81:1159–1171.

4. Maki DG, Mermel LA. Infections due to infusion therapy. In: Bennett JV, Brachman PS, eds. *Hospital Infections*. Philadelphia: Lippincott-Raven; 1998:689–724.

5. Jarvis WR, Edwards JE, Culver DH, et al. Nosocomial infection rates in adult and pediatric intensive care units in the United States. *Am J Med* 1991;91 (Suppl 3B):185S–191S.

6. Maki DG. Pathogenesis, prevention and management of infections due to intravascular devices used for infusion therapy. In: Bison AL, Waldvogel F, eds. *Infections Associated With Indwelling Medical Devices*. Washington, DC: American Society for Microbiology; 1989:161–177.

7. Banerjee SN, Emori TG, Culver DH, et al. Secular trends in nosocomial bloodstream infections in the United States, 1980–1989. National Nosocomial Surveillance System. *Am J Med* 1991;91(3B):86S–89S.

8. Heiselman D. Nosocomial bloodstream infection in the critically ill. *JAMA* 1994;272:1819–1820.

9. Maki DG. Nosocomial bacteremia. *Am J Med* 1981;70:183–196.

10. Mermel LA. Prevention of intravascular catheter-related infections. *Ann Intern Med* 2000;132: 391–401.

11. Raad I. Intravascular catheter-related infections. *Lancet* 1998;351:893–898.

12. Arnow PM, Quimosing EM, Beach M. Consequences of intravascular catheter sepsis. *Clin Infect Dis* 1993;16:778–784.

13. Pittet D, Tarara D, Wenzel RP. Nosocomial bloodstream infection in critically ill patients: excess length of stay, extra costs and attributable mortality. *JAMA* 1994;271:1598–1601.

14. Smith RL, Meixler SM, Simberkoff MS. Excess mortality in critically ill patients with nosocomial bloodstream infections. *Chest* 1991;100:164–167.

15. Warren DK, Quadir WW, Hollenbeak CS, et al. Attributable costs of catheter-associated bloodstream infections among intensive care patients in a nonteaching hospital. *Crit Care Med* 2006;34:2084–2089.

16. Di Giovine B, Chenoweth C, Watts C, Higgins M. The attributable mortality and costs of primary nosocomial bloodstream infections in the intensive care unit. *Am J Respir Crit Care Med* 1999;160:976–981.

17. Haley RW, Culver DH, White JW, et al. The efficacy of infection surveillance and control programs in preventing nosocomial infections in US hospitals. *Am J Epidemiol* 1985;121:182–205.

18. Curran ET, Coia JE, Gilmour H, McNamee S, Hood J. Multicentre research surveillance project to reduce infections/phlebitis associated with peripheral vascular catheters. *J Hosp Infect* 2000;46:194–202.

19. Curran ET, Booth M, Hood J. Central venous catheters and infection. Surveillance is effective in reducing catheter related sepsis. *BMJ* 1998;317:683.

20. Vandenberghe A, Laterre PF, Goenen M, et al. Surveillance of hospital-acquired infections in an intensive care department—the benefit of the full time presence of an infection control nurse. *J Hosp Infect* 2002;52:56–59.

21. CDC. Monitoring hospital-acquired infections to promote patient safety—United States, 1990–1999. *MMWR Morb Mortal Wkly Rep* 2000;49:149–153.

22. Pittet D, Tarara D, Wenzel RP. Nosocomial bloodstream infection in critically ill patients: excess length of stay, extra costs, and attributable mortality. *JAMA* 1994;271(120): 1819–1820.

23. Di Giovine B, Chenoweth C, Watts C, et al. The attributable mortality and costs of primary nosocomial bloodstream infection in the intensive care unit. *Am J Resp Crit Care Med* 1999;160:976–981.

24. Dimick JB, Pelz RK, Consunji R, et al. Increased resource use associated with catheter-related bloodstream infection in the surgical intensive care unit. *Arch Surg* 2001;136:229–234.

25. Slater F. Cost-effective infection control success story: a case presentation. *Emerg Infect Dis* 2001;7:293–294.

26. Mahieu LM, Buitenweg N, Beutels Ph, De Dooy JJ. Additional hospital stay and charges due to hospital-acquired infections in a neonatal intensive care unit. *J Hosp Infect* 2001;47: 223–229.

27. The Joint Commission. 2009 National Patient Safety Goals (NPSGs). http://www.jointcommission.org/patientsafety/nationalpatientsafetygoals/. Accessed July 27, 2009.

28. Perl TM. Surveillance, reporting, and the use of computers. In: Wenzel, ed. *Prevention and Control of Nosocomial Infections*. 3rd ed. Baltimore, MD: Williams & Wilkins; 1997:127–161.

29. German RR, Lee LM, Horan JM, Milstein RL, Pertowski CA, Waller MN; Guidelines Working Group Centers for Disease Control and Prevention (CDC). Updated guidelines for evaluating public health surveillance systems: recommendations from the Guidelines Working Group. *MMWR Recomm Rep* 2001;50(RR-13):1–35.

30. Freeman J, McGowan JE Jr. Methodologic issues in hospital epidemiology. I. Rates, case-finding, and interpretation. *Rev Infect Dis* 1981;3:658–667.

31. Wenzel RP, Osterman CA, Hunting KJ, Gwaltney JM. Hospital-acquired infections. I. Surveillance in a university hospital. *Am J Epidemiol* 1976;103:251–260.

32. Emori TG, Culver DH, Horan TC, et al. National nosocomial infections surveillance methods. *Am J Infect Control* 1991; 19:19–35.

33. Haley RW, Aber RC, Bennett JV. Surveillance of nosocomial infections. In: Bennett JV, Brachman PS, eds. *Hospital Infections*. Boston: Little, Brown & Co; 1986:51–71.

34. Hambraeus A, Malmborg A. Surveillance of hospital infections: at the bedside or at the bacteriology laboratory? *Scand J Infect Dis* 1977;9:289–292.

35. Wenzel RP, Thompson RL, Landry SM, et al. Hospital acquired infections in intensive care unit patients: an overview with emphasis on epidemics. *Infect Control* 1983;4(5):371–375.

36. Pittet D, Wenzel RP. Nosocomial bloodstream infections. *Arch Intern Med* 1995;155:1177–1184.

37. Mermel LA, Allon M, Bouza E, et al. Clinical practice guidelines for the diagnosis and management of intravascular catheter-related infection: 2009 update by the Infectious Diseases Society of America. *Clin Infect Dis* 2009;49:1–45.

38. Owings MF, Kozak LJ. Ambulatory and inpatient procedures in the United States, 1996. National Center for Health Statistics. *Vital Health Stat* 1998;13:19.

39. Pokras R, Kozak LJ, McCarthy E, et al. Trends in hospital utilization: United States, 1965–86. National Center for Health Statistics. *Vital Health Stat* 1998:13:28.

40. Haupt BJ, Jones A. National Home and Hospice Care Survey: annual summary, 1996. National Center for Health Statistics. *Vital Health Stat* 1999;13:8.

41. Popovic JR, Hall MJ. National Hospital Discharge Survey. National Center for Health Statistics. *Advance Data* 2001; 319:10.

42. Sands K, Vineyard G, Platt R. Surgical site infections occurring after hospital discharge. *J Infect Dis* 1996;173:963–970.

43. Gaynes RP, Culver DH, Horan TC, Edwards JR, Richards C, Tolson JS. Surgical site infection (SSI) rates in the United States, 1992–1998: the National Nosocomial Surveillance System, basic SSI risk index. *Clin Infect Dis* 2001;33(Suppl 2):69–77.

44. Moureau N, Poole S, Murdock MA, Gray SM, Semba CP. Central venous catheters in home infusion care: outcomes analysis in 50,470 patients. *J Vasc Interv Radiol* 2002;13: 1009–1016.

45. Hugonnet S, Eggimann P, Touveneau S, et al. ICU-acquired nosocomial infections: is post-discharge surveillance worth? In: Program and abstracts of the 41st Intersciences Conference on Antimicrobial Agents and Chemotherapy (ICAAC); Decmber 16–19, 2001; Chicago, IL. Abstract K1128.

46. Tokars JI, Miller ER, Stein G. New national surveillance system for hemodialysis-associated infections: initial results. *Am J Infect Control* 2002;30:288–295.

47. Finelli L, Miller JT, Tokars JI, Alter MJ, Arduino MJ. National surveillance of dialysis-associated diseases in the United States, 2002. *Semin Dial* 2005;18:52–61.

48. Mayhall CG. Diagnosis and management of infections of implantable devices used for prolonged venous access. *Curr Clin Top Infect Dis* 1992;12:83–110.

49. Kiehn TE, Armstrong D. Changes in the spectrum of organisms causing bacteremia and fungemia in immunocompromised patients due to venous access devices. *Eur J Clin Microbiol Infect Dis* 1990;9(7):869–872.

50. Maki DG, Weise CE, Sarafin HW. A semi-quantitative culture method for identifying intravenous-catheter-related infection. *N Engl J Med* 1977;296:1305–1309.

51. Maki DG, Cobb L, Garman JK, et al. An attachable silver-impregnated cuff for prevention of infection with central venous catheters: a prospective randomized multicenter trial. *Am J Med* 1988;85:307–314.

52. Sherertz RJ, Raad II, Belani A, et al. Three-year experience with sonicated vascular catheter cultures in a clinical microbiology laboratory. *J Clin Microbiol* 1990;28:76–82.

53. Horan TC, Andrus M, Dudeck MA. CDC/NHSN surveillance definition of health care–associated infection and criteria for specific types of infections in the acute care setting. *Am J Infect Control* 2008;36:309–32.

54. Gaur AH, Flynn PM, Heine DJ, et al. Diagnosis of catheter-related bloodstream infections among pediatric oncology patients lacking a peripheral culture, using differential time to detection. *Pediatr Infect Dis J* 2005;24:445–449.

55. National Committee for Clinical Laboratory Standards. http://www.nccls.org. Accessed July 27, 2009.

56. Brachman PS. Nosocomial infections surveillance. *Infect Control Hosp Epidemiol* 1993;14:194–196.

57. Rothman KJ, Greenland S. Precision and validity in epidemiologic studies. In: Rothman KJ, Greenland S, eds. *Modern Epidemiology*. 2nd ed. Philadelphia: Lippincott-Raven; 1998:115–134.

58. Marschall J, Mermel LA, Classen D, et al. Strategies to prevent central line-associated bloodstream infections in acute care hospitals. *Infect Control Hosp Epidemiol* 2008;29(Suppl 1): S22–S30.

59. Pronovost P, Needham D, Berenholtz S, et al. An intervention to decrease catheter-related bloodstream infections in the ICU. *N Engl J Med* 2006;355:2725–2732.

60. Wenzel RP, Edmond MB. The evolving technology of venous access. *N Engl J Med* 1999;340:48–50.

61. Eggiman P, Pittet D. Overview of catheter-related infections with special emphasis on prevention based on educational programs. *Clin Microbiol Infect* 2002;8:295–309.

62. Sherertz RJ, Wesley E, Westbrokk DM, et al. Education of physicians-in-training can decrease the risk for vascular catheter infection. *Ann Intern Med* 2000;132:641–648.

Antimicrobial-Resistant Organisms

Chapter 17 Control of Gram-Positive Multidrug-Resistant Pathogens

Trevor Van Schooneveld, MD, and Mark E. Rupp, MD

Antimicrobial-resistant pathogens are a significant and increasingly important threat to human health. More than half of healthcare-associated infections are caused by antibiotic-resistant pathogens.[1] Patients with infections due to antimicrobial-resistant pathogens have healthcare costs that are $6,000–$30,000 higher than those for patients infected with antimicrobial-susceptible organisms.[2] Healthcare settings are crucial pivot points in the initial development of antimicrobial resistance traits and the clonal expansion of antibiotic-resistant pathogens via person-to-person transmission. Healthcare epidemiologists are increasingly involved in programs to reinforce prudent use of antimicrobial agents and to control epidemic and endemic transmission of multidrug-resistant organisms (MDROs). This chapter is intended as a brief overview of the major issues regarding control of transmission of gram-positive MDROs.

Definition

MDROs are often defined as organisms that are resistant to more than 1 class of antimicrobial agents. Although the names of the most common MDROs, such as methicillin-resistant *Staphylococcus aureus* (MRSA) or vancomycin-resistant enterococci (VRE), imply resistance to only 1 antibiotic, these pathogens are often resistant to all but a few available antimicrobial agents.

Prevalence and Significance

Unfortunately, the prevalence of MDROs is increasing dramatically. For example, MRSA was observed in Europe approximately 40 years ago, concomitant with the introduction of antistaphylococcal penicillins. During the 1970s and 1980s, outbreaks of MRSA occurred in hospitals throughout the world. The prevalence of methicillin resistance among *S. aureus* isolates has been steadily increasing, with the Centers for Disease Control and Prevention (CDC) reporting that, from 1998 through 2002, 51.3% of *S. aureus* isolates recovered from patients in intensive care unit (ICUs) were resistant to methicillin.[3] For the same period, the prevalence of methicillin resistance among *S. aureus* isolates recovered from patients in non-ICU inpatient areas was 41.4%.[3] In 2004, the CDC noted that 63% of *S. aureus* infections were due to MRSA.[4] A survey of 463 hospitals from 2006–2007 noted that 56.2% of device-associated and healthcare-associated *S. aureus* infections were due to MRSA.[3] Recently, an explosive growth of MRSA infections in the outpatient setting has been noted, with more than half of invasive MRSA infections now occurring in outpatients.[5] The number of hospitalizations for treatment of VRE infections doubled between 2003 and 2006, and VRE isolates make up 12.8% of enterococcal isolates recovered from ICU patients, 12% recovered from non-ICU patients, 23% recovered from patients with surgical site infections, and 4.7% recovered from outpatient areas.[1,3,6] Although penicillin-resistant pneumococci may be considered gram-positive MDROs, they rarely result in nosocomial infection and will not be considered further in this chapter. Finally, resistance traits are also prevalent among gram-negative pathogens. Gram-negative MDROs are discussed in detail in Chapter 18.

Antibiotic resistance is associated with less favorable clinical outcomes. Kollef et al.[7] found that infection-related mortality was 2.37 times as likely among ICU patients for whom antimicrobial treatment was inadequate, most commonly because the causal pathogen was antibiotic resistant, than among patients who received antibiotics to which the causative pathogen was susceptible (P < .001). With regard to gram-positive MDROs, numerous investigators have documented their clinical significance. Compared with patients with infections due to methicillin-susceptible *S. aureus* (MSSA), patients with infections due to MRSA have significantly greater mortality, length of hospital stay, and hospital costs.[8-12] For example, Engemann et al.[8] studied staphylococcal surgical site infections and found that patients infected with MRSA were 3.4 times as likely to die than patients infected with MSSA. Excess hospital charges attributed to MRSA infections were $13,901 per infection.[8] However, it remains unclear as to whether these differences are due to intrinsic differences in the virulence of the microbes, differences in the underlying host issues, or variation in antimicrobial agent efficacy. Although similar observations have been made regarding the significance of VRE, conclusions drawn from these findings are even less clear cut, owing to multiple confounding variables that often exist among patients infected with VRE. Edmond et al.[13] observed an attributable mortality of 37% and a risk ratio for mortality of 2.3 in a comparison between patients with VRE bacteremia and matched control subjects. Bhavnani et al.[14] noted that VRE bacteremia, when compared with vancomycin-susceptible enterococcal bacteremia, was associated with an increased clinical failure rate (60% vs 40% of patients; P < .001) and all-cause mortality (52% vs 27% of patients; P < .001). In a prospective, multicenter study, Vergis et al.[15] noted that vancomycin resistance was an independent predictor of mortality among patients with enterococcal bacteremia. However, investigators in several similar studies involving various patient cohorts noted that vancomycin resistance was not associated with differences in outcomes.[16-19] Despite these conflicting findings, all agree that antimicrobial-resistant pathogens are problematic because they limit the number of therapeutic choices, require more-costly and potentially more-toxic antimicrobial agents, and increase the costs associated with performance of surveillance cultures and placement of patients in isolation.

Mechanism of Resistance and Reservoir for Transmission

MRSA

Methicillin resistance in *S. aureus* is due to the production of an alternate penicillin-binding protein, PBP2a, that has a low affinity for all β-lactam antibiotics and generates stable peptidoglycan products in the presence of inhibitory concentrations of β-lactam antibiotics.[20] The genetic element encoding methicillin resistance is carried on the staphylococcal chromosome cassette mec (SCCmec), which is a large chromosomal element containing the *mecA* gene, regulators, and usually a variety of other resistance-conferring genes. Until recently, genetic transfer of SCCmec from strain to strain has been a very rare event, and thus the worldwide spread of MRSA was almost exclusively due to clonal expansion of a few strains with this genetic background via person-to-person spread. Transmission of MRSA has traditionally been associated with the healthcare system, and previously almost all cases of colonization or infection with MRSA could be traced back to the subject's treatment at an inpatient care facility, receipt of hemodialysis, stay at a long-term care facility, or receipt of home infusion therapy.[21]

However, in more-recent years, community-associated MRSA (CA-MRSA) strains have been recovered from persons without other risk factors for acquisition of MRSA. These strains carry a smaller, more mobile and less physiologically burdensome chromosomal element, termed SCCmec type IV. This genetic element usually carries only the *mecA* gene, with no other resistance determinants differentiating it from genetic elements traditionally found in nosocomial strains of MRSA, which are usually multidrug resistant.[22,23] However, multidrug-resistant CA-MRSA strains have been described.[24] The most prevalent strain of CA-MRSA is the USA300 strain. This strain of CA-MRSA has radically altered the epidemiologic characteristics of MRSA and has even replaced MSSA as the most common cause of purulent skin and soft-tissue infections.[25] The rapid expansion in the prevalence of the USA300 strain has blurred the line between community and nosocomial strains. In San Francisco, 43.5% of hospital-onset *S. aureus* infections during 2004 and 2005 were due to the USA300 strain.[26] From 2000 through 2006, investigators at a large inner-city hospital observed that the percentage of hospital-onset MRSA bloodstream infections due to genotypic CA-MRSA increased from 24% to 49%.[27]

Despite the emergence of CA-MRSA, the major MRSA reservoir still consists of patients with significant contact with the healthcare system, and the organism is usually spread from patient to patient via contact with contaminated healthcare workers and, to a lesser extent, with medical fomites, such as stethoscopes, blood-pressure cuffs, and thermometers, and with environmental surfaces, such as bed rails and tables. It should be emphasized that the majority of carriers of MRSA are patients with asymptomatic colonization. The most common site of MRSA colonization is the anterior nares, but other sites, such as axillae, the rectum, the throat, wounds, and implanted devices, may become colonized.[28] There is evidence that transmission of CA-MRSA strains might be less dependent on nasal colonization and more dependent on fomite or skin-to-skin contact.

Other concerning developments involving *S. aureus* are increases in the minimum inhibitory concentration (MIC) of vancomycin (sometimes referred to as "MIC creep"), the emergence of vancomycin-intermediate strains (VISA) and vancomycin-heteroresistant strains (hVISA), and the detection of vancomycin-resistant *S. aureus* (VRSA). Evidence has accumulated that the vancomcyin MIC for MRSA strains has increased since 2000.[29,30] The cause and clinical significance of this finding are a topic of much debate, but isolates with vancomycin MICs of less than 1 µg/mL have been associated with increased mortality.[31] Subpopulations of *S. aureus* cells with vancomycin susceptibilities in the intermediate range have also been described. These strains have been designated as hVISA, and their detection by standard microbiologic methods is difficult.[32] hVISA strains may be associated with treatment failure and prolonged bacteremia. This prompted a change in the criteria for interpreting the vancomycin MIC, to improve detection of hVISA strains.[32,33] The mechanism behind this decreased susceptibility is a thickening of the bacterial cell wall and biomatrix.[34] Complete resistance to vancomycin in *S. aureus* isolates occurs via acquisition of the *vanA* gene from VRE species. These isolates have been surprisingly rare, and their transmission to other patients has not been documented; however, with the continued heavy use of vancomycin, resistance is expected to increase in all forms.[35]

VRE

Vancomycin resistance in *Enterococcus faecalis* and *Enterococcus faecium* is primarily due to the acquisition of *vanA* or *vanB* gene clusters, which encode enzymes responsible for the production of peptidoglycan precursors that bind to glycopeptides with reduced affinity. The resistance genes are carried on mobile genetic elements that are readily transferable between enterococcal strains.

In the United States, VRE is almost always linked to persons with significant contact with the healthcare system. The transmission of VRE in the United States typically occurs via healthcare workers and medical fomites. In Europe, the prevalence of VRE varies by country, and the epidemiology of VRE is somewhat different than in the United States. Until 1997, avoparcin, a glycopeptide, was widely used as a growth promoter in farm animals, and transmission by food products played a significant role in VRE acquisition. However, this mode of acquisition appears to have diminished dramatically in response to the prohibition of avoparcin as a growth promoter, and VRE transmission in Europe is now similar to transmission in the United States.[36,37] The natural ecologic niche of enterococci is the gut, and VRE can be readily recovered from cultures of rectal swab specimens or stool specimens from colonized persons.

Control Measures

Infection control efforts to limit the spread of gram-positive MDROs must be considered in a larger context and should be part of a comprehensive, system-wide program directed at antimicrobial resistance. Such programs should be strongly supported by hospital administration and should include educational efforts with facility-wide and unit-specific scopes. It must also be recognized that the major driving factors in the emergence of antimicrobial resistance are overuse and inappropriate use of antimicrobial agents.[38] Control of antibiotic use and antimicrobial stewardship is discussed in detail in Chapter 20. Efforts to reduce selective pressure through more-prudent use of antimicrobials should be coupled with primary measures to prevent infection, such as vaccination programs and campaigns to prevent nosocomial infections.[39] Finally, the chain of contagion must be broken through rigorous use of standard infection control precautions and contact-isolation procedures to prevent transmission.

Comprehensive statements from the Society for Healthcare Epidemiology of America (SHEA) and the CDC regarding control of gram-positive MDROs have been promulgated.[40,41] A 2008 compendium on infection prevention in acute-care hospitals, jointly authored by SHEA and the Infectious Diseases Society of America, includes a section on preventing transmission of MRSA.[42] The CDC Hospital Infection Control Practices Advisory Committee (HICPAC) authored a document that covers infection control considerations in a variety of healthcare settings for a broad range of potential pathogens, including gram-positive MDROs.[43] Although these publications differ in a number of areas, both the SHEA MRSA guidelines and the CDC MDRO guidelines advocate a tiered approach to gram-positive MDRO control. This process initially involves the use of basic practices, such as hand hygiene, contact isolation, and proper disinfection of equipment and environmental surfaces, coupled with monitoring of MDROs. If these basic control measures are ineffective in decreasing the prevalence of a target MDRO or an MDRO that has been identified in a highly vulnerable patient population or unit (eg, the neonatal ICU and the burn unit), more-intensive interventions should be used. These measures include the implementation of an active surveillance culture program or decolonization procedures, with continual assessment of their effectiveness until control of the target organism is achieved. Figure 17-1 outlines the major steps in controlling the transmission

Review published guidelines

Institute basic practices
- Conduct an MRSA risk assessment
- Ensure compliance with hand hygiene recommendations
- Ensure compliance with contact precautions for MRSA-colonized and -infected patients
- Ensure proper disinfection with equipment and environment
- Educate healthcare personnel regarding MRSA
- Implement an MRSA monitoring program
  - Implement an MRSA line list
  - Implement a laboratory-based alert system so that new cases of MRSA colonization or infection are immediately identified by IC program
  - Implement an alert system that identified readmitted or transferred MRSA-colonized or -infected patients

Continue to monitor MRSA rates
- Develop a regular reporting system to relevant stakeholders, physicians, nurses, staff, and other hospital leaders
- Hold relevant individuals and groups accountable for implementing and complying with basic prevention measures

Determine if MRSA has been effectively controlled

MRSA NOT effectively controlled

Ensure compliance with basic practices

MRSA effectively controlled
- Continue basic practices
- Continue to monitor MRSA rates
- Continue MRSA reporting and accountability system

Determine if MRSA has been effectively controlled

MRSA NOT effectively controlled

MRSA effectively controlled
- Continue basic practices
- Continue to monitor MRSA rates
- Continue MRSA reporting and accountability system

Institute one or more special approaches
- Conduct active surveillance testing for MRSA colonization among patients
  - Ensure compliance with active surveillance testing program
- Provide decolonization therapy to MRSA-colonized patients in conjunction with an active surveillance testing program
- Bathe adult intensive care unit patients with chlorhexidine

Determine if MRSA has been effectively controlled

MRSA NOT effectively controlled

MRSA effectively controlled
- Continue basic practices
- Continue to monitor MRSA rates
- Continue MRSA reporting and accountability system

- Ensure compliance with special approaches
- Assess need to intensify active surveillance testing program
- Consider additional special approaches
- Continue to monitor MRSA rates
- Continue MRSA reporting and accountability system

Figure 17-1. Major steps in controlling the transmission of methicillin-resistant *Staphylococcus aureus* (MRSA) transmission. IC, infection control. Reproduced from Calfee et al.[40]

of MRSA. Key issues in a comprehensive program to limit the spread of gram-positive MDROs are discussed below.

Basic Practices

Surveillance

Surveillance is critical to the control of MDROs because it allows for identification of new emerging pathogens, monitoring of epidemiologic trends, and assessment of the effectiveness of interventions. Facility-wide antimicrobial susceptibility results ("antibiograms") are one of the simplest ways to monitor MDROs and should be both facility and unit specific. A specific program for MRSA monitoring that can track and identify patients colonized or infected with MRSA is recommended in the SHEA guidelines.[40] It is also recommended that a system be in place to notify infection control personnel when a culture positive for MRSA is reported and to identify patients transferred or admitted who have previously been colonized or infected with MRSA.[40] A full description of laboratory methods to detect gram-positive MDROs and the role of the clinical microbiology laboratory in infection control efforts is beyond the scope of this chapter, and the interested reader is referred to recent reviews.[44,45] In recent years, several molecular-based assays and more-efficient chromogenic agar culture-based methods to detect MRSA have been introduced to the market. The cost-effectiveness of these laboratory techniques and their impact on surveillance and prevention programs is an area of intense study.

Isolation Precautions

Standard Precautions

Standard precautions should be used during all encounters with patient. The CDC HICPAC guidelines recognize that colonization with MDROs is frequently undetected and emphasize the role of standard precautions.[41] Hand hygiene is a cornerstone of standard precautions. Healthcare workers should be encouraged to use an approved alcohol-based hand rub for routine hand disinfection and to wash their hands with soap and water whenever their hands are visibly soiled with blood or body fluids.[46]

Contact Isolation Precautions

Patients known to be or strongly suspected of harboring a gram-positive MDRO should be cared for under contact isolation precautions. Patients should be housed to provide spatial separation as a means to reduce the risk of transmission. The most effective means of accomplishing this goal is to mandate use of private rooms for persons infected or colonized with gram-positive MDROs. When this is not practical, patients harboring the same species of gram-positive MDRO may be cohorted with one another to provide physical barriers between colonized or infected patients and patients who do not harbor MDROs. Limited data suggest that there may be detrimental effects on a patient's mental and physical well being when they are placed in contact isolation, and efforts should be made to monitor for and counteract potential adverse effects.[47,48]

Barrier Precautions: Gloves, Gowns, and Masks

Gloves should be worn as part of standard precautions whenever it can be reasonably anticipated that contact with blood, mucous membranes, potentially infectious material, or colonized skin will occur.[49] The use of gloves is uniformly recommended when caring for a person infected or colonized with a gram-positive MDRO. It should be stressed to healthcare workers that gloves should be changed between contact with different patients and, for a single patient, between performance of a contamination-prone task (eg, repositioning a patient, changing diapers, and emptying a bedpan) and a task involving a clean site (eg, manipulation of an intravenous catheter and performance of an intramuscular injection).[46] In addition, use of gloves does not obviate the need for hand hygiene, and hands should be disinfected following removal of gloves.[46]

Gowns should be worn as part of standard precautions to protect uncovered skin and prevent soiling of clothing during patient-care activities that are likely to generate splashes or sprays of blood or body fluids.[49] However, many healthcare workers question the need for gowns in the routine care of patients asymptomatically colonized with gram-positive MDROs. Several issues should be noted in this regard. First, colonization or infection with gram-positive MDROs often results in widespread contamination of the patient's environment.[50-52] Furthermore, it has been demonstrated that healthcare workers readily contaminate their clothing in the routine care of patients colonized with gram-positive MDROs and that gowns prevent such contamination.[50,53,54] Last, most studies examining the role of gowns in the prevention of transmission of gram-positive MDROs have indicated better control of transmission when gowns are in use.[39,55,56]

The SHEA MRSA and CDC MDRO guidelines differ from previously published guidelines and recommend against the use of masks in the control of MRSA. The CDC document does relate that masks should be used as part of standard precautions during any splash-generating procedure, care of an open tracheostomy, and when transmission from a heavily colonized source (eg, burn wounds) is likely.[41]

Equipment

Numerous studies have documented that devices such as stethoscopes, thermometers, tourniquets, and glucose monitors become contaminated with gram-positive MDROs during patient-care activities.[57-60] Furthermore, some investigators have linked contaminated equipment with transmission of gram-positive MDROs to patients.[59,60] Therefore, noncritical patient care equipment should be dedicated to a single patient. If use of nondedicated equipment is unavoidable, items should be carefully cleaned and disinfected between use involving different patients.

Environmental Measures

As previously mentioned, harboring of gram-positive MDROs can result in widespread contamination of the patient-care environment, including bed clothes, linens, bed rails, wheelchairs, bedside tables, patient-care equipment, doorknobs, faucet handles, telephone handsets, and computer keyboards.[40,41,51,55,61-63] In addition, gram-positive MDROs are quite hardy, resist desiccation, and remain viable on inanimate surfaces for days to months, and their presence in the environment predisposes patients to colonization.[52,64-66] Therefore, it is important to include the environmental-services department in a comprehensive program to combat the spread of gram-positive MDROs. Environmental-services workers should be educated, and procedures should be implemented to ensure consistent cleaning and disinfection, particularly of surfaces most likely to be touched, such as bed rails, doorknobs, and faucet handles.[40,41] Lack of adherence to prescribed facility procedures is associated with continued environmental contamination, and to combat this, a system that monitors adherence to protocols is desirable.[40,41] The use of education, feedback, and enforcement of standard cleaning policies can result in significant reductions in environmental contamination without significant financial burden.[67,68] MRSA and VRE are rapidly killed by standard, low-level disinfectants.[69] Cleaning and disinfection must be performed with careful attention to the adequacy of cleaning, the dilution of the disinfectant, and the duration of disinfectant contact with environmental surfaces.[70] Environmental cleaning can be enhanced by ensuring that the same individuals perform cleaning and disinfection services.[41] Environmental cultures are recommended only when there is epidemiologic evidence suggesting that an environmental source is responsible for transmission.[41]

Discontinuation of Contact Isolation Precautions

Indications for the discontinuation of contact isolation precautions are controversial, and both the CDC guidelines and the SHEA guidelines consider this an unresolved issue awaiting more-definitive studies.[40,41] Previous guidelines from the CDC recommend discontinuing contact isolation for VRE when cultures of 3 stool specimens obtained at weekly intervals have negative results.[71] More recent experience has indicated that such screening may not detect persons with low-level colonization and that VRE may reemerge when patients are exposed to antimicrobials.[72-74] Similarly, colonization with MRSA may not be detected by screening of only 1 anatomic site, may be persistent or intermittent, and may be difficult to detect in persons with low-level colonization or intermittent shedding.[75-78] Both guidelines suggest that patients colonized with an MDRO should be considered colonized until results of 3 surveillance cultures performed over the 1–2-week period after completion of antimicrobial therapy are negative.[40,41] Whether an assessment for continued colonization should be done is debatable, but an interval of at least 3–4 months is suggested.

Intensive Interventions

Active Surveillance

As previously mentioned, the majority of patients harboring MRSA or VRE are asymptomatically colonized. Therefore, case finding based solely on detection of gram-positive MDROs from routinely submitted clinical specimens will not detect the majority of asymptomatic carriers.[40,79] For example, Girou et al.[80] noted that approximately one-half of cases of MRSA colonization and MRSA infection among patients admitted to an ICU were discovered only by screening cultures. Similarly, Huang et al.[81] found that routine surveillance cultures for VRE detected up to 15 times as many colonized patients than did routine clinical cultures. This is the impetus behind the use of active surveillance cultures.

Both the SHEA and CDC guidelines recognize the significance of patients who are asymptomatically colonized with gram-positive MDROs. The current guidelines from these organizations[40,41] differ somewhat from previous SHEA guidelines, which advocated the routine use of surveillance cultures throughout the healthcare system, with the goal of identifying all colonized patients.[82] Both of the more recent guidelines recommend consideration of active surveillance cultures as part of a multifaceted program targeted at the control of MDRO, when adherence to basic practices has been unsuccessful at controlling MDRO spread.[40,41] The exact circumstances in which active surveillance cultures should be used are not well-defined but should be specifically tailored to individual facilities. Recent studies have produced conflicting results regarding the use of active surveillance cultures. An observational cohort study performed in 3 hospitals found that use of active surveillance cultures reduced MRSA infections by nearly 70%.[83] Conversely, a crossover cohort trial involving surgical patients at a single, large institution found no reduction in the incidence of nosocomial MRSA infections.[84] Preliminary analysis of a multicenter, cluster-randomized trial of active surveillance cultures in critical care units noted no decrease in the acquisition of either MRSA or VRE.[85] These results make the universal application of active surveillance cultures a continuing uncertainty. The use of active surveillance cultures should be considered in institutions that continue to experience unacceptably high rates of MDRO acquisition despite use of basic practices for MDRO control.[40,41]

When beginning an active surveillance program, many factors must be taken into account. The first consideration is the additional support needed to implement the program: both personnel to collect samples for culture and to process them in the microbiology laboratory, and also the means to communicate the findings and measure compliance with the screening procedures. Other important considerations are where to implement the program, which patients to screen (eg, patients in the ICU, patients with a high risk of infection, or all patients), when to perform the screening (on admission to the prescribed unit is considered the minimum), and which anatomic sites to screen. The anterior nares is the most frequent culture-positive site, but screening only the nares has been noted to miss up to 27% of carriers.[86] The anterior nares should always be included in screening, and other sites, such as open wounds, perirectal areas, throat, or foreign bodies, may also be included to increase yield.[40,41] It is imperative that the screening program is part of a multifaceted effort to control the transmission of MDROs. Simply identifying colonized patients without adherence to isolation, hand hygiene, and environmental disinfection procedures is unlikely to be effective.

Multiple laboratory methods are available for screening, including culture-based techniques and molecular assays. The major disadvantage of using a culture-based method is the time required for results. Traditional culture methods require an interval of at least 48 hours, but newer chromogenic media yield findings in 24 hours.[87] Molecular assays are very rapid (2 hours or less, if tests are run continuously and not batched) and highly sensitive and specific, but they are limited by their cost (at least $10 per assay, compared with less than $1 for a swab culture) and recently pub-lished findings showing that their use may not decrease the frequency of MRSA transmission.[88] The preferred test for a given facility should be determined by consideration of a number of factors, including the performance characteristics of the test, the turnaround time, the laboratory's capabilities, the number of specimens anticipated, and the cost.[40] How patients should be managed while they await screening results should be determined, and it is reasonable to manage patients empirically by placing them in contact isolation until negative results of active surveillance cultures are available.[40,41] Finally, compliance with screening recommendations, isolation precautions, and communication of results should be monitored.

Decolonization

Decolonization is not routinely recommended but can be used in concert with active surveillance cultures as a component of intensive interventions to control the spread of MRSA.[40,41] The use of decolonization regimens for MRSA has been shown by some studies to decrease the spread of MRSA and has been a component of the control of MRSA infection outbreaks.[89,90] A variety of decolonization regimens have been described, and the optimal regimen is unknown. A recent systematic review noted decolonization regimens that use 2% mupirocin ointment intranasally for 4–7 days were 90% effective at eliminating colonization at 1 week and approximately 60% successful at longer follow-up durations.[91] The addition of daily chlorhexidine bathing to a regimen of mupirocin therapy has also shown excellent efficacy.[89,92,93] Oral or systemic antibiotic therapy is less successful than topical mupirocin therapy.[91] Unfortunately, suppression of MRSA colonization is often transient and, at times, is complicated by the emergence of resistance to agents used in the decolonization scheme.[91,94,95] After decolonization has been attempted, surveillance cultures should continue to be performed to document clearance and monitor for resistance. Decolonization of healthcare workers colonized with MRSA should be limited to instances in which healthcare workers have been implicated epidemiologically to transmission. The efficacy of decolonization regimens of VRE has not been established, and they should not, at this time, be routinely used.

Chlorhexidine Bathing

The bathing of patients with chlorhexidine should be considered when basic measures are not successful in interrupting transmission of MDROs.[40] Chlorhexidine bathing in the ICU may decrease the acquisition of

MRSA and VRE and may decrease the incidence of catheter-associated bloodstream infections.[89,96,97] A variety of chlorhexidine products are available for use, but it should be noted that the use of 4% chlorhexidine was associated with adverse skin reactions in one study.[98] The use of chlorhexidine bathing outside of the ICU has not been sufficiently studied.

Practice Settings

Consensus exists that patients in acute care settings (eg, ICUs, burn units, and inpatient wards) are at high risk for the development of nosocomial infections and that comprehensive measures should be implemented to prevent nosocomial acquisition and transmission of gram-positive MDROs.[40,41,82] However, it is less clear what measures should be practiced in ambulatory care, long-term care, or home care settings. One view, espoused in the 2003 SHEA guideline, advocates that transmission of gram-positive MDROs can occur in any healthcare setting and that, for control of gram-positive MDROs to have the greatest effectiveness and long-term impact, widespread performance of surveillance cultures and strict adherence to contact isolation precautions should be implemented throughout the healthcare system.[82] Alternatively, the CDC guideline emphasizes standard precautions and points out that risk factors for infection differ markedly across care settings and patient populations.[41] In long-term care facilities, the CDC recommends that contact precautions be used for patients who are colonized with gram-positive MDROs and totally dependent on healthcare personnel and that standard precautions are acceptable for individual patients unless contact is expected with uncontrolled respiratory secretions, pressure ulcers, draining wounds, stool, or ostomy bags.[41] Similar recommendations should be followed in the ambulatory care setting, although this issue is unresolved.

Conclusion

Antimicrobial resistance among gram-positive pathogens is a significant and growing problem. The fearful specter of a "postantibiotic era" in which drug development does not keep pace with the emergence of antimicrobial resistance is a realistic possibility. Therefore, efforts to control the spread of gram-positive MDROs are of paramount importance. Infection control efforts should be part of a comprehensive program that includes antimicrobial stewardship and primary infection prevention measures.

The foundation of infection control programs directed at the prevention of transmission of gram-positive MDROs is the uniform use of standard precautions and hand hygiene. When these practices, along with contact isolation and appropriate environmental disinfection, are inadequate to control MDROs, more-intensive practices should be implemented. These practices should be tailored to each hospital, unit, and patient group and may include performance of active surveillance cultures to detect patients asymptomatically colonized with VRE or MRSA, decolonization of patients identified by these cultures, or routine implementation of chlorhexidine bathing. These measures are particularly important to prevent infection in high-risk populations but should be used as broadly as possible and practical.

Challenges for the future will include adaptation to the changing epidemiologic characteristics of MRSA, particularly in the form of the rapidly expanding reservoir of persons colonized with community-acquired MRSA. Because of the biological characteristics of staphylococci, the easily transmissible nature of the *mecA* gene, and the expanding patient population colonized with CA-MRSA, it is likely that the prevalence of MRSA will continue to increase in healthcare settings, despite use of appropriate infection control measures. Despite this probable expansion in the prevalence of MRSA acquisition, the aggressive use of multiple infection control modalities might prevent infections from occurring. A recent analysis of data reported to the CDC from hospitals around the United States noted that the prevalence of MRSA acquisition increased from 48% in 1997 to 65% in 2007.[99] Despite this, the incidence of MRSA central-line associated bloodstream infections decreased almost 50% during the same period.[99] Although the reasons for this decrease are not addressed, it is reasonable to speculate that it was due to efforts to prevent central line–associated bloodstream infections, such as implementing use of full sterile barrier precautions, chlorhexidine skin disinfection, and a "checklist" approach. This highlights the continued need for emphasis on the basics of infection control, such as hand hygiene, environmental disinfection, limiting the number of fomites, and infection prevention efforts aimed at preventing device-associated infections and surgical site infections. These practices will limit transmission and infection due to a variety of pathogens, whereas measures that are specific to a single pathogen may detract from such general efforts. There remain major gaps in our knowledge of the pathogenesis of disease, factors that influence colonization and infection by gram-positive MDROs, how best to apply new techniques in rapid MDRO detection, and the most effective and cost-effective means to eliminate or block MDRO colonization and transmission.

References

1. Hidron AI, Edwards JR, Patel J, et al. Antimicrobial-resistant pathogens associated with healthcare-associated infections: annual summary of data reported to the National Healthcare Safety Network at the Centers for Disease Control and Prevention, 2006–2007. *Infect Control Hosp Epidemiol* 2008; 29:996–1011.

2. Cosgrove SE. The relationship between antimicrobial resistance and patient outcomes: Mortality, length of hospital stay and health care costs. *Clin Infect Dis* 2006;42:S82–S89.

3. National Nosocomial Infections Surveillance (NNIS) system report, data summary from January 1992 to June 2002, issued August 2002. *Am J Infect Control* 2002;30:458–475.

4. Centers for Disease Control and Prevention. *MRSA in Healthcare Settings*. Modified October 3, 2007. http://www.cdc.gov/ncidod/dhqp/ar_MRSA_spotlight_2006.html. Accessed August 5, 2009.

5. Klevens RM, Morrison MA, Nadle J, et al. Invasive methicillin-resistant *Staphylococcus aureus* infections in the United States. *JAMA* 2007;298:1763–1771.

6. Ramsey AM, Zilberberg MD. Secular trends in hopitalizations with vancomycin-resistant *Enterococcus* infection in the United States, 2000–2006. *Infect Control Hosp Epidemiol* 2009;30:184–186.

7. Kollef MH, Sherman G, Ward S, Fraser VJ. Inadequate antimicrobial treatment of infections: a risk factor for hospital mortality among critically ill patients. *Chest* 1999;115:462–474.

8. Engemann JJ, Carmeli Y, Cosgrove SE, et al. Adverse clinical and economic outcomes attributable to methicillin resistance among patients with *Staphylococcus aureus* surgical site infection. *Clin Infect Dis* 2003;36:592–598.

9. Cosgrove SE, Qi Y, Kaye KS, et al. The impact of methicillin-resistance in *Staphylococcus aureus* bacteremia on patient outcomes: mortality, length of stay, and hospital charges. *Infect Control Hosp Epidemiol* 2005;26:166–174.

10. Cosgrove SE, Sakoulas G, Perencevich EN, Schwaber J, Karchmer AW, Carmeli Y. Comparison of mortality associated with methicillin-resistant and methicillin-susceptible *Staphylococcus aureus* bacteremia: a meta-analysis. *Clin Infect Dis* 2003;36:53–59.

11. Romero-Vivas J, Ruio M, Fernandez C, Picazo JJ. Mortality associated with nosocomial bacteremia due to methicillin-resistant *Staphylococcus aureus*. *Clin Infect Dis* 1995;21:1417–1423.

12. Rubin RJ, Harrington CA, Poon A, Dietrich K, Greene JA, Moiduddin A. The economic impact of *Staphylococcus aureus* infection in New York city hospitals. *Emerg Infect Dis* 1999;5:9–17.

13. Edmond MB, Ober JF, Dawson JD, Weinbaum DL, Wenzel RP. Vancomycin-resistant enterococcal bacteremia: natural history and attributable mortality. *Clin Infect Dis* 1996;23:1234–1239.

14. Bhavnani SM, Drake JA, Forrest A, et al. A nationwide, multicenter, case-control study comparing risk factors, treatment, and outcome for vancomycin-resistant and –susceptible enterococcal bacteremia. *Diag Microbiol Infect Dis* 2000;36:145–158.

15. Vergis EN, Hayden MK, Chow JW, et al. Determinants of vancomycin resistance and mortality rates in enterococcal bacteremia. *Ann Intern Med* 2001;135:484–492.

16. Shay DK, Maloney SA, Montecalvo M, et al. Epidemiology and mortality risk of vancomycin-resistant enterococcal bloodstream infections. *J Infect Dis* 1995;172:993–1000.

17. Stroud L, Edwards J, Danzig L, Culver D, Gaynes R. Risk factors for mortality associated with enterococcal bloodstream infections. *Infect Control Hosp Epidemiol* 1996;17:576–580.

18. Garbutt JM, Ventrapragada M, Littenberg B, Mundy LM. Association between resistance to vancomycin and death in cases of *Enterococcus faecium* bacteremia. *Clin Infect Dis* 2000;30:466–472.

19. Erlandson KM, Sun J, Iwen PC, Rupp ME. Impact of more-potent antibiotics quinopristin-dalfopristin and linezolid on outcome measure of patients with vancomycin-resistant *Enterococcus* bacteremia. *Clin Infect Dis* 2008; 46:30–36.

20. Chambers HF. Methicillin-resistant staphylococci. *Clin Microbiol Rev* 1988;1:173–186.

21. Salgado CD, Farr BM, Calfee DP. Community-acquired methicillin-resistant *Staphylococcus aureus:* a meta-analysis of prevalence and risk factors. *Clin Infect Dis* 2003;36:131–139.

22. Naimi TS, LeDell KH, Como-Sabetti K, et al. Comparison of community- and healthcare-associated methicillin-resistant *Staphylococcus aureus* infection. *JAMA* 2003;290:2976–2984.

23. Said-Salim B, Mathema B, Kreiswirth BN. Community-acquired methicillin-resistant *Staphylococcus aureus:* an emerging pathogen. *Infect Control Hosp Epidemiol* 2003;24:451–455.

24. Diep BA, Chambers HF, Graber CJ, et al. Emergence of multidrug-resistant, community-associated methicillin-resistant *Staphylococcus aureus* clone USA300 in men who have sex with men. *Ann Intern Med* 2008;148:249–257.

25. Kaplan SL, Hulten KG, Gonzalez BE, et al. Three-year surveillance of community-acquired *Staphylococcus aurues* infections in children. *Clin Infect Dis* 2005;40:1785–1791.

26. Lui C, Graber CJ, Karr M, et al. A population-based study of the incidence and molecular epidemiology of methicillin-resistant *Staphylococcus aureus* disease in San Francisco, 2004–2005. *Clin Infect Dis* 2008;46:1637–1646.

27. Popovich KJ, Weinstein RA, Hota B. Are community-associated methicillin-resistant *Staphylococcus aureus* strains replacing traditional nosocomial MRSA strains. *Clin Infect Dis* 2008;46:787–794.

28. Currie A, Davis L, Odrobina E, et al. Sensitivities of nasal and rectal swabs for detection of methicillin-resistant *Staphylococcus aureus* colonization in an active surveillance program. *J Clin Microbiol* 2008;46:3101–3103.

29. Steinkraus G, White R, Friedrich L. Vancomycin MIC creep in non-vancomycin-intermediate *Staphylococcus aureus* (VISA), vancomycin-susceptible clinical methicillin-resistant *S. aureus* (MRSA) blood isolates from 2001–2005. *J Antimicrob Chemother* 2007;60:788–794.

30. Wang G, Hindler JF, Ward KW, et al. Increased vancomycin MICs for *Staphylococcus aureus* clinical isolates from a university hospital during a 5-year period. *J Clin Microbiol* 2006;44:3883–3886.

31. Soriano A, Marco F, Martinez JA, et al. Influence of vancomycin minimum inhibitory concentration of the treatment of methicillin-resistant *Staphylococcus aureus* bacteremia. *Clin Infect Dis* 2008;46:193–200.

32. Tenover FC, Moellering RC. The rationale for revising the clinical and laboratory standards institute vancmycin minimal inhibitory concentration interpretive criteria for *Staphylococcus aureus*. *Clin Infect Dis* 2007;44:1208–1215.

33. Maor Y, Hagin M, Belausov N, et al. Clinical features of heteroresistant vancomycin-intermediate *Staphylococcus aureus* bacteremia versus those of methicillin-resistant *S. aureus* bacteremia. *J Infect Dis* 2009;199:619–624.

34. Cui L, Ma X, Sato K, et al. Cell wall thickening is a common feature of vancomycin resistance in *Staphylococcus aureus*. *J Clin Microbiol* 2003;41:5–14.

35. Sievert DM, Rudrik JT, Patel JB, et al. Vancomycin-resistant *Staphylococcus aureus* in the United States, 2002–2006. *Clin Infect Dis* 2008;46:668–674.

36. Wegener HK. Ending the use of antimicrobial growth promoters is making a difference. *ASM News* 2003;69:443–448.

37. Werner G, Coque TM, Hammerum AM, et al. Emergence and spread of vancomycin resistance among enterococci in Europe. *Euro Surveill* 2008;13(47):pii=19046.

38. World Health Organization. WHO global strategy for containment of antimicrobial resistance. http://www.who.int/csr/resources/publications/drugresist/EGlobal_Strat.pdf. Accessed August 5, 2009.

39. Centers for Disease Control and Prevention. Campaign to prevent antimicrobial resistance in healthcare settings: 12 steps to prevent antimicrobial resistance among hospitalized adults. http://www.cdc.gov/drugresistance/healthcare/ha/12steps_HA.htm. Accessed August 5, 2009.

40. Calfee DP, Salgado CD, Classen D, et al. Strategies to prevent transmission of methicillin-resistant *Staphylococcus aureus* in acute care hospitals. *Infect Control Hosp Epidemiol* 2008;29:S62-S80.

41. Siegel JD, Rhinehart E, Jackson M, et al. Management of multidrug-resistant organisms in healthcare settings, 2006. *Am J Infect Control* 2007;35:S165-S193.

42. Society for Healthcare Epidemiologist of America, Infectious Diseases Society of America. *Compendium of Strategies to Prevent Healthcare-Associated Infections in Acute Care Hospitals*. 2008. http://www.shea-online.org. Accessed August 5, 2009.

43. Centers for Disease Control and Prevention. Infection Control in Healthcare Settings. Updated July 23, 2009. http://www.cdc.gov/NCIDOD/DHQP/. Accessed August 5, 2009.

44. Struelens MJ, Hawkey PM, French GL, et al. Laboratory tools and strategies for methicillin-resistant *Staphylococcus aureus* screening, surveillance and typing: state of the art and unmet needs. *Clin Microbiol Infect* 2009;15:112–119.

45. Malhotra-Kumar S, Haccuria K, Michiels M, et al. Current trends in rapid diagnostics for methicillin-resistant *Staphylococcus aureus* and glycopeptide-resistant *Enterococcus* species. *J Clin Microbiol* 2008;46:1577–1587.

46. Centers for Disease Control and Prevention. Guideline for hand hygiene in health-care settings: recommendations of the healthcare infection control practices advisory committee and the HICPAC/SHEA/APIC/IDSA hand hygiene task force. *MMWR Recomm Rep* 2002;51(RR-16):1–45.

47. Stelfox HT, Bates DW, Redelmeier DA. Safety of patients isolated for infection control. *JAMA* 2003;290:1899–1905.

48. Kirkland KB. Taking off the gloves: toward a less dogmatic approach to the use of contact isolation. *Clin Infect Dis* 2009;48:766–771.

49. Occupational exposure to bloodborne pathogens—OSHA: final rule. *Fed Regist* 1991;56:64004–64182.

50. Boyce JM, Potter-Bynoe G, Chenevert C, King T. Environmental contamination due to methicillin-resistant *Staphylococcus aureus*: possible infection control implications. *Infect Control Hosp Epidemiol* 1997;18:622–627.

51. Boyce JM, Havill NL, Otter JA, et al. Widespread environmental contamination associated with patients with diarrhea and methicillin-resistant *Staphylococcus aureus* colonization of the gastrointestinal tract. *Infect Control Hosp Epidemiol* 2007;28:1142–1147.

52. Falk PS, Winnike J, Woodmansee C, Desai M, Mayhall CG. Outbreak due to vancomycin-resistant enterococci (VRE) in a burn unit. *Infect Control Hosp Epidemiol* 2000; 21:575–582.

53. Smith TL, Iwen PC, Olson SB, Rupp ME. Environmental contamination with vancomycin-resistant enterococci in an outpatient setting. *Infect Control Hosp Epidemiol* 1998;19:515–518.

54. Snyder GM, Thom KA, Furuno JP, et al. Detection of methicillin-resistant *Staphylococcus aureus* and vancomycin-resistant enterococci on the gowns and gloves of healthcare workers. *Infect Control Hosp Epidemiol* 2008;29:583–589.

55. Puzniak LA, Leet T, Mayfield J, Kollef M, Mundy LM. To gown or not to gown: the effect on acquisition of vancomycin resistant enterococci. *Clin Infect Dis* 2002;35:18–25.

56. Srinivasan A, Song X, Bower R, et al. A prospective study to determine whether cover gowns in addition to gloves decrease nosocomial transmission of vancomycin-resistant enterococci in an ICU. *Infect Control Hosp Epidemiol* 2002;23:424–428.

57. Bernard L, Kereveur A, Durand D, et al. Bacterial contamination of hospital physicians' stethoscopes. *Infect Control Hosp Epidemiol* 1999;20:626–628.

58. de Gialluly C, Morange V, de Gialluly E, et al. Blood pressure cuff as a potential vector of pathogenic microorganisms: a prospective study in a teaching hospital. *Infect Control Hosp Epidemiol* 2006;27:940–943.

59. Livornese LL, Dias S, Romanowski B, et al. Hospital-acquired infection with vancomycin-resistant *Enterococcus faecium* transmitted by electronic thermometers. *Ann Intern Med* 1992;117:112–116.

60. Brooks S, Khan A, Stoica D, Griffith J. Reduction in vancomycin-resistant *Enterococcus* and *Clostridium difficile* infections following change to tympanic thermometers. *Infect Control Hosp Epidemiol* 1998;19:333–336.

61. Boyce JM, Potter-Bynoe G, Chenevert C, King T. Environmental contamination due to methicillin-resistant *Staphylococcus aureus*: possible infection control implications. *Infect Control Hosp Epidemiol* 1997;18:622–627.

62. Noskin GA, Peterson L, Warren J. *Enterococcus faecium* and *Enterococcus faecalis* bacteremia: acquisition and outcome. *Clin Infect Dis* 1995;20:296–301.

63. Rutula W, Katz E, Sherertz R, Sarubbi F. Environmental study of methicillin-resistant *Staphylococcus aureus* epidemic in a burn unit. *J Clin Microbiol* 1983;18:683–688.

64. Neely AN, Maley MP. Survival of enterococci and staphylococci on hospital fabrics and plastics. *J Clin Microbiol* 2000;38:724–726.

65. Wendt C, Wiesenthal B, Dietz E, Ruden H. Survival of vancomycin-resistant and vancomycin-susceptible enterococci on dry surfaces. *J Clin Microbiol* 1998;36:3734–3736.

66. Drees M, Snydman DR, Schmid CH, et al. Prior environmental contamination increases the risk of acquisition of vancomycin-resistant enterococci. *Clin Infect Dis* 2008;46:678–685.

67. Carling PC, Parry MM, Rupp ME, et al. Improving cleaning of the environment surrounding patients in 36 acute care hospitals. *Infect Control Hosp Epidemiol* 2008;29: 1035–1041.

68. Hayden MK, Bonten MJ, Blom DW, et al. Reduction in acquisition of vancomycin-resistant enterococcus after enforcement of routine environmental cleaning measures. *Clin Infect Dis* 2006;42:1552–1560.

69. Ayliffe GAJ. Control of *Staphylococcus aureus* and enterococcal infections. In: Block SS, ed. *Disinfection, Sterilization, and Preservation*. 5th ed. Philadelphia, PA: Lippincott Williams & Wilkins;2001:491–504.

70. Sehulster L, Chinn RY. Guidelines for environmental infection control in health-care facilities: recommendations of CDC and the Healthcare Infection Control Practices Advisory Committee (HICPAC). Centers for Disease Control and Prevention, HICPAC. *MMWR Recomm Rep* 2003;52(RR-10):1–42.

71. Centers for Disease Control and Prevention. Healthcare Infection Control Practices Advisory Committee (HICPAC). Recommendations for preventing the spread of vancomycin resistance. *MMWR Recomm Rep* 1995;44(RR-12):1–13.

72. Byers KE, Anglim AM, Anneski CJ, et al. Duration of colonization with vancomycin-resistant *Enterococcus*. *Infect Control Hosp Epidemiol* 2002;23:207–211.

73. Donskey CJ, Hoyen CK, Das SM, Helfand MS, Hecker MT. Recurrence of vancomycin-resistant *Enterococcus* stool colonization during antibiotic therapy. *Infect Control Hosp Epidemiol* 2002;23:436–440.

74. D'Agata EM. Antimicrobial-resistant gram-positive bacteria among patients undergoing chronic hemodialysis. *Clin Infect Dis* 2002;35:1212–1218.

75. Scanvic A, Denic L, Gaillon S, Giry P, Andremont A, Lucet JC. Duration of colonization by methicillin-resistant *Staphylococcus aureus* after hospital discharge and risk factors for prolonged carriage. *Clin Infect Dis* 2001;32:1393–1298.

76. Hachem R, Raad I. Failure of oral antimicrobial agents in eradicating gastrointestinal colonization with vancomycin-resistant enterococci. *Infect Control Hosp Epidemiol* 2002;23:43–44.

77. Harbarth S, Liassine N, Dharan S, Herrault P, Auckenthaler R, Pittet D. Risk factors for persistent carriage of methicillin-resistant *Staphylococcus aureus*. *Clin Infect Dis* 2000;31: 1380–1385.

78. Marschall J, Muhlemann K. Duration of methicillin-resistant *Staphylococcus aureus* carriage, according to risk factors for acquisition. *Infect Control Hosp Epidemiol* 2006;27:1206–1212.

79. Warren DK, Fraser VJ. Infection control measures to limit antimicrobial resistance. *Crit Care Med* 2001;29(Suppl 4): N128–N134.

80. Girou E, Pujade G, Legrand P, Cizeau F, Brun-Buisson C. Selective screening of carriers for control of methicillin-resistant *Staphylococcus aureus* (MRSA) in high-risk hospital areas with a high level of endemic MRSA. *Clin Infect Dis* 1998;27:543–550.

81. Huang SS, Rifas-Shiman SL, Pottinger JM, et al. Improving the assessment of vancomycin-resistant enterococci by routine screening. *J Infect Dis* 2007;195:339–346.

82. Muto CA, Jernigan JA, Ostrowsky BE, et al. SHEA guideline for preventing nosocomial transmission of multidrug-resistant strains of *Staphylococcus aureus* and *Enterococcus*. *Infect Control Hosp Epidemiol* 2003;24:362–386.

83. Robiscsek A, Beaumont JL, Paule SM, et al. Universal surveillance for methicillin-resistant *Staphylococcus aureus* in 2 affiliated hospitals. *Ann Intern Med* 2008;148:409–418.

84. Harbarth S, Fankhauser C, Schrenzel J, et al. Universal screening for methicillin-resistant *Staphylococcus aureus* at hospital admission and nosocomial infection in surgical patients. *JAMA* 2008;299:1149–1157.

85. Huskins C. New insights into MRSA screening and reporting: the results of the strategies to reduce transmission of antimicrobial resistant bacteria in ICUs (STAR*ICU). Oral presentation at the 17th Annual Scientific Meeting of the Society for Healthcare Epidemiology of America; April 14–17, 2007; Baltimore, MD. Symposium 16-16.

86. Eveillard M, de Lassence A, Lancien E, et al. Evaluation of a strategy of screening multiple anatomical sites for methicillin-resistant *Staphylococcus aureus* at admission to a teaching hospital. *Infect Control Hosp Epidemiol* 2006;27:181–184.

87. Cherkaoui A, Renzi G, Francios P, Schrenzel J. Comparison of four chromogenic media for culture-based screening of methicillin-resistant *Staphylococcus aureus*. *J Med Microbiol* 2007;56:500–3.

88. Jeyaratnam D, Whitty CJ, Phillips K, et al. Impact of rapid screening tests on acquisition of methicillin-resistant *Staphylococcus aureus:* cluster randomized crossover trial. *BMJ* 2008;336:927–330.

89. Ridenour G, Lampen R, Federspiel J, et al. Selective use of intranasal mupirocin and chlorhexidine bathing and the incidence of methicillin-resistant *Staphylococcus aureus* colonization and infection among intensive care unit patients. *Infect Control Hosp Epidemiol* 2007;28:1155–1161.

90. Saima L, Cronquist A, Wu F, et al. An outbreak of methicillin-resistant *Staphylococcus aureus* in a neonatal intensive care unit. *Infect Control Hosp Epidemiol* 2003;24:317–321.

91. Ammerlaan HS, Kluytmans JA, Wertheim HF, et al. Eradication of methicillin-resistant *Staphylococcus aureus* carriage: a systematic review. *Clin Infect Dis* 2009;48:922–930.

92. van Rijen M, Bonten M, Wenzel R, et al. Mupirocin ointment for preventing *Staphylococcus aureus* infections in nasal carriers. *Cochrane Database Syst Rev* 2008;CD006216.

93. Falagas ME, Bliziotis IA, Fragoulis KN. Oral rifampin for eradication of *Staphylococcus aureus* carriage from healthy and sick populations: a systematic review of the evidence from comparative trials. *Am J Infect Control* 2007;35:106–114.

94. Mody L, Kauffman CA, McNeil SA, et al. Mupirocin-based decolonization of *Staphylococcus aureus* carries in residents of 2 long-term care facilities: a randomized, double-blind, placebo-controlled trial. *Clin Infect Dis* 2003;37:1467–1474.

95. Miller MA, Dascal A, Portnory J, Medelson J. Development of mupirocin resistance among methicillin-resistant *Staphylococcus aureus* after widespread use of nasal mupirocin ointment. *Infect Control Hosp Epidemiol* 1996;17:811–813.

96. Vernon MO, Hayden MK, Trick WE, et al. Chlorhexidine gluconate to cleanse patients in a medical intensive care unit: the effectiveness of source control to reduce bioburden of vancomycin-resistant enterococci. *Arch Intern Med* 2006; 166:306–312.

97. Bleasdale SC, Trick WE, Gonzalez IM, et al. Effectiveness of chlorhexidine bathing to reduce catheter-associated bloodstream infections in medical intensive care units. *Arch Intern Med* 2007;167:2073–2079.
98. Wendt C, Schinke S, Wurttemberger M, et al. Value of whole-body washing with chlorhexidine for the eradication of methicillin-resistant *Staphylococcus aureus:* a randomized, placebo-controlled, double-blind clinical trial. *Infect Control Hosp Epidemiol* 2007;28:1036–1043.
99. Burton DC, Edwards JR, Horan TC, et al. Methicillin-resistant *Staphylococcus aureus* central line-associated bloodstream infections in US intensive care units, 1997–2007. *JAMA* 2009;301:727–736.

Chapter 18 Control of Antibiotic-Resistant Gram-Negative Pathogens

Anthony D. Harris, MD, MPH, and Kerri A. Thom, MD, MS

Over the last 15 years, considerable attention has been given to antibiotic resistance in gram-positive cocci. Methicillin-resistant *Staphylococcus aureus* (MRSA) and vancomycin-resistant *Enterococcus faecium* (VRE) have become well-known to hospital epidemiologists and infection control practitioners. In response to the threat from these organisms, pharmaceutical manufacturers have developed a considerable armamentarium of drugs with activity against drug-resistant gram-positive cocci. Linezolid, quinupristin-dalfopristin, tigecycline, ceftobiprole, and daptomycin are the first agents in a significant pipeline of antibiotics active against multidrug-resistant gram-positive cocci (which are discussed in Chapter 17).

In contrast, virtually no new antibiotics active against antibiotic-resistant gram-negative bacilli (GNB) have become available.[1,2] Furthermore, trends in drug resistance among GNB have been toward resistance to multiple antibiotic agents and, in some instances, resistance to all commercially available antibiotics.[3-7] Given this development, control of the spread of antibiotic-resistant GNB has become a significant infection control problem. The purpose of this chapter is to present a brief review of the mechanisms of antibiotic resistance in GNB and a detailed discussion of potential infection control strategies for the control of these organisms in the hospital environment.

Microbiologic and Epidemiologic Characteristics of GNB Typically Found in Hospitalized Patients

The most clinically relevant classification of the common GNB involves distinguishing between Enterobacteriaceae (which ferment rather than oxidize D-glucose and other sugars) and nonfermentative GNB, which usually degrade glucose oxidatively (Table 18-1). Although this distinction does not adequately cover some pathogenic species of GNB, it does address the vast majority of aerobic GNB encountered in hospitals.

Enterobacteriaceae

All members of the Enterobacteriaceae family are GNB and usually reside in the gastrointestinal tract. Examples of such organisms include *Escherichia coli*, *Klebsiella pneumoniae*, *Enterobacter cloacae*, and *Citrobacter freundii* (Table 18-1). In the National Healthcare Safety Network (NHSN; formerly known as the National Nosocomial Infections Surveillance System) of US hospitals, *E. coli*, *Klebsiella* species, and *Enterobacter* species accounted for 21.3% of all healthcare-associated infections, defined as central line–associated bloodstream infection (CLABSI), catheter-associated urinary tract infection, ventilator-associated pneumonia, and surgical site infection.[8] The risk of Enterobacteriaceae infection is high among patients with a pathologic intra-abdominal condition. Spillage of enteric organisms into the peritoneal cavity in such patients may lead to formation of an intra-abdominal abscess. Urinary tract infection due to Enterobacteriaceae organisms may occur in both catheterized and noncatheterized patients because of the proximity of the urethral meatus to the anus. Finally, patients who have been in the hospital for a prolonged period may develop skin and upper respiratory-tract colonization with gastrointestinal-tract flora,[9,10] leading to possible central venous catheter–related infection

Table 18-1. Gram-negative bacteria commonly encountered in the hospital setting

Enterobacteriaceae (glucose fermenters)

Klebsiella species
   K. pneumoniae
   K. oxytoca

Escherichia coli

Enterobacter species
   E. cloacae
   E. aerogenes

Serratia marcescens

Morganella morganii

Citrobacter species
   C. freundii
   C. diversus

Proteus species
   P. mirabilis
   P. vulgaris

Providencia stuartii

Nonfermentating bacteria (non–glucose fermenters)

Pseudomonas aeruginosa

Acinetobacter baumannii

Stenotrophomonas maltophilia

and ventilator-associated pneumonia due to Enterobacteriaceae.

Unfortunately, over the past decade, an increasing prevalence of antibiotic resistance has been observed among Enterobacteriaceae isolates. The NHSN has provided useful data on the extent of antibiotic resistance in the United States for more than 10 years.[8,11] In 2008, the NHSN reported that 27.1% of *K. pneumoniae* isolates from persons with CLABSI were resistant to ceftriaxone or ceftazidime and that 10.8% were resistant to imipenem, meropenem, or ertapenem.[8] In addition, 8.1% of *E. coli* isolates associated with CLABSI were resistant to ceftriaxone or ceftazidime, 0.9% were resistant to imipenem, meropenem, or ertapenem, and 30.8% were resistant to fluoroquinolones.[8] In Europe, a 1999 study showed that 23% to 51% (depending on the country) of *Enterobacter* isolates recovered from ICU patients were resistant to piperacillin-tazobactam and that more than 30% of *Enterobacter* isolates in some countries were resistant to ciprofloxacin.[1,12] A 2004 worldwide study of GNB that caused intra-abdominal infections showed that extended-spectrum β-lactamases (ESBLs) were detected phenotypically in 10% of *E. coli* isolates, 17% of *Klebsiella* isolates, and 22% of *Enterobacter* isolates.[13]

In general, rates of antibiotic resistance among Enterobacteriaceae organisms are much higher in intensive care units (ICUs) than in other areas of the hospital. However, fluoroquinolone resistance may be one case in which these divergences are less pronounced. NNIS data from 1998 through 2002 showed that 5.8% of *E. coli* isolates recovered from ICUs were resistant to fluoroquinolones, whereas 5.3% of *E. coli* isolates from non-ICU inpatient areas were fluoroquinolone resistant.[11] Additionally, in long-term care facilities and certain specialized units of acute-care hospitals, rates of resistance to many antibiotics may parallel rates seen in ICUs.[14-17]

Carbapenems have long been the last line of defense for treatment of infection with antibiotic-resistant Enterobacteriaceae. However, this has changed dramatically in the United States during the past 10 years. Carbapenem resistance has emerged, particularly in *K. pneumoniae* isolates, although it can be found in other Enterobacteriaceae species. Resistance is due to the production of a carbapenemase, often known as a *Klebsiella pneumoniae* carbapenemase (KPC). Carbapenem-resistant *Klebsiella* strains have been reported in 24 states and are endemic in parts of New York and New Jersey.[18] The NHSN reported that 8% of all *Klebsiella* isolates recovered in 2007 were carbapenem resistant, compared with slightly less than 1% in 2000.[8] This is most likely due to the emergence of KPCs in the United States.

Proper microbiologic detection of KPC is critical and often difficult, requiring additional microbiologic testing, including the modified Hodge test.[18,19] For *K. pneumoniae*, ertapenem susceptibility testing is preferred over imipenem susceptibility testing, because of the risk of false-positive susceptibility results for imipenem in the presence of a KPC.

Nonfermentative GNB

Examples of nonfermentative GNB include *Pseudomonas aeruginosa*, *Acinetobacter* species, and *Stenotrophomonas maltophilia*. According to the NHSN, *P. aeruginosa* accounted for 7.9% of all healthcare-associated infections.[8] Of note, *P. aeruginosa* accounted for 16.3% of all cases of ventilator-associated pneumonia.

Acinetobacter baumannii, a species whose pathogenicity was once questionable, has emerged as an infectious agent of importance to hospitals worldwide. The incidences of both antibiotic-susceptible and antibiotic-resistant *A. baumannii* infections have increased globally in the past decade.[8,20] The reasons for the emergence of this pathogen are uncertain. *A. baumannii* accounted for 2.7% of all healthcare-associated infections reported to the NHSN.[8]

Antibiotic resistance among nonfermentative GNB is also a considerable concern. NHSN data on *P. aeruginosa* isolates associated with CLABSI revealed the following mean prevalences of resistance to common antipseudomonal agents: fluoroquinolones, 30.5%, piperacillin or piperacillin-tazobactam, 20.2%; imipenem or meropenem, 23.0%; and ceftazidime 18.7%.[8] Among *A. baumannii* isolates from persons with CLABSI, 29.2% were resistant to imipenem or meropenem.[8] As with Enterobacteriaceae organisms, resistance was more frequent among isolates from ICUs than among isolates from other areas of the hospital. A most disturbing problem involving nonfermentative GNB has been the emergence of resistance to multiple antibiotics in the same strain. Given the lack of availability of new antibiotics active against organisms such as *P. aeruginosa* and *A. baumannii*, it is likely that multidrug resistance will continue to be a major problem among nonfermentative GNB.

Mechanisms of Resistance Among Common GNB

Resistance to Cephalosporins

The predominant mechanism of resistance to cephalosporins among common GNB is β-lactamase production. More than 300 different β-lactamases have been described.[21] They have been classified in a variety of different ways. However, the most commonly used classifications are based on molecular structure (ie, classes A through D) and functional characteristics, such as substrate and inhibitor profile (ie, groups 1 through 4).[22,23] The β-lactamases inactivate cephalosporins by splitting the amide bond in the β-lactam ring of the antibiotics.

Most strains of *P. aeruginosa, Enterobacter* species, *Citrobacter* species, *Providencia* species, *Morganella morganii,* and *Serratia* species resistant to third-generation cephalosporins produce functional group 1 (ie, molecular class C) β-lactamases. A representative of the group 1 β-lactamases is AmpC. Characteristically, this β-lactamase can inactivate first-generation cephalosporins, second-generation cephalosporins (including cephamycins such as cefoxitin and cefotetan), and third-generation cephalosporins (Table 18-2). These β-lactamases are not inhibited by β-lactamase inhibitors, such as clavulanic acid. An important characteristic of group 1 β-lactamases is that their production can be increased if the bacteria producing them are exposed to certain antibiotics. This phenomenon is known as induction. The magnitude of the increased β-lactamase production depends on the concentration of the antibiotic and the duration of exposure.

In most populations of organisms, such as *Enterobacter* species, mutants exist that permanently hyperproduce group 1 β-lactamases.[22] These population

Table 18-2. Common mechanisms of resistance to antimicrobials in gram-negative bacilli, by agent

Third-generation cephalosporins
Extended-spectrum β-lactamase production
AmpC-type β-lactamase production
Carbapenemase production
Efflux pumps

Piperacillin-tazobactam, other β-lactam–β-lactamase inhibitor combinations
β-lactamase hyperproduction
Outer-membrane protein deficiencies
Efflux pumps

Carbapenems
Outer-membrane protein deficiencies
Carbapenemase production
Efflux pumps

Quinolones
Target mutations
Efflux pumps

frequency of such mutants usually ranges from 10^{-5} to 10^{-8}. Their presence at this low frequency is not enough to result in frank resistance to antibiotics such as third-generation cephalosporins. However, these mutants have important clinical implications. Antibiotic agents that do not induce transient β-lactamase production (eg, third-generation cephalosporins) will kill all organisms in the colony except mutants that permanently hyperproduce β-lactamase. These mutants therefore become the dominant population at the site of infection, leading to frank resistance to third-generation cephalosporins. This can result in the emergence of resistance during therapy.[24,25]

Organisms such as *K. pneumoniae, E. coli,* and *Proteus mirabilis* do not characteristically hyperproduce group 1 β-lactamases. On occasion, they may acquire plasmid-mediated group 1 β-lactamases. However, much more commonly, they may acquire plasmid-mediated group 2be β-lactamases, known as ESBLs. ESBLs differ from group 1 β-lactamases, and these differences have significant clinical implications.

ESBLs were first described in 1983.[26] Importantly, these β-lactamases were not found before the introduction of third-generation cephalosporins into clinical practice. The gene encoding the first identified ESBL was demonstrated to have a mutation that changed a single nucleotide in the gene encoding SHV-1, a β-lactamase produced by more than 95% of *K. pneumoniae* isolates.[2,26] Production of SHV-1 produces resistance to ampicillin and first-generation cephalosporins but not to third-generation cephalosporins. The ESBL derivative of SHV-1 is resistant to the third-generation cephalosporins. Other ESBLs were soon discovered that were closely related to TEM-1 and TEM-2, β-lactamases which were commonly produced by E. coli, resulting in resistance of the organism to ampicillin. Again, the ESBL derivative of TEM-1 or TEM-2 was able to confer resistance to extended-spectrum cephalosporins.[27]

Because these new β-lactamases had a spectrum of activity that was greater than that of their parent enzymes, they were termed "ESBLs." It is not appropriate to designate group 1 β-lactamases as ESBLs, because their spectrum of activity is not greater than that of any parent enzyme. The ESBLs also differ from group 1 β-lactamases in that they are not able to inactivate the cephamycins (eg, cefoxitin and cefotetan) and because they are inactivated in vitro by the β-lactamase inhibitor clavulanic acid.

Worse still, the genes encoding ESBLs are carried on plasmids. This enables transfer of genetic material from organism to organism. Furthermore, genes encoding factors involved in mechanisms of resistance to other antibiotics (eg, aminoglycosides and trimethoprim-sulfamethoxazole) are carried on the same plasmids as ESBLs. This means that ESBL-producing organisms are truly multidrug resistant. ESBLs received a lot of national attention because, similar to KPCs, there was a time when microbiology laboratories were unable to identify ESBLs and were thus reporting false clinical susceptibility profiles of cephalosporins to clinicians.[27-29]

Resistance to β-Lactam–β-Lactamase Inhibitor Combinations

Piperacillin-tazobactam, ticarcillin-clavulanate, and ampicillin-sulbactam are examples of formulations in which a β-lactam is combined with a β-lactamase inhibitor. The theoretical advantage of adding a β-lactamase inhibitor is that it protects the β-lactam from the destructive effects of β-lactamase. The β-lactamase inhibitors in current use are active against the β-lactamases produced by anaerobic organisms such as Bacteroides fragilis, the TEM-1 and SHV-1 β-lactamases commonly produced by E. coli and K. pneumoniae, and the penicillinases produced by S. aureus.

Unfortunately, multiple mechanisms of resistance may work together in producing resistance to the β-lactam–β-lactamase inhibitor combinations. Entry of any β-lactam into the bacterial cell occurs via outer membrane proteins, which function as channels through which the antibiotics pass. These proteins may be lost, contributing to decreased antibiotic levels in the cell and decreased antimicrobial activity. For some infection sites and types in which the organism load is high (eg, intra-abdominal

abscesses and severe ventilator-associated pneumonia), the sheer magnitude of β-lactamase production by a high inoculum of organisms may overcome the effects of β-lactamase inhibitors.[30] Finally, as noted above, group 1 β-lactamases (produced by organisms such as Enterobacter species) are not susceptible to the effects of β-lactamase inhibitors and therefore may be inherently resistant to β-lactam–β-lactamase inhibitor combinations.

Resistance to Fluoroquinolones

For some organisms of the same species, the probability of resistance to fluoroquinolones among ESBL producers is greater than that among non–ESBL producers.[31,32] The reasons for this coresistance are not entirely clear. Plasmid-mediated fluoroquinolone resistance, although reported in the literature,[3] does not yet appear to be widespread. It appears more likely that seriously ill patients are subject to multiple antibiotic exposures. If third-generation cephalosporins and quinolones are used in the same patient, ESBL production and fluoroquinolone resistance may be selected by independent mechanisms.

One mechanism of resistance to fluoroquinolones is mutation of the genes that encode the target enzymes DNA gyrase and topoisomerase IV. A stepwise increase in resistance occurs after mutations in one and then both genes encoding these enzymes.

Alterations in outer membrane proteins coupled with active efflux pumps (which pump antibiotics out of the bacterial cell) appear to be important additional mechanisms of bacterial resistance to fluoroquinolones, especially in P. aeruginosa. Most of the common pump systems in P. aeruginosa remove drugs belonging to multiple antibiotic classes, such as fluoroquinolones, penicillins, carbapenems, and cephalosporins. This makes pump-associated alterations potent causes of multidrug resistance.[4,33]

Resistance to Carbapenems

Carbapenems are often used as drugs of last resort in the treatment of serious infections due to GNB, but they are increasingly used empirically in hospitals where multi-drug resistance abounds. In P. aeruginosa, imipenem resistance may arise via loss of OprD, an outer membrane protein that is accessible to carbapenems but not to other antibiotics. Loss of OprD is the major mechanism of resistance for imipenem-resistant P. aeruginosa. Some efflux mechanisms result in reduced susceptibility to meropenem and doripenem.[4,33]

S. maltophilia has been known for many years to be carbapenem resistant by way of production of class B β-lactamases, known as metalloenzymes. A variety of metalloenzymes have also been detected in P. aeruginosa,

Acinetobacter species, and even members of the Entero-bacteriaceae family. Of concern, many of these are transferable from one bacterial genus to others. To this point, metalloenzymes have not been as much of a problem in the United States as they have been in Asia.[34-36]

Carbapenem resistance among *K. pneumoniae* is becoming a major problem, as discussed earlier in this chapter. Resistance is due to the production of a KPC. KPCs are broad-spectrum β-lactamases that can confer resistance to all β-lactams, β-lactam–β-lactamase inhibitor combinations, and carbapenems. To date, there are 6 subtypes of KPCs that have either been reported in the literature or had sequences submitted to the GenBank database.[35] Unfortunately, the enzyme that confers resistance in KPCs is carried on a plasmid. This has the potential to greatly increase the chance of widespread dissemination of KPCs.

Proper microbiologic detection of KPC is critical and is often difficult, requiring additional microbiologic testing, including the modified Hodge test.[18,19] Testing of *K. pneumoniae* for susceptibility to ertapenem is preferred over testing for susceptibility to imipenem, owing to false susceptibility findings for imipenem in the presence of a KPC.

Resistance to Tigecycline

Tigecycline is a relatively new antibiotic. It is a glycylcycline with an expanded broad spectrum of in vitro activity that is approved for the treatment of complicated skin and soft-tissue infections and complicated intra-abdominal infections.[37] Tigecycline resistance is already being reported among *A. baumannii* and is likely to occur in other GNB.[38,39] The resistance mechanism is believed to involve an efflux pump.[40]

Infection Control Options for Control of Multidrug-Resistant GNB in the Hospital

Causes and Methods of Spread of Antibiotic-Resistant GNB

There is a long list of causal factors that may be driving the spread of gram-negative antibiotic-resistant bacteria. Figure 18-1 illustrates the complicated interaction among many of these factors. The 2 most important parameters for deciding what infection control options should be implemented for the control of multidrug-resistant GNB are the organism-specific proportion of antibiotic resistance attributable to antibiotic use (ie, the attributable fraction) and the organism-specific attributable fraction due to patient-to-patient transmission. With these 2 parameters, relatively simple cost-effectiveness studies could be performed and yield answers that are useful to hospital epidemiologists and, potentially, are beneficial to society as a whole. The ability to

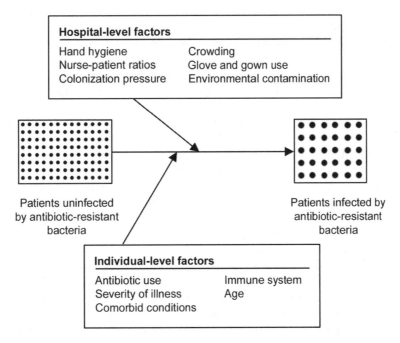

Figure 18-1. Interaction among causal factors contributing to antibiotic resistance in gram-negative bacteria.

quantify both parameters could lead to answers to the following questions:

1. Should I implement contact precautions for patients colonized with antibiotic-resistant GNB?
2. Should I obtain active surveillance cultures to detect patients colonized with multidrug-resistant GNB?
3. Should I cohort patients colonized with multidrug-resistant GNB?
4. Should I focus on infection control interventions aimed at decreasing the amount of patient-to-patient transmission or on antimicrobial stewardship interventions?

However, at present, accurate estimates of these parameters are lacking for many multidrug-resistant GNB in the nonoutbreak setting. Many factors have contributed to this current lack of knowledge. These include a lack of studies measuring these parameters in the nonoutbreak setting, limited sample size of existing studies, variable and sometimes poor molecular epidemiologic techniques to quantify patient-to-patient transmission, lack of basic science information on the impact of antibiotic resistance on virulence and gut ecology, and poor epidemiologic and statistical methods to assess the causal role of antibiotic therapy. However, even with the best epidemiologic intentions, the complicated interplay and interaction of causal factors makes ascertainment and quantification of the difference between the attributable fraction of antibiotic therapy and the attributable fraction of patient-to-patient transmission difficult (Figure 18-1).[41]

In the past, there was a widespread belief that multidrug-resistant GNB are selected primarily by antibiotic use rather than by spread from person to person. A number of studies have changed this belief. A review of more than 50 studies that used molecular epidemiologic techniques to analyze outbreaks of ESBL-producing organisms found that, in all cases, at least 2 patients in each hospital shared genotypically similar strains, implying person-to-person spread of the organism.[5,42] Environmental foci were discovered in less than 10% of these outbreaks. Several studies, most of which were done in an outbreak setting, have demonstrated clonal spread of carbapenemase-producing *K. pneumoniae* or *P. aeruginosa*.[6-8,43-45]

Efforts have been made to determine the proportion of antibiotic-resistant gram-negative bacteria due to patient-to-patient-transmission in the nonoutbreak setting. These studies are difficult to perform because they require a large cohort of patients, owing to the fact that acquisition of antibiotic resistance is often a rare event. Harris et al.[46-48] have studied ESBL-producing *E. coli*, ESBL-producing *K. pneumoniae* and *Klebsiella oxytoca*,

and imipenem-resistant *P. aeruginosa* in a large cohort. Methods used by Harris and colleagues included active surveillance cultures, to determine bacterial acquisition; pulsed-field gel electrophoresis (PFGE), to determine the genetic relatedness of isolates; and analysis of data sets that overlap in time, to help determine epidemiologic relatedness. For ESBL-producing *E. coli*, it was determined that 13% of cases in which this pathogen was acquired in the ICU setting were due to patient-to-patient transmission. For ESBL-producing *Klebsiella* organisms, it was determined that 52% of acquisitions in the ICU setting were due to patient-to-patient transmission. For imipenem-resistant *P. aeruginosa*, the results were a little less clear in that many patients with isolates that had the same PFGE profile did not have a temporal overlap in their hospital stay, in contrast to results involving other GNB. Thirty-one percent of cases in which this GNB was acquired in the ICU setting were due to patient-to-patient transmission, based on the similarity of the PFGE profiles among isolates recovered from these individuals, but only 11% of the acquisitions also involved a temporal overlap in hospitalization.[48] No longer should multidrug-resistant GNB be regarded as selected by antibiotic use to the exclusion of the possibility of person-to-person transmission by the hands of healthcare workers.

Role of the Healthcare Worker

When patient-to-patient transmission is believed to be a cause of the acquisition of antibiotic-resistant GNB, the healthcare worker, as a transient vector, is often believed to be the major mechanism of transmission. The hypothesis is that the healthcare worker touches the patient and/or the patient's environment, becomes colonized with the bacteria, and then transmits it to a subsequent patient in that healthcare workers' care.

This hypothesis has been supported by a number of studies that demonstrate that healthcare workers get colonized on a frequent basis with antibiotic-resistant bacteria such as MRSA and VRE.[49,50] Less work has been done to determine the role of gloves and gowns in protecting patients harboring antibiotic-resistant GNB. Gloves and/or gowns get colonized with antibiotic-resistant *P. aeruginosa* on 8% of physician contacts with patients harboring this pathogen. Gloves and/or gowns get colonized with antibiotic-resistant *A. baumannii* on 38% of physician contacts with patients harboring this pathogen. These results suggest that contact precautions should decrease the frequency of patient-to-patient acquisition of antibiotic-resistant GNB. They also demonstrate the need for good compliance with contact precautions, owing to the frequency at which routine encounters between healthcare workers and patients leads to healthcare worker colonization.

Role of the Environment

The environment of patients harboring antibiotic-resistant bacteria often becomes colonized with the same antibiotic-resistant bacteria. However, the role of this colonization in the transmission of antibiotic-resistant GNB is less clear and has been hotly debated.[51,52]

Studies involving patients colonized with VRE and MRSA have demonstrated that their environment is frequently colonized with the same bacteria.[50] Other studies have observed similar findings for GNB.[51,53]

For *Acinetobacter* species, there is a concern that the environment may have a more important role. Although most GNB remain viable on inanimate surfaces for only short periods, *A. baumannii* is relatively resistant to desiccation and may remain viable for extended periods.[51] Under experimental conditions, *A. baumannii* was shown to persist on dry surfaces for up to 32 days in one study and for 4 months in another.[54] *A. baumannii* has been isolated from a variety of environmental surfaces throughout the hospital, including patient beds, bed rails, bedside tables, curtains, sinks, counter tops, and floors.[51,55-57] One study recovered *A. baumannii* from a bed rail 9 days after a patient infected with the same strain was discharged from that room.[56] Furthermore, several outbreaks due to *A. baumannii* have been linked to an environmental reservoir.[56,57-59]

Nonfermentative bacteria in particular (but not exclusively) have been linked to water reservoirs in the hospital.[9,60] For *Pseudomonas* organisms, the role of water as a potential source is a controversial topic.[61] It is clear that *Pseudomonas* species can be found in hospital water and that it often exists in a biofilm in faucets and faucet aerators. These organisms can exist on many inanimate surfaces. Potable water itself may be a source, although *P. aeruginosa* isolates recovered from potable water in a healthcare setting are rarely multidrug resistant unless faucets or faucet aerators are contaminated. However, the relative importance of faucets and hospital water supplies as modes of patient-to-patient transmission in the GNB-endemic setting is still not clear.

Role of Antibiotics

To most physicians, it is clear that increasing the use of antibiotics leads to an increase in antibiotic resistance.[61] This has been demonstrated by means of numerous study designs, including ecological studies, case-control studies, and cohort studies. However, it is often difficult in studies occurring in the hospital setting to properly perform risk factor analyses that clearly identify a particular antibiotic or antibiotic class as a major causal factor in the emergence of antibiotic resistance.[62] In addition, owing to the rareness of the outcome of antibiotic resistance and to multiple causal variables (demonstrated in

Figure 18-1), it is often difficult to determine whether an antimicrobial intervention led to a decrease in the emergence of antibiotic resistance. This has led to tremendous controversy about which antibiotic stewardship interventions are most cost-effective and which are most likely to have an impact on the control of antibiotic-resistant GNB.

There are numerous different potential antibiotic interventions that may decrease the emergence of antibiotic-resistant GNB. Major antibiotic interventions discussed in the literature include decreasing the duration of therapy, de-escalation of therapy, use of more-appropriate empirical therapy, use of prediction rules to guide antibiotic therapy, front-end restriction of the availability of certain antibiotics, and selective decolonization of the oral or digestive tract.[63-65] Other interventions include the use of more-rapid diagnostic measures to better guide antibiotic therapy.[66,67] An intervention that was once popular and has now fallen out of vogue is antibiotic cycling.[68-70] Many of these possible control measures may be beneficial from biological plausibility and construct validity points of view. However, very few of these antibiotic interventions have been studied by use of upper-level study designs, such as randomized controlled trials, or well-performed quasi-experimental designs.[71,72] More upper-level studies are needed to assess the cost-effectiveness and impact of interventions on antibiotic-resistant GNB. The remaining portion of this section will focus the antibiotic-control discussion on a few antibiotic stewardship interventions that are believed to hold the most promise.

Elimination of overly broad antibiotic therapy, once culture results are available, is a practical way of reducing the burden of antibiotic use on the selection of multidrug-resistant organisms. This method also would seemingly have very few potential negative consequences. The value of such streamlining or narrowing of coverage may seem intuitive to infectious diseases physicians and infection control practitioners, but for many other physicians an attitude of "don't change a winning team" is pervasive. Furthermore, in some situations culture results are negative, yet an infection is clinically present, whereas in other situations culture results are positive but indicate the presence of colonization rather than true infection. Thus, there are some practical issues to the practice of streamlining.

There is currently little information on the optimal duration of antibiotic therapy for many commonly encountered conditions. Exceptions in which the duration of therapy is reasonably well-defined include urinary tract infection, right-sided endocarditis, and ventilator-associated pneumonia.[73] Studies comparing different durations of antibiotic therapy are desperately needed.

However, these studies are often costly, are not deemed beneficial to industry, and thus are often not performed.

Use of invasive testing strategies (such as quantitative culture of bronchoscopically obtained respiratory specimens) or more-rapid tests to rule in or rule out certain pathogens may allow differentiation between colonization from true infection and narrow the spectrum of potential pathogens, thus leading to more-appropriate use of antibiotics.[10,66,67]

Current Infection Control Practices and Guidelines Relative to Antibiotic-Resistant GNB

The Healthcare Infection Control Practices Advisory Committee (HICPAC) published a guideline in 2007 on the use of isolation precautions to prevent transmission of infectious agents in healthcare settings.[74] There are also recommendations published by the HICPAC on the management of multidrug-resistant organisms in healthcare settings. The recommendations for multidrug-resistant organisms (including antibiotic-resistant GNB) are as follows:

1. Preventing the emergence and transmission of these pathogens requires a comprehensive approach that includes administrative involvement and measures (eg, nurse staffing, communication systems, and performance improvement processes to ensure adherence to recommended infection control measures), education and training of medical and other healthcare workers, judicious antibiotic use, comprehensive surveillance for targeted MDROs, application of infection control precautions during patient care, environmental measures (eg, cleaning and disinfection of the patient care environment and equipment and dedicated single-patient use of noncritical equipment), and decolonization therapy when appropriate.
2. Contact precautions are recommended in settings with evidence of ongoing transmission, in acute care settings with an increased risk for transmission, and during contact with patients who have wounds that cannot be contained by dressings.

The Centers for Disease Control and Prevention and the HICPAC have published a guideline for environmental infection control in healthcare facilities.[75] The executive summary of the guidelines includes the following recommendations for controlling the incidence of healthcare-associated infections and pseudo-outbreaks: appropriate use of cleaners and disinfectants, appropriate maintenance of equipment including endoscopes, adherence to water-quality standards for hemodialysis,

and adherence to ventilation standards for specialized-care environments. The guidelines also note that routine environmental sampling is not usually advised.

Conclusions

It appears safe to say that multidrug-resistant GNB will be with us for many years to come. The likely shortage of new types of antibiotics to cope with these organisms will place a spotlight on hospital epidemiologists' efforts to control their spread. This will likely require a multi-pronged approach that includes novel infection control interventions and novel antimicrobial stewardship interventions. The discovery of effective interventions will likely require a better understanding of resistance mechanisms, gut ecology, and the effect of antibiotics on gut ecology; improved molecular epidemiologic techniques; improved rapid diagnostic modalities; and increased funding of upper-level study designs, such as randomized cluster trials, to study infection control and antimicrobial stewardship interventions.

References

1. Talbot GH, Bradley J, Edwards JE Jr, Gilbert D, Scheld M, Bartlett JG. Bad bugs need drugs: an update on the development pipeline from the antimicrobial availability task force of the infectious diseases society of america. Antimicrobial Task Force of the Infectious Diseases Society of America. *Clin Infect Dis* 2006;42:657–668.
2. Boucher HW, Talbot GH, Bradley JS, et al. Bad bugs, no drugs: no ESKAPE! an update from the Infectious Diseases Society of America. *Clin Infect Dis* 2009;48:1–12.
3. Paterson DL, Doi Y. A step closer to extreme drug resistance (XDR) in gram-negative bacilli. *Clin Infect Dis* 2007;45:1179–1181.
4. Tato M, Coque TM, Ruiz-Garbajosa P, et al. Complex clonal and plasmid epidemiology in the first outbreak of Enterobacteriaceae infection involving VIM-1 metallo-beta-lactamase in spain: toward endemicity? *Clin Infect Dis* 2007;45:1171–1178.
5. Bratu S, Landman D, Haag R, et al. Rapid spread of carbapenem-resistant *Klebsiella pneumoniae* in new york city: a new threat to our antibiotic armamentarium. *Arch Intern Med* 2005;165:1430–1435.
6. Michalopoulos A, Falagas ME. Colistin and polymyxin B in critical care. *Crit Care Clin* 2008;24:377–391, x.
7. Falagas ME, Kasiakou SK. Colistin: the revival of polymyxins for the management of multidrug-resistant gram-negative bacterial infections. *Clin Infect Dis* 2005;40:1333–1341.
8. Hidron AI, Edwards JR, Patel J, et al. Antimicrobial-resistant pathogens associated with healthcare-associated infections: annual summary of data reported to the National Healthcare Safety Network at the Centers for Disease Control and

Prevention, 2006–2007. *Infect Control Hosp Epidemiol* 2008;29:996–1011.

9. Garrouste-Orgeas M, Chevret S, Arlet G, et al. Oropharyngeal or gastric colonization and nosocomial pneumonia in adult intensive care unit patients: a prospective study based on genomic DNA analysis. *Am J Respir Crit Care Med* 1997; 156:1647–1655.

10. Johanson WG Jr, Pierce AK, Sanford JP, Thomas GD. Nosocomial respiratory infections with gram-negative bacilli: the significance of colonization of the respiratory tract. *Ann Intern Med* 1972;77:701–706.

11. National Nosocomial Infections Surveillance (NNIS) System report, data summary from January 1992 through June 2004, issued October 2004. *Am J Infect Control* 2004;32: 470–485.

12. Hanberger H, Garcia-Rodriguez JA, Gobernado M, Goossens H, Nilsson LE, Struelens MJ. Antibiotic susceptibility among aerobic gram-negative bacilli in intensive care units in 5 European countries. French and Portuguese ICU Study Groups. *JAMA* 1999;281:67–71.

13. Rossi F, Baquero F, Hsueh PR,, et al. In vitro susceptibilities of aerobic and facultatively anaerobic gram-negative bacilli isolated from patients with intra–abdominal infections worldwide: 2004 results from SMART (Study for Monitoring Antimicrobial Resistance Trends). *J Antimicrob Chemother* 2006;58:205–210.

14. Furuno JP, Hebden JN, Standiford HC, et al. Prevalence of methicillin-resistant *Staphylococcus aureus* and *Acinetobacter baumannii* in a long-term acute care facility. *Am J Infect Control* 2008;36:468–471.

15. Manzur A, Gavalda L, Ruiz de Gopegui E, et al. Prevalence of methicillin-resistant *Staphylococcus aureus* and factors associated with colonization among residents in community long-term-care facilities in spain. *Clin Microbiol Infect* 2008;14: 867–872.

16. Bonomo RA. Multiple antibiotic-resistant bacteria in long-term-care facilities: an emerging problem in the practice of infectious diseases. *Clin Infect Dis* 2000;31:1414–1422.

17. Maslow JN, Lee B, Lautenbach E. Fluoroquinolone-resistant *Escherichia coli* carriage in long-term care facility. *Emerg Infect Dis* 2005;11:889–894.

18. Srinivasan A, Patel JB. *Klebsiella pneumoniae* carbapenemase-producing organisms: an ounce of prevention really is worth a pound of cure. *Infect Control Hosp Epidemiol* 2008;29: 1107–1109.

19. Anderson KF, Lonsway DR, Rasheed JK, et al. Evaluation of methods to identify the *Klebsiella pneumoniae* carbapenemase in Enterobacteriaceae. *J Clin Microbiol* 2007;45: 2723–2725.

20. Gaynes R, Edwards JR. Overview of nosocomial infections caused by gram-negative bacilli. *Clin Infect Dis* 2005;41: 848–854.

21. Jacoby G, Bush K. Amino acid sequences for TEM, SHV and OXA extended-spectrum and inhibitor resistant β-lactamases. Updated August 4, 2009. http://www.lahey.org/Studies/. Accessed August 9, 2009.

22. Livermore DM. Beta-lactamases in laboratory and clinical resistance. *Clin Microbiol Rev* 1995;8:557–584.

23. Bush K, Jacoby GA, Medeiros AA. A functional classification scheme for beta-lactamases and its correlation with molecular structure. *Antimicrob Agents Chemother* 1995;39:1211–1233.

24. Chow JW, Fine MJ, Shlaes DM, et al. *Enterobacter* bacteremia: clinical features and emergence of antibiotic resistance during therapy. *Ann Intern Med* 1991;115:585–590.

25. Kaye KS, Cosgrove S, Harris A, Eliopoulos GM, Carmeli Y. Risk factors for emergence of resistance to broad-spectrum cephalosporins among *Enterobacter* spp. *Antimicrob Agents Chemother* 2001;45:2628–2630.

26. Knothe H, Shah P, Krcmery V, Antal M, Mitsuhashi S. Transferable resistance to cefotaxime, cefoxitin, cefamandole and cefuroxime in clinical isolates of *Klebsiella pneumoniae* and *Serratia marcescens*. *Infection* 1983;11:315–317.

27. Bradford PA. Extended-spectrum beta-lactamases in the 21st century: characterization, epidemiology, and detection of this important resistance threat. *Clin Microbiol Rev* 2001;14: 933–951.

28. Tenover FC, Mohammed MJ, Gorton TS, Dembek ZF. Detection and reporting of organisms producing extended-spectrum beta-lactamases: survey of laboratories in connecticut. *J Clin Microbiol* 1999;37:4065–4070.

29. Yang K, Guglielmo BJ. Diagnosis and treatment of extended-spectrum and AmpC beta-lactamase-producing organisms. *Ann Pharmacother* 2007;41:1427–1435.

30. Thomson KS, Moland ES. Cefepime, piperacillin-tazobactam, and the inoculum effect in tests with extended-spectrum beta-lactamase-producing Enterobacteriaceae. *Antimicrob Agents Chemother* 2001;45:3548–3554.

31. Lautenbach E, Strom BL, Bilker WB, Patel JB, Edelstein PH, Fishman NO. Epidemiological investigation of fluoroquinolone resistance in infections due to extended-spectrum beta-lactamase-producing *Escherichia coli* and *Klebsiella pneumoniae*. *Clin Infect Dis* 2001;33:1288–1294.

32. Paterson DL, Mulazimoglu L, Casellas JM, et al. Epidemiology of ciprofloxacin resistance and its relationship to extended-spectrum beta-lactamase production in *Klebsiella pneumoniae* isolates causing bacteremia. *Clin Infect Dis* 2000;30:473–478.

33. Livermore DM. Multiple mechanisms of antimicrobial resistance in *Pseudomonas aeruginosa*: our worst nightmare? *Clin Infect Dis* 2002;34:634–640.

34. Lolans K, Queenan AM, Bush K, Sahud A, Quinn JP. First nosocomial outbreak of *Pseudomonas aeruginosa* producing an integron-borne metallo-beta-lactamase (VIM-2) in the United States. *Antimicrob Agents Chemother* 2005;49: 3538–3540.

35. Queenan AM, Bush K. Carbapenemases: the versatile beta-lactamases. *Clin Microbiol Rev* 2007;20:440–458.

36. Yan JJ, Hsueh PR, Lu JJ, Chang FY, Ko WC, Wu JJ. Characterization of acquired beta-lactamases and their genetic support in multidrug-resistant *Pseudomonas aeruginosa* isolates in Taiwan: the prevalence of unusual integrons. *J Antimicrob Chemother* 2006;58:530–536.

37. Peterson LR. A review of tigecycline—the first glycylcycline. *Int J Antimicrob Agents* 2008;32(Suppl 4):S215–S222.

38. Navon-Venezia S, Leavitt A, Carmeli Y. High tigecycline resistance in multidrug-resistant *Acinetobacter baumannii*. *J Antimicrob Chemother* 2007;59:772–774.

39. Dowzicky MJ, Park CH. Update on antimicrobial susceptibility rates among gram-negative and gram-positive organisms in the United States: results from the tigecycline evaluation and surveillance trial (TEST) 2005 to 2007. *Clin Ther* 2008;30: 2040–2050.

40. Peleg AY, Adams J, Paterson DL. Tigecycline efflux as a mechanism for nonsusceptibility in *Acinetobacter baumannii*. *Antimicrob Agents Chemother* 2007;51:2065–2069.

41. Harris AD, McGregor JC, Furuno JP. What infection control interventions should be undertaken to control multidrug-resistant gram-negative bacteria? *Clin Infect Dis* 2006;43 (Suppl 2):S57–S61.

42. Paterson DL, Yu VL. Extended-spectrum beta-lactamases: a call for improved detection and control. *Clin Infect Dis* 1999;29:1419–1422.

43. Cornaglia G, Mazzariol A, Lauretti L, Rossolini GM, Fontana R. Hospital outbreak of carbapenem-resistant *Pseudomonas aeruginosa* producing VIM-1, a novel transferable metallo-beta-lactamase. *Clin Infect Dis* 2000;31:1119–1125.

44. Yan JJ, Ko WC, Tsai SH, Wu HM, Wu JJ. Outbreak of infection with multidrug-resistant *Klebsiella pneumoniae* carrying bla(IMP-8) in a university medical center in Taiwan. *J Clin Microbiol* 2001;39:4433–4439.

45. Giakkoupi P, Xanthaki A, Kanelopoulou M, et al. VIM-1 metallo-beta-lactamase-producing *Klebsiella pneumoniae* strains in Greek hospitals. *J Clin Microbiol* 2003;41: 3893–3896.

46. Harris AD, Kotetishvili M, Shurland S, et al. How important is patient-to-patient transmission in extended-spectrum beta-lactamase *Escherichia coli* acquisition. *Am J Infect Control* 2007;35:97–101.

47. Harris AD, Perencevich EN, Johnson JK, et al. Patient-to-patient transmission is important in extended-spectrum beta-lactamase-producing *Klebsiella pneumoniae* acquisition. *Clin Infect Dis* 2007;45:1347–1350.

48. Johnson JK, Smith G, Warren M, et al. The role of patient-to-patient transmission in the acquisition of imipenem-resistant *Pseudomonas aeruginosa* colonization in the intensive care unit. *J Infect Dis* 2009 (in press).

49. Snyder GM, Thom KA, Furuno JP, et al. Detection of methicillin-resistant *Staphylococcus aureus* and vancomycin-resistant enterococci on the gowns and gloves of healthcare workers. *Infect Control Hosp Epidemiol* 2008;29:583–589.

50. Hayden MK, Blom DW, Lyle EA, Moore CG, Weinstein RA. Risk of hand or glove contamination after contact with patients colonized with vancomycin-resistant *Enterococcus* or the colonized patients' environment. *Infect Control Hosp Epidemiol* 2008;29:149–154.

51. Hota B. Contamination, disinfection, and cross-colonization: are hospital surfaces reservoirs for nosocomial infection? *Clin Infect Dis* 2004;39:1182–1189.

52. Boyce JM. Environmental contamination makes an important contribution to hospital infection. *J Hosp Infect* 2007; 65(Suppl 2):50–54.

53. Simor AE, Lee M, Vearncombe M, et al. An outbreak due to multiresistant *Acinetobacter baumannii* in a burn unit: risk factors for acquisition and management. *Infect Control Hosp Epidemiol* 2002;23:261–267.

54. Jawad A, Seifert H, Snelling AM, Heritage J, Hawkey PM. Survival of *Acinetobacter baumannii* on dry surfaces: comparison of outbreak and sporadic isolates. *J Clin Microbiol* 1998;36:1938–1941.

55. Levin AS, Gobara S, Mendes CM, Cursino MR, Sinto S. Environmental contamination by multidrug-resistant *Acinetobacter baumannii* in an intensive care unit. *Infect Control Hosp Epidemiol* 2001;22:717–720.

56. Catalano M, Quelle LS, Jeric PE, Di Martino A, Maimone SM. Survival of *Acinetobacter baumannii* on bed rails during an outbreak and during sporadic cases. *J Hosp Infect* 1999;42: 27–35.

57. Das I, Lambert P, Hill D, Noy M, Bion J, Elliott T. Carbapenem-resistant *Acinetobacter* and role of curtains in an outbreak in intensive care units. *J Hosp Infect* 2002;50: 110–114.

58. Aygun G, Demirkiran O, Utku T, et al. Environmental contamination during a carbapenem-resistant *Acinetobacter baumannii* outbreak in an intensive care unit. *J Hosp Infect* 2002;52:259–262.

59. Denton M, Wilcox MH, Parnell P, et al. Role of environmental cleaning in controlling an outbreak of *Acinetobacter baumannii* on a neurosurgical intensive care unit. *Intensive Crit Care Nurs* 2005;21:94–98.

60. Anaissie EJ, Penzak SR, Dignani MC. The hospital water supply as a source of nosocomial infections: a plea for action. *Arch Intern Med* 2002;162:1483–1492.

61. Bonten MJ, Weinstein RA. Transmission pathways of *Pseudomonas aeruginosa* in intensive care units: don't go near the water. *Crit Care Med* 2002;30:2384–2385.

62. Harris AD, Karchmer TB, Carmeli Y, Samore MH. Methodological principles of case-control studies that analyzed risk factors for antibiotic resistance: a systematic review. *Clin Infect Dis* 2001;32:1055–1061.

63. Owens RC Jr. Antimicrobial stewardship: concepts and strategies in the 21st century. *Diagn Microbiol Infect Dis* 2008;61:110–128.

64. Owens RC Jr, Ambrose PG. Antimicrobial stewardship and the role of pharmacokinetics-pharmacodynamics in the modern antibiotic era. *Diagn Microbiol Infect Dis* 2007;57 (3 Suppl):77S-83S.

65. de Smet AM, Kluytmans JA, Cooper BS, et al. Decontamination of the digestive tract and oropharynx in ICU patients. *N Engl J Med* 2009;360:20–31.

66. Fagon JY, Chastre J, Wolff M, et al. Invasive and noninvasive strategies for management of suspected ventilator-associated pneumonia: a randomized trial. *Ann Intern Med* 2000; 132:621–630.

67. Forrest GN, Roghmann MC, Toombs LS, et al. Peptide nucleic acid fluorescent in situ hybridization for hospital-acquired enterococcal bacteremia: delivering earlier effective antimicrobial therapy. *Antimicrob Agents Chemother* 2008;52:3558–3563.

68. Merz LR, Warren DK, Kollef MH, Fridkin SK, Fraser VJ. The impact of an antibiotic cycling program on empirical therapy for gram-negative infections. *Chest* 2006;130:1672–1678.

69. Warren DK, Hill HA, Merz LR, et al. Cycling empirical antimicrobial agents to prevent emergence of antimicrobial-resistant gram-negative bacteria among intensive care unit patients. *Crit Care Med* 2004;32:2450–2456.

70. Kollef MH. Is antibiotic cycling the answer to preventing the emergence of bacterial resistance in the intensive care unit? *Clin Infect Dis* 2006;43(Suppl 2):S82–S88.

71. Harris AD, Bradham DD, Baumgarten M, Zuckerman IH, Fink JC, Perencevich EN. The use and interpretation of quasi-experimental studies in infectious diseases. *Clin Infect Dis* 2004;38:1586–1591.

72. Cooper BS, Cookson BD, Davey PG, Stone SP. Introducing the ORION statement, a CONSORT equivalent for infection control studies. *J Hosp Infect* 2007;65(Suppl 2):85–87.

73. Chastre J, Wolff M, Fagon JY, et al. Comparison of 8 vs 15 days of antibiotic therapy for ventilator-associated pneumonia in adults: a randomized trial. *JAMA* 2003;290:2588–2598.

74. Siegel JD, Rhinehart E, Jackson M, Chiarello L. *2007 Guidelines for Isolation Precautions: Preventing Transmission of Infectious Agents in Healthcare Settings*. http://www.cdc.gov/ncidod/dhqp/pdf/guidelines/Isolation2007.pdf. Accessed August 9, 2009.

75. Sehulster L, Chinn RY. Guidelines for environmental infection control in health-care facilities. recommendations of CDC and the healthcare infection control practices advisory committee (HICPAC). *MMWR Recomm Rep* 2003;52(RR-10):1–42.

Chapter 19 *Clostridium difficile* Infection

Natasha Bagdasarian, MD, MPH, and
Preeti N. Malani, MD, MSJ

When *Clostridium difficile* was first identified as the major cause of pseudomembranous colitis in the 1970s, the infection was recognized as an easily managed iatrogenic complication. In the past decade, *C. difficile* infection (CDI) has evolved and become more frequent, more severe, and more difficult to manage.[1] CDI is now responsible for increased hospital costs and length of stay, and the cost of CDI in the United States is estimated at $1.1 billion per year.[2] The growing CDI problem is related in large part to the nosocomial spread of *C. difficile*, which has been impacted by the emergence of the epidemic *C. difficile* strain BI/NAP1. In addition to being associated with outbreaks of clinically severe disease, some strains have also been shown to hypersporulate, thus increasing the ease of transmission.[3,4] Targeted infection control measures aimed at curtailing the healthcare-associated transmission of *C. difficile* remain absolutely vital.

Risk Factors

Factors associated with increased risk of developing CDI include the following: age greater than 65 years, prior hospitalization, increased length of stay in an intensive care unit (ICU), nonsurgical admission to the hospital, exposure to broad-spectrum antimicrobials, longer duration of antimicrobial use, exposure to multiple antimicrobials, exposure to acid-suppressive therapy, receipt of cancer chemotherapy, renal insufficiency, hemodialysis, and presence of a nasogastric tube.[5-9] These risk factors may help us to identify patients at greatest risk of developing CDI and also highlight the importance of antimi-crobial stewardship in the approach to prevention. Additionally, those patients residing in a room previously occupied by a *C. difficile*–positive patient or in a bed adjacent to or down the hall from a patient with CDI appear to be at greater risk for the development of CDI.[10] These observations reflect the critical role of the environment and healthcare workers' hands in the transmission of *C. difficile,* especially in light of the increased incidence of CDI during the past decade.

Epidemiology

Three- to 4-fold increases in the incidence of CDI have been reported during outbreaks, with similar increases in severe outcomes.[11] The annual incidence of CDI has also been on the rise and, in the United States, the portion of hospital discharges listing CDI as a diagnosis doubled between 1996 and 2003.[12] Such increases in incidence and severity of CDI have been reported across the United States, Canada, and countries in Europe.[12-16] The increase in severity of CDI has been closely linked to the emergence of the epidemic BI/NAP1 strain of *C. difficile*[13,15] but may also be related to the increasing age of hospitalized patients and the increasing prevalence of comorbidities among them, because patients over the age of 65 years have been disproportionately affected.[12] A review of US children's hospitals in the years 2001–2006 found that the incidence of CDI has also increased among hospitalized children, although the associated colectomy and mortality rates have not increased as they have among adults.[17]

Treatment

Severe CDI can often progress to fulminant disease requiring colectomy, despite medical therapy. The BI/NAP1 epidemic strain of *C. difficile* produces increased levels of toxins A and B, the major virulence determinants of *C. difficile,* as well as binary toxin.[18,19] Although resistance to certain classes of antibiotics has been noted (notably fluoroquinolone resistance in the BI/NAP1 epidemic strain), resistance to the primary antibiotics used in the treatment of CDI (metronidazole and vancomycin) does not appear to play a significant role in CDI treatment failures.[18,20] In contrast, rifampin resistance rates of up to 30% have been seen in some settings, mostly in BI/NAP1 isolates.[21] Rifamycins are sometimes used as an adjunctive therapy, especially for refractory or recurrent cases of CDI. However, the development of resistance during therapy (particularly resistance to rifaximin) can limit the utility of this class of drugs.

Diagnosis

A case of CDI should be defined as a case of diarrhea or toxic megacolon without other known etiology in a patient who has a stool sample positive for *C. difficile* (either by an assay for toxins A and/or B, by culture, or by other means) and/or evidence of pseudomembranous colitis detected during endoscopy or surgery or on histopathologic examination.[22] Both a high index of clinical suspicion and use of a sensitive test are essential for early isolation of infected patients and prevention of nosocomial transmission of CDI. Use of a highly specific confirmatory test is critical for accurately monitoring the CDI rate and identifying true outbreaks of infection.

Although there is no clear criterion standard ("gold standard") test for detection of *C. difficile,* toxigenic culture and cell culture cytotoxicity neutralization assay (CCNA) have long been considered the best microbiological tests for detection of CDI[23-25] (Table 19-1). That being said, *C. difficile* toxin detection enzyme immunoassays (CDT-EIA), rather than either toxigenic culture or CCNA, remain the primary testing strategy at more than 90% of laboratories in the United States.[24] Earlier assays detected the presence of *C. difficile* toxin A alone, and newer tests detect the presence of both toxin A and toxin B. CDT-EIAs test directly for the presence of toxin; however, a recent systematic review of several CDT-EIAs demonstrated that they had a low positive predictive value (less than 50% in some circumstances) when the prevalence of CDI was relatively low (positive results for fewer than 10% of stool samples tested).[26] The sensitivities of these CDT-EIAs have also been variable (69%–99%).[26] Many healthcare institutions have attempted

to overcome the low or variable sensitivities by sending multiple stool samples for testing. However, automatic repeat testing should be strongly discouraged. Instead, repeat testing should be done on the basis of the clinical index of suspicion. For example, a patient who has mild diarrhea and no changes in temperature, white blood cell count, abdominal pain, and stool output and who has a negative result for an initial test could be monitored closely, instead of automatically retested.

Several authorities have proposed 2-stage testing strategies that combine a highly sensitive rapid screening test with a more specific confirmatory test. The screening tests are usually CDT-EIAs or glutamate dehydrogenase antigen EIAs (GDH-EIAs). A recent study compared a standard CDT-EIA with a 2-step testing method that used a GDH-EIA followed by a confirmatory CCNA.[27] Use of the CDT-EIA screening test reduced the workload and costs, compared with use of CCNA alone; however, the 3 days required to obtain CCNA results remained a major limitation. Another multiple-step strategy utilizes a screening GDH-EIA followed by a confirmatory CDT-EIA for specimens that test positive, and toxigenic culture only for specimens that test negative.[28] This strategy allowed final test results to be obtained within 4 hours for most specimens submitted.[28] However, some investigators have reported low sensitivity (76%) with GDH-EIAs. If this estimate is true, any potential benefit from screening with these assays would be negated.

Polymerase chain reaction (PCR)–based techniques to detect *C. difficile* have been investigated as well. One such assay is a real-time PCR that amplifies the *tcdB* gene (which encodes toxin B). The test result can be obtained in 3 hours and, compared with toxigenic culture as the gold standard, it has a high sensitivity and specificity (83.6% and 98.2%, respectively).[25] One concern regarding use of PCR is the emergence of new genotypes of toxigenic *C. difficile*; however, use of PCR coupled with periodic toxigenic culture for surveillance could potentially overcome this obstacle. One group of investigators has attempted to avoid this problem by developing a multiplex PCR that amplifies 4 genes that encode *C. difficile* toxin (*tcdA, tcdB, cdtA,* and *cdtB*).[29] Although the optimal diagnostic paradigm remains unclear, it is likely that more centers will switch to PCR-based approaches for CDI diagnosis in the future.

Surveillance

Surveillance for CDI is essential in order to recognize outbreaks and to monitor the effectiveness of infection control practices. Healthcare facility–onset, healthcare facility–associated CDI (HO-CDI) is defined as a case with symptom onset that occurs more than 48 hours

Table 19-1. Comparison of laboratory tests for *Clostridium difficile*

Test	Comments	Advantages	Disadvantages
Toxigenic culture	• Combines stool culture with CCNA or EIA	• Highly sensitive and specific	• Technically demanding • Time consuming (>72 h for results)
Cell culture cytotoxicity neutralization assay (CCNA)	• Stool sample is added to cell culture, exerting a cytopathic effect if *C. difficile* toxin is present	• Sensitive and specific	• Technically demanding • Time consuming (24–48 h for results)
Anaerobic stool culture	• Stool culture on selective, cycloserine-cefoxitin-fructose agar	• Highly sensitive • Essential for molecular typing and performing antibiotic susceptibility	• Technically demanding • Time consuming (48–72 h for results) • Not specific for toxin-producing strains
Toxin detection EIA (CDT-EIA)	• Direct detection of toxin • Older tests only tested for toxin A; newer tests look for both toxin A and B	• Rapid (1–2 h for results) • Relatively inexpensive	• Variable sensitivity and specificity
Glutamate dehydrogenase antigen EIA (GDH-EIA)	• EIA for glutamate dehydrogenase (GDH), which is produced by all *C. difficile* strains, as well as some other organisms	• Rapid (15–45 min for results) • Sensitive • Inexpensive • Possibly good screening test	• Not specific for toxin-producing strains
Multiple-step testing strategies	• Combinations of various of the tests listed above • Rapid screening tests are combined with slower confirmatory tests	• Attempt to increase sensitivity and specificity while reducing costs • Can be relatively inexpensive and rapid depending on methods chosen	• Discordant results may be difficult to interpret • Depends on sensitivity of GDH-EIA and sensitivity of confirmatory test
PCR-based techniques	• Uses primers targeting specific genes encoding toxins	• Rapid (<3 h for results) • Sensitive and specific	• Would not detect emergence of virulent strains with new genotypes • Require trained personnel and specialized equipment

NOTE: Data are from several studies.[23-25] EIA, enzyme immunoassay; PCR, polymerase chain reaction.

after the patient is admitted to a healthcare facility. Community-onset, healthcare facility–associated CDI (CO/HCFA-CDI) is defined as a case with symptom onset that occurs in the community or within 48 hours after the patient is admitted, provided that symptom onset occurs less than 4 weeks after the patient's last discharge from a healthcare facility[22] (Table 19-2). At a minimum, healthcare facilities should track the rate of HO-CDI (reported as the number of CDI cases per 10,000 patient-days). Although the HO-CDI rate does not capture all cases of healthcare-associated CDI, it acts as a good surrogate marker and allows healthcare epidemiologists to effectively gauge responses to CDI control measures.[30,31] Because of the time and expense required for case ascertainment that uses surveillance definitions, some institutions have looked at using administrative data, including *International Classification of Diseases, Ninth Revision* codes, as surrogate measures for the CDI burden.[32,33] Large-scale validation of these methods is not yet available.

Preventing Iatrogenic Spread

C. difficile is shed in the feces and is a common environmental contaminant in rooms of patients with CDI. This organism has been found on the floors, toilets, bedding, and furniture in rooms of patients with CDI.[34] When *C. difficile*, an obligate anaerobe, is exposed to the air, the vegetative cells die and the spore form of the organism predominates. Although many cleaning agents are effective against the vegetative form of *C. difficile*, only a

Table 19-2. Surveillance definitions for types of *Clostridium difficile* infection (CDI)

CDI type	Definition
Healthcare facility onset, healthcare facility associated	Symptom onset occurred >48 h after admission to a healthcare facility
Community onset, healthcare facility associated	Symptom onset occurred in the community or ≤48 h after admission, provided that symptom onset occurred ≤4 weeks after the patient's last discharge from a healthcare facility
Community associated	Symptom onset occurred in the community or ≤48 h after admission to a healthcare facility, provided that symptom onset occurred >12 weeks after the patient's last discharge from a healthcare facility
Indeterminate onset	Case does not fit any of the above criteria for exposure setting (eg, onset occurred in the community >4 weeks but <2 weeks after the patient's last discharge from a healthcare facility)
Unknown	Exposure setting cannot be determined because of a lack of available data
Recurrent	Episode occurred ≤8 weeks after the onset of a previous episode, provided that CDI symptoms from the earlier episode resolved

NOTE: Adapted from Dubberke et al.[22]

few are sporicidal. *C. difficile* spores can resist desiccation for months and have been found on hard surfaces for up to 5 months following inoculation, although it has been proposed that the spores have decreased survival on surfaces composed of certain compounds (copper alloys).[35] There is a strong association between the level of environmental contamination and nosocomial transmission,[36] and the epidemic BI/NAP1 strain of *C. difficile* has been found to hypersporulate, thus potentially increasing the level of environmental contamination.[3,4]

Infection control measures targeting CDI typically focus on 2 issues: preventing patient exposure to *C. difficile* and reducing the chance of CDI development if a patient is exposed to the organism.[22,37] Strategies to reduce the exposure of patients to *C. difficile* include use of contact precautions, use of special environmental cleaning practices, and hand hygiene policies. Strategies to reduce the development of CDI, once exposure has occurred, center on antimicrobial stewardship.

Many investigators have documented the utility of infection control measures in preventing the spread of

HO-CDI, even since emergence of the epidemic BI/NAP1 strain.[11,38] These "bundles" of infection control practices have emphasized expansion of existing measures, such as use of contact precautions for a longer duration (expanded from the duration of illness to the duration of hospitalization) and daily, enhanced cleaning of CDI patient rooms using sodium hypochlorite; however, it is generally difficult to assess the efficacy of individual components of a *C. difficile* infection control bundle.[11,38]

Although most acute care hospitals utilize infection control policies to prevent the spread of CDI, this does not mean that practices are being implemented uniformly. A study of hospitals in the Canadian Nosocomial Infection Surveillance Program found that, although some form of *C. difficile* infection control precautions was used in all the hospitals surveyed, there was considerable variation in terms of testing strategies, cleaning and disinfection protocols and products, and isolation practices.[39] The practice recommendations regarding CDI in the "Compendium of Strategies to Prevent Healthcare-Associated Infections" compiled in 2008 by the Society for Healthcare Epidemiology of America (SHEA) and Infectious Diseases Society of America (IDSA)[22] offer a comprehensive approach to implementing prevention. They highlight several interventions to prevent patient exposure to CDI; these include (1) use of contact precautions for patients with CDI, including the use of gowns and gloves; (2) meticulous adherence to hand hygiene practices, in compliance with Centers for Disease Control and Prevention or World Health Organization recommendations; and (3) proper cleaning and disinfection of equipment and environment.[22] Each of these strategies is discussed in further detail below.

Contact Precautions

Patients with CDI should be placed under contact precautions. When private rooms are available in the acute care setting, a patient with CDI should be placed in such a room. Many hospitals continue to use semiprivate rather than private rooms for hospitalized patients, although single-patient rooms are the standard for newly constructed inpatient facilities in the United States. Semiprivate rooms may be more difficult to clean and decontaminate, and compliance with hand hygiene may be negatively impacted in multiple-bed rooms.[40] Cohorting of patients is another approach to consider if private rooms are not available, although it is important not to cohort patients discordant for infection with other epidemiologically important organisms (such as vancomycin-resistant *Enterococcus* or methicillin-resistant *Staphylococcus aureus*).

In addition, barrier precautions (the practice of donning a gown and gloves prior to entering the infected patient's room) should be implemented. The use of vinyl gloves has been shown to decrease the incidence of CDI,[41] and the use of gowns is a logical extension of this idea, given the association between environmental contamination, carriage by healthcare workers, and nosocomial transmission.[36]

Early initiation of contact precautions is one potential way to decrease nosocomial transmission. During outbreaks or in environments where C. difficile is hyperendemic, patients at high risk who have diarrhea can be placed under contact precautions preemptively, before results of C. difficile testing are available, or even if initial test results are negative, if clinical suspicion is high and the screening tests being used have low sensitivity. Several risk factors are associated with development of CDI (as discussed above), and these factors may be used to identify patients at greatest risk for CDI and to guide empiric implementation of contact precautions at the onset of diarrhea. However, we must keep in mind that CDI has also been described in groups of patients at low risk, including young peripartum women without a history of antibiotic use or prolonged hospitalization[42] (although CDI in these low-risk groups generally has occurred without healthcare exposures). If rates of CDI are unacceptably high, another infection control strategy involves extending the duration of contact precautions until time of discharge from the hospital, since C. difficile spores may be shed in the stool even after diarrhea resolves.[22]

Hand Hygiene

C. difficile has been isolated from the hands of healthcare workers, indicating that this is likely an important mode of nosocomial spread[34] (see Chapter 21, on hand hygiene). Adherence to hand hygiene practices has long been touted as a mechanism to prevent nosocomial infections; however, rates of adherence to hand hygiene using traditional soap and water have been poor. For this reason, many hospitals have switched to alcohol-based hand rubs to improve rates of hand hygiene. Although alcohol-based hand rubs have activity against the vegetative form of C. difficile, they are not sporicidal. Many healthcare facilities have continued to use alcohol-based hand rubs despite their lack of activity against C. difficile spores, as it is believed that this deficit may be offset by improved hand hygiene compliance. Evidence indicates that hand hygiene adherence improves dramatically when alcohol-based hand cleansers are available and that routine use of alcohol-based hand rubs does not significantly increase rates of CDI.[43-45] The 2008 SHEA/IDSA Compendium notes that the use of alcohol-based hand rubs for prevention of CDI spread is an area of controversy.[22] A new aloe vera–based hand gel (Xgel; Remedy Research) that contains CuAL42, a copper-based biocidal formulation, has shown in vitro activity against C. difficile spores[46]; however, this has yet to be tested in a clinical setting.

Environmental Decontamination

Quaternary ammonium–based detergents and other commonly used hospital cleaning agents do not kill C. difficile spores.[3,4] Sodium hypochlorite (bleach) solutions have been shown to kill spores and to decrease levels of environmental contamination with C. difficile and have been used successfully to control outbreaks of CDI.[47,48] The SHEA/IDSA Compendium[22] recommends the use of a solution of sodium hypochlorite (diluted 1:10 in water), with a contact time of 10 minutes, for environmental disinfection during outbreaks or in settings of hyperendemicity. However, the use of sodium hypochlorite solution has several disadvantages, including the strong odor, the possibility of respiratory distress after exposure, the risk of corrosion or pitting of equipment, and the need for the solution to be mixed daily.[22,37] Moreover, use of sodium hypochlorite has not been found to be effective at reducing CDI rates in non-outbreak circumstances. Despite terminal room-cleaning policies that specify use of sodium hypochlorite, nosocomial transmission of CDI in acute care settings has continued to be an ongoing issue, likely because of poor compliance with cleaning guidelines.[48] It is important to note that spores are physically removed if proper cleaning technique is used; therefore, it may not be necessary to kill spores if proper cleaning is done. Conversely, if proper cleaning is not done, sporicidal agents will also not work. Hospital epidemiologists should work closely with environmental services staff to ensure consistent, effective cleaning procedures.

Other cleaning strategies are under investigation, including hydrogen peroxide vapor cleaning systems. Such systems have been shown to decrease environmental contamination with C. difficile significantly after a single 3-hour cleaning cycle, even reducing contamination in rooms that had previously been cleaned with sodium hypochlorite solution.[49] However, these systems also have limitations, including the need for staff training, the greater length of time needed for room decontamination, and higher cost.[50] Their advantages include that they decontaminate hard-to-reach surfaces and that they potentially reduce variation in cleaning practices.[50] These newer hydrogen peroxide vapor methods have yet to be studied on a broad scale during CDI outbreaks or in settings of hyperendemicity, and the ultimate role of this technology in the control of CDI remains unclear. (See also Chapter 7, on disinfection and sterilization.)

Antimicrobial Stewardship

Prior exposure to antimicrobials is a strong risk factor for the development of CDI, especially when multiple agents are used, or when treatment is given for a prolonged period of time. Clindamycin has long been identified as a major culprit in the development of CDI, and fluoroquinolones have also been implicated after emergence of the fluoroquinolone-resistant epidemic strain BI/NAP1.[51] However, virtually all antimicrobials can increase the propensity for development of CDI.[51] For this reason, controlling the use of antimicrobials (antimicrobial stewardship) is a cornerstone of *C. difficile* infection control and has been effective in reducing the rate of CDI in the inpatient setting.[52,53] However, antimicrobial stewardship is sometimes difficult to evaluate as a single infection control strategy, given that it is often implemented as part of an infection control bundle that includes enhanced contact precautions and specialized cleaning methods.[11,38]

A successful antimicrobial stewardship program aims to limit inappropriate use of antimicrobials while optimizing the choice of agent and the dose, route, and duration of therapy to maximize clinical utility and limit the adverse consequences of therapy[54] (see Chapter 20, on improving antimicrobial usage). In the case of *C. difficile*, most antimicrobial stewardship programs have focused on restricting the use of high-risk antibiotics (second- and third-generation cephalosporins, fluoroquinolones, and clindamycin); however, antimicrobial stewardship programs can also reduce use of these agents through nonrestrictive means, such as clinician education and provision of antimicrobial guidelines rather than restrictions.[53]

Antimicrobial stewardship impacts both the risk for the individual patient and the global risk for other patients; fewer patients receiving antimicrobials results in fewer patients with CDI, and fewer patients with CDI means fewer patients contributing to the spread of infection.

Other Preventative Strategies

Practices such as testing asymptomatic patients for *C. difficile* toxin and repeating *C. difficile* testing at the end of treatment for CDI are not recommended as routine components of a CDI prevention strategy.[22] Prophylactic use of vancomycin or metronidazole to prevent the development of CDI in patients receiving antibiotics or in asymptomatic *C. difficile* carriers is also not recommended.[22] In fact, asymptomatic colonization with *C. difficile* has a protective effect against development of CDI.[37] Patients treated for asymptomatic colonization may be at higher risk for developing CDI.

Proton-pump inhibitors are now one of the most-prescribed groups of drugs in the United States. Several studies have noted an association between receipt of acid-suppressive therapy and development of CDI, likely because *C. difficile* spore forms have increased survival in higher-pH gastric acid.[7,9] The evidence has been mixed, but one study found that hospitalized patients exposed to proton-pump inhibitors were at 3.6 times higher risk of developing CDI, controlling for other factors.[7] This issue remains somewhat controversial, but stopping unnecessary proton-pump inhibitor therapy on admission may be an adjunctive strategy to reduce the rate of nosocomial transmission of *C. difficile*.

Summary

- In the past decade, CDI has become more frequent, more severe, and more difficult to manage.
- Nosocomial spread of *C. difficile* has been exacerbated by the emergence of the epidemic *C. difficile* strain BI/NAP1.
- CDT-EIA is the primary test for CDI at most hospitals; however, some facilities are switching to multiple-step testing strategies or PCR-based tests.
- Rates of HO-CDI should be tracked for surveillance purposes.
- There is a strong association between environmental contamination with *C. difficile*, carriage by healthcare workers, and nosocomial transmission of CDI.
- Patients with CDI should be placed in private rooms whenever possible, and barrier precautions should be used.
- Antimicrobial stewardship programs should focus on reducing unnecessary use of high-risk agents (including second- and third-generation cephalosporins, fluoroquinolones, and clindamycin).
- Testing asymptomatic patients for *C. difficile* toxin and repeating *C. difficile* testing at the end of treatment for CDI are not recommended.
- Curtailing the unnecessary use of proton-pump inhibitors may help prevent the transmission of *C. difficile*.

Acknowledgement

We thank Erik R. Dubberke, MD, for his thoughtful review of this chapter.

References

1. Bartlett JG. Narrative review: the new epidemic of *Clostridium difficile*–associated enteric disease. *Ann Intern Med* 2006;145: 758–764.

2. Kyne L, Hamel MB, Polavaram R, Kelly CP. Health care costs and mortality associated with nosocomial diarrhea due to *Clostridium difficile*. *Clin Infect Dis* 2002;34:346–353.

3. Wilcox MH, Fawley WN. Hospital disinfectants and spore formation by *Clostridium difficile*. *Lancet* 2000;356:1324.

4. Fawley WN, Underwood S, Freeman J, et al. Efficacy of hospital cleaning agents and germicides against epidemic *Clostridium difficile* strains. *Infect Control Hosp Epidemiol* 2007;28: 920–925.

5. Brown E, Talbot GH, Axelrod P, Provencher M, Hoegg C. Risk factors for *Clostridium difficile* toxin–associated diarrhea. *Infect Control Hosp Epidemiol* 1990;11:283–290.

6. McFarland LV, Surawicz CM, Stamm WE. Risk factors for *Clostridium difficile* carriage and C. *difficile*–associated diarrhea in a cohort of hospitalized patients. *J Infect Dis* 1990; 162:678–684.

7. Aseeri M, Schroeder T, Kramer J, Zackula R. Gastric acid suppression by proton pump inhibitors as a risk factor for *Clostridium difficile*–associated diarrhea in hospitalized patients. *Am J Gastroenterol* 2008;103:2308–2313.

8. Garey KW, Dao-Tran TK, Jiang ZD, Price MP, Gentry LO, Dupont HL. A clinical risk index for *Clostridium difficile* infection in hospitalised patients receiving broad-spectrum antibiotics. *J Hosp Infect* 2008;70:142–147.

9. Dial S, Delaney JA, Barkun AN, Suissa S. Use of gastric acid-suppressive agents and the risk of community-acquired *Clostridium difficile*–associated disease. *JAMA* 2005;294: 2989–2995.

10. Howitt JR, Grace JW, Schaefer MG, Dolder C, Cannella C, Schaefer RS. *Clostridium difficile*–positive stools: a retrospective identification of risk factors. *Am J Infect Control* 2008;36: 488–491.

11. Muto CA, Blank MK, Marsh JW, et al. Control of an outbreak of infection with the hypervirulent *Clostridium difficile* BI strain in a university hospital using a comprehensive "bundle" approach. *Clin Infect Dis* 2007;45:1266–1273.

12. McDonald LC, Owings M, Jernigan DB. *Clostridium difficile* infection in patients discharged from US short-stay hospitals, 1996–2003. *Emerg Infect Dis* 2006;12:409–415.

13. Dubberke ER, Wertheimer AI. Review of current literature on the economic burden of *Clostridium difficile* infection. *Infect Control Hosp Epidemiol* 2009;30:57–66.

14. Akerlund T, Persson I, Unemo M, et al. Increased sporulation rate of epidemic *Clostridium difficile* type 027/NAP1. *J Clin Microbiol* 2008;46:1530–1533.

15. McDonald LC, Killgore GE, Thompson A, et al. An epidemic, toxin gene-variant strain of *Clostridium difficile*. *N Engl J Med* 2005;353:2433–2441.

16. Pepin J, Valiquette L, Alary ME, et al. *Clostridium difficile*–associated diarrhea in a region of Quebec from 1991 to 2003: a changing pattern of disease severity. *CMAJ* 2004;171: 466–472.

17. Kim J, Smathers SA, Prasad P, Leckerman KH, Coffin S, Zaoutis T. Epidemiological features of *Clostridium difficile*–associated disease among inpatients at children's hospitals in the United States, 2001–2006. *Pediatrics* 2008;122:1266–1270.

18. Kelly CP, LaMont JT. *Clostridium difficile*—more difficult than ever. *N Engl J Med* 2008;359:1932–1940.

19. Warny M, Pepin J, Fang A, et al. Toxin production by an emerging strain of *Clostridium difficile* associated with outbreaks of severe disease in North America and Europe. *Lancet* 2005;366:1079–1084.

20. Brazier JS, Raybould R, Patel B, et al. Distribution and antimicrobial susceptibility patterns of *Clostridium difficile* PCR ribotypes in English hospitals, 2007–8. *Euro Surveill* 2008; 13(41):601–604.

21. Curry SR, Marsh JW, Shutt KA, et al. High frequency of rifampin resistance identified in an epidemic *Clostridium difficile* clone from a large teaching hospital. *Clin Infect Dis* 2009; 48(4):425–429.

22. Dubberke ER, Gerding DN, Classen D, et al. Strategies to prevent *Clostridium difficile* infections in acute care hospitals. *Infect Control Hosp Epidemiol* 2008;29(Suppl 1): S81–S92.

23. Wilkins TD, Lyerly DM. *Clostridium difficile* testing: after 20 years, still challenging. *J Clin Microbiol* 2003;41:531–534.

24. Bartlett JG, Gerding DN. Clinical recognition and diagnosis of *Clostridium difficile* infection. *Clin Infect Dis* 2008;46(Suppl 1): S12–S18.

25. Stamper PD, Alcabasa R, Aird D, et al. Comparison of a commercial real-time PCR assay for tcdB detection to a cell culture cytotoxicity assay and toxigenic culture for direct detection of toxin-producing *Clostridium difficile* in clinical samples. *J Clin Microbiol* 2009;47:373–378.

26. Planche T, Aghaizu A, Holliman R, et al. Diagnosis of *Clostridium difficile* infection by toxin detection kits: a systematic review. *Lancet Infect Dis* 2008;8:777–784.

27. Ticehurst JR, Aird DZ, Dam LM, Borek AP, Hargrove JT, Carroll KC. Effective detection of toxigenic *Clostridium difficile* by a two-step algorithm including tests for antigen and cytotoxin. *J Clin Microbiol* 2006;44:1145–1149.

28. Fenner L, Widmer AF, Goy G, Rudin S, Frei R. Rapid and reliable diagnostic algorithm for detection of *Clostridium difficile*. *J Clin Microbiol* 2008;46:328–330.

29. Persson S, Torpdahl M, Olsen KE. New multiplex PCR method for the detection of *Clostridium difficile* toxin A (tcdA) and toxin B (tcdB) and the binary toxin (cdtA/cdtB) genes applied to a Danish strain collection. *Clin Microbiol Infect* 2008;14:1057–1064.

30. Dubberke ER, McMullen KM, Mayfield JL, et al. Hospital-associated *Clostridium difficile* infection: is it necessary to track community-onset disease? *Infect Control Hosp Epidemiol* 2009;30:332–337.

31. Dubberke ER, Butler AM, Hota B, et al. Multicenter study of the impact of community-onset *Clostridium difficile* infection on surveillance for C. *difficile* infection. *Infect Control Hosp Epidemiol* 2009;30:518–525.

32. Dubberke ER, Reske KA, McDonald LC, Fraser VJ. *ICD-9* codes and surveillance for *Clostridium difficile*–associated disease. *Emerg Infect Dis* 2006;12:1576–1579.

33. Scheurer DB, Hicks LS, Cook EF, Schnipper JL. Accuracy of *ICD-9* coding for *Clostridium difficile* infections: a retrospective cohort. *Epidemiol Infect* 2007;135:1010–1013.

34. Fekety R, Kim KH, Brown D, Batts DH, Cudmore M, Silva J Jr. Epidemiology of antibiotic-associated colitis; isolation of *Clostridium difficile* from the hospital environment. *Am J Med* 1981;70:906–908.

35. Weaver L, Michels HT, Keevil CW. Survival of *Clostridium difficile* on copper and steel: futuristic options for hospital hygiene. *J Hosp Infect* 2008;68:145–151.

36. Samore MH, Venkataraman L, DeGirolami PC, Arbeit RD, Karchmer AW. Clinical and molecular epidemiology of sporadic and clustered cases of nosocomial *Clostridium difficile* diarrhea. *Am J Med* 1996;100:32–40.

37. Gerding DN, Muto CA, Owens RC Jr. Measures to control and prevent *Clostridium difficile* infection. *Clin Infect Dis* 2008;46(Suppl 1):S43–S49.

38. Weiss K, Boisvert A, Chagnon M, et al. Multipronged intervention strategy to control an outbreak of *Clostridium difficile* infection (CDI) and its impact on the rates of CDI from 2002 to 2007. *Infect Control Hosp Epidemiol* 2009;30:156–162.

39. Gravel D, Gardam M, Taylor G, et al. Infection control practices related to *Clostridium difficile* infection in acute care hospitals in Canada. *Am J Infect Control* 2009;37:9–14.

40. Detsky ME, Etchells E. Single-patient rooms for safe patient-centered hospitals. *JAMA* 2008;300:954–956.

41. Johnson S, Gerding DN, Olson MM, et al. Prospective, controlled study of vinyl glove use to interrupt *Clostridium difficile* nosocomial transmission. *Am J Med* 1990;88:137–140.

42. Garey KW, Jiang ZD, Yadav Y, Mullins B, Wong K, Dupont HL. Peripartum *Clostridium difficile* infection: case series and review of the literature. *Am J Obstet Gynecol* 2008;199: 332–337.

43. Vernaz N, Sax H, Pittet D, Bonnabry P, Schrenzel J, Harbarth S. Temporal effects of antibiotic use and hand rub consumption on the incidence of MRSA and *Clostridium difficile*. *J Antimicrob Chemother* 2008;62:601–607.

44. Boyce JM, Ligi C, Kohan C, Dumigan D, Havill NL. Lack of association between the increased incidence of *Clostridium difficile*–associated disease and the increasing use of alcohol-based hand rubs. *Infect Control Hosp Epidemiol* 2006;27: 479–483.

45. Rupp ME, Fitzgerald T, Puumala S, et al. Prospective, controlled, cross-over trial of alcohol-based hand gel in critical care units. *Infect Control Hosp Epidemiol* 2008;29:8–15.

46. Hall TJ, Wren MW, Jeanes A, Gant VA. A comparison of the antibacterial efficacy and cytotoxicity to cultured human skin cells of 7 commercial hand rubs and Xgel, a new copper-based biocidal hand rub. *Am J Infect Control* 2009;37: 322–326.

47. McMullen KM, Zack J, Coopersmith CM, Kollef M, Dubberke E, Warren DK. Use of hypochlorite solution to decrease rates of *Clostridium difficile*–associated diarrhea. *Infect Control Hosp Epidemiol* 2007;28:205–207.

48. Eckstein BC, Adams DA, Eckstein EC, et al. Reduction of *Clostridium difficile* and vancomycin-resistant *Enterococcus* contamination of environmental surfaces after an intervention to improve cleaning methods. *BMC Infect Dis* 2007;7:61.

49. Shapey S, Machin K, Levi K, Boswell TC. Activity of a dry mist hydrogen peroxide system against environmental *Clostridium difficile* contamination in elderly care wards. *J Hosp Infect* 2008;70:136–141.

50. Boyce JM, Havill NL, Otter JA, et al. Impact of hydrogen peroxide vapor room decontamination on *Clostridium difficile* environmental contamination and transmission in a healthcare setting. *Infect Control Hosp Epidemiol* 2008; 29:723–729.

51. Owens RC Jr, Donskey CJ, Gaynes RP, Loo VG, Muto CA. Antimicrobial-associated risk factors for *Clostridium difficile* infection. *Clin Infect Dis* 2008;46(Suppl 1):S19–S31.

52. Nuila F, Cadle RM, Logan N, Musher DM. Antibiotic stewardship and *Clostridium difficile*–associated disease. *Infect Control Hosp Epidemiol* 2008;29:1096–1097.

53. Valiquette L, Cossette B, Garant MP, Diab H, Pepin J. Impact of a reduction in the use of high-risk antibiotics on the course of an epidemic of *Clostridium difficile*–associated disease caused by the hypervirulent NAP1/027 strain. *Clin Infect Dis* 2007;45(Suppl 2):S112–S121.

54. Dellit TH, Owens RC, McGowan JE Jr, et al. Infectious Diseases Society of America and the Society for Healthcare Epidemiology of America guidelines for developing an institutional program to enhance antimicrobial stewardship. *Clin Infect Dis* 2007;44:159–177.

Chapter 20 Improving Use of Antimicrobial Agents

Robert A. Duncan, MD, MPH, and
Kenneth R. Lawrence, PharmD

Antibiotics comprise the second most commonly used class of drugs in hospital formularies: 30%–65% of hospitalized patients receive antimicrobial agents,[1,2] and expenditures for these drugs may comprise 10%–40% of the hospital pharmacy budget.[1,3] They are unique among pharmaceuticals in that they affect not only individual patients, but the larger microbiological environment as well. Nearly all practicing physicians readily prescribe antibiotics, with varying levels of competence.[4] It is estimated that 37% (range, 25%–50%) of antibiotic use in hospitals is inappropriate,[5,6,7,8] and hospitals are often where patterns are set for outpatient practice.[9] Physicians commonly focus on the individual patient and are rarely aware of the ecologic effects of antimicrobial agents on the patient, the hospital, long-term care facilities, the community, or the world at large.[10] Reflecting these concerns, there is an expanding literature regarding "adequacy of therapy."[11] These studies are typically set in intensive care units and determine, in retrospect, whether a pathogen isolated from a patient with sepsis was sensitive to one or more components of the chosen therapeutic regimen. "Inadequate" therapy is associated with increased mortality. This is often interpreted to mean that an empiric regimen of the newest, broadest-spectrum, most potent agents would have provided a better outcome. Caution must be used when interpreting these results, as they refer not to whether the empiric regimen was a wise choice for the circumstances, but rather whether the choice actually paid off. Retrospective analyses of adequacy do not substitute for prospective trials comparing expert advice or decision support to usual practice. The common "broader is better" conclusion

"panders simultaneously to the physician's ego and fears,"[12(p575)] a well-established tactic of the pharmaceutical industry.

A recent contribution to this field introduces the "likelihood of inadequate therapy."[13] This concept brings the potential prescriber to the treatment choice most appropriate for the circumstances, rather than to an impossible prediction of what regimen would have been active against the eventually isolated pathogen.

Excessive use of antibiotics is linked not only to the emergence and spread of resistance[14,15,16] but also to adverse drug reactions, the added cost of high-end treatment, and increased infection control efforts required to prevent the spread of pathogens.[17,18,19,20] Consequently, there is a growing awareness among hospital administrators and insurers of the clinical and financial benefits of containing antibiotic use and the emergence of resistant organisms.

Resistant organisms continue to blossom.[21,22] Discovery of glycopeptide-resistant gram-positive cocci,[23] multidrug-resistant gram-negative bacilli,[24] and carbapenem-resistant Enterobacteriaciae,[25] nearly untreatable pathogens, brings new challenges for both treatment and infection control. Compounding these problems is what has been described as a "perfect storm" of sicker patients and a decimated process of development of new drugs to fight the worsening resistance.[26] The Infectious Diseases Society of America (IDSA) has addressed this in its position paper, "Bad Bugs, No Drugs,"[27] an effort to induce the healthcare industry, pharmaceutical companies, legislatures, and the public to join in combating the growing crisis of antimicrobial resistance, as McGowan put it, "as we try to mount our counteroffensive against the wily and ingenious resistant microorganisms of the 1990s."[12(p576)]

These organisms have only become more crafty in the decades that followed. Inappropriate use of antimicrobial agents and the resultant increase in the number of resistant organisms have gained national attention. In 1997 a joint committee of the Society for Healthcare Epidemiology of America (SHEA) and the IDSA published guidelines for the prevention of antimicrobial resistance in hospitals.[28] The Centers for Disease Control and Prevention (CDC), the Food and Drug Administration, and the World Health Organization (WHO) have since identified antimicrobial stewardship as an essential component of efforts to prevent and reduce resistance.[29,30,31] In addition, the Joint Commission has identified prevention of healthcare-acquired infection as a primary target of its evolving National Patient Safety Goals.[32] The Centers for Medicare and Medicaid Services and the US Congress were not far behind, resulting in a burgeoning list of healthcare-associated conditions for which the Centers for Medicare and Medicaid Services will no longer pay.[33] This combination of therapeutic, economic, and regulatory incentives provides a driving force for clinical and ecological reforms that previously concerned a more limited audience.

Antimicrobial stewardship lies at the intersection of infectious diseases, infection control, safety and quality improvement, and cost containment. It has been defined as the "appropriate selection, dosing, and duration of antimicrobial therapy to achieve optimal efficacy in managing infections."[28(p590)] It is thus an important tool in the effort to reduce the inappropriate use of antimicrobials and the subsequent development of both resistant microorganisms and drug-related adverse events.[34] However, in a review of top stewardship practices in the United States and in Massachusetts, Barlam and DiVall[35] found that use of a restricted formulary and an antibiotic approval process, measures active in the vast majority of academic centers, were implemented in only 49% and 29% of community hospitals, respectively. The IDSA and SHEA have since collaborated to provide comprehensive guidelines for developing institutional programs to enhance antimicrobial stewardship.[36]

Effective antimicrobial stewardship requires the collaboration of many disciplines, including the medical, surgical, pediatric, and critical care services; nursing; infection control; the laboratory; and the pharmacy. Often ignored are emergency services, which may operate with relative independence. Monitoring antimicrobial use is an important component of stewardship because of the close relationship between use of antimicrobial agents and the emergence of bacterial resistance.[15,16,37,-40] For example, excessive use of third-generation cephalosporins and of vancomycin has been linked to emergence of several resistant organisms, including methicillin-resistant *Staphylococcus aureus* (MRSA), vancomycin-resistant enterococci (VRE), and extended-spectrum β-lactamase (ESBL)–producing *E. coli* and *Klebsiella* species.[38-40]

This chapter describes how to organize an antimicrobial stewardship program, including options for hospitals in a range of settings. The available resources have grown dramatically. A Medline search of the term "antibiotic stewardship" since 1995 identified 122 articles; 71 (58%) were published in the past 2 years (data from R.A.D.; search performed on April 23, 2009). In addition to the IDSA/SHEA guidelines, reviews of the experiences at large institutions have been published.[41,44-45] Internet resources regarding both antimicrobial stewardship[46] and infection control[47] are available, as is a "Compendium of Strategies to Prevent Healthcare-Associated Infections in Acute Care Hospitals,"[48] strategies which include forming antimicrobial stewardship programs. Although few carefully designed, randomized, controlled trials have been done (70% of articles published from 1980 through 2003 were deemed to have inadequate methodology[49,50]), newer studies show promise. Developing a solid evidence base that corroborates the myriad benefits of stewardship will be essential when soliciting financial and staffing support, especially in challenging economic environments.

Organizing a Program

Organization and Personnel

An effective program of antibiotic utilization reform requires a team approach. This starts with formation of an antibiotic utilization committee, chaired by a member of the infectious diseases staff, a clinical pharmacist with infectious diseases training, or someone with sophisticated knowledge of infectious disease treatment. These leaders should function under the auspices of the quality and safety administration and receive appropriate compensation for their efforts[36,44] (see Chapter 2), as well as adequate training and support.[35] There should be representatives from the departments of pharmacy, microbiology, and infection control (including the hospital epidemiologist and infection preventionists), and a data/informatics manager. Depending on the size and structure of the hospital, representation from several other departments may be helpful (Table 20-1). Residency staff should be included, since they often prescribe most (or all) antibiotics in training hospitals. Alternatively, community hospitals should recruit major physician opinion leaders. Direct involvement of a key administrator provides liaison to higher levels of hospital leadership, as

Table 20-1. Composition of an antimicrobial utilization committee

Infectious disease staff (Co-chair)
Infectious disease pharmacist (Co-chair)
Pharmacy director
Microbiology director or supervisor
Infection control practitioner(s)
Data manager
Surgeon
General internist
Pediatrician
Intensivist
Emergency department staff
House staff
Staff nurse
Intensive care unit nurse
Quality improvement staff
Administrator

well as vital support when reform efforts must be promoted among the medical staff.[51] In addition, collaboration with quality improvement teams may bring substantial administrative resources to bear and facilitates innovation and change. A small working group that includes the chair, an infectious diseases pharmacist, and a data manager provides most of the effort, with advice from the larger committee. A small group can still provide significant incremental benefit in hospitals with limited resources, whether because of financial, geographic, or staffing constraints.[52,53]

Data Collection

Data for monitoring antimicrobial prescribing come from many sources, ranging from pharmacy purchasing records or individual patient records to computerized clinical databases. Pharmacy purchasing data are easily obtainable and may be useful when tracking trends over a period of time, yet have limited utility for epidemiologic analysis. Retrospective or prospective review of a patient's medical record yields actual dispensed doses but is laborious. Computerized systems can provide the time, date, location, prescriber, and dosage of each drug. These data allow calculation of defined drug densities and cost per hospitalization or per patient-day, as well as density of drug use within specific areas of the hospital (eg, vancomycin use per patient-day in an intensive care unit). This helps to pinpoint areas of misuse and provides targets for educating clinicians. Finally, summary data from health maintenance organizations and Medicare allow researchers to analyze utilization of antimicrobial agents on a broader

scale. Despite availability, national utilization data are difficult to link to practices in local hospitals. Commercial databases used by the pharmaceutical industry may provide surprisingly detailed prescriber-specific information.

Expert data management and medical informatics resources are becoming increasingly essential elements of antimicrobial stewardship programs[36] (see Chapter 10). Enhanced communication between pharmacy, chemistry and microbiology laboratories, infection control, and administrative databases facilitates clinical care and infection prevention, as well as complying with the exploding requirements of state and national reporting of processes, outcomes, and adverse events. The future is likely to demand improved communication as patients move among disparate healthcare systems, often with incompatible information technologies. Medical informaticists will be increasingly valuable.

Benchmarking studies, utilizing standardized days of therapy,[2] have been suggested as a means to compare the use of antimicrobials between hospitals of similar size, acuity, and function, aiming to identify both "best practice" and significant variation in usage patterns.[54] In a study of 22 university teaching hospitals from 2002 to 2006, Pacyz et al.[2] found that 63.5% of patients received an antibacterial drug, increasing from 798 days of therapy per 1000 patient-days in 2002 to 855 days of therapy per 1000 patient-days in 2006 ($P <$ .001). Clearly, the clinical areas in which each hospital specializes will affect the types and patterns of prescribed agents and must be considered when examining the results from such studies. But the strongest predictor of broad-spectrum antibiotic use was variation in duration of therapy between different hospitals,[2] confirming findings by Carling et al.[55] The National Healthcare Surveillance Network, rapidly growing as states mandate public reporting of hospital-associated infections, is increasingly valuable for comparative, severity-adjusted data.[22]

Drug utilization evaluations identify usage patterns and trends, according to hospital service or unit.[56] This process focuses on a particular drug or class of drugs, assessing optimal prescribing, as defined by local experts (eg, parenteral fluoroquinolone use). Other projects may examine antibiotic therapy for a specific disease, such as community-acquired pneumonia. A drug utilization evaluation should examine a few questions, rather than attempt a comprehensive review. It may also detect variability in practice or identify unrecognized medication errors. Follow-up drug utilization evaluations give essential feedback to clinicians and document whether reforms were effective. Initial projects might examine the role of antibiotics in emergence of a resistant pathogen (eg, imipenem-resistant *Pseudomonas aeruginosa*), over-utilization of antimicrobial agents, dosing regimens that are ineffective or toxic, or surgical prophylaxis. A list of the pharmacy's "top 200" expenditures reveals crude patterns of drug use and helps identify targets for intervention[45] (eg, if purchases of parenteral formulations of a fluoroquinolone are outpacing purchases of oral formulations, an aggressive campaign to switch therapy from intravenous to oral formulations ["intravenous-to-oral conversion"] may be warranted).

The microbiology laboratory is of vital importance in any antimicrobial stewardship program. It supplies hospital-wide antimicrobial susceptibility results ("antibiograms") to help in choosing empiric antibiotic therapy, as well as providing separate profiles for intensive care units; each must be updated regularly. The laboratory also must inform the committee about emerging problems with resistant organisms and document changes in susceptibility patterns that may be attributable to changes in use of antimicrobial agents.[57]

Improving diagnosis is an essential component of antimicrobial stewardship. Much of the waste and abuse of antibiotics results from attempts to "cover everything." Quantitative and semiquantitative culture techniques may aid in diagnosis. This has been standard practice for decades for urinary tract infection but can also improve the diagnostic accuracy in ventilator-associated pneumonia.[58,59]

Rapid molecular diagnostic testing is dramatically altering the pace of laboratory identification of invasive pathogens and resistance (see Chapter 9). Not only are polymerase chain reaction (PCR)–based batteries of tests now able to detect any one of the protean causes of meningitis, but "real-time" rapid PCR methods can indicate the presence of a *mec*-A or *KPC* resistance mechanism, thus identifying MRSA or carbapenemase-producing *Klebsiella* species within a few hours, rather than days.[60] Rapid and definitive diagnosis speeds detection and isolation of patients carrying resistant organisms and assists in identifying clonal spread, but it also shortens the duration of empiric broad-spectrum therapy, because the regimen can be tailored to an optimal, narrow-spectrum agent. This immediately ensures that therapy is "adequate" and obviates expanding empiric treatment.

Interventions

Development of Antimicrobial Management Programs

After establishing baseline characteristics of current antimicrobial use and local hospital resistance patterns, the committee should determine what they wish to accomplish and then develop a mission statement.[45,61] A good starting point is succinctly stated in the IDSA/SHEA guidelines: "The ultimate goal of antimicrobial stewardship is to improve patient care and health care outcomes."[36(p162)]

When developing programs, the committee should consider several factors. These include hospital size, physician make-up (private practice model or training program), special patient populations (eg, solid organ or bone marrow transplantation, pediatrics, trauma, or burns), referral resources (eg, rehabilitation or long-term care facilities), and hospital politics.

Initial recommendations and interventions should be limited in scope and should have strong potential to improve patient care, reduce medication errors, and decrease costs, but they should also have a high likelihood of success and physician acceptance. Medical staff support is usually strong: in a survey of 490 internal medicine physicians at 4 Chicago hospitals, almost 90% considered antibiotic resistance to be an important problem, and still more felt that inappropriate antimicrobial prescribing was an important contributor.[62] The committee and administrators should agree on formulas to measure the results of its initiatives.

After initial success, the committee may tackle more challenging programs. These could include local adaptations of treatment guidelines, such as a ventilator-associated pneumonia management pathway,[63] or joining a national collaborative, such as the Surgical Care Improvement Project (discussed below).[64]

Dellit et al.[36] describe 2 core strategies to improve antimicrobial use: a prospective audit with intervention and feedback (rating of supporting evidence, A-I) and formulary restriction and a requirement for preauthorization (rating of supporting evidence, B-II). These are supplemented by several other techniques. Interventions to improve antimicrobial prescribing, reduce resistance, and decrease costs are listed in Table 20-2 and discussed below.[1,10,45,65] Table 20-3 lists several specific "low-hanging fruit" targets for intervention. Caveats are noted in Table 20-4.

Revising the Formulary

Formulary revision is a common and easily implemented method of modulating antimicrobial use.[66,67] Open formularies impose few controls on availability or prescribing of medications by physicians, whereas closed formularies utilize a systematic process to choose the best 1 or 2 agents among existing drugs in a class and to evaluate new medications. The plethora of available oral and parenteral antimicrobial agents has made selection of an appropriate agent difficult for clinicians and the committee can help to simplify these choices. Inclusion in

Table 20-2. Components of antimicrobial stewardship and cost-containment programs

Data collection and target identification

Routine monitoring of pharmacy purchasing volume and costs
Measure days of therapy or defined daily doses
Focused, problem-oriented drug utilization evaluations for individual drugs or prescribing practices
Review of "top 200–300" high-cost formulary agents
Comprehensive review of patterns of antimicrobial usage, with feedback from infectious diseases physicians or pharmacists
Computerized drug utilization evaluations that integrate data from the departments of pharmacy, microbiology, chemistry, and radiology

Formulary revision

Expert committee selection of formulary and limitation of pharmacy stocks to one or a few optimal drugs within a therapeutic class
Substitution of generic agents for proprietary agents
Selection of newly off-patent agents
Competitive contract bidding for similar drugs with equivalent efficacy

Microbiology testing and reporting

Routine susceptibility testing done only for formulary agents
Graded susceptibility reporting, based on level of resistance and cost-effectiveness
Regular reporting of susceptibility patterns and empiric drugs of choice, according to intensive care unit, ward, or outpatient isolates
Reporting may include level of restriction, usual dosing regimens, renal dose adjustments, and costs

Education

Direct education of healthcare providers by physicians or pharmacists, one-on-one or by group
Use of clinical management pathways and guidelines
"Counter-detailing" of drug information
Concurrent review and advice provided by infectious diseases physicians or pharmacists
Computerized decision support of prescribing choices
Feedback given to providers
Education and recruitment of senior department heads and opinion leaders
Use the institution's internal guideline on antimicrobial therapy, which may be made available on the internal computer network

Restriction policies

Policies ranked by increasing restriction: open formulary; unrestricted but closed formulary; monitored drugs; infectious diseases telephone
  approval required; infectious diseases consultation required
Limitation on use of drugs by clinical scenario (eg, diabetic foot infection), location (eg, intensive care units), or hospital service
Removal or restriction of specific problem agents (eg, habitual antibiotic choices)
Use of computerized approval system on internal computer network

Ordering policies

Use of antimicrobial agent order forms, including common dosing parameters and educational information
Specification of duration of therapy and prophylaxis
Use of surgical prophylaxis standing orders with specified doses and durations
Use of automatic stop orders for prophylaxis, empiric therapy, and specific therapy, with duration of use limited unless approved
Use of computerized physician order entry incorporating decision-support tools

Drug administration

Establishment of an infectious diseases pharmacist clinical intervention program
Use of pharmacokinetic consultation
Revision of standard dosing regimens, based on new pharmacodynamic data
Streamlining (optimizing) use of broad-spectrum agents or multiple-drug antimicrobial regimens
Use of once-daily aminoglycoside dosing
Campaign to increase intravenous-to-oral conversion and/or step-down conversion
Use of home-based intravenous therapy

Limiting contact with pharmaceutical representatives

Restriction of access of pharmaceutical representatives to clinical care areas
Review (or elimination) of "detailing" information and coordination with representatives
Therapeutic "partnering" with pharmaceutical firms to coordinate education, clinical treatment pathways, and drug selection according to
  institutionally selected criteria
Restriction of the use of pharmeceutical samples

Table 20-3. Common targets of antimicrobial management programs

Intravenous-to-oral conversion
Vancomycin restriction
Fluoroquinolone use and MRSA and *Clostridium difficile*
Carbapenem use and resistant *Pseudomonas aeruginosa*
Third-generation cephalosporin use and ESBL producers and VRE
Proton pump inhibitors and risk of pneumonia and *C. difficile*
Weight-based and renal dose adjustment
Selection of off-patent alternatives within a drug class
Standardized orders for surgical prophylaxis

NOTE: ESBL, extended-spectrum β-lactamase; MRSA, methicillin-resistant *Staphylococcus aureus*; VRE, vancomycin-resistant *Enterococcus*.

the formulary should be based upon clinical efficacy, safety, pharmacokinetic and pharmacodynamic profiles, and dosing convenience. In addition, certain agents may have a greater propensity to induce resistance or alteration in hospital flora—for instance, receipt of fluoroquinolones may promote acquisition of *Clostridium difficile*.[68,69] Communication with the microbiology laboratory is important to ensure that susceptibility testing for a new agent not included in the current susceptibility testing panels does not incur unnecessary cost.

If 2 or more drugs within a pharmacologic class are considered therapeutically equivalent, cost often becomes an important factor in the final choice. Negotiation with competing pharmaceutical companies for the best contract can result in substantial savings. However, close review is required to ensure that market-share agreements do not adversely affect appropriate use of antibiotics. The committee should also consider the formularies of managed healthcare organizations to ensure continuity of therapy on hospital discharge but should ultimately make decisions in the best interest of the hospital, its patients, and its budget. Periodic review of the formulary avoids therapeutic duplication and eliminates little-used agents or those associated with newly reported adverse events. Similarly, agents with expiring patent protection can offer substantial saving opportunities, given the (usually) lower costs of equivalent generic drugs.

One recent but important variable is a manufacturer's shortage of medication.[70] Drug shortages occur because of multiple factors, including shortages of raw materials and unsafe manufacturing practices, that cause a product to be temporarily unavailable. Medication shortages increasingly affect patient care[70] and may force compensatory alterations of the formulary (eg, substitution of cefazolin plus metronidazole for cefoxitin or cefotetan).

Polk et al.[71] challenged the benefit of a closed antimicrobial formulary. They demonstrated that centers employing a narrow, closed antibiotic formulary experienced an associated increase in bacterial resistance, compared with facilities with a more varied formulary. These data suggest that heterogeneous antimicrobial prescribing may slow the development of antimicrobial resistance[72] and raise concerns about the consequences of relatively uniform prescribing in clinical management pathways or guidelines. Similarly, mathematical models suggest that, rather than use focusing on a rotating script of a few agents, mixed use of various antibiotics may be more effective in slowing the spread of resistance.[73] The benefits of narrowly constricted formularies are uncertain, as choosing one agent over another may lead to trade-offs in adverse effects.[36]

Reporting Laboratory Data

Susceptibility reporting should include results only for agents available in the formulary. Laboratories can influence antibiotic use by reporting susceptibility results for expensive, broad-spectrum agents only when organisms are resistant to narrower-spectrum and less expensive drugs. Use of new Clinical and Laboratory Standards Institute breakpoints is also critical to ensure that broad-spectrum agents are prescribed only when necessary—for example, revised breakpoints for pneumococcal species suggest that penicillin remains the drug of choice for many patients with pneumococcal pneumonia or bacteremia.[57]

Educational Programs

To assist physicians in prescribing antimicrobial therapy appropriately, a comprehensive educational program should be developed. This should include formal presentations, such as grand rounds and staff conferences, as well as more informal conferences for residents and for patient care rounds. Presentations should incorporate evidence-based national guidelines, current resistance trends in the hospital, and the importance of appropriate antimicrobial prescribing, with an emphasis on specific

Table 20-4. Caveats to bear in mind when implementing an antimicrobial stewardship program

Provide stewardship; don't become a policeman or zealot
Avoid formulary changes for only short-term gain
Beware the "package deal" from supply-chain managers
Financial concerns should not supersede clinical efficacy and safety
Substitution may alienate devoted users of a specific agent
Formulary alterations may necessitate changes in automated susceptibility testing
Don't add to clinicians' paperwork or burden them with collecting your data

problems identified within the hospital. Hospitals with training programs should regularly review basic antimicrobial therapeutics and clinical microbiology with house staff. An audience response system is a useful teaching tool, providing anonymous assessment, comparison with peers, and immediate expert feedback. Newsletters may include discussion of emerging resistance issues, lessons from antimicrobial drug utilization evaluations, and reviews of new antimicrobials or additions to the hospital formulary.

"Home-grown" institutional pocket-sized guides that contain information on antimicrobial therapy are popular with house staff and attending physicians. The Lahey Clinic "Guideline to Antimicrobial Therapy" has evolved from a 16-page booklet to a 76-page manual and is accessible on the institution's internal computer network. Information in the guide may include the hospital antibiogram, usual antimicrobial dosages, drug-level monitoring recommendations, dosage adjustments for impaired renal and hepatic function, common drug interactions, antibiotic cost, and approved treatment guidelines. A list of restricted agents and criteria for use of specific antibiotics should be included and updated regularly. A hospital with an internal computer network can use it to post educational information; a "question of the week" program can promote simple but useful advice for prescribers. Classic resources such as *The Sanford Guide to Antimicrobial Therapy*[74] and *The Johns Hopkins Abx Guide*[75] are now also available on the Internet and/or in formats for hand-held computers.[46]

An effective but labor-intensive form of education is one-on-one or small-group discussion, often referred to as "academic counter-detailing."[76] This method uses an evidence-based approach to educate clinicians about appropriate drug therapy in a specific clinical scenario and is often implemented in response to advice from manufacturers' representatives, advertising campaigns, or promotional literature that may be inconsistent with institutional goals.

One-time educational programs have little long-term impact but, when applied continuously, are essential strategic components of every antibiotic stewardship program. It is also clear that education is most effective in synergy with other multifaceted interventions, including systematizing best practice.

Restriction and Preauthorization

One of the most widely used active antibiotic containment strategies is restriction of use. Usage policies range from open, unrestricted formularies that allow any available pharmaceutical to be used to heavily restricted formularies that disallow prescription of specific drugs without consultation and approval by the infectious dis-

eases service. Options are delineated below, in order of increasing restriction.[45]

1. *Open formulary.* Physicians may prescribe any available pharmaceutical agent, without restriction.
2. *Unrestricted but closed formulary.* The formulary is limited to agents approved by the hospital's Pharmacy and Therapeutics Committee, but physicians may prescribe any drug in the formulary, without restriction.
3. *Monitored drugs.* The hospital's Antimicrobial Management Team prospectively monitors use of particular agents and assesses the appropriateness of use, then provides direct feedback to the providers.
4. *Limited (criteria-based) drugs.* Use of some drugs is limited to specific clinical scenarios (eg, diabetic foot infections), locations (eg, intensive care units), or services. Any other use is subject to approval by the infectious diseases service or the Antimicrobial Management Team.
5. *Approval required from the infectious diseases service or Antimicrobial Management Team.* Physicians who wish to use specified antimicrobial agents must discuss the case with the infectious diseases service or the Antimicrobial Management Team to obtain verbal approval. An initial grace period (eg, 24 hours) is commonly allowed before approval must be obtained.
6. *Infectious diseases consultation required.* Prescribers who wish to use a restricted drug must obtain on-site consultation by the infectious diseases service before the pharmacy will release the specified drug. An initial grace period may be offered.

Antimicrobial restriction programs have evolved over the decades since their introduction by McGowan and Finland,[77] shifting from simple approval programs administered by infectious diseases fellows or Antimicrobial Management Teams to educational efforts combined with preestablished criteria for use of broad-spectrum agents.[78-80] In one dramatic example, White et al.[81] describe instituting a prior-approval system for restricted antibiotics in response to an outbreak of multidrug-resistant *Acinetobacter* infection. This was administered by infectious diseases faculty who were available 24 hours per day, and it resulted in increased antimicrobial susceptibility (especially in intensive care units), a 32% reduction in expenditures for parenteral antimicrobials, and an estimated annual pharmacy budget savings of $863,100, at a cost of less than $150,000. Programs administered by infectious diseases fellows may not perform as well as those staffed by infectious diseases consultants or infectious diseases pharmacists.[82] A successful program that utilized the hospital's internal computer network in Australia has been described, which yielded reductions in the use of targeted antibiotics and improvement in resistance patterns.[83] Other, more complex programs are described below.

Although commonly used, antimicrobial restriction programs are controversial. Unrestricted drug access exposes staff to complex choices for which they have neither the time, information, nor expertise required to select drugs appropriately. Indeed, Kunin[84] has argued that "use of high-cost, specialized antimicrobial agents should be a privilege of infectious disease consultants and others trained in their use, just as performance of invasive procedures is limited to those who are qualified."[84] However, excessive antibiotic restriction fosters an adversarial relationship between infectious diseases consultants and medical, surgical, and house-staff services and can interfere with timely antibiotic administration. Thus, most hospitals maintain graduated levels of monitoring and restriction, depending on the severity of prescribing and resistance problems, potential toxicity, and costs. The most restrictive policies are usually reserved for the most toxic, unfamiliar, or expensive agents (eg, antifungal agents).

Intensive care units may be the most appropriate sites for active stewardship because of the concentration of severe illness, widespread antimicrobial use, and opportunities for spread of infection and antimicrobial resistance.[85-87] A recent survey of 276 US intensive care units concluded that mortality rates in those that did not employ clinical pharmacists in the direct care of patients with infections were 4.8% higher for patients with sepsis, 16.2% higher for patients with community-acquired infections, and 23.6% higher for patients with hospital-acquired infections. Furthermore, the length of stay in the intensive care unit was longer, and billing charges were 12%–13% greater.[88] These investigators estimated savings of $24.81 in total charges for every dollar spent on salary and benefits for an intensive care unit pharmacist.

Tightly restricted formularies are rarely practical outside of large teaching hospitals, where resistance problems are severe enough to provide an impetus and infectious diseases fellows can staff 24-hour approval programs. Experience shows, however, that many telephone communications (39%) contain inaccurate information[89] and lead to inappropriate recommendations from the Antimicrobial Management Team.[90] Some programs have also noted episodes of "stealth" dosing and other efforts to circumvent restrictions.[91] Narrowing, rather than restricting, options, along with personal interaction with prescribers, may achieve many of the same goals while maintaining collegial relationships.

Antimicrobial Order Forms

Antimicrobial order forms may incorporate formalized criteria for use of antimicrobial agents, suggest dosing regimens, and define the duration of prophylactic, empiric, or specific therapy.[92] Pharmacies often provide time limits for empiric therapy (eg, for the initial 72 hours). However, such policies must include measures to prevent inadvertent lapses in appropriate therapy. Use of antimicrobial order forms can facilitate audits of antimicrobial use, yet the quality of information is dependent on those filling out the forms and must be viewed with caution.

Surgical Prophylaxis

In the past, approximately 1 of every 3 antibiotic prescriptions in hospitals was for perioperative prophylaxis,[12] most of them unnecessary. In 1981, Durbin et al.[93] showed that requiring physicians to indicate on a form whether antimicrobial use was prophylactic, empiric, or therapeutic increased the number of patients receiving appropriate prophylactic antibiotics and reduced the duration of surgical prophylaxis by 2 days (from 4.9 to 2.9 days). However, controversy remains concerning the duration of prophylaxis. In a systematic review of 28 studies, McDonald et al.[94] concluded that there was no clear advantage to either multiple-dose or single-dose prophylaxis regimens for preventing surgical site infection (OR, 1.06 [95% confidence interval, 0.89–1.25]). They recommended single-dose prophylaxis. Professional societies have conducted similar reviews, reaching similar conclusions, yet left open the option to continue postoperative prophylaxis for up to 24–48 hours, largely for prosthetic implant procedures.[95]

Nearly 3 decades later, many centers now use standing order forms for selected surgical procedures, which incorporate recommended prophylactic dosing regimens and limit duration to a single preoperative dose or to include a 24-hour postoperative period. This standardizes many aspects of perioperative antimicrobial orders, minimizes excessive prophylaxis, and reduces errors and toxicity. Such orders best originate from within the department, directed by a surgeon who is an "influence leader."[12] Input from an infectious diseases pharmacist is invaluable and facilitates subsequent passage through the Pharmacy and Therapeutics Committee, which maintains ultimate oversight. At the Lahey Clinic, we formed subcommittees of the Pharmacy and Therapeutics Committee for antibiotics, standing orders, and chemotherapy to provide specialized expertise, preserving the full committee from detailed review.

The Surgical Care Improvement Project is a national collaborative effort to optimize the numerous processes surrounding operative care.[96] This began with a directive on choosing and then administering appropriate antibiotics for prophylaxis within 1 hour after first incision[97] and has further evolved from a voluntary pilot program to more inclusive directives for a broadening scope of procedures.[98] The Joint Commission, the Centers for Medicare and Medicaid Services, and the US Congress now encourage "voluntary" public reporting of compliance with the component parts of the Surgical Care Improvement Project by withholding 2% of payment to those who do not participate.[99] Notable successes have been associated with use of coordinated order sets, administration of prophylaxis by anesthesiologists, and use of "time-outs" before incision to ensure that each measure has been implemented. As it moves into the ambulatory surgery arena, the Surgical Care Improvement Project has the potential to further improve the use of prophylaxis. For example, new guidelines from the Urology Society,[100] implemented through the Surgical Care Improvement Project, should alter the heretofore common practice of prescribing a fluoroquinolone for 1–2 days before and 3–5 days after transrectal prostate biopsy, changing this prophylaxis to a single preoperative oral dose. Guidance is needed in surgery of the eye and ear as well; for example, prophylaxis lasting 1–2 weeks is not uncommon following stapedectomy (R.A.D., personal observation).

Computerized Physician Order Entry

Computerized physician order entry offers an automated and more versatile version of the paper order form for those hospitals with adequate computer resources.[101] Software can generate educational "pop-up" screens in response to requests for restricted or monitored agents or may require prescribers to justify choices before therapy can be dispensed. When designed adequately, these systems also yield prospective, provider-specific utilization data for review without being burdensome and may increase the success of antimicrobial stewardship programs.[101]

Computerized Decision Support

Researchers at LDS Hospital (Salt Lake City, Utah) have written extensively on the capabilities of their computer system, which features decision support.[102,103] They developed a computer-assisted management program that provides integrated, real-time data pertaining to treatment of infections, with multiple advantages.

First, the program can evaluate clinical data and alert clinicians that antibiotic therapy is necessary. It then suggests appropriate empiric or therapeutic antimicrobial regimens for individual patients, after reviewing diagnoses, vital signs, renal function, and microbiology reports, as well as other data. In an intensive care unit trial, the program significantly reduced medication

errors, inappropriate use of antimicrobial therapy, and total hospital and antimicrobial costs.[102] At present, this methodology offers integration of clinical, pharmacologic, and epidemiologic data, generating informed clinical advice at the bedside; it is now available as a commercial product on the Internet.[104]

In a recent Cochrane review, computerized advice for drug dosing was shown to provide significant benefits by increasing the initial dose and its subsequent serum concentration and reducing the time to therapeutic stabilization, the risk of a toxic drug level, and the length of hospital stay, although it had no effect on adverse reactions.[105]

Antimicrobial Cycling

Several attempts to prevent antimicrobial resistance have utilized different classes of antimicrobial agents in a sequential schedule,[72] injecting heterogeneity by scheduling changes of empiric antimicrobial regimens.[106] Antibiotic cycling programs typically rotate empiric use of β-lactam/β-lactamase inhibitor combinations, fluoroquinolones, carbapenems, and cephalosporins, often in an intensive care unit setting, changing every few months. Pilot studies claimed to have reduced the emergence of antimicrobial resistance, the incidence of infections due to drug-resistant pathogens, and the mortality rate, compared with historical controls, among critically ill intensive care unit patients.[107,108] However, these studies were often confounded by other concurrent interventions, and well-designed, multicenter studies are lacking. In addition, some studies have shown that up to half of patients receive off-cycle antimicrobial agents, because of prescriber concerns.[106] These programs do little to foster familiarity or expertise with a selected antimicrobial armamentarium and have an unclear long-term effect on prescribing. In addition, antimicrobial resistance mechanisms may linger for many years after removing the selective pressures exerted by the target drugs.[109] In a systematic review, Dellit et al.[36] concluded that there were insufficient data to recommend the routine use of antibiotic cycling as a means of preventing or reducing resistance over a prolonged period of time (rating of supporting evidence, C-II).

Intravenous-to-Oral Conversion

Many antimicrobial agents have excellent oral bioavailability, including fluoroquinolones, most azole antifungal agents, linezolid, rifampin, metronidazole, clindamycin, trimethoprim-sulfamethoxazole, azithromycin, doxycycline, valacyclovir, and valganciclovir. Because oral formulations of these agents offer clinical efficacy similar to that of parenteral formulations, eliminate risks of catheter-associated complications, and enable earlier patient discharge, oral therapy should be recommended in place of parenteral therapy for patients who are improving.[110] Furthermore, the cost of oral formulations is often 5- to 10-fold less than that of the parenteral forms, which reaps substantial savings.

In "switch therapy" a parenteral formulation is replaced with an oral form of the same drug (eg, intravenous ciprofloxacin is replaced with oral ciprofloxacin); in "step-down therapy" one intravenous agent is replaced by a different oral agent with a similar in vitro spectrum (eg, intravenous ceftriaxone is replaced with oral cefpodoxime or cefdinir). Several controlled trials have demonstrated the safety, efficacy, and significant cost savings associated with intravenous-to-oral conversion programs.[110,111] In our experience, automatic conversion for selected agents is both appropriate and more efficient than making recommendations that must then be implemented by the clinical services; this program alone results in over $200,000 in savings annually at Lahey Clinic.

Antimicrobial Optimization (Streamlining)

Programs that seek to optimize antimicrobial therapy by selecting the most appropriate drug and dosage are increasingly popular. New Food and Drug Administration labeling reflects the time-honored dictum that culture and drug-susceptibility information be considered when selecting and then modifying antibacterial therapy.[30] Antimicrobial therapy should be tailored to provide a narrow spectrum of activity for a specific pathogen, based on in vitro susceptibility data.[112] However, it is often difficult to convince clinicians to switch from broad-spectrum to targeted, narrow-spectrum therapy when the patient is responding clinically. In a careful multinational study of empiric therapy in intensive care units, Aarts et al.[113] found that only 20% of patients suspected to have a hospital-acquired infection were actually infected and that prolonged empiric therapy was both common and associated with a trend toward increased mortality for patients who received it, compared with patients whose empiric therapy was discontinued (multivariate odds ratio for 28-day mortality, 3.8 [95% confidence interval, 0.9–15.5]; $P = .07$).

Many of the strategies discussed here have been advocated as ways to minimize selection of resistant organisms while maximizing individual outcomes.[114]

The science of antibacterial pharmacodynamics describes the relationship between serum and tissue concentrations and the effect of the drug, allowing clinicians to optimize antimicrobial therapy, improve clinical efficacy, and reduce the development of antimicrobial resistance.[115] Investigators at Hartford Hospital have used pharmacodynamic modeling to recommend lower doses

of antibiotics that maintain adequate serum and tissue concentrations, providing clinical outcomes similar to those with Food and Drug Administration–approved dosage regimens.[116] Use of these principles also allows less frequent administration of a drug while providing an unchanged total daily dose.[117] Continuous infusion and once-daily aminoglycoside dosing protocols offer improved efficacy but also reduce pharmacy preparation time and nursing administration time, thus reducing cost.

Antimicrobial Management Teams

In 1988 the IDSA suggested development of antimicrobial advisory teams as a way to improve antibiotic prescribing,[56] and the 2007 IDSA/SHEA guidelines on enhancing antimicrobial stewardship provide a systematic review of the topic.[36] Prospective review of antimicrobial regimens by multidisciplinary teams has since proven to be an effective way to improve use of antibiotics.[118,119] Armed with pharmacokinetic and pharmacodynamic data, teams staffed by pharmacists and/or physicians trained in infectious diseases recommend changes in dosing amounts and frequency, intravenous-to-oral conversion, and discontinuance or streamlining of antibiotic therapy. Physicians are notified of team recommendations by direct communication or by notes left in the medical record. Recommendations from these teams have an acceptance rate, in most studies, of greater than 80%.

Antimicrobial prescribing advice may be helpful in long-term care facilities as well.[120] Recent experience with the spread of KPC-producing organisms (a variety of gram-negative bacilli expressing carbapenemases) in Chicago,[120] New York,[121] and in New Jersey, Delaware, and Pennsylvania[122] suggests that long-term care facilities can be a source of these organisms and a wide palette of other resistant organisms. D'Agata et al.[123] described extensive and aggressive antibiotic use in patients with end-stage dementia, which was associated with high levels of resistance. The value of antimicrobial therapy in such settings should be strongly weighed, considering the broader public health and ethical ramifications of unfettered use.[124] Collaborative communication between referring facilities is essential, warning of the presence of such organisms at the time of patient transfer.

Comprehensive Antimicrobial Management Programs

A multifaceted program at the University of Pennsylvania Hospital encompasses many of the methods used to alter antimicrobial prescribing.[82] Antimicrobial management began at this hospital in 1993 with the goal of improving the quality of patient care by ensuring the effective use of antibiotics. In collaboration with the infection control and infectious diseases departments, the formulary was reviewed and modified, use of vancomycin and broad-spectrum antibiotics was restricted, and empiric treatment guidelines were initiated. In addition, they developed dosing recommendations based on disease state, pharmacokinetics, and pharmacodynamics, along with antibiotic streamlining, continuous education, and monitoring of antibiotic usage. An infectious diseases–trained pharmacist and an infectious diseases physician were responsible for approving restricted drugs prior to use. This program increased the appropriate use of restricted antibiotics and demonstrated significant improvements in patient outcomes, as well as reducing antibiotic and total hospital costs, compared with a control group program that received advice from infectious diseases fellows.[82] This study did not address use of unrestricted antibiotics.

In further work these researchers studied the effect of antibiotic restriction on resistance. Similar methods were successful in reducing third-generation cephalosporin use by 85.8% but had little durable effect on vancomycin use.[125] Despite dramatic overall changes in antibiotic utilization, the prevalence of vancomycin-resistant enterococci continued to increase. Although disappointing, these efforts may simply have been inadequate to counter widespread burgeoning rates of colonization and infection with vancomycin-resistant enterococci. Given the numerous risk factors selecting for drug resistance and the complexity of "bug-drug" combinations,[38] successful elimination of multidrug-resistant organisms will likely require this type of multidisciplinary effort (involving surveillance, infection control, and antimicrobial stewardship), as well as collaborative interventions among broad networks of institutions.[51,126] The gains of smaller, local programs may be more difficult to document, despite their intrinsic value.

Outcomes

Most studies of antimicrobial stewardship programs report reduction of antibiotic costs, and some report significant declines in adverse drug events and errors,[102,103] but few have systematically studied the impact of such programs on bacterial resistance or patient outcomes.[66,78,81] In a program instituted in response to increasing bacterial resistance, Rahal and colleagues[127] observed that restricting use of cephalosporins reduced the incidence of colonization and infection with ceftazidime-resistant *Klebsiella* species by almost 45%. In this study an 80% reduction in cephalosporin use was observed after restrictions were implemented. However, during the same time period, use of imipenem increased 140%, and there was a hospital-wide increase in the number of imipenem-resistant *P. aeruginosa* isolates recovered,[127] followed by an outbreak of infections with multidrug-resistant *Acinetobacter baumannii*.[128] These organisms can now be found in most New York hospitals; most recently, *Klebsiella pneumoniae* that produce carbapenemases have been detected.[129]

These results, examples of "squeezing the balloon,"[130] highlight the hospitals' capacity to alter resistance problems by changing antimicrobial usage, as well as the complexity of hospital ecologies and the need for careful monitoring of antimicrobial use and bacterial resistance after implementing restriction programs. This is compounded by the emergence of community-associated strains of MRSA as hospital-acquired pathogens. Demonstrating an effective intervention is made more difficult as concurrent interventions, such as hand hygiene campaigns, are mandated.[131] Carefully designed multicenter studies will be needed to assess the relative value of various interventions. As Ramsay[49] has commented, 70% of relevant studies from 1980 through 2003 had inadequate methodology.

Buising et al.[83] describe a computerized antibiotic approval program, based on an internal computer network, that was implemented at their tertiary hospital in Melbourne, Australia. Using time-series analysis, they reported a decline in use of third- and fourth-generation cephalosporins, glycopeptides, carbapenems, aminoglycosides, and quinolones, while use of extended-spectrum penicillins increased. They also noted trends toward improved susceptibility of *S. aureus* and *P. aeruginosa*.

Funding

It is important to enumerate the many benefits that antimicrobial improvement programs offer to patients and hospitals.[132] Patients receive more efficient antimicrobial therapy, experience fewer adverse events and drug-resistant infections, and often have shorter hospitalizations.[36] Prescribers gain knowledge of appropriate antibiotic use, and the institution frequently reduces its pharmacy and other expenditures, all while achieving better patient outcomes. Third-party payers, contractors, and other administrators are interested in these improvements as well and may base referral contracts on these indicators of high-quality health care. As Kunin[133] noted, "These programs can save lives as well as money." Yet the expectation that antimicrobial stewardship (or infection control) must save money is unreasonable. Perencevich et al.[134] argue that society should be willing to pay for programs that provide significant value. Regardless, many antimicrobial stewardship programs seem to operate well into the black (and are expected to do so).

In 1987 Woodward and colleagues[66] reported saving $24,620 per month by restricting the formulary and requiring justification for use of restricted agents, spawning numerous subsequent endeavors. In a comprehensive review of cost-containment programs, John and Fishman[65] found that a focus on a single drug could save several thousand dollars, but multidisciplinary interventional strategies could save as much as $500,000 per year. Estimated costs for salary and benefits for an infectious diseases pharmacist and a part-time infectious diseases physician were $104,810, suggesting a 5:1 return on investment. As an example, the program described in John and Fishman[65] improved antibiotic selection and the cure rate while reducing clinical, microbiological, and other failure rates by 80%, compared with the rates with usual prescribing practice, and saved $302,400 per year in antibiotic costs, $533,000 in infection-related costs, and more than $4.25 million in total costs.[43]

Our experience at Lahey Clinic is similar. Our Antimicrobial Management Program is a multidisciplinary effort similar to the University of Pennsylvania Hospital program and is funded through an at-risk arrangement with the hospital, based on projected cost savings. In the 15 years since the advent of our program in 1993,[135] pharmacy expenditures per patient-day for antimicrobial agents, adjusted for inflation and case-mix index, decreased by 30%, whereas the rest of the budget increased by 171%. The proportion of the budget spent on antimicrobials decreased from 13.7% to 4.6%, despite substantial increases in the complexity of services provided and in the case-mix index. Over time, the infectious diseases pharmacist position was intermittently diverted to other duties; antimicrobial expenditures decreased while the stewardship program was active but climbed rapidly during inactive periods. Using 2 alternate methods to predict antimicrobial expenditures with and without a stewardship program, adjusting for inflation and case-mix index, we estimate savings ranging from $2.4 to $8.8 million over 15 years, or $162,250–$588,970 per year, at a cost of less than $110,000 annually. Because the infectious diseases pharmacist position was funded and active for about half the study period, this suggests a roughly 3- to 11-fold return on the dollar, based on pharmacy expenditures alone (R.A.D., unpublished data).

Maintaining and Evaluating the Success of Your Program

It is essential to make administrators aware of the significance of antimicrobial stewardship efforts[132,136] and to periodically reinforce that impression. In addition to pharmacy budget savings, the volume and scope of interventions can be tallied, as well as rates of medication errors and adverse events. Antimicrobial usage and resistance patterns should be monitored to document successes and to provide alerts to new problems. Costs associated with specific diagnosis-related groups may drop if targeted programs result in shorter length of hospitalization. These data should be shared with hospital administrators to reinforce the financial and clinical value of the program.

Benefits can be further expressed in measures of prevented morbidity, mortality, drug resistance, errors, adverse events, and wasted medications, as well as the "added value" of decreased adverse publicity and liability. Shortened length of hospitalization translates to greater opportunity to use those beds for more-productive care. Estimates of savings associated with these parameters may be based on published data, if cost-measuring databases are unavailable. Use of economic models and collaboration with the institution's financial officers can add sophistication to a business plan and can bolster credibility with other members of administration.[134]

Conclusions

Antimicrobial stewardship is a prominent and growing part of local and national efforts to contain and reverse the emergence of antimicrobial resistance. A range of intervention options is available to institutions with varying depths of resources and can yield substantial improvements in morbidity, mortality, safety and quality of care, and cost. The cost of delivering such programs is dwarfed by the benefits and provides an opportunity for hospital epidemiologists to garner support. This suggests that antimicrobial management programs belong to the rarefied group of truly cost-saving quality improvement initiatives; well-designed, multicenter studies evaluating this would be welcome.

Future informatics systems promise greater integration and analysis of data, facilitated delivery of information to the clinician, and rapid and expert decision support that will optimize patient outcomes while minimizing antimicrobial resistance. They may also offer our best hope for avoiding an "antibiotic Armageddon."[133]

References

1. Bryan CS. Strategies to improve antibiotic use. *Infect Dis Clin North Am* 1989;3(4):723–734.
2. Pakyz AL, MacDougall C, Oinonen M, Polk RE. Trends in antibacterial use in US academic health centers, 2002 to 2006. *Arch Intern Med* 2008; 168:2254–2260.

3. Salama S, Rotstein C, Mandell L. A multidisciplinary hospital-based antimicrobial use program: impact on hospital pharmacy expenditures and drug use. *Can J Infect Dis* 1996; 7:104–109.

4. Neu HC, Howrey SP. Testing the physician's knowledge of antibiotic use: self-assessment and learning via videotape. *N Engl J Med* 1975;293:1291–1295.

5. Willemson I, Groenhuijzen A, Bogaers D, Stuurman A, van Keulen P, Kluytmans J. Appropriateness of antimicrobial therapy measured by repeated prevalence surveys. *Antimicrob Agents Chemother* 2007;51:864–867.

6. Maki DG, Schuna A. A study of antimicrobial misuse in a university hospital. *Am J Med Sci* 1978;275:271–282.

7. Hecker MT, Aron DC, Patel NP, et al. Unnecessary use of antimicrobials in hospitalized patients: current patterns of misuse with emphasis on the antianaerobic spectrum of activity. *Arch Intern Med* 2003;163:972–978.

8. Dunagan WC, Woodward RS, Medoff G, et al. Antimicrobial misuse in patients with positive blood cultures. *Am J Med* 1989;87:253–259.

9. Owens RC Jr, Prato BS, Lucas FL, Bachman D, Glenski SL. Impact of a hospital's formulary on outpatient antimicrobial prescribing. In: Program and abstracts of the 41st annual meeting of the Infectious Disease Society of America; October 9–12, 2003; San Diego, CA. Abstract 162.

10. Kunin CM. Problems in antibiotic usage. In: Mandell GL, Douglas RG, Bennett JE, eds. *Principles and Practice of Infectious Diseases*. 3rd ed. New York: Churchill Livingstone; 1990:427–434.

11. Ibrahim EH, Sherman G, Ward S, Fraser VJ, Kollef MH. The influence of inadequate antimicrobial treatment of bloodstream infections on patient outcomes in the ICU setting. *Chest* 2000;118:146–155.

12. McGowan JE Jr. Improving antibiotic use has become essential—can surgery lead the way? *Infect Control Hosp Epidemiol* 1990;11:575–577.

13. Burgmann H, Stoiser B, Heinz G, et al. Likelihood of inadequate treatment: a novel approach to evaluating drug-resistance patterns. *Infect Control Hosp Epidemiol* 2009; 30:672–677.

14. Neu HC. The crisis in antibiotic resistance. *Science* 1992; 257:1064–1073.

15. McGowan JE Jr. Antimicrobial resistance in hospital organisms and its relation to antibiotic use. *Rev Infect Dis* 1983; 5:1033–1048.

16. Rogues AM, Dumartin C, Amadéo B, et al. Relationship between rates of antimicrobial consumption and the incidence of antimicrobial resistance in *Staphylococcus aureus* and *Pseudomonas aeruginosa* isolates from 47 French hospitals. *Infect Control Hosp Epidemiol* 2007;28: 1389–1395.

17. Roberts RR, Scott RD, Cordell R, et al. The use of economic modeling to determine the hospital cost associated with nosocomial infections. *Clin Infect Dis* 2003;36: 1424–1432.

18. Burke JP. Infection control: a problem for patient safety. *N Engl J Med* 2003;348:651–656.

19. Engemann JJ, Carmeli Y, Cosgrove SE, et al. Adverse clinical and economic outcomes attributable to methicillin resistance among patients with *Staphylococcus aureus* surgical site infection. *Clin Infect Dis* 2003;36:592–598.

20. Cosgrove S, Carmeli Y. The impact of antimicrobial resistance on health and economic outcomes. *Clin Infect Dis* 2003; 36:1433–1437.

21. McDonald LC. Trends in antimicrobial resistance in healthcare-associated pathogens and effect on treatment. *Clin Infect Dis* 2006;42(suppl):S65–S71.

22. Hidron AI, Edwards JR, Patel J, et al. Antimicrobial-resistant pathogens associated with healthcare-associated infections: annual summary of data reported to the National Healthcare Safety Network at the Centers for Disease Control and Prevention, 2006–2007. *Infect Control Hosp Epidemiol* 2008;29:996–1011.

23. Centers for Disease Control and Prevention (CDC). *Staphylococcus aureus* resistant to vancomycin—United States, 2002. *MMWR Morb Mortal Wkly Rep* 2002; 51:565–567.

24. Lautenbach E, Polk RE. Resistant gram-negative bacilli: a neglected healthcare crisis? *Am J Health Syst Pharm* 2007; 64(suppl):S3–S21.

25. Schwaber MJ, Carmeli Y. Carbapenem-resistant Enterobacteriaceae: a potential threat. *JAMA* 2008;300:2911–2913.

26. Spellberg B, Guidos R, Gilbert D, et al. The epidemic of antibiotic-resistant infections: a call to action for the medical community from the Infectious Diseases Society of America. *Clin Infect Dis* 2008;46:155–164.

27. Talbot GH, Bradley J, Edwards JE Jr, Gilbert D, Scheld M, Bartlett JG. Bad bugs need drugs: an update on the development pipeline from the Antimicrobial Availability Task Force of the Infectious Diseases Society of America. *Clin Infect Dis* 2006;42:657–668.

28. Shlaes DM, Gerding DN, John JF, et al. Society for Healthcare Epidemiology of America and Infectious Diseases Society of America Joint Committee on the Prevention of Antimicrobial Resistance. Guidelines for the prevention of antimicrobial resistance in hospitals. *Clin Infect Dis* 1997; 25:584–599.

29. Centers for Disease Control and Prevention (CDC). Campaign to prevent antimicrobial resistance in healthcare settings. http://www.cdc.gov/drugresistance/healthcare/overview.htm. Accessed July 5, 2009.

30. Food and Drug Administration, Department of Health and Human Services. Labeling requirements for systemic antibacterial drug products for human use. *Fed Regist* 2003; 68(25):6062–6081.

31. World Health Organization (WHO). Global strategy for containment of antimicrobial resistance. http://www.who.int/csr/resources/publications/drugresist/EGlobal_Strat.pdf. Accessed July 1, 2003.

32. The Joint Commission. 2009 National Patient Safety Goals. http://www.jointcommission.org/patientsafety/nationalpatientsafetygoals/. Accessed July 6, 2009.

33. Rosenthal MB. Nonpayment for performance? Medicare's new reimbursement rule. *N Engl J Med* 2007;357: 1573–1575.

34. Gerding DN. Good antimicrobial stewardship in the hospital setting: fitting, but flagrantly flagging. *Infect Control Hosp Epidemiol* 2000;21(4):253–235.

35. Barlam TF, DiVall M. Antibiotic-stewardship practices at top academic centers throughout the United States and at hospital throughout Massachusetts. *Infect Control Hosp Epidemiol* 2006;27:695–703.

36. Dellit TH, Owens RC, McGowan JE Jr, et al. Infectious Diseases Society of America and the Society for Healthcare Epidemiology of America guidelines for developing an institutional program to enhance antimicrobial stewardship. *Clin Infect Dis* 2007;44:159–177.

37. Austin DJ, Kristinsson KG, Anderson RM. The relationship between the volume of antimicrobial consumption in human communities and the frequency of resistance. *Proc Natl Acad Sci USA* 1999;96:1152–1156.

38. Safdar N, Maki DG. The commonality of risk factors for nosocomial colonization and infection with antimicrobial-resistant *Staphylococcus aureus, Enterococcus,* gram-negative bacilli, *Clostridium difficile,* and *Candida. Ann Intern Med* 2002;136:834–844.

39. Fridkin SK, Edwards JR, Courval JM, et al. The effect of vancomycin and third-generation cephalosporins on prevalence of vancomycin-resistant enterococci in 126 US adult intensive care units. *Ann Intern Med* 2001;135: 175–183.

40. Monnet DL. Methicillin-resistant *Staphylococcus aureus* and its relationship to antimicrobial use: possible implications for control. *Infect Control Hosp Epidemiol* 1998;19:552–559.

41. Pestotnik SL, Classen DL, Evans RS, Burke JP. Implementing antibiotic practice guidelines through computer-assisted decision support: clinical and financial outcomes. *Ann Intern Med* 1996;124:884–890.

42. Lautenbach E, LaRosa LA, Marr AM, Nachamkin I, Bilker WB, Fishman NO. Changes in the prevalence of vancomycin-resistant enterococci in response to antimicrobial formulary interventions: impact of progressive restrictions on use of vancomycin and third-generation cephalosporins. *Clin Infect Dis* 2003;36:440–446.

43. Fishman N. Antimicrobial stewardship. *Am J Med* 2006; 119(suppl):S53–S61.

44. Owens RC Jr. Antimicrobial stewardship: programmatic efforts to optimize antimicrobial use. In: Jarvis WR, ed. *Bennett and Brachman's Hospital Infections.* 5th ed. Philadelphia: Lippincott Williams & Wilkins; 2007:179–192.

45. Duncan RA. Controlling use of antimicrobial agents. *Infect Control Hosp Epidemiol* 1997;18:260–266.

46. Pagani L, Gyssens IC, Huttner B, Nathwani D, Harbarth S. Navigating the Web in search of resources on antimicrobial stewardship in health care institutions. *Clin Infect Dis* 2009; 48:626–632.

47. Johnson LE, Reyes K, Zervos MJ. Resources for infection prevention and control on the World Wide Web. *Clin Infect Dis* 2009;48:1585–1595.

48. Yokoe DS, Mermel LA, Anderson DJ, et al. Executive summary: a compendium of strategies to prevent healthcare-associated infection in acute care hospitals. *Infect Control Hosp Epidemiol* 2008;29(suppl):S12–S21.

49. Ramsay C, Brown E, Hartman G, Davey P. Room for improvement: a systematic review of the quality of evaluations of interventions to improve hospital antibiotic prescribing. *J Antimicrob Chemother* 2003;52:764–771.

50. Davey P, Brown E, Fenelon L, et al. Systematic review of antimicrobial drug prescribing in hospitals. *Emerg Infect Dis* 2006;12(2):211–216.

51. Goldmann DA, Weinstein RA, Wenzel RP, et al. Strategies to prevent and control the emergence and spread of antimicrobial-resistant microorganisms in hospitals: a challenge to hospital leadership. *JAMA* 1996;275:234–240.

52. Rosenthal VD, Maki DG, Salomao R, et al. Device-associated nosocomial infections in 55 intensive care units of 8 developing countries. *Ann Intern Med* 2006;145:582–591.

53. Pittet D, Donaldson L. Clean care is safer care: the first global challenge of the who world alliance for patient safety. *Infect Control Hosp Epidemiol* 2005;26:891–894.

54. Rifenburg RP, Paladine JA, Hanson SC, et al. Benchmark analysis of strategies hospitals use to control antimicrobial expenditures. *Am J Health Syst Pharm* 1996;53:2054–2062.

55. Carling P, Fung T, Killion A, Terrin N, Barza M. Favorable impact of a multidisciplinary antibiotic management program conducted during 7 years. *Infect Control Hosp Epidemiol* 2003;24:699–706.

56. Marr JJ, Moffet HL, Kunin CM. Guidelines for improving the use of antimicrobial agents in hospitals: a statement by the Infectious Disease Society of America. *J Infect Dis* 1988;157:869–876.

57. Centers for Disease Control and Prevention (CDC). Effects of new penicillin susceptibility breakpoints for *Streptococcus pneumoniae*—United States, 2006—2007. *MMWR Morb Mortal Wkly Rep* 2008;57:1353–1355.

58. Craven DE, Duncan RA. Preventing ventilator-associated pneumonia: tiptoeing through a minefield. *Am J Resp Crit Care Med* 2006;173:1297–1299.

59. Chastre J, Wolff M, Fagon JY, et al. Comparison of 8 vs 15 days of antibiotic therapy for ventilator-associated pneumonia in adults: a randomized trial. *JAMA* 2003;290: 2588–2598.

60. Zhanel GG, Decorby M, Nichol KA, et al. Molecular characterization of methicillin-resistant *Staphylococcus aureus,* vancomycin-resistant enterococci and extended-spectrum β-lactamase–producing *Escherichia coli* in intensive care units in Canada: results of the Canadian National Intensive Care Unit (CAN-ICU) study (2005–2006). *Can J Infect Dis Med Microbiol* 2008;19:243–249.

61. Fishman NO. Antimicrobial management and cost containment. In: Mandell GL, Bennett JE, Dolin R, eds. *Principles and Practice of Infectious Diseases.* 5th ed. New York: Churchill Livingstone; 2000:539–546.

62. Wester CW. Antibiotic resistance: a survey of physician perceptions. *Arch Intern Med* 2002;162:2210.

63. American Thoracic Society and the Infectious Diseases Society of America Guideline Committee. Guidelines for the management of adults with hospital-acquired, ventilator-associated, and health care–associated pneumonia. *Am J Respir Crit Care Med* 2005;171:388–416.

64. Surgical Care Improvement Project (SCIP) Web site. http://www.SCIPpartnership@okqio.sdps.org. Accessed July 6, 2009.

65. John JF, Fishman NO. Programmatic role of the infectious disease physician in controlling antimicrobial cost in the hospital. *Clin Infect Dis* 1997;24:471–485.

66. Woodward RS, Medoff G, Smith MD, Gray JL 3rd. Antibiotic cost savings from formulary restrictions and physician monitoring in a medical school–affiliated hospital. *Am J Med* 1987;83(5):817–823.

67. Klapp DL, Ramphal R. Antibiotic restrictions in hospitals associated with medical schools. *Am J Hosp Pharm* 1983;40: 1957–1960.

68. Pépin J, Saheb N, Coulombe MA, et al. Emergence of fluoroquinolones as the predominant risk factor for *Clostridium difficile*–associated diarrhea: a cohort study during an epidemic in Quebec. *Clin Infect Dis* 2005;41:1254–1260.

69. Valiquette L, Cossette B, Garant M-P, Pepin J. Impact of a reduction in the use of high-risk antibiotics on the course of an epidemic of *Clostridium difficile*–associated disease caused by the hypervirulent NAP1/027 strain. *Clin Infect Dis* 2007; 45(suppl):S112–S121.

70. Strausbaugh LJ, Jernigan DB, Liedtke LA, et al. National shortages of antimicrobial agents: results of 2 surveys from the Infectious Disease Society of America Emerging Infections Network. *Clin Infect Dis* 2001;33:1495–1501.

71. Polk RE, Nichols M, Johnson CK. Hospitals with "open" antibiotic formularies may be associated with lower rates of bacterial resistance: from the SCOPE-MMIT Antimicrobial Surveillance Network. In: Program and abstracts of the 42nd Interscience Conference on Antimicrobial Agents and Chemotherapy, American Society for Microbiology; September 27–30, 2002; San Diego, CA. Abstract K-1352.

72. Bonhoeffer S, Lipsitch M, Levin BR. Evaluating treatment protocols to prevent antibiotic resistance. *Proc Natl Acad Sci USA* 1997;94:12106–12111.

73. Bergstrom CT, Lo M, Lipsitch M. Ecological theory suggests that antimicrobial cycling will not reduce antimicrobial resistance in hospitals. *Proc Natl Acad Sci USA* 2004;101: 13285–13290.

74. Gilbert DN, Moellering RC Jr, Eliopoulos GM, Chambers HF, Saag MS, eds. *The Sanford Guide to Antimicrobial Therapy*. 39th ed. Sperryville, VA: Antimicrobial Therapy; 2009.

75. The Johns Hopkins ABx Guide Web page. http://www. hopkins-abxguide.org/. Accessed July 7, 2009.

76. Avorn J, Solomon DH. Cultural and economic factors that (mis)shape antibiotic use: the non-pharmacologic basis of therapeutics. *Ann Intern Med* 2000;133:128–135.

77. McGowan JE Jr, Finland M. Usage of antibiotics in a general hospital: effect of requiring justification. *J Infect Dis* 1974; 130:165–168.

78. Coleman RW, Rodondi LC, Kaubisch S, et al. Cost-effectiveness of prospective and continuous parenteral antibiotic control: experience at the Palo Alto Veterans Affairs Medical Center from 1987 to 1989. *Am J Med* 1991; 90:439–444.

79. Ahern JW, Grace CJ. Effectiveness of a criteria-based educational program for appropriate use of antibiotic. *Infect Med* 2002;19:364–374.

80. Regal RE, DePestel DD, VandenBussche HL. The effect of an antimicrobial restriction program on *Pseudomonas aeruginosa* resistance to β-lactams in a large teaching hospital. *Pharmacotherapy* 2003;23(5):618–624.

81. White AC, Atmar RL, Wilson J, et al. Effects of requiring prior authorization for selected antimicrobials: expenditures, susceptibilities, and clinical outcomes. *Clin Infect Dis* 1997; 25:230–239.

82. Gross R, Morgan AS, Kinky DE, et al. Impact of a hospital-based antimicrobial management program on clinical and economic outcomes. *Clin Infect Dis* 2001; 33:289–295.

83. Buising KL, Thursky KA, Robertson MB, et al. Electronic antibiotic stewardship—reduced consumption of broad-spectrum antibiotics using a computerized antimicrobial approval system in a hospital setting. *J Antimicrob Chemother* 2008;62:608–616.

84. Kunin CM. The responsibility of the infectious disease community for the optimal use of antimicrobial agents. *J Infect Dis* 1985;151(3):388–398.

85. Paterson DL. Restrictive antibiotic policies are appropriate in intensive care units. *Crit Care Med* 2003;31(suppl 1): S25–S28.

86. Lawrence KL, Kollef MH. Antimicrobial stewardship in the intensive care unit: advances and obstacles. *Am J Respir Crit Care Med* 2009;179:434–438.

87. Horn E, Jacobi J. The critical care clinical pharmacist: evolution of an essential team member. *Crit Care Med* 2006; 34(suppl):S46–S51.

88. MacLaren R, Bond CA, Martin SJ, Fike D. Clinical and economic outcomes of involving pharmacists in the direct care of critically ill patients with infections. *Crit Care Med* 2008; 36:3184–3189.

89. Linkin DR, Paris S, Fishman NO, Metlay JP, Lautenbach E. Inaccurate communications in telephone calls to an antimicrobial stewardship program. *Infect Control Hosp Epidemiol* 2006;27:688–694.

90. Linkin DR, Fishman NO, Landis JR, et al. Effect of communication errors during calls to an antimicrobial stewardship program. *Infect Control Hosp Epidemiol* 2007;28: 1374–1381.

91. LaRosa LA, Fishman NO, Lautenbach E, Koppel RJ, Morales KH, Linkin DR. Evaluation of antimicrobial therapy orders circumventing an antimicrobial stewardship program: investigating the strategy of "stealth dosing". *Infect Control Hosp Epidemiol* 2007;28:531–536.

92. Lesar TS, Briceland LL. Survey of antibiotic control policies in university-affiliated teaching institutions. *Ann Pharmacother* 1996;30(1):31–34.

93. Durbin WA, Lapidas B, Goldmann DA. Improved antibiotic usage following introduction of a novel prescription system. *JAMA* 1981;246:1796–1800.

94. McDonald M, Grabsch E, Marshall C, Forbes A. Single- versus multiple-dose antimicrobial prophylaxis for major surgery: a systematic review. *Aust N Z J Surg* 1998;68: 388–396.

95. Engelman R, Shahian D, Shemin R, et al. The Society of Thoracic Surgeons practice guideline series: antibiotic prophylaxis in cardiac surgery, part II: antibiotic choice. *Ann Thorac Surg* 2007;83:1569–1576.

96. Bratzler DW, Houck PM. Antimicrobial prophylaxis for surgery: an advisory statement from the National Surgical Infection Prevention Project. *Clin Infect Dis* 2004;38: 1706–1715.

97. Classen DC, Evans RS, Pestotnik SL, Horn SD, Menlove RL, Burke JP. The timing of prophylactic administration of antibiotics and the risk of surgical-wound infection. *N Engl J Med* 1992;326:281–286.

98. Centers for Medicare and Medicaid Services. Fact sheet: proposals to improve quality of care in inpatient stays in acute care hospitals in FY 2010. May 1, 2009. http://www.cms. hhs.gov/apps/media/fact_sheets.asp. Accessed July 1, 2009.

99. The Joint Commission. Performance Measurement. http:// www.jointcommission.org/PerformanceMeasurement. Accessed July 1, 2009.

100. Wolf LS Jr, Bennett CJ, Dmochowski RR, Hollenbeck BK, Pearle MS, Schaeffer AJ. Best practice policy statement on urologic surgery antimicrobial prophylaxis. *J Urol* 2008; 179:1379–1390.

101. Kaushal R, Shojania K, Bates DW. Effects of computerized physician order entry and clinical decision support systems on medication safety. *Arch Intern Med* 2003;163:1409–1416.

102. Evans RS, Pestotnik SL, Classen DC, et al. A computer-assisted management program for antibiotics and other anti-infective agents. *N Engl J Med* 1998;338:232–238.

103. Pestotnik SL, Classen DL, Evans RS, Burke JP. Implementing antibiotic practice guidelines through computer-assisted decision support: clinical and financial outcomes. *Ann Intern Med* 1996;124:884–890.

104. TheraDoc Web page. http://www.theradoc.com. Accessed Spetember 1, 2009.

105. Durieux P, Trinquart L, Colombet I, et al. Computerized advice on drug dosage to improve prescribing practice. *Cochrane Database of Systematic Reviews* 2009, Issue 2. http://www.thecochranelibrary.com. Accessed September 1, 2009.

106. Fridkin SK. Routine cycling of antimicrobial agents as an infection control measure. *Clin Infect Dis* 2003;36:1438–1444.

107. Kollef MH, Vlasnik J, Sharpless L, et al. Scheduled change of antibiotic classes: a strategy to decrease the incidence of ventilator-associated pneumonia. *Am J Respir Crit Care Med* 1997;156:1040–1048.

108. Gruson D, Hilbert G, Vargas F, et al. Rotation and restricted use of antibiotics in a medical intensive care unit. *Am J Respir Crit Care Med* 2000;162:837–843.

109. Johnsen PJ, Townsend JP, Bohn T, Simonsen GS, Sundsfjord A, Nielsen KM. Factors affecting the reversal of antimicrobial-drug resistance. *Lancet Infect Dis* 2009;9:357–364.

110. Ramirez JA, Vargas S, Ritter GW, et al. Early switch from intravenous to oral antibiotics and early hospital discharge: a prospective observational study of 200 consecutive patients with community acquired pneumonia. *Arch Intern Med* 1999;159:2449–2454.

111. Kuti JL, Le TN, Nightingale CH, et al. Pharmacoeconomics of a pharmacist-managed program for automatically converting levofloxacin route from IV to oral. *Am J Health Syst Pharm* 2002;59(22):2209–2215.

112. Briceland LL, Nightingale CH, Quintiliani R, et al. Antibiotic streamlining from combination therapy to monotherapy utilizing an interdisciplinary approach. *Arch Intern Med* 1988;148:2019–2022.

113. Aarts M-A W, Brun-Buisson C, Cook DJ, et al. Antibiotic management of suspected nosocomial ICU-acquired infection: does prolonged empiric therapy improve outcome? *Intensive Care Med* 2007;33:1369–1378.

114. Paterson DL, Rice LB. Empirical antibiotic choice for the seriously ill patient: are minimization of selection of resistant organisms and maximization of individual outcome mutually exclusive? *Clin Infect Dis* 2003;36:1006–1012.

115. Craig WA. Pharmacokinetic/pharmacodynamic parameters: rationale for antibacterial dosing of mice and men. *Clin Infect Dis* 1998;26:1–12.

116. Kuti JL, Maglio D, Nightingale CH, et al. Economic benefit of a meropenem dosing strategy based on pharmacodynamic concepts. *Am J Health Syst Pharm* 2003;60:565–568.

117. Kim MK, Capitano B, Mattoes HM, et al. Pharmacokinetic and pharmacodynamic evaluations of two dosing regimens for piperacillin-tazobactam. *Pharmacotherapy* 2002;22(5):569–577.

118. Fraser GL, Stogsdill P, Dickens JD, et al. Antibiotic optimization: an evaluation of patient safety and economic outcomes. *Arch Intern Med* 1997;157:1689–1694.

119. Gums JG, Yancey RW, Hamilton CA, et al. A randomized, prospective study measuring outcomes after antibiotic therapy intervention by a multidisciplinary consult team. *Pharmacotherapy* 1999;19(12):1369–1377.

120. Munoz-Price LS. Long-term acute care hospitals. *Clin Infect Dis* 2009;49:438–443.

121. Urban C, Bradford PA, Tuckman M, et al. Carbapenem-resistant *Escherichia coli* harboring *Klebsiella pneumoniae* carbapenemase β-lactamases associated with long-term care facilities. *Clin Infect Dis* 2008;46:e127–e130.

122. Lautenbach E, Marsicano R, Tolomeo P, Heard M, Serrano S, Stieritz DD. Epidemiology of antimicrobial resistance among gram-negative organisms recovered from patients in a multistate network of long-term care facilities. *Infect Control Hosp Epidemiol* 2009;30:790–793.

123. D'Agata E, Mitchell SL. Patterns of antimicrobial use among nursing home residents with advanced dementia. *Arch Intern Med* 2008;168(4):357–362.

124. Schwaber MJ, Carmeli Y. Antibiotic therapy in the demented elderly population: redefining the ethical dilemma. *Arch Intern Med* 2008;168:349–350.

125. Lautenbach E, LaRosa LA, Marr AM, Nachamkin I, Bilker WB, Fishman NO. Changes in the prevalence of vancomycin-resistant enterococci in response to antimicrobial formulary interventions: impact of progressive restrictions on use of vancomycin and third-generation cephalosporins. *Clin Infect Dis* 2003;36:440–446.

126. Weinstein RA. Controlling antimicrobial resistance in hospitals: infection control and use of antibiotics. *Emerg Infect Dis* 2001;7(2):188–192.

127. Rahal JJ, Urban C, Horn D, et al. Class restriction of cephalosporin use to control total cephalosporin resistance in nosocomial *Klebsiella*. *JAMA* 1998;280:133–137.

128. Rahal JJ, Urban C, Segal-Maurer S. Nosocomial antibiotic resistance in multiple gram-negative species: experience at one hospital with squeezing the resistance balloon at multiple sites. *Clin Infect Dis* 2002;34:499–503.

129. Bratu S, Landman D, Haag R, et al. Rapid spread of carbapenem-resistant *Klebsiella pneumoniae* in New York City: a new threat to our antibiotic armamentarium. *Arch Intern Med* 2005;165:1430–1435.

130. Burke JP. Antibiotic resistance—squeezing the balloon? *JAMA* 1998;280:1270.

131. Hota B. Infection control or formulary control: what is the best tool to reduce nosocomial infections due to methicillin-resistant *Staphylococcus aureus*? *Clin Infect Dis* 2006;42:785–787.

132. McQuillen DP, Petrak RM, Wasserman RB, Nahass RG, Scull JA, Martinelli LP. The value of infectious diseases specialists: non–patient care activities. *Clin Infect Dis* 2008;47:1051–1063.

133. Kunin CM. Antibiotic armageddon. *Clin Infect Dis* 1997;25:240–241.

134. Perencevich EN, Stone PW, Wright SB, Carmeli Y, Fisman DN, Cosgrove SE. Raising standards while watching the bottom line: making a business case for infection control. *Infect Control Hosp Epidemiol* 2007;28:1121–1133.

135. Duncan RA, Segarra M, Anderson ER, Chow LS, Needham C, Jacoby GA. A comprehensive approach to reforming therapeutic antibiotic (Abx) use. *Infect Control Hosp Epidemiol* 1995;16(suppl):37. Abstract M7.

136. Petrak RM, Sexton DJ, Butera ML, et al. The value of an infectious diseases specialist. *Clin Infect Dis* 2003;36:1013–1017.

Special Topics

Chapter 21 Improving Hand Hygiene in Healthcare Settings

John M. Boyce, MD, and Didier Pittet, MD, MS

In the mid-1800s, studies by Ignaz Semmelweis in Vienna and by Oliver Wendell Holmes in Boston established that hospital-acquired diseases, now known to be caused by infectious agents, were transmitted by way of the hands of healthcare workers (HCWs). A prospective controlled trial conducted in a hospital nursery and investigations conducted during the last 40 years have confirmed the important role that contaminated HCWs' hands play in the transmission of healthcare-associated pathogens.[1,2] As a result, handwashing has for many years been considered one of the most important measures for preventing the spread of pathogens in healthcare settings.[3]

Despite the fact that handwashing guidelines were published by the Centers for Disease Control and Prevention (CDC) in 1985 and by the Association for Professionals in Infection Control and Epidemiology (APIC) in 1988 and 1995, HCWs' rate of adherence to recommended handwashing procedures remained unacceptably low for decades. Among 34 published observational surveys of hand hygiene adherence among HCWs published between 1981 and 2000, mean rates of adherence ranged from 5% to 81% of indicated opportunities, with an average of about 40% (Figure 21-1). These low rates of compliance persisted despite the fact that multidrug-resistant pathogens, such as methicillin-resistant *Staphylococcus aureus*, vancomycin-resistant enterococci (VRE) and resistant gram-negative bacilli, are transmitted primarily on the hands of HCWs. Continued poor handwashing adherence among HCWs and the increasing spread of multidrug-resistant healthcare-associated pathogens suggested the need to study the causes of poor

handwashing practices among HCWs and to develop new strategies for improving them.

In the largest study of its kind, Pittet and colleagues[5] observed more than 2,800 opportunities for hand hygiene in a large hospital and utilized multivariate analysis to determine factors associated with poor compliance among HCWs. Factors associated with poor compliance included a high intensity of care (ie, many opportunities for hand hygiene per hour of patient care), it being a weekday (ie, Monday through Friday), work location in an intensive care unit, and performance of a procedure associated with a high risk of contamination. Poor adherence associated with a high intensity of care suggested that HCWs did not have enough time to wash their hands as frequently as recommended, a complaint registered by nurses in previous studies.

In fact, a study by Voss and Widmer[6] revealed that it took intensive care unit nurses an average of 62 seconds to leave a patient's bedside, find a sink, wash and dry their hands, and return to patient care activities. These investigators estimated that in an intensive care unit with 12 nurses on duty each shift, achieving 100% compliance with recommended handwashing practices would require 16 hours of nursing time per shift. In contrast, if bedside dispensers of an alcohol-based hand rub were available, achieving 100% compliance with recommended hand hygiene practices would require only 4 hours of nursing time per shift. Other factors associated with poor adherence to recommended handwashing practices include irritant contact dermatitis caused by frequent exposure to soap and water, lack of convenient access to handwashing facilities (eg, sinks, paper towels),

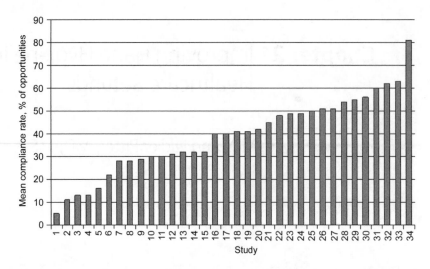

Figure 21-1. Handwashing compliance rates among healthcare workers in 34 observational surveys, 1981–2000. Adapted from the Hand Hygiene Resource Center,[4] with permission.

lack of awareness of recommended practices by HCWs, and lack of administrative concern.[7-10]

Alcohol-based hand rubs (whether gel, rinse, or foam) have characteristics that address a number of the problems associated with washing hands with soap and water (Table 21-1). At least 20 published studies have shown that use of an alcohol solution reduces bacterial counts on the hands of volunteers to a greater degree than washing hands with non-antimicrobial soap and water, and, in all but 2 studies, use of an alcohol product was more effective than washing hands with anti-microbial soap and water.[11] Several clinical trials involving HCWs have documented that cleaning hands with well-formulated alcohol-based hand rubs containing emollients causes less skin irritation and dryness with repeated use than does washing hands with soap and water.[12-14] Also, because alcohol-based hand rubs do not require the use of water, they can be made available at many more locations than handwashing facilities. Finally, making such products readily accessible (near each patient's bedside or in pocket-sized bottles) has been shown to improve HCWs' adherence to recommended hand hygiene practices.[15] In intensive care units, where the workload is high, with corresponding elevated numbers of indications for hand hygiene per hour of patient care,[5] the preferential use of hand rubs can reduce the impact of time constraints and subsequently improve compliance.[16]

However, making alcohol-based hand rub products available in patient care areas, by itself, is not sufficient to improve hand hygiene practices among HCWs.[17] Several studies have demonstrated that multimodal, multidisciplinary promotional campaigns are needed to achieve long-lasting improvements in hand hygiene compliance. Such programs must include education, performance feedback to HCWs, innovative motivational material, and appropriate administrative support.[10,11,15,18-20]

To address the issue of poor handwashing compliance, the Healthcare Infection Control Practices Advisory Committee (HICPAC) of the CDC, in conjunction with the Society for Healthcare Epidemiology of America (SHEA), APIC, and the Infectious Diseases Society of America (IDSA) developed and published the "Guideline for Hand Hygiene in Healthcare Settings" in 2002.[11] The guideline was intended for adoption primarily by healthcare institutions located in the United States and other countries with similar resources.

Subsequently, the World Health Organization (WHO) identified the promotion of hand hygiene practices in health care as a priority measure for reducing healthcare-associated infections worldwide, and it launched the First Global Patient Safety Challenge in October 2005. With the assistance of more than 100 international experts, the WHO World Alliance for Patient Safety conducted systematic reviews of the literature using PubMed, Ovid, MEDLINE, Embase, and the Cochrane Library, reviewed national and international

Table 21-1. Advantages of using alcohol-based hand rub, compared with traditional soap-and-water handwashing

Takes less time to use

Can be made much more accessible because it does not require the use of sinks: can be made available at many locations in patient care areas, and can even be supplied in pocket-sized bottles

Causes less skin irritation and dryness; virtually all current products designed for use by health care workers contain emollients that help prevent dry, irritated skin

Reduces bacterial counts on hands to a greater degree

Does not require the use of water or paper towels

guidelines and textbooks, formed task forces dedicated to specific topics, and held 3 consultations of a core group of experts at WHO headquarters in Geneva, Switzerland. In April 2006, the WHO World Alliance for Patient Safety issued the advanced draft of the WHO guidelines on hand hygiene in health care.[21]

In conjunction with the preparation of the Advanced Draft of the WHO guidelines,[21] an multimodal implementation strategy was developed,[22] together with a variety of tools (Pilot Implementation Pack) designed to help healthcare settings translate the guidelines into practice. A key element of the implementation strategy is the concept developed by Sax et al.,[23] "My 5 moments for hand hygiene" (Figure 21-2), a graphic that facilitates the understanding of the indications for hand hygiene during a patient care episode. In accordance with WHO policies regarding guideline preparation, the recommendations of the WHO guidelines Advanced Draft[21] were tested together with their implementation strategy and related tools in 8 pilot healthcare settings in 7 countries representing all WHO regions. These tests had the following objectives: to provide local data on the resources required to carry out the recommendations; to generate information on the feasibility, validity, reliability, and cost-effectiveness of the interventions; and to adapt and refine proposed implementation strategies. In 2008, updated evidence, perspectives gained from pilot test sites, and input from 2 additional expert consultations were incorporated into the guidelines. Following input from external and internal reviewers, the final version of the WHO guidelines was published in 2009.[24]

The WHO guidelines[24] provide HCWs, hospital administrators, and health authorities with a thorough review of evidence on hand hygiene in health care and specific recommendations to improve practices and reduce the transmission of healthcare-associated pathogens. The guidelines are intended to be implemented in any situation in which health care is delivered either to a patient or to a specific group in a population and in all settings where health care is continuously or occasionally performed, including during home care by birth attendants. They expand on earlier international and national guidelines: they bring a global perspective; they aim to bridge the gap between developing and developed countries, irrespective of resources available; and their feasibility has been tested in settings with different cultural backgrounds and levels of development. The WHO guidelines[24] address a number of unique aspects of hand hygiene, including religious and cultural aspects, promotion on a national scale, social marketing, safety issues, infrastructures required for hand hygiene, and new strategies for improvement.[25] Work related to the production of these guidelines and related tools and results from interventions has been summarized in more than 50 peer-reviewed articles published between 2005 and 2009.[26]

The WHO guidelines[24] and the associated WHO Multimodal Hand Hygiene Improvement Strategy and Implementation Toolkit are designed to offer health-care facilities in member states a conceptual framework and practical tools for implementing recommendations in all types of healthcare settings. Each of the consensus recommendations was categorized on the basis of existing scientific data, theoretical rationale, applicability, and economic impact, using criteria developed by HICPAC. The WHO guidelines and the accompanying implementation toolkit were issued by WHO Patient Safety in May 2009 on the occasion of the launch of the "Save Lives: Clean Your Hands" initiative.[27] The consensus recommendations are listed in the final section of this chapter.

Implementing a Hand Hygiene Improvement Program

Administrative Support

In order to implement an effective hand hygiene promotional campaign, it is imperative to seek support from high-level administrators who can make the necessary resources available. In the past, many administrators and purchasing staff sought to acquire the least-expensive soap for use by HCWs, despite the fact that frequent use of such products sometimes causes considerable irritant contact dermatitis. In order to convince administrators that purchasing alcohol-based hand rub products is worthwhile, it is often helpful to make a presentation to the administration regarding the high cost of healthcare-associated infections.[28] There is no single original scientific study that has dealt specifically with the subject of the cost-effectiveness of different hand hygiene products or the exact financial benefit of increased hand hygiene compliance in the hospital setting. A few recent articles roughly estimate some of the financial effects of hand hygiene and describe the potential benefits of using alcohol-based hand antisepsis compared with soap and water. Not surprisingly, the current consensus is that the modest increases in costs for alcohol-based hand hygiene products are tiny in comparison to the excess hospital costs and years of life lost that are associated with severe nosocomial infections. According to Boyce,[29] if the frequent use of an alcohol-based hand rub by HCWs prevents only a few surgical site infections or bloodstream infections per year, the savings accrued by preventing such infections will be greater than the institution's entire annual budget for hand hygiene products. Webster and colleagues[30] reported a cost saving of approximately US$17,000 resulting from the reduction

Your 5 Moments
for Hand Hygiene

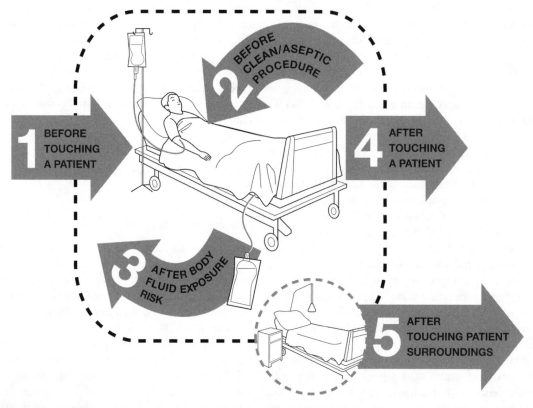

1	BEFORE TOUCHING A PATIENT	WHEN?	Clean your hands before touching a patient when approaching him/her.
		WHY?	To protect the patient against harmful germs carried on your hands.
2	BEFORE CLEAN/ ASEPTIC PROCEDURE	WHEN?	Clean your hands immediately before performing a clean/aseptic procedure.
		WHY?	To protect the patient against harmful germs, including the patient's own, from entering his/her body.
3	AFTER BODY FLUID EXPOSURE RISK	WHEN?	Clean your hands immediately after an exposure risk to body fluids (and after glove removal).
		WHY?	To protect yourself and the health-care environment from harmful patient germs.
4	AFTER TOUCHING A PATIENT	WHEN?	Clean your hands after touching a patient and her/his immediate surroundings, when leaving the patient's side.
		WHY?	To protect yourself and the health-care environment from harmful patient germs.
5	AFTER TOUCHING PATIENT SURROUNDINGS	WHEN?	Clean your hands after touching any object or furniture in the patient's immediate surroundings, when leaving – even if the patient has not been touched.
		WHY?	To protect yourself and the health-care environment from harmful patient germs.

World Health Organization

Patient Safety
A World Alliance for Safer Health Care

SAVE LIVES
Clean **Your** Hands

May 2009

Figure 2. Poster "My 5 moments for hand hygiene." Reproduced with permission from Sax et al.[23]

in the use of vancomycin after the observed decrease in the incidence of colonization or infection with methicillin-resistant *S. aureus* in a neonatal intensive care unit during a 7-month period.

The hand hygiene promotion campaign at the University of Geneva Hospitals (Geneva, Switzerland) constitutes the first reported experience of a sustained improvement in compliance with hand hygiene, coinciding with a reduction of nosocomial infections with and transmission of methicillin-resistant *S. aureus*.[15] Including both direct costs associated with the intervention and indirect costs associated with healthcare personnel time, costs of the promotion campaign were less than US$2.30 per patient admitted. The total costs of hand hygiene promotion accounted for less than 1% of the costs associated with nosocomial infection over a 7-year period.[31] More studies need to be conducted, ideally in the form of randomized clinical trials that include prospectively collected cost information and concurrent surveillance of nosocomial infections, to provide more evidence of the cost benefit of hand hygiene promotion strategies. Although refined cost-effectiveness analyses comparing the costs of alternative strategies for achieving a given outcome are needed, it is clear that improvement in hand hygiene compliance is cost-effective in most instances, seen from a societal perspective. Finally, preventing healthcare-associated infections through improved HCW hand hygiene practices can be incorporated into the institution's overall plan to improve patient safety.

Form a Multidisciplinary Committee

The WHO guidelines[24] strongly recommend that institutions develop a multimodal, multidisciplinary promotional campaign to improve hand hygiene practices. Such a committee should be comprised of HCWs from various departments and with different job descriptions.[15] To stimulate HCW interest in and commitment to improving hand hygiene practices, it is helpful to engage nurses, physicians, and other caregivers in planning various aspects of a promotional campaign.[15,32,33] This can be done by involving HCWs in the selection of a slogan for the promotional campaign, by having them participate in design of motivational materials (brochures and visual reminders of the importance of hand hygiene), and in the selection of hand hygiene agents to be used in the institution.[15,32] Another strategy that may be helpful is to make the improvement in hand hygiene adherence a nursing unit–based quality improvement project.

Selecting Hand Hygiene Agents for Use in the Facility

Non-antimicrobial soaps, antimicrobial soaps, and alcohol-based hand rubs possess widely differing characteristics that can have a major impact on the acceptance of products by HCWs. Such characteristics include their scent, consistency (feel), ease of application, and propensity to cause irritant contact dermatitis with frequent use.[34,35] In addition, alcohol-based hand rubs are available as rinses (with low viscosity), gels, and foams, which vary in how they feel when applied, in the extent to which they cause a sticky sensation during or after application (sometimes a major cause of decreased compliance), in their drying time, in the likelihood they will cause a build-up of emollients with repeated use, in their in vivo antimicrobial activity, in their dispenser design, and in their tendency to interact with the powder used in some types of gloves.[33] As a result, it is imperative to involve HCWs from various departments and job descriptions when evaluating the hand hygiene products being considered for use in a facility. Selection of a product that has one or more undesirable characteristics or that has poorly-functioning dispensers can adversely affect the frequency of use by HCWs[36] (see Chapter 33 on product evaluation).

If alcohol-based hand rub products are used for preoperative surgical hand antisepsis, it is important to determine that the product has in vivo activity for an appropriate time period after application. In the United States, the Food and Drug Administration requires that such products maintain bacterial counts on the hands at or below baseline levels for 6 hours after application.[37] In Europe, if it is claimed that a product has sustained activity, it must be demonstrated that it maintains a bacterial count at 3 hours after application that is significantly lower than a reference standard.[38,39] A more in-depth discussion of the selection of alcohol-based hand rub products for use in healthcare institutions can be found in the "Educational Tools" section of the Hand Hygiene Resource Center Web page.[4]

Education of HCWs

Educating HCWs regarding the types of patient care activities that result in hand contamination and regarding the advantages and disadvantages of soap and water handwashing and use of alcohol-based hand rubs can be achieved in various ways. It is often helpful to use one or more of the following approaches:

- Conduct small educational sessions on each nursing unit or in each clinical department.
- Give conferences utilizing audience participation technology.
- Make interactive, computer-based hand hygiene educational modules available on the institution's internal computer network (intranet).

- Have HCWs place their hand on an agar plate and showing them the subsequent growth of bacteria on the plate.
- Use fluorescent powder or liquid to demonstrate the efficacy (or lack thereof) of handwashing or hand antisepsis technique.

Motivational Materials

Successful promotional campaigns have also developed innovative motivational material stressing the importance of good hand hygiene. For example, Pittet and colleagues[15] enlisted the help of a professional cartoonist, who met with ward personnel and developed a collection of colorful reminders of various aspects of hand hygiene.[40] The cartoonist's posters were placed in specific, designated locations on all wards and in public areas of the hospital and were changed by housekeeping personnel every week—a poster that remains on a ward for many months or for an entire year is not likely to be noticed by personnel after a few weeks. Other motivational strategies that might be considered include rewarding HCWs on wards or services with the highest levels of adherence to recommended hand hygiene policies, or printing hand hygiene reminders on objects such as ballpoint pens or coffee cups. In one of our facilities, HCWs on wards with the best adherence rates were given coupons that allowed them to receive complimentary food items at the hospital's cafeteria.[32]

In addition to the motivational strategies listed above, attempts have been made to involve patients in hand hygiene promotion, so-called "patient empowerment" programs.[41,42] The WHO has recognized the importance of encouraging patients to play an active role in their care, and has outlined the fundamental elements that are necessary for patient empowerment to be effective.[24] They include patients' understanding of their role, patients having sufficient knowledge to interact with healthcare providers, patients being comfortable with asking healthcare workers about hand hygiene, and the presence of a facilitating environment. Several preliminary studies have suggested that patient empowerment can lead to improved hand hygiene practices among HCWs,[41,42] but additional research is needed to identify the key factors necessary for successful patient participation in improving patient safety.[43]

Measuring Hand Hygiene Performance

Monitoring hand hygiene performance is an essential element of successful hand hygiene promotion campaigns. However, a recent review of strategies for monitoring hand hygiene revealed that there is no standard method for measuring hand hygiene performance, and each method has advantages and disadvantages.[44] Conduct-

ing periodic observational surveys of HCWs engaged in patient care activities is currently the "gold standard" for determining hand hygiene adherence rates among personnel. Although such surveys are somewhat time-consuming, they are currently the most reliable method of assessing the adequacy of hand hygiene practices. Observations should be carried out using a structured data collection sheet that includes entries for the categories of HCWs observed; whether the HCWs cleaned their hands before patient contact, after various types of patient care activities and after removing gloves; and whether hands were cleaned by washing with soap and water or by using an alcohol-based hand rub. In general, we do not record the identity of HCWs being observed. The most widely used monitoring method is the one proposed by the WHO,[45] which is in line with the "My 5 moments for hand hygiene" concept (Figure 21-2).[23]

Monitoring the amount of hand hygiene products used (alcohol-based hand rub and/or soap) is not as time-consuming as performing an observational survey, and it can provide information regarding trends in the frequency that hand hygiene is performed, but it cannot measure actual hand hygiene adherence rates or technique.[11,15] The amount of product used can be expressed as the number of liters of product used per 1,000 patient-days, or it can be used to *estimate* the number of hand hygiene episodes that have occurred.[15,41] Electronic counting devices placed in dispensers of alcohol-based hand rub have also been used to monitor the frequency that hand hygiene is performed.[46] Electronic monitoring systems designed to measure product usage in relation to specific events (eg, entering or leaving a patient room), and in some instances to record hand hygiene episodes by individual HCWs, are under development but have not yet been shown to result in sustained improvement in hand hygiene adherence rates.

Questionnaire surveys of HCWs, patients, and family members can provide data on perceptions and attitudes toward hand hygiene, but they often yield inaccurate information regarding hand hygiene adherence rates[44] and cannot be recommended as a sole method of measuring hand hygiene adherence.

Providing HCWs With Feedback

Providing HCWs with feedback regarding how well they adhere to recommended patient care practices, including hand hygiene, has been one of the most effective methods for modifying patterns of behavior.[11,15,33] Feedback can be provided during ward-based educational sessions or larger conferences with HCWs, or by posting adherence rate data for wards or services in highly visible areas of the facility. We favor giving feedback to groups of HCWs on each ward or service; no attempt should be made to single out individuals with the lowest adherence rates.

Importance of Dispenser Design and Placement

Before selecting alcohol-based hand rub products or soaps for use by HCWs, institutions should evaluate the design and reliability of dispensers that will be provided by the manufacturer or vendor. One institution installed dispensers of a viscous alcohol-based hand rinse hospital-wide on the basis of a short trial utilizing table-top pump bottles.[36] Within months after the installation of wall-mounted dispensers, HCWs noticed that dispensers often became partly or totally plugged. Partly plugged dispensers often squirted product onto the wall or floor instead of onto the HCW's hand. Constant problems with dispensers probably accounted in part for the low rate of utilization of the product by personnel.[36]

Placing alcohol-based hand rub dispensers adjacent to patients' beds or making pocket-sized bottles available can assure that the product is readily available to HCWs.[6,15] Some experts also favor placing dispensers in hallways adjacent to patient room doors, but placement of alcohol-based hand rub dispensers at such locations may not comply with some local fire safety codes.[47] Therefore, before installing dispensers in hallways of healthcare institutions, it is important to check with fire safety officials regarding local regulations.

Surgical Hand Antisepsis

Surgical hand antisepsis has traditionally been performed by using a brush or a sponge to scrub hands with an antimicrobial soap for 2-5 minutes or even longer. Frequently used, such a regimen causes irritant contact dermatitis in some individuals, and it is time-consuming if longer scrub times are utilized. Although both the HICPAC guidelines[11] and the WHO guidelines[24] recommend that surgical hand antisepsis be performed by using either an antimicrobial soap or an alcohol-based hand rub product with persistent activity, some surgeons have questioned whether the incidence of surgical site infections is comparable with the 2 regimens. A prospective controlled trial involving several thousand surgical patients revealed that the incidence of surgical site infections among patients whose surgeons used an alcohol-based hand rub for surgical hand antisepsis was nearly identical to the incidence among patients whose surgeons performed a traditional scrub using an antimicrobial soap.[48] Making surgical personnel aware of this study may allay concerns of the efficacy of alcohol-based hand rubs for surgical hand antisepsis.

Other Logistical Issues

Because alcohol-based hand rubs are flammable, it is imperative that large stocks of such products be stored in approved areas of the facility, and that they be protected from high temperatures or flames. Specific individuals should be assigned the responsibility for transporting the product from bulk storage areas to clinical areas and for checking dispensers on a regular basis to make sure they are not empty. In some institutions, one or more of these individuals can assist with monitoring the amount of product that is used or the amount that is delivered to each ward.

If providing hand hygiene products (soaps and alcohol-based hand rubs) is part of housekeeping functions that are outsourced to contractors, it is important that managers of such contractors understand the importance of hand hygiene and comply with policies of the healthcare facility. Some facilities belong to large healthcare-associated buying groups that dictate the hand hygiene products that are available to the institution. If evaluation by HCWs of alcohol-based hand rub products available through the buying group reveals that none are acceptable to a majority of personnel, then the institution should examine buying group contracts to determine if preferred products can be purchased from other vendors.

Depending on the consistency of the alcohol-based hand rub product in use and the type of dispensers used, alcohol-containing rinses or gels may sometimes drip from the hands of HCWs during application. In some institutions, this has affected the wax finish on floors underneath dispensers and has required a change in floor care procedures.

Any difficulties in implementing a hand hygiene promotional campaign, as well as evidence of improved hand hygiene compliance that is achieved, should be discussed with high-level administrators and be presented at appropriate committee meetings and conferences.

WHO Guidelines Consensus Recommendations, 2009

The consensus recommendations in this section are reproduced with permission from the WHO guidelines, Part II, Sections 1–9.[24] The strength of the supporting evidence, categorized according to the CDC/HICPAC system (Table 21-2), is given in parentheses for each recommendation.

1. **Indications for hand hygiene**
   A. Wash hands with soap and water when visibly dirty or visibly soiled with blood or other body fluids (IB) or after using the toilet (II).
   B. If exposure to potential spore-forming pathogens is strongly suspected or proven, including

Table 21-2. Ranking system for evidence supporting recommendations in the World Health Organization guidelines on hand hygiene

Category	Description
IA	Strongly recommended for implementation and strongly supported by well-designed experimental, clinical, or epidemiological studies
IB	Strongly recommended for implementation and supported by some experimental, clinical, or epidemiological studies and a strong theoretical rationale
IC	Required for implementation, as mandated by federal and/or state regulation or standard
II	Suggested for implementation and supported by suggestive clinical or epidemiological studies or a theoretical rationale or a consensus by a panel of experts

NOTE: Adapted from the World Health Organization guidelines.[24]

outbreaks of *Clostridium difficile,* handwashing with soap and water is the preferred means (IB).

C. Use an alcohol-based handrub as the preferred means for routine hand antisepsis in all other clinical situations described in items D(a) to D(f) listed below, if hands are not visibly soiled (IA). If alcohol-based handrub is not obtainable, wash hands with soap and water (IB).

D. Perform hand hygiene:
   a. before and after touching the patient (IB);
   b. before handling an invasive device for patient care, regardless of whether or not gloves are used (IB);
   c. after contact with body fluids or excretions, mucous membranes, non-intact skin, or wound dressings (IA);
   d. if moving from a contaminated body site to another body site during care of the same patient (IB);
   e. after contact with inanimate surfaces and objects (including medical equipment) in the immediate vicinity of the patient (IB);
   f. after removing sterile (II) or non-sterile (IB) gloves.

E. Before handling medication or preparing food, perform hand hygiene using an alcohol-based handrub or wash hands with either plain or antimicrobial soap and water (IB).

F. Soap and alcohol-based handrub should not be used concomitantly (II).

2. **Hand hygiene technique**
   A. Apply a palmful of alcohol-based handrub and cover all surfaces of the hands. Rub hands until dry (IB).
   B. When washing hands with soap and water, wet hands with water and apply the amount of product necessary to cover all surfaces. Rinse hands with water and dry thoroughly with a single-use towel. Use clean, running water whenever possible. Avoid using hot water, as repeated exposure to hot water may increase the risk of dermatitis (IB). Use towel to turn off tap/faucet (IB). Dry hands thoroughly using a method that does not recontaminate hands. Make sure towels are not used multiple times or by multiple people (IB).
   C. Liquid, bar, leaf, or powdered forms of soap are acceptable. When bar soap is used, small bars of soap in racks that facilitate drainage should be used to allow the bars to dry (II).

3. **Recommendations for surgical hand preparation**
   A. Remove rings, wrist-watch, and bracelets before beginning surgical hand preparation (II). Artificial nails are prohibited (IB).
   B. Sinks should be designed to reduce the risk of splashes (II).
   C. If hands are visibly soiled, wash hands with plain soap before surgical hand preparation (II). Remove debris from underneath fingernails using a nail cleaner, preferably under running water (II).
   D. Brushes are not recommended for surgical hand preparation (IB).
   E. Surgical hand antisepsis should be performed using either a suitable antimicrobial soap or suitable alcohol-based handrub, preferably with a product ensuring sustained activity, before donning sterile gloves (IB).
   F. If quality of water is not assured in the operating theatre, surgical hand antisepsis using an alcohol-based handrub is recommended before donning sterile gloves when performing surgical procedures (II).
   G. When performing surgical hand antisepsis using an antimicrobial soap, scrub hands and forearms for the length of time recommended by the manufacturer, typically 2–5 minutes. Long scrub times (eg, 10 minutes) are not necessary (IB).
   H. When using an alcohol-based surgical handrub product with sustained activity, follow the manufacturer's instructions for application times. Apply the product to dry hands only (IB). Do not combine surgical hand scrub and surgical handrub with alcohol-based products sequentially (II).

I. When using an alcohol-based handrub, use sufficient product to keep hands and forearms wet with the handrub throughout the surgical hand preparation procedure (IB).

J. After application of the alcohol-based handrub as recommended, allow hands and forearms to dry thoroughly before donning sterile gloves (IB).

4. **Selection and handling of hand hygiene agents**

   A. Provide HCWs with efficacious hand hygiene products that have low irritancy potential (IB).

   B. To maximize acceptance of hand hygiene products by HCWs, solicit their input regarding the skin tolerance, feel, and fragrance of any products under consideration (IB). Comparative evaluations may greatly help in this process.

   C. When selecting hand hygiene products

   a. determine any known interaction between products used to clean hands, skin care products, and the types of glove used in the institution (II);

   b. solicit information from manufacturers about the risk of product contamination (IB);

   c. ensure that dispensers are accessible at the point of care (IB);

   d. ensure that dispensers function adequately and reliably and deliver an appropriate volume of the product (II);

   e. ensure that the dispenser system for alcohol-based handrubs is approved for flammable materials (IC);

   f. solicit and evaluate information from manufacturers regarding any effect that hand lotions, creams, or alcohol-based handrubs may have on the effects of antimicrobial soaps being used in the institution (IB);

   g. cost comparisons should only be made for products that meet requirements for efficacy, skin tolerance, and acceptability (II).

   D. Do not add soap (IA) or alcohol-based formulations (II) to a partially empty dispenser. If soap dispensers are reused, follow recommended procedures for cleansing.

5. **Skin care**

   A. Include information regarding hand-care practices designed to reduce the risk of irritant contact dermatitis and other skin damage in education programmes for HCWs (IB).

   B. Provide alternative hand hygiene products for HCWs with confirmed allergies or adverse reactions to standard products used in the healthcare setting (II).

   C. Provide HCWs with hand lotions or creams to minimize the occurrence of irritant contact dermatitis associated with hand antisepsis or handwashing (IA).

   D. When alcohol-based handrub is available in the healthcare facility for hygienic hand antisepsis, the use of antimicrobial soap is not recommended (II).

   E. Soap and alcohol-based handrub should not be used concomitantly (II).

6. **Use of gloves**

   A. The use of gloves does not replace the need for hand hygiene by either handrubbing or handwashing (IB).

   B. Wear gloves when it can be reasonably anticipated that contact with blood or other potentially infectious materials, mucous membranes, or non-intact skin will occur (IC).

   C. Remove gloves after caring for a patient. Do not wear the same pair of gloves for the care of more than one patient (IB).

   D. When wearing gloves, change or remove gloves during patient care if moving from a contaminated body site to either another body site (including non-intact skin, mucous membrane or medical device) within the same patient or the environment (II).

   E. The reuse of gloves is not recommended (IB). In the case of glove reuse, implement the safest reprocessing method (II).

7. **Other aspects of hand hygiene**

   A. Do not wear artificial fingernails or extenders when having direct contact with patients (IA).

   B. Keep natural nails short (tips less than 0.5 cm long or approximately 1/4 inch) (II).

8. **Educational and motivational programmes for healthcare workers**

   A. In hand hygiene promotion programmes for HCWs, focus specifically on factors currently found to have a significant influence on behavior, and not solely on the type of hand hygiene products. The strategy should be multifaceted and multimodal and include education and senior executive support for implementation (IA).

   B. Educate HCWs about the type of patient-care activities that can result in hand contamination and about the advantages and disadvantages of various methods used to clean their hands (II).

C. Monitor HCWs' adherence to recommended hand hygiene practices and provide them with performance feedback (IA).

D. Encourage partnerships between patients, their families, and HCWs to promote hand hygiene in healthcare settings (II).

9. **Governmental and institutional responsibilities**

9.1. For healthcare administrators

A. It is essential that administrators ensure conditions are conducive to the promotion of a multifaceted, multimodal hand hygiene strategy and an approach that promotes a patient safety culture by implementation of points B–I below.

B. Provide HCWs with access to a safe, continuous water supply at all outlets and access to the necessary facilities to perform handwashing (IB).

C. Provide HCWs with a readily accessible alcohol-based handrub at the point of patient care (IA).

D. Make improved hand hygiene adherence (compliance) an institutional priority and provide appropriate leadership, administrative support, financial resources, and support for hand hygiene and other infection prevention and control activities (IB).

E. Ensure HCWs have dedicated time for infection control training, including sessions on hand hygiene (II).

F. Implement a multidisciplinary, multifaceted and multimodal programme designed to improve adherence of HCWs to recommended hand hygiene practices (IB).

G. With regard to hand hygiene, ensure that the water supply is physically separated from drainage and sewerage within the healthcare setting, and provide routine system monitoring and management (IB).

H. Provide strong leadership and support for hand hygiene and other infection prevention and control activities (II).

I. Alcohol-based handrub production and storage must adhere to the national safety guidelines and local legal requirements (II).

9.2. For national governments

A. Make improved hand hygiene adherence a national priority and consider provision of a funded, coordinated implementation programme, while ensuring monitoring and long-term sustainability (II).

B. Support strengthening of infection control capacities within healthcare settings (II).

C. Promote hand hygiene at the community level to strengthen both self-protection and the protection of others (II).

D. Encourage healthcare settings to use hand hygiene as a quality indicator (Australia, Belgium, France, Scotland, USA) (II).

References

1. Mortimer EA, Lipsitz PJ, Wolinsky E, et al. Transmission of staphylococci between newborns. *Am J Dis Child* 1962;104:289–295.
2. Ehrenkranz NJ, Alfonso BC. Failure of bland soap handwash to prevent hand transfer of patient bacteria to urethral catheters. *Infect Control Hosp Epidemiol* 1991;12:654–662.
3. Larson E. A causal link between handwashing and risk of infection? Examination of the evidence. *Infect Control Hosp Epidemiol* 1988;9:28–36.
4. Hand Hygiene Resource Center Web page. Educational tools. http://www.handhygiene.org/educational_tools.asp. Accessed September 14, 2009.
5. Pittet D, Mourouga P, Perneger TV. Compliance with handwashing in a teaching hospital. *Ann Intern Med* 1999;130:126–130.
6. Voss A, Widmer AF. No time for handwashing!? Handwashing versus alcoholic rub: can we afford 100% compliance? *Infect Control Hosp Epidemiol* 1997;18:205–208.
7. Larson E, Killien M. Factors influencing handwashing behavior of patient care personnel. *Am J Infect Control* 1982;10:93–99.
8. Larson E, McGeer A, Quraishi ZA, et al. Effect of an automated sink on handwashing practices and attitudes in high-risk units. *Infect Control Hosp Epidemiol* 1991;12:422–428.
9. Larson E, Friedman C, Cohran J, Treston-Aurand J, Green S. Prevalence and correlates of skin damage on the hands of nurses. *Heart Lung* 1997;26:404–412.
10. Pittet D, Boyce JM. Hand hygiene and patient care: pursuing the Semmelweis legacy. *Lancet Infect Dis* 2001;1:9–20.
11. Boyce JM, Pittet D. Guideline for hand hygiene in health-care settings. Recommendations of the Healthcare Infection Control Practices Advisory Committee and the HICPAC/SHEA/APIC/IDSA Hand Hygiene Task Force. Society for Healthcare Epidemiology of America/Association for Professionals in Infection Control/Infectious Diseases Society of America. *MMWR Recomm Rep* 2002;51(RR-16):1–45.
12. Boyce JM, Kelliher S, Vallande N. Skin irritation and dryness associated with two hand-hygiene regimens: soap-and-water hand washing versus hand antisepsis with an alcoholic hand gel. *Infect Control Hosp Epidemiol* 2000;21:442–448.
13. Winnefeld M, Richard MA, Drancourt M, Grob JJ. Skin tolerance and effectiveness of two hand decontamination procedures in everyday hospital use. *Br J Dermatol* 2000;143:546–550.
14. Larson EL, Aiello AE, Heilman JM, et al. Comparison of different regimens for surgical hand preparation. *AORN J* 2001;73:412–418, 420.
15. Pittet D, Hugonnet S, Harbarth S, et al. Effectiveness of a hospital-wide programme to improve compliance with hand hygiene. *Lancet* 2000;356:1307–1312.

16. Hugonnet S, Perneger TV, Pittet D. Alcohol-based handrub improves compliance with hand hygiene in intensive care units. *Arch Intern Med* 2002;162:1037–1043.

17. Muto CA, Sistrom MG, Farr BM. Hand hygiene rates unaffected by installation of dispensers of a rapidly acting hand antiseptic. *Am J Infect Control* 2000;28:273–276.

18. Larson EL, Early E, Cloonan P, Sugrue S, Parides M. An organizational climate intervention associated with increased handwashing and decreased nosocomial infections. *Behav Med* 2000;26:14–22.

19. Pittet D. Improving compliance with hand hygiene in hospitals. *Infect Control Hosp Epidemiol* 2000;21:381–386.

20. Grayson ML, Jarvie LJ, Martin R, et al. Significant reductions in methicillin-resistant *Staphylococcus aureus* bacteraemia and clinical isolates associated with a multisite, hand hygiene culture-change program and subsequent successful statewide roll-out. *Med J Aust* 2008;188:633–640.

21. World Health Organization (WHO). World Health Organization guidelines on hand hygiene in health care (advanced draft). Geneva, Switzerland: WHO, 2006:7–702.

22. World Health Organization (WHO). WHO Multimodal Hand Hygiene Improvement Strategy. http://www.who.int/gpsc/en/. Accessed September 14, 2009.

23. Sax H, Allegranzi B, Uckay I, Larson E, Boyce J, Pittet D. "My 5 moments for hand hygiene": a user-centred design approach to understand, train, monitor and report hand hygiene. *J Hosp Infect* 2007;67:9–21.

24. World Health Organization (WHO). World Health Organization guidelines on hand hygiene in health care. Geneva: WHO; 2009. http://www.who.int/gpsc/5may/en/. Accessed September 14, 2009.

25. Ahmed QA, Memish ZA, Allegranzi B, Pittet D. Muslim health-care workers and alcohol-based handrubs. *Lancet* 2006;367:1025–1027.

26. World Health Organization (WHO). Key articles published on the First Global Patient Safety Challenge [Web page]. http://www.who.int/gpsc/information_centre/key_articles/en/index.html. Accessed September 14, 2009.

27. World Health Organization (WHO). Save Lives: Clean Your Hands initiative [Web page]. http://www.sho.int/gpsc/en/. Accessed September 14, 2009.

28. Jarvis WR. Selected aspects of the socioeconomic impact of nosocomial infections: morbidity, mortality, cost, and prevention. *Infect Control Hosp Epidemiol* 1996;17:552–557.

29. Boyce JM. Antiseptic technology: access, affordability and acceptance. *Emerg Infect Dis* 2001;7:231–233.

30. Webster J, Faoagali JL, Cartwright D. Elimination of methicillin-resistant *Staphylococcus aureus* from a neonatal intensive care unit after hand washing with triclosan. *J Paediatr Child Health* 1994;30:59–64.

31. Pittet D, Sax H, Hugonnet S, Harbarth S. Cost implications of successful hand hygiene promotion. *Infect Control Hosp Epidemiol* 2004;25:264–266.

32. Ligi CE, Kohan CA, Dumigan DG, Havill NL, Pittet D, Boyce JM. A multifaceted approach to improving hand hygiene practices among healthcare workers using an alcohol-based hand gel. In: Program and abstracts of the 13th Annual Meeting of the Society for Healthcare Epidemiology of America; Arlington, VA; 5–8 April 2003. Abstract 160.

33. Harbarth S, Pittet D, Grady L, et al. Interventional study to evaluate the impact of an alcohol-based hand gel in improving hand hygiene compliance. *Pediatr Infect Dis J* 2002;21: 489–495.

34. Larson E, Leyden JJ, McGinley KJ, Grove GL, Talbot GH. Physiologic and microbiologic changes in skin related to frequent handwashing. *Infect Control* 1986;7:59–63.

35. Scott D, Barnes A, Lister M, Arkell P. An evaluation of the user acceptability of chlorhexidine handwash formulations. *J Hosp Infect* 1991;18:51–55.

36. Kohan C, Ligi C, Dumigan DG, Boyce JM. The importance of evaluating product dispensers when selecting alcohol-based handrubs. *Am J Infect Control* 2002;30:373–375.

37. Food and Drug Administration. Tentative final monograph for healthcare antiseptic drug products; proposed rule. *Fed Reg* 1994; 31441–31452.

38. European Committee for Standardization. European standard (pr)EN 12791. Chemical disinfectants and antiseptics: surgical hand disinfection—test method and requirements. 2004. http://www.cen.eu/esearch/. Accessed October 19, 2009.

39. Rotter M, Sattar S, Dharan B, Allegranzi B, Mathai E, Pittet D. Methods to evaluate the microbiocidal activities of handrub and handwash agents. *J Hosp Infect* 2009;73(3): 191–199.

40. Hospisafe.ch: A New Approach to Prevent Infections in Hospital [Web site]. www.hopisafe.ch. Accessed September 14, 2009.

41. McGuckin M, Waterman R, Porten L, et al. Patient education model for increasing handwashing compliance. *Am J Infect Control* 1999;27:309–314.

42. McGuckin M, Taylor A, Martin V, Porten L, Salcido R. Evaluation of a patient education model for increasing hand hygiene compliance in an inpatient rehabilitation unit. *Am J Infect Control* 2004;32:235–238.

43. Longtin Y, Sax H, Leape LL, Sheridan SE, Donaldson L, Pittet D. Patient participation: current knowledge and applicability to patient safety. *Mayo Clinic Proceedings* 2009 (in press).

44. Joint Commission. Measuring hand hygiene adherence: overcoming the challenges. 2009. http://www.jointcommission.org/PatientSafety/InfectionControl/hh_monograph.htm. Accessed Sep 14, 2009.

45. World Health Organization (WHO). Save Lives: Clean Your Hands initiative: tools for evaluation and feedback [Web site]. http://www.who.int/gpsc/5may/tools/evaluation_feedback/en/index.html. Accessed Sep 14, 2009.

46. Boyce JM, Cooper T, Dolan MJ. Evaluation of an electronic device for real-time measurement of alcohol-based hand rub use. *Infect Control Hosp Epidemiol* 2009;30:1090–1095.

47. Boyce JM, Pearson M. Low frequency of fires from alcohol-based hand rub dispensers in healthcare facilities. *Infect Control Hosp Epidemiol* 2003;24:618–619.

48. Parienti JJ, Thibon P, Heller R, et al., Group amotACdMS. Hand-rubbing with an aqueous alcoholic solution vs. traditional surgical hand-scrubbing and 30-day surgical site infection rates. *JAMA* 2002;288:722–727.

Chapter 22 Biological Disaster Preparedness

Sandro Cinti, MD, and Eden Wells, MD, MPH

Biological disasters, including bioterrorism, severe acute respiratory syndrome (SARS), and pandemic influenza present a unique set of challenges for the healthcare epidemiologist. These events appear suddenly, cause high morbidity and mortality, and produce high levels of fear and, potentially, panic. The response to a biological disaster requires coordination of many local, state, and federal entities. Accordingly, it is important for the healthcare epidemiologist to be familiar not only with the relevant biological agents and diseases but also with his or her role in responding to a biological disaster. This chapter reviews the National Incident Management Structure (NIMS), the role of the healthcare epidemiologist in a biological disaster, the early characteristics of a biological disaster, and, finally, the diseases and agents associated with biological disaster: the Centers for Disease Control and Prevention (CDC) Category A agents and diseases of bioterrorism (variola virus [smallpox], *Bacillus anthracis* [anthrax], *Francisella tularensis* [tularemia], *Clostridium botulinum* toxin [botulism], viral hemorrhagic fever viruses, and *Yersinia pestis* [plague]),[1] SARS, and pandemic influenza.

The National Incident Management Structure

Overall Structure

The healthcare epidemiologist will have an important role within the healthcare facilities' disaster response plan to a biological disaster. In the event of a large-scale infectious disease outbreak, as may be seen with a bioter-

rorist attack or with an influenza pandemic, hospitals and their emergency departments will be on the front line for response. The hospital planners must collaborate with other community responders so that the overall response is delivered in a coordinated fashion. Interagency, hospital, and community response during a disaster event can become extremely complex; in 2003, Homeland Security Presidential Directive 5 (HSPD-5) created NIMS,[2] which enhances the ability of the United States to manage any domestic incident by establishing a single, comprehensive national incident management system.[2]

NIMS provides a national template for all government and private organizations to work together in a coordinated fashion to prepare for, prevent, respond to, and recover from domestic incidents, including acts of terrorism.[2] In 2006 the NIMS Integration Center and the US Department of Health and Human Services released a guide to NIMS implementation activities for hospitals and healthcare systems,[3] which contains 17 elements for hospital planning (Table 22-1). While the American Hospital Association strongly encourages all hospitals to become NIMS compliant, only hospitals that receive federal preparedness and response funds are required to do so.[4,5] Regardless, emergency responders from surrounding agencies are themselves operating under NIMS-compliant response systems, and it is important for hospitals to coordinate planning and response with these agencies.[5] Hospitals will need to interact with numerous response partners in a bioterrorism event, including prehospital emergency services, the local emergency operations center, other hospitals, the regional hospital

Table 22-1. Elements for hospital planning from the 2006 National Incident Management System (NIMS) Integration Center and Department of Health and Human Services guide

1. Adopt NIMS at all organizational levels
2. Manage incidents in accordance with the Incident Command System
3. Establish multiagency coordination (MAC)
4. Establish timely and accurate communication through Joint Information Center (JIC)
5. Perform NIMS implementation tracking annually to enhance the Emergency Management Program
6. Receive, implement, and document preparedness funding
7. Revise and update plans to incorporate NIMS
8. Participate in and promote mutual aid agreements
9. Complete instructional course IS 700: Introduction to NIMS
10. Complete instructional course IS 800: Introduction to National Response Plan
11. Complete instructional course ICS 100 HC: Introduction to ICS and ICS 200 HC—Basics of ICS
12. Incorporate NIMS into all trainings and exercises
13. Participate in "all hazards" exercise program, involving multiple partners
14. Incorporate corrective actions into plans and procedures
15. Maintain inventory of hospital response assets
16. Establish procurement that follows national standards and guidance, to the extent possible, to achieve interoperability (resource typing)
17. Apply standard and consistent terminology—use "plain English" and no acronyms

NOTE: ICS, Incident Command System; IS, independent study [course]. Adapted from NIMS documentation.[3]

coordination center, and public safety and public health agencies, to name just a few (Figure 22-1). Hospitals and hospital systems that are recipients of federal hospital preparedness funds work together in local and regional groups to coordinate their disaster and bioterrorism response plans, regardless of whether the healthcare epidemiologist's facility is a recipient of these funds. It will be important for the healthcare epidemiologist to identify these preparedness activities within his or her community and healthcare region.

The Hospital Incident Command System

A critical element for hospital response in a disaster such as a bioterrorism event is the incident command structure. The Hospital Incident Command System (HICS) is one example of a NIMS-compliant incident command system methodology for healthcare systems. HICS was adapted from the Hospital Emergency Incident Command System (HEICS), which was developed in the 1980s and was used by more than 6,000 US hospitals to

prepare and respond to a variety of disasters.[6] HICS, however, can be implemented for both emergent and nonemergent incidents. A sample HICS command structure is outlined in Figure 22-2.

The healthcare epidemiologist is part of the hospital's incident command system response, either in developing the hospital response plan, in surveillance for the detection of a bioterrorism or pandemic event, or in the control and management of the disease agent within the hospital facility. Therefore, the healthcare epidemiologist may be asked to assume a role within the operation section or the planning section of the incident command system (Figure 22-1). These roles may be different from the healthcare epidemiologist's usual roles, and regular training and exercise in the hospital facility's incident command system or HICS is important for all involved employees. Furthermore, as a subject matter expert in infection control and infectious diseases, the healthcare epidemiologist may be asked to provide information to the incident commander or section chiefs in the Emergency Operation Center during any serious disaster event in the facility or community.

Role of the Healthcare Epidemiologist in Biological Disasters

The healthcare epidemiologist will play a large role in the preparation for and response to any biological disaster. Within the HICS, the healthcare epidemiologist will be a key consultant to the incident commander. Particular duties that will fall on the healthcare epidemiologist before and during a biological disaster include surveillance for infection, infection control, and communication of infectious risk to the staff.

Surveillance for Infection

The healthcare epidemiologist is responsible for developing surveillance protocols within the hospital but must also monitor external sources for up-to-date information on potential biological threats. During the SARS outbreak of 2003, many US hospitals screened patients presenting to the emergency department for clinical symptoms and exposure history based on the case definition of the CDC.[7] Had SARS become more widespread in the United States, more stringent surveillance may have been necessary. In Toronto, Canada, where SARS was more widespread, hospitals instituted strict temperature checks for both hospital staff and patients.[8] Regarding laboratory surveillance, the healthcare epidemiologist must guide clinicians and hospital

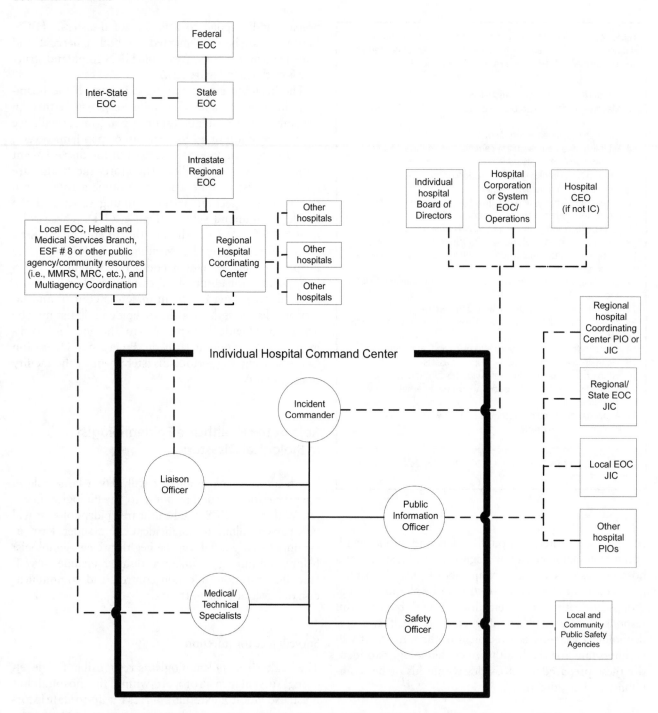

Figure 22-1. Hospital community response partners. CEO, Chief Operating Officer; EOC, Emergency Operations Center; ESF, Emergency Support Function; IC, Incident Commander; JIC, Joint Information Center; MMRS, Metropolitan Medical Respopnse Team; PIO, Public Information Officer.

laboratories on which laboratory tests should be used for surveillance and diagnosis of a biological threat. As an example, early in an influenza pandemic, results of rapid influenza tests may help to signal the arrival of influenza in a particular hospital. However, given the relative insensitivity of rapid influenza tests,[9] it may not make sense to use this test later in the pandemic or as a

means of identifying infected patients who should be treated.

Regarding external surveillance, the healthcare epidemiologist will be called upon to interpret information coming from multiple sources, including the community, the state, the federal government, and the World Health Organization (WHO). Rapid decisions will have to be

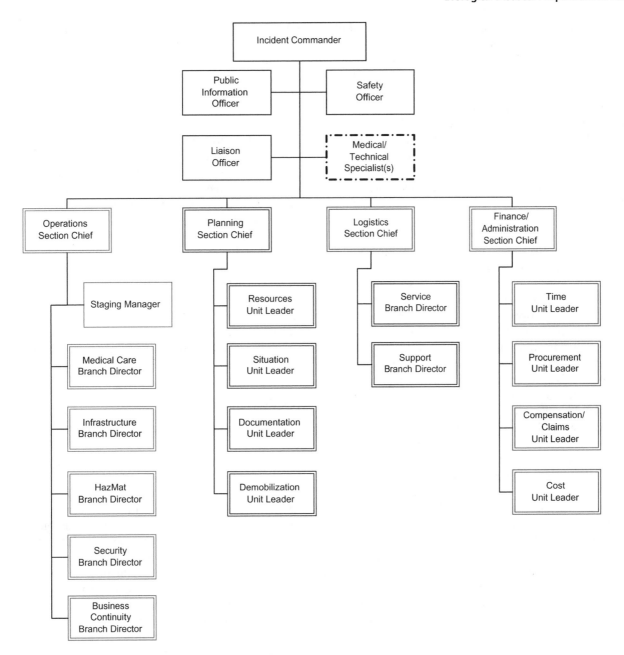

Figure 22-2. Hospital incident command system (HICS) sample organizational chart (E5).

made about activating biological disaster plans within the hospital. As an example, an increase in the WHO pandemic alert phase from 3 to 4 would indicate increased human-to-human transmission of a pandemic strain of influenza.[10] Even if this increased transmission is not occurring in the United States, certain actions— such as stockpiling vaccine, personal protective equipment, antibiotics, and antivirals—might be important for a hospital. If the pandemic progresses, strategies to improve hospital capacity—such as curtailing elective procedures and doctors appointments, as well as reassigning staff to influenza wards—might need to be insti-

tuted. Gathering accurate information on disease activity locally, statewide, nationwide, and worldwide will be crucial to a successful hospital response.

Infection Control

Although infection control is the daily task of the healthcare epidemiologist, biological disasters present unique challenges. Pre-event duties of the healthcare epidemiologist in planning for a biological disaster include the following: (1) helping to develop stockpiles of personal protective equipment (PPE), vaccines, antivirals, and antibiotics; (2) establishing protocols for the use of PPE, isolation procedures, and cohorting of patients to prevent disease transmission; (3) helping to establish priority groups and develop mechanisms for the distribution of vaccines, antivirals, and antibiotics; and (4) helping to develop treatment, prophylaxis, and immunization protocols that will be used within the hospital.

Stockpiling of antibiotics, antivirals, PPE, and, possibly, vaccine will be very important, as these will be scarce resources during a biological disaster. During the SARS outbreak in Toronto, N-95 masks became scarce even in the United States, where very few cases and no fatalities had occurred.[11] There has been enough concern about bioterrorism and pandemic influenza that the states and federal government have stockpiled antibiotics and antivirals.[12,13] In anticipation of a biological event, the healthcare epidemiologist must work with the departments of material services, pharmacy, and infectious diseases and the hospital leadership, to determine which resources and how much of each resource should be stockpiled. The input of the healthcare epidemiologist is particularly crucial when it comes to stockpiling PPE, as protocols for fit-testing of N-95 masks and proper use of PPE will be crucial for determining the numbers of items that should be held on-site.

Hospital leadership will look to the healthcare epidemiologist to develop protocols for PPE use, isolation procedures, and cohorting strategies in all areas of the hospital. Each biological threat has different requirements for infection control (these are summarized in Table 22-2), and many biological agents and diseases, such as anthrax, plague, SARS, and influenza virus A subtype H5N1, have rarely or never been seen in most hospitals. Thus the healthcare epidemiologist must develop strategies for infection control in a mass-casualty biological event. Although SARS requires airborne infection isolation precautions, a huge influx of patients might force a hospital to cohort SARS-infected patients to a certain ward, section of the hospital, or even off-site.[24] Plans for cohorting, therefore, must be in place well ahead of a biological event. Protocols for PPE use must also take into account that shortages will occur during a biological disaster, and provisions must be put in place by the healthcare epidemiologist to allow for alternate use of PPE. Generally, an N-95 mask must be discarded after one use.[25] However, during a biological event such as SARS or pandemic influenza, N-95 masks will become scarce, and protocols that allow for reuse of masks will need to be activated.

The healthcare epidemiologist will need to be involved in developing distribution plans for antivirals, antibiotics, and vaccines during a biological disaster. Distribution plans must be established for both patients and staff with input from the departments of employee health, in-

fectious diseases, and pharmacy and the hospital leadership. Although priority lists have been established for the distribution of vaccines and antivirals during an influenza pandemic,[14] they do not currently exist for other biological disasters. The healthcare epidemiologist must be involved in prioritizing these valuable resources as they become scarce.

Treatment, immunization, and prophylaxis protocols have been developed for most biological agents and diseases. However, as a biological event progresses and resources become scarce, these protocols might change. As an example, current guidelines for using antivirals for pandemic influenza recommend that, if enough medication is available, healthcare workers at high risk should get outbreak (or preexposure) prophylaxis to prevent disease.[14] However, as antivirals become scarce, this strategy will have to change, and the healthcare epidemiologist will have to work with the departments of infectious diseases, employee health, and pharmacy and the hospital leadership to modify antiviral use protocols. Mechanisms for modifying the distribution of scarce resources must be worked out before a biological disaster occurs.

Risk Communication

One of the most important tasks for the healthcare epidemiologist during a biological disaster will be communication of risk to hospital employees and patients. Pre-event preparation includes maintaining contact with local, state, and federal public health entities. Involvement with local, regional, and state biopreparedness groups is an excellent way for the healthcare epidemiologist to stay familiar with key public health, emergency management, law enforcement, and hospital personnel. Also, participating in local, regional, and state tabletop or functional exercises is an excellent way to understand how public health, emergency management, law enforcement, and hospital personnel will work together during a biological disaster.[26] The healthcare epidemiologist will also participate in educating hospital personnel on biopreparedness plans within the hospital. Certain biological disasters, particularly pandemic influenza, will alter the standards of medical care and staffing ratios within a hospital.[27] It is crucial that hospital personnel are aware of their roles and the role of the hospital in responding to a biological disaster. The healthcare epidemiologist will be called upon to explain the risks and possible consequences of a biological disaster.

During a biological disaster, the healthcare epidemiologist will be a key advisor to the incident commander (see the section on NIMS, above), but he or she must also be available to the staff to answer questions and contribute to changes in preestablished protocols. Accordingly, the healthcare epidemiologist must closely monitor how

Table 22-2. Infection control issues related to biological disaster–associated diseases and agents

Disease (causative agent)	Person-to-person spread	Infection control precautions[a]	Decontamination	Postmortem care
Anthrax (*Bacillus anthracis*)	No	• Contact precautions	• For exposed persons, thoroughly wash exposed areas with soap and water • Clean contaminated surfaces with 1:100 dilution of household bleach in water • Contaminated clothing should be placed in a biohazard bag	• Cremation is preferred • Autopsy instruments should be autoclaved or incinerated
Smallpox (variola virus)	Yes, via droplets and from skin lesions	• Contact precautions • Droplet precautions • Airborne precautions: place the patient in a negative air-pressure room	**Personnel** Not applicable: if it is an aerosolized agent exposure event, detection of cases would occur long after viable virus was present in the environment (also see Vaccination in Table 4) **Environment** • Clean contaminated surfaces with manufacturer-recommended concentrations of EPA-registered germicides • Use all FDA-approved sterilization methods for medical instruments and devices • Clothing and bedding should be bagged or contained at the point of use, with minimal agitation to avoid spread of fomites, in accordance with OSHA regulations • Laundry should be autoclaved or laundered in hot water with bleach added	• Cremation whenever possible • Mortuary workers should be vaccinated
Plague (*Yersinia pestis*)	Yes, via droplets	• Droplet precautions for 48 hours after treatment and until clinical improvement • Infected patient should wear a mask (a respirator is not necessary) when they are outside the hospital room • Cohorting of plague patients may be used if patient numbers are high • If buboes are aspirated, contact precautions and droplet precautions should be followed	• Environmental decontamination is not generally recommended; *Y. pestis* is infectious for only ~1 hour • If cleaning of environmental surfaces is performed, commercially available bleach or 0.5% hypochlorite solution (1:10 dilution of household bleach) is considered adequate for cleaning.	• Standard precautions are required at autopsy • N-95 masks should be used at autopsy for aerosol-generating procedures (eg, bone sawing) • Disinfect patient clothing according to standard precautions protocols
Tularemia (*Francisella tularensis*)	No	• Standard precautions • Microbiology laboratory personnel should be alerted when tularemia is suspected, and diagnostic procedures should be performed under Biological Safety Level 2 conditions	• Surface decontamination: spray the suspected contaminated area with a 10% bleach solution (1 part household bleach and 9 parts water) • Persons with direct exposure to powder or liquid aerosols containing *F. tularensis* should wash body surfaces and clothing with soap and water	• Standard precautions are required at autopsy • Aerosolizing procedures (eg, bone sawing) should be avoided • Disinfect patient clothing according to standard precautions protocols *(continued)*

Table 22-2. (continued)

Disease (causative agent)	Person-to-person spread	Infection control precautions[a]	Decontamination	Postmortem care
Botulism (*Clostridium botulinum* toxin)	No	• Standard precautions only	• Clothing and skin exposed to botulinum toxin should be washed with soap and water • Contaminated objects or surfaces should be cleaned with 0.1% hypochlorite bleach solution • Botulinum toxin degrades 2 days after aerosolization	• Standard precautions are required at autopsy

VHF (Ebola, Marburg, Lassa fever, Rift Valley fever, yellow fever viruses)	Yes: Ebola, Marburg, and Lassa fever viruses can be spread though contact with blood or body fluids; cadavers can spread virus in fluids	• Contact precautions
		• Airborne precautions: although airborne transmission is rare, it has occurred
		• HCWs coming into contact with a suspected VHF patient should wear the following PPE
		  – N-95 respirator or powered air-purifying respirator
		  – Double (leak-proof) gloves
		  – Impermeable gown
		  – Face shield
		  – Goggles for eye protection
		  – Leg and shoe coverings
		• Observation for 21 days of all persons with high-risk contact, including mucous membrane contact; percutaneous injury involving contact with secretions, excretions, or blood from a patient with VHF; or close contact (persons who live with, shake hands with, hug, process laboratory specimens from, or care for a patient with VHF)
			• Environmental surfaces, inanimate contaminated objects, or contaminated equipment should be disinfected with a 1:100 dilution of household bleach using standard procedures
			• Contaminated linens should be incinerated, autoclaved, or placed in double (ie, leak-proof) bags at the site of use and washed without sorting in a normal hot water cycle with bleach
			• Hospital housekeeping staff and linen handlers should wear appropriate PPE when handling or cleaning potentially contaminated material or surfaces
			• Decontaminate stool, fluids, and secretions before disposal: such fluids should be autoclaved, processed in a chemical toilet, or treated with several ounces of household bleach for 5 minutes or more before flushing or disposal
				• Standard precautions should be used at autopsy
				• N-95 masks should be used at autopsy
				• Avoid aerosol-generating procedures (eg, bone sawing)
				• Burial
				  – Limit contact to trained personnel
				  – Cremation is preferable

SARS (SARS CoV)	Yes, predominantly via close contact or droplets, however, airborne spread has not been excluded and may occur in some cases (via "superspreaders")	**Healthcare facility**
		• Emergency Department and ambulatory screening respiratory protocols during a SARS outbreak
		  – Negative air-pressure room or cohorting for patients who fit case definition
		  – PPE for HCWs entering room of suspected SARS patient
		  – Promote respiratory hygiene and hand washing
		• Hospitalized patients
		  – Place SARS patients in negative air-pressure room if possible or cohort during an epidemic outbreak
		  – PPE should be donned on entering the room and doffed before exiting the room
			• Designate certain personnel in hospitals to clean and disinfect rooms
			• Cleaners should wear PPE
			• Use any EPA-registered hospital detergent-disinfectant
			• Clean and disinfect SARS patient's room at least daily and more often when there is visible soiling and/or contamination
			• Clean and disinfect room after patient discharge or transfer
				• Wear PPE for autopsy
				  – Surgical mask if no aerosols
				• Use biosafety cabinets for specimens

	– Hand hygiene before and after seeing patient – Limit patient transport; if needed, patient should wear surgical mask and clean gown • Aerosol-generating procedures (eg, intubation or bronchoscopy) – Limit use – Perform in an airborne-infection isolation room – Wear PPE during procedure • HCWs who care for SARS patients – Should be monitored for fever and respiratory symptoms – Should be kept from work if they get sick **SARS patient's home** – SARS patient should remain at home until 10 days after symptoms resolve – Separate patient from others and limit caretakers – Rigorous hand hygiene by patient and household – Patient should wear a surgical mask, if possible – Contacts should wear a surgical mask – Laundry, utensils and other household items should not be shared with the patient – Disinfect surfaces, wash clothes in warm water, and bag and dispose of gloves, tissues, and other waste – Caregivers should be monitored for fever and respiratory symptoms		• For exposed persons, thoroughly wash exposed areas, especially hands, with soap and water • Use EPA-registered hospital detergent-disinfectant • Follow standard facility procedures for cleaning and disinfection of environmental surfaces • Emphasize cleaning and disinfection of frequently touched surfaces	• Follow standard facility practices for care of deceased • Standard precautions should be used for any contact with blood or body fluids
Pandemic influenza (influenza virus)	Yes: the transmission dynamics of pandemic influenza may differ from that of seasonal influenza and/or the novel "avian influenza" (influenza A virus subtype H5N1), and will depend on the characteristics of the pandemic strain	• Contact precautions • Droplet precautions • Airborne precautions[b] – Use for all personnel involved in direct care activities and when performing high-risk aerosol-generating procedures – If negative air-pressure rooms are used up, put patients in single rooms, or cohort patients		

NOTE: Table based on information from multiple studies.[14-23] EPA, Environmental Protection Agency; FDA, Food and Drug Administration; HCW, healthcare worker; OSHA, Occupational Safety and Health Administration; PPE, personal protective equipment (N-95 mask, gloves, gown, and face shield); SARS, severe acute respiratory syndrome; SARS CoV, SARS-associated coronavirus; VHF, viral hemorrhagic fever.

a Standard precautions apply to all diseases and agents. The requirements for the different types of precautions are summarized as follows. *Standard precautions:* hand hygiene; use of mask, gown, gloves, mask, face shield, or eye protection depending on exposure; safe injection practices. *Contact precautions:* wear gown and gloves for interactions that involve contact with the patient or contaminated areas around the patient. *Droplet precautions:* patient should be in single room; persons entering the room must wear a mask (a respirator is not necessary) or maintain a distance of more than 1 m (3 feet) from the patient. *Airborne precautions:* place patient in a negative air-pressure room with an anteroom; N-95 mask use required for people entering the room.

b Negative air-pressure isolation is not required for routine patient care of individuals with pandemic influenza. If possible, airborne infection isolation rooms should be used when performing high-risk aerosol-generating procedures.

the biological event is evolving internationally, nationally, and at the community level.

Early Identification of a Biological Disaster

The CDC Category A bioterrorism agents and diseases[1] and most of the recently identified emerging infections, such as SARS, "avian influenza" (caused by influenza virus A subtype H5N1), and West Nile fever,[28,29,30] present with nonspecific symptoms, especially early in the disease course. Thus, clues to a specific illness may not be evident and may not raise concern. Attention to syndrome complexes may be a better way to quickly identify unusual diseases. While much work has been done on syndromic surveillance of populations (eg, emergency department patients),[31] there has been little training of clinicians in syndrome-complex surveillance for bioterrorism or other biological disaster agents and diseases. In addressing biological disaster agents and diseases, 4 syndrome complexes emerge: (1) febrile pulmonary syndromes, (2) febrile rash syndromes, (3) febrile and nonfebrile neurologic syndromes, and (4) viral hemorrhagic fever syndromes.

Febrile Pulmonary Syndromes

Febrile pulmonary syndromes are any constellation of acute symptoms that include fevers *and* shortness of breath, cough, or dyspnea. Examples of febrile pulmonary syndromes associated with bioterrorism include anthrax,[15] plague,[32] and tularemia.[33] Biological disaster diseases that present as febrile pulmonary syndromes include SARS[28] and influenza A (H5N1).[30] At initial presentation, febrile pulmonary syndromes may not raise much concern. However, several "red flag" findings and features warrant further testing and precautions. They are as follows:

- Rapid progression of symptoms in a previously healthy young patient
- Widened mediastinum visible on chest x-ray (suggests anthrax)[15]
- Gram-positive rods growing in cultures of blood samples within 24 hours after being drawn (suggests anthrax)[15]
- A febrile pulmonary syndrome with meningeal signs and symptoms (suggests anthrax or plague)[15,32]
- Recent travel overseas (suggests SARS, influenza A [H5N1] infection, anthrax, or plague)[15,28,30,32]
- Patient is a healthcare worker
- Recent exposure to a person hospitalized for a febrile pulmonary syndrome

If the initial diagnostic workup of a patient with a febrile pulmonary syndrome does not lead to suspicion of a biological disaster disease, subsequent "red flag" features might include the following:

- No or minimal response to empiric antibiotic therapy (suggests anthrax or plague)[32]
- Enlarged mediastinal lymph nodes visible on a computed tomography scan of the chest (suggests anthrax)[15]
- Additional patients, especially contacts of the initial patient, presenting with a similar febrile pulmonary syndromes over a brief period

The presence of any febrile pulmonary syndrome and one or more of the "red flag" findings listed above should prompt immediate action, including efforts to protect staff and other patients if a contagious agent is suspected and reporting of the case to other clinicians and, especially, to the public health department. In addition, aggressive efforts to diagnose the disease and treat the patient(s) should obviously continue.

Febrile Rash Syndrome

Febrile rash syndrome is defined as a constellation of acute symptoms that include fever *and* a bodily rash. Biological disaster diseases that can produce febrile rash syndrome include smallpox, hemorrhagic fevers (Ebola fever), and anthrax[15,16,34] Recently emerging infections that present in this fashion include monkeypox and West Nile fever[29,35] Although the presence of fever and a rash is not uncommon and can be seen with common infections, cancers, drug reactions and rheumatologic conditions, the following features warrant increased suspicion of a biological attack or emerging infection:

- Vesicular or pustular rash in the same stage of development on the face and extremities and more than on the trunk (suggests smallpox or monkeypox)[16,35]
- Rash presenting several days after fever (suggests smallpox or monkeypox)[16,35]
- Diffuse rash in a toxic-appearing patient (suggests smallpox, monkeypox, hemorrhagic fevers, or West Nile fever)[16,29,34,35]
- Fever and rash and recent foreign travel (suggests anthrax, smallpox, monkeypox, or hemorrhagic fevers)[15,16,34,35]
- Fever and rash and recent contact with a patient with febrile rash syndrome (suggests smallpox, monkeypox, or hemorrhagic fevers)[16,34,35]
- Painless, edematous lesion with an eschar in a febrile patient (suggests anthrax)[15]

The presence of any febrile rash syndrome and one or more of the "red flag" findings listed above should prompt immediate action, including efforts to protect staff and other patients, if a contagious agent is suspected, and reporting of the case to other clinicians and, especially, to the public health department. In addition, aggressive efforts to diagnose the disease and treat the patient(s) should continue.

Neurological Syndromes

Neurological syndromes are any constellation of acute symptoms involving diffuse or focal weakness and/or symptoms of meningitis (nuchal rigidity, headache, photophobia, and/or lethargy) or encephalitis (altered mental status, motor and sensory deficits, and/or speech disorder). Neurological syndromes occur with several bioterrorism-related agents and diseases, including anthrax, botulism, and plague. Also, like West Nile fever, emerging infections may present with an encephalitic clinical picture. Clinicians are familiar with neurological complaints among their patients. Strokes occur 500,000 times per year in the United States,[36] and there are thousands of cases of viral and bacterial meningitis per year.[37] However, there are some presentations of neurological syndromes that should raise suspicion of a biological disaster–associated disease or an emerging infection. If a patient presents with a neurological syndrome consisting of no fever and a descending, symmetric, flaccid paralysis that begins in the bulbar muscles, a diagnosis of botulism should be considered.[38] If the patient has no history of a recent wound or obvious food exposure, then the possibility of bioterrorism should be entertained.[38] If a patient presents with a febrile neurological syndrome, the following features should provoke suspicion:

- Encephalitis in a young previously healthy adult (suggests West Nile fever, St. Louis encephalitis, Venezuelan equine encephalitis, or Eastern equine encephalitis)[29]
- Meningitis due to gram-positive rods in an immunocompetent host (suggests anthrax)[15]
- Meningoencephalitis with diffuse weakness, blindness, or other focal deficits (suggests West Nile fever)[29]

Viral Hemorrhagic Fever Syndrome

Viral hemorrhagic fever syndrome (VHFS) results from infection with one of several viruses, including filoviruses (Ebola and Marburg viruses), arenaviruses (Lassa fever virus), bunyaviruses (Rift Valley fever virus), and flaviviruses (yellow fever virus).[34] In the early stages, VHFS symptoms are nonspecific: patients present with fever, headache, myalgias, nausea, and nonbloody diarrhea.[34] Patients may also have conjunctivitis, pharyngitis, and a nonspecific rash. As the disease progresses, a hemorrhagic diathesis ensues and may include petichiae, mucous membrane and conjunctival hemorrhage, hematuria, hematemesis, and melena.[34] VHFS in any patient warrants increased suspicion of bioterrorism or an emerging infection, especially if the patient has not recently traveled. Some hemorrhagic fever viruses (eg, yellow fever and dengue fever viruses) are common in parts of Africa, Southeast Asia, and Central and South America, and travelers to these areas may present with VHFS.[34] However, hemorrhagic fever viruses are uncommon in North America and Europe. Because certain hemorrhagic fever viruses (Ebola, Marburg, and Lassa fever viruses and New World arenaviruses) are transmissible from person to person,[34] immediate measures should be undertaken to protect others from any patient thought to have VHFS. Also, such cases should be reported to the health department as soon as possible. Rarely, smallpox can present in a hemorrhagic form that is rapidly fatal and highly contagious.[16]

Other Clues to a Biological Disaster

The following are early clues to a biological disaster that may not fit into the 4 categories discussed above:[39,40]

- Animals and humans ill at the same time (suggests bioterrorism agents or West Nile virus)
- Two or more previously healthy young patients presenting with rapid sepsis
- Clusters of patients presenting with a similar presentation (suggests SARS, influenza A [H5N1] infection, or a bioterrorism-associated disease)
- Unusual temporal or geographic clustering of illness (eg, a severe influenza outbreak in the summer, or Ebola hemorrhagic fever in a region where it is not endemic, such as the United States)
- Unusual presentation of an illness (inhalational anthrax vs cutaneous anthrax)
- Higher morbidity and mortality than expected with a common disease or syndrome
- Multiple unusual or unexplained disease entities coexisting in the same patient without other explanation
- A causative agent that is unusual, atypical, genetically engineered, or is an antiquated strain
- Simultaneous clusters of similar illness in noncontiguous areas, domestic or foreign
- Multiple atypical presentations of diseases or agents

Public Health Surveillance Systems

Since the terrorist attacks on the United States on September 11, 2001, the US public health sector has been developing surveillance systems for the detection of infectious disease outbreaks, including bioterrorism events. Traditional surveillance relies upon the provider, the infection preventionist, the healthcare epidemiologist, or the laboratory to report any cases of disease of public health significance to the local public health department, whether the cases are laboratory diagnosed or clinically suspected.[41] At this point, all state public health departments have incorporated electronic disease-reporting systems, many of which are accessible by hospital-based providers. This traditional disease reporting system can alert public health authorities at any time that the patient has already entered the healthcare system with clinical signs and systems, which, unfortunately, may be long after the possibly undetected exposure event.

The purpose of a syndrome surveillance system is to attempt to provide an earlier warning of an infectious disease outbreak than traditional systems can.[31,42] Many of these systems utilize hospital-based information, such as the surveillance of the chief complaints of emergency department patients, registered at presentation to the emergency department, which can detect specific syndrome complexes (see the previous section). Numerous states have implemented syndromic surveillance systems, and there are several national initiatives, which are described below.

BioSense

BioSense, provided by the CDC and in operation in the United States since November 2003, is an Internet-based syndromic surveillance application for the early detection of both intentional and natural infectious disease outbreaks.[43] BioSense receives *International Classification of Diseases, Ninth Revision, Clinical Modification (ICD-9-CM)* diagnosis and procedure codes from US Department of Defense and the Department of Veterans Affairs ambulatory care visits; the system also receives pharmacy over-the-counter sales information from various retail outlets, and the Laboratory Centers of America provides information on laboratory tests ordered. These data are analyzed daily at the CDC.[44]

Real-Time Outbreak and Disease Surveillance System

Software has been developed by the Real-Time Outbreak and Disease Surveillance (RODS) Laboratory, a collaboration of the University of Pittsburgh and Carnegie Mellon University. This software can collect and analyze many types of clinical data, such as emergency department chief complaints or orders for particular laboratory tests, and is used by a number of state public health departments for early detection of infectious disease or bioterrorism events.[43]

Electronic Surveillance System for the Early Notification of Community-Based Epidemics

This is a syndromic surveillance system, abbreviated ESSENCE, provided by the US Department of Defense that automatically collects *ICD-9* codes at participating military treatment facilities. Analysis of the frequency and distribution of 7 syndrome types is provided: respiratory, gastrointestinal, fever, neurologic, dermatologic (infectious), and dermatologic (hemorrhagic) syndromes, and coma or sudden death.[43,45]

Disease Reporting Requirements

Hospitals and their emergency departments may be where a victim of a bioterrorism event may be suspected and subsequently identified. Healthcare epidemiologists, along with healthcare providers, infection preventionists, and laboratorians must immediately report any *suspected* case to their local or state public health department.[41] In many states, this important reporting requirement is encoded in legal statute, yet the list of diseases required to be reported can vary between states and territories.[41] Because of the public health importance of early and rapid reporting of diseases of public health importance, especially of those that may indicate natural or intentional release of potential bioterrorism agents, healthcare professionals are integral to public health efforts at the local, state, and federal levels.[41] This early reporting of even suspected cases allows the immediate public health investigation and response that may be required, and reporting should not be deferred for confirmatory testing.

In fact, confirmatory testing for any biological disaster–associated agent or disease should be performed under specific biosafety conditions within the state public health laboratory. At the time of reporting, public health workers will provide information to the provider regarding specific and safe specimen collection and transportation to the closest diagnostic public health laboratory. The healthcare epidemiologist will need to ensure the hospital laboratory's capabilities for securely obtaining, packaging, and transporting any samples suspected of harboring bioterrorism agents to the directed public health laboratory, so that diagnostic testing can be performed. Again, involvement by the healthcare epidemiologist with local and state public health departments, as well as local, regional, and state biopreparedness groups, will allow familiarity with the response plans of public health, emergency management, law enforcement, and surrounding hospitals.

Biological Disaster Agents and Diseases

Many diseases and agents are capable of causing a widespread biological disaster, and the CDC has identified more than 60 agents or diseases that might be used as a biological weapon.[40] This chapter will focus on the 6 CDC Category A bioterrorism agents and diseases (anthrax, smallpox, tularemia, plague, botulinum toxin, and viral hemorrhagic fevers), SARS, and influenza A (H5N1) infection (avian influenza). These agents and diseases represent a wide range of presentations and modes of spread and, thus, provide an excellent framework for preparedness activities and education.

Bioterrorism Agents and Diseases

Bioterrorism is the malevolent use of bacteria, viruses, or toxins against humans, animals, or plants in an attempt to cause harm and to create fear. Although concern about bioterrorism has always existed, the 2001 anthrax attack in the United States that killed 5 people and prompted preemptive treatment of more than 10,000 others has focused attention on this form of terrorism as an imminent threat to national security.[46,47] To understand this concern, one must be familiar with the pathogens that are most likely to be used in any future attack and must be aware of the available methods for detecting and responding to such an event.

The use of biological agents for warfare has a centuries-old history, predating even the concept of germ theory.[48] The first documented case was in 1346 when the Tartars, frustrated after years of laying siege to the Black Sea city of Kaffa, catapulted plague victims over the unassailable city walls.[48] The Black Plague epidemic that followed and eventually spread from Kaffa wiped out almost half of Europe. In 1763, Sir Jeffrey Amherst, the Commander of British troops in America, sanctioned the use of blankets contaminated with smallpox virus as germ warfare implements against the American Indians, who were highly susceptible to this deadly virus.[48]

In World War I, the Germans infected cattle destined for consumption by Allied forces with anthrax and glanders. This act resulted in the 1925 Geneva Protocol that prohibited the use of biological weapons. However, in spite of this agreement and the 1972 Biological and Chemical Weapons Convention, several nations continued to produce biological weapons. Biopreparat, the biological weapons program of the Soviet Union, was the largest in the world, with 10,000 scientists working in 50 production facilities. In 1979 an accidental release of weaponized anthrax from a production plant in Sverdlovsk resulted in 66 deaths downwind of the facility. The program was dismantled in 1992 after Russian president Boris Yeltsin finally admitted that it existed.[48]

Aum Shinrikyo, a Japanese cult, attempted several unsuccessful biological attacks with anthrax and botulinum toxin before releasing sarin gas in the Tokyo subway in 1995.[49] The most successful biological attack in the United States was perpetrated by a religious cult, the Rajneeshees. In 1984, in an attempt to affect elections in a small Oregon town, the cult poisoned 10 restaurant salad bars with *Salmonella enterica* serotype Typhimurium and sickened more than 700 people.[48]

Biological weapons, unlike other weapons of mass destruction, are inexpensive to make. Nuclear and chemical weapons programs are 800 times and 600 times more costly, respectively, than a comparable bioweapons program. The pathogens are relatively available, and the materials and equipment for producing biological weapons are the same as those used for peaceful purposes. For example, the organism that causes anthrax, *B. anthracis,* is present in the soil in many countries, and the organism can be grown in standard laboratory culture medium. The *S.* Typhimurium used in the Oregon attack was easily obtained by a member of the Rajneeshee cult. The necessary culture media, incubators, and milling equipment are available for purchase, and information on cultivating the organisms and on generating antibiotic-resistant strains is available in the scientific literature and from the internet.

Biological weapons produce fear and panic, as was apparent in the 2001 anthrax attack.[47] The weapon can be released covertly, and its effects become apparent only days later, when the terrorist is gone. Unlike the situation with chemical and nuclear events, the point of release of a biological agent may not be apparent. If it is a contagious agent, the uncertainty about who has been exposed could lead to widespread panic. Finally, the lethality of a biological attack could exceed that of other weapons of mass destruction. By one governmental estimate, if 50 kg of powdered anthrax similar to that used in the 2001 attack in the United States was released over a city of 500,000 people, it would cause death in 95,000 and incapacity in 125,000.[50]

The characteristics of an ideal agent for bioterrorism include accessibility, durability, infectiousness, and communicability. Although hundreds of agents and toxins may qualify, the CDC has developed categories (A, B, and C) on the basis of the perceived threat. Category A agents pose the greatest threat, and these agents are discussed individually in this chapter.

Category A agents can be released by aerosolization, contamination of food, contamination of water, person-to-person transmission of a contagious agent, or release of infected insect vectors. Aerosolization—the dispersion of organisms into the environment—is the most feared and potentially lethal method of releasing biological agents. As mentioned above, a very small amount of weaponized anthrax (50 kg) could result in very high casualties (95,000 dead).[50] The Russian bioweapons program spent considerable resources developing effective methods of aerosolizing biological agents, especially the pathogens that cause anthrax, tularemia, and smallpox. Both the 1979 Sverdlovsk outbreak of anthrax and the 2001 anthrax attack in the United States involved the release of finely powdered *B. anthracis* into the environment. Means of aerosolizing a biological agent include spraying devices (crop dusters), air-handling systems in buildings, incendiary devices (bombs), and the postal system (infected mail).[15] However, aerosolization of a biological agent is not easy. Infectious particles must be precisely the right size (0.5–5 microns) to enter and to infect the lungs.[15] Furthermore, radiation from sunlight, shear forces from sprayers, or explosions from incendiary devices would likely destroy most of the released organism.

Contamination of the food or water supply with biological agents is another method of dissemination. The contamination of salad bars with *S.* Typhimurium in Oregon is an example of a successful attack. The agents most likely to be disseminated in this fashion include botulinum toxin, *B. anthracis,* and diarrheal agents, including *S.* Typhimurium. However, contamination of food and water would be difficult because of food inspection criteria, the use of water purification and filtering systems, and increased security at water reservoirs and food distribution centers.

Dissemination through person-to-person transmission is a likely means of dispersing a contagious agent, such as variola virus (smallpox). As few as 100 smallpox-infected people could start an epidemic in a nonimmune population. The fear, panic, and quarantine measures that would result from such an epidemic could rapidly overwhelm healthcare resources and destabilize the affected country or countries.

Finally, zoonotic delivery, or the use of insect vectors to disperse a biological agent, is an unlikely method of bioterrorism. It is inefficient and unpredictable. The agents most likely to be delivered in this way include *Y. pestis* (plague) or viral hemorrhagic fever viruses (Ebola hemorrhagic fever virus and Lassa hemorrhagic fever virus).

Category A Biological Agents and Diseases

The CDC defines Category A bioterrorism agents as those that can be easily disseminated or transmitted from person to person, that result in high mortality rates and have the potential for major public health impact, that might cause public panic and social disruption, and that require special action for public health preparedness.[1] The 6 agents and diseases on the list are *B. anthracis*

(anthrax), *C. botulinum* toxin (botulism), *Y. pestis* (plague), variola virus (smallpox), *Francisella tularensis* (tularemia), and hemorrhagic fever viruses.[1] The clinical presentation and diagnosis of the diseases caused by Category A agents are summarized in Table 22-3; treatment, prophylaxis, and vaccination strategies are summarized in Table 22-4.

Anthrax

B. anthracis is an ideal bioterrorism agent because (1) the organism forms a hardy spore that can exist in the environment for years and can survive aerosolization, (2) it is readily available in the soil in many countries, (3) it produces a severe inhalational form of disease with a high mortality rate, and (4) the organism can be manipulated to create antibiotic-resistant strains.[15] Several countries have developed *B. anthracis* into a biological weapon, and in 2001 this agent was used in a biological attack in the United States that created widespread panic and 5 deaths.[15,47]

B. anthracis causes 3 distinct diseases; cutaneous anthrax, gastrointestinal anthrax, and inhalational anthrax (Table 22-3).[15] Anthrax meningitis is associated with disseminated anthrax and occurs in 50% of people with inhalational anthrax. While a bioterrorism attack could manifest as any of these forms, a case of inhalational anthrax should immediately raise suspicion of a biological attack. Aerosolized forms of *B. anthracis* have been developed by several countries, and a finely milled form (Ames strain) was used in the 2001 attack.[15] Anthrax is not transmissible from person to person and, therefore, requires no isolation of infected patients (Table 22-2).[15]

Cutaneous anthrax is quite common in areas around the world where it is endemic, and the mortality rate is very low (less than 1%) if treatment is prompt.[15] Gastrointestinal, meningeal, and inhalational anthrax have much higher mortality rates, but the 2001 anthrax attack revealed that prompt therapy with a multiple-antibiotic regimen reduced the mortality rate from the 90% seen previously to less than 50%.[15]

Treatment and prophylactic regimens are detailed in Table 22-4. Generally, inhalational, gastrointestinal, and meningeal anthrax require multiple antibiotics to achieve cure, whereas cutaneous anthrax can be treated with a single antibiotic.[15] Prophylaxis is administered to exposed and potentially exposed persons. Anthrax vaccine is available and is administered both before exposure (for researchers or military personnel) and after exposure to aerosolized anthrax (Table 22-4).[15]

Smallpox

Smallpox, or variola major, is an effective biological weapon because it is highly contagious and has a high mortality rate and because there is no treatment for the

Table 22-3. Clinical presentation and diagnosis of biological disaster–associated diseases and agents

Disease (causative agent)	Clinical presentation	Incubation period	Diagnosis
Anthrax (*Bacillus anthracis*)	**Cutaneous** • Papule to vesicle to eschar, painless, extensive surrounding edema • Lymphangitis may occur with painful lymphadenopathy • Fevers and chills may occur with dissemination • Eschar drys and falls off over 1–2 weeks **Inhalational** • Stage 1: Fever, dyspnea, cough, headache, vomiting, chills, weakness, abdominal pain, and chest pain. May improve slightly after a few days • Stage 2: Abrupt fever, dyspnea, stridor, cyanosis, diaphoresis, and shock with rapid death; lymphadenopathy may occur, and 50% of patients get meningitis **Gastrointestinal** • Upper symptoms: sore throat, dysphagia, oral or esophageal ulcers, nausea, vomiting, fevers, and hemoptysis • Lower symptoms: fevers, abdominal pain, bloody stools, lymphadenopathy, ascites, acute abdomen, and sepsis **Meningitis** • Headache, neck stiffness, photophobia, fever, chills, and altered mental status	**Cutaneous:** 1–10 days **Inhalational:** 1–6 days (residual spores may cause disease up to 60 days later) **Gastrointestinal:** 3–7 days (residual spores may cause disease up to 60 days later) **Meningitis:** 1–6 days (residual spores may cause disease up to 60 days later)	**Cutaneous** • Culture of vesicular fluid, skin biopsy sample, or blood may grow *B. anthracis* • Gram staining of vesicular fluid, wound exudate, or punch biopsy may show gram-positive bacilli • Perform PCR or IHS of punch biopsy of the skin (send to the CDC) **Inhalational** • Gram staining of blood buffy coat and/or blood showing gram-positive rods may be the earliest sign of infection • Blood culture positive for a gram-positive rod within 6–24 hours after onset of infection (unless antibiotics given before) and positive for *Bacillus* species within 48 hours — Antimicrobial susceptibility testing should be performed • Perform PCR or IHS of blood, pleural fluid, and spinal fluid (send to the CDC) • Bloody pleural fluid • Send suspect samples to the CDC for confirmation of diagnosis • Sputum culture and Gram stain are rarely helpful • Radiologic findings: chest radiograph showing widened mediastinum and/or pleural effusion and/or infiltrates; chest CT showing mediastinal lymphadenopathy, pleural effusion and/or pulmonary infiltrates • Nasal swab samples and blood antibody levels are useful only for epidemiologic purposes **Gastrointestinal** • Same laboratory tests as for inhalational • Stool cultures are not helpful • Radiology: abdominal x-ray with bowel obstruction; abdominal CT showing lymphadenopathy and thickened bowel **Meningitis** • Bloody CSF • Gram stain showing gram-positive rods • Culture of CSF positive for a gram-positive rod within 6–24 hours after onset of infection (unless antibiotics given before) and positive for *Bacillus* species within 48 hours **Postmortem** • Hemorrhagic necrotizing lymphadenitis, meningitis, and mediastinitis
Smallpox (variola virus)	**Prodrome** (duration, 2–4 days) • Sometimes contagious • Fever: temperature in the range of 38.3°C–40°C (101°F–104°F)	12–14 days (range, 7–17 days)	• Use the CDC rash protocol • Laboratory diagnostic testing for variola virus should be conducted in a CDC LRN laboratory using LRN-approved PCR tests and protocols for variola virus

(continued)

Table 22-3. (continued)

Disease (causative agent)	Clinical presentation	Incubation period	Diagnosis

	• Malaise
	• Head and body aches and prostration
	• Abdominal pain and delirium may be present
	• Occasionally, vomiting
	Early rash (duration, ~4 days)
	• Most contagious period
	• Rash appears as small red spots on the tongue and in the mouth (enanthem)
	• Rash develops into sores that break open and spread large amounts of the virus into mouth and throat
	• Rash then appears on the skin, starting on the face and spreading to the arms and legs and then to the hands and feet
	• By the third day of the rash, the rash becomes small bumps
	• By the fourth day, the vesicles become pustular and fill with a thick, opaque fluid, and often will be umbilicated
	• Fever often will increase again at this time and remain high until scabs form over vesicles
	• Patient may appear critically ill
	Pustular rash (duration, ~5 days)
	• Contagious
	• Rash becomes more vesicular, round, tense, and deeply embedded in the dermis
	• Patient appears critically ill
	Pustules and scabs (duration, ~5 days)
	• Contagious
	• Crusts begin to form on about day 8 or day 9 after rash onset
	Resolving scabs (duration, ~6 days)
	• Contagious
	• The scabs begin to fall off, leaving pitted scars
	• Most scabs will have fallen off by 3 weeks after rash onset
	• The patient is contagious until all the scabs have fallen off
	Scabs resolved
	• Scabs have fallen off
	• Patient is no longer contagious
	Other forms
	• Hemorrhagic, flat smallpox
	• Difficult to diagnose, with a high mortality rate
			• Perform culture of vesicular or pustular fluid specimen(s)
			  – Specimen should be collected by someone who has recently been vaccinated and who is wearing appropriate PPE
			  – To obtain vesicular or pustular fluid, lesions may need to be opened with the blunt edge of a scalpel. The fluid can then be harvested on a cotton swab
			• Diagnostic laboratory criteria
			  – PCR identification of variola virus DNA in a clinical specimen
			  – Isolation of variola virus from a clinical specimen with PCR confirmation
			• Initial confirmation of a smallpox outbreak requires additional confirmatory testing at the CDC

Plague (*Yersinia pestis*)	**Foodborne**
	• Gastrointestinal
	  – Initial: symptoms can include nausea, vomiting, abdominal cramps, or diarrhea
	  – Constipation (after neurological symptom onset)
		Pneumonic, septicemic, pharyngeal, meningitis: 1–6 days	**Pneumonic**
			• Gram stain of sputum, blood, or lymph node aspirate showing gram-negative bacilli or coccobacilli with bipolar ("safety pin") staining

Bubonic (from flea bite): 2–8 days

- Neurological
  - Initial: dry mouth, blurred vision, and diplopia
  - Dysphonia, dysarthria, dysphagia, and peripheral muscle weakness
  - Characteristic proximal-to-distal pattern: symmetric descending paralysis, begins with the cranial nerves, affects upper extremities, then respiratory muscles, and then the lower extremities

Wound

- Gastrointestinal: none
- Neurological: same as above

Infant

- Gastrointestinal
  - Constipation
  - Poor feeding
- Neurological
  - Weak cry
  - Poor muscle tone and floppy head
  - Decreased sucking and lethargy

Intentional, or bioterrorism, event

- Gastrointestinal: similar to those for *Foodborne* illness, above
- Neurological: similar to those for *Foodborne* illness, above
- *Inhalational* events may not be associated with gastrointestinal symptoms, just neurological

---

- Culture of sputum, blood, or lymph node aspirate that yields gram-negative bacilli or coccobacilli after 24–48 hours of growth, with later identification as *Yersinia* species
  - Antimicrobial susceptibility testing should be performed
- Antigen detection, IgM EIA, immunostaining, and PCR are available only from the CDC and state laboratories
- Leukocytosis, coagulation abnormalities, elevated aminotransferase levels, azotemia, and other evidence of multiorgan failure
- Radiologic findings: chest radiograph showing lobar pneumonia or bilateral infiltrates and/or pleural effusions

Septicemic

- Same as for *Pneumonic* except chest radiograph may not show pneumonia, and there may be no buboes

Bubonic

- Gram stain of lymph node aspirate showing gram-negative bacilli or coccobacilli with bipolar ("safety pin") staining
- Culture of lymph node aspirate that yields gram-negative bacilli or coccobacilli after 24–48 hours of growth, with later identification as *Yersinia* species

Pharyngeal

- Culture of throat specimen that yields gram-negative bacilli or coccobacilli
- Culture of lymph node aspirate that yields gram-negative bacilli or coccobacilli after 24–48 hours of growth, with later identification as *Yersinia* species

Meningitis

- Culture of CSF and Gram stain of CSF that show gram-negative bacilli or coccobacilli with bipolar ("safety pin") staining

Postmortem

- Lobular exudation, bacillary aggregation, and areas of necrosis in pulmonary parenchyma

---

Tularemia (*Francisella tularensis*)

All forms: 1–14 days

Pneumonic

- Abrupt onset of fever, headache, chills, myalgias, and sore throat
- Dry or slightly productive cough, substernal chest pain, or tightness
- Nausea, vomiting, and diarrhea
- Symptoms can last weeks to months

Typhoidal

- Fevers, chills, myalgias, abdominal pain, and diarrhea
- No regional lymphadenopathy or skin lesions

Oropharyngeal

- Exudative pharyngitis, tonsillitis, and oral ulcerations
- Cervical lymphadenopathy
- Fevers, headache, chills, and myalgias

Pneumonic

- High index of suspicion, given nonspecific symptoms
- Gram stain of blood, sputum, or lymph node aspirate: *F. tularensis* is a gram-negative coccobacillus that stains poorly
- Culture of sputum, pharyngeal swabs, lymph node aspirate, and rarely blood: *F. tularensis* is difficult to culture and requires cysteine-enriched agar
  - Highly infectious, *so microbiology laboratory must be notified if tularemia is suspected*
  - Antimicrobial susceptibility testing should be performed in special laboratories
- Fluorescent antibody staining is performed in special laboratories
- Antigen detection assays, PCR, ELISAs, immunoblotting, and pulsed-field gel electrophoresis are done only in special laboratories

(continued)

273

Table 22-3. (continued)

Disease (causative agent)	Clinical presentation	Incubation period	Diagnosis
	Glandular • Regional lymphadenopathy without skin or ocular lesions • Fevers, headache, chills, and myalgias **Ulceroglandular and oculoglandular** • Less likely to occur from a biological attack • Ulcerative skin or ocular lesions with regional lymphadenopathy • Fevers, headache, chills, and myalgias		• Serum antibody levels: 4-fold increase in titer or single IgG titer of 1:160 (these findings only appear after 10 days of illness) • Leukocytosis, coagulation abnormalities, elevated aminotransferase levels, azotemia, and other evidence of multiorgan failure • Radiologic findings: chest radiograph and/or chest CT showing bronchopneumonia in 1 or more lobes, hilar adenopathy, pleural effusion • Pathology and/or postmortem: acute necrosis in involved tissues (lungs, lymph nodes, spleen, liver, and kidney) with a granulomatous reaction **Typhoidal, glandular, ulceroglandular, pharyngeal, oculoglandular** • Same as above, without pulmonary tests and including skin and/or ocular specimens
Botulism (*Clostridium botulinum* toxin)	**Ingestion or inhalation of pure toxin** • Bulbar weakness: blurred vision, double vision, dysarthria, dysphagia, and sore throat – The "4 Ds": diplopia, dysarthria, dysphonia, and dysphagia • Symmetric descending paralysis • No fever • Clear sensorium (patient is alert) • Fatigue and constipation • Physical examination findings: ptosis, gaze paralysis, dilated pupils, facial palsy, diminished gag reflex, tongue weakness, decreased reflexes, and ataxia	2 hours to 8 days, depending on the toxin dose absorbed; the *Inhalational* form has effect by 3 days	• Clinical presentation is the way to diagnose • Results of routine testing are unremarkable • Serum, stool, gastric aspirates, vomitus, and suspected food or liquid should be sent to the CDC or state laboratory for testing for mouse bioassay (detects 0.03 ng of botulinum toxin in 1–2 days); also, *C. botulinum* will grow in anaerobic culture in 7–10 days • Electromyogram findings: normal nerve conduction velocity, normal sensory nerve function, a pattern of brief small-amplitude motor potentials, and an incremental response (facilitation) to repetitive stimulation often seen only at 50 Hz • CSF findings: normal • Rule out Guillain-Barré syndrome and myasthenia gravis: Guillain-Barré syndrome has ascending paralysis, and both it and myasthenia gravis have characteristic electromyogram findings
VHF (Ebola, Marburg, Lassa fever, Rift Valley fever, yellow fever viruses)	**Filovirus** (Marburg and Ebola hemorrhagic fever) • Fever, headache, myalgias, abdominal pain, nausea, and vomiting • Maculopapular rash 5 days after onset of symptoms and jaundice • Hemorrhagic complications: petichiae, hematemesis, hemoptysis, purpura, bloody diarrhea, bleeding at puncture sites, shock, and DIC • Delerium, coma, and seizures **Arenavirus** (Lassa fever) • Fever, headache, malaise, arthralgias, back pain, and nonproductive cough • Maculopapular rash, prostration, and exudative pharyngitis • Hemorrhagic complications: see *Filovirus*, above	**Filovirus** (Marburg and Ebola hemorrhagic fever): 2–21 days **Arenavirus** (Lassa fever): 5–16 days **Bunyavirus** (Rift Valley fever): 2–6 days **Flavivirus** (yellow fever): 3–6 days	• World Health Organization surveillance standards for VHF (all must be present) – Patient has acute onset of fever of less than 3 weeks' duration and is severely ill – Patient has no known predisposing host factors for hemorrhagic manifestations – Patient has any 2 of the following symptoms: hemorrhagic or purpuric rash, epistaxis, hematemesis, hemoptysis, blood in stool, or other hemorrhagic symptom • Organism identification of any VHF agent must be performed at the CDC, and specimens should be collected and sent in consultation with local public health department and the CDC • Serologic testing (ELISA), PCR, and viral isolation (requires Biological Safety Level 4 facility) **Filovirus** (Marburg and Ebola hemorrhagic fever)

Bunyavirus (Rift Valley fever)
- Fever, headache, photophobia, and retro-orbital pain
- Hemorrhagic complications: see *Filovirus*, above
- Encephalitis: confusion, lethargy, tremors, ataxia, coma, seizures, meningismus, vertigo, and choreiform movements
- Retinitis: bilateral; hemorrhages, exudates, and cotton wool spots may be visible on macula; retinal detachment may occur
- Hepatitis

Flavivirus (yellow fever)
- Fever, headache, myalgias, facial flushing, conjunctival injection, and relative bradycardia
- Hemorrhagic complications: see *Filovirus*, above
- Malignant disease: severe hepatic involvement, bleeding manifestations, renal failure, shock, and death

- Laboratory findings: leukopenia, thrombocytopenia, elevated levels of liver enzyme and amylase, DIC

Arenevirus (Lassa fever)
- Laboratory findings: elevated levels of liver enzymes and hemoconcentration

Bunyavirus (Rift Valley fever)
- Laboratory findings: thrombocytopenia, leukocytosis initially followed by leukocytopenia, DIC, and elevated levels of hepatic enzymes

Flavivirus (yellow fever)
- Laboratory findings: leukopenia, thrombocytopenia, elevated levels of liver enzymes, and hyperbilirubinemia

SARS (SARS CoV)

Prodrome — **Prodrome:** 2–7 days
- Influenza-like symptoms, including fever, myalgias, headache, and diarrhea

Early respiratory phase — **Early respiratory phase:** 1–5 days
- Dry, nonproductive cough and mild dyspnea

Late respiratory phase — **Late respiratory phase:** 7 days
- Progressive hypoxia and dyspnea on exertion (10%–20% of patients require mechanical ventilation)

Case definition
- Based on clinical, epidemiological, and laboratory criteria[a]

RT-PCR
- Positive result must be confirmed by testing for another region of the SARS virus genome
- Test nasopharyngeal swab specimens (high sensitivity even in the first 5 days), and pharyngeal and stool specimens
- Serum specimens do not have as high sensitivity but should be sent
- Viral quantification may be helpful in infection control

Antibody assays
- Testing for IgM will not be positive until 7 days after symptoms begin; testing for IgG is positive after 20–26 days, so serologic testing in the acute period is not helpful
- Serologic testing can be done by immunofluorescent assays, ELISAs, and Western blot assays

Antigen detection
- Serum EIA for SARS-CoV N protein is good for detection of early disease (<7 days)
- Sensitivity decreases after 7 days
- For nonserum specimens, antigen EIA not as sensitive as RT-PCR

General laboratory tests are not helpful

Radiologic findings
- Chest radiograph and chest CT with nonspecific bilateral infiltrates

(continued)

Table 22-3. (continued)

Disease (causative agent)	Clinical presentation	Incubation period	Diagnosis

Pandemic influenza	**Influenza-like illness**
	• Fever (temperature >38°C)
	• Sore throat
	• Cough
	
	Other possible signs and symptoms
	• Headache
	• Fatigue
	• Rhinorrhea and nasal congestion
	• Myalgias
	• Gastrointestinal symptoms may occur (more commonly in children), such as nausea, vomiting, and diarrhea
	• More severe presentations may occur, such as shortness of breath and/or dyspnea, hypotension, pneumonia, "cytokine storm," shock, or death
	• Presentation may also be altered by secondary or underlying infections
		1–7 days (E24)	• Respiratory-tract specimens are best collected within the first 3 days after illness onset: specimens of (1) nasopharyngeal wash and/or aspirate, (2) nasopharyngeal swab, (3) oropharyngeal swab, (4) broncheoalveolar lavage, (5) tracheal aspirate, (6) pleural fluid tap, and (7) sputum, and (8) autopsy specimens
			• Nasopharyngeal wash and/or aspirate samples are the specimen of choice for respiratory virus detection
			• Other types of specimens may be requested, depending on the pandemic strain
			• Viral isolation is usually the "gold standard" for diagnosis of influenza
			• Real-time RT-PCR is the preferred assay to detect influenza A (H5N1)
			  – Biological Safety Level 2 conditions required: the CDC has made H5-specific primers and probes available to state health department laboratories (E24)
			• Use immunofluorescence antibody staining after virus isolation to identify influenza type (A, B) and influenza A hemagglutinin subtypes using a specific panel of antisera
			• Some rapid diagnostic tests may be able to detect the pandemic strain with adequate sensitivity and specificity during a pandemic
			• Testing of serum with ELISA, hemagglutination inhibition, and microneutralization assays can be used to confirm influenza infection retrospectively
			• Postmortem specimens (the state health department will arrange with the CDC if deemed necessary for novel or pandemic strain influenza case): a minimum total of 8 blocks or fixed-tissue specimens representing samples from each of the following sites should be obtained
			  – Central (hilar) lung with segmental bronchi
			  – Right and left primary bronchi
			  – Trachea (proximal and distal)
			  – Representative pulmonary parenchyma from right and left lung
			• Radiologic findings: chest radiograph showing pleural effusion and/or infiltrates; chest CT showing pleural effusion and/or pulmonary infiltrates

NOTE: Table based on information from multiple studies.[14-22] CDC, Centers for Disease Control and Prevention; CT, computed tomography; CSF, cerebrospinal fluid; DIC, disseminated intravascular coagulation (prolonged bleeding time, prothrombin time, and activated partial thromboplastin time; elevated levels of fibrin degradation products; decreased level of fibrinogen); EIA, enzyme immunoassay; ELISA, enzyme-linked immunosorbent assay; IHS, immunohistochemical staining; LRN, Laboratory Response Network; PCR, polymerase chain reaction; RT-PCR, reverse-transcriptase PCR; PPE, personal protective equipment (N-95 mask, gloves, gown, and face shield); SARS, severe acute respiratory syndrome; SARS CoV, SARS-associated coronavirus; VHF, viral hemorrhagic fever.
a See the CDC guide.[51]

Table 22-4. Treatment, prophylaxis, and vaccination for biological disaster–associated diseases and agents

Disease (causative agent)	Treatment	Prophylaxis	Duration	Vaccination
Anthrax (*Bacillus anthracis*)	**Cutaneous** ● Adults (including pregnant women[a]): ciprofloxacin, 500 mg by mouth twice per day, **or** doxycycline, 100 mg by mouth twice per day ● Children: ciprofloxacin, 10–15 mg/kg every 12 h (not to exceed 1 g per day), **or** doxycycline in age- and weight-dependent dosage: aged >8 y and weight >45 kg, 100 mg every 12 h; aged >8 y and weight ≤45 kg, 2.2 mg/kg every 12 h; aged ≤8 y, 2.2 mg/kg every 12 h **Inhalational or gastrointestinal**[b] ● Adults (including pregnant women[a]): ciprofloxacin, 400 mg every 12 h, **or** doxycycline, 100 mg every 12 h PLUS 1 or 2 additional antimicrobials[c] ● Children[a]: ciprofloxacin, 10–15 mg/kg every 12 h, **or** doxycycline in age- and weight-dependent dosage: aged >8 y and weight >45 kg, 100 mg every 12 h; aged >8 y and weight ≤45 kg, 2.2 mg/kg every 12 h; aged <8 y, 2.2 mg/kg every 12 h PLUS 1 or 2 additional antimicrobials[c] **Meningitis** ● Should be suspected and treated in all cases of systemic anthrax ● Same treatment as for inhalational or gastrointestinal forms ● Include treatment with 1 or more antibiotics that penetrate the central nervous system (ampicillin or penicillin, meropenem, rifampin, or vancomycin) ● Consider steroid therapy as an adjunct to antibiotics **Immunotherapeutics:** Human-derived anthrax immune globulin has been successfully used, but studies are lacking; this drug must be obtained from the CDC	**PEP with antibiotics** ● Adults: *initial PEP*, ciprofloxacin, 500 mg by mouth twice per day; *alternative PEP*, doxycycline, 100 mg by mouth twice per day, **or** amoxicillin, 500 mg by mouth 3 times per day ● Children: *initial PEP*, ciprofloxacin, 20–30 mg/kg per day by mouth in 2 doses per day (not to exceed 1 g per day); *alternative PEP*, in weight-dependent dosage: weight ≥20 kg, amoxicillin, 500 mg by mouth every 8 h; weight <20 kg, amoxicillin, 40 mg/kg by mouth in 3 doses every 8 h ● Pregnant women: *initial PEP*, ciprofloxacin, 500 mg by mouth every 12 h; *alternative PEP*, amoxicillin, 500 mg by mouth every 8 h **PEP with vaccine:** see column *Vaccination*, at right	**Treatment and PEP** **Cutaneous:** 60 days; disease has occurred in humans up to 58 days after exposure through inhalation **Inhalational, gastrointestinal, or meningitis:** 60 days; intravenous treatment initially and oral therapy when clinically appropriate	**Anthrax vaccine adsorbed** ● FDA licensed for 6-dose preexposure vaccination but not for postexposure vaccination ● Preexposure vaccination: 6 doses of 0.5 mL subcutaneously at weeks 0, 2, and 4 and at months 6, 12, and 18 ● Postexposure vaccination: 3 doses of 0.5 mL subcutaneously on days 0, 14, and 28, with PEP for 60 days

(continued)

Table 22-4. (continued)

Disease (causative agent)	Treatment	Prophylaxis	Duration	Vaccination
Smallpox (Variola minor)	under emergency as an investigatory new drug Currently, there is no proven treatment for smallpox **Supportive care** ●Isolation of the patient to prevent transmission of variola virus to nonimmune persons ●Monitoring and maintaining fluid and electrolyte balance ●Monitoring for and treatment of complications ●Skin care	**Vaccination:** see column *Vaccination*, at right ●ACAM2000 Smallpox Vaccine **Pre-event** ●Civilian laboratory personnel ●Military personnel ●State public health preparedness programs	Not applicable	**Vaccination, postevent** ●Use ACAM2000 Smallpox Vaccine **PEP:** ●Vaccinate within 3 days after exposure to prevent or significantly modify smallpox disease ●Vaccination within 4–7 days may offer some protection from disease, or modify disease severity **Vaccine adverse effects** ●Live vaccine can cause eczema vaccinatum, generalized vaccinia, and immune reactions, particularly in persons with immunocompromise or eczema ●Mortality rate is 1–2 deaths per million
Plague (*Yersinia pestis*)	**All forms** ●Adults, preferred therapy[d]: streptomycin, 1 g intramuscularly twice per day, **or** gentamicin, 5 mg/kg intramuscularly or IV once per day or a 2 mg/kg loading dose followed by 1.7 mg/kg intramuscularly or IV 3 times per day ●Adults, alternative choices[a]: doxycycline, 100 mg IV twice per day or 200 mg IV once per day, **or** ciprofloxacin, 400 mg IV twice per day, **or** chloramphenicol, 25 mg/kg IV 4 times per day ●Children, preferred therapy[d]: streptomycin, 15 mg/kg intramuscularly twice per day (not to exceed 2 g per day), **or** gentamicin, 2.5 mg/kg intramuscularly or IV 3 times per day ●Children, alternative choices[a]: doxycycline in weight-adjusted dosage: weight >45 kg, administer adult dosage;	**PEP** ●Adults, preferred therapy[a]: doxycycline, 100 mg by mouth twice per day, **or** ciprofloxacin, 500 mg by mouth twice per day ●Adults, alternative choices: chloramphenicol, 25 mg/kg by mouth 4 times per day ●Children, preferred therapy[a]: doxycycline in a weight-dependent dosage: weight ≥45 kg, administer adult dosage; weight <45 kg, administer 2.2 mg/kg by mouth twice per day **or** ciprofloxacin, 20 mg/kg by mouth twice per day (not to exceed 1 g per day) ●Children, alternative choices[e]: chloramphenicol, 25 mg/kg by mouth 4 times per day (not to exceed 4 g per day)	**Treatment for all forms:** 10–14 days (intravenous therapy initially, then oral therapy) **PEP:** 7 days	*Currently, there is no FDA-approved vaccine:* killed whole-cell plague vaccines are effective against bubonic plague in animal models but do not appear effective in combating pneumonic plague

weight <45 kg, 2.2 mg/kg IV twice per day (not to exceed 200 mg per day) **or** ciprofloxacin, 15 mg/kg IV twice per day (not to exceed 1 g per day) **or** chloramphenicol,[e] 25 mg/kg IV 4 times per day (not to exceed 4 g per day)
• Pregnant women[a,d]: same treatment as for adults, except avoid streptomycin and chloramphenicol because of fetal toxicity

• Pregnant women: same as adults except avoid chloramphenicol because of fetal toxicity

Tularemia (*Francisella tularensis*)

PEP
• Adults, preferred therapy[a]: doxycycline, 100 mg by mouth twice per day, **or** ciprofloxacin, 500 mg by mouth twice per day
• Children, preferred therapy[a]: doxycycline in a weight-dependent dosage: weight ≥45 kg, administer adult dosage; weight <45 kg, administer 2.2 mg/kg by mouth twice per day **or** ciprofloxacin, 20 mg/kg by mouth twice per day (not to exceed 1 g per day)
• Pregnant women[a]: same as for adults

All forms
• Adults, preferred therapy[d]: streptomycin, 1 g intramuscularly twice per day, **or** Gentamicin, 5 mg/kg intramuscularly or IV once per day, or a 2 mg/kg loading dose followed by 1.7 mg/kg intramuscularly or IV 3 times per day
• Adults, alternative choices[a]: doxycycline, 100 mg IV twice per day, or 200 mg IV once per day, **or** ciprofloxacin, 400 mg IV twice per day, **or** chloramphenicol, 25 mg/kg IV 4 times per day
• Children, preferred therapy[d]: streptomycin, 15 mg/kg intramuscularly twice per day (not to exceed 2 g per day) **or** gentamicin, 2.5 mg/kg intramuscularly or IV 3 times per day
• Children, alternative choices[a]: doxycycline in a weight-dependent dosage: weight >45 kg, administer adult dosage; weight <45 kg, administer 2.2 mg/kg IV twice per day (not to exceed 200 mg per day), **or** ciprofloxacin, 15 mg/kg IV twice per day (not to exceed 1 g per day), **or** chloramphenicol,[e] 25 mg/kg IV 4 times per day (not to exceed 4 g per day)
• Pregnant women[a,d]: same treatment as for adults, except avoid streptomycin and chloramphenicol because of fetal toxicity

Treatment: 10–14 days (intravenous initially, then oral therapy)
PEP: 14 days

Currently, there is FDA-approved vaccine

(continued)

Table 22-4. (continued)

Disease (causative agent)	Treatment	Prophylaxis	Duration	Vaccination
Botulism (*Clostridium botulinum* toxin)	**Supportive care** • Mechanical ventilation, intensive care unit stay • Nutrition: administer parenterally or through enteral tube • Treatment of secondary infections (pulmonary) • Trenedelenburg position (20° tilt) for optimal breathing **Antitoxin** • Equine antitoxin: 10-mL vial per patient, diluted 1:10 in 0.9% saline solution, administered by slow IV infusion • Obtain antitoxin from the CDC: telephone 404-639-2206 or, after regular business hours, 404-639-2888	**PEP:** Equine antitoxin can be given as PEP, but the supply is limited, and it may be best to monitor potentially exposed persons carefully and treat them if symptoms appear	**Duration of illness:** weeks to months **Duration of treatment with antitoxin:** 1 dose	*Currently, there is FDA-approved vaccine* **Investigational pentavalent toxoid vaccine** can be used for laboratory workers at high risk of exposure to botulinum toxin
VHF (Ebola, Marburg, Lassa fever, Rift Valley fever, yellow fever viruses)	**Supportive care** • Blood pressure, fluids, and electrolytes • Mechanical ventilation • Renal dialysis **Ribavirin**[f] has activity against Lassa fever virus, other arenaviruses, and Rift Valley fever. It lacks FDA approval but may be used against VHF in a mass-casualty scenario • Adults: — Contained casualty scenario: administer a loading dose of 30 mg/kg IV once (maximum dose, 2 g), then 16 mg/kg IV every 6 h for 4 days (maximum dose, 1 g), then 8 mg/kg IV every 8 h for 6 days (maximum dose, 500 mg) — Mass-casualty scenario: administer a loading dose of 2,000 mg by mouth once, then a weight-dependant dosage: weight >75 kg, 1,200 mg per day by mouth in 2 divided doses for 10 days; weight ≤75 kg, 1,000 mg per day by mouth in divided doses (400 mg in the morning and 600 mg in the afternoon) for 10 days	None	**Treatment:** 10 days	*The only VHF agent vaccine is for yellow fever:* it is administered to people traveling to areas of endemicity; it is not useful after exposure, because the incubation period for yellow fever is 3–6 days, and vaccine takes 10 days to work *There are no vaccines for other VHF agents*

Disease	Treatment	Prophylaxis	Duration	Vaccination / Notes
	• Children: – Contained casualty scenario: administer a loading dose of 30 mg/kg IV once (maximum dose, 2 g), then 16 mg/kg IV every 6 h for 4 days (maximum dose, 1 g), then 8 mg/kg every 8 h for 6 days (maximum dose, 500 mg IV) – Mass casualty scenario: administer a loading dose of 30 mg/kg by mouth once, then 15 mg/kg per day by mouth in 2 divided doses for 10 days			*There is currently no FDA-approved vaccine, although there are ongoing trials of candidate vaccines*
SARS (SARS CoV)	**Supportive Care** • Mechanical ventilation • Antibiotic therapy for secondary pneumonia • Anti-inflammatory agents: corticosteroids (high dose) **Antivirals:** none shown to be effective, although studies are limited	None	Not applicable	
Pandemic influenza (influenza virus)	**Oseltamivir**** (treatment for persons aged ≥1 y) • Adults and children aged ≥13 y: 75 mg by mouth twice per day • Children: weight <15 kg, 30 mg twice per day; weight 15–23 kg, 45 mg twice per day; weight >23 and ≤40 kg, 60 mg twice per day; weight >40 kg, 75 mg twice per day **Zanamivir**** (treatment for persons aged ≥7 y) • Adults and children aged ≥7 y: 10 mg (2 inhalations) twice per day **Amantadine:** (consult package insert for persons with creatinine clearance of ≤50 mL/min/1.73 m²) • Adults and children aged ≥13 y: 100 mg twice per day (aged ≥65 y, ≤100 mg per day) • Children: 5 mg/kg per day, up to 150 mg, in 2 divided doses • Children aged ≥10 y and weight <40 kg, 5 mg/kg per day	**Oseltamivir** (prophylaxis for persons aged ≥1 y) • Adults and children aged ≥13 y: 75 mg by mouth once per day • Children: weight <15 kg, 30 mg once per day; weight 15–23 kg, 45 mg once per day; weight >23 and ≤40 kg, 60 mg once per day; weight >40 kg, 75 mg once per day **Zanamivir** (prophylaxis for persons aged ≥5 y) • Adults and children aged ≥5 y: 10 mg (2 inhalations) once per day **Amantadine** (consult package insert for persons with creatinine clearance ≤50 mL/min/1.73 m²) • Adults: 100 mg twice per day (aged ≥65 years, ≤100 mg per day) • Children: 5 mg/kg per day, up to 150 mg, in 2 divided doses • Children aged ≥10 y and weight <40 kg, 5 mg/kg per day	**Oseltamivir** • Treatment: 5 days • PEP: 10 days **Zanamivir** • Treatment: 5 days • PEP: 10 days **Amantadine:** • Treatment: 7 days • PEP: 10 days **Rimantidine:** Treatment: 7 days PEP: 10 days	**Prepandemic vaccination** • The US government is currently stockpiling vaccine for highly pathogenic influenza A (H5N1) virus in the event that this strain mutates into a human pandemic influenza strain **Pandemic vaccination** • Pandemic vaccine production will begin immediately upon identification of the pandemic influenza virus strain • Pandemic vaccine: may be 4–6 months before it is available for use

(continued)

281

Table 22-4. (continued)

Disease (causative agent)	Treatment	Prophylaxis	Duration	Vaccination
	Rimantadine Note: A reduction in dosage to 100 mg per day of rimantadine is recommended for persons who have severe hepatic dysfunction or those with creatinine clearance <10 mL/min. Other persons with less severe hepatic or renal dysfunction taking 100 mg per day of rimantadine should be observed closely, and the dosage should be reduced or the drug discontinued, if necessary • Adults and children ≥13 y: 100 mg twice per day (aged ≥65 y, 100 mg per day) • Children: not FDA-approved for treatment of children Rimantadine is approved by FDA for treatment of adults. However, certain specialists in the management of influenza consider rimantadine appropriate for treatment of children. Studies evaluating the efficacy of amantadine and rimantadine in children are limited, but they indicate that treatment with either drug diminishes the severity of influenza A infection when administered within 48 h after illness onset	**Rimantidine** • Adults children aged ≥13 years: 100 mg twice per day (aged ≥65 y, 100 mg per day) • Children: 5 mg/kg per day, up to 150 mg, in 2 divided doses		

NOTE: Table based on information from multiple studies.[14-22,52] CDC, Centers for Disease Control and Prevention; EPA, Environmental Protection Agency; FDA, Food and Drug Administration; h, hour; HCW, healthcare worker; IV, intravenous; PEP, postexposure prophylaxis; PPE, personal protective equipment (N-95 mask, gloves, gown, and face shield); SARS, severe acute respiratory syndrome; SARS CoV, SARS-associated coronavirus; VHF, viral hemorrhagic fever; y, year.

a The use of ciprofloxacin and doxycycline in pregnant women and children is justified for life-threatening illness, since the risk of death outweighs the adverse events. The ciprofloxacin dose should not exceed 1 g per day in children.

b Inhalational, gastrointestinal, and meningeal anthrax should be initially treated with intravenous antibiotics, until the patient's condition improves, and then therapy can be completed with 1 or 2 oral agents.

c Additional antibiotics with activity against *B. anthracis* include rifampin, vancomycin, penicillin, ampicillin, chloramphenicol, imipenem, clindamycin, and clarithromycin. *B. anthracis* resistance to β-lactam antibiotics has been described, and these medications should not be used alone to treat inhalational, gastrointestinal, or meningeal anthrax. Antibiotic susceptibility testing should be performed.

d Aminoglycoside dosages must be adjusted according to the patient's renal function, and the patient's renal function and aminoglycoside levels should be monitored during therapy.

e Children younger than 2 years should not receive chloramphenicol.

f Ribavirin is teratogenic in animal models and should not be used in pregnant women unless the risk to the patient outweighs the risk to the fetus.

disease. Furthermore, because vaccination efforts against smallpox ceased in 1982, few people are adequately protected against infection.[16,53] Currently, the smallpox virus, variola, is held in only 2 WHO laboratories in the world, which include the CDC in Atlanta, Georgia, and the Institute of Virus Preparations in Moscow, Russia. However, there is concern that stocks of smallpox virus may exist in other countries.[16,53]

Smallpox could be used as a biological weapon either by aerosolizing the virus or by direct person-to-person spread from an infected bioterrorist or unsuspecting infected individual.[16,53] A single case of smallpox anywhere in the world would be considered a bioterrorist event, given that smallpox has been eradicated.[16,53] A suspected case of smallpox *must* be reported immediately to local public health authorities.[53]

There are 2 clinical forms of smallpox, variola major and variola minor. Variola major is the most common form of smallpox, and also the most severe, and cases present with a more extensive rash and high fevers (Table 22-3). There are 4 types of variola major smallpox: ordinary, which accounts for 90% of cases or more; modified, which is milder and occurs in previously vaccinated persons; flat; and hemorrhagic. The 2 latter forms of smallpox include a malignant smallpox characterized by severe toxemia and flat, confluent lesions that do not progress to pustules and a hemorrhagic-type smallpox characterized by a severe prodrome, toxemia, and a hemorrhagic rash. These 2 less common forms of smallpox have mortality rates greater than 95%.[16,53] Variola minor is another form of smallpox and presents much less commonly and with less severe disease; historically, death rates have been 1% or less.[53]

Smallpox is transmitted person to person by means of respiratory droplets or contact with infected skin lesions.[16,53] Symptoms begin approximately 12–14 days after exposure; the clinical presentation and diagnosis are outlined in Table 22-3. Smallpox patients are most infectious during the first week of the rash, when the oral mucosa lesions ulcerate and release large amounts of virus into the saliva.[53] The patient is no longer infectious once all scabs have separated from the skin, which usually occurs 3–4 weeks after the onset of rash.[16,53]

The CDC case definition for smallpox identifies it as an illness with an acute onset of fever with a temperature of 38.3°C (101°F) or higher, followed by a rash characterized by firm, deep-seated vesicles or pustules in the same stage of development without other apparent cause.[53] The laboratory criteria for confirmation require identification of variola virus DNA in a clinical specimen by polymerase chain reaction (PCR) or isolation of the smallpox virus from a clinical specimen with confirmation by variola virus PCR. The healthcare epidemiologist must be aware that the laboratory diagnostic testing for variola virus should be conducted *only* in a CDC Laboratory Response Network laboratory, such as the state public health department's laboratory, utilizing Laboratory Response Network–approved PCR tests and protocols.[53] Initial confirmation of a smallpox outbreak requires additional testing at the CDC. Because few physicians have seen smallpox, the CDC has developed a vesicular/pustular rash algorithm that helps differentiate smallpox from other common diseases, particularly chickenpox[54] (Figure 22-3). When the healthcare epidemiologist is faced with a patient with a febrile rash syndrome, he or she and other clinical staff should be familiar with this information and be able to access this diagnostic aid quickly.

There is currently no treatment for smallpox, although cidofovir, an antiviral drug, has shown some effect in vitro.[16] Smallpox vaccine is a live vaccinia vaccine produced by Acambis.[55] It can be administered before exposure, as was done in 2003 for 40,000 healthcare workers and approximately 500,000 military personnel.[55] The vaccine can also be given after exposure; it prevents disease if given within 4 days and prevents death if given within 8 days.[16,55] Smallpox vaccine, being live, is contraindicated for immunocompromised patients and persons with certain skin conditions, including eczema (Table 22-4).[16,53,55] Smallpox vaccine can only be procured from the CDC or, possibly, state public health departments.[16,55]

The healthcare epidemiologist will need to implement appropriate isolation and disease control measures within the facility. If a patient with an acute generalized vesicular or pustular rash illness presents to the emergency department, a clinic, or a healthcare provider's office, actions should be taken to decrease the risk of transmission. The patient should not be left in any common waiting areas and should be immediately triaged into a private room where he or she can be assessed quickly to determine the actual risk of smallpox, using the CDC algorithm "Evaluating Patients for Smallpox: Acute, Generalized Vesicular or Pustular Rash Illness Protocol"[53] (Figure 22-3). If in a clinic, the door to the examination room must remain closed until the risk for smallpox is determined. If in a healthcare facility, immediate and appropriate airborne-transmission and contact precautions should be instituted, and the healthcare epidemiologist and/or infection control department should be immediately alerted, if that has not been done already. The healthcare epidemiologist should ensure that the patient is placed in a private, negative air-pressure room. The door should be kept closed at all times, except when staff or the patient must enter or exit. Staff and visitors should wear properly fitted N-95 respirators, gloves, and gowns, and the patient should wear a surgical mask whenever he or she must be outside their negative

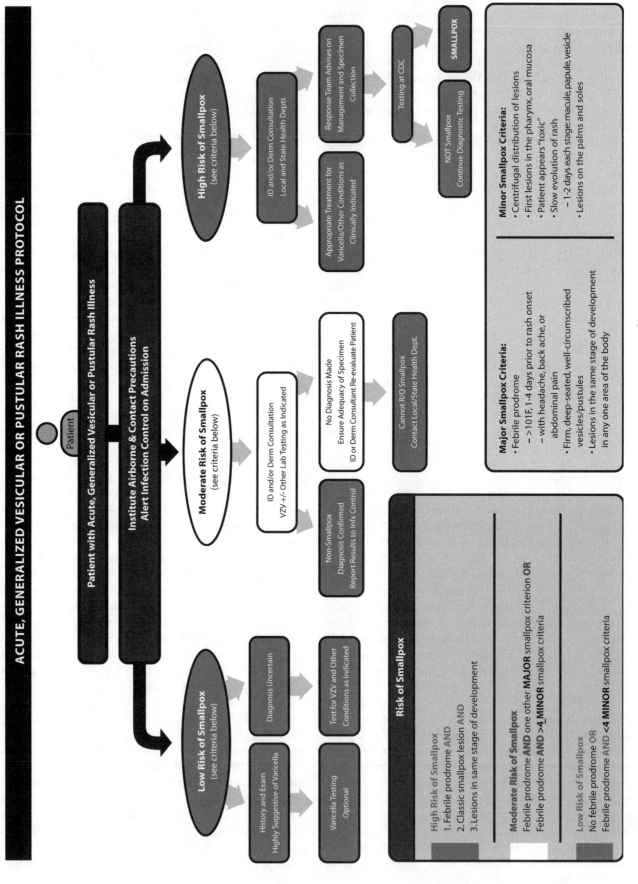

Figure 22-3. Algorithm for evaluating patients for smallpox, from the US Department of Health and Human Services.[54]

air-pressure isolation room. The patient's rash should be fully covered with a gown and/or sheet, if the patient is transported out of the isolation room.[53]

Plague

Y. pestis, the gram-negative bacterium that causes plague, historically has been transmitted to humans through a flea bite.[17] Although it is reported that the Japanese attempted to disseminate plague-infected fleas in China during World War II, the most likely method of release in a biological attack would be in an aerosolized form.[17] Plague was developed as a biological weapon in the Soviet Union, and it is considered an attractive disease for bioterrorism because of its high mortality rate, human-to-human transmission, high infectivity (requiring 100–500 Y. pestis organisms), and high panic factor.[17]

Naturally occurring plague has 3 main forms—bubonic, pneumonic, and septicemic—but bubonic plague presenting as painful lymph nodes and fever after a flea bite is the most common form and occurs frequently throughout the world.[17] A bioterrorist release of aerosolized Y. pestis would most likely present as pneumonic or septicemic plague (Table 22-3). Pneumonic plague can be transmitted from person to person, and infected patients must be isolated (Table 22-2).[17]

Bubonic plague is relatively easy to identify, and in the antibiotic era, prompt treatment has decreased mortality to 5%–14%.[17] Pneumonic and septicemic plague, however, present with nonspecific signs and symptoms (Table 22-3), and delayed diagnosis leads to mortality rates of more than 50% if antibiotic therapy is not started within 24 hours after the onset of symptoms.[17] Initial treatment of pneumonic and septicemic plague requires intravenous aminoglycosides (Table 22-4), antibiotics not typically given as empiric treatment for pneumonia. Therefore a high level of suspicion must exist for appropriate therapy to be initiated.[17] Prophylactic therapy should be given to close contacts of pneumonic plague patients and those exposed to aerosolized Y. pestis (Table 22-4).[17] There is currently no vaccine available for plague (Table 22-4).[17]

Tularemia

F. tularensis is the gram-negative bacterium that causes tularemia.[18] Humans are naturally infected with tularemia through the bites of arthropods (ticks), contact with infected animal tissue (from rabbits), ingestion of contaminated food or water, and, rarely, through aerosol inhalation (laboratory accident).[18] F. tularensis has been developed as a biological agent because of its high infectivity; 1 or 2 organisms can cause infection.[18] Japan, the Soviet Union, and the United States have all developed F. tularensis as a biological weapon in the past.[18]

There are 6 naturally occurring forms of tularemia; ulceroglandular, glandular, oculoglandular, oropharyngeal, pneumonic, typhoidal, and septic tularemia.[18] The most likely form of tularemia from a biological attack would be pneumonic disease, but typhoidal, septic, and oropharyngeal infection might also occur (Table 22-3).[18] Tularemia is not transmitted from person to person, and thus the patient need not be isolated (Table 22-2).[18]

Pneumonic tularemia would present as a nonspecific pneumonia and might be difficult to diagnose (Table 22-3).[18] The F. tularensis organism is fastidious and does not grow well in standard cultures. Furthermore, because it is highly infectious, it must be handled very carefully in the laboratory.[18]

Without use of antibiotics, the mortality rate is 30%–60% for pneumonic tularemia and 5%–15% for other forms of tularemia.[18] The prompt administration of antibiotics—including aminoglycosides (gentamicin or streptomycin), doxycycline, flouroquinolones, or chloramphenicol—decreases mortality to less than 2% (Table 22-4).[18] Prophylactic antibiotic therapy should be given to persons exposed to aerosolized F. tularensis after a biological attack (Table 22-4).[18] There is no Food and Drug Administration (FDA)–approved vaccine for tularemia, but a live attenuated vaccine is currently undergoing clinical trials.[56]

Botulism

Botulism is a disease caused by botulinum toxin, one of the most lethal toxins known to man.[19] An inhaled dose of 0.70 µg is enough to kill a 70 kg human. Botulinum toxin is produced by C. botulinum, an anaerobic bacterium. The disease occurs naturally in 3 forms: foodborne, wound, and intestinal botulism.[19] Botulinum toxin was used as a biological weapon by Aum Shinrikyo in Japan in the early 1990s, and it was produced as a biological weapon in the United States during World War II.[19] It is also believed that the Soviet Union and Iraq developed botulinum toxin as a weapon.[19]

A biological attack with botulinum toxin would most likely involve an aerosol release or poisoning of the food or water supply.[19] Although there is some concern about terrorists using commercial formulations of botulinum toxin, these aliquots do not contain enough toxin to be lethal (0.005%–0.3% of the estimated lethal dose).[19] Botulinum toxin blocks acetylcholine release and thereby prevents normal muscle contraction.[19] Patients affected by botulinum toxin have a classic triad of symptoms: (1) symmetric, descending flaccid paralysis with bulbar signs; (2) absence of fever; and (3) clear sensorium (Table 22-3).[19]

Diagnosis of botulism is difficult, if it is not clinically suspected, and laboratory testing can be delayed and difficult (Table 22-3).[19] The mortality rate is 25% or more for patients not given treatment and ventilatory support.[19]

Care of a patient with botulism is generally supportive (eg, ventilatory care); however, antitoxin is available through the CDC via the local or state public health department.[19] There is currently no FDA-approved vaccine for botulism; however, pentavalent botulinum toxoid is available for special populations (eg, researchers) through the CDC as an investigational new drug.[19]

Viral Hemorrhagic Fever

Viral hemorrhagic fever (VHF) agents include a variety of viruses in 4 families: Filoviridae, Arenaviridae, Bunyaviridae, and Flaviviridae. Specific viruses of concern include Ebola, Marburg, Lassa fever, yellow fever, and Rift Valley fever viruses.[20] Several countries, including the United States, the Soviet Union, and Russia, have developed VHF viruses as biological weapons in the past.[20] The Aum Shinrikyo cult unsuccessfully attempted to acquire Ebola virus and develop this agent as a biological weapon.[20]

As biological weapon, a VHF virus would most likely be released as an aerosol.[20] After an incubation period of 2–21 days, illness would most likely manifest as a nonspecific viral syndrome that included myalgias, headache, fever, nausea, and vomiting (Table 22-3).[20] As the disease progresses, the patient might develop hemorrhagic manifestations, including conjunctival petichiae or hemorrhages, gastrointestinal bleeding, hemoptysis, hematemesis, and petichiae or purpura of the skin.[20] Patients with yellow fever might develop jaundice and/or scleral icterus.[20] Laboratory diagnosis of these organisms is difficult, and a high level of suspicion is warranted in patients presenting with nonspecific viral symptoms with hemorrhagic complications.[20] The mortality rate for VHF varies from less than 1% for Rift Valley fever to 50%–90% for Ebola hemorrhagic fever.[20]

Treatment is generally supportive and includes administration of intravenous fluids and/or blood products, mechanical ventilation, and administration of nonsteroidal anti-inflammatory drugs.[20] (Table 22-4). Ribavirin, an antiviral medication, has activity against arenaviruses and bunyaviruses and should also be used against these viruses or when the identity of the virus in a case of VHF is not known.[20] Postexposure prophylaxis is not recommended, and only yellow fever virus has an FDA-approved vaccine.[20]

Severe Acute Respiratory Syndrome (SARS)

SARS is caused by SARS-associated coronavirus (SARS-CoV), first isolated in the pandemic outbreak of 2003.[21,22] Cases of SARS first appeared in south China (Guandong province) in November 2002. In February 2003 an ill physician from Guandong transmitted the virus to 10 other people at a conference in Hong Kong.[21]

These 10 individuals then spread the virus to other countries around the world, and on March 13, 2003, the WHO issued a warning about the worldwide spread of this virus.[21] The global outbreak ended in July 2003, after 8,422 cases had been diagnosed and 916 patients had died.[21] There were only 29 cases of SARS in the United States and no deaths.[21] Only 9 other cases of SARS have been diagnosed since July 2003, and these were all related to a researcher infected in the laboratory.[21]

The reservoirs for SARS include bats and palm civets, and humans are infected through contact with these animals.[21] Person-to-person transmission of SARS-CoV occurs by means of respiratory droplets or contact with bodily secretions, such as stool.[21] Certain persons seemed to transmit SARS-CoV at a high rate ("superspreaders"), particularly when they were very ill and hospitalized. As a result, much of the early transmission of SARS-CoV occurred in the hospital setting.[21] Because of the possibility of airborne spread, SARS patients should be placed in respiratory isolation in a negative air-pressure room (Table 22-2).[21]

Persons infected with SARS-CoV present with nonspecific symptoms of fever, chills, myalgias, headache, and diarrhea (Table 22-3).[21] After 2 days, the patient develops a nonproductive cough and dyspnea. Respiratory symptoms progress, and 10%–20% of patients require mechanical ventilation.[21] Patients are often hospitalized for more than 2 weeks, and the case-fatality rate approaches 11%.[21] The diagnosis of SARS is based on a case definition that depends on exposure history, clinical symptoms, and laboratory confirmation of infection.[57] Serologic assays and a reverse-transcriptase PCR assay have been developed for the diagnosis of SARS (Table 22-4).[21]

Treatment for SARS is supportive, and there is no specific antiviral medication that has been demonstrated to be effective (Table 22-4).[21] Steroid therapy has been used to decrease the immune-mediated lung damage, but no prospective studies have been done.[21] Infection control precautions must be taken in order to protect family members, the community, healthcare workers, and other patients (Table 22-2).[21,22]

Pandemic Influenza

In 2005 the US Government released its first version of a national pandemic influenza plan, the Department of Health and Human Services Pandemic Influenza Plan.[14] Globally, nations are developing response plans in response to an unprecedented spread of a novel, highly pathogenic strain, influenza virus A subtype H5N1 or "avian influenza virus," which has raised concerns about an imminent influenza pandemic.[26,58] Even if influenza virus A (H5N1) does not mutate into a pandemic human strain, 3 or 4 pandemics of influenza occur each century; although the pandemics vary in severity, world governments have understood the necessity for the development of pandemic influenza response plans. Pandemic planning in US hospitals has now been incorporated into current hospital biological disaster preparedness programs, again in coordination with local, state, and federal groups.

Impact

Pandemic influenza, like any other pandemic disease, encompasses multiple countries or the whole globe and has the potential to have a significantly larger impact across the healthcare system than a potentially localized bioterrorism event. A bioterrorism event could have a devastating impact but may be circumscribed in the geographic area affected and in the time during which it directly or indirectly impacts the population. Pandemic influenza, however, once a novel strain has developed the ability to be transmitted from person to person, has the potential to encompass the globe rapidly—perhaps more quickly than have past pandemics, given the global nature of air travel and trade. The Department of Health and Human Services Pandemic Influenza Plan estimates that approximately 75–105 million US citizens may develop influenza, and 209,000–1,903,000 of them may die.[14,26] The influenza season in the United States is associated with an average of 40,000 deaths annually.[14]

Because of the global nature of pandemic influenza, hospital preparedness plans that are adequate to address the release of a biological disaster agent could be insufficient for a pandemic influenza event. Surge-capacity plans that may rely on surrounding hospitals or surrounding communities may not be sufficient, as those hospitals and communities, too, will need to activate their response plans. Further, although a bioterrorism event may be localized temporally, pandemic influenza may require hospitals to address surge capacity issues for a period as long as 6–8 weeks.[14,59,60] Surge capacity needs may be compromised by a limited ability to replenish stores of PPE, basic medical supplies, pharmaceuticals (including antivirals for treatment or prophylaxis), and critical care equipment.[59,60] Because of impacts on healthcare capacity and provisions, the healthcare epidemiologist and clinical providers may need to implement altered standards of care, which will lead to difficult decisions regarding allocation of scarce supplies. The healthcare epidemiologist's role, as explained earlier in this chapter, will be vital for prioritizing the use of PPE and antivirals and vaccines according to national and state public health guidance.[14,25,60,61] Furthermore, the healthcare epidemiologist can assist the planning team in identifying distributors of vital supplies and ensuring that the hospital's potential

needs are part of the distributor's contingency plans for a pandemic of influenza as well.[59]

Hospital Pandemic Preparedness

Table 22-5 outlines the basic activities a hospital should undertake for pandemic influenza preparedness developed by the Center for Biosecurity of the University of Pittsburgh Medical Center.[11,60] A pandemic preparedness committee should involve all key responders within the hospital organization and have a dedicated leader, a role that may be best served by the bioterrorism preparedness planner or disaster/emergency coordinator.

Table 22-5. Basic activities a hospital should undertake for pandemic influenza preparedness, as developed by the Center for Biosecurity of University of Pittsburgh Medical Center

---

Form a pandemic preparedness committee

> Have a full-time, dedicated disaster coordinator, and coordinate with regional hospitals and public health officials
> The CDC's FluSurge version 2.0 (E18) can be used to guide planning
> Plan to make 30% of beds available within 1 week
> Double the licensed bed capacity within 2 weeks
> Anticipate supply needs (including medications)

Focus on methods to limit nosocomial spread

> Stockpile a 3-week supply of PPE
> Have surgical masks for use by everyone in the facility
> Stockpile N-95 respirators for use by healthcare workers
> Prevent infected staff from working
> Limit exposures by cohorting patients
> Improve hospital surveillance

Maintain and augment the healthcare workforce

> Make rapid influenza testing readily available
> Treat with antiviral agents within 6 h after symptom onset
> Organize in-home child care in the event of school closures
> Maintain open and honest communication
> Expand occupational health services
> Provide training to staff
> Develop plans to augment clinical staff
> Coordinate activities with other regional hospitals

Allocate resources efficiently

> Prioritize services
> Establish clinical care guidelines
> Address legal and ethical issues
> Allocate resources and healthcare
> Prioritize use of antivirals, vaccine, ventilators, and beds
> Develop a framework for fair and efficient allocation of antivirals and vaccines
> Provide credible communication to health care personnel and the public

---

NOTE: Data from Toner and Waldhorn[11] and Bartlett and Borio.[60] CDC, Centers for Disease Control and Prevention; PPE, personal protective equipment (N-95 mask, gloves, gown, and face shield).

Anticipated surge capacity needs should be documented, and hospitals should prepare to make 30% of their beds available quickly (within 1 week). The CDC's FluSurge software is a valuable tool the healthcare epidemiologist can use to assist in estimating the potential impacts on surge capacity based on various pandemic influenza scenarios.[62] Toner and Waldhorn[11] note the potential ability of hospital facilities to quickly make approximately 10%–20% of a hospital's bed capacity available within a few hours simply by accelerating discharges and utilizing discharge holding areas, as well as by converting single rooms to double rooms and opening previously closed areas.

As mentioned previously in this chapter, hospital involvement in community and regional preparedness programs is vital. An important feature of influenza pandemics is that they impact all sectors of society and all institutions within the community. Hospitals and their healthcare epidemiologists should work together in a regional hospital coordinating group that includes neighboring hospitals, local public health officials, and emergency management personnel to address the surge capacity and scarce resource issues that will arise in a pandemic influenza event.[11,59,60] If faced with an impending influenza pandemic like that of 1918, regional hospital planners should be able to address a potential need to increase licensed bed capacity by 200% within the region in a period as short as 2 weeks.[11]

The healthcare epidemiologist will need to limit the transmissibility of the pandemic virus strain within the healthcare facility.[11,60] Respiratory etiquette protocols and guidelines should be implemented, and everyone entering the facility, including staff, patients, and visitors, should utilize simple surgical masks. For this capacity, hospitals should stockpile enough of these surgical masks to cover their needs for at least 3 weeks.[11]

Fit-tested N-95 respirators should be used during aerosol-generating procedures, during cardiopulmonary resuscitation, and in situations that call for repeated direct contact with patients who have influenza or pneumonia, and it is prudent to use them for other direct patient-care activities. Powered air-purifying respirators should be available for use in high-risk aerosol-generating procedures if the provider is unable to use an N-95 mask. The healthcare epidemiologist can refer to the Occupational Safety and Health Administration document "Guidance on Preparing Workplaces for an Influenza Pandemic"[63] to assess risk for pandemic influenza exposure among all hospital personnel, not just direct patient-care personnel. Healthcare personnel performing aerosol-generating procedures on patients known to have or suspected of having pandemic influenza (such as cough-induction procedures, bronchoscopy, some dental procedures, or invasive specimen collection) and laboratory personnel collecting or handling specimens are within the very highest risk category for exposure, as determined by the Occupational Safety and Health Administration.[63]

N-95 masks will likely be in short supply, and resupply of these within a pandemic period could be difficult. They should be stockpiled, and their use should be continuously monitored.[11] If no other masks are available, surgical masks, which do provide adequate droplet protection, should be used.[11,64] The healthcare epidemiologist should be aware of the capability of their suppliers to deliver this PPE during a potential influenza pandemic and may need to stockpile even more supplies, depending on their distributor's continuity of business plan and response during this type of disaster.

The healthcare epidemiologist will need to assist the administration in the cohorting of patients, which will assist in limiting the number of staff working with patients with this contagious disease. Limit the number of staff who are exposed to patients with influenza. Adjustment of schedules may be needed to limit the number of dedicated direct-care staff. Employees who are either vaccinated for the pandemic virus strain or who have recovered from pandemic influenza illness, and thus considered immune, would be candidates to work with these cohorts. Despite strain on healthcare resources—especially on staff—infected staff or those potentially infected and requiring quarantine should be excluded from working, although they could work with influenza patients, and the healthcare epidemiologist and infection control staff will need to track and monitor staff accordingly.[11]

The Center for Biosecurity of the University of Pittsburgh Medical Center also recommends that hospitals be able to minimize infection by utilizing rapid influenza testing, if it is available for the pandemic strain circulating in the community, and by making antiviral treatment available to staff within 6 hours after onset of influenza symptoms.[11] On-site day care, medical care for family members, and the utilization of medical or support volunteers all should be incorporated in the hospital pandemic preparedness plans. Many regional bioterrorism preparedness planning efforts incorporate volunteer registries that could be utilized.[11,60]

The allocation of scarce resources will be a significant logistical, clinical, and ethical issue. Although there are currently national guidelines for the allocation of antivirals and pandemic influenza vaccine,[14,25,61] the Agency for Health Research and Quality noted in 2004 that many healthcare system preparations do not provide sufficient planning and guidance concerning the altered standards of care that would be required to respond to a mass-casualty event.[65] Alterations in normal care routines may need to occur if routine care cannot be delivered despite all efforts. Importantly, legal and ethical templates for this action should be developed prior to a mass-casualty event.[11] The healthcare epidemiologist, in regional collaboration with other hospitals, will be able to assist in criteria and clinical guideline development for the use of resources, such as mechanical ventilation, based on evolving national guidelines, such as those now being developed by the Agency for Health Research and Quality.[11,59,60,65]

Conclusion

The healthcare epidemiologist plays an important role in the response to a biological disaster. The healthcare epidemiologist's responsibilities will include surveillance for infection, infection control, and communication of infectious risk to the staff. In addition, the healthcare epidemiologist will be a key consultant to the hospital incident commander during the event. Finally, as new diseases, such as SARS, emerge and as old diseases, such as influenza, reemerge, the healthcare epidemiologist must remain up to date on the latest developments.

References

1. Darling RG, Catlett CL, Huebner KD, Jarrett DG. Threats in bioterrorism. I: CDC category A agents. *Emerg Med Clin North Am* 2002;20:273–309.
2. Department of Homeland Security. National Incident Management System. FEMA Publication P-501, December 2008:1–156.
3. National Incident Management System (NIMS). NIMS Implementation Activities for Hospitals and Healthcare Systems. http://www.fema.gov/pdf/emergency/nims/imp_hos.pdf. Accessed March 28, 2009.
4. American Hospital Association. Disaster Readiness Advisory. National Incident Management System Hospital Frequently Asked Questions. May 23, 2007. http://www.gnyha.org/6852/File.aspx. Accessed March 28, 2009.
5. McLaughlin S. Ready for anything. *Health Facilities Management* 2007;20:39–40,42.
6. Hospital Incident Command System Guidebook. California Emergency Medical Services Authority (EMSA), 2006. http://www.emsa.ca.gov/HICS/default.asp. Accessed March 28, 2009.
7. Centers for Disease Control and Prevention (CDC). Revised US surveillance case definition for severe acute respiratory syndrome (SARS) and update on SARS cases—United States and worldwide, December 2003. *MMWR Morb Mortal Wkly Rep* 2003;52:1202–1206.
8. Nam WW. Role of SARS screening clinic in the ED. *CJEM* 2004;6:78–79.
9. McGeer AJ. Diagnostic testing or empirical therapy for patients hospitalized with suspected influenza: what to do? *Clin Infect Dis* 2009;48(Suppl 1):S14–S19.
10. World Health Organization (WHO). WHO pandemic alerts. http://www.who.int/csr/disease/avian_influenza/phase/en/index.html. Accessed August 26, 2009.
11. Toner E, Waldhorn R. What hospitals should do to prepare for an influenza pandemic. *Biosecur Bioterror* 2006;4(4):397–402.
12. Esbitt D. The Strategic National Stockpile: roles and responsibilities of health care professionals for receiving the stockpile assets. *Disaster Manag Response* 2003;1:68–70
13. Hota S, McGeer A. Antivirals and the control of influenza outbreaks. *Clin Infect Dis* 2007;45:1362–1368.
14. Department of Health and Human Services (HHS). HHS Pandemic Influenza Plan http://www.hhs.gov/pandemicflu/plan/. Accessed August 26, 2009.
15. Inglesby TV, O'Toole T, Henderson DA, et al. Anthrax as a biological weapon, 2002: updated recommendations for management. *JAMA* 2002;287:2236–2252.
16. Henderson DA, Inglesby TV, Bartlett JG, et al. Smallpox as a biological weapon: medical and public health management. Working Group on Civilian Biodefense. *JAMA* 1999;281:2127–2137.
17. Inglesby TV, Dennis DT, Henderson DA, et al. Plague as a biological weapon: medical and public health management. Working Group on Civilian Biodefense. *JAMA* 2000;283(17):2281–2290.
18. Dennis DT, Inglesby TV, Henderson DA, et al. Tularemia as a biological weapon: medical and public health management. *JAMA* 2001;285:2763–2773.
19. Arnon SS, Schechter R, Inglesby TV, et al. Botulinum toxin as a biological weapon: medical and public health management. *JAMA* 2001;285:1059–1070.
20. Borio L, Inglesby T, Peters CJ, et al. Hemorrhagic fever viruses as biological weapons: medical and public health management. *JAMA* 2002;287:2391–2405.
21. Christian MD, Poutanen SM, Loutfy MR, et al. Severe acute respiratory syndrome. *Clin Infect Dis* 2004;38:1420.
22. Cheng VCC, Lau SKP, Woo PCY, et al. Severe acute respiratory syndrome coronavirus as an agent of emerging and reemerging infection. *Clin Microbiol Rev* 2007;20:660–694.
23. Siegel JD, Rhinehart E, Jackson M, et al. 2007 Guideline for isolation precautions: preventing transmission of infectious agents in health care settings. *Am J Infect Control* 2007;35:S65–S164.
24. Public Health Guidance for Community-Level Preparedness and Response to Severe Acute Respiratory Syndrome (SARS), Supplement I: Infection Control in Healthcare, Home, and Community Settings, CDC. http://www.cdc.gov/ncidod/sars/guidance/I/healthcare.htm. Accessed August 26, 2009.
25. Department of Health and Human Services (HHS). Interim Guidance on Planning for the Use of Surgical Masks and Respirators in Health Care Settings during an Influenza Pandemic. http://www.pandemicflu.gov/plan/healthcare/maskguidancehc.html. Accessed August 26, 2009.
26. Cinti SK, Wilkerson W, Holmes JG, et al. Pandemic influenza and acute care centers: taking care of sick patients in a nonhospital setting. *Biosecur Bioterror* 2008;6:335–348.
27. Barnitz L, Berkwits M. The health care response to pandemic influenza. *Ann Intern Med* 2006;145:135–137.
28. Hui DS, Chan MC, Wu AK, Ng PC. Severe acute respiratory syndrome (SARS): epidemiology and clinical features. *Postgrad Med J* 2004;80:373–381.
29. Sampathkumar P. West Nile virus: epidemiology, clinical presentation, diagnosis, and prevention. *Mayo Clin Proc* 2003;78:1137–1143.
30. Hien ND, Ha NH, Van NT, et al. Human infection with highly pathogenic avian influenza virus (H5N1) in northern Vietnam, 2004–2005. *Emerg Infect Dis* 2009;15(1):19–23.

31. Lewis MD, Pavlin JA, Mansfield JL, et al. Disease outbreak detection system using syndromic data in the greater Washington DC area. *Am J Prev Med* 2002;23:180–186.

32. Inglesby TV, Dennis DT, Henderson DA, et al. Plague as a biological weapon: medical and public health management. Working Group on Civilian Biodefense. *JAMA* 2000;283: 2281–90.

33. Dennis DT, Inglesby TV, Henderson DA, et al. Tularemia as a biological weapon: medical and public health management. *JAMA* 2001;285:2763–2773.

34. Borio L, Inglesby T, Peters CJ, et al. Hemorrhagic fever viruses as biological weapons: medical and public health management. *JAMA* 2002;287:2391–2405.

35. Sejvar JJ, Chowdary Y, Schomogyi M, et al. Human monkeypox infection: a family cluster in the midwestern United States. *J Infect Dis* 2004;190:1833–1840.

36. State-specific mortality from stroke and distribution of place of death—United States, 1999. *MMWR Morb Mortal Wkly Rep* 2002;51:429–433.

37. Schlech WF III, Ward JI, Band JD. Bacterial meningitis in the United States, 1978 through 1981. The National Bacterial Meningitis Surveillance Study. *JAMA* 1985;253: 1749–1754.

38. Arnon SS, Schechter R, Inglesby TV, et al. Botulinum toxin as a biological weapon: medical and public health management. *JAMA* 2001;285:2081.

39. Centers for Disease Control and Prevention (CDC). Recognition of illness associated with the intentional release of a biologic agent. *MMWR Morb Mortal Wkly Rep* 2001;50(41): 893–897.

40. Buehler JW, Berkelman RL, Hartley DM, Peters CJ. Syndromic surveillance and bioterrorism-related epidemics. *Emerg Infect Dis* 2003;9:1197–1204.

41. Roush S, Birkhead G, Koo D, Cobb A, Fleming D. Mandatory reporting of diseases and conditions by health care professionals and laboratories. *JAMA* 1999;282(2):164–170.

42. Buehler JW, Berkelman RL, Hartley DM, et al. Syndromic surveillance and bioterrorism related epidemics. *Emerg Infect Dis* 2003;9:1197–1204.

43. Borchardt SM, Ritger KA, Dworkin MS. Categorization, prioritization, and surveillance of potential bioterrorism agents. *Infect Dis Clin North Am* 2006;20:213–225.

44. Sokolow LZ, Grady N, Rolka H, et al. Deciphering data anomalies in BioSense. *MMWR Morb Mortal Wkly Rep* 2005;54(Suppl):133–139.

45. Rotz LD, Hughes JM. Advances in detecting and responding to threats from bioterrorism and emerging infectious disease. *Nat Med* 2004;10:S130–S136.

46. Blank S, Moskin LC, Zucker JR. An ounce of prevention is a ton of work: mass antibiotic prophylaxis for anthrax, New York City, 2001. *Emerg Infect Dis* 2003;9:615–622.

47. Dewan PK, Fry AM, Laserson K, et al. Inhalational anthrax outbreak among postal workers, Washington, D.C., 2001. *Emerg Infect Dis* 2002;8:1066–1072.

48. Christopher GW, Cieslak TJ, Pavlin JA, et al. Biological warfare: a historical perspective. *JAMA* 1997;278:412–417.

49. Sugishima M. Aum Shinrikyo and the Japanese law on bioterrorism. *Prehosp Disaster Med* 2003;18:179–183.

50. Office of Technology Assessment, US Congress. Proliferation of Weapons of Mass Destruction. Washington, DC: US Government Printing Office; 1993. Publication OTA-ISC-559.

51. Centers for Disease Control and Prevention. Supplement B: SARS Surveillance Appendix B1: Revised CSTE SARS surveillance case definition. http://www.cdc.gov/ncidod/sars/guidance/b/app1.htm. Accessed March 29, 2009.

52. Centers for Disease Control and Prevention. E25: Recommended Daily Dosage of Seasonal Influenza Antiviral Medications for Treatment and Chemoprophylaxis for the 2008–09 Season—United States. http://www.cdc.gov/flu/professionals/antivirals/dosagetable.htm#table. Accessed April 20, 2009.

53. Centers for Disease Control and Prevention. Smallpox Response Plan and Guidelines, Version 3.0. 2002. http://www.bt.cdc.gov/agent/smallpox/response-plan/. Accessed March 29, 2009.

54. Centers for Disease Control and Prevention: Acute, Generalized Vesicular or Pustular Rash Illness Testing Protocol in the United States. http://www.bt.cdc.gov/agent/smallpox/diagnosis/. Accessed November 24, 2009.

55. Division of Bioterrorism Preparedness and Response (DBPR), National Center for Preparedness, Detection, and Control of Infectious Diseases (NCPDCID), Coordinating Center for Infectious Diseases (CCID). CDC Interim Guidance for Revaccination of Eligible Persons who Participated in the US Civilian Smallpox Preparedness and Response Program, October 2008. http://www.bt.cdc.gov/agent/smallpox/revaxmemo.asp. Accessed March 20, 2009.

56. Pasetti MF, Cuberos L, Horn TL, et al. An improved *Francisella tularensis* live vaccine strain (LVS) is well tolerated and highly immunogenic when administered to rabbits in escalating doses using various immunization routes. *Vaccine* 2008;26:1773–1785.

57. World Health Organization. Alert, verification and public health management of SARS in the post-outbreak period. August 14, 2003. http://www.who.int/csr/sars/postoutbreak/en/. Accessed April 26, 2004.

58. Oshitani H, Kamigaki T, Suzuki A. Major issues and challenges of influenza pandemic preparedness in developing countries. *Emerging Infect Dis* 2008;14(6):875–880.

59. Eyck RP. Ability of regional hospitals to meet projected avian flu pandemic surge capacity requirement. *Prehosp Disaster Med* 2008;23(2):102–112.

60. Bartlett JG, Borio L. The current status of planning for pandemic influenza and implications for health care planning in the United States. *Clin Infect Dis* 2008;46:919–925.

61. Department of Health and Human Services (DHHS). Guidance on Allocating and Targeting Pandemic Influenza Vaccine. DHHS, 2007. http://www.pandemicflu.gov/vaccine/allocationguidance.pdf. Accessed April 11, 2009.

62. Centers for Disease Control and Prevention. FluSurge, version 2.0. http://www.cdc.gov/flu/tools/flusurge/. Accessed April 11, 2009.

63. Occupational Safety and Health Administration (OSHA). Guidance on Preparing Workplaces for an Influenza Pandemic. OSHA publication 3327–02N 2007. http://www.osha.gov/Publications/influenza_pandemic.html. Accessed August 26, 2009.

64. Hick JL, Hanfling D, Burstein JL, et al. Health care facility and community strategies for patient care surge capacity. *Ann Emerg Med* 2004;44(3):253–261.

65. Agency for Healthcare Research and Quality (AHRQ). Bioterrorism and Other Public Health Emergencies: Altered Standards of Care in Mass Casualty Events. AHRQ publication 05–0043. April 2005. http://www.ahrq.gov/research/altstand/. Accessed April 11, 2009.

Chapter 23 Exposure Workups

Louise-Marie Dembry, MD, MS, MBA,
Stephanie Holley, RN, CIC, BSN, Hitoshi Honda, MD,
Loreen A. Herwaldt, MD, and Jean M. Pottinger, RN, MA, CIC

There seem to be several axioms about exposures in the hospital. First, they always come at inconvenient times. There is no good time for an exposure even if it's not Friday. The corollary to this axiom is that epidemiologic workups for exposures always interrupt other important infection prevention activities. The second axiom is that exposures usually involve more than one department and that at least one of the affected areas will be a large open room in which many persons—who may be very difficult to identify—congregate. The third axiom is that exposures almost always involve the most vulnerable patients or healthcare workers. The corollary to this axiom is that exposures are guaranteed to cause great anxiety among patients and staff.

Infection prevention personnel define exposures as events in which persons were exposed to infectious microorganisms or to ectoparasites. The goals of an epidemiologic workup for an exposure are to prevent disease, if possible, in the person or persons who were exposed to the etiologic agent and to prevent further transmission if exposed persons become ill. To achieve these goals, infection prevention personnel must identify all patients, visitors, and staff who might have been exposed and then must determine whether these persons are susceptible or immune. If the exposed persons are immune to the etiologic agent, they do not require further investigations or interventions. If exposed persons are not immune to the etiologic agent or do not know their immune status, infection prevention personnel may need to obtain further data, prescribe prophylactic treatment, and institute work restrictions. In this chapter, we describe exposure workups for a number of important pathogens.

Many healthcare-associated exposures do not cause secondary cases of infection, or, if secondary cases occur, they are often mild. However, on occasion, patients, visitors, or healthcare workers acquire infections that cause serious short-term or long-term consequences, including prolonged absence from work, exposure to toxic treatments, incurable chronic illness, irreversible disability, or death.[1] Regardless of the ultimate consequences, exposure workups consume considerable time, money, and other resources.[2-4] Therefore, infection prevention staff should strive to prevent exposures using the following measures:

- Implement policies that reduce the number of susceptible persons exposed (eg, by requiring all healthcare workers to be immune to measles, mumps, and rubella or by requiring all outpatient-treatment areas to screen patients for symptoms consistent with communicable diseases).
- Teach healthcare workers to recognize when they should stay home to prevent the spread of infectious agents.
- Teach healthcare workers to apply standard precautions and transmission-based precautions properly.
- Implement respiratory etiquette and cough hygiene (ie, use of masks for persons with signs or symptoms of a respiratory infection, such as fever and cough) for patients and the persons accompanying them in outpatient-treatment areas and emergency departments.

Given the serious consequences that can result from exposures, healthcare facilities must manage exposures in a systematic and consistent manner. Many healthcare facilities assign this responsibility to infection prevention personnel. In this chapter, we describe the steps in exposure investigations. We have summarized our recommendations in a generic algorithm for the epidemiologic workup of exposures (Figure 23-1) and in a table (Table 23-1).

In developing the recommendations in this chapter, we used the 2009 Red Book,[5] the *Control of Communicable Diseases Manual*,[6] the guidelines published by the Hospital Infection Control Practices Advisory Committee (HICPAC),[7] the guidelines published by the Advisory Committee on Immunization Practices (ACIP),[8] the Web sites of the Centers for Disease Control and Prevention (CDC)[9] and the World Health Organization (WHO),[10] other published studies, and our own experience. We also consulted *Principles and Practice of Infectious Diseases*.[11] Infection prevention personnel who prefer the recommendations published in other references will need to modify our recommendations to suit their needs. We excluded bloodborne pathogen exposures from this discussion (they are discussed in Chapter 24 on occupational health), and we included only the etiologic agents that cause most exposures in hospitals. Hospitals vary; thus, some facilities may have numerous exposures to infectious agents that we have not discussed. In addition, newer agents, such as the coronavirus that causes severe acute respiratory syndrome (SARS), monkeypox virus, and novel influenza A virus subtype H1N1 ("swine flu" virus) can and will continue to afflict hospitals. Therefore, infection prevention personnel need to know what is happening in their communities and around the world. One way infection prevention personnel can keep abreast of what is happening is to join Internet list services ("listservs"), such as the Emerging Infections Network sponsored by the Infectious Diseases Society of America.[12] In addition, some state or local health departments inform infection prevention personnel of important developments through e-mail or faxed bulletins.

General Recommendations Regarding Exposure Workups

Obtain a Mandate from the Administration

If the hospital administration assigns the responsibility for doing exposure workups to the infection prevention program, the administrators also must define the scope of that responsibility and must delegate the authority for the associated activities to staff in the infection prevention program. The hospital administration must define prospectively what tests and prophylactic treatments the hospital will provide. In addition, the hospital administration must specify if exposed healthcare workers will be granted administrative leave, leave with pay, or leave without pay or if they will be allowed to work in non–patient-care areas during the period in which they might be infectious.[13,14]

Develop Policies and Procedures

Once the infection prevention program has been given the authority to do exposure workups, the staff must develop specific policies that define exposures to various bacterial and viral pathogens and to ectoparasites and must describe the investigative and preventive measures that should be undertaken for exposure to each agent. The staff also should develop general policies and procedures that define what tasks should be undertaken and who will do them.

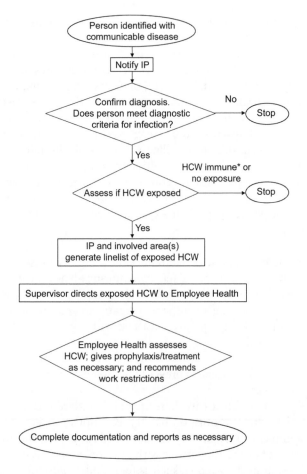

Figure 23-1. Flow chart summarizing the management of communicable disease exposures in the University of Iowa Hospitals and Clinics. HCW, healthcare worker; IP, infection preventionist. *See Table 23-1 for details.

Table 23-1. Basic Information Regarding the Etiologic Agents of the Most Common Healthcare-Associated Exposures

Organism	Incubation period	Diagnostic criteria	Exposure criteria	Period of communicability	Occupational health actions and issues	Work restrictions, by exposure status	Prophylaxis
Varicella-zoster virus	• Usually 14–16 days (range, 10–21 days); up to 28 days in persons who received VZIG	**Chickenpox** • Fever and vesicular rash **Shingles** • Grouped vesicular lesions • May need to consult dermatology dept.	**Chickenpox or disseminated zoster** • Continuous household contact • Face-to-face contact with infected person for 5 minutes or more without wearing a respirator • Direct contact with vesicle fluid without wearing gloves **Shingles** • Direct contact with vesicle fluid without wearing gloves	**Chickenpox** • Most contagious 1–2 days before and shortly after rash appears • Transmission can occur until all lesions are crusted • Immunocompromised persons may be contagious as long as new lesions are appearing **Shingles** • Contagious for 24 hours before the first lesion appears and until all lesions are crusted	• Assess exposed person's immunity • HCW is susceptible unless: – HCW has serologic evidence of immunity – HCW has documentation proving receipt of 2 doses of varicella vaccine • Consider obtaining varicella IgG antibody titer to determine immune status before HCW is exposed	**Exposed** • Day 1–7, no restrictions • Days 8–21 after a single exposure, or day 8 after first exposure through day 21 after last exposure, restrictions are: – HCW must not work, **or** – HCW must not have direct patient contact and must work only with immune persons away from patient-care areas • For an HCW who received VZIG, restrict work through day 28 **Infected** • HCW may return to work after all lesions are crusted	• For nonimmune, immunocompromised person, consider giving VZIG within 96 hours after exposure • For susceptible HCW, consider giving varicella virus vaccine within 3 days after exposure to prevent or modify infection; giving the vaccine does not change the work restrictions
Rubeolla (measles) virus	• Usually 8–12 days (range, 7–18 days)	• Prodromal symptoms including conjunctivitis, coryza, and cough • Fever and rash with positive measles IgM antibody titer • May need to consult dermatology dept.	• Spent time, without wearing a respirator, in a room with an infected person • If air is recirculated, spent time in the area supplied by the air-handling system while infected person was present or within 1 hour after the person's departure	• 3–5 days before rash to 4–7 days after rash appears, but transmission is minimal 2–4 days after rash appears • Immunocompromised persons may be contagious for the duration of the illness	• Assess exposed person's immunity • HCW is susceptible unless: – HCW has serologic evidence of immunity – HCW has documentation proving receipt of 1 dose of measles vaccine, if born before 1957,a or 2 doses of vaccine, if born after	**Exposed** • Days 1–4, no restrictions • Days 5–21 after a single exposure, or day 5 after first exposure through day 21 after last exposure, restriction is: – HCW must be excluded from work setting	• For staff who have not received 2 doses of measles vaccine, consider giving MMR vaccine within 3 days after exposure, to modify infection • Vaccine or immune globulin given after exposure does not change work restrictions

(continued)

Table 23-1. (continued)

Organism	Incubation period	Diagnostic criteria	Exposure criteria	Period of communicability	Occupational health actions and issues	Work restrictions, by exposure status	Prophylaxis
			• Contact, without wearing gloves, with nasal or oral secretions from an infected person or items contaminated with these secretions		January 1, 1968, and on or after first birthday, if born after 1957 • Obtain blood samples to determine IgG antibody titer, as needed	**Infected** • HCW may return to work 4 days after developing rash	
Rubella virus	• Usually 14–17 days (range, 12–23 days)	• Mild febrile exanthem and positive rubella IgM antibody titer • May need to consult dermatology dept.	• Contact, without wearing a mask, within 1 m (3 feet) of infected person • Contact, without wearing gloves, with nasopharyngeal secretions from an infected person or items contaminated with these secretions • Contact, without wearing gloves, with nasopharyngeal secretions or urine from infant with congenital rubella	• 7 days before rash to 7 days after rash appears • Up to 1 year for infants with congenital rubella	• Assess exposed person's immunity • HCW is susceptible unless: – HCW can provide serologic evidence of immunity – HCW has documentation proving receipt of 1 dose of rubella vaccine after Jan 1, 1970 • Obtain blood samples to determine IgG antibody titer, as needed	**Exposed** • Day 1–6, no restrictions • Day 7–21 for a single exposure or day 5 of first exposure through day 21 of last exposure, restrictions are: • HCW must not work or • HCW must have no direct patient contact and work only with immune persons away from patient-care areas **Infected** • HCW may return to work 7 days after developing rash	• None • Rubella vaccine does not prevent infection after exposure • Immune globulin does not prevent infection
Mumps virus	• Usually 16–18 days (range, 12–25 days)	• Fever with swelling and tenderness of the salivary glands or testes and positive mumps IgM antibody titer	• Contact, without wearing a mask, within 1 m (3 feet) of infected person • Contact, without wearing gloves, with saliva or	• Most communicable 48 hours before onset of illness, but may begin as early as 7 days before onset of overt parotitis and/or	• Assess exposed person's immunity • HCW is susceptible unless: – HCW can provide serologic	**Exposed** • Day 1–10, no restrictions • Day 11–26 for a single exposure, or day 11 after first exposure through day 26 after last	• None • Mumps vaccine is not proven to prevent infection after exposure • Mumps immune globulin does not prevent infection

Infectious agent	Incubation period / clinical features	Diagnostic criteria	Transmission / infectivity	Management of exposure	Work restrictions
(mumps, continued)	orchitis and continue 5–9 days thereafter (average, 5 days)		items contaminated with saliva from an infected person	evidence of immunity – HCW can document receipt of 1 dose of mumps vaccine, if born before 1957, during nonepidemic periods, or receipt of 2 doses, during epidemic periods; or receipt of 2 doses, if born after 1957 • Obtain blood samples to determine IgG antibody titer, as needed	exposure, restrictions are: – HCW must not work, **or** – HCW must have no direct patient contact and work only with immune persons away from patient-care areas **Infected** • HCW may return to work 9 days after onset of parotid gland swelling
Parvovirus B19	• Usually 4–14 days (range, up to 21 days) • Rash and joint symptoms occur 2–3 weeks after infection	• "Slapped cheek" rash and positive serum parvovirus B-19 IgM antibody titer	• Criteria have not been defined but probably include the following: – Contact, without wearing a mask, within 1 m (3 feet) of infected person – Contact, without wearing gloves, with respiratory secretions from an infected person or items contaminated with these secretions • Infected persons are unlikely to be infectious after the onset of rash • Immunocompromised persons can have chronic infections and can shed virus for prolonged periods	• Assess exposed person's immunity • Obtain blood samples to determine IgG antibody titer, as needed • Describe signs and symptoms and inform exposed HCWs that they should not work if these symptoms occur • Refer a pregnant HCW to her obstetrician	• Not necessary
Hepatitis A virus	• Usually 28–30 days (range, 15–50 days)	• Positive hepatitis A IgM antibody titer	• Contact, without wearing gloves, with stool of infected person • Consuming uncooked food prepared by an infected person • Virus is shed in stool for 1–3 weeks • Highest viral titers are found in stool 1–2 weeks before onset of symptoms	• Assess exposed person's immunity • Obtain blood samples to determine IgG antibody titer, as needed	**Exposed** • None **Infected** • May return to work on day 7 after onset of jaundice or other clinical symptoms • Consider giving immune globulin to exposed HCW within 2 weeks after exposure

(continued)

Table 23-1. (continued)

Organism	Incubation period	Diagnostic criteria	Exposure criteria	Period of communicability	Occupational health actions and issues	Work restrictions, by exposure status	Prophylaxis
				• Risk of transmission is minimal 1 week after onset of symptoms	• Describe signs and symptoms and ask exposed HCW to return to employee health dept. if these occur		
Influenza virus	• Usually 1–4 days	• Influenza-like illness during influenza season (northern hemisphere, Oct–Apr; southern hemisphere, May–Aug; tropics, all months) • Positive result of diagnostic test for influenza	• Contact, without wearing a mask, within 1 m (3 feet) of infected person • Direct contact, without wearing gloves, with respiratory tract secretions of infected person or items contaminated with these secretions	• Infected persons are most infectious 24 hours before onset of symptoms • Viral shedding usually ceases within 7 days but can persist longer in children	• Assess immunization status • Discuss risks and benefits of chemoprophylaxis • Describe signs and symptoms and inform HCWs that they should not work if these symptoms occur	**Exposed** • Restrictions have not been defined for exposed nonimmune HCWs **Infected** • HCW who is ill should not work	• Consider vaccinating exposed nonimmune HCWs • Amantidine or rimantadine, 100 mg twice per day, for adults exposed to influenza A only • Oseltamivir, 75 mg twice per day, for adults exposed to influenza A or B • Zanamivir, 10 mg (2 inhalations) once per day, for adults exposed to influenza A or B • Chemoprophylaxis may vary by location, season, and in vitro drug susceptibility
CJD agent (prion)[b]	• 15 months to more than 30 years	• Progressive dementia • Receipt of cadaver-derived pituitary hormones • Family history of prion disease	• Criteria have not been defined but probably include the following, if the source patient has a diagnosis of CJD or has risk factors for CJD:[c] – Puncture or cut with instruments contaminated with patient's blood or CSF	• Period of communicability is unknown, but transmission probably occurs during symptomatic illness and for an undetermined period before symptoms appear	• Educate employee about CJD and risk of transmission • Counsel employee using data from the literature indicating that the risk of transmission is very low	• None	• None

			– Handling, without gloves, CSF or tissue from brain, spinal cord, or eye – Working in the operating room, autopsy suite, ophthalmology department, pathology laboratory, or microbiology laboratory poses the highest risk of exposure for HCWs – Routine patient care poses a very low risk for HCWs			
Mycobacterium tuberculosis	• 2–10 weeks from exposure to positive TST result • Risk of developing active disease is greatest in first 2 years after infection	• *M. tuberculosis* or AFB found in respiratory secretions or wound drainage	• Spending time, without wearing a respirator, in a room with a person who has active disease • Packing or irrigating wounds infected with *M. tuberculosis* without wearing a respirator	• Persons are considered infectious if they meet the following criteria: – Person is coughing – Person is undergoing cough-inducing or aerosol-generating procedures – Person has sputum smears positive for AFB – Person is not receiving therapy – Person has just started therapy – Person has poor clinical response to therapy • Persons are considered infectious until they are receiving effective antituberculosis chemotherapy,	• Obtain baseline TST result within 2 weeks after exposure, if the HCW previously had a negative TST result • Perform postexposure TST at 12 weeks • Prescribe treatment if postexposure TST result is positive	**Exposed** • None for persons whose TST result becomes positive **Infected** • Restrict HCWs with active tuberculosis until they are receiving effective antituberculosis chemotherapy, respond to therapy, and have 3 consecutive sputum smears negative for AFB	• Isoniazid, 300 mg daily, for 9 months • Pyridoxine, 25–50 mg daily, may be added to the treatment regimen for persons with conditions in which neuropathy is common • Refer to the CDC Core Curriculum on Tuberculosis[218] for alternative treatment options

(continued)

Table 23-1. (continued)

Organism	Incubation period	Diagnostic criteria	Exposure criteria	Period of communicability	Occupational health actions and issues	Work restrictions, by exposure status	Prophylaxis
				respond to therapy, and have 3 consecutive sputum smears negative for AFB • Children with primary pulmonary tuberculosis are rarely contagious			
Neisseria meningitidis	• Usually 4 days or less (range, 1–10 days)	• Clinical signs of sepsis, meningitis, or pneumonia, and gram-negative diplococci in blood, CSF, sputum, synovial fluid, pericardial fluid, or skin scraping	• Extensive contact, without wearing a mask, with respiratory secretions from an infected person, particularly during these procedures: – Suctioning – Resuscitation – Intubation – Extensive oral or pharyngeal examination	• Persons are infectious until they have received effective antibiotic therapy for 24 hours	• Prescribe prophylaxis • Educate exposed HCW about signs and symptoms of meningitis	**Exposed** • None	• Ciprofloxacin, 20 mg/kg (maximum, 500 mg) in a single dose (contraindicated in pregnancy), **or** • Rifampin, 10 mg/kg (maximum, 600 mg) every 12 hours for 2 days (contraindicated in pregnancy) **or** • Ceftriaxone, 250 mg intramuscularly, in a single dose (safe during pregnancy)
Bordetella pertussis	• Usually 7–10 days (range, 6–20 days)	• Paroxysmal cough, other respiratory symptoms, or inspiratory whoop with positive result of DFA, culture, PCR, or serologic test for *B. pertussis*	• Contact, without wearing a mask, within 1 m (3 feet) of infected person • Direct contact, without wearing gloves, with respiratory tract secretions from infected persons or items contaminated with these secretions	• Infected persons are most contagious during the catarrhal state • Communicability diminishes rapidly after onset of cough, but can persist as long as 3 weeks	• If not symptomatic, HCW should begin prophylaxis and can return to work • If symptomatic, HCW should begin therapy and be relieved from work	**Exposed** • No restrictions **Infected** • HCW may return to work after receiving effective therapy for at least 5 days	• Azithromycin, 500 mg per day for 5 days • Erythromycin, 40 mg/kg per day in 4 divided doses (maximum, 2 g per day) for 14 days (estolate preparation preferred) • Tdap should be considered for a nonimmune HCW

Organism	Incubation/Life cycle	Clinical features/Diagnosis	Transmission	Period of communicability	Management	Work restrictions	Treatment
Lice (*Pediculus humanus capitis, Pediculus humanus corporis, Phthirus pubis*)	**Head lice** • 7–12 days (from egg to hatching) • 9–12 days (from adulthood to reproduction by egg) **Body lice** • 1–2 weeks (from egg to hatching) • 9–19 days (from adulthood to reproduction by egg) **Pubic lice** • 6–10 days (from egg to hatching) • 2–3 weeks (from adult lice to reproduction by egg)	**Head lice** • Live lice or nits (eggs) on hair shaft 0.5 inch or less from skin • May need to consult dermatology dept.	**Head lice** • Hair-to-hair contact with infested person **Body lice** • Contact with linen or clothes of infested person without wearing gloves **Pubic lice** • Sexual contact	• As long as lice or eggs remain alive on infested person, clothing, and/or personal items • Survival times for lice away from the host: – Head lice: 1–2 days (adult lice), 7–10 days (egg) – Body lice: 5–7 days (adult lice), up to 1 month (egg) – Pubic lice: 2 days (adult lice) • Nits 10 mm or more from scalp have been present 2 weeks or more and may not be viable	• Treat HCW only if infested	**Exposed** • No restrictions **Infested** • Immediate restriction until 24 hours after start of treatment	• Not recommended
Scabies (*Sarcoptes scabiei*)	• 4–5 weeks, if person has no history of previous infestation • 1–4 days, if person has history of previous infestation	• Burrows or papular lesions in classic body sites and intense itching at night • May need to consult dermatology dept.	• Prolonged, close personal contact • Minimal direct contact with crusted (Norwegian) scabies can result in transmission	• Transmission can occur before the onset of symptoms • Person remains contagious until treated	• Prescribe scabicide for all HCWs exposed to persons with crusted scabies • Pregnant women should not use lindane	**Exposed** • No restriction **Infested** • Immediate restriction until 24 hours after start of treatment	• Drug of choice: 5% permethrin • Alternative drugs: lindane or 10% crotamiton

NOTE: Data are drawn from multiple studies.[5,6,11,15,119] AFB, acid-fast bacilli; CJD, Creutzfeldt-Jakob disease; CSF, cerebrospinal fluid; DFA, direct fluorescent antibody test; HCW, healthcare worker; MMR, measles, mumps, and rubella vaccine; PCR, polymerase chain reaction; Tdap, tetanus-diphtheria-acellular pertussis vaccine; TST, tuberculin skin test; VZIG, varicella-zoster immune globulin.

a A small percentage of HCWs born before 1957 will not be immune to measles.[49] Infection prevention personnel should determine whether this criterion is appropriate for the staff in their hospital.

b These precautions should be used for persons who have CJD or progressive dementia or who have a family history of prion disease, CJD, Gerstmann-Sträussler-Scheinker syndrome, or fatal familial insomnia.

c Data from a study with mice suggest that prions might be transmitted by contact of infected blood or CSF with mucous membranes.[219] No data on humans are available to support or refute the data on mice.

Collaborate With Occupational Health

In many institutions, the infection prevention staff initiates the exposure workup and recommends the prophylaxis and work restrictions for exposed healthcare workers. However, staff in the occupational health service actually evaluate whether the employee was exposed and susceptible, examine the healthcare worker, enforce work restrictions, and give permission for healthcare workers to return to work. Thus, as they develop policies and procedures, infection prevention personnel must collaborate extensively with staff in occupational health and must clearly delineate responsibilities.

Develop a Database on the Immune Status of Healthcare Workers

Infection prevention personnel will save countless hours if they have a database in which they store information on the immune status of all healthcare workers. The most important data are the employee's immune status with respect to chicken pox, measles, mumps, rubella, and hepatitis B. However, some hospitals might find it useful to test employees for antibody to parvovirus B19, if they work in antepartum clinics or with patients who are immunocompromised or have hemolytic anemia. For each employee, tuberculin skin test results and the results of respirator fit-testing should be recorded in the database. The database also could store information on each employee's immunity to diphtheria, tetanus, and hepatitis A. Baseline data should be obtained from all new healthcare workers before they start working in the institution. If the hospital is establishing a new database, the same information should be obtained from all current employees.

The database should be computerized; it may be as simple as a spreadsheet format that can be easily managed in smaller hospitals (see Chapter 10 on informatics). The persons who develop and maintain the database could be in the hospital's information management group, in the infection prevention program, or in the occupational health service. Regardless of who manages the database, the persons investigating exposures must have unobstructed access to the database so that they can use the data regardless of when the exposure occurs, while ensuring employee confidentiality. Some programs have found that storing the database on a shared computer drive accessible to clinicians, infection prevention staff, and employee health staff is useful to facilitate the data sharing necessary for complete, prompt exposure evaluations.

Develop a Data Collection Form

Infection prevention staff must investigate exposures in a consistent fashion. Therefore, in addition to developing policies and procedures, infection prevention staff should design a standardized form (preferably as an electronic file) with which they can collect the necessary data for each exposure. A list of healthcare workers who were in the affected areas should be generated that indicates either the immune status of these employees or the date of their last tuberculin skin test, as appropriate. This list can usually be generated from staffing records but may also require additional medical-record review to capture data on healthcare workers who were in the affected area and are not hospital employees (eg, physicians) or who were assigned to a particular area or unit. However the list is generated and whatever form it is in (ie, on paper or in an electronic form), it should be shared as soon as possible with occupational health personnel so that they are prepared when exposed staff come to them for follow-up.

Educate Staff

Healthcare workers should know the modes of transmission for common communicable pathogens and the basic infection prevention practices that limit the spread of microorganisms. In addition, epidemiologic workups after exposures will go more smoothly if infection prevention personnel prospectively educate healthcare workers about exposure workups in general and about the specific steps taken during common exposure workups. Infection prevention staff also will need to educate and reassure the staff while conducting an exposure workup. It is not uncommon for staff to be anxious regarding an exposure, and they may panic and act irrationally; this is a common response when staff think they have been exposed to *Neisseria meningitidis* or to lice, for example.

Collect and Evaluate Data on Exposures

Infection prevention personnel should collect data on the exposure workups that they conduct. At least once per year, infection prevention staff should assess the following points:

- The number of exposures
- The etiologic agents
- The affected locations
- The number of susceptible healthcare workers, patients, and visitors exposed
- The number of secondary cases
- The number of healthcare workers who were placed on leave
- The number of leave-days
- The failures in infection prevention technique that led to the exposures

- The prophylactic treatments given
- The cost in time and money

Infection prevention personnel should report these data to the infection control committee and should use these data to do the following actions:

- Document their effort to the administration
- Identify topics for in-service educational programs
- Identify interventions (eg, offer influenza vaccine free of charge to all employees)
- Identify areas for collaboration with other departments (eg, work with staff in other departments to develop methods for screening and triage of potentially infectious patients)
- Identify areas for improvement
- Document quality improvement efforts required for accreditation

Disease Agent–Specific Recommendations Regarding Exposure Workups

Viruses

Varicella-Zoster Virus

Varicella-zoster virus causes a primary infection, chicken pox, and a recrudescent infection, herpes zoster or shingles. Varicella-zoster virus can be transmitted through the air by persons with chicken pox or through direct contact with fresh chicken pox or herpes zoster lesions. Thus, patients with chicken pox or disseminated zoster should be placed under airborne transmission precautions and contact precautions until all lesions are crusted, to prevent exposures within hospitals.[15] Nonimmune patients who have been exposed to chicken pox should be placed under airborne precautions for the period from day 10 through day 21 after their exposure (or through day 28, if the person is immunocompromised or received varicella-zoster immune globulin). Varicella-zoster virus rarely is spread through the air from persons with localized herpes zoster, so patients, visitors, and healthcare workers with this disease entity do not need to be restricted if their lesions can be covered.

Approximately 4%–15% of susceptible healthcare workers will develop chicken pox each year.[16] Currently, 2%–5% of all healthcare workers are not immune to varicella-zoster virus,[16,17] and 28% of those with no known history of chicken pox are susceptible to this virus.[16] Nonimmune healthcare workers who have been exposed to a person with chicken pox could be incubating the infection. To prevent the spread of varicella-zoster virus, infection prevention personnel must identify those healthcare workers and must restrict their work during the incubation period. Most healthcare facilities do not allow susceptible, exposed healthcare workers to continue their patient-care duties during the incubation period. Some healthcare facilities place such staff on leave,[13,14] and other facilities reassign exposed susceptible staff to non–patient-care areas, if all the employees in that area are immune.[18] Exposed staff who are permitted to work must take care not to expose persons as they enter and exit the building. Staff members who develop active disease must not work until all lesions are crusted. Exposed visitors who are not immune should not be allowed to enter the hospital during the incubation period. Exposed visitors who do not know their immune status should not enter the hospital during the incubation period until they have antibody levels tested and are documented to be immune.

Infection prevention personnel should work with occupational health staff, expert clinicians, pharmacists, and hospital administrators to determine whether exposed persons will be offered the chicken pox vaccine (Varivax; Merck) or varicella-zoster immune globulin. If the healthcare facility provides one or both of these agents, this group should decide prospectively which persons will be offered which agent. In addition, this group should decide whether their healthcare facility will offer the chicken pox vaccine to all nonimmune healthcare workers. This decision may not be a simple one, for the following reasons:

- Five percent of healthcare workers who receive the vaccine will develop a varicella-like rash that will require them to miss work because transmission of the vaccine virus has been documented.
- For approximately 6 weeks after receiving the vaccination, healthcare workers should not care for susceptible, high-risk persons, including immunocompromised persons, pregnant women who do not have a history of chicken pox or detectable antibody to varicella-zoster virus, and newborns of such women.
- According to the vaccine package insert, up to 27% of vaccinated persons will have subclinical or breakthrough varicella infection after close exposure to a person with chicken pox.[19]

Vaccinated persons who acquire chicken pox often have milder disease than do nonvaccinated persons.[20-23] Some healthcare facilities have experienced substantial rates of transmission of varicella-zoster virus,[2,3,24] and several investigators found that varicella vaccine would be cost-effective, given the costs associated with secondary cases.[25-27]

Table 23-1 outlines an approach to managing healthcare workers who have been exposed to varicella-zoster virus.[5,14,23] Infection preventionists who want additional information about varicella-zoster virus exposures should consult the appropriate references.[6,11,16-18,28]

Measles Virus

Measles is a febrile illness that is characterized by Koplik's spots on the buccal mucosa and by an erythematous rash. The measles rash starts on the face and spreads to the trunk and extremities and also progresses from maculopapular to confluent. Measles virus, which is highly communicable, is spread by airborne transmission. Despite sensitivity to acid, strong light, and drying, the measles virus can remain viable in airborne droplets for hours, especially if the relative humidity is low. Consequently, outbreaks have occurred in healthcare facilities when the index patient was no longer present.[29,30] To prevent the spread of measles virus within healthcare facilities, patients with measles should be placed under airborne precautions.[15]

Before 1963, when the measles vaccine was licensed, 500,000 cases of measles occurred in the United States each year. Subsequently, the number of measles cases in the United States declined dramatically, reaching a nadir in 1983. Thereafter, the incidence of measles increased for several years. More recently, increased immunization rates and routine use of 2 doses of the vaccine have helped decrease the number of measles cases. Measles now occurs most frequently in preschool children, many of whom are too young to be vaccinated. Seo et al.[31] reported that measles seropositivity rates among healthcare workers in their twenties was lower for those hired between 1998 and 1999 than for those hired between 1983 and 1988. Thus, healthcare workers who were born after 1989 should have their antibody levels checked when they are hired.

Despite the declining incidence of measles and despite recommendations that all persons receive 2 doses of measles vaccine and that all healthcare workers should be immune to measles, outbreaks and healthcare-associated transmission continue to occur.[29,30,32-49] Steingart et al.[48] reported that 8 of 31 persons in Clark County, Washington, who acquired measles in 1996 were healthcare workers, and 5 were patients or visitors in healthcare facilities. Healthcare workers who acquired measles worked in facilities that did not require proof of measles immunity. Compared with adults in Clark County, the relative risk of measles in healthcare workers was 18.6 (95% confidence interval, 7.4–45.8; $P < .001$). Only 47% of facilities surveyed by these investigators had measles immunization policies, and only 21% met the ACIP recommendations[8] and enforced their policies. Kelly et al.[50] described several outbreaks of measles in Australian healthcare facilities. They concluded that the outbreaks occurred because published guidelines for

preventing healthcare-associated measles were not followed. They stated that transmission of measles in a healthcare facility could be considered "a sentinel sign of system failure."[50] At present, 5%–10% of healthcare workers are susceptible to measles,[16] including 4.7% of those born before 1957, 16% of those born in the 1960s, and 34% of those born in the 1970s.[45] In fact, healthcare workers are the source of 5%–10% of all measles cases, and they account for 28% of measles cases acquired in medical settings in the United States.[16]

Although measles exposures are infrequent, infection prevention personnel still must develop policies that will limit the spread of measles if it is introduced into the hospital. A study conducted by Enguidanos et al.[45] suggests that infection prevention programs may be ignoring measles because the incidence is low. These investigators noted that 74 adults employed in acute-care hospitals acquired measles during a community-wide outbreak in 1987 through 1989. They surveyed all 102 infection preventionists in the acute-care hospitals in Los Angeles County to determine whether infection prevention policies were adequate. Only 17% of the hospitals required healthcare workers to document immunity to measles, and only 4% had policies that covered students or volunteers. The investigators also surveyed the healthcare workers who became ill. Of these 74 persons, 46% worked in hospitals that did not have measles infection prevention policies, 43% were born before 1957, and 31% were working in jobs that have not been considered to increase the risk of measles exposure.[45]

During outbreaks in 2008, some persons who were thought to be immune on the basis of their age acquired measles. Thus, the CDC has proposed new criteria for measles immunity among healthcare workers. The draft proposal suggests that only documentation of measles, mumps, and rubella vaccination or serologic evidence of immunity be considered evidence of measles immunity. Thus, birth date and documentation of physician diagnosed disease in the past would no longer be considered evidence of immunity.

As discussed previously, infection prevention personnel should work with occupational health personnel to develop a database that has each healthcare worker's history of measles vaccination. If such a database is not available and a person with measles comes to the hospital, the infection prevention staff must identify exposed personnel and then determine whether these persons are immune. Nonimmune patients who have been exposed to measles should be placed under airborne precautions from day 5 through day 21 after their exposure.[15] Nonimmune family members and friends who have been exposed to a person with measles should not come to the hospital during the incubation period. Table 23–1 provides information necessary for managing a measles exposure.[5]

Rubella Virus

Rubella (German measles) is an acute exanthematous viral infection that affects children and adults. Postnatal rubella, which resembles a mild case of measles, is characterized by rash, fever, and lymphadenopathy. In contrast, rubella acquired in pregnancy can cause fetal death, premature labor, and severe congenital defects. Consequently, it is very important to prevent the spread of rubella in healthcare facilities. However, the mild clinical symptoms associated with rubella have, at times, facilitated nosocomial spread of rubella virus, because healthcare workers have continued to work while they were ill. The literature documents numerous outbreaks of rubella in medical facilities, some of which affected many susceptible pregnant women.[51-59] Furthermore, these institutions had to invest large amounts of time and money to control the outbreaks.[55,57,59]

The epidemiology of rubella has been changing. Data from the CDC indicate that the incidence of rubella has been decreasing among children less than 15 years old but has been increasing among adults, primarily those born outside the United States.[60,61] In fact, 21 of 23 infants with congenital rubella syndrome reported to the CDC between 1997 and 1999 were born to foreign-born women, most of whom were Hispanic. Sheridan et al.,[62] in the United Kingdom, reported a case of healthcare-associated transmission of rubella from one neonate to another whose bed was nearby. The mother of the index patient was from Bangladesh and apparently had a mild influenza-like illness without a rash when she was 10 weeks pregnant. The infant was not recognized as having congenital rubella.

Rubella virus is spread in droplets that are shed from the respiratory secretions of infected persons. Persons with rubella are most contagious when the rash is erupting. In addition, persons with subclinical illness also may transmit the virus. To prevent healthcare-associated spread, patients with rubella should be placed under droplet precautions until day 7 after the onset of the rash. Infants with congenital rubella shed large quantities of virus for many months, despite having high titers of neutralizing antibody. Such patients should be placed under droplet precautions each time they are admitted during the first year of life, unless nasopharyngeal and urine cultures, after 3 months of age, are negative for rubella virus.

Nonimmune patients who have been exposed to rubella virus should be placed under droplet precautions from day 7 through day 21 after exposure.[15] Nonimmune family members and friends who have been exposed to a person with rubella should not come to the hospital during the incubation period.

Despite vaccination campaigns, 10%–20% of hospital personnel are susceptible to rubella.[16] Given the adverse effects of rubella virus on the fetus, many healthcare facilities require employees, especially those working in obstetrics, to be immune to rubella.[34,51] Table 23-1 provides information that infection prevention personnel need for evaluating exposures to persons with rubella.[5]

Mumps Virus

Mumps is characterized by fever and parotitis. In postpubertal men, mumps virus also can cause orchitis, which can be the primary manifestation of the infection. The mumps virus is transmitted through direct contact with contaminated respiratory secretions, through inhalation of droplet nuclei, or through contact with fomites contaminated by respiratory secretions. Transmission of mumps virus requires more intimate contact with the infected person than does transmission of either measles virus or varicella-zoster virus. To prevent exposures in healthcare facilities, persons with mumps should be placed under droplet precautions until day 9 after parotid (or other glandular) swelling began.[15]

The incidence of mumps decreased substantially in the United States after the vaccine was licensed in 1967.[5,34] Consequently, exposures to persons with mumps and healthcare-associated transmission of mumps are rare.[63,64] In 1996, Fischer et al.[64] published a report of an outbreak of mumps in which a 3-year-old patient, a nurse, and a physical therapist (who had been vaccinated) acquired mumps after a 12-year-old Mexican girl who was incubating mumps was admitted to the hospital. Neither the patient who acquired mumps nor the physical therapist had direct contact with the index patient.

In 2006, the United States experienced a large epidemic of mumps, which was centered in the Midwest.[65,66] This community-based outbreak caused numerous exposures in healthcare facilities. At the University of Iowa Hospitals and Clinics alone, more than 500 staff members were exposed to 28 patients or staff members who had either confirmed or probable mumps during April and May 2006. During this outbreak, the ACIP changed its requirements for evidence of mumps immunity among healthcare workers.[67] Before the outbreak, ACIP's criteria were as follows: (1) documentation of adequate vaccination, (2) serologic evidence of mumps immunity, (3) birth date before 1957, or (4) documentation of physician-diagnosed mumps. Subsequently, ACIP recommended that hospitals no long accept birth before 1957 and a doctor's diagnosis as evidence of immunity. Rather, the ACIP recommended that, during nonepidemic periods, healthcare workers without other evidence of immunity be given 2 doses of mumps vaccine if they were born during or after 1957 and 1 dose of mumps vaccine if they were born before 1957. In addition, the ACIP recommended that healthcare workers born before 1957 who do not have other evidence of immunity should receive 2 doses of live mumps virus vaccine during outbreaks.

Polgreen et al.[68] studied the duration of shedding of the mumps virus during the outbreak in Iowa and found that the probability of mumps virus shedding decreased rapidly after the onset of symptoms. However, they estimated that 8%–15% of patients are still shedding the virus 5 days after the onset of symptoms and, thus, may still be contagious during this period. They concluded that their statistical model and the absence of positive culture results more than 9 days after the onset of symptoms may support excluding healthcare workers from work for up to 9 days after the onset of symptoms.

Most adults are immune to mumps, and approximately 90% of adults who have no history of mumps have antibody to the virus.[5] Thus, only a small proportion of healthcare workers will be susceptible to mumps. For example, Nichol and Olson[69] found that 6.7% of the medical students they studied were nonimmune. Consequently, it might be cost beneficial to assess antibody titers of exposed persons who have no history of mumps and who have not received the mumps vaccine. Only those who are seronegative would be excluded from patient care during the incubation period (day 11 through day 26). Table 23-1 provides information necessary to evaluate an exposure to a person with mumps. Nonimmune patients who have been exposed to mumps should be placed under droplet precautions from day 11 through day 26 after exposure.[15] Nonimmune family members and friends who have been exposed to a person with mumps should not come to the hospital during the incubation period.

Parvovirus B19

Erythema infectiosum, or fifth disease, is a common manifestation of acute parvovirus B19 infection. Fifth disease acquired its name because common childhood exanthems were numbered, in the late 19th century. The first 3 illnesses were scarlet fever, rubeola, and rubella, and the fourth was a variation of scarlet fever known as Filatov-Dukes disease. Erythema infectiosum was the fifth disease, and roseola infantum was the sixth. Erythema infectiosum is characterized by mild systemic symptoms (fever in 15%–30% of patients), followed in 1–4 days by an erythematous rash on the cheeks—the "slapped cheek" appearance. Subsequently, an asymmetric macular or maculopapular lace-like erythematous rash can involve the trunk and extremities.

Parvovirus B19 has a predilection for infecting rapidly dividing cells, especially rapidly dividing red blood cells. Thus, persons with sickle cell disease, hereditary spherocytosis, pyruvate kinase deficiency, and other hemolytic anemias can develop transient hemolytic crises. Parvovirus B19 can cause severe chronic anemia associated with red cell aplasia in persons who are receiving maintenance chemotherapy for acute lymphocytic leukemia, who have congenital immunodeficiencies, or who have human immunodeficiency virus infection or acquired immunodeficiency syndrome. Parvovirus B19 also can cause hydrops fetalis. However, most parvovirus B19 infections during pregnancy do not affect the fetus adversely. Several studies indicate that the risk of fetal death is less than 10% in infected fetuses.[5]

Parvovirus B19 DNA has been found in the respiratory secretions of patients with viremia, but most persons are no longer viremic when the rash appears. In general, healthcare workers with parvovirus B19 infection do not need to be removed from patient care, because the infection is usually not diagnosed until after the rash appears. Some hospitals might choose to restrict healthcare workers from caring for patients at high risk of complications until the healthcare worker's symptoms have resolved. Persons with transient hemolytic crises and babies with hydrops fetalis can remain viremic for prolonged periods. These patients can be the source of infection for susceptible patients or healthcare workers and thus should be placed under droplet precautions while they are hospitalized, to prevent spread of parvovirus B19.[15] Lui et al.[70] documented healthcare-associated patient-to-patient transmission of this virus from a renal transplant patient, who apparently transmitted the virus many weeks after the onset of symptoms.

Transmission of parvovirus B19 is common in the community.[71] Outbreaks have occurred in day-care centers and in elementary and junior high schools. Secondary spread to susceptible household contacts also is frequent. Documented transmission within hospitals has been uncommon.[72-77] However, when transmission occurs, a high proportion (13%–50%) of susceptible persons may be infected.[73-75,78]

Transmission of parvovirus B19 usually requires prolonged, frequent, close contact. Adler et al.[79] investigated the rate of seroconversion to this virus among people employed either in schools or hospitals during an endemic period. These investigators found that the risk of seroconversion for persons who had daily contact with school-aged children at home (aged 5–11 years) or at work (aged 5–18 years) was 5 times higher than that for other study participants. The overall rate of seroconversion was 5.2% for primary-school employees, 2.4% for other school employees, and 0%–0.5% for hospital employees.

In general, routine infection prevention precautions should minimize nosocomial transmission of this virus.[80] A study by Cartter et al.[81] of risk factors for parvovirus B19 infection in pregnant women demonstrated that the rate of infection was highest among nurses who cared for patients before the patients were placed in isolation. These results suggest that isolation precautions can prevent healthcare-associated spread of this virus from infected patients. Ray et al.[74] obtained serologic test results for parvovirus B19 infection from 32 nonimmune healthcare workers who cared for 2 patients with transient aplastic crisis before they were put in isolation and from 37 nonimmune healthcare workers who were not exposed. Serologic evidence of recent parvovirus B19 infection was present for 3.1% of the exposed healthcare workers and 8.1% of the healthcare workers in the comparison group ($P = .06$). On the basis of their data, Ray et al.[74] concluded that the risk of healthcare-associated transmission was low even when isolation precautions are not implemented.

Table 23-1 provides information about how infection prevention personnel could evaluate an exposure to parvovirus B19.[5] In addition, the article by Crowcroft and colleagues[77] reviews relevant literature and provides recommendations for protecting "at-risk seronegative healthcare workers" and "at-risk patients."

Hepatitis A Virus

Hepatitis A is transmitted primarily by the fecal-oral route, but, in hospitals, hepatitis A virus also can be transmitted by blood transfusions. Infected persons excrete the highest concentration of virus in their stools during the 2 weeks before their symptoms begin. Most persons are no longer shedding the virus 1 week after they become jaundiced. However, infants can shed the virus in their stools for months.

Healthcare-associated transmission of hepatitis A is relatively uncommon. Most healthcare-associated outbreaks have occurred after an infant or a young child has received blood from a viremic but asymptomatic donor. The child often has an asymptomatic infection.[82-88] Occasionally, healthcare-associated outbreaks have occurred when healthcare workers cared for an older child or an adult who had vomiting, diarrhea, or fecal incontinence.[89-98]

Healthcare workers who are exposed to the stool of infected patients are at greatest risk for acquiring hepatitis A virus infection. Occasionally, patients, visitors, and healthcare workers could be at risk of acquiring hepatitis A if they eat uncooked food prepared by a food handler who is shedding the virus. Several food-related healthcare-associated epidemics have been reported.[99,100]

To prevent healthcare-associated transmission of hepatitis A virus, healthcare workers should follow standard precautions (ie, wear gowns and gloves whenever they might contaminate their hands or clothes with a patient's stool). Healthcare workers must perform hand hygiene after doing any patient-care activities and after removing their gloves. Adult patients with hepatitis A who are continent do not require private rooms, but diapered or incontinent persons should be placed in private rooms.[15] Healthcare workers who cared for patients with hepatitis A do not need to be restricted from working unless they develop hepatitis, because the risk of acquiring hepatitis from a patient is low, and the risk of transmission from infected healthcare workers to patients also is low. Healthcare workers with hepatitis A virus infection should not work during the first 7 days of their symptomatic illnesses. Table 23-1 provides information that infection prevention personnel need when evaluating an exposure to hepatitis A.[5]

The HICPAC guideline for infection control in healthcare personnel states that "Immune globulin given within 2 weeks after an HAV [hepatitis A virus] exposure is more than 85% effective in preventing HAV infection and may be advisable in some outbreak situations."[7(p427)] The usual dosage of immune globulin is 0.02 mg/kg administered intramuscularly, when given as postexposure prophylaxis. The hepatitis A vaccine has helped terminate outbreaks in the community, but its role in hospitals has not been determined.

Influenza Virus

Healthcare workers tend to think that transmission of influenza virus occurs primarily in the community, not in the hospital. However, Evans et al.[101] identified 17 reports of healthcare-associated influenza transmission that were published between 1959 and 1994. In 5 of these outbreaks, healthcare workers were implicated in transmitting the virus, and in 12 outbreaks, healthcare workers became infected with influenza virus. There have subsequently been additional reports of healthcare-associated influenza outbreaks, at least 4 of which clearly stated that healthcare workers were affected. The affected units included neonatal intensive care units (3 reports), a pediatric unit (1), a solid-organ transplant unit (1), an adult bone marrow transplant unit (1), a unit for cancer patients (1), and an adult pulmonary unit (1). Thus, patients, visitors, and healthcare workers can spread this virus in healthcare facilities.[101-117] In the outbreak described by Pachucki et al.,[103] 118 workers were affected, including 8% of the nurses and 3%–6% of the doctors. Everts[117] described 2 outbreaks of influenza A affecting wards that treated and rehabilitated elderly patients. The attack rate among patients was 48% on one ward and 58% on the other; 46% of the ill patients had lower respiratory tract involvement, and 7% died. The attack rate among staff was 69% on one ward and 36% on the other.

Healthcare-associated influenza probably goes unrecognized in many instances. Clinicians and infection prevention personnel should consider this diagnosis when staff or hospitalized patients develop symptoms of influenza during the appropriate season. Table 23-1 describes how infection prevention personnel could manage healthcare workers who were exposed to influenza.[5] The role of prophylaxis with amantidine, rimantidine, zanamivir, or oseltamivir has not been defined for healthcare workers who are exposed to influenza in acute care facilities.[118]

Influenza increases absenteeism among staff and increases the costs associated with sick leave. In addition, the ACIP has recommended that physicians, nurses, and other personnel in both inpatient-care and outpatient-care settings who have contact with persons at high risk receive the influenza vaccine.[119] Thus, many healthcare facilities offer their employees the influenza vaccine free of charge to protect the staff and to prevent spread of influenza within the healthcare facility. More recently, a live influenza virus vaccine (FluMist; MedImmune) has become available, but there have been concerns about its use in acute care facilities. The package insert states that vaccinated persons should not have close contact with immunocompromised patients for at least 21 days after receiving the vaccine. In addition, the vaccine should be administered before exposure to influenza, and the safety of the vaccine has not been demonstrated for the persons at highest risk of complications from influenza. Many institutions use this live influenza virus vaccine for their nonclinical staff, particularly in times of shortage of the injectable form of influenza vaccine, given the stated concerns.

Even though influenza vaccination is offered to healthcare workers at no cost, the acceptance rate remains low. There are many suggested ways to increase participation (eg, roving vaccination teams, vaccination "fairs," financial incentives, and pandemic preparedness drills[120]), but none has been consistently effective in improving the rate of acceptance of vaccination. Thus, there is much discussion now about requiring influenza vaccination for healthcare workers. Among the benefits of influenza vaccination are the reduction of influenza transmission in healthcare settings and decreases in rates of staff illness and absenteeism. However, these benefits need to be weighed against the ethical arguments against mandatory vaccination, which are focused on maintaining healthcare workers' rights and respecting their autonomy.[121]

The Creutzfeldt-Jakob Disease Prion

The agent that causes Creutzfeldt-Jakob disease, a prion, has been transmitted in the healthcare setting by brain-to-brain inoculation (eg, through contaminated instruments) and by contaminated tissues or tissue extracts. To date, there have been no documented instances of transmission to healthcare workers, and the incidence of

Creutzfeldt-Jakob disease is not higher in healthcare workers than it is in the general population.[122,123] Berger and Noble[124] reported that 24 healthcare workers have been identified as having Creutzfeldt-Jakob disease. These authors provided 5 case reports of healthcare workers (1 neurosurgeon, 1 pathologist, 1 internist who did autopsies for 1 year during his training, and 2 histopathology technicians) who developed Creutzfeldt-Jakob disease.[124] However, none of these persons had documented exposures to the agent.

Criteria for defining exposures to the Creutzfeldt-Jakob agent have not been developed. The WHO categorizes the tissues with high infectivity as brain, spinal cord, and eye tissues,[125,126] and categorizes the tissues or fluids with low infectivity as cerebrospinal fluid and kidney, liver, lung, lymph nodes, spleen, and placenta tissues.[125,126] The WHO categorizes the following tissues or fluids as having no detectable infectivity: adipose, adrenal gland, gingival, heart muscle, intestine, peripheral nerve, prostate, skeletal muscle, testis, thyroid gland tissues and tears, nasal mucosa, saliva, sweat, serous exudates, milk, semen, urine, and feces.[125,126] The WHO also classifies blood as having "no detectable infectivity," despite the fact that blood and its components have been found to have very low levels of infectivity in experimental models. The WHO classified blood in this way because the epidemiologic evidence indicates that blood has never transmitted the Creutzfeldt-Jakob disease prion to humans.[125,126]

The highest-risk injuries involve high-risk tissues and needlestick injuries with inoculation. Exposures via mucous membranes have the "theoretical risk" of transmitting the Creutzfeldt-Jakob disease prion. The WHO recommends the following procedures if an exposure occurs[125]:

- Wash exposed unbroken skin with detergent and abundant quantities of warm water (avoid scrubbing). Then rinse and dry the affected area. Brief (1 minute) exposure to 0.1 N NaOH or a 1:10 bleach solution can be used for maximum safety.
- After a needle stick or laceration, gently encourage bleeding, wash (avoid scrubbing) as described above, rinse, dry, and cover with a dressing.
- After splashing an eye or mouth, irrigate the affected area with saline (the eye) or water (the mouth).
- Report any exposures to the appropriate department.

If an exposure occurs, infection prevention personnel should create a list of all exposed staff, which should be saved indefinitely in case anyone develops the disease. Hospital epidemiology and occupational health service staff also should counsel healthcare workers. In addition, infection prevention staff should work with staff from the operating suite and central sterile supply to ensure that, if possible, the reusable surgical instruments used on the index patient are recalled and reprocessed properly and that all contaminated equipment in other departments (eg, the pathology department) is properly cleaned and disinfected.

The best exposure management for the Creutzfeldt-Jakob prion is to prevent exposures from occurring. Therefore, infection prevention staff would be wise to work with persons from the operating suite, the neurosurgery department, the ophthalmology department, the pathology department, the laboratory, central sterile supply, and the morgue to develop policies that prevent exposures. These precautions should be used for all persons who undergo invasive procedures or ophthalmologic examinations and who are known to have Creutzfeldt-Jakob disease or a progressive dementia or who have a family history of prion disease, Creutzfeldt-Jakob disease, fatal familial insomnia, or Gerstmann-Sträussler-Scheinker syndrome.[125-127] The precautions also should be used for patients who have received gonadotropin or human growth hormone extracted from cadaveric pituitary glands.[127]

Infection prevention personnel who are developing these policies should review the recommendations of Steelman[127,128] and HICPAC.[129] These documents recommend methods for protecting staff from exposure to potentially infectious tissues, limiting contamination of equipment and the environment, and effectively eradicating the Creutzfeldt-Jakob prion from surgical equipment. The guidelines on the care of surgical equipment are extremely important, because the Creutzfeldt-Jakob agent is not killed by routine chemical and physical means of sterilization (including routine steam sterilization, ethylene oxide sterilization, and dry heat sterilization; processes using peracetic acid, hydrogen peroxide, UV light, radiation, freezing, drying, or hot bead glass; and any level of cleaning and disinfection with glutaraldehyde, dry heat radiation, detergents, or formaldehyde). Of note, some of the recommendations differ between the documents developed by Steelman and those by HICPAC (effective sterilization methods are summarized in Chapter 7).

Variant Creutzfeldt-Jakob disease has become an important issue in the United Kingdom and Europe.[130-132] It is thought to be transmitted from beef infected with the prion that causes bovine spongiform encephalopathy. The United States has not identified bovine spongiform encephalopathy as a problem, and thus many people in this country are not concerned about variant Creutzfeldt-Jakob disease. However, given the ease with which people travel, the presence of chronic wasting disease (another spongiform encephalopathy) in cervids in the United States, and the lax regulation of the animal-products rendering industry, infection prevention personnel should not ignore variant Creutzfeldt-Jakob disease.

Unlike the prion that causes Creutzfeldt-Jakob disease, the prion that causes variant Creutzfeldt-Jakob disease infects the lymphoreticular tissues. A tonsilar biopsy is the preferred diagnostic test. Thus, a wider variety of tissues may be able to transmit this agent. The Department of Health in the United Kingdom has mandated that decontamination facilities be upgraded and requires that all adenotonsillectomy procedures be performed using disposable instruments.[132] In addition, decontamination and sterilization of equipment is different for variant Creutzfeldt-Jakob disease than for Creutzfeldt-Jakob disease.[133,134] Because lymphoreticular tissue is affected and infected persons may not show symptoms or signs of the disease for years, many hospitals in Europe have changed their general decontamination and sterilization procedures to ensure that the variant Creutzfeldt-Jakob prion will be inactivated.

Bacteria

Mycobacterium tuberculosis

M. tuberculosis is an acid-fast bacillus that is spread through the air. This organism causes a primary infection, which in normal hosts usually is not manifested as clinical disease, a recrudescent pulmonary disease, or disseminated disease. Persons who are infected with *M. tuberculosis* have positive tuberculin skin test results but are not contagious. Those who have active pulmonary disease are infectious and are the persons who cause most healthcare-associated exposures. On occasion, patients who have active infections at other sites also can cause exposures. For example, a patient with a large soft-tissue abscess underwent incision, drainage, and irrigation in an operating suite.[135] Because he continued to have copious drainage from the wound, it was cleaned with a pressurized irrigation system. Subsequently, 59 employees were identified who had tuberculin skin test conversion, and 9 persons acquired active tuberculosis (5 employees, 2 patients, and 2 family members of the index patient). Matlow et al.[136] reported that 111 healthcare workers were exposed to tuberculosis while caring for an infant with peritoneal tuberculosis; 2 (5%) of the primary-care nurses but no doctors or housekeepers had tuberculin skin test conversion.

Persons are considered to have been exposed to *M. tuberculosis* if they shared air space with a patient who had active pulmonary tuberculosis or who had an extrapulmonary site of infection from which *M. tuberculosis* was aerosolized and the healthcare worker was not wearing a respirator rated N-95 or higher. During outbreaks, a large proportion (3.6%–100%) of exposed persons may have their tuberculin skin test result convert to positive.[135,137-139] In general, approximately 30% of persons will become infected when they are exposed to a patient whose sputum contains acid-fast bacilli, whereas only 10% of persons will become infected when they are exposed to an infected patient whose sputum does not contain visible acid-fast bacilli.[140]

As with the other airborne infections—measles and chicken pox—it is best to prevent exposures by screening patients in clinics and on admission for symptoms and signs of tuberculosis. However, screening can be difficult, because some patients present with atypical signs or symptoms and others do not answer truthfully to screening questions designed to identify patients who might have tuberculosis so that they can be isolated before they expose persons in the healthcare setting. Moreover, some patients present with tuberculosis at unusual sites, and immunocompromised patients can have atypical signs and symptoms.

Healthcare workers continue to acquire *M. tuberculosis* through occupational exposures.[141-151] In countries where tuberculosis is common, healthcare workers may be at considerable risk of acquiring tuberculosis.[148-152] Persons who move from a country with a high incidence of tuberculosis to a country with a low incidence can cause substantial exposures in healthcare facilities.[153] In addition, studies done in Canada indicate that delays in diagnosis, inadequate ventilation (fewer than 2 air exchanges per hour) in general patient rooms, involvement in certain types of work (nursing, respiratory therapy, physical therapy, and housekeeping), and the duration of work all increase the risk of transmission.[142,154]

The goal of an epidemiological workup after an *M. tuberculosis* exposure is to identify all patients, visitors, and healthcare workers who were exposed, so that those who become infected can be treated with antimycobacterial drugs. This task can be very difficult if, before the diagnosis is made, the infectious person visited many clinics and diagnostic laboratories or was hospitalized in an open bay of an intensive care unit. All persons who meet the criteria for exposure should have a baseline tuberculin skin test performed, if they have not had one recently (within the previous 6 weeks in a high-prevalence area or within 1 year in a low-prevalence area). Healthcare workers should be evaluated by occupational health service staff. Patients and visitors should be notified about the exposure and told to contact their own physician or should be offered the opportunity to have a skin test done at the medical facility where the exposure occurred. In addition, the patients' primary physicians should receive letters informing them of the exposure. Twelve weeks after the exposure, exposed persons should have another skin test. If the result of that skin test is positive, they should be encouraged to take prophylaxis.[138]

N. meningitidis

N. meningitidis is a gram-negative diplococcus that causes meningitis and septicemia. Household contacts of persons with invasive meningococcal disease are at 500–800 times greater risk of acquiring meningococcal infection than are members of the general public.[155] Other semiclosed or closed populations, such as persons living in college dormitories, long-term care hospitals, or military barracks or attending nursery schools, also are at high risk of infection.[156] Despite caring for patients with meningococcal infection, healthcare workers are not at higher risk for acquiring this infection than are members of the general population.[16]

N. meningitidis is transmitted by respiratory droplets. Thus, patients with meningococcal infections should be placed under droplet precautions for the first 24 hours of treatment.[15] Nosocomial transmission of *N. meningitidis*, which has occurred rarely, may be more likely to occur from patients who have meningococcal pneumonia than from patients with meningitis or septicemia.[157,158] Persons are considered exposed to *N. meningitidis* if they did not wear a mask and either had prolonged close contact with a person who had meningococcal disease or had contact with the patient's respiratory secretions while not wearing appropriate personal protective equipment. Exposed persons should begin prophylactic treatment within 24 hours after their exposure.[5] Thus, immediately upon identifying a patient with meningococcal disease, infection prevention personnel must determine whether any healthcare workers meet the criteria for exposure. Staff who meet the criteria for exposure must be sent to the occupational health service at once to receive a prescription for an appropriate antimicrobial agent.

Despite the very low risk of transmission to healthcare workers, staff often are very anxious when they learn that a patient with meningococcal disease has been admitted to their unit. Staff who do not meet the criteria for exposure frequently demand prescriptions for antimicrobial prophylaxis. If infection prevention personnel refuse to oblige them, these staff members often have other physicians write prescriptions for them. However, prophylactic treatment is not without complications (eg, allergic reactions, side effects of the medications, and development of *Clostridium difficile* colitis), and its use should be discouraged when the criteria for exposure are not met.

Bordetella pertussis

The whole-cell pertussis vaccines dramatically altered the epidemiology of pertussis. Before the vaccines were introduced, most adults were immune to pertussis because they had the disease during childhood, and their immunity was probably boosted by frequent exposures to infected persons. However, most adults are now susceptible to pertussis, because vaccine-induced immunity disappears within 12 years after the last vaccination.[159] Consequently, the incidence of pertussis in adults is now increasing,[160-164] and adolescents and adults have become the primary source of infection for susceptible young children.[163] In the United States the proportion of persons with pertussis who are over 10 years of age increased from 7.2% during 1992–1994 to more than 50% during 1997–2000.[165]

Pertussis may be transmitted in the hospital by patients, visitors, and healthcare workers.[166-175] Outbreaks also have occurred in other healthcare institutions, including homes for handicapped persons[176-178] and a nursing home.[164] During the outbreak in the nursing home, 11 (10%) of 107 residents and 17 (14%) of 116 employees developed clinical or laboratory-confirmed pertussis infection.[164] The mean age of persons with clinical infection was 75 years for residents and 34 years for employees. Recently, Wright et al.[175] reported the results of a study in which they observed 106 resident physicians and 39 emergency department physicians over time to see if their levels of antibody to pertussis toxin and filamentous hemagglutinin increased by 50% (a finding diagnostic of pertussis infection) during a 1–3 year follow-up period. Two residents (1.3% [95% confidence interval, 0%–3.5%]) and 3 emergency physicians (3.6% [95% confidence interval, 0%–9.6%]) had serologic evidence of recent pertussis infection. Only 2 of these 5 physicians had symptomatic illnesses.

Most adults with pertussis have persistent and sometimes severe cough. These adults frequently receive a diagnosis of bronchitis. Thus, many exposures are not identified. Several studies indicate that erythromycin treatment early in the course of illness decreases the frequency of secondary spread.[177-179] However, physicians rarely see adult patients early in their illness.

Several communities have experienced outbreaks of pertussis in the past few years.[178,180] Patients involved in these outbreaks have caused exposures when they were evaluated in clinics or were admitted to a hospital. The ACIP recommended administration of the newer tetanus-diphtheria-acellular pertussis (Tdap) vaccine, which has been generally well tolerated, to boost healthcare workers' immunity.[181] Table 23-1 illustrates how infection prevention personnel could evaluate an exposure to a person with pertussis.[5,182]

Group A Streptococcus

Although group A *Streptococcus* infection is not typically considered a disease warranting postexposure intervention, there have been reports of outbreaks that affected healthcare workers.[183-185] These outbreaks demonstrate that group A streptococci can spread quickly to both patients and healthcare workers. To our knowledge, there are no guidelines for treating healthcare workers who have been exposed to patients with streptococcal infection. However, given that group A *Streptococcus* is on occasion transmitted to healthcare workers, prophylaxis may be appropriate under some circumstances.

Ectoparasites

Infection prevention personnel also must investigate exposures to ectoparasites, such as lice and scabies.[186] Healthcare workers often react more irrationally to these exposures than they do to exposures involving infectious agents and expect prophylactic treatment when it is not necessary.

Lice

Pediculus humanus capitis, *Pediculus humanus corporis*, and *Phthirus pubis* are found not infrequently on patients admitted to healthcare facilities. These ectoparasites are transmitted by direct contact with infested persons or their clothing. Persons infested with lice should be placed under contact precautions until they have been treated.[15,187] All clothing, bedding, hats, and other personal-care items should be washed in hot water and machine dried at a hot temperature setting; lice and their eggs cannot survive temperatures above 53.5°C.[11] Clothes that cannot be washed should be dry cleaned or placed in a sealed plastic bag for 2 weeks.[11] Brushes and combs should be soaked in a pediculicide shampoo.[5] Healthcare workers who have had direct contact with the patient's head (head lice) or clothes (body lice) should be evaluated by the occupational health service. Because the risk of acquiring lice in a healthcare facility is very low, only staff who become infested should be treated with a pediculicide.

Scabies

In contrast to lice, the mite *Sarcoptes scabiei* can be transmitted easily within healthcare facilities, especially if the index patient has crusted (Norwegian) scabies.[188-203] Of note, several outbreaks have occurred because patients with human immunodeficiency virus infection and unrecognized crusted or keratotic (Norwegian) scabies were admitted without the necessary precautions.[192,197,200,202]

Such exposures can be quite expensive. For example, an outbreak of scabies occurred in an extended-care unit that was attached to an acute care hospital. To terminate the outbreak, 78 residents and more than 100 staff and family members were treated, at a cost of more than

$20,000.[194] Scabies spread within the unit, in part because the protocol for control of this ectoparasite was inadequate. The policy was based on the assumption that staff had previous experience with scabies exposures and would know what to do. The outbreak described by Obasanjo et al.[198] was enormous (773 healthcare workers and 204 patients were exposed) and was not terminated until precautions beyond those recommended by the CDC[7] were implemented. This included the following precautions: (1) early identification of infested patients, (2) administration of prophylactic topical treatment to all exposed healthcare workers, (3) administration of 2 courses of treatment for patients with Norwegian scabies, (4) use of barrier isolation precautions until 24 hours after administration of the second course of treatment, (5) and administration of oral ivermectin treatment to patients for whom conventional therapy failed.

Van Vliet et al.[195] identified 6 reasons for spread of scabies in healthcare facilities: (1) many patients who have scabies are at risk of developing Norwegian scabies, (2) many people have contact with these patients, (3) diagnosis is often delayed, (4) the epidemiologic evaluation is often inadequate, (5) treatment failures occur, and (6) follow-up is often inadequate.

Persons with scabies should be placed under contact precautions until they are treated.[15] Personnel who have cared for patients with Norwegian scabies or cared for patients during outbreaks of scabies when transmission continues to occur should be evaluated by the occupational health service, and those who had contact with an infested patient's skin should be treated. In "routine" cases of scabies (ie, cases that are noncrusted scabies and that occur in a nonoutbreak situation), exposed healthcare workers should be treated only if they acquire scabies. If 2 or more persons who live or work in a long-term care facility acquire scabies, all residents and employees should be treated, to prevent further spread. Persons receiving effective therapy may have pruritus for up to 2 weeks after therapy ends. Thus, infection prevention personnel should not interpret pruritus occurring during this time period as treatment failure.

The index patient's bedding and clothes that contacted the index patient's skin should be washed in hot water and machine dried at a hot temperature setting.[5] Clothes that cannot be washed can be stored in a sealed plastic bag for several days to a week, because the *S. scabiei* mite cannot survive more than 3–4 days in the environment.[5]

Emerging Pathogens

Just when it seems that we in hospital epidemiology have survived one crisis and are ready to restore some normalcy, a new disease emerges and upsets our fragile equilibrium. We anticipate that more organisms of epidemiologic import within healthcare facilities will emerge as the global population continues to grow and as world travel remains rapid and common. We chose to discuss 3 viral diseases in this category that infection prevention staff in the United States have had to spend considerable time dealing with in the past several years.

Smallpox Virus

Smallpox (variola) is a serious, contagious, and sometimes fatal infectious disease. Although smallpox was declared globally eradicated in 1980, there is concern that smallpox virus may be used for bioterrorism (see Chapter 22 on biological disaster preparedness). There is no specific treatment for smallpox disease, and the only prevention is vaccination. The smallpox vaccine, which was routinely administered to Americans until 1972, is highly effective in protecting against the disease when given before or shortly after exposure to the virus. Though protection by the live vaccinia virus persists for a long time and may prevent death from illness in individuals who were vaccinated more than 2 decades ago, all children and most adults are now considered susceptible unless they were recently vaccinated.[204,205] Because of concerns that smallpox could be used as a bioweapon, a pre-event vaccination program was undertaken at many hospitals in early 2003.[206-209] This would allow for recently vaccinated personnel to care for patients with smallpox (or patients with suspected smallpox) and to vaccinate other healthcare workers.

If smallpox virus was released into the community, one would expect transmission to occur as an infected person's fever peaks and the skin rash starts. Persons with smallpox are occasionally contagious during the prodrome phase, but they become more contagious with the onset of the rash. Fever usually begins 10–14 days after the initial infection (range, 7–19 days), and the rash typically occurs approximately 2–4 days later.[204] Infectious particles are released during the sloughing of oropharyngeal lesions (duration, approximately 1 week). Transmission by way of contact with material from the smallpox pustules or crusted scabs can also occur; however, scabs are much less infectious than respiratory secretions are. Generally, direct and fairly prolonged face-to-face contact is required to spread smallpox virus from one person to another. Smallpox also can be spread through direct contact with infected bodily fluids or contaminated objects, such as bedding or clothing. Rarely, smallpox has been spread by the airborne route in enclosed settings, such as buildings, buses, and trains.

A healthcare worker would be considered exposed to smallpox if he or she had unprotected contact with an infected patient (ie, approach within 2.1 m [7 feet] without wearing an N-95 respirator and/or contact with lesions without use of gloves). Follow-up would include monitoring the healthcare worker's temperature twice daily for 17 days after the last exposure date (including vaccination days). It is likely that healthcare workers exposed to a person with smallpox would be quarantined and that infection prevention guidelines created by public health officials at the time will direct management of smallpox exposures. At present, infection prevention and public health personnel disagree about whether healthcare workers who have recently received the smallpox vaccine should be restricted from patient care. The official guidelines do not recommend special precautions,[204] but many infection prevention experts disagree.

SARS-Associated Coronavirus

Worldwide, numerous healthcare workers and patients acquired SARS in healthcare facilities in 2003.[210] In fact, transmission in healthcare facilities was amplified beyond that in the community. In general, transmission appears to have occurred if there was close contact with symptomatic individuals before infection prevention measures were implemented or if breaches in infection prevention practices occurred. Studies indicate that appropriate use of masks or respirators, gloves and gowns, and hand hygiene significantly decreased the risk of acquiring the SARS coronavirus while caring for patients with SARS[211]; however, some healthcare workers did acquire the SARS coronavirus despite wearing appropriate protective equipment (gown, mask, goggles or face shield, and gloves) while helping intubate patients with SARS.[212]

The incubation period for SARS ranges from 2 to 10 days, but most patients develop symptoms around day 4 or 5.[213,214] To manage SARS exposures, infection prevention staff need mechanisms for monitoring healthcare personnel for fever and respiratory symptoms and for managing asymptomatic exposed healthcare workers, symptomatic exposed healthcare workers, and symptomatic exposed visitors.[215] The definition of a SARS exposure will likely continue to change. Thus, infection prevention personnel should check the CDC Web site[9] for current definitions should SARS reemerge and should they think an exposure may have occurred.

Asymptomatic exposed persons. During the first SARS outbreak in 2003, the CDC did not recommend work restrictions for asymptomatic exposed persons unless they had unprotected high-risk exposures. The CDC did recommend that exposed healthcare workers be monitored for respiratory symptoms and fever (ie, check temperature twice daily) for 10 days after their last exposure. If fever or respiratory symptoms develop, the person should notify their healthcare provider, restrict their movements outside their home, and reassess the

situation in 72 hours. However, a number of hospitals took a more restrictive approach, such as placing healthcare workers who had unprotected exposure to patients with SARS on leave for 10 days from the last date of exposure. The CDC recommended that healthcare workers who have unprotected high-risk exposures should be excluded from duty for 10 days following the exposure. An unprotected high-risk exposure is defined as being present in the room when a probable or confirmed SARS patient underwent an aerosol-generating procedure without compliance with the recommended infection prevention precautions.

Symptomatic exposed healthcare workers. An exposed healthcare worker who develops either fever or respiratory symptoms within 10 days after exposure should be excluded from duty and should be evaluated in a manner that does not expose other persons to the SARS coronavirus. If symptoms improve or resolve in 72 hours after onset of symptoms, the person may be allowed to return to duty after consultation with infection prevention and local public health staff. For persons whose condition progresses to meet the case definition of SARS, infection prevention precautions should be continued until 10 days after fever and respiratory symptoms have resolved.

Symptomatic exposed visitors. To prevent exposures within healthcare facilities, infection prevention staff should consider designing a process for screening healthcare workers who traveled to areas where the SARS virus is being transmitted. In addition, symptomatic exposed visitors should not be allowed to visit their family member or friend but should be evaluated to determine whether they may have SARS. Thus, infection prevention personnel must design a way to identify visitors who might have been exposed and to screen them for symptoms and signs of SARS. To prevent transmission within exposed healthcare worker's homes, infection prevention personnel should counsel exposed persons to avoid contact with members of their household members (ie, avoid physical contact, stay in a separate part of the house, avoid eating together, and use separate bathrooms) or to find alternative living arrangements for household members during the 10 days following exposure.

Many of the lessons learned from the 2003 SARS experience, including the use of respiratory hygiene and cough etiquette, are being applied to planning for pandemic influenza and to managing the 2009 pandemic of influenza due to the novel influenza A (H1N1) virus ("swine flu").

Monkeypox Virus

Monkeypox is a rare viral disease that occurs mostly in central and western Africa. The monkeypox virus belongs to a group of viruses that includes the smallpox virus (variola), the virus used in the smallpox vaccine (vaccinia), and the cowpox virus. In early June 2003, monkeypox was identified in the United States among persons who had contact with ill pet prairie dogs.[216,217] This was the first time that there has been an outbreak of monkeypox in the United States. People can get monkeypox from an animal with monkeypox if they are bitten or if they touch the animal's blood, body fluids, or skin rash. Person-to-person transmission is believed to occur primarily through direct contact with lesions and also by way of respiratory droplets.

The current recommendations from the CDC do not restrict the activities of healthcare workers who have unprotected exposure to patients with monkeypox (ie, contact when they were not wearing personal protective equipment).[217] However, such healthcare workers should measure their temperature at least twice daily for 21 days following the exposure. Before reporting for duty each day, exposed healthcare workers should be interviewed to determine whether they have fever or rash. Healthcare workers who have cared for patients with monkeypox and who adhered to recommended infection prevention precautions do not need to be monitored. Healthcare workers who cared for a patient with monkeypox should monitor themselves for symptoms suggestive of monkeypox for 21 days after they last had contact with that patient. Healthcare workers who note symptoms of concern should notify infection prevention and/or occupational health staff and should be evaluated. Interested infection preventionists should consult the CDC Web site[9,217] for the most current recommendations.

Conclusion

Exposure workups are an important responsibility for infection prevention personnel. If they evaluate exposures promptly and effectively, infection prevention staff can prevent transmission of infectious agents or ectoparasites to numerous healthcare workers, patients, and visitors. Exposure workups consume resources, such as time and money, that could be used for other infection prevention activities. In addition, many exposures could be averted if healthcare workers were immune to vaccine-preventable infections, if staff used isolation precautions appropriately and consistently, and if healthcare workers did not come to work when they have communicable illnesses. Thus, wise infection preventionists learn from their own experience and develop policies and procedures to limit the number of exposures in their institutions.

References

1. Weltman AC, DiFerdinando GT, Washko R, Lipsky WM. A death associated with therapy for nosocomially acquired multidrug-resistant tuberculosis. *Chest* 1996;110:279–281.
2. Weber DJ, Rutala WA, Parham C. Impact and costs of varicella prevention in a university hospital. *Am J Public Health* 1988;78:19–23.
3. Faoagali JL, Darcy R. Chicken pox outbreak among the staff of a large, urban adult hospital: costs of monitoring and control. *Am J Infect Control* 1995;23:247–250.
4. Christie CDC, Glover AM, Willke MJ, Marx ML, Reising SF, Hutchinson NM. Containment of pertussis in the regional pediatric hospital during the greater Cincinnati epidemic of 1993. *Infect Control Hosp Epidemiol* 1995;16:556–563.
5. Committee on Infectious Diseases, American Academy of Pediatrics. *2009 Red Book: Report of the Committee on Infectious Diseases*. 28th ed. Elk Grove Village, IL: American Academy of Pediatrics; 2009.
6. Heymann DL. *Control of Communicable Diseases Manual.* 19th ed. Washington, DC: American Public Health Association; 2008
7. Bolyard EA, Tablan OC, Williams WW, Pearson ML, Shapiro CN, Deitchmann SD. Guideline for infection control in health care personnel, 1998. *Infect Control Hosp Epidemiol* 1998;19:407–463.
8. Centers for Disease Control and Prevention (CDC) Advisory Committee on Immunization Practices. Recommended adult immunization schedule: United States, 2009. *Ann Intern Med* 2009;150(1):40–44
9. Centers for Disease Control and Prevention Web site. http://www.cdc.gov. Accessed October 1, 2009.
10. World Health Organization Web site. http://www.who.int. Accessed October 1, 2009.
11. Mandell GL, Bennett JE, Dolin R, eds. *Principles and Practice of Infectious Diseases*. 6th ed. New York: Churchill Livingstone; 2005.
12. Emerging Infections Network Web site. http://ein.idsociety.org/. Accessed October 1, 2009.
13. Valenti WM. Employee work restrictions for infection control. *Infect Control* 1984;5:583–584.
14. Meyers MG, Rasley DA, Hierholzer WJ. Hospital infection control for varicella-zoster virus infection. *Pediatrics* 1982;70:199–202.
15. Siegel JD, Rhinehart E, Jackson M, Chiarello L, Healthcare Infection Control Practices Advisory Committee. 2007 Guideline for isolation precautions: preventing transmission of infectious agents in healthcare settings. http://www.guideline.gov/summary/summary.aspx?ss=15&doc_id=10984&nbr=5764. Accessed October 2, 2009.
16. Sepkowitz KA. Occupationally acquired infections in healthcare workers. Part I. *Ann Intern Med* 1996;125:826–834.
17. McKinney WP, Horowitz MM, Battiola RJ. Susceptibility of hospital-based health care personnel to varicella-zoster virus infections. *Am J Infect Control* 1989;17:26–30.
18. Hayden GF, Meyers JD, Dixon RE. Nosocomial varicella. Part II: suggested guidelines for management. *West J Med* 1979;130:300–303.
19. Varivax [package insert]. Whitehouse Station, NJ: Merck; 2005.

20. Galil K, Lee B, Strine T, et al. Outbreak of varicella at a day-care center despite vaccination. *N Engl J Med* 2002; 347:1909–1915.

21. Galil K, Fair E, Mountcastle N, Britz P, Seward J. Younger age at vaccination may increase risk of varicella vaccine failure. *J Infect Dis* 2002;186:102–105.

22. Saiman L, LaRussa P, Steinberg SP, et al. Persistence of immunity to varicella-zoster virus after vaccination of healthcare workers. *Infect Control Hosp Epidemiol* 2001;22: 279–283.

23. Wurtz R, Check IJ. Breakthrough varicella infection in a healthcare worker despite immunity after varicella vaccination. *Infect Control Hosp Epidemiol* 1999;20:561–562.

24. Richard VS, John TJ, Kenneth J, Ramaprabha P, Kuruvilla PJ, Chandy GM. Should health care workers in the tropics be immunized against varicella? *J Hosp Infect* 2001;47:243–245.

25. Weinstock DM, Rogers M, Lim S, Eagan J, Sepkowitz KA. Seroconversion rates in healthcare workers using a latex agglutination assay after varicella virus vaccination. *Infect Control Hosp Epidemiol* 1999;20:504–507.

26. Nettleman MD, Schmid M. Controlling varicella in the healthcare setting: the cost effectiveness of using varicella vaccine in healthcare workers. *Infect Control Hosp Epidemiol* 1997;18:504–508.

27. Tennenberg AM, Brassard JE, Van Lieu J, Drusin LM. Varicella vaccination for healthcare workers at a university hospital: an analysis of costs and benefits. *Infect Control Hosp Epidemiol* 1997;18:405–411.

28. Weitekamp MR, Schan P, Aber RC. An algorithm for the control of nosocomial varicella-zoster virus infection. *Am J Infect Control* 1985;13:193–198.

29. Bloch AB, Orenstein WA, Ewing WM, et al. Measles outbreak in a pediatric practice: airborne transmission in an office setting. *Pediatrics* 1985;75:676–683.

30. Remington PL, Hall WN, Davis IH, Herald A, Gunn RA. Airborne transmission of measles in a physician's office. *JAMA* 1985;253:1574–1577.

31. Seo SK, Malak SF, Lim S, Eagan J, Sepkowitz KA. Prevalence of measles antibody among young adult healthcare workers in a cancer hospital: 1980s versus 1998–1999. *Infect Control Hosp Epidemiol* 2002;23:276–278.

32. Davis RM, Orenstein WA, Frank JA, et al. Transmission of measles in medical settings 1980 through 1984. *JAMA* 1986;255:1295–1298.

33. Istre GR, McKee PA, West GR, et al. Measles spread in medical settings: an important focus of disease transmission? *Pediatrics* 1987;79:356–358.

34. Papania M, Reef S, Jumaan A, Lingappa JR, Williams WW. Nosocomial measles, mumps, rubella, and other viral infections. In: Mayhall CG, ed. *Hospital Epidemiology and Infection Control.* 3rd ed. Philadelphia, PA: Lippincott, Williams & Wilkins; 2004:829–849.

35. Edmonson MB, Addiss DG, McPherson Berg JL, Circo SR, Davis JP. Mild measles and secondary vaccine failure during a sustained outbreak in a highly vaccinated population. *JAMA* 1990;263:2467–2471.

36. Atkinson WL, Markowitz LE, Adams NC, Seastrom GR. Transmission of measles in medical settings—United States, 1985–1989. *Am J Med* 1991;91:320S–324S.

37. Raad II, Sherertz RJ, Rains CS, et al. The importance of nosocomial transmission of measles in the propagation of a community outbreak. *Infect Control Hosp Epidemiol* 1989;10:161–166.

38. Sienko DG, Friedman C, McGee, et al. A measles outbreak at university medical settings involving health care providers. *Am J Public Health* 1987;77:1222–1224.

39. Rivera ME, Mason WH, Ross LA, Wright HT. Nosocomial measles infection in a pediatric hospital during a community-wide epidemic. *J Pediatr* 1991;119:183–186.

40. Rank EL, Brettman L, Katz-Pollack H, DeHertogh D, Neville D. Chronology of a hospital-wide measles outbreak: lessons learned and shared from an extraordinary week in late March 1989. *Am J Infect Control* 1992;20:315–318.

41. Farizo KM, Stehr-Green PA, Simpson DM, Markowitz LE. Pediatric emergency room visits: a risk factor for acquiring measles. *Pediatrics* 1991;87:74–79.

42. Weber DJ, Rutala WA, Orenstein WA. Prevention of mumps, measles and rubella among hospital personnel. *J Pediatr* 1991;119:322–326.

43. Subbarao EK, Andrews-Mann L, Amin S, Greenberg J, Kumar ML. Postexposure prophylaxis for measles in a neonatal intensive care unit. *J Pediatr* 1990;117:782–785.

44. Ammari LK, Bell LM, Hodinka RL. Secondary measles vaccine failure in healthcare workers exposed to infected patients. *Infect Control Hosp Epidemiol* 1993;14:81–86.

45. Enguidanos R, Mascola L, Frederick P. A survey of hospital infection control policies and employee measles cases during Los Angeles County's measles epidemic, 1987 to 1989. *Am J Infect Control* 1992;20:301–304.

46. Wright LJ, Carlquist JF. Measles immunity in employees of a multihospital healthcare provider. *Infect Control Hosp Epidemiol* 1994;15:8–11.

47. de Swart RL, Wertheim-van Dillen PM, van Binnendijk RS, Muller CP, Frenkel J, Osterhaus AD. Measles in a Dutch hospital introduced by an immuno-compromised infant from Indonesia infected with a new virus genotype. *Lancet* 2000;355:201–202.

48. Steingart KR, Thomas AR, Dykewicz CA, Redd SC. Transmission of measles virus in healthcare settings during a community-wide outbreak. *Infect Control Hosp Epidemiol* 1999;20:115–119.

49. Marshall TM, Hlatswayo D, Schoub B. Nosocomial outbreaks—a potential threat to the elimination of measles? *J Infect Dis* 2003;187:S97–S101.

50. Kelly HA, Riddell MA, Andrews RM. Measles transmission in healthcare settings in Australia. *Med J Aust* 2002;176:50–51.

51. Greaves WL, Orenstein WA, Stetler HC, Preblud SR, Hinnman AR, Bart KJ. Prevention of rubella transmission in medical facilities. *JAMA* 1982;248:861–864.

52. Polk BF, White JA, DeGirolami PC, Modlin JF. An outbreak of rubella among hospital personnel. *N Engl J Med* 1980; 303:541–545.

53. Centers for Disease Control and Prevention (CDC). Rubella in hospitals—California. *MMWR Morb Mortal Wkly Rep* 1983;32:37–39.

54. Poland GA, Nichol KL. Medical students as sources of rubella and measles outbreaks. *Arch Intern Med* 1990; 150:44–46.

55. Storch GA, Gruber C, Benz B, Beaudoin J, Hayes J. A rubella outbreak among dental students: description of the outbreak and analysis of control measures. *Infect Control* 1985; 6:150–156.

56. Strassburg MA, Stephenson TG, Habel LA, Fannin SL. Rubella in hospital employees. *Infect Control* 1984;5: 123–126.

57. Fliegel PE, Weinstein WM. Rubella outbreak in a prenatal clinic: management and prevention. *Am J Infect Control* 1982;10:29–33.

58. Strassburg MA, Imagawa DT, Fannin SL, et al. Rubella outbreak among hospital employees. *Obstet Gynecol* 1981; 57:283–288.

59. Gladstone JL, Millian SJ. Rubella exposure in an obstetric clinic. *Obstet Gynecol* 1981;57:182–186.

60. Reef SE, Frey TK, Theall K, et al. The changing epidemiology of rubella in the 1990s: on the verge of elimination and new challenges for control and prevention. *JAMA* 2002;287: 464–472.

61. Centers for Disease Control and Prevention (CDC). Control and prevention of rubella: evaluation and management of suspected outbreaks, rubella in pregnant women, and surveillance for congenital rubella syndrome. *MMWR Morb Mortal Wkly Rep* 2001;50:1–23.

62. Sheridan E, Aitken C, Jeffries D, Hird M, Thayalasekaran P. Congenital rubella syndrome: a risk in immigrant populations. *Lancet* 2002;359:674–675.

63. Wharton M, Cochi SL, Hutcheson RH, Schaffner W. Mumps transmission in hospitals. *Arch Intern Med* 1990;150:47–49.

64. Fischer PR, Brunetti C, Welch V, Christenson JC. Nosocomial mumps: report of an outbreak and its control. *Am J Infect Control* 1996;24:13–18.

65. Centers for Disease Control and Prevention (CDC). Mumps epidemic—Iowa, 2006. *MMWR Morb Mortal Wkly Rep* 2006;55:366–368.

66. Dayan GH, Quinlisk MP, Parker AA, et al. Recent resurgence of mumps in the United States. *N Engl J Med* 2008; 358:1580–1589.

67. Centers for Disease Control and Prevention (CDC). Notice to readers: updated recommendations of the Advisory Committee on Immunization Practices (ACIP) for the control and elimination of mumps. *MMWR Morb Mortal Wkly Rep* 2009;55(22):629–630.

68. Polgreen PM, Bohnett LC, Cavanaugh JE, et al. The duration of mumps virus shedding after the onset of symptoms. *Clin Infect Dis* 2008;46:1447–1449.

69. Nichol KL, Olson R. Medical students' exposure and immunity to vaccine-preventable diseases. *Arch Intern Med* 1993;153:1913–1916.

70. Lui SL, Luk WK, Cheung CY, Chan TM, Lai KN, Peiris JS. Nosomial outbreak of parvovirus B19 infection in a renal transplant unit. *Transplantation* 2001;71:59–64.

71. Dowell SF, Torok TJ, Thorp JA, et al. Parvovirus B19 infection in hospital workers: community or hospital acquisition? *J Infect Dis* 1995;172:1076–1079.

72. Bell LM, Naides SJ, Stoffman P, Hodinka RL, Plotkin SA. Human parvovirus B19 infection among hospital staff members after contact with infected patients. *N Engl J Med* 1989;321:485–491.

73. Seng C, Watkins P, Morse D, et al. Parvovirus B19 outbreak on an adult ward. *Epidemiol Infect* 1994;113:345–353.

74. Ray SM, Erdman DD, Berschling JD, Cooper JE, Torok TJ, Blumberg HM. Nosocomial exposure to parvovirus B19: low risk of transmission to healthcare workers. *Infect Control Hosp Epidemiol* 1997;18:109–114.

75. Shishiba T, Matsunaga Y. An outbreak of erythema infectiosum among hospital staff members including a patient with pleural fluid and pericardial effusion. *J Am Acad Dermatol* 1993;29:265–267.

76. Miyamoto K, Ogami M, Takahashi Y, et al. Outbreak of human parvovirus B19 in hospital workers. *J Hosp Infect* 2000;45:238–241.

77. Crowcroft NS, Roth CE, Cohen BJ, Miller E. Guidance for control of parvovirus B19 infection in healthcare settings and the community. *J Public Health Med* 1999;21: 439–446.

78. Lohiya GS, Stewart K, Perot K, Widman R. Parvovirus B19 outbreak in a developmental center. *Am J Infect Control* 1995;23:373–376.

79. Adler SP, Manganello AM, Koch WC, Hempfling SH, Best AM. Risk of human parvovirus B19 infections among school and hospital employees during endemic periods. *J Infect Dis* 1993;168:361–368.

80. Centers for Disease Control and Prevention (CDC). Risks associated with human parvovirus B19 infection. *MMWR Morb Mortal Wkly Rep* 1989;38:81–88, 93–97.

81. Cartter ML, Farley TA, Rosengren S, et al. Occupational risk factors for infection with parvovirus B19 among pregnant women. *J Infect Dis* 1991;163:282–285.

82. Noble RC, Kane MA, Reeves SA, Roeckel I. Posttransfusion hepatitis A in a neonatal intensive care unit. *JAMA* 1984; 252:2711–2715.

83. Azimi PH, Roberto RR, Guralnik J, et al. Transfusion-acquired hepatitis A in a premature infant with secondary nosocomial spread in an intensive care nursery. *Am J Dis Child* 1986;140:23–27.

84. Giacoia GP, Kasprisin DO. Transfusion-acquired hepatitis A. *South Med J* 1989;82:1357–1360.

85. Seeberg S, Brandberg A, Hermodsson S, Larsson P, Lundgren S. Hospital outbreak of hepatitis A secondary to blood exchange in a baby. *Lancet* 1981;1:1155–1156.

86. Klein BS, Michaels JA, Rytel MW, Berg KG, Davis JP. Nosocomial hepatitis A. A multinursery outbreak in Wisconsin. *JAMA* 1984;252:2716–2721.

87. Rosenblum LS, Villarino ME, Nainan OV, et al. Hepatitis A outbreak in a neonatal intensive care unit: risk factors for transmission and evidence of prolonged viral excretion among preterm infants. *J Infect Dis* 1991;164:476–482.

88. Burkholder BT, Coronado VG, Brown J, et al. Nosocomial transmission of hepatitis A in a pediatric hospital traced to an anti-hepatitis A virus-negative patient with immunodeficiency. *Pediatr Infect Dis J* 1995;14:261–266.

89. Drusin LM, Sohmer M, Groshen SL, Spiritos MD, Senterfit LB, Christenson WN. Nosocomial hepatitis A infection in a pediatric intensive care unit. *Arch Dis Child* 1987;62: 690–695.

90. Reed CM, Gustafson TL, Siegel J, Duer P. Nosocomial transmission of hepatitis A from a hospital-acquired case. *Pediatr Infect Dis J* 1984;3:300–303.

91. Krober MS, Bass JW, Brown JD, Lemon SM, Rupert KJ. Hospital outbreak of hepatitis A: risk factors for spread. *Pediatr Infect Dis J* 1984;3:296–299.

92. Orenstein WA, Wu E, Wilkins J, et al. Hospital-acquired hepatitis A: report of an outbreak. *Pediatrics* 1981;67:494–497.

93. Edgar WM, Campbell AD. Nosocomial infection with hepatitis A. *J Infect* 1985;10:43–47.

94. Goodman RA, Carder CC, Allen JR, Orenstein WA, Finton RJ. Nosocomial hepatitis A transmission by an adult patient with diarrhea. *Am J Med* 1982;73:220–226.

95. Petrosillo N, Raffaele B, Martini L, et al. A nosocomial and occupational cluster of hepatitis A virus infection in a pediatric ward. *Infect Control Hosp Epidemiol* 2002;23:343–345.

96. Jensenius M, Ringertz SH, Berild D, Bell H, Espinoza R, Grinde B. Prolonged nosomial outbreak of hepatitis A arising from an alcoholic with pneumonia. *Scand J Infect Dis* 1998; 30:119–123.

97. Hanna JN, Loewenthal MR, Negel P, Wenck DJ. An outbreak of hepatitis A in an intensive care unit. *Anaesth Intensive Care* 1996;24:440–444.

98. Doebbeling BN, Li N, Wenzel RP. An outbreak of hepatitis A among health care workers: risk factors for transmission. *Am J Public Health* 1993;83:1679–1684.

99. Eisenstein AB, Aach RD, Jacobsohn W, Goldman A. An epidemic of infectious hepatitis in a general hospital: probable transmission by contaminated orange juice. *JAMA* 1963; 185:171–174.

100. Meyers JD, Romm FJ, Tihen WS, Bryan JA. Food-borne hepatitis A in a general hospital: epidemiologic study of an outbreak attributed to sandwiches. *JAMA* 1975;231: 1049–1053.

101. Evans ME, Hall KL, Berry SE. Influenza control in acute care hospitals. *Am J Infect Control* 1997;25:357–362.

102. Weingarten S, Friedlander M, Rascon D, Ault M, Morgan M, Meyer RD. Influenza surveillance in an acute-care hospital. *Arch Intern Med* 1988;148:113–116.

103. Pachucki CT, Pappas SA, Fuller GF, Krause SL, Lentino JR, Schaaff DM. Influenza A among hospital personnel and patients. Implications for recognition, prevention, and control. *Arch Intern Med* 1989;149:77–80.

104. Berlinberg CD, Weingarten SR, Bolton LB, Waterman SH. Occupational exposure to influenza—introduction of an index case to a hospital. *Infect Control Hosp Epidemiol* 1989;10:70–73.

105. Centers for Disease Control and Prevention (CDC). Suspected nosocomial influenza cases in an intensive care unit. *MMWR Morb Mortal Wkly Rep* 1988;37:3–4, 9.

106. Adal KA, Flowers RH, Anglim AM. Prevention of nosocomial influenza. *Infect Control Hosp Epidemiol* 1996; 17:641–648.

107. Horcajada JP, Pumarola T, Martinez JA, et al. A nosocomial outbreak of influenza during a period without influenza epidemic activity. *Eur Respir J* 2003;21:303–307.

108. Berg HF, Van Gendt J, Rimmelzwaan GF, Peeters MF, Van Keulen P. Nosocomial influenza infection among post-influenza-vaccinated patients with severe pulmonary diseases. *J Infect* 2003;46:129–132.

109. Slinger R, Dennis P. Nosocomial influenza at a Canadian pediatric hospital from 1995 to 1999: opportunities for prevention. *Infect Control Hosp Epidemiol* 2002;23: 627–629.

110. Hirji Z, O'Grady S, Bonham J, et al. Utility of zanamivir for chemoprophylaxis of concomitant influenza A and B in a complex continuing care population. *Infect Control Hosp Epidemiol* 2002;23:604–608.

111. Sagrera X, Ginovart G, Raspall F, et al. Outbreaks of influenza A virus infection in neonatal intensive care units. *Pediatr Infect Dis J* 2002;21:196–200.

112. Malavaud S, Malavaud B, Sandres K, et al. Nosocomial outbreak of influenza virus A (H3N2) infection in a solid organ transplant department. *Transplantation* 2001;72:535–537.

113. Weinstock DM, Eagan J, Malak SA, et al. Control of influenza A on a bone marrow transplant unit. *Infect Control Hosp Epidemiol* 2000;21:730–732.

114. Cunney RJ, Bialachowski A, Thornley D, Smaill FM, Pennie RA. An outbreak of influenza A in a neonatal intensive care unit. *Infect Control Hosp Epidemiol* 2000;21:449–454.

115. Munoz FM, Campbell JR, Atmar RL, et al. Influenza A virus outbreak in a neonatal intensive care unit. *Pediatr Infect Dis J* 1999;18:811–815.

116. Schepetiuk S, Papanaoum K, Qiao M. Spread of influenza A virus infection in hospitalised patients with cancer. *Aust N Z J Med* 1998;28:475–476.

117. Everts RJ, Hanger HC, Jennings LC, Hawkins A, Sainsbury R. Outbreaks of influenza A among elderly hospital inpatients. *N Z Med J* 1996;109:272–274.

118. Salgado CD, Farr BM, Hall KK, Hayden FG. Influenza in the acute hospital setting. *Lancet Infect Dis* 2002;2:145–155.

119. Fiore AE, Shay DK, Broder K, et al.; Centers for Disease Control and Prevention (CDC); Advisory Committee on Immunization Practices (ACIP). Prevention and control of influenza: recommendations of the Advisory Committee on Immunization Practices (ACIP), 2008. *MMWR Recomm Rep* 2008;57(RR-7):1–60.

120. Kuntz JL, Holley S, Helms CM, et al. Use of a pandemic preparedness drill to increase influenza vaccination among healthcare workers. *Infect Control Hosp Epidemiol* 2008;29:111–115.

121. Anikeeva O, Braunack-Mayer A, Rogers W. Requiring influenza vaccination for health care workers. *Am J Public Health* 2009;99:24–29.

122. Will RG. Epidemiology of Creutzfeldt-Jakob disease. *Br Med Bull* 1993;49:960–970.

123. Harries-Jones R, Knight R, Will RG, Cousens S, Smith PG, Matthews WB. Creutzfeldt-Jakob disease in England and Wales, 1980–1984: a case-control study of potential risk factors. *J Neurol Neurosurg Psychiatry* 1988;51:1113–1119.

124. Berger JR, Noble JD. Creutzfeldt-Jakob disease in a physician: a review of the disorder in health care workers. *Neurology* 1993;43:205–206.

125. World Health Organization (WHO). WHO infection control guidelines for transmissible spongiform encephalopathies: report of a WHO consultation, Geneva, Switzerland, 23–26 March 1999. http://www.who.int/csr/resources/publications/bse/WHO_CDS_CSR_APH_2000_3/en/. Accessed October 2, 2009.

126. World Health Organization (WHO). WHO guidelines on transmissible spongiform encephalopathies in relation to biological and pharmaceutical products. Geneva, Switzerland: WHO; 2003. http://www.who.int/biologicals/publications/trs/areas/vaccines/tses/en/index.html. Accessed October 2, 2009.

127. Steelman VM. Creutzfeldt-Jakob disease: recommendations for infection control. *Am J Infect Control* 1994;22:312–318.

128. Steelman VM. Creutzfeldt-Jakob disease: decontamination issues. *Infect Control Sterilization Technology* 1996; 2:32–39.

129. Centers for Disease Control and Prevention, Healthcare Infection Control Practices Advisory Committee (HICPAC).

Guideline for disinfection and sterilization in healthcare facilities. Atlanta, GA: CDC; 2008. http://www.cdc.gov/ncidod/dhqp/sterile.html. Accessed October 2, 2009.

130. Irani DN, Johnson RT. Diagnosis and prevention of bovine spongiform encephalopathy and variant Creutzfeldt-Jakab disease. *Annu Rev Med* 2003;54:305–319.

131. Centers for Disease Control and Prevention (CDC). Probable variant Creutzfeldt-Jakob disease in a U.S. resident—Florida, 2002. *MMWR Morb Mortal Wkly Rep* 2002;51:927–929.

132. Frosh A, Joyce R, Johnson A. Iatrogenic vCJD from surgical instruments: the risk is unknown, but improved decontamination will help reduce the risk. *BMJ* 2001;322:1558–1559.

133. Spencer RC, Ridgway GL, vCJD Consensus Group. Sterilization issues: vCJD—towards a consensus: meeting between the Central Sterilizing Club and Hospital Infection Society, September 12, 2000. *J Hosp Infect* 2002;51:168–174.

134. Axon AT, Beilenhoff U, Bramble MG, et al. Variant Creutzfeldt-Jakob disease (vCJD) and gastrointestinal endoscopy. *Endoscopy* 2001;33:1070–1080.

135. Hutton MD, Stead WW, Cauthen GM, Block AB, Ewig WM. Nosocomial transmission of tuberculosis associated with a draining abscess. *J Infect Dis* 1990;161:286–295.

136. Matlow AG, Harrison A, Monteath A, Roach P, Balfe JW. Nosocomial transmission of tuberculosis (TB) associated with care of an infant with peritoneal TB. *Infect Control Hosp Epidemiol* 2000;21:222–223.

137. Bowden KM, McDiarmid MA. Occupationally acquired tuberculosis: what's known. *J Occup Med* 1994;36:320–325.

138. Stead WW. Management of health care workers after inadvertent exposure to tuberculosis: a guide for the use of preventive therapy. *Ann Intern Med* 1995;122:906–912.

139. Templeton GL, Illing LA, Young L, Cave D, Stead WW, Bates JH. The risk for transmission of *Mycobacterium tuberculosis* at the bedside and during autopsy. *Ann Intern Med* 1995;122:922–925.

140. Sepkowitz KA, Raffalli J, Riley L, Kiehn TE, Armstrong D. Tuberculosis in the AIDS era. *Clin Microbiol Rev* 1995;8:180–199.

141. Conover C, Ridzon R, Valway S, et al. Outbreak of multidrug-resistant tuberculosis at a methadone treatment program. *Int J Tuberc Lung Dis* 2001;5:59–64.

142. Menzies D, Fanning A, Yuan L, FitzGerald JM. Hospital ventilation and risk for tuberculous infection in Canadian health care workers. Canadian Collaborative Group in Nosocomial Transmission of TB. *Ann Intern Med* 2000;133:779–789.

143. Suzuki K, Onozaki I, Shimura A. Tuberculosis infection control practice in hospitals from the viewpoint of occupational health. *Kekkaku* 1999;74:413–420.

144. Nakasone T. Tuberculosis among health care workers in Okinawa Prefecture. *Kekkaku* 1999;74:389–395.

145. Schoch OD, Graf-Deuel E, Knoblauch A. Tuberculin testing of hospital personnel: large investment with little impact. *Schweiz Med Wochenschr* 1999;129:217–224.

146. Ridzon R, Kenyon T, Luskin-Hawk R, Schultz C, Valway S, Onorato IM. Nosocomial transmission of human immunodeficiency virus and subsequent transmission of multidrug-resistant tuberculosis in a healthcare worker. *Infect Control Hosp Epidemiol* 1997;18:422–423.

147. Centers for Disease Control and Prevention (CDC). Multidrug-resistant tuberculosis outbreak on an HIV ward—Madrid, Spain, 1991–1995. *MMWR Morb Mortal Wkly Rep* 1996;45:330–333.

148. Yanai H, Limpakarnjanarat K, Uthaivoravit W, Mastro TD, Mori T, Tappero JW. Risk of *Mycobacterium tuberculosis* infection and disease among health care workers, Chiang Rai, Thailand. *Int J Tuberc Lung Dis* 2003;7:36–45.

149. Uyamadu N, Ahkee S, Carrico R, Tolentino A, Wojda B, Ramirez J. Reduction in tuberculin skin-test conversion rate after improved adherence to tuberculosis isolation. *Infect Control Hosp Epidemiol* 1997;18:575–579.

150. Silva VM, Cunha AJ, Oliveira JR, et al. Medical students at risk of nosocomial transmission of *Mycobacterium tuberculosis*. *Int J Tuberc Lung Dis* 2000;4:420–426.

151. Kilinc O, Ucan ES, Cakan MD, et al. Risk of tuberculosis among healthcare workers: can tuberculosis be considered as an occupational disease? *Respir Med* 2002;96:506–510.

152. Eyob G, Gebeyhu M, Goshu S, Girma M, Lemma E, Fontanet A. Increase in tuberculosis incidence among the staff working at the Tuberculosis Demonstration and Training Centre in Addis Ababa, Ethiopia: a retrospective cohort study (1989–1998). *Int J Tuberc Lung Dis* 2002;6:85–88.

153. Moore DA, Lightstone L, Javid B, Friedland JS. High rates of tuberculosis in end-stage renal failure: the impact of international migration. *Emerg Infect Dis* 2002;8:77–78.

154. Greenaway C, Menzies D, Fanning A, et al. Delay in diagnosis among hospitalized patients with active tuberculosis—predictors and outcomes. *Am J Respir Crit Care Med* 2002;165:927–933.

155. The Meningococcal Disease Surveillance Group. Analysis of endemic meningococcal disease by serogroup and evaluation of chemoprophylaxis. *J Infect Dis* 1976;134:201–204.

156. Apicella MA. *Neisseria meningitidis*. In: Mandell GL, Bennett JE, Dolin R, eds. *Principles and Practice of Infectious Diseases*. 6th ed. New York: Churchill Livingstone; 2005:2498–2513.

157. Rose HD, Lenz IE, Sheth NK. Meningococcal pneumonia. A source of nosocomial infection. *Arch Intern Med* 1981;141:575–577.

158. Cohen MS, Steere AC, Baltimore R, et al. Possible nosocomial transmission of group Y *Neisseria meningitidis* among oncology patients. *Ann Intern Med* 1979;91:7–12.

159. Lambert HJ. Epidemiology of a small pertussis outbreak in Kent County, Michigan. *Public Health Rep* 1965;80:365–369.

160. Bass JW, Stephenson SR. The return of pertussis. *Pediatr Infect Dis J* 1987;6:141–144.

161. Mortimer EA Jr. Pertussis and its prevention: a family affair. *J Infect Dis* 1990;161:473–479.

162. Mink CM, Cherry J, Christenson P. A search for *Bordetella pertussis* infection in college students. *Clin Infect Dis* 1992;14:464–471.

163. Nelson JD. The changing epidemiology of pertussis in young infants: the role of adults as reservoirs of infection. *Am J Dis Child* 1978;132:371–373.

164. Addiss DG, Davis JP, Meade BD, et al. A pertussis outbreak in a nursing home. *J Infect Dis* 1991;164:704–710.

165. Robbins JB. Pertussis in adults: introduction. *Clin Infect Dis* 1999;28(Suppl 2):S91–S93.

166. Linnemann CC Jr, Nasenbeny J. Pertussis in the adult. *Annu Rev Med* 1977;28:179–185.

167. Kurt TL, Yeager AS, Guenette S, Dunlop S. Spread of pertussis by hospital staff. *JAMA* 1972;221:264–267.

168. Linnemann CC Jr, Ramundo N, Perlstein PH, et al. Use of pertussis vaccine in an epidemic involving hospital staff. *Lancet* 1975;2:540–543.

169. Valenti WM, Pincus PH, Messner MK. Nosocomial pertussis: possible spread by a hospital visitor. *Am J Dis Child* 1980;134:520–521.

170. Martinez SM, Kemper CA, Haiduven D, Cody SH, Deresinski SC. Azithromycin prophylaxis during a hospitalwide outbreak of a pertussis-like illness. *Infect Control Hosp Epidemiol* 2001;22:781–783.

171. Spearing NM, Horvath RL, McCormack JG. Pertussis: adults as a source in healthcare settings. *Med J Aust* 2002;177:568–569.

172. Karino T, Osaki K, Nakano E, Okimoto N. A pertussis outbreak in a ward for severely retarded. *Kansenshogaku Zasshi* 2001;75:916–922.

173. Gehanno JF, Pestel-Caron M, Nouvellon M, Caillard JF. Nosocomial pertussis in healthcare workers from a pediatric emergency unit in France. *Infect Control Hosp Epidemiol* 1999;20:549–552.

174. Matlow AG, Nelson S, Wray R, Cox P. Nosocomial acquisition of pertussis diagnosed by polymerase chain reaction. *Infect Control Hosp Epidemiol* 1997;18:715–716.

175. Wright SW, Decker MD, Edwards KM. Incidence of pertussis infection in healthcare workers. *Infect Control Hosp Epidemiol* 1999;20:120–123.

176. Steketee RW, Burstyn DG, Wassilak SGF, et al. A comparison of laboratory and clinical methods for diagnosing pertussis in an outbreak in a facility for the developmentally disabled. *J Infect Dis* 1988;157:441–449.

177. Steketee RW, Wassilak SGF, Adkins WN, et al. Evidence for a high attack rate and efficacy of erythromycin prophylaxis in a pertussis outbreak in a facility for the developmentally disabled. *J Infect Dis* 1988;157:434–440.

178. Fisher MC, Long SS, McGowan KL, Kaselis E, Smith DG. Outbreak of pertussis in a residential facility for handicapped people. *J Pediatr* 1989;114:934–939.

179. Biellik RJ, Patriarca PA, Mullen JR, et al. Risk factors for community- and household-acquired pertussis during a large-scale outbreak in central Wisconsin. *J Infect Dis* 1988;157:1134–1141.

180. Christie CD, Marx ML, Marchant CD, Reising SF. The 1993 epidemic of pertussis in Cincinnati. Resurgence of disease in a highly immunized population of children. *N Engl J Med* 1994;331:16–21.

181. Sandora TJ, Pfoh E, Lee GM. Adverse events after administration of tetanus-diphtheria-acellular pertussis vaccine to healthcare workers. *Infect Control Hosp Epidemiol* 2009;30:389–391.

182. Weber DJ, Rutala WA. Management of healthcare workers exposed to pertussis. *Infect Control Hosp Epidemiol* 1994;15:411–415.

183. Kakis A, Gibbs L, Eguia J, et al. An outbreak of group A streptococcal infection among health care workers. *Clin Infect Dis* 2002;35:1353–1359.

184. Ramage L, Green K, Pyskir D, Simor AE. An outbreak of fatal nosocomial infections due to group A streptococcus on a medical ward. *Infect Control Hosp Epidemiol* 1996; 17:429–431.

185. Hagberg C, Radulescu A, Rex JH. Necrotizing fasciitis due to group A streptococcus after an accidental needle-stick injury. *N Engl J Med* 1997;337:1699.

186. Lettau LA. Nosocomial transmission and infection control aspects of parasitic and ectoparasitic diseases, part III: ectoparasites/summary and conclusions. *Infect Control Hosp Epidemiol* 1991;12:179–185.

187. Meinking TL, Taplin D, Kalter DC, Eberle MW. Comparative efficacy of treatments for *Pediculosis capitis* infestations. *Arch Dermatol* 1986;122:267–271.

188. Degelau J. Scabies in long-term care facilities. *Infect Control Hosp Epidemiol* 1992;13:421–425.

189. Pasternak J, Richtmann R, Ganme APP, et al. Scabies epidemic: price and prejudice. *Infect Control Hosp Epidemiol* 1994;15:540–542.

190. Yonkosky D, Ladia L, Gackenheimer L, Schultz MW. Scabies in nursing homes: an eradication program with permethrin 5% cream. *J Am Acad Dermatol* 1990;23:1133–1136.

191. Clark J, Friesen DL, Williams WA. Management of an outbreak of Norwegian scabies. *Am J Infect Control* 1992;20:217–220.

192. Corbett EL, Crossley I, Holton J, Levell N, Miller R, De Cock KM. Crusted ("Norwegian") scabies in a specialist HIV unit: successful use of ivermectin and failure to prevent nosocomial transmission. *Genitourin Med* 1996;72:115–117.

193. Sirera G, Rius F, Romeu J, et al. Hospital outbreak of scabies stemming from two AIDS patients with Norwegian scabies. *Lancet* 1990;335:1227.

194. Jack M. Scabies outbreak in an extended care unit—a positive outcome. *Can J Infect Control* 1993;8:11–13.

195. van Vliet JA, Samsom M, van Steenbergen JE. Causes of spread and return of scabies in health care institutions: literature analysis of 44 epidemics. *Ned Tijdschr Geneeskd* 1998;142:354–357.

196. Meltzer E. Ivermectin and the treatment of outbreaks of scabies in medical facilities. *Harefuah* 2002;141:948–952, 1011.

197. Portu JJ, Santamaria JM, Zubero Z, Almeida-Llamas MV, Aldamiz-Etxebarria San Sebastian M, Gutierrez AR. Atypical scabies in HIV-positive patients. *J Am Acad Dermatol* 1996;34:915–917.

198. Obasanjo OO, Wu P, Conlon M, et al. An outbreak of scabies in a teaching hospital: lessons learned. *Infect Control Hosp Epidemiol* 2001;22:13–18.

199. Deabate MC, Calitri V, Licata C, et al. Scabies in a dialysis unit: mystery and prejudice. *Minerva Urol Nefrol* 2001; 53:69–73.

200. Zafar AB, Beidas SO, Sylvester LK. Control of transmission of Norwegian scabies. *Infect Control Hosp Epidemiol* 2002;23:278–279.

201. Robles García M, de la Lama López-Areal J, Avellaneda Martínez C, Gimenez García R, Cortejoso Gonzalo B, Vaquero Puerta JL. Nosocomial scabies outbreak [in Spanish]. *Revista Clinica Espanola* 2000;200:538–542.

202. Boix V, Sanchez-Paya J, Portilla J, Merino E. Nosocomial outbreak of scabies clinically resistant to lindane. *Infect Control Hosp Epidemiol* 1997;18:677.

203. Nandwani R, Pozniak AL, Fuller LC, Wade J. Crusted ("Norwegian") scabies in a specialist HIV unit. *Genitourin Med* 1996;72:453.

204. Centers for Disease Control and Prevention (CDC). Emergency preparedness and response: smallpox. http://www.bt.cdc.gov/agent/smallpox/index.asp. Accessed October 2, 2009.

205. Centers for Disease Control and Prevention (CDC). Notice to readers: smallpox: what every clinician should know—a self-study course [Web broadcast]. *MMWR Morb Mortal Wkly Rep* 2002;51(16):352. http://www.cdc.gov/mmwr/preview/mmwrhtml/mm5116a5.htm. Accessed October 2, 2009.

206. Wharton M, Strikas RA, Harpaz R, et al.; Advisory Committee on Immunization Practices; Healthcare Infection Control Practices Advisory Committee. Recommendations for using smallpox vaccine in a pre-event vaccination program: supplemental recommendations of the Advisory Committee on Immunization Practices (ACIP) and the Healthcare Infection Control Practices Advisory Committee (HICPAC). *MMWR Morb Mortal Wkly Rep* 2003;52(RR-7):1–16.

207. Committee on Smallpox Vaccination Program Implementation, Board on Health Promotion and Disease Prevention, Institute of Medicine. Review of the Centers for Disease Control and Prevention's smallpox vaccination program implementation, letter report #4. Institute of Medicine; 2003. http://www.iom.edu/CMS/3793/4781/14631.aspx. Accessed October 2, 2009.

208. Centers for Disease Control and Prevention (CDC). Notice to readers: supplemental recommendations on adverse events following smallpox vaccine in the pre-event vaccination program: recommendations of the Advisory Committee on Immunization Practices. *MMWR Morb Mortal Wkly Rep* 2003;52:282–284.

209. Centers for Disease Control and Prevention (CDC). Update: adverse events following civilian smallpox vaccination—United States, 2003. *MMWR Morb Mortal Wkly Rep* 2003;52:819–820.

210. Centers for Disease Control and Prevention (CDC). Update: severe acute respiratory syndrome—worldwide and United States, 2003. *MMWR Morb Mortal Wkly Rep* 2003;52:664–665.

211. Seto WH, Tsang D, Yung RW, et al. Effectiveness of precautions against droplets and contact in prevention of nosocomial transmission of severe acute respiratory syndrome (SARS). *Lancet* 2003;361:1519–1520.

212. Centers for Disease Control and Prevention (CDC). Cluster of severe acute respiratory syndrome cases among protected health-care workers—Toronto, Canada, April 2003. *MMWR Morb Mortal Wkly Rep* 2003;52:433–436.

213. Centers for Disease Control and Prevention (CDC). Severe Acute Respiratory Syndrome (SARS) [Web page]. http://www.cdc.gov/ncidod/sars. Accessed October 2, 2009.

214. World Health Organization (WHO). Global Alert and Response: Severe Acute Respiratory Syndrome (SARS) [Web page]. http://www.who.int/csr/sars/en. Accessed October 2, 2009.

215. Centers for Disease Control and Prevention (CDC). Use of Quarantine to Prevent Transmission of Severe Acute Respiratory Syndrome—Taiwan, 2003. *MMWR Morb Mortal Wkly Rep* 2003;52:680–683.

216. Centers for Disease Control and Prevention (CDC). Update: multi-state outbreak of monkeypox—Illinois, Indiana, Kansas, Missouri, Ohio, and Wisconsin, 2003. *MMWR Morb Mortal Wkly Rep* 2003;52(27):642–646.

217. Centers for Disease Control and Prevention (CDC). Monkeypox [Web page]. http://www.cdc.gov/ncidod/monkeypox. Accessed October 2, 2009.

218. Centers for Disease Control and Prevention (CDC). Core Curriculum on Tuberculosis: What the Clinician Should Know. 3rd ed. Atlanta, GA: CDC; 1994. (This print version is in revision; the most up-to-date information is available as an interactive teaching course, at http://www.cdc.gov/tb/webcourses/corecurr/index.htm. Accessed October 2, 2009.)

219. Scott JR, Foster JD, Fraser H. Conjunctival instillation of scrapie in mice can produce disease. *Vet Microbiol* 1993;34:305–309.

Chapter 24 Employee Health and Occupational Medicine

Tara N. Palmore, MD, and David K. Henderson, MD

An institution's occupational medicine program is 1 of 3 vital programs that provide occupational medical and safety support for its healthcare workers. The other 2 programs are the institution's biosafety division and its hospital epidemiology service. These 3 programs work in concert to ensure the health and safety of workers and patients in healthcare institutions. This chapter reviews the intersection of these 3 programs in screening workers for infectious diseases, providing preexposure education and immunoprophylaxis, and in ensuring an adequate infrastructure for the safe provision of care.

Serologic Screening and Immunization

Preplacement Examination

Prior to entry into the workplace, hospital personnel should be evaluated by the occupational medicine service to ensure their fitness for duty. The most important aspect of this evaluation is the employee's medical history. From the perspective of the hospital epidemiologist, the critical aspects of the employee's medical history are his or her communicable disease history (including immunization history) and the presence of underlying conditions that place the employee at elevated risk for occupational infection in the healthcare workplace. An "entry-onto-duty" focused physical examination is performed by some occupational medical services; others require that the employee's personal physician provide the findings from a recent examination.

Laboratory Evaluation

Routine, unfocused laboratory testing is of limited value in screening potential employees. Screening should be limited to diseases and organisms that present significant or unique risks to patients and other staff (eg, varicella zoster virus in children's hospitals or oncology centers; rubella in obstetrical and gynecology settings). Some institutions that provide care primarily to children or immunocompromised patients screen all employees for varicella immunity, whereas others consider a history of chickenpox to be sufficiently reliable evidence of immunity. Routine serologic screening for hepatitis B virus (HBV) is generally not cost-effective, unless the prevalence of HBV infection is high enough in the population of healthcare workers being screened that it is possible to avoid vaccination of enough employees to pay for the screening process. Routine screening for susceptibility to measles, mumps, and rubella is not advisable. Individuals who cannot document that they were adequately vaccinated or that they had physician-diagnosed infection may be given the measles-mumps-rubella (MMR) vaccine. If an institution wishes to ensure that all employees are immune to these childhood illnesses, immunity to these infections should be made a condition of employment. History alone is not adequate to determine rubella immunity; serological examination should be required.

Screening for Prior Tuberculosis Infection

The occupational medicine, hospital epidemiology, and institutional biosafety programs should work together to establish an effective ongoing tuberculosis prevention program, including a tuberculosis surveillance system. This scope of the program should be based on a detailed risk assessment, including the annual tuberculosis case load, history of tuberculosis transmission in the institution, and the prevalence of tuberculosis in the hospital's catchment area. Each new employee should receive intradermal Mantoux skin testing, using the 2-step approach, during his or her initial evaluation, unless the employee has documentation of prior active tuberculosis, a previous positive skin test result, and/or completion of therapy for infection. Use of "control" skin tests (eg, for the presence of *Candida* species or mumps virus) is not advisable. Occupational medicine staff should evaluate tuberculin skin test results using stringent, consistent criteria. Alternatively, the whole-blood interferon-γ release assay (Quanti-FERON TB Gold; Cellestis) has been approved by the Food and Drug Administration and may be used in place of skin testing for screening healthcare workers.[1] This is an in vitro assay that detects cell-mediated immune reactivity to *Mycobacterium tuberculosis* by detecting the release of interferon-γ. Some investigators have reported substantially discordant results between skin tests and the cytokine-based test.[2,3] The significance of these discordant results is difficult to ascertain, since there is no adequate criterion standard test ("gold standard"). Healthcare workers with positive skin test or cytokine test results should be managed according to existing guidelines.[4]

Pre-event Immunoprophylaxis

As noted above, vaccination against certain infectious diseases is advisable for healthcare providers. Table 24-1 provides a summary list of immunizations recommended for healthcare workers. Certainly, any healthcare worker whose job entails potential exposure to blood or blood-containing body fluids should have immunity against HBV. The occupational medicine and hospital epidemiology services should work together to ensure that an efficient program is in place to educate staff about the occupational risk of bloodborne pathogen infection and to provide HBV vaccination. As noted above, depending on the institution, immunization against measles, mumps, and rubella and varicella may be appropriate. If the risk for transmission of these childhood exanthems is present in your institution, or, conversely, if transmission of these infections may present life-threatening risks to immunosuppressed patients in your institution, you may wish to consider establishing ongoing immunization programs for your healthcare workers.

For most institutions, particularly those serving elderly and/or immunocompromised patients, the risks of transmission of influenza A virus and influenza B virus are substantial. For this reason, influenza vaccination deserves special emphasis. Institutions and their occupational medicine services should, at a minimum, offer voluntary influenza vaccination to all providers who have patient contact. From a hospital epidemiology perspective, annual vaccination of healthcare workers is the single most efficacious strategy for reducing the risk for influenza transmission to hospitalized patients. Unfortunately, voluntary influenza vaccination programs for healthcare workers historically have produced less than optimal immunization rates.[6] The national vaccination rate among healthcare workers is approximately 40%.[7] Low healthcare worker influenza vaccination rates have been associated with nosocomial outbreaks,[8] whereas higher vaccination rates have been associated with both a reduced incidence of influenza-like illness[9,10] and reduced mortality[6,11-13] among patients. For these reasons, we view influenza vaccination of healthcare providers as a patient safety issue. Several national organizations, as well as the Centers for Disease Control and Prevention (CDC), recommend mandatory influenza vaccination of healthcare workers.[6,14-19] In light of the increasing concern about pandemic influenza, the development of effective vaccination strategies for healthcare workers has become even more important. In our own institution, implementation of a new hospital policy requiring healthcare workers to accept or decline vaccination, as well as the simultaneous implementation of a streamlined vaccination process and vaccination tracking program, resulted in vaccination of 89% of all healthcare workers who have patient contact in our facility.[20] Additional beneficial effects of a successful influenza immunization program include a decrease in worker absenteeism during influenza epidemics and a decrease in healthcare costs.

The only major change relating to healthcare worker vaccination since the previous edition of this textbook is the recommendation that all providers receive a booster vaccination for pertussis immunity (a single dose of tetanus toxoid, reduced diphtheria toxoid, and acellular pertussis vaccine [Tdap]), if they have not previously received Tdap and if their last tetanus vaccination was more than 2 years previously.[21] The acellular pertussis vaccine has been approved for use in adults, and both the US Department of Health and Human Services' Healthcare Infection Control Practices Advisory Committee (HICPAC) and the CDC's Advisory Committee on Immunization Practices have endorsed the use of the 3-component vaccine in healthcare workers.[21]

Infectious Disease Surveillance for Employees

The one infectious disease for which active surveillance is almost uniformly recommended for healthcare

Table 24-1. Vaccinations recommended for hospital employees

Disease or pathogen	Indication	Vaccine and dosage	Cautions
Diphtheria	In an outbreak or following documented exposures for employees who have not been vaccinated in the past 10 years or those who lack serological evidence of immunity	Td, 0.5 mL intramuscularly (or Tdap if not boosted previously for pertussis; see below)	Known hypersensitivity to thimerosal or any component of the vaccine
Hepatitis A	Employees working in high-risk areas (eg, dietary service, cafeteria, or hepatitis ward) who do not have serologic evidence of previous hepatitis A virus infection	Hepatitis A vaccine, 1.0 mL intramuscularly at months 0 and 6–12	Known hypersensitivity to any component of the vaccine
Hepatitis B	All employees at risk for occupational exposure to blood or body fluids	Hepatitis B vaccine, 1.0 mL intramuscularly (in the deltoid muscle) at months 0, 1, and 6	Yeast sensitivity or allergy
Influenza	All hospital employees	Influenza vaccine, 0.5 mL intramuscularly annually	History of anaphylactic reaction to eggs or to prior doses of influenza vaccine
Measles	Employees who have never had physician-diagnosed measles or laboratory evidence of immunity[a]	Trivalent MMR, 0.5 mL subcutaneously	Pregnancy, history of anaphylactic reaction to eggs or neomycin, severe febrile illness, immunosuppression, recent receipt of intravenous immunoglobulin
Mumps	Employees who have never had physician-diagnosed mumps or do not have laboratory evidence of immunity or proof of vaccination on or after their first birthday[a]	Trivalent MMR, 0.5 mL subcutaneously	Pregnancy, history of anaphylactic reaction to eggs or neomycin, severe febrile illness, immunosuppression, recent receipt of intravenous immunoglobulin
Meningococcus	In there is an institutional outbreak caused by the A, C, Y, or W-135 strains, vaccination may be useful	Reconstituted meningococcal vaccine, 0.5 mL	Safety in pregnancy uncertain; sensitivity to thimerosal or any other component of the vaccine
Pertussis	HCWs who have direct patient contact[b]	Tdap, 0.5 mL	Known hypersensitivity to any component of the vaccine, prior encephalopathy associated with primary immunization
Pneumococcus	Employees who are >65 years of age or have underlying cardiac, pulmonary, liver, renal, or immunocompromising disease	Pneumococcal vaccine, 0.5 mL subcutaneously or intramuscularly; booster dose	Safety in pregnancy undocumented
Rubella	Employees who have never received the live vaccine on or after their first birthday or who do not have serological proof of immunity	Trivalent MMR, 0.5 mL subcutaneously	Pregnancy, history of anaphylactic reaction to eggs or neomycin, severe febrile illness, immunosuppression, recent receipt of intravenous immunoglobulin
Tetanus	Employees who sustain tetanus-prone wounds, those who never completed the initial vaccination series, and those who have not received a booster dose within the past 10 years[b]	*Initial series:* Td, 0.5 mL intramuscularly at months 0, 1, and 6–12; *booster:* 0.5 mL intramuscularly	History of neurological or hypersensitivity reaction following a previous dose; first trimester of pregnancy
Varicella	Employees with patient contact who have no history of chickenpox and have a negative varicella titer[b]	Varicella vaccine, 0.5 mL at weeks 0 and 4–8	Hypersensitivity to vaccine, gelatin, or neomycin; immunosuppression or immunodeficiency; active tuberculosis; febrile illness; pregnancy

NOTE: HCW, healthcare worker; MMR, measles, mumps, and rubella vaccine; Td, tetanus and diphtheria toxoids; Tdap, tetanus toxoid, reduced diphtheria toxoid, and acellular pertussis vaccine.
a HCWs who have never received measles vaccine and/or have no history of immunity to measles require 2 doses of MMR, administered no less than 1 month apart.
b Current recommendation from the Advisory Committee on Immunization Practices, Centers for Disease Control and Prevention.[5]

employees is tuberculosis. In addition to ensuring that the tuberculosis control program is tailored to the unique aspects of risk in their own environments, institutions should develop programs that address the variable risks for exposure of individuals in differing job categories. At a minimum, healthcare workers who have prior negative tuberculin skin test results should be retested at appropriate intervals, on the basis of the institutional risk assessment, as recommended by the most recent guidelines from the CDC and the US Public Health Service.[4] Staff members who, on the basis of their job categories, are at higher risk for occupational exposure to tuberculosis (eg, critical care physicians and nurses, pulmonologists, anesthesiologists, and respiratory therapists) should be tested more frequently. Employees with underlying immunodeficiencies that place them at high risk of developing active tuberculosis may be discouraged from caring for patients with tuberculosis.

The most recent guidelines from the CDC and the US Public Health Service on preventing the transmission of *M. tuberculosis* in healthcare settings included several modifications from the previous edition of this book.[4] Among several substantial changes in the guideline, the following significant changes are directly relevant to occupational medicine and healthcare epidemiology:

- The risk assessment process for institutions includes the assessment of additional aspects of infection control.
- As noted above, the whole-blood interferon-γ release assay (QuantiFERON TB Gold; Cellestis) is accepted as a possible alternative to skin testing.
- The frequency of tuberculosis screening recommended for healthcare workers in various settings has decreased.
- The criteria for determining frequency of screening have been modified.
- Criteria for serial testing of healthcare workers for *M. tuberculosis* infection are more clearly defined, resulting in a decrease in the number of healthcare workers who require serial testing.
- The guidelines include new recommendations for annual respirator training, initial and periodic respirator-fit testing, as well as the scientific evidence and federal regulations supporting the need for fit testing.
- The guidelines provide more detailed information about use of UV germicidal irradiation and room-air recirculation units.

Additional details can be found in Chapter 25, on control of tuberculosis.

Postexposure Prophylaxis for Occupational Exposures to Bloodborne Pathogens

The occupational medical service is the first stop for any employee with a blood or body fluid exposure. Assisted by published guidelines, occupational medicine specialists evaluate the nature of the exposure, the clinical status of the source patient (if known), and the underlying health of the employee to determine the risk of infectious disease transmission and the need for postexposure prophylaxis. For HBV, influenza virus, and some other viral pathogens, postexposure prophylaxis is simple and effective. For others, such as hepatitis C virus (HCV), there is no known safe, reliable postexposure prophylaxis. Occupational medicine providers work closely with infectious diseases specialists to administer antiretroviral drugs as postexposure prophylaxis to healthcare workers who have an exposure at high risk of transmitting human immunodeficiency virus (HIV). Postexposure management strategies for occupational exposures to the major bloodborne pathogens (ie, HBV, HCV, and HIV) are discussed below. The management of other occupational exposures and the issues relating to their postexposure management are discussed in Chapter 23, on exposure workups.

HBV Exposure

Historically, prior to the development of the HBV vaccine, hepatitis B represented one of the most significant workplace risks for healthcare providers (particularly those who had occupational exposure to blood). As a result of exposures to HBV in the workplace, healthcare workers were at significantly increased risk for HBV infection when compared with the population at-large.

Transmission
HBV is transmitted parenterally, with percutaneous exposure to infected blood the most important mode of occupational transmission. Mucous membrane exposure also may result in infection. HBV also can be transmitted sexually and perinatally. Both acutely and chronically infected individuals transmit infection; infected individuals who have high viral loads and high levels of hepatitis B e antigen (HBeAg) represent the greatest risk for infection. The risk for transmission of HBV following parenteral exposure to blood from an HBeAg-positive patient is 27%–43% per exposure, whereas the risk associated with exposure to blood from a patient who tests positive for hepatitis B surface antigen (HBsAg) but negative for HBeAg is approximately 6%–10%.[22]

Criteria for Exposure

Any worker who sustains a percutaneous, mucous membrane, or nonintact skin exposure to blood or body fluids that may contain blood from an HBsAg-positive patient (or a patient whose HBV serologic status cannot be determined) should be considered exposed. Source patients with unknown serologic status should be tested as soon as possible after the exposure.

Postexposure Prophylaxis

If a healthcare worker who has not been immunized with the HBV vaccine sustains an occupational exposure to HBV, the worker should be given 0.06 mL/kg of hepatitis B immune globulin intramuscularly. Ideally, this first dose should be administered within 24 hours after the exposure. The first dose of the HBV vaccine series should be administered at the same time, followed by additional doses 1 month later and 6 months later. Prior to treatment, a baseline level of antibody to HBs (anti-HBs) should be determined. If the test result is positive, no additional treatment is indicated. For previously vaccinated healthcare workers, the anti-HBs level should be determined. Workers whose anti-HBs levels are greater than 10 mIU/mL are protected. Although routine administration of a booster dose of HBV vaccine is not recommended by the US Public Health Service, we administer it to employees who are known to have had protective antibody levels but whose levels have fallen below 10 mIU/mL. Employees who have anti-HBs levels below 10 mIU/mL and who were never demonstrated to have had an adequate vaccine response should be treated as if they had no response (ie, they should be given both a single dose of hepatitis B immune globulin and a vaccine booster). For employees who for some reason cannot be vaccinated, a second dose of hepatitis B immune globulin should be administered 1 month after the first dose. More detailed discussion of issues relating to the management of occupational exposures to HBV can be found in Beekmann and Henderson,[22] the CDC guidelines on occupational exposures,[23] and Panlilio et al.[24]

Control Measures

Universal vaccination of healthcare workers against HBV should be a primary goal of the occupational medicine service, and education and vaccination campaigns should focus on achieving that goal. Vaccination of healthcare workers, which provides immunity in more than 93% of vaccinees, should mitigate the risk of HBV transmission from patients to healthcare workers. Whereas patient-to-healthcare worker transmission occurs far more frequently than does healthcare worker-to-patient transmission, the latter type of transmission does occur, particularly when the healthcare worker is HBeAg positive and conducts invasive procedures. More than

400 instances of iatrogenic HBV transmission have been reported in the literature. Current US Public Health Service guidelines (last updated in 1991) recommend restricting the practices of healthcare workers who are HBeAg positive.[25] In the United Kingdom, HBV-infected providers who are positive or negative for HBeAg but have HBV DNA levels greater than 10^3 genome equivalents (GE) per milliliter may not conduct what are termed "exposure-prone" invasive procedures. HBV-infected providers who are HBeAg negative and have circulating HBV DNA levels of less than 10^3 GE/mL are permitted to perform exposure-prone invasive procedures but must be retested at least every 12 months to ensure that the viral load remains below 10^3 GE/mL.[26] United Kingdom guidelines[27] also permit HBV-infected healthcare workers who are HBeAg negative and have pretreatment HBV DNA levels between 10^3 and 10^5 GE/mL to perform exposure-prone procedures if they are receiving suppressive oral antiviral therapy and if their viral load has decreased to less than 10^3 GE/mL. The major challenge associated with this latter recommendation is the development of an effective monitoring strategy to make certain that the circulating viral burden remains less than 10^3 GE/mL.[27] The variability of various testing systems further complicates monitoring.

A European consortium developed a set of guidelines that do not permit HBV-infected healthcare workers who are HBeAg positive to conduct exposure-prone procedures[28] but that do permit HBV-infected healthcare workers who are HBeAg negative but have HBV DNA levels of less than 10^4 GE/mL to perform such procedures. Such individuals must be retested at least annually to make certain that the circulating viral burden remains below 10^4 GE/mL.[28] These guidelines also do not allow healthcare workers who are identified as having transmitted HBV to perform exposure-prone procedures and permit HBV-infected healthcare workers who have been treated and have posttreatment DNA levels that have fallen to less than 10^4 GE/mL to conduct exposure-prone procedures, so long as the healthcare worker is retested every 3 months to ensure that the viral burden remains below 10^4 GE/mL.[28]

Authorities from the Netherlands published a third guideline that allows HBV-infected healthcare workers who have HBV DNA levels of less than 10^5 GE/mL to conduct exposure-prone invasive procedures but requires them to be retested at least annually.[29]

The US Public Health Service guidelines[25] were last updated in 1991. The Society for Healthcare Epidemiology of America (SHEA) has issued 2 sets of guidelines for infected healthcare workers.[30,31] SHEA is currently revising these guidelines, and updated recommendations should be published in 2010. For a more detailed discussion of issues related to the risk of iatrogenic transmission of HBV, see Beekmann and Henderson[22] and the CDC guidelines on occupational exposures.[23]

HCV Exposure

Healthcare workers are at risk for HCV infection as a result of parenteral or mucous membrane exposures to blood from patients infected with HCV. For a detailed discussion of the risks for occupational HCV infection, the changing understanding of the pathogenesis and immunopathogenesis of HCV infection, and approaches to the management of HCV infection, see Henderson.[32] A long standing concern has been the fact that a significant proportion of the individuals who develop productive HCV infection develop chronic infection and are at risk for serious sequelae of this infection, such as chronic active hepatitis, cirrhosis, hepatocellular carcinoma, and death. Whereas new information about the immunopathogenesis of HCV infection suggests this risk may not be as high as it was once thought to be, a substantial fraction of infected patients develop serious sequelae of infection.

Transmission
Occupational risk for HCV transmission is likely linked to the same routes of transmission as those for HBV. Occupational HCV infection has been most frequently associated with parenteral exposures. A few instances of mucous membrane transmission have been reported, and nonapparent parenteral transmission (caused by exposure of nonintact skin to blood of an HCV-infected individual) likely also occurs, albeit at a substantially lower rate than is the case for HBV. The risk for occupational infection associated with a single parenteral exposure has been estimated to be approximately 2%.[32]

Criteria for Exposure
Occupational medicine staff should consider any healthcare worker who has sustained a percutaneous, mucous membrane, or nonintact skin exposure to blood, or a body fluid potentially containing blood, from an HCV-infected patient as having been exposed. As is the case for HBV infection, in instances in which the source patient for an exposure is unknown, cannot be tested, or is known to have epidemiological risk factors associated with HCV infection, the worker also should be considered exposed.

Postexposure Management
All HCV-exposed individuals should be tested to determine their levels of aminotransferases (alanine and aspartate) and should have their HCV antibody level determined with a sensitive and specific antibody test at baseline and again at least 15 weeks after exposure. The healthcare worker also should be encouraged to seek prompt medical attention for any symptoms suggestive of systemic illness or acute hepatitis. Immune serum globulin should not be administered for occupational exposures to HCV. Neutralizing antibodies to HCV are highly strain specific and have not been shown to afford protection against reinfection. Additionally, even if antibody were protective, donors for current immune globulin preparations are screened for hepatitis C antibody (anti-HCV) and are eliminated from the donor pool if found to be positive for anti-HCV. HCV infection also has been associated with the administration of HCV-contaminated intravenous immunoglobulin in 2 clusters. Finally, the theoretical risk that intramuscularly administered immune serum globulin might result in HCV transmission was not identified in the one study we are aware of that attempted to look for it. Most academic centers now monitor exposed healthcare workers with HCV RNA polymerase chain reaction tests and then use immunomodulators to treat the healthcare workers who become infected. Both "preemptive therapy" and "watchful waiting" strategies have been proposed as reasonable strategies for postexposure management.[32] As yet, data from studies of healthcare workers treated with these approaches are too preliminary to provide the basis for a formal recommendation of an optimal management strategy. At our own institution, we use the "watchful waiting" strategy.[32]

Control Measures
Avoidance of exposures through the routine use of universal or standard precautions (ie, primary prevention) is the only effective preventive strategy currently available. Iatrogenic transmission of HCV from healthcare workers to their patients has been uncommon, particularly in the United States. The past several years have seen several reports of healthcare worker-to-patient HCV transmission. The experience in the United Kingdom has been quite distinctive, in that HCV transmission from healthcare workers to patients seems to be occurring at higher rates than in the United States. Although the limited data available effectively preclude identifying factors associated with a risk of iatrogenic transmission, the fact that 2 gynecologists, 3 cardiac or thoracic surgeons, and 1 orthopedic surgeon are involved in reported cases suggest that risk factors for transmission are likely to be similar to those identified for HBV transmission. A number of cases of iatrogenic HCV transmission have been linked to healthcare workers injecting drugs that were intended for patients into themselves and then reusing the needle to inject the patients. The magnitude of risk for iatrogenic HCV transmission has not been measured and is likely to be so small that precise measurement will be

difficult, if not impossible. Because there have been several clusters of healthcare worker-to-patient transmission of HCV in the United Kingdom, public health authorities there have issued practice restrictions for HCV-infected healthcare workers. In the United States, neither the US Public Health Service nor professional organizations have, at least as yet, recommended restricting the practices of HCV-infected healthcare workers. If the experience in the United Kingdom is shown to be representative, the existing US guidelines for managing HCV-infected healthcare workers[25] will clearly need to be revised. As we have noted, SHEA is currently revising its guideline for management of healthcare workers infected with bloodborne pathogens.

HIV Exposure

The magnitude of the risk of infection from each exposure to HIV is approximately 0.3% per parenteral exposure.[33] Whereas this risk is substantially smaller than that for other bloodborne infections, the consequences of infection are life altering. For a thorough discussion of the issues relating to nosocomial transmission of HIV, see Henderson.[34]

Transmission

HIV is transmitted parenterally, sexually, and vertically between mother and child (ie, across the placenta, perinatally, or through breast-feeding). Occupational

Table 24-2. Centers for Disease Control and Prevention recommendations for postexposure prophylaxis (PEP) for different types of exposure to human immunodeficiency virus (HIV)

Exposure type	HIV-positive source patient[a]		HIV status of source patient unknown or source unknown	HIV-negative source patient
	Class 1: asymptomatic HIV infection or known low viral load (eg, <1,500 GE/mL)	Class 2: symptomatic HIV infection, AIDS, acute sero-conversion, or known high viral load		
Percutaneous				
Less severe (eg, solid needle or superficial injury)	Recommend basic PEP	Recommend expanded PEP	Generally, no PEP warranted; however, consider basic PEP[b] if the source patient has risk factors for HIV infection[c]	No PEP warranted
More severe (eg, large-bore hollow needle, deep puncture, visible blood on device, or needle used in patient's artery or vein)	Recommend expanded PEP	Recommend expanded PEP	Generally, no PEP warranted; however, consider basic PEP[b] if the source patient has risk factors for HIV infection[c]	No PEP warranted
Mucous membrane or nonintact skin				
Small volume (eg, few drops or brief contact)	Consider basic PEP[b]	Recommend basic PEP	Generally, no PEP warranted; however, consider basic PEP[b] if the source patient has risk factors for HIV infection[c]	No PEP warranted
Large volume (eg, major blood splash or prolonged contact)	Recommend basic PEP	Recommend expanded PEP	Generally, no PEP warranted; however, consider basic PEP[b] if the source patient has risk factors for HIV infection[c]	No PEP warranted

NOTE: "Basic" and "expanded" PEP are defined in Table 24-3. GE, genome equivalents. Modified from the Centers for Disease Control and Prevention guidelines on occupational exposure to HIV.[23]
a If drug resistance is a concern, obtain expert consultation. Initiation of PEP should not be delayed pending expert consultation, and, because expert consultation alone cannot substitute for face-to-face counseling, resources should be available to provide immediate evaluation and follow-up care for all exposed healthcare workers.
b The designation "consider PEP" indicates that PEP is optional and should be based on an individualized decision between the exposed healthcare worker and the treating clinician.
c If PEP is offered and taken, and the source patient is later determined to be HIV negative, PEP should be discontinued.

transmission has been reported after percutaneous, mucous membrane, and nonintact skin exposure to HIV-infected blood. HIV is present in much lower amounts in other blood cell-containing body fluids, including inflammatory exudates, amniotic fluid, saliva, and vaginal secretions. The risk of seroconversion following mucous membrane or nonintact skin exposure is too low to be estimated with precision.

Criteria for Exposure
As for HBV and HCV exposures, any healthcare worker who has sustained a percutaneous, mucous membrane, or nonintact skin exposure to the blood or body fluid potentially containing blood from an HIV-infected patient should be considered exposed. The risk for infection associated with any discrete exposure depends on a number of variables, including the inoculum size, the exposure severity, and the stage of the source patient's illness (ie, circulating viral burden).

Postexposure Management and Postexposure Antiretroviral Prophylaxis
The efficacy of antiretroviral chemoprophylaxis for occupational HIV exposure will likely never be definitively established in a prospective clinical trial. Nonetheless, a variety of types of studies provide indirect evidence of the efficacy of postexposure prophylaxis, including the efficacy of antiretroviral agents in preventing retroviral infection in animal models, the efficacy of antiretrovirals in preventing maternal-fetal transmission, the results of the CDC's retrospective case-control study of occupational HIV infection, and our clinical experience using antiretroviral agents for postexposure prophylaxis at our institution since 1988 (discussed in more detail in Henderson[34]). The CDC's recommendations regarding postexposure management[23] are summarized in Table 24-2, and the current US Public Health Service recommendations regarding antiretroviral regimens for postexposure prophylaxis[24] are summarized in Table 24-3. At approximately 3–5 year intervals, the Public Health Service revises its recommendations for chemoprophylaxis after occupational HIV exposure, taking into account new information about the toxicity of established regimens, the ability of exposed healthcare workers to adhere to the established regimens, the development and marketing of new antiretroviral agents, and patterns of antiretroviral resistance. Updated recommendations should be published in the next few years. These recommendations will likely take the same shape as those outlined in Table 24-2 and Table 24-3 but will likely offer more agents as options for postexposure prophylaxis.

The basic principles for offering chemoprophylaxis are listed in Table 24-4. Some of these are worthy of a bit of additional discussion. The fact that treatment is immediately accessible should be widely publicized throughout the healthcare institution. All employees should be aware of the postexposure prophylaxis program and how it works. At our institution, we have distributed posters and flyers, emphasize the program, and also include information about the program in mandatory annual education sessions for clinical staff. Occupational medicine staff have to be very familiar with what constitutes an exposure and must make certain that they do not over-prescribe antiretrovirals. Prescribing physicians should carefully choose a regimen that can be taken by the healthcare worker. More is not necessarily better, especially if all of the drugs are vomited up. Prescribing physicians also need to be cognizant of the source patient's therapy and viral burden (if this information is immediately available) and should use this information in developing an optimal regimen. If the prescriber is not familiar with the primary or alternative agents, she or he should get assistance from infectious diseases physicians

Table 24-3. US Public Health Service recommendations for postexposure chemoprophylaxis for occupational exposures to human immunodeficiency virus (HIV)

Class of exposure, type of regimen	Antiretroviral agent(s)
Exposure associated with a recognized transmission risk	
Basic regimen A	Zidovudine plus lamivudine or emtricitibine
Basic regimen B	Tenofovir plus lamivudine or emtricitibine
Alternative basic regimen A	Stavudine plus lamivudine
Alternative basic regimen B	Stavudine plus emtricitibine[a]
Exposure of a nature suggesting elevated transmission risk[b]	
Expanded regimen	Basic regimen plus Lopinavir-ritonavir
Alternative expanded regimens	Basic regimen plus one or more of these agents:[c] Atazanavir Fosamprenavir[d] Indinavir[e] plus ritonavir Saquinavir plus ritonavir Efavirenz[b]

NOTE: Modified from Panlilio et al.[24]
a Agent not recommended for use in pregnant women.
b Elevated risk is associated with "deep" injury, injury with a device that has been used in an HIV-infected patient's artery or vein, and injuries associated with "larger" volumes of blood and/or blood containing a high titer of HIV.
c The role of newer agents (eg, integrase inhibitors and CCR5 inhibitors) remains undetermined.
d Agent has been reported to have some toxicity.[37]
e Agent should be taken on an empty stomach, with increased fluid consumption (eg, six 8-ounce [0.237-L] glasses of water per day).

who are used to prescribing antiretrovirals. To the extent that it is possible, prescribers should become familiar with the antiretroviral agents, their side effects, and the appropriate strategies to manage toxicity. Prescribers should anticipate and prophylactically treat side effects (eg, providing antiemetics for nausea and antispasmodics for diarrhea). Prescribing staff also should carefully monitor all healthcare workers who are taking antiretrovirals for the development of signs of toxicity, as well as for adherence to the regimen. The occupational medical service should also have ready access to (and frequently should seek the advice of) expert consultants. If no experts are immediately available, expert guidance is available around the clock from PEPline, the National Clinicians' Postexposure Prophylaxis Hotline (sponsored by the CDC and the University of California), either by telephone (at 888-448–4911 or via the Internet (at http://www.nccc.ucsf.edu/Hotlines/PEPline.html).

Other Considerations
Several additional issues must be taken into consideration when considering HIV postexposure chemoprophylaxis. Counseling and serologic testing of exposed personnel should be performed as soon as possible after the exposure. These services should be available 24 hours per day. All personnel involved in postexposure evaluation and counseling, including emergency department workers, must be trained in and familiar with institutional protocols. Follow-up serologic testing should be performed at 6 weeks, 3 months, and 6 months after the exposure. Some institutions (including our own) offer additional testing at 1 year after the exposure. Exposed workers should be instructed to return immediately for clinical evaluation if they develop signs or symptoms of either drug toxicity or of acute retroviral infection (eg, fever, rash, and lymphadenopathy). Occupational HIV exposure can cause severe psychological symptoms, including depression, anxiety, anger, fear, sleep disturbances, conversion symptoms, suicidal ideation, and psychosis. Postexposure counselors should be alert to these possibilities and be quick to refer the employee to specialists in crisis intervention and counseling, if necessary.

Healthcare Worker–to-Patient Transmission
Transmission of HIV from healthcare worker to patient occurs extremely uncommonly. Nonetheless, a few such cases have been described in the literature.[32,35] Current US Public Health guidelines recommend that individual states either adopt the 1991 CDC guidelines or construct guidelines that are certified by the states as equivalent to the CDC guidelines.[25] Although the individual state guidelines vary substantially, some state guidelines do include restricting the practices of HIV-infected providers

Table 24-4. Basic principles for human immunodeficiency virus postexposure chemoprophylaxis programs

Treatment must be immediately accessible
Make certain an exposure has occurred
Choose a regimen that the exposed person can tolerate
Be cognizant of source-patient's therapy and viral burden (if available)
Be familiar with the antiretroviral agents, their side effects, and the management of toxicity
Anticipate and prophylactically treat side effects
Monitor for toxicity and for adherence
Provide access to (and use) expert consultants

who perform what the CDC terms "exposure-prone" procedures. The CDC guidelines clearly recommend that HIV-infected healthcare workers who do not perform such procedures need not have their practices restricted. As noted above, SHEA will soon publish its revised guideline for managing healthcare workers infected with HIV or other bloodborne pathogens.

Education and Orientation

Employee education is another area in which the biosafety, hospital epidemiology, and occupational medicine programs should work together. New employee orientation should contain basic information about infection control and prevention and should provide a detailed list of resources for additional information. Educational efforts should focus on the routine use of universal or standard precautions and handwashing to reduce the risk of transmitting infection. Ongoing educational programs for staff should emphasize the basic tenets of infection control, such as use of universal or standard precautions, optimal use of personal protective equipment, handwashing, vaccine safety and efficacy, and identification of healthcare worker illnesses that require prompt evaluation by occupational medicine staff (eg, conjunctivitis, varicella, skin and soft-tissue infections, herpes zoster, other childhood illnesses, jaundice, and diarrhea).

Outbreak Investigation

The occurrence of clusters of infections caused by the same organism is another instance in which cooperation and collaboration among the occupational medicine, hospital epidemiology, and biosafety programs are essential. Depending on the type of epidemic, the hospital epidemiology team will likely conduct the "shoe-leather" studies, identify healthcare workers at risk, and refer them to the occupational medicine service.

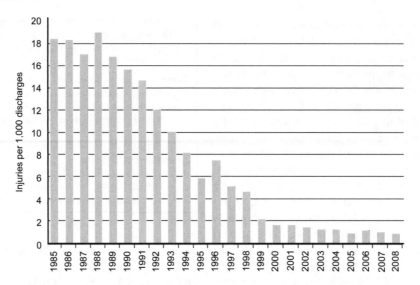

Figure 24-1. Graph showing the number of parenteral injuries per 1,000 discharges sustained by healthcare workers at the National Institutes of Health Clinical Center, 1985–2008.

Occupational medicine staff will conduct careful interviews, provide postexposure treatment as appropriate, and (in collaboration with hospital epidemiology and biosafety personnel) try to identify factors that are associated with a risk for transmission. During an outbreak, effective communication and daily interaction among these 3 groups is essential to effective interdiction (see Chapter 12, on outbreak investigations).

Noninfectious Adverse Events Among Hospital Staff

Hospital epidemiology should also work closely with both the biosafety team and the occupational medicine service to evaluate noninfectious adverse events occurring among hospital staff. Clusters of certain kinds of events (eg, needlestick injuries associated with similar circumstances of exposure) can become the stimulus for institutional performance improvement activities that can ultimately reduce the risk of these events substantially. Following an occupational HIV infection in a staff member that occurred in 1988, we developed a working group to evaluate the circumstance of every occupational exposure to blood that occurred in our institution, with the expectation that a more complete understanding of the circumstances of these exposures might provide a path to performance improvement, through the continued education and training of our staff about risks and risk reduction, instruction of staff about the appropriate use of infection control precautions, the use of intrinsically safer devices, and the modification of work practices associated with exposures. Figure 24-1 demonstrates the efficacy of this approach in reducing parenteral exposures to blood in our institution. Thus, as is the case for healthcare-associated infections, noninfectious adverse events in the hospital also have healthcare-associated epidemiologies that, when delineated, may provide insight into appropriate interventions and risk-reduction strategies. The hospital epidemiologist is ideally placed in the organization to facilitate the identification of factors contributing risk for such adverse events and for leading the team to design, implement, and evaluate the success of interventions designed to mitigate those risks.

References

1. Mazurek GH, Jereb J, Lobue P, Iademarco MF, Metchock B, Vernon A. Guidelines for using the QuantiFERON-TB Gold test for detecting *Mycobacterium tuberculosis* infection, United States. *MMWR Recomm Rep* 2005;54(RR-15):49–55.
2. Pollock NR, Campos-Neto A, Kashino S, et al. Discordant QuantiFERON-TB Gold test results among US healthcare workers with increased risk of latent tuberculosis infection: a problem or solution? *Infect Control Hosp Epidemiol* 2008; 29(9):878–886.
3. Carvalho AC, Crotti N, Crippa M, et al. QuantiFERON-TB Gold test for healthcare workers. *J Hosp Infect* 2008;69(1): 91–92.
4. Jensen PA, Lambert LA, Iademarco MF, Ridzon R. Guidelines for preventing the transmission of *Mycobacterium tuberculosis* in health-care settings, 2005. *MMWR Recomm Rep* 2005;54(RR-17):1–141.
5. Advisory Committee on Immunization Practices, Centers for Disease Control and Prevention. Immunization Action

Coalition Web page. http://immunize.org/acip/. Accessed October 6, 2009.

6. Pearson ML, Bridges CB, Harper SA. Influenza vaccination of health-care personnel: recommendations of the Healthcare Infection Control Practices Advisory Committee (HICPAC) and the Advisory Committee on Immunization Practices (ACIP). *MMWR Recomm Rep* 2006;55(RR-2):1–16.

7. Fiore AE, Shay DK, Broder K, et al. Prevention and control of influenza: recommendations of the Advisory Committee on Immunization Practices (ACIP), 2008. *MMWR Recomm Rep* 2008;57(RR-7):1–60.

8. Cunney RJ, Bialachowski A, Thornley D, Smaill FM, Pennie RA. An outbreak of influenza A in a neonatal intensive care unit. *Infect Control Hosp Epidemiol* 2000;21(7):449–454.

9. Saito R, Suzuki H, Oshitani H, Sakai T, Seki N, Tanabe N. The effectiveness of influenza vaccine against influenza A (H3N2) virus infections in nursing homes in Niigata, Japan, during the 1998–1999 and 1999–2000 seasons. *Infect Control Hosp Epidemiol* 2002;23(2):82–86.

10. Potter J, Stott DJ, Roberts MA, et al. Influenza vaccination of health care workers in long-term-care hospitals reduces the mortality of elderly patients. *J Infect Dis* 1997;175(1):1–6.

11. Carman WF, Elder AG, Wallace LA, et al. Effects of influenza vaccination of health-care workers on mortality of elderly people in long-term care: a randomised controlled trial. *Lancet* 2000;355(9198):93–97.

12. Hayward AC, Harling R, Wetten S, et al. Effectiveness of an influenza vaccine programme for care home staff to prevent death, morbidity, and health service use among residents: cluster randomised controlled trial. *BMJ* 2006;333(7581):1241.

13. Thomas RE, Jefferson TO, Demicheli V, Rivetti D. Influenza vaccination for health-care workers who work with elderly people in institutions: a systematic review. *Lancet Infect Dis* 2006;6(5):273–279.

14. Talbot TR, Bradley SE, Cosgrove SE, Ruef C, Siegel JD, Weber DJ. Influenza vaccination of healthcare workers and vaccine allocation for healthcare workers during vaccine shortages. *Infect Control Hosp Epidemiol* 2005;26(11):882–890.

15. Infectious Diseases Society of America. Flu Experts Call for Mandatory Shots for Health Care Workers. 2007. http://www.idsociety.org/Content.aspx?id=3708. Accessed October 6, 2009.

16. American College of Physicians. ACP Policy on Influenza Vaccination of Health Care Workers. 2008. http://www.acponline.org/clinical_information/resources/adult_immunization/. Accessed October 6, 2009.

17. National Foundation for Infectious Diseases. Improving influenza vaccination rates in healthcare workers: strategies to increase protection for workers and patients. 2008. http://www.nfid.org/publications/. Accessed October 6, 2009.

18. Dash GP, Fauerbach L, Pfeiffer J, et al. APIC position paper: improving health care worker influenza immunization rates. *Am J Infect Control* 2004;32(3):123–125.

19. American Nurses Association. Report to the Board of Directors on seasonal influenza vaccination for registered nurses: consent action report. 2008. http://www.preventinfluenza.com/profs_workers.asp. Accessed October 6, 2009.

20. Palmore TN, Vandersluis JP, Morris J, et al. A successful mandatory influenza vaccination campaign using an innovative electronic tracking system. *Infect Control Hosp Epidemiol* 2009 (in press).

21. Kretsinger K, Broder KR, Cortese MM, et al. Preventing tetanus, diphtheria, and pertussis among adults: use of tetanus toxoid, reduced diphtheria toxoid and acellular pertussis vaccine recommendations of the Advisory Committee on Immunization Practices (ACIP) and recommendation of ACIP, supported by the Healthcare Infection Control Practices Advisory Committee (HICPAC), for use of Tdap among healthcare personnel. *MMWR Recomm Rep* 2006;55(RR-17):1–37.

22. Beekmann SE, Henderson DK. Nosocomial viral hepatitis in healthcare workers, Chapter 78. In: Mayhall CG, ed. *Hospital Epidemiology and Infection Control*. 3rd ed. Philadelphia: Lippincott, Williams and Wilkins; 2004:1337–6130.

23. Centers for Disease Control and Prevention. Updated US Public Health Service Guidelines for the management of occupational exposures to HBV, HCV, and HIV and recommendations for postexposure prophylaxis. *MMWR Morb Mortal Wkly Rep* 2001;50(RR-11):1–52.

24. Panlilio AL, Cardo DM, Grohskopf LA, Heneine W, Ross CS. Updated U.S. Public Health Service guidelines for the management of occupational exposures to HIV and recommendations for postexposure prophylaxis. *MMWR Recomm Rep* 2005; 54(RR-9):1–17.

25. Centers for Disease Control and Prevention. Recommendations for preventing transmission of human immunodeficiency virus and hepatitis B virus to patients during exposure-prone invasive procedures. *MMWR Morb Mortal Wkly Rep* 1991;40(RR-8):1–9.

26. Department of Health (UK). Hepatitis B Infected Health Care Workers: Guidance on Implementation of Health Service Circular 2000/020. London; 2000. http://www.dh.gov.uk/en/Publicationsandstatistics/Publications/index.htm. Accessed October 6, 2009.

27. Department of Health (UK). Hepatitis B infected healthcare workers and antiviral therapy. London; 2007. Updated March 2007. http://www.dh.gov.uk/en/Publicationsandstatistics/Publications/index.htm. Accessed October 6, 2009.

28. Gunson RN, Shouval D, Roggendorf M, et al. Hepatitis B virus (HBV) and hepatitis C virus (HCV) infections in health care workers (HCWs): guidelines for prevention of transmission of HBV and HCV from HCW to patients. *J Clin Virol* 2003;27(3):213–230.

29. Buster EH, van der Eijk AA, Schalm SW. Doctor to patient transmission of hepatitis B virus: implications of HBV DNA levels and potential new solutions. *Antiviral Res* 2003;60(2): 79–85.

30. AIDS/Tuberculosis Subcommittee of the Society for Healthcare Epidemiology of America. Management of healthcare workers infected with hepatitis B virus, hepatitis C virus, human immunodeficiency virus, or other bloodborne pathogens. *Infect Control Hosp Epidemiol* 1997;18(5):349–363.

31. Rhame FS, Pitt H, Tapper ML, et al. Position paper: the HIV-infected health care worker. *Infect Control Hosp Epidemiol* 1990;11:647–656.

32. Henderson DK. Managing occupational risks for hepatitis C transmission in the healthcare setting. *Clin Microbiol Rev* 2003;16(3):546–568.

33. Henderson DK, Fahey BJ, Willy ME, Schmitt JM, Carey K, Koziol DE. Risk for occupational transmission of human immunodeficiency virus type 1 (HIV-1) associated with clinical exposures: a prospective evaluation. *Ann Intern Med* 1990; 113:740–746.

34. Henderson DK. Human immunodeficiency virus in the health-care setting. In: Mandell GL, Bennett JE, Dolin R, eds. *Principles and Practice of Infectious Diseases.* 7th ed. New York: Elsevier; 2009 (in press).

35. Henderson DK, Gerberding JL. Healthcare worker issues, including occupational and nonoccupational postexposure management. In: Dolin R, Masur H, Saag MS, eds. *AIDS Therapy.* 2nd ed. New York: Churchill Livingstone; 2002:327–346.

36. Centers for Disease Control and Prevention. Guidelines for preventing the transmission of *Mycobacterium tuberculosis* in health-care facilities. *MMWR Morb Mortal Wkly Rep* 1994;43(RR-13):1–132.

37. Pavel S, Burty C, Alcaraz I, et al. Severe liver toxicity in post-exposure prophylaxis for HIV infection with a zidovudine, lamivudine and fosamprenavir/ritonavir regimen. *AIDS* 2007;21(2):268–269.

Suggested Reading

Alexander EM, Travis S, Booms C, et al. Pertussis outbreak on a neonatal unit: identification of a healthcare worker as the likely source. *J Hosp Infect* 2008;69(2):131–134.

American College of Occupational and Environmental Medicine (ACOEM). Pertussis vaccination of health care workers. *J Occup Environ Med* 2007;49(6):700–702.

Atkinson WL, Markowitz LE, Adams NC, Seastrom GR. Transmission of measles in medical settings—United States, 1985–1989. *Am J Med* 1991;91(3B):320S–324S.

American College of Occupational and Environmental Medicine (ACOEM). ACOEM guidelines for protecting health care workers against tuberculosis. *J Occup Environ Med* 1998; 40(9):765–767.

Barnhart S, Sheppard L, Beaudet N, Stover B, Balmes J. Tuberculosis in health care settings and the estimated benefits of engineering controls and respiratory protection. *J Occup Environ Med* 1997;39(9):849–854.

Benenson AS, ed. *Control of Communicable Diseases in Man.* 16th ed. Washington, DC: American Public Health Association; 1995.

Christini AB, Shutt KA, Byers KE. Influenza vaccination rates and motivators among healthcare worker groups. *Infect Control Hosp Epidemiol* 2007;28(2):171–177.

Croft DR, Sotir MJ, Williams CJ, et al. Occupational risks during a monkeypox outbreak, Wisconsin, 2003. *Emerg Infect Dis* 2007;13(8):1150–1157.

Daskalaki I, Hennessey P, Hubler R, Long SS. Resource consumption in the infection control management of pertussis exposure among healthcare workers in pediatrics. *Infect Control Hosp Epidemiol* 2007;28(4):412–417.

Diodati C. Mandatory vaccination of health care workers. *CMAJ* 2002;166(3):301–302.

Dworsky ME, Welch K, Cassady G, Stagno S. Occupational risk for primary cytomegalovirus infection among pediatric health-care workers. *N Engl J Med* 1983;309(16):950–953.

Gundlach DC. Protecting health care workers from the occupational risk of disease. *QRB Qual Rev Bull* 1988;14(5): 144–146.

Hallak KM, Schenk M, Neale AV. Evaluation of the two-step tuberculin skin test in health care workers at an inner-city medical center. *J Occup Environ Med* 1999;41(5):393–396.

Jarvis WR, Bolyard EA, Bozzi CJ, et al. Respirators, recommendations, and regulations: the controversy surrounding protection of health care workers from tuberculosis. *Ann Intern Med* 1995;122(2):142–146.

Larsen NM, Biddle CL, Sotir MJ, White N, Parrott P, Blumberg HM. Risk of tuberculin skin test conversion among health care workers: occupational versus community exposure and infection. *Clin Infect Dis* 2002;35(7):796–801.

Lewy R. Organization and conduct of a hospital occupational health service. *Occup Med* 1987;2(3):617–638.

Maunder R, Hunter J, Vincent L, et al. The immediate psychological and occupational impact of the 2003 SARS outbreak in a teaching hospital. *CMAJ* 2003;168(10):1245–1251.

Moore RM Jr, Kaczmarek RG. Occupational hazards to health care workers: diverse, ill-defined, and not fully appreciated. *Am J Infect Control* 1990;18(5):316–327.

Nardell EA. Issues facing TB control (4.1). Nosocomial tuberculosis transmission—problems of health care workers. *Scott Med J* 2000;45(5 Suppl):34–37.

O'Reilly FW, Stevens AB. Sickness absence due to influenza. *Occup Med (Lond)* 2002;52(5):265–269.

Polder JA, Tablan OC, Williams WW. Personnel health services. In: Bennett JV, Brachman PS, eds. *Hospital Infections.* 3rd ed. Boston: Little, Brown; 1992:31–61.

Sepkowitz KA. Occupationally acquired infections in health care workers. Part I. *Ann Intern Med* 1996;125(10):826–834.

Sepkowitz KA. Occupationally acquired infections in health care workers. Part II. *Ann Intern Med* 1996;125(11): 917–928.

Simard EP, Miller JT, George PA, et al. Hepatitis B vaccination coverage levels among healthcare workers in the United States, 2002–2003. *Infect Control Hosp Epidemiol* 2007;28(7): 783–790.

Swinker M. Occupational infections in health care workers: prevention and intervention. *Am Fam Physician* 1997;56(9): 2291–2300, 2303–2306.

Thomas DL, Factor SH, Kelen GD, Washington AS, Taylor E Jr, Quinn TC. Viral hepatitis in health care personnel at The Johns Hopkins Hospital: the seroprevalence of and risk factors for hepatitis B virus and hepatitis C virus infection. *Arch Intern Med* 1993;153(14):1705–1712.

Vagholkar S, Ng J, Chan RC, Bunker JM, Zwar NA. Healthcare workers and immunity to infectious diseases. *Aust N Z J Public Health* 2008;32(4):367–371.

Valenti WM. Infection control and the pregnant health care worker. *Am J Infect Control* 1986;14(1):20–27.

Weber DJ, Rutala WA. Risks and prevention of nosocomial transmission of rare zoonotic diseases. *Clin Infect Dis* 2001; 32(3):446–456.

Wharton M, Cochi SL, Hutcheson RH, Schaffner W. Mumps transmission in hospitals. *Arch Intern Med* 1990; 150(1): 47–49.

Wicker S, Rabenau HF, Gottschalk R, Doerr HW, Allwinn R. Seroprevalence of vaccine preventable and blood transmissible viral infections (measles, mumps, rubella, polio, HBV, HCV and HIV) in medical students. *Med Microbiol Immunol* 2007;196(3):145–150.

Zimmerman RK, Middleton DB, Smith NJ. Vaccines for persons at high risk due to medical conditions, occupation, environment, or lifestyle, 2003. *J Fam Pract* 2003;52(1 Suppl): S22–S35.

Chapter 25 Tuberculosis Infection Control in Healthcare Settings

Henry M. Blumberg, MD

Tuberculosis has emerged as an enormous global public health problem. The World Health Organization (WHO) estimates that there are more than 9 million new tuberculosis cases each year and nearly 2 million deaths each year due to tuberculosis.[1] The vast majority of cases occur in low- and middle-income countries; only approximately 3% of all tuberculosis cases occur in the Americas. Tuberculosis is spread from person to person by means of airborne droplet nuclei. Interestingly, tuberculosis has been recognized and accepted by the medical community as a potential occupational hazard only for several decades.[2] The risk of transmission of *Mycobacterium tuberculosis* from patients with tuberculosis disease to other hospitalized patients and healthcare workers (HCWs) was established by the 1950s,[2] when, as noted by Myers et al.,[3] "a rapid decline of tuberculosis in the general population made the disease among physicians more conspicuous." With the introduction of effective chemotherapy for tuberculosis in the 1950s and a progressive decline in the incidence of tuberculosis in the United States until the mid-1980s, the risk of occupational infection and clinical tuberculosis declined among US HCWs. There were only scattered reports of hospital outbreaks in the 1960s, 1970s, and early 1980s.[4-7] With this decline, less and less attention was paid toward tuberculosis infection control measures in hospitals. Thus few healthcare facilities were prepared for the changing epidemiology of tuberculosis in the mid-1980s and early 1990s.

Between 1985 and 1992, there was a resurgence of tuberculosis in the United States, with a 20% increase in the number of reported cases. This resurgence was fueled by the decay of the public health infrastructure (because of underfunding) and the epidemic of human immunodeficiency virus (HIV) infection.[8-10] The surge of cases, combined with neglect of tuberculosis infection control activities and ineffective control measures, led to a number of reports of nosocomial transmission of tuberculosis in the late 1980s and early 1990s.[11-22] A number of these explosive and devastating outbreaks involved transmission of multidrug-resistant (MDR) strains of *M. tuberculosis* (with resistance to at least both isoniazid and rifampin) to patients and HCWs that was associated with significant morbidity and mortality, especially among persons infected with HIV and other immunocompromised persons. Devastating outbreaks of MDR tuberculosis disease (MDR-TB) and extensively drug-resistant tuberculosis (XDR-TB) have also been reported from lower- and middle-income countries.[17,22-24] XDR-TB is defined as disease caused by a strain of *M. tuberculosis* that has resistance to isoniazid and rifampin, to a fluoroquinolone, and to a second-line injectable agent (such as amikacin, capreomycin, or kanamycin). In many if not most lower- and middle-income countries, there has been little attention paid to tuberculosis infection control measures.[1,25] The reports of the emergence of XDR-TB, nosocomial transmission, and high mortality due to XDR-TB among HIV-infected persons have emphasized the need for healthcare institutions throughout the world to implement effective tuberculosis infection control measures. High rates of latent tuberculosis infection (LTBI) among HCWs in lower- and middle-income countries suggest that nosocomial transmission is ongoing and that tuberculosis remains a very important occupational risk for HCWs in these regions.[26-29]

Several factors contributed to outbreaks of tuberculosis in hospitals (Table 25-1 and Table 25-2). Breaks in basic tuberculosis infection control strategies, such as delays in suspecting and diagnosing tuberculosis, delays in identification of drug resistance, and delays in initiation of appropriate therapy, have postponed the proper isolation of patients and prolonged the infectiousness of patients. In addition, clustering of patients with unsuspected tuberculosis (or for many lower- and middle-income countries with known infectious pulmonary tuberculosis) with susceptible immunocompromised patients (most often HIV-infected persons) has facilitated nosocomial transmission of tuberculosis. Second, engineering controls are often inadequate. For example, studies have found airborne infection (respiratory) isolation rooms with positive rather than negative air pressure, that have air recirculated from them to other areas, and that have their doors left open, as well as isolation precautions that are discontinued too soon and HCWs who do not wear adequate respiratory protection.[31]

Implementation of effective tuberculosis infection control measures recommended in the Centers for Disease Control and Prevention (CDC) guidelines for preventing the transmission of *M. tuberculosis*[32] and the decreasing incidence of tuberculosis in the community since 1992 led to a dramatic decrease in the risk of nosocomial transmission of tuberculosis in healthcare facilities in the United States[9,33-35] The control of tuberculosis in healthcare facilities contributed to enhanced control in the

community, which was also facilitated by rebuilding of the public health infrastructure and the greatly expanded use of directly observed therapy. A hierarchy of tuberculosis infection control measures, including administrative controls (which are the most important), engineering controls, and use of respiratory protection, has reduced the rate of nosocomial transmission significantly and has prevented transmission of tuberculosis in healthcare settings.[9,35] In 2005, the CDC published updated guidance based on the changing epidemiology of tuberculosis since the 1990s in the United States.[35]

Despite the decreasing incidence of tuberculosis in the United States since 1992[36] and the lower risk of nosocomial transmission of tuberculosis in US healthcare facilities, tuberculosis infection control remains an important responsibility of healthcare personnel. Many reports have highlighted delays in the diagnosis of patients with tuberculosis.[37] Failure to be vigilant and recognize undiagnosed tuberculosis in patients continues to be reported and has resulted in nosocomial transmission of tuberculosis in the United States, even in an era when tuberculosis control has improved, as highlighted in reports from the CDC and others.[38,39] Unfortunately, despite guidelines by the WHO about prevention of healthcare-associated transmission of tuberculosis, including an addendum focusing on HIV infection and tuberculosis,[40,41] in the past little or no attention has been given to tuberculosis infection control measures in most lower- and middle-income countries. The WHO indicates that tuberculosis infection control is at an early stage of development in most of these countries.[1] Acquisition of tuberculosis by HCWs in lower- and middle-income countries has too often been accepted as an occupational hazard.[25] However, given the recent report of the emergence and spread of extensively drug-resistant strains of *M. tuberculosis*, including a recent outbreak in South Africa that affected both patients and HCWs and that had a 98% mortality rate among persons coinfected with HIV, tuberculosis infection control measures can no longer be ignored.[23] Control of XDR-TB will require implementation of measures to prevent institutional spread of tuberculosis.

This chapter is not intended to provide an exhaustive summary of tuberculosis infection control measures; CDC tuberculosis infection control guidelines published in 2005 outline recommendations for US healthcare facilities,[35] and the WHO has published updated recommendations that are particularly focused on resource-limited countries.[42] Rather, this chapter outlines the basic framework for developing a tuberculosis control program, including how to assess the risk of tuberculosis in a healthcare delivery setting, how to prioritize control measures on the basis of their effectiveness, and how to meet current US regulatory requirements.

Table 25-1. Factors facilitating nosocomial transmission of tuberculosis

Inefficient infection control procedures

Delayed suspicion and diagnosis
- Clustering of patients who have unsuspected tuberculosis with susceptible immunocompromised patients
- Delayed recognition of tuberculosis in HIV-infected patients because of atypical presentation or low clinical suspicion, leading to misdiagnosis
- Failure to recognize and isolate patients with active pulmonary disease

Failure to recognize ongoing infectiousness of patients

Laboratory delays in identification and susceptibility testing of *Mycobacterium tuberculosis* isolates

Inadequate airborne infection (respiratory) isolation facilities and engineering controls

Lack of airborne infection (respiratory) isolation rooms
Recirculation of air from airborne infection isolation rooms to other parts of the hospital

Delayed initiation of effective antituberculosis therapy

NOTE: HIV, human immunodeficiency virus. Adapted from Blumberg.[30]

Table 25-2. Factors that may facilitate nosocomial transmission of tuberculosis in low- and middle-income countries

Factors that increase risk for nosocomial exposure

Overwhelming numbers of tuberculosis patients and repeated exposure to AFB smear–positive tuberculosis patients

Unnecessary or prolonged hospitalization of AFB smear–positive tuberculosis patients

Delays in initiating antituberculosis treatment for those with active tuberculosis disease

Poor adherence to treatment, use of suboptimal treatment regimens, and lack of adequate patient support to improve adherence

Interruptions in the supply of tuberculosis medications in healthcare facilities

Lack of effective infection control procedures

Failure to recognize and isolate patients with active pulmonary tuberculosis

Laboratory delays in identification of tuberculosis, and poor use of tests, such as sputum microscopy, to identify infectious tuberculosis cases

Clustering of patients who have tuberculosis with susceptible and vulnerable patients (eg, HIV-infected patients)

Lack of HIV testing services and delayed recognition of tuberculosis in HIV-infected patients because of atypical presentation or low level of clinical suspicion

Inadequate airborne infection (respiratory) isolation facilities and engineering controls

Overcrowded hospital wards and outpatient departments

Poorly ventilated wards and rooms

Lack of airborne infection isolation rooms

Lack of personal protection equipment (eg, respirators)

Lack of screening programs to detect and treat latent tuberculosis infection and active tuberculosis among healthcare workers

Lack of commitment on the part of hospitals to invest in infection control programs

Lack of national guidelines on nosocomial tuberculosis tailored to the local country's healthcare environment

Gaps in knowledge and awareness

Lack of awareness about nosocomial tuberculosis transmission in healthcare settings

Healthcare workers' belief that nosocomial infection is an occupational hazard that cannot be avoided

Lack of educational programs on occupational safety and hygiene

Poor patient education regarding cough etiquette and sputum disposal

NOTE: AFB, acid-fast bacilli; HIV, human immunodeficiency virus. Adapted from Pai et al.[25]

Institutional Controls for the Prevention of Nosocomial Tuberculosis

Nosocomial tuberculosis is driven by the occurrence of disease in the community served by the hospital or healthcare system[34] and the efficacy of tuberculosis infection control measures instituted by a healthcare institution. Rates of tuberculosis in the United States have decreased significantly since 1992[36] but still vary widely by geographic area. Frequently, inner-city areas[43] and areas that have large numbers of foreign-born persons from areas where tuberculosis is endemic have the highest rates of tuberculosis disease. There are large disparities among US-born persons with tuberculosis; rates are 8 times higher among US-born African Americans than among white persons.[36] The global epidemic of tuberculosis has had a significant impact upon the United States: as nearly 60% of cases in the United States occurred among foreign-born persons in 2008; in addition, foreign-born persons accounted for more than 80% of MDR-TB cases in the United States in 2007.[36]

An effective tuberculosis infection control program requires early identification, use of airborne infection isolation precautions, and effective treatment of persons with active tuberculosis disease.[32] The importance of an effective tuberculosis infection control program is highlighted by the devastating outbreaks of tuberculosis, including outbreaks of infection with MDR-TB, that occurred in the late 1980s and early 1990s in the United States and elsewhere and that were associated with high rates of morbidity and mortality, especially among HIV-infected and immunocompromised patients and HCWs.[11-22] The devastating nature of the outbreak of XDR-TB infection in South Africa[23] again reinforces the importance of tuberculosis infection control measures throughout the world, including in lower- and middle-income countries, which have the highest incidence of tuberculosis disease. The termination of outbreaks in the United States and the prevention of nosocomial transmission of tuberculosis followed implementation of effective infection control programs.[9,33-35,44,45] Policies and procedures regarding tuberculosis infection control that reflect the risk and the patient population served should be developed by all healthcare facilities. All healthcare settings need a tuberculosis infection control program designed to detect disease early and to isolate or geographically separate patients with known or suspected tuberculosis and promptly refer or treat those who have tuberculosis disease. The major goals of a tuberculosis infection control program are outlined in Table 25-3.

Assignment of Responsibility

The first step in establishing an effective tuberculosis infection control program is for the institution to assign responsibility to a specific person or persons and to ensure they have the authority and support to implement such a program. The person or persons should have expertise or access to expertise in the areas of infection

Table 25-3. Major goals in the control and prevention of nosocomial tuberculosis

Place patients under airborne infection isolation precautions as soon as tuberculosis is suspected, whether during emergency care or on admission to the institution

Start empirical antituberculosis therapy as soon as tuberculosis is suspected, with an appropriate regimen that includes at least 2 drugs to which the organism is likely to be susceptible (generally, a 4-drug initial regimen of rifampin, isoniazid, pyrazinamide, and ethambutol)

Comply with isolation procedures during the patient's hospitalization until laboratory and clinical evidence eliminates the possibility of tuberculosis or the risk of transmission

Conduct laboratory studies as soon as possible to confirm or exclude the presence of tuberculosis and to identify multidrug-resistant strains of *Mycobacterium tuberculosis*

Enhance occupational health services to monitor for tuberculosis infection and disease in healthcare workers

Discharge tuberculosis patients from acute care only when they are no longer infectious or when arrangements have been made for appropriate isolation from contact with susceptible individuals (eg, in a stable home or another stable location where no new persons will be exposed)

Cooperate closely with public health and other community agencies to provide resources that ensure the completion of therapy (eg, directly observed therapy)

Provide education to healthcare workers on tuberculosis to support the goals stated above

NOTE: Adapted from McGowan[34] with additional information from Blumberg.[30]

control and healthcare epidemiology, public health, employee health, engineering, and clinical microbiology. Frequently this responsibility is given to an institution's Infection Control Committee. The group should develop written tuberculosis infection control policies on the basis of the institution's risk assessment. Policies and procedures should be reviewed at least annually and updated as indicated. At large institutions located in urban areas that care for sizable numbers of patients with tuberculosis, it has been very useful to designate an individual (eg, one of the infection preventionists) to serve as the coordinator of tuberculosis infection control activities.

Risk Assessment

Tuberculosis is not evenly distributed among the population; it is more common among foreign-born persons, inner-city residents, ethnic and racial minorities, and homeless and indigent persons. Additionally, the incidence of tuberculosis can vary widely by geographic area. Even within a defined metropolitan area, tuberculosis case rates can vary widely.[43] Therefore a uniform, "one-size-fits-all" approach is not appropriate, and the measures implemented should reflect the institution's risk assessment.[34] All healthcare facilities should conduct regular, periodic (at least annual) tuberculosis risk assessments, regardless of whether patients with suspected or confirmed tuberculosis disease will receive care at the institution.

The tuberculosis risk assessment determines the risk of nosocomial transmission of *M. tuberculosis* in the healthcare setting by examining a number of factors, including the following: (1) the incidence of tuberculosis disease in the community; (2) the number of patients with tuberculosis presenting for care at the healthcare facility, regardless of whether they receive care in the facility or are transferred to another healthcare facility; (3) the timeliness of the recognition, isolation, and evaluation of patients with suspected or confirmed tuberculosis; and (4) evidence of transmission of *M. tuberculosis* in the facility. Local or state public health departments can help infection control personnel obtain information about their community's tuberculosis profile. Other sources of information on tuberculosis cases include extended-care facilities, schools, homeless shelters, and prisons. Even if there are no reported cases of tuberculosis in a community, infection control staff still should determine if patients with tuberculosis may have been admitted to or treated in the facility. Good sources for this information are the microbiology laboratory's database, infection control records, and medical records databases that contain discharge diagnoses, autopsy, and surgical pathology reports.

The CDC has recommended using a risk classification system for healthcare facilities that is based on the institution and the number of persons with active tuberculosis disease seen at the institution; this system has categories for low risk, medium risk, and ongoing transmission.[35] In general, a risk classification is determined for the entire institution, although in certain circumstances, such as a large healthcare organization that encompasses several sites, specific areas can be defined according to geography, functional units, or location. For hospitals with 200 or more beds, those that provide care for fewer than 6 patients with tuberculosis per year are considered to be at low risk, and those that care for 6 or more patients with tuberculosis per year are considered to be at medium risk. For inpatient settings with fewer than

200 beds, those that provide care for fewer than 3 patients with tuberculosis per year are considered to be at low risk, and those that care for 3 or more patients with tuberculosis per year are considered to be at medium risk. For outpatient clinics, outreach programs, and home healthcare settings, those that provide care to fewer than 3 patients with tuberculosis per year are considered to be at low risk, and those that provide care for 3 or more patients with tuberculosis are considered to be at medium risk. Tuberculosis clinics and outreach programs, as well as other outpatient settings where care of persons with tuberculosis is provided, should be considered to be at medium risk. Any institution, clinic, or setting with evidence of transmission of *M. tuberculosis* from patient to patient or from patient to HCW or evidence of ongoing nosocomial transmission of tuberculosis should be considered to have potential ongoing transmission, until appropriate infection control measures have been implemented and transmission has been demonstrated to have been stopped. "Potential ongoing transmission" should be a temporary classification only. When nosocomial transmission of tuberculosis is suspected, an immediate investigation should be undertaken, and active and corrective steps should be implemented. This may include consultation with public health officials or experts in healthcare epidemiology and infection control. Evidence of potential nosocomial transmission of tuberculosis includes clusters of new positive results of tests for latent tuberculosis infection (tuberculin skin test [TST] or interferon-γ release assays [IGRA], including the QuantiFERON-TB Gold In-Tube test [Cellestis] or the TSPOT.TB test [Oxford Immunotec]) among HCWs, increased rates of TST or IGRA conversions for HCWs, an HCW with potentially infectious tuberculosis, unrecognized disease in patients or HCWs, or recognition of an identical strain of *M. tuberculosis* in more than 1 patient or HCW with tuberculosis disease.

Based on the finding of the risk assessment, the appropriate administrative, environmental, and respiratory protection policies to prevent occupational exposure to and nosocomial transmission of tuberculosis can be determined. The frequency of diagnostic testing for latent tuberculosis infection (with either the TST or an IGRA) among HCWs is also based on the finding of the risk assessment; it is discussed in additional detail below.

Hierarchy of Tuberculosis Infection Control Measures

A "hierarchy of controls" is recommended by the CDC to prevent nosocomial transmission of tuberculosis,[35] which includes administrative controls, engineering controls, and respiratory protection (Table 25-4). Implemen-

Table 25-4. Hierarchy of tuberculosis infection control measures

1. Administrative controls (most essential component)

Careful screening of patients, isolation, early diagnosis, and treatment
Healthcare worker–directed measures
Comprehensive tuberculin skin testing program for healthcare workers
Healthcare worker education

2. Environmental controls

Airborne infection isolation (ie, negative air-pressure) rooms; a single-pass ventilation system is preferred; use of high-efficiency particulate air filtration, if recirculation of air is necessary
UV germicidal irradiation (eg, in selected locations)

3. Personal respiratory protection equipment (including use of N-95 respirators)

NOTE: From the Centers for Disease Control and Prevention.[35]

tation of this hierarchy of controls has been noted to be effective in terminating outbreaks and in preventing nosocomial transmission of tuberculosis.[9,33] An infection control program should achieve the following goals: early identification of patients with tuberculosis disease, prompt implementation of airborne infection isolation (AII), and prompt diagnosis and effective treatment of persons with active disease (or rapid transfer of the patient to another facility that treats patients with tuberculosis, if the admitting facility does not). The specific control measures can be prioritized on the basis of their relative effectiveness in reducing risk of transmission and are discussed below.

Administrative Controls

Administrative controls are the most important tuberculosis infection control measures and consist of measures to reduce the risk of exposure to persons with infectious tuberculosis (Table 25-4). A healthcare facility should implement administrative controls first, because these controls most effectively reduce the risk of nosocomial transmission.[9,33,46,47] Early identification and implementation of airborne infection isolation is the key to controlling and preventing nosocomial tuberculosis. As noted above, nosocomial transmission has occurred primarily because of the failure to recognize and isolate patients with infectious tuberculosis. Administrative controls include development and implementation of effective policies and protocols, to ensure that persons likely to have tuberculosis disease are identified rapidly, isolated properly, evaluated clinically, and treated appropriately. It requires that HCWs carefully evaluate patients at their initial encounter and promptly isolate any patient who they suspect may have tuberculosis until laboratory and clinical evidence eliminates this diagnosis. Hospitals can implement an early identification and isolation protocol more efficiently by authorizing both nurses and physicians to isolate patients with suspected tuberculosis and by developing policies that allow staff to isolate certain patients automatically (eg, patients for whom tuberculosis is in the differential diagnosis or from whom specimens are ordered for acid-fast bacilli [AFB] smear or culture).[33] Many institutions have implemented policies that include mandatory use of airborne infection (respiratory) isolation for certain patients, to facilitate the success of administrative controls.[2,33,48] Moreover, because patients with HIV infection may present with atypical signs and symptoms, some facilities isolate all patients with HIV infection who have clinical symptoms suggestive of tuberculosis (eg, fever, cough, and/or abnormal chest radiograph findings) until AFB smears or cultures of appropriate specimens yield negative results. For example, at Grady Memorial Hospital (Atlanta, Georgia), which cares for large numbers of patients with tuberculosis, including those who are HIV infected, the respiratory isolation policy requires that all patients admitted to the hospital with known tuberculosis, all patients who have tuberculosis in the differential diagnosis or for whom sputum or respiratory specimens for AFB are ordered, and all patients who are HIV infected and have an abnormal chest radiograph finding be placed under airborne infection isolation precautions until the diagnosis of tuberculosis is excluded or "ruled out." Generally, the diagnosis is excluded if there are negative results for 3 AFB smears of sputum or other respiratory specimens. Airborne infection isolation precautions policies and procedures should be developed on the basis of the local epidemiology of the disease in the community served by a particular facility.

The protocol for early identification of patients with tuberculosis and the definition of "suspected case" will determine the number of isolation rooms required. It should be anticipated that some patients who do not have disease will be isolated (ie, there will be overisolation) to prevent nosocomial transmission. The degree of overisolation will depend on the institution's policy and the prevalence of tuberculosis disease in the community and in the patient population served by the institution. At our institution, the "rule out" ratio of patients isolated to patients found to have tuberculosis disease is approximately 10:1.[48] In Iowa, a Midwestern state with a low prevalence of tuberculosis disease, a group of investigators predicted that as many as 93 patients would be isolated without tuberculosis for every patient isolated with tuberculosis diagnosed.[49] The expected "rule out" ratio is not well defined and likely varies by

geographic area on the basis of the prevalence of tuberculosis in the community served and at the facility. However, because there is little or no margin of error in the detection of persons with tuberculosis disease, in that undiagnosed infection in a single person can lead to an outbreak,[32,35,38] a high sensitivity is required, and therefore some degree of overisolation is to be expected. At large institutions, increased efficiency in the evaluation of patients who subsequently have tuberculosis "ruled out" has been demonstrated by clustering airborne infection isolation rooms on a respiratory isolation ward.[48] This enhanced efficiency can provide significant cost savings to the institution and better use of airborne infection isolation rooms, which are often in limited supply. In addition, it should be noted that individuals with suspected or known infectious tuberculosis should wear a surgical mask when not in an airborne infection isolation (negative air-pressure) room or a local exhaust ventilation enclosure (eg, when transported to undergo a procedure or diagnostic test).[50] The purpose of the surgical mask is to block aerosols produced by coughing, talking, and breathing. In general, the time out of an airborne infection isolation room should be minimized, and when infectious or potentially infectious patients are required to leave the room, they should be monitored to ensure compliance with wearing a mask.

Surveillance for LTBI

Surveillance for LTBI in HCWs is a component of the administrative controls. The appropriate frequency of performing diagnostic tests (either TST or IGRA) for HCWs is determined by the risk assessment described above. Given the low positive predictive value of diagnostic tests when testing populations at low risk for LTBI and with a low prevalence of LTBI,[9,35,51] frequent testing of HCWs in low-incidence and low-risk settings is not recommended, because it will lead to false-positive results. In fact, testing of persons at very low risk can result in the majority of positive test results being falsely positive. Institutions also need to recognize that false-positive TST results have occurred when institutions have switched brands of purified protein derivative (PPD) reagent, for example from Tubersol (Sanofi Pasteur) to Aplisol (JHP Pharmaceuticals).[52] All HCWs should undergo baseline testing with a test diagnostic for LTBI (TST or IGRA). For those who undergo TST, 2-step testing is recommended at the time of employment if the HCW has not been previously tested in the preceding year.[35] Two-step baseline TST can help infection control staff identify LTBI in new personnel who otherwise would be classified as having recent TST conversion. Two-step testing is not required if an IGRA is used.

It is not recommended that HCWs in low-risk settings (as determined by the risk assessment) undergo routine periodic follow-up testing; follow-up testing is recommended only if there is an exposure to a patient with active tuberculosis (ie, a patient not initially isolated but later found to have laryngeal or pulmonary tuberculosis). HCWs working at medium-risk settings should undergo baseline and annual testing, as well as testing after a tuberculosis exposure episode. Institutions with ongoing nosocomial transmission should perform diagnostic testing for LTBI for at-risk HCWs every 3 months until it is documented that transmission has been terminated.[35] Surveillance for TST (or IGRA) conversions is one way to assess the efficacy of an infection control program (eg, in medium-risk settings); a cluster of conversions may be the first indication of ongoing nosocomial transmission. In addition, such surveillance is a mechanism to demonstrate termination of transmission in situations where there has been ongoing nosocomial transmission.

When tuberculin skin testing of HCWs is performed, the Mantoux method should be used. PPD is injected intradermally (0.1 mL of 5 tuberculin units), and the diameter of the induration is recorded in millimeters 48–72 hours after placement.[51] HCWs with a positive TST or IGRA result (either at baseline or at follow-up testing) should have a chest radiograph performed to exclude the diagnosis of active disease. If there is an abnormal chest radiograph finding, the HCW should be removed from the work setting until the diagnosis of active tuberculosis disease is excluded. If an HCW with normal chest radiograph findings is determined to have LTBI and is at increased risk for progression to active disease (eg, has recent conversion or an underlying medical condition),[51] the HCW should be strongly encouraged to take and complete therapy for LTBI (see the section on treatment, below).

The infection control staff, working closely with employee health clinic staff, should consider a number of important issues when developing a program for diagnostic testing for LTBI among HCWs. Institutions should assume responsibility for surveillance and should mandate testing of all HCWs working at a particular institution (eg, all paid and unpaid staff, including students, agency nurses, residents, attending physicians, volunteers, and others), not just employees. This is particularly important in the current era of "outsourcing," when many HCWs may not be employees of the institution they are working at. For institutions where routine follow-up testing is warranted (eg, medium-risk facilities), diagnostic test results should be recorded in the individual HCW's health record and in an aggregate database of all results. TST (and/or IGRA) conversion rates should be calculated for the facility as a whole and, if appropriate, for specific areas of the facility and for occupational groups. Conversion rates should be calculated by dividing the number of HCWs who have TST (or IGRA) conversion in each area of the facility or group (ie, the numerator) by the total number of HCWs who previously had negative TST (or IGRA) results in each area of the facility or group (ie, the denominator). In collaboration with employee health clinic staff, infection control personnel should interpret TST (or IGRA) conversion rates. If the number of workers in a particular area is small, the conversion rate may be high, although the actual risk may not be higher than it is in other areas of the facility. In contrast, statistical analysis may miss significant problems if the number of workers is small. If HCWs become TST or IGRA positive (ie, have conversion), infection control staff should investigate to determine whether the likely source is in the facility or in the community. Of note, HCWs in some facilities are more likely to be exposed to tuberculosis in the community than in the hospital[53-55]; this may particularly be the case following implementation of effective tuberculosis infection control measures. One challenge of TST programs is to ensure that staff report to the employee health staff for TST placement and for follow-up assessment. Some facilities have improved compliance by offering TST testing at the work site, thereby removing the time and distance barriers and increasing peer pressure. Others have tied undergoing TST to issuance of employee identification badges required to work at the facility and to credentialling of physicians.

For more than 100 years, the TST had been the only diagnostic test available for LTBI. Given the limitations of the TST,[9,51] the development of more effective diagnostic tests is urgently needed.[56] Recently, there has been the development of T cell–based IGRAs that measure T-cell release of interferon-γ in response to stimulation with relatively tuberculosis-specific antigens, such as early secreted antigenic target 6 (ESAT6) and culture filtrate protein 10 (CFP10), using methods such as the enzyme-linked immunosorbent assay (ELISA) or the enzyme-linked immunospot (ELISPOT) assay.[57] Two IGRAs approved by the Food and Drug Administration are available in the United States and also are currently registered for use in Canada: the QuantiFERON-TB Gold In-Tube assay (Cellestis), and the T-SPOT.TB assay (Oxford Immunotec). The CDC has published guidelines for use of a previous version of the QuantiFERON test,[58] and it is anticipated that updated recommendations on the 2 IGRAs approved by the Food and Drug Administration will be published before the end of 2009. The Canadian Tuberculosis Committee published guidelines on the use of these 2 IGRAs in October 2008.[59] The CDC has recommended that the IGRAs can be used in place of the TST, including for contact investigations and testing of HCWs.[58] A major advantage of the IGRAs is that they do not cross-react with bacillus Calmette-Guérin and most nontuberculous mycobacteria. The specificity of the

IGRAs is much greater than the TST among foreign-born persons who have had BCG vaccination,[60,61] a group frequently encountered among HCWs. However, there are very limited data on the use of IGRAs for serial testing of HCWs and no definition of what constitutes a conversion using the IGRAs (only a static cutoff value to distinguish a positive from a negative result). This has the potential to lead to conversions or reversions for persons who have a result near the cut-off value for a positive IGRA result.[61,62] This has led the Canadian Tuberculosis Committee to recommend against the routine use of IGRAs for serial testing of HCWs,[59] although they suggest that IGRAs may be used as a confirmatory test if a TST result for an HCW at low risk is suspected of being falsely positive. Clearly, more data are needed on the utility of IGRAs for serial testing, as well as definitions for what constitutions an IGRA conversion.

Education

HCW education is an important component of an effective tuberculosis infection control program.[35] HCWs should receive training and education on the variety of components of an effective tuberculosis infection control program and their responsibilities in implementing and carrying out the institution's infection control plan. HCWs need to appreciate the risk of occupational exposure to patients with tuberculosis, as well as the measures (eg, the hierarchy of controls) and policies adopted by the healthcare facilities to prevent nosocomial transmission of tuberculosis. Tuberculosis education should be provided at the time of employment and then subsequently each year. Basic information should be provided to all HCWs, and more in-depth education and training can be targeted to HCWs working in areas or settings where patients who are at risk or who have tuberculosis may receive care. The US Occupational Health and Safety Administration (OSHA) requires that US healthcare facilities provide annual training, and a number of institutions have incorporated tuberculosis education into OSHA-mandated training on bloodborne pathogens (see Chapter 24 on occupational health).

Long-Term Care Facilities

Many of the considerations for control of tuberculosis in hospitals apply to extended care facilities, including the risk assessment recommendations. Elderly persons who reside in a nursing home are at a higher risk of developing active tuberculosis than are those who live at home in the community.[64,65] As is the case for hospitals, effective tuberculosis control measures for extended care facilities include a high index of suspicion, prompt detection of residents with active cases, isolation of residents with infectious cases, initiation of appropriate therapy, identification and evaluation of contacts, and, when appropriate,

targeted testing for and treatment of LTBI. Generally, long-term care facilities do not have airborne infection isolation rooms, and therefore patients with suspected tuberculosis should be referred to acute care hospitals. Patients found to have tuberculosis should not be infectious at the time of admission or readmission to a long-term care facility. All residents entering long-term care facilities should have a baseline diagnostic test for LTBI performed (either a TST or an IGRA) unless they are documented to have had a positive result previously.[66] If a TST is used, 2-step testing should be performed, unless the newly admitted resident has undergone TST during the previous 12 months. Persons found to have a positive TST or IGRA result should have a chest radiograph performed, and if findings are negative, they should be evaluated for treatment of LTBI.[51] Stead[67] has published data suggesting that treating LTBI in elderly residents of nursing homes has value as a means to prevent future outbreaks.

Environmental Controls

The second level of tuberculosis controls are environmental controls that reduce or eliminate M. tuberculosis–laden droplet nuclei in the air. These controls include local exhaust ventilation, general or central ventilation, air filtration with high-efficiency particulate air (HEPA) filters, and air disinfection with UV germicidal irradiation (UVGI).

Local Exhaust Ventilation

Local exhaust ventilation is a source control method for capturing airborne contaminants, including infectious droplet nuclei or other infectious particles, before they are dispersed into the general environment. Local exhaust ventilation that uses a booth, hood, or tent can be an efficient engineering control technique, because it captures a contaminant at its source. Local exhaust ventilation should be used for cough-inducing procedures (eg, a sputum induction booth) and aerosol-generating procedures. If local exhaust ventilation is not feasible, cough-inducing and aerosol-generating procedures (eg, bronchoscopy) should be performed in a room that meets the requirements of an airborne infection isolation room.

General Ventilation

General ventilation includes mechanisms that dilute and remove contaminated air and control the direction of airflow to prevent an infectious source from contaminating the air in nearby areas. These mechanisms include maintaining negative air-pressure and circulating air to dilute and remove infectious droplet nuclei (eg, exchanging room air). Air flow should be from more clean areas to more contaminated (or less clean) areas[68,69]; thus, air should flow from corridors into airborne infection isolation rooms to prevent the spread of tuberculosis. Airborne infection isolation rooms are used to house a patient with suspected or confirmed tuberculosis who is being cared for at a healthcare facility. Airborne infection isolation rooms should have negative air-pressure to prevent the escape of droplet nuclei, and the CDC recommends a minimum of 6 air exchanges per hour (and 12 air exchanges per hour, if feasible), to decrease the concentration of infectious particles. For newly constructed or renovated facilities, a minimum of 12 air exchanges per hour for airborne infection isolation rooms is recommended by the CDC.[35] A single-pass ventilation system is the preferred choice for airborne infection isolation rooms; in such a system, after air passes through the room or area, 100% of that air is exhausted to the outside. If this is not possible, HEPA filtration must be employed to filter air from an airborne infection isolation room that is recirculated into the general ventilation system. HEPA filtration must also be used when discharging air from local exhaust ventilation booths or enclosures (eg, sputum induction booths).

The number of airborne infection isolation rooms and the location of these rooms (eg, in the wards, the emergency department, and the intensive care unit) should be determined on the basis of the risk assessment. Grouping of airborne infection isolation rooms in one area of the facility (eg, in the respiratory isolation ward) may facilitate the care of patients with suspected or proven tuberculosis[48] and the installation and maintenance of optimal environmental controls. Airborne infection isolation rooms should be checked regularly to ensure they are under negative air-pressure, which is done using smoke tubes or other devices. The CDC recommends that these rooms be checked before occupancy and daily while occupied by a patient with suspected or confirmed tuberculosis. When negative air-pressure is required, the CDC and American Institute of Architects recommend that the pressure differential be greater than 0.01 inch (0.25 mm) of water gauge, relative to adjacent areas. Detailed recommendations for designing and operating ventilation systems have been published in recent years.[35,68-70] A maintenance plan that outlines the responsibility and authority for maintenance of the environmental controls and that addresses staff training needs should be part of the written tuberculosis control plan. Standard operating procedures should include the notification of infection control personnel before performing maintenance on ventilation systems serving tuberculosis patient care areas.

Portable Air Filtration Units

Portable room-air recirculation units (which are often referred to as portable air filtration units or portable HEPA filters) have been shown to be effective in removing bioaerosols and aerosolized particles from room air,[71,72] and therefore may be helpful in reducing airborne transmission of disease. If portable devices are used, units with relatively high volumetric airflow rates that provide maximum flow through the HEPA filter are preferred. Portable HEPA units should be designed to achieve 12 or more equivalent air exchanges per hour and to ensure adequate air mixing in all areas of the rooms, and they should be compatible with the ventilation system.[35] Placement of the units is important and should be selected to optimize the recirculation of air from the airborne infection isolation room through the HEPA filter. These portable units may be useful as an interim engineering control measure. These units enable hospitals to establish tuberculosis isolation rooms in outpatient departments and in patient care areas when other tuberculosis isolation rooms are in use. In addition, facilities that do not have isolation rooms can use these units to convert general patient rooms to tuberculosis isolation rooms. The effectiveness of these portable units is affected by the room's configuration, the furniture and persons in the room, and the placement of the HEPA filtration unit relative to the supply air vent and exhaust grilles. Portable air filtration units may also include UVGI, as discussed below.

UVGI

UVGI is an air cleaning technology that can be installed in a room or corridor to irradiate the air in the upper portion of the room (upper air irradiation), installed in a duct to irradiate air passing through the duct (duct irradiation), or incorporated into room air-recirculation units. The effective use of UVGI is associated with exposure of M. tuberculosis organisms contained in a droplet to a sufficient dose of light in the UV-C range to ensure they are inactivated.[35] Germicidal lamps used in upperroom UVGI systems consist of low-pressure mercury vapor lamps enclosed in special UV transmitting glass tubes. Approximately 95% of the energy from these lamps is radiated at 253.7 nm, in the UV-C range.[73] The CDC considers UVGI to be a supplementary measure for tuberculosis control and recommends that UVGI not be used as a substitute negative air-pressure or HEPA filtration.[35] Others have advocated more vigorously to expand the role of UVGI for tuberculosis infection control.[74] A recent report from Peru indicated that upperroom UV irradiation, combined with adequate air mixing, prevented most airborne tuberculosis transmission to guinea pigs exposed to hospital room air from rooms with patients with active tuberculosis disease.[75] Further investigations involving humans are needed, but this approach might provide a relatively low-cost intervention for possible use in lower- and middle-income countries, as well as settings where other types of environmental controls are hard to implement, such as waiting rooms and other overcrowded areas in healthcare facilities.

Air-cleaning technologies, such as UVGI and HEPA filtration, can be used to increase equivalent air changes per hour in waiting areas and airborne infection isolation rooms. Air mixing, air velocity, relative humidity, UVGI intensity, and lamp configuration affect the efficacy of UVGI systems. In practical terms, it can be difficult to achieve the desired effects unless the system is properly designed. It is strongly recommended that healthcare facility managers consult a UVGI system designer to address safety and efficacy considerations before such a system is procured and installed.[35,73] Experts who can be consulted include industrial hygienists, engineers, and health physicists.

In upper-room air irradiation, UVGI lamps are suspended from the ceiling or mounted on the wall with a shield at the bottom of the lamp that directs the rays upward. As the air circulates, nonirradiated air moves from the lower part to the upper part of the room, and irradiated air moves from the upper part to the lower part of the room. For upper-air systems, airborne microorganisms in the lower, occupied areas of the room must move to the upper part of the room to be killed or inactivated by upper-room UVGI. For optimal efficacy of upper-room UVGI, relative humidity should be maintained at 60% or lower, a level that is consistent with current recommendations for providing acceptable indoor air quality and minimizing environmental microbial contamination in indoor environments.[76] The most useful places to consider for use of UVGI include areas in the facility with a high tuberculosis prevalence that are difficult to control with ventilation measures alone, such as waiting rooms, the emergency department, corridors, and other central areas of a facility where patients with undiagnosed tuberculosis could contaminate the air— including operating rooms and adjacent corridors where procedures are performed on patients with tuberculosis disease. Details about the types of UVGI, their applications, and limitations can be found in guidelines from the CDC and the National Institute for Occupational Safety and Health (NIOSH) and in other resources.[35,73]

UVGI-containing portable room air cleaners are another way that UVGI has been used in healthcare facilities. In portable room air-recirculation units that incorporate UVGI, a fan moves a volume of room air across UVGI lamps to disinfect the air before it is recirculated back to the room. Some portable units contain both a HEPA filter (or other high-efficiency filter) and UVGI lamps. One study has reported that portable room air cleaners with UVGI lamps are effective in inactivating or killing more than 99% of airborne vegetative bacteria.[60] Potential locations to use portable room air cleaners with UVGI include airborne infection isolation rooms, as a supplemental method of air cleaning, as well as waiting rooms, emergency departments, corridors, central areas, or other large areas where individuals with undiagnosed tuberculosis could potentially contaminate the air.

There are a number of health and safety issues related to the use of UVGI. For example, short-term overexposure to UV radiation can cause erythema, photokeratitis, and conjunctivitis. If UVGI is used (eg, in upper-room UVGI systems), it is important that the UVGI fixtures be designed and installed to ensure that room occupants' exposure to UV radiation is below current safe exposure levels. Health-hazard evaluations by the CDC and NIOSH have identified potential problems at some facilities using UVGI systems.[35,73] These include overexposure of HCWs to UV light and inadequate maintenance, training, labeling, and use of personal protective equipment. It is believed that, in most instances, UVGI fixtures that are properly designed, installed, and maintained provide protection from most, if not all, of the direct UV radiation in the lower part of the room.[74] When UVGI is used, it is important that these systems be monitored appropriately, as would be expected with other types of engineering controls: that responsible individuals maintain them and that HCWs receive appropriate education about issues related to UVGI safety.[35]

Personal Respiratory Protection

Personal respiratory protection is the last step in the hierarchy of tuberculosis infection control measures. It is recommended that personal respiratory equipment (eg, N-95 respirators) be used by HCWs when they enter high-risk areas where exposure to airborne *M. tuberculosis* may occur (eg, airborne infection isolation rooms and rooms where cough-producing or aerosol-producing procedures are performed, including the bronchoscopy suite where procedures are performed on patients with suspected or proven tuberculosis).[35] The most controversial area of tuberculosis infection control has involved personal respiratory protection, because OSHA has issued federal mandates regarding respirator fit-testing and because data are lacking; the precise level of effectiveness of personal respiratory protection in protecting HCWs from *M. tuberculosis* transmission in healthcare settings has not been determined. Prior to 1996, OSHA had mandated the use of HEPA respirators in healthcare facilities. Two cost-effectiveness analyses performed at the University of Virginia suggested that HEPA respirators would offer negligible additional efficacy at a great cost (eg, $7 million per case of tuberculosis prevented).[77,78] However, all US federal agencies involved in this issue (NIOSH, OSHA, and the CDC) are in agreement that the minimal acceptable respiratory protection is a NIOSH-certified N-95 respirator.[79]

In October 1997, OSHA published a proposed standard for occupational exposure to tuberculosis.[80] The Institute of Medicine was subsequently asked by the US Congress to evaluate the risk of tuberculosis among HCWs and the impact of the proposed OSHA tuberculosis standard. The Institute of Medicine report,[9] published in 2001, questioned the validity of the OSHA risk assessment that the standard was based on and noted that the risk of occupational exposure to tuberculosis and HCWs' risk of occupationally acquired infection had decreased significantly following the implementation of CDC-recommended tuberculosis infection control guidelines and the decreasing incidence of tuberculosis in the community. The Institute of Medicine report[9] also concluded that implementation of the CDC's 1994 tuberculosis infection control guidelines[32] was effective in terminating outbreaks and in preventing nosocomial infection of tuberculosis. A survey by the CDC and the American Hospital Association has noted that most hospitals had implemented the CDC recommended tuberculosis infection control guidelines by the mid-1990s.[81] In 2003, OSHA announced that it had decided to withdraw its 1997 proposal, because "it does not believe a standard would substantially reduce the occupational risk of tuberculosis infection."[82] Even though it did not issue a separate tuberculosis standard, OSHA maintains regulatory control over tuberculosis in healthcare settings under the Code of Federal Regulations (CFR) Title 29, Part 1910.134, and Section 5(a)(1) of the OSH Act, often referred to as "the General Duty clause."[83] The impact of this decision is that healthcare facilities are now required by OSHA to perform respirator fit-testing annually, rather than just at the time of employment, as had been the case previously.

Respirator fit-testing has been an extremely contentious issue. Observational studies have demonstrated that tuberculosis outbreaks in the United States were terminated prior to the availability or use of N-95 or HEPA respirators or the use of fit-testing.[34] Fit-testing is time consuming and logistically difficult, and it can be expensive at large institutions, which may have thousands of HCWs. There are no definitive data demonstrating the benefit of fit-testing, and recent studies by NIOSH have demonstrated a variety of problems with fit-testing. Coffey and colleagues[84] at NIOSH reported that when the most rigorous criterion of fit-testing was used (the 1% pass/fail criterion recommended by the American National Standards Institute and required by OSHA), a substantial majority of tested individuals failed the fit-test for 17 of 21 brands of N-95 respirators tested; thus most individuals could not be successfully fitted.[9] There are a number of different methodologies available for fit-testing, although in healthcare facilities, the qualitative fit method is most commonly used. In an additional investigation, Coffey et al.[85] compared 5 methods for fit-testing N-95 respirators, using both qualitative and quantitative methods. They found that there was a wide variation in results between these fit-testing methods and that none of the 5 methods met criteria for determining whether a fit-test adequately screened out poorly fitting respirators. They concluded that the accuracy of fit-testing methods and the fitting characteristics of N-95 respirators need to be improved.

In a more recent investigation, Coffey and colleagues[86] reported on the fit characteristics of 18 different models of N-95 respirators using 4 different analytical methods used to measure the performance of N-95 respirators. They found that the most important characteristic in providing protection was the inherent fitting characteristics of the N-95 respirators. Only 3 of the 18 N-95 respirators had good fitting characteristics and met the expected level of protection without fit-testing. Passing a fit-test, however, did not guarantee the wearer an adequately fitting respirator. There was little or no additional benefit of fit-testing for those models of respirators with good fitting characteristics. Use of poorly fitting respirators with fit-testing continued to be inferior to use of well-fitting respirators without fit-testing. Thus, those respirators with good fitting characteristics provided better protection "out of the box," with no fit-testing, than did respirators with poor fitting characteristics after fit-testing. These findings led the NIOSH investigators to conclude that given the "current state of fit-testing, it may be of more benefit to the user to wear a respirator model with good-fitting characteristics without fit-testing than to wear a respirator model with poor-fitting characteristics after passing a fit-test."[86(p271)] In 1995, NIOSH published new certification regulations for particulate respirators.[87] Unfortunately, there is no provision requiring good fit characteristics as part of the certification process.

OSHA requires healthcare settings in which HCWs use personal respiratory protection to develop, implement, and maintain a respiratory-protection program.[35] OSHA permits a HCW to reuse a respirator as long as it maintains its structural and functional integrity and the filter material is not damaged or soiled. Each facility should include in its tuberculosis control program policy a protocol that defines when a disposable respirator must be discarded (eg, if it becomes contaminated with blood or other body fluids). Healthcare facilities should strongly consider selecting a brand of N-95 respirator on the basis of its fit characteristics (ie, whether it has good fitting characteristics), as outlined by Coffey et al.[86] In addition to selecting N-95 respirators, each healthcare facility needs a complete respiratory protection program. The OSHA respiratory protection standard has the following requirements[35]:

- The institution must assign responsibility for the program to a specific person or group.
- The institution must write procedures for all aspects of the program.
- The institution must screen all employees for medical conditions that prevent them from wearing respirators.
- The institution must train and educate employees about respiratory protocols (and tuberculosis infection control measures).
- The institution must fit-test the respirators on each employee (annually) and have employees check the fit each time they use a respirator.
- The institution must develop policies and procedures that describe how to inspect, maintain, and reuse respirators, and must define when respirators are contaminated and must be discarded.
- The institution must evaluate the program periodically.

Despite the limitations of fit-testing,[9,84-86,88] OSHA requires that it be performed annually. A qualitative fit-testing method is generally used for disposable N-95 respirators at most healthcare facilities. This method involves exposing the employee to saccharin. It has been recommended that healthcare facilities should follow the manufacturer's instructions and recommendations for fit-testing.[35] OSHA requires that healthcare facilities screen employees to determine whether they can wear respirators. Other than severe cardiac or pulmonary disease, few medical conditions should preclude the use of disposable respirators. Many facilities use a general questionnaire to screen employees for medical conditions and to determine whether an employee should be evaluated further. Personal respiratory protection (eg, N-95 respirators) should be used by persons entering rooms in which patients with suspected or confirmed infectious tuberculosis are being isolated (eg, airborne infection isolation rooms); should be used by persons present during cough-inducing or aerosol-generating procedures performed on patients with suspected or confirmed infectious tuberculosis; and should be used by persons in other areas and circumstances in which administrative and environmental controls are not likely to protect them from inhaling infectious airborne droplet nuclei. This includes emergency medical technicians and other persons who transport patients who might have infectious tuberculosis in ambulances or other vehicles and persons who provide urgent surgical or dental care to patients who might have infectious tuberculosis. In addition, laboratory workers conducting aerosol-producing procedures involving specimens that might contain *M. tuberculosis* should also use personal respiratory protection. Detailed recommendations about the environment (including use of a biosafety cabinet and other biosafety procedures) used for performing such procedures have been published by the CDC and the National Institutes of Health.[89] It is recommended that visitors to airborne infection isolation rooms or other areas where patients with suspected or confirmed infectious tuberculosis are present should wear an N-95 respirator. Visitors can be given N-95 respirators and instructed in their use but do not need to be fit-tested.

As discussed above, OSHA's minimum requirement for respiratory protection is the N-95 respirator. However, particular situations may warrant more-protective respirators. Modeling studies have suggested that the benefits of personal respiratory protection are directly proportional to the degree of risk.[90] For example, personnel who perform extremely high-risk procedures, such as bronchoscopy on patients with known or suspected MDR-TB, may need additional respiratory protection. One example of a more-protective respirator is a powered air-purifying respirator. NIOSH has published a guide on respirators for tuberculosis that describes the types of respirators that are available.[91]

Laboratory Diagnosis

Laboratory tests (eg, AFB smear and culture) are necessary to confirm or exclude the diagnosis of tuberculosis and to identify drug-resistant isolates of *M. tuberculosis*.[92,93] If a clinical laboratory cannot perform the most rapid tests, the hospital may need to send specimens to a referral laboratory. This will become increasingly the case in areas with a low incidence.[94] The healthcare facility must ensure that arrangements comply with the CDC's guidelines for transporting specimens and reporting results (eg, AFB smear results should be reported within 24 hours after specimen collection).[95] The use of nucleic acid amplification tests may be quite useful when caring for patients coinfected with HIV, from whom it is common to recover nontuberculous mycobacteria. Given the relatively low positive predictive value of a positive AFB smear of sputum from an HIV-infected person, nucleic acid amplification tests that determine whether *M. tuberculosis* is present can help facilitate more appropriate and efficient care of the patients.[96] The CDC has published updated guidelines on the use of nucleic acid amplification tests, and reports from institutions in the United States have indicated that this test has a high sensitivity and specificity for specimens positive for tuberculosis organisms by AFB smear and culture.[96-98] Those patients who have a positive AFB smear result for a respiratory specimen but who are found to not have tuberculosis on the basis of nucleic acid amplification test results could have isolation and therapy discontinued in an expeditious fashion rather than having to wait several weeks for results of a culture.

Treatment of Tuberculosis Disease and Latent Tuberculosis Infection

Clinicians should start empirical therapy as soon as they suspect that a patient has tuberculosis disease. The current recommendation is to begin empirical therapy with a 4-drug regimen (rifampin, isoniazid, pyrazinamide, and ethambutol).[93,99] Definitive therapy depends on drug-susceptibility test results. The American Thoracic Society, the Infectious Diseases Society of America, and the CDC have published guidelines on the treatment of tuberculosis disease that provide detailed guidance.[93] Directly observed therapy is an important component of therapy and has been demonstrated to improve therapy completion rates and outcomes.[100]

Treatment of LTBI has been demonstrated to be effective in reducing the risk of progression to active disease and is recommended for those individuals at increased risk of progression, including HCWs. Recommendations for the treatment of LTBI have been published and updated[51,99,101] (Table 25-5). Isoniazid therapy for 9 months is generally the preferred regimen, and rifampin therapy for 4 months is an alternative regimen for the treatment of LTBI.

Despite the benefits of treatment for LTBI, HCWs have historically had poor rates of initiation and completion of LTBI therapy, with the majority of HCWs not initiating or completing therapy.[102-104] However, in facilities that have a comprehensive tuberculosis infection control program[105] and/or a program that has focused efforts on delivering LTBI therapy to HCWs,[106] much higher rates of initiation and completion have been reported. For example, investigators in St. Louis reported that 98% of HCWs with LTBI initiated isoniazid therapy and 82% completed therapy at Barnes-Jewish Hospital.[106] The authors of this study attributed the high initiation and completion rates to active follow-up, consisting of physician counseling and monthly telephone consultations by nurses at the institution's occupational health department, along with the provision of free services and medication. These investigators found that foreign-born HCWs who had received BCG vaccination were less likely to complete LTBI therapy and recommended addressing cultural barriers that may lead HCWs to refuse therapy and not adhere to therapy.[106] Preliminary data suggest that use of an IGRA test (which does not cross-react with bacillus Calmette-Guérin) in lieu of the TST at the time of employment may increase acceptance of therapy for LTBI by foreign-born HCWs who have a positive test result.[107] Further investigations are warranted.

Improved infection control measures and the decreasing incidence of tuberculosis since 1992 in the United States has led to a significant reduction in HCW risk.[9] After the establishment of effective tuberculosis

---

Table 25-5. Abbreviated guidelines for the treatment of latent tuberculosis infection

Drug, dosing interval and duration	Comments[a]
Isoniazid	
Once per day for 9 months[b,c]	Preferred regimen. For HIV-infected persons, isoniazid may be administered concurrently with NRTIs, NNRTIs, or protease inhibitors
Twice per week for 9 months[b,c]	DOT must be used with twice-weekly dosing
Once per day for 6 months[c]	Not indicated for HIV-infected persons, patients with fibrotic lesions on chest radiographs, or children
Twice per week for 6 months[c]	DOT must be used with twice-weekly dosing
Rifampin[d]	
Once per day for 4 months	Used for persons who are contacts of tuberculosis patients infected with an isoniazid-resistant, rifampin-susceptible strain. For HIV-infected persons, do not administer most protease inhibitors or delavirdine concurrently with rifampin. Rifabutin, with appropriate dose adjustments, can be used with some protease inhibitors and NNRTIs (except delavirdine). Consult updates on the Internet for the latest specific recommendations.

NOTE: Adapted from the Centers for Disease Control and Prevention.[101] A 6-month regimen of isoniazid is an alternative for patients who will not or cannot complete 9 months of treatment. DOT, directly observed therapy; HIV, human immunodeficiency virus; NRTI, nucleoside reverse-transcriptase inhibitor; NNRTI, nonnucleoside reverse-transcriptase inhibitor.
a Interactions with antiretroviral drugs are updated frequently and are available at http://www.aidsinfo.nih.gov/guidelines.
b Recommended regimen for persons aged less than 18 years.
c Recommended regimen for pregnant women.
d The substitution of rifapentine for rifampin is not recommended, because rifapentine's safety and effectiveness have not been established for patients with latent tuberculosis infection.

control measures in hospitals, it is the case for many HCWs that community factors pose a greater risk for infection than does occupational exposure.[53-55] At many institutions, a large proportion of HCWs are foreign born and may be found to have LTBI at the time of employment, presumably in large part because infection was acquired in their home country where the incidence of disease is high. Thus, in part, surveillance for LTBI among HCWs is part of a public health strategy for treating individuals with LTBI who may be at increased risk for progression to active disease (eg, persons who have immigrated to the United States within the past 5 years).

Discharge Planning and Collaboration With Public Health

Healthcare facilities and local and state public health officials have responsibilities to work closely with each other to further tuberculosis control in the community and state. Public health officials can provide important data to healthcare facilities regarding incidence of tuberculosis in the community, which is needed for the institution's risk assessment. It is important for healthcare facilities to establish contact with local public health authorities and report tuberculosis cases to them. All US states require that tuberculosis cases be reported, and often the physician caring for the patient is responsible for this. Frequently, infection control departments have assumed this responsibility for their facility, to ensure the reporting occurs in a timely fashion. Healthcare facilities and public health officials also need to work closely on discharge planning to ensure a seamless transition of care from an inpatient setting to an outpatient clinic (eg, the tuberculosis clinic at the patient's local health department) and to help ensure that patients are not lost to follow-up after discharge. A written policy or critical-pathway management of tuberculosis patient discharges that provides guidance as to what constitutes an appropriate transfer (for programs that do not provide care to patients with proven or suspected tuberculosis but refer them to other sites) or discharge (for sites that do provide care) should be established and included as part of a tuberculosis infection control program.[34] For example, these measures may include (1) ensuring that patients are discharged while receiving an appropriate antituberculosis regimen, (2) ensuring that there is close follow-up of the patient after discharge (eg, the patient is contacted in the hospital by the public health outreach worker who will provide directly observed therapy after hospital discharge), and/or (3) ensuring that patients meet the appropriate criteria for discharge (eg, the patient is medically ready for discharge and has a stable home or other stable location to go to, if the patient is potentially infectious).

Summary

Much progress has been made over the past 2 decades in greatly reducing the risk of occupational exposure to tuberculosis and occupationally acquired infection due to *M. tuberculosis*. The CDC recommended guidelines (ie, the hierarchy of controls) have been shown to be effective in preventing and terminating outbreaks and in preventing nosocomial transmission of tuberculosis.[9,35] The improved safety for HCWs (and patients) has resulted from a combination of improved infection control measures implemented in hospitals and a decrease in the incidence of tuberculosis in the community. The annual risk of TST conversion among HCWs has been reported to be in the range of 2–4 conversions per 1,000 person-years worked, even in high-prevalence areas in the United States.[54,108] Recommendations made in this chapter focus on tuberculosis infection control for the United States (and would be applicable to other industrialized countries), and the CDC's most recent detailed tuberculosis infection control guidelines from 2005.[35] The WHO has published guidelines for limited-resource areas.[42] The emergence of XDR-TB and devastating outbreaks among HIV-infected persons have raised awareness about the importance of infection control measures throughout the world. However, fundamental changes in how patients are cared for are needed in lower- and middle-income countries in order to adequately address tuberculosis infection control. Despite progress made in the United States over the past 2 decades, a number of controversial areas remain, especially regarding personal respiratory protection and fit-testing. It is essential that guidelines and regulatory requirements be evidence based, as much as possible, and that research continue into unresolved scientific issues. Finally, HCWs must remain vigilant. Even in an era when the incidence of tuberculosis is decreasing in the United States, failure to consider the diagnosis and take appropriate infection control measures can lead to nosocomial transmission of tuberculosis in healthcare facilities.

References

1. World Health Organization (WHO). Global tuberculosis control 2009: epidemiology, strategy, financing. Geneva: WHO; 2009. http://www.who.int/tb/publications/global_report/en/. Accessed October 14, 2009.
2. Sepkowitz KA. Tuberculosis and the health care worker: a historical perspective. *Ann Intern Med* 1994;120:71–79.
3. Myers JA, Diehl HS, Boynton RE, Horns HL. Tuberculosis in physicians. *JAMA* 1955;158:1–8.
4. Lincoln EM. Epidemic of tuberculosis. *Adv Tuberc Res* 1965;14:157–201.

5. Alpert ME, Levison ME. An epidemic of tuberculosis in medical school. *N Engl J Med* 1965;332:92–98.

6. Ehrenkranz NJ, Kirklighter JL. Tuberculosis outbreak in a general hospital: evidence for airborne spread of infection. *Ann Intern Med* 1972;77:377–382.

7. Catanzaro A. Nosocomial tuberculosis. *Am Rev Respir Dis* 1982;125:559–562.

8. Menzies D, Fanning A, Yuan L, Fitzgerald M. Tuberculosis among health care workers. *N Engl J Med* 1995;332: 92–98.

9. Institute of Medicine Committee on Regulating Occupational Exposure to Tuberculosis. Field MJ, ed. *Tuberculosis in the Workplace*. Washington, DC: National Academy Press; 2001.

10. Snider DE Jr, Roper WL. The new tuberculosis. *N Engl J Med* 1992;326:703–705.

11. Jarvis WR. Nosocomial transmission of multidrug-resistant *Mycobacterium tuberculosis*. *Res Microbiol* 1993;144: 117–122.

12. Fischl MA, Uttamchandani RB, Caikos GL, et al. An outbreak of tuberculosis caused by multidrug-resistant tubercle bacilli among patients with HIV infection. *Ann Intern Med* 1992;117:177–183.

13. Beck-Sague C, Dooley SW, Hutton MD, et al. Hospital outbreak of multidrug-resistant *Mycobacterium tuberculosis* infections: factors in transmission to staff and HIV-infected patients. *JAMA* 1992;268:1280–1286.

14. Edlin BR, Tokars JI, Grieco MH, et al. An outbreak of multidrug-resistant tuberculosis among hospitalized patients with the acquired immunodeficiency syndrome. *N Engl J Med* 1992;326:1514–1521.

15. Dooley SW, Villarino ME, Lawrence M, et al. Nosocomial transmission of tuberculosis in a hospital unit for HIV-infected patients. *JAMA* 1992;267:2632–2634.

16. Zaza S, Blumberg HM, Beck-Sague C, et al. Nosocomial transmission of *Mycobacterium tuberculosis*: role of health care workers in outbreak propagation. *J Infect Dis* 1995; 172:1542–1549.

17. Aita J, Barrera L, Reniero A, et al. Hospital transmission of multidrug-resistant *Mycobacterium tuberculosis* in Rosario, Argentina. *Medicina (B Aires)* 1996;56:48–50.

18. Jereb JA, Klevens RM, Privett TD, et al. Tuberculosis in health care workers at a hospital with an outbreak of multidrug-resistant *Mycobacterium tuberculosis*. *Arch Intern Med* 1995;155:854–859.

19. Coronado VG, Beck-Sague CM, Hutton MD, et al. Transmission of multidrug-resistant *Mycobacterium tuberculosis* among persons with human immunodeficiency virus infection in an urban hospital: epidemiologic and restriction fragment length polymorphism analysis. *J Infect Dis* 1993; 168:1052–1055.

20. Pearson ML, Jereb JA, Frieden TR, et al. Nosocomial transmission of multidrug-resistant *Mycobacterium tuberculosis*: a risk to patients and health care workers. *Ann Intern Med* 1992;117:191–196.

21. Ikeda RM, Birkhead GS, DiFerdinando GT Jr, et al. Nosocomial tuberculosis: an outbreak of a strain resistant to seven drugs. *Infect Control Hosp Epidemiol* 1995;16:152–159.

22. Ritacco V, Di Lonardo M, Reniero A, et al. Nosocomial spread of human immunodeficiency virus-related multidrug-resistant tuberculosis in Buenos Aires. *J Infect Dis* 1997;176: 637–642.

23. Gandhi NR, Moll A, Sturm AW, et al. Extensively drug-resistant tuberculosis as a cause of death in patients co-infected with tuberculosis and HIV in a rural area of South Africa. *Lancet* 2006;368:1575–1580.

24. Bock NN, Jensen PA, Miller B, Nardell E. Tuberculosis infection control in resource-limited settings in the era of expanding HIV care and treatment. *J Infect Dis* 2007;196(Suppl 1): S108–S113.

25. Pai M, Kalantri S, Aggarwal AN, Menzies D, Blumberg HM. Nosocomial tuberculosis in India. *Emerg Infect Dis* 2006;12: 1311–1318.

26. Galgalo T, Dalal S, Cain KP, et al. Tuberculosis risk among staff of a large public hospital in Kenya. *Int J Tuberc Lung Dis* 2008;12:949–954.

27. Menzies D, Joshi R, Pai M. Risk of tuberculosis infection and disease associated with work in health care settings. *Int J Tuberc Lung Dis* 2007;11:593–605.

28. Joshi R, Reingold AL, Menzies D, Pai M. Tuberculosis among health-care workers in low- and middle-income countries: a systematic review. *PLoS Med* 2006;3(12):e494.

29. Mirtskhulava V, Kempker R, Shields KL, et al. Prevalence and risk factors for latent tuberculosis infection among health care workers in Georgia. *Int J Tuberc Lung Dis* 2008;12: 513–519.

30. Blumberg HM. Tuberculosis infection control. In: Reichman LB, Hershfeld ES, eds. *Tuberculosis: a comprehensive international approach*. 2nd ed. New York: Marcel Dekker; 2000.

31. Jarvis WR. Nosocomial transmission of multidrug-resistant *Mycobacterium tuberculosis*. *Am J Infect Control* 1995;23: 146–151.

32. Centers for Disease Control and Prevention. Guidelines for preventing the transmission of *Mycobacterium tuberculosis* in health-care settings, 1994. *MMWR Recomm Rep* 1994; 43(RR-13):1–132.

33. Blumberg HM, Watkins DL, Berschling JD, et al. Preventing the nosocomial transmission of tuberculosis. *Ann Intern Med* 1995;122:658–663.

34. McGowan JE. Nosocomial tuberculosis: new progress in control and prevention. *Clin Infect Dis* 1995;21:489–505.

35. Jensen PA, Lambert LA, Iademarco MF, Ridzon R; Centers for Disease Control and Prevention. Guidelines for preventing the transmission of *Mycobacterium tuberculosis* in health-care settings, 2005. *MMWR Recomm Rep* 2005; 54(RR-17):1–141.

36. Centers for Disease Control and Prevention. Trends in tuberculosis—United States, 2008. *MMWR Morb Mortal Wkly Rep* 2009;58(10):249–253.

37. Sreeramareddy CT, Kishore PV, Menten J, Van den Ende J. Time delays in diagnosis of pulmonary tuberculosis: a systematic review of literature. *BMC Infect Dis* 2009;9(1):91.

38. Centers for Disease Control and Prevention. Tuberculosis outbreak in a community hospital–District of Columbia, 2002. *MMWR Morb Mortal Wkly Rep* 2004;53:214–216.

39. Lee EH, Graham PL 3rd, O'Keefe M, Fuentes L, Saiman L. Nosocomial transmission of *Mycobacterium tuberculosis* in a children's hospital. *Int J Tuberc Lung Dis* 2005;9:689–692.

40. World Health Organization (WHO). Guidelines for the prevention of tuberculosis in health care facilities in resource-limited settings. Geneva: WHO; 1999. http://www.who.int/tb/publications/who_tb_99_269/en/index.html. Accessed October 14, 2009.

41. World Health Organization (WHO), Centers for Disease Control and Prevention (CDC). Tuberculosis infection control in the era of expanding HIV care and treatment: an addendum to WHO guidelines for the prevention of tuberculosis in heath care facilities in resource-limited settings, 1999. Atlanta: CDC, 2006. http://www.who.int/tb/challenges/hiv/en/index.html. Accessed October 14, 2009.

42. World Health Organization (WHO). WHO Policy on TB Infection Control in Health-Care Facilities, Congregate Settings and Households. WHO/HTM/TB/2009.419. Geneva, Switzerland: WHO, 2009. http://www.who.int/tb/publications/2009/en/index.html. Accessed November 23, 2009.

43. Sotir MJ, Parrott P, Metchock B, et al. Tuberculosis in the inner city: impact of a continuing epidemic in the 1990's. *Clin Infect Dis* 1999;29:1138–1144.

44. Wenger PN, Otten J, Breeden A, Orfas D, Beck-Sague CM, Jarvis WR. Control of nosocomial transmission of multidrug-resistant *Mycobacterium tuberculosis* among healthcare workers and HIV-infected patients. *Lancet* 1995;345: 235–240.

45. Maloney SA, Pearson ML, Gordon MT, del Castillo R, Boyle JF, Jarvis WR. Efficacy of control measures in preventing nosocomial transmission of multidrug-resistant tuberculosis to patients and health care workers. *Ann Intern Med* 1995; 122:90–95.

46. Welbel SF, French AL, Bush P, Deguzman D, Weinstein RA. Protecting health care workers from tuberculosis: a 10-year experience. *Am J Infect Control* 2009;37(8):668–673.

47. da Costa PA, Trajman A, Mello FC, et al. Administrative measures for preventing *Mycobacterium tuberculosis* infection among healthcare workers in a teaching hospital in Rio de Janeiro, Brazil. *J Hosp Infect* 2009;72:57–64.

48. Leonard MK, Egan KB, Kourbatova E, et al. Increased efficiency in evaluating patients with suspected tuberculosis by use of a dedicated airborne infection isolation unit. *Am J Infect Control* 2006;34:69–72.

49. Scott B, Schmid M, Nettleman M. Early identification of and isolation of inpatients at high risk for tuberculosis. *Arch Intern Med* 1994;154:326–330.

50. Francis J. *Tuberculosis Infection Control: A Practical Manual for Preventing TB*. San Francisco, CA: Francis J. Curry National Tuberculosis Center;2007. http://www.national-tbcenter.edu/TB_IC/. Accessed October 14, 2009.

51. American Thoracic Society, Centers for Disease Control and Prevention. Targeted tuberculin testing and treatment of latent tuberculosis infection. *Am J Resp Crit Care Med* 2000;161:S221–S247.

52. Blumberg HM, White N, Parrott P, Gordon W, Hunter M, Ray S. False-positive tuberculin skin test results among health care workers. *JAMA* 2000;283:2793.

53. Bailey TC, Fraser VJ, Spitznagel EL, Dunagan WC. Risk factors for a positive tuberculin skin test among employees of an urban, midwestern teaching hospital. *Ann Intern Med* 1995;122:580–585.

54. Larsen NM, Biddle CL, Sotir MJ, White N, Parrott P, Blumberg HM. Risk of tuberculin skin test conversion among healthcare workers: occupational versus community exposure and infection. *Clin Infect Dis* 2002;35:796–801.

55. Driver CR, Stricof RL, Granville K, et al. Tuberculosis in health care workers during declining tuberculosis incidence in New York State. *Am J Infect Control* 2005;33:519–526.

56. Institute of Medicine. *Ending Neglect: The Elimination of Tuberculosis in the United States*. Washington, DC: National Academy Press; 2000.

57. Pai M, Dheda K, Cunningham J, Scano F, O'Brien R. T-cell assays for the diagnosis of latent tuberculosis infection: moving the research agenda forward. *Lancet Infect Dis* 2007;7: 428–438.

58. Mazurek GH, Jereb J, Lobue P, Iademarco MF, Metchock B, Vernon A; Division of Tuberculosis Elimination, National Center for HIV, STD, and TB Prevention, Centers for Disease Control and Prevention (CDC). Guidelines for using the QuantiFERON-TB Gold test for detecting *Mycobacterium tuberculosis* infection, United States. *MMWR Recomm Rep* 2005;54(RR-15):49–55

59. Canadian Tuberculosis Committee (CTC). Updated recommendations on interferon gamma release assays for latent tuberculosis infection: an Advisory Committee Statement (ACS-6). *Can Commun Dis Rep* 2008;34: 1–13.

60. Menzies D, Pai M, Comstock G. Meta-analysis: new tests for the diagnosis of latent tuberculosis infection: areas of uncertainty and recommendations for research. *Ann Intern Med* 2007;146:340–354.

61. Pai M, Zwerling A, Menzies D. Systematic review: T-cell-based assays for the diagnosis of latent tuberculosis infection: an update. *Ann Intern Med* 2008;149:177–184.

62. Pai M, O'Brien R. Serial testing for tuberculosis: can we make sense of T cell assay conversions and reversions? *PLoS Med* 2007;4(6):e208.

63. Perry S, Sanchez L, Yang S, Agarwal Z, Hurst P, Parsonnet J. Reproducibility of QuantiFERON-TB Gold In-Tube assay. *Clin Vaccine Immunol* 2008;15:425–432.

64. Centers for Disease Control and Prevention. Prevention and control of tuberculosis in facilities providing long term care to the elderly: recommendations of the Advisory Council for the Elimination of Tuberculosis. *MMWR Morb Mortal Wkly Rep* 1990;39(RR-10):7–13.

65. Ijaz K, Dillaha JA, Yang Z, Cave MD, Bates JH. Unrecognized tuberculosis in a nursing home causing death with spread of tuberculosis to the community. *J Am Geriatr Soc* 2002;50:1213–1218.

66. Thrupp L, Bradley S, Smith P, et al.; SHEA Long-Term Care Committee. Tuberculosis prevention and control in long-term care facilities for older adults. *Infect Control Hosp Epidemiol* 2004;25:1097–1108.

67. Stead WW. Tuberculosis among elderly persons, as observed among nursing home residents. *Int J Tuberc Lung Dis* 1998;2(Suppl 1):S64–S70.

68. American Society of Heating, Refrigerating and Air-Conditioning Engineers (ASHRAE). Health care facilities. In: *2003 ASHRAE Handbook: HVAC Applications*. Atlanta, GA: ASHRAE; 2003.

69. American Institute of Architects (AIA). Guidelines for design and construction of hospital and health care facilities. Washington, DC: AIA; 2001.

70. American Conference of Governmental Industrial Hygienists (ACGIH). Industrial ventilation: a manual of recommended practice. 24th ed. Cincinnati: ACGIH; 2001.

71. Rutala WA, Jones SM, Worthington JM, Reist PC, Weber DJ. Efficacy of portable filtration units in reducing aerosolized particles in the size range of *Mycobacterium*

tuberculosis. Infect Control Hosp Epidemiol 1995;16: 391–398.

72. Miller SL, Hernandez M. Evaluating portable air cleaner removal efficiencies for bioaerosols. Boulder, CO: University of Colorado at Boulder; 2002. NIOSH contract report PO-36755-R-00077B5D.

73. National Institute for Occupational Safety and Health (NIOSH). Environmental control for tuberculosis: basic upper-room ultraviolet germicidal irradiation guidelines for healthcare settings, March 2009. DHHS (NIOSH) publication no. 2009–105. http://www.cdc.gov/niosh/docs/2009-105/. Accessed October 14, 2009.

74. Nardell EA. Environmental infection control of tuberculosis. *Semin Respir Infect* 2003;18:307–319.

75. Escombe AR, Moore DA, Gilman RH, et al. Upper-room ultraviolet light and negative air ionization to prevent tuberculosis transmission. *PLoS Med* 2009;6(3):e43.

76. American National Standards Institute, American Society of Heating, Refrigerating and Air-Conditioning Engineers (ASHRAE). Standard 55-2004: Thermal environmental conditions for human occupancy. Atlanta: ASHRAE; 2004.

77. Adal KA, Anglim AM, Palumbo CL, Titus MG, Coyner BJ, Farr BM. The use of high-efficiency particulate air-filter respirators to protect hospital workers from tuberculosis. A cost-effectiveness analysis. *N Engl J Med* 1994;331: 169–173.

78. Nettleman MD, Fredrickson M, Good NL, Hunter SA. Tuberculosis control strategies: the cost of particulate respirators. *Ann Intern Med* 1994;121:37–40.

79. Jarvis WR, Bolyard EA, Bozzi CJ, et al. Respirators, recommendations, and regulations: the controversy surrounding protection of healthcare workers from tuberculosis. *Ann Intern Med* 1995;122:142–146.

80. Department of Labor, Occupational Safety and Health Administration. Occupational exposure to tuberculosis: proposed rule. *Fed Regist* 1997;62:54159–54308.

81. Managan LP, Bennett CL, Tablan N, et al. Nosocomial tuberculosis prevention measures among two groups of U.S. hospitals, 1992 to 1996. *Chest* 2000;117:380–384.

82. Department of Labor, Occupational Safety and Health Administration (OSHA). Occupational exposure to tuberculosis; proposed rule; termination of rulemaking respiratory protection for *M. tuberculosis;* final rule; revocation. *Fed Regist* 2003;68:75767–75775.

83. Department of Labor, Occupational Safety and Health Administration (OSHA). Respiratory protection for *M. tuberculosis.* 29 CFR part 1910 [docket no. H–371]. *Fed Regist* 2003;68:75776–75780.

84. Coffey CC, Campbell DL, Zhuang Z. Simulated workplace performance of N95 respirators. *Am Ind Hyg Assoc J* 1999; 60(5):618–624.

85. Coffey CC, Lawrence RB, Zhuang Z, Campbell DL, Jensen PA, Myers WR. Comparison of five methods for fit-testing N95 filtering-facepiece respirators. *Appl Occup Environ Hyg* 2002;17:723–730.

86. Coffey CC, Lawrence RB, Campbell DL, Zhuang Z, Calvert CA, Jensen PA. Fitting characteristics of eighteen N95 filtering-facepiece respirators. *J Occup Environ Hyg* 2004;1:262–271.

87. National Institute for Occupational Safety and Health (NIOSH). Respiratory protective devices: final rules and notice. *Fed Regist* 1995;60(110):30336–30398.

88. Coffey CC, Lawrence RB, Zhuang Z, Duling MG, Campbell DL. Errors associated with three methods of assessing respirator fit. *J Occup Environ Hyg* 2006;3:44–52.

89. Department of Health and Human Services, Centers for Disease Control and Prevention, National Institutes of Health. Richmond JY, McKinney RW, eds. *Biosafety in Microbiological and Biomedical Laboratories.* 4th ed. Washington, DC: US Government Printing Office; 1999.

90. Fennelly K, Nardell E. The relative efficacy of respirators and room ventilation in preventing occupational tuberculosis. *Infect Control Hosp Epidemiol* 1998;19:754–759.

91. Centers for Disease Control and Prevention. TB respiratory protection program in health care facilities: administrator's guide. Cincinnati, OH: US Department of Health and Human Services, Public Health Service, CDC, National Institute for Occupational Safety and Health; 1999. NIOSH publication 99–143. http://www.cdc.gov/niosh/docs/99-143/. Accessed October 14, 2009.

92. American Thoracic Society, Centers for Disease Control and Prevention. Diagnostic standards and classification of tuberculosis in adults and children. *Am J Respir Crit Care Med* 2000;161:1376–1395.

93. Blumberg HM, Burman WJ, Chaisson RE, et al.; American Thoracic Society, Centers for Disease Control and Prevention, Infectious Diseases Society. Treatment of tuberculosis. *Am J Respir Crit Care Med* 2003;167:603–662.

94. Advisory Council for the Elimination of Tuberculosis (ACET). Tuberculosis elimination revisited: obstacles, opportunities, and a renewed commitment. *MMWR Recomm Rep* 1999;48(RR-9):1–13.

95. Taylor Z, Nolan CM, Blumberg HM; American Thoracic Society; Centers for Disease Control and Prevention; Infectious Diseases Society of America. Controlling tuberculosis in the United States: recommendations from the American Thoracic Society, CDC, and the Infectious Diseases Society of America. *MMWR Recomm Rep* 2005;54(RR-12):1–81

96. Kourbatova E, Wang YF, Leonard MK, White N, McFarland D, Blumberg HM. Cost-effectiveness of the nucleic acid amplification test in the setting of a high prevalence of TB and HIV infection. In: Program and abstracts of the 40th Union World Conference on Lung Health; Cancun, Mexico; December 2009.

97. Centers for Disease Control and Prevention. Updated guidelines for the use of nucleic acid amplification tests in the diagnosis of tuberculosis. *MMWR Morb Mortal Wkly Rep* 2009;58:7–10.

98. Laraque F, Griggs A, Slopen M, Munsiff SS. Performance of nucleic acid amplification tests for diagnosis of tuberculosis in a large urban setting. *Clin Infect Dis* 2009;49:46–54.

99. Blumberg HM, Leonard MK Jr, Jasmer RM. Update on the treatment of tuberculosis and latent tuberculosis infection. *JAMA* 2005;293:2776–2784.

100. Chaulk CP, Kazandjian VA. Directly observed therapy for treatment completion of pulmonary tuberculosis: consensus statement of the Public Health Tuberculosis Guidelines Panel. *JAMA* 1998;279:943–948.

101. Centers for Disease Control and Prevention. Update: adverse event data and revised American Thoracic Society/CDC recommendations against the use of rifampin and pyrazinamide for treatment of latent tuberculosis infection—United States, 2003. *MMWR Morb Mortal Wkly Rep* 2003;52:735–739.

102. Barrett-Connor E. The epidemiology of tuberculosis in physicians. *JAMA* 1979;241:33–38.

103. Fraser VJ, Kilo CM, Bailey TC, Medoff G, Dunagan WC. Screening of physicians for tuberculosis. *Infect Control Hosp Epidemiol* 1994;15:95–100.

104. Gieseler PJ, Nelson KE, Crispen RG. Tuberculosis in physicians: compliance with preventive measures. *Am Rev Respir Dis* 1987;135:3–9.

105. Camins BC, Bock N, Watkins DL, Blumberg HM. Acceptance of isoniazid therapy by health care workers after tuberculin skin test conversion. *JAMA* 1996;275:1013–1015.

106. Shukla SJ, Warren DK, Woeltje KF, Gruber CA, Fraser VJ. Factors associated with the treatment of latent tuberculosis infection among health-care workers at a midwestern teaching hospital. *Chest* 2002;122:1609–1614.

107. Sahni R, Miranda C, Yen-Lieberman B, et al. Does the implementation of an interferon-gamma release assay in lieu of a tuberculin skin test increase acceptance of preventive therapy for latent tuberculosis among healthcare workers? *Infect Control Hosp Epidemiol* 2009;30:197–199.

108. Wilson JCE, Blumberg HM. Low risk of house staff tuberculin skin test conversion at an inner-city hospital in a high endemic area. In: Program and abstracts of the 40th Annual Meeting of the Infectious Diseases Society of America; October 24–27, 2002; Chicago, IL.

Chapter 26 Infection Control in Long-Term Care Facilities

Lindsay E. Nicolle, MD

Patients in long-term care facilities (LTCFs) are at increased risk of infection because of aging-associated changes, comorbidities, and functional impairment, as well as institutional residence. Effective infection control practices can minimize infection in this setting. This chapter reviews types of and risk factors for infection in LTCFs, the components of an infection control program, and recommendations for prevention of specific infections of importance. Special characteristics of LTCFs that challenge personnel charged with conducting effective infection control programs are also discussed.

Background

Many individuals in developed countries reside for extended periods in long-term care institutions. While different types of institutions provide a wide variety of services to diverse groups of patients, the majority of residents in these facilities are elderly persons who reside in nursing homes. Approximately 43% of Americans who became 65 years old in 1990 will have resided in a nursing home for some time before they die.[1] Infection control programs in these facilities contribute to optimal resident quality of life by limiting morbidity and mortality from infections and supporting optimal use of resources.

Infections are common in LTCFs. Reported rates of infection in nursing homes have varied from 1.8 to 9.4 infections per 1,000 resident-days (Table 26-1). The prevalence of infection has varied from 1.6% to 14% of residents.[2] This wide variation in reported infection rates reflects differences in patient populations, definitions of infection, and surveillance methods used. A recent report of infections in 17 Idaho nursing homes that used standard training, case ascertainment, and definitions for surveillance reported rates varying from 1.45 to 5.95 infections per 1,000 resident-days in individual facilities, with an average infection rate of 3.64 infections per 1,000 resident-days.[3]

The most common endemic infections in nursing homes are urinary tract infections, upper and lower respiratory tract infections, gastrointestinal tract infections, and skin and mucous membrane infections (ie, infected pressure ulcers, cellulitis, infected vascular ulcers, conjunctivitis, and candidiasis). Surgical site infections, the second most common cause of nosocomial infections in acute care facilities, are uncommon. Outbreaks of infection, usually respiratory or gastrointestinal infection, occur frequently.[2] Etiologic agents that have been repeatedly reported as causes of outbreaks are listed in Table 26-2.

Infections among residents of LTCFs may cause patient discomfort, accelerated functional decline, or death.[4] In addition, infections are costly to the facility. Residents of LTCFs have many associated comorbidities and impaired functional status, and the usual annual mortality rate is 10%–30% of residents. The contribution of infection to mortality may be difficult to ascertain, given this high expected mortality. The case-fatality rate for pneumonia is 6%–23%; this is the only infection that contributes substantially to overall mortality in these facilities. The case-fatality rate for bacteremia ranges from 10% to 25%, but bacteremia is relatively uncommon.[2] There is little information describing the

Table 26-1. Reported incidence and prevalence of common infections in nursing homes

Infection	Incidence, cases per 1,000 resident-days	Prevalence, %
All types of infection	1.8–9.4	1.6–13.9
Urinary tract infection		
Symptomatic	0.19–2.2	2.6–3.5
Asymptomatic	1.1	15–50
Respiratory tract infection		
Lower (eg, pneumonia or bronchitis)	0.3–4.7	0.3–5.8
Sinusitis, otitis, or pharyngitis	0.003–2.3	1.5
Skin and soft-tissue infection		
All types	0.14–1.1	5.6–8.4
Infected pressure ulcers	0.1–0.3	2.6–24
Cellulitis or cutaneous abscesses	0.19–0.23	7.2–8.7
Conjunctivitis	0.17–1.0	5–13
Candida infection	0.28	33–47
Bacteremia	0.2–0.36	…
Gastrointestinal infection	0–2.5	0.5–1.3

costs of infections in LTCFs. Some factors that may contribute to increased costs include the need for more frequent evaluation by nursing and medical staff, laboratory and radiologic testing, antimicrobial therapy and other treatment, intensified nursing care, infection control interventions if necessary, and transfer of residents to acute care facilities.

Reasons for Increased Risk of Infection

Changes in body organ systems that are a normal accompaniment of aging may contribute to an increased risk of infection. Aging-associated alterations in immune function appear to contribute little to the increased occurrence of infection, with the exception of reactivation of latent infections, such as tuberculosis or varicella zoster virus (shingles), as cell-mediated immunity declines in older individuals. However, nonimmunologic aging-associated changes that enhance an LTCF resident's susceptibility to infection occur in virtually all body systems (Table 26-3).

Institutionalized elderly persons have numerous comorbidities that substantially increase their risk of infection. For instance, urologic abnormalities, such as

prostatic hypertrophy, are common and are associated with urinary tract infections. Chronic obstructive lung disease and congestive heart failure increase a patient's risk of developing pneumonia, and prior vascular surgery or leg edema is a risk factor for recurrent lower limb erysipelas. Diabetes or vascular insufficiency may lead to more-frequent and more-severe skin infections. Functional impairment, including decreased mobility and incontinence, further increases the resident's risk of infection. Not only are residents who are more functionally impaired at greater risk of infection, but when infection occurs it accelerates further decline in functional status.[4]

Interventions necessary for resident care may also increase the risk of infection for individual patients. Approximately 5% of residents in nursing homes will have long-term indwelling urinary catheters, and these patients will always have bacteriuria.[2] The use of other invasive devices, such as central lines for hemodialysis and tracheostomy tubes for long-term respirator use, is increasing in LTCFs, contributing to bloodstream and pulmonary infections in residents in whom these devices

Table 26-2. Etiologic agents identified as causes of outbreaks of infection in long-term care facilities

Viruses

Rhinovirus
Influenza virus
Respiratory syncytial virus
Parainfluenza virus
Metapneumovirus
Rotavirus
Noroviruses
Astroviruses (rare)

Bacteria

Streptococcus pneumoniae
Haemophilus influenzae
Mycobacterium tuberculosis
Bordetella pertussis (rare)
Group A *Streptococcus*
Legionella species
Chlamydia pneumoniae
Salmonella species
Escherichia coli 0157:H7
Shigella species
Clostridium difficile
Staphylococcus aureus
Staphylococcus aureus (food poisoning)
Clostridium perfringens (food poisoning)
Bacillus cereus (food poisoning)
Aeromonas hydrophilia (rare)
Campylobacter jejuni (rare)

Other

Giardia lamblia
Scabies

Table 26-3. Aging-associated changes that may promote infection in residents of long-term care facilities

Factor	Change
Skin	Epidermal thinning; decreased elasticity, subcutaneous tissue, vascularity, and wound healing
Respiratory tract	Decreased cough reflex, elastic tissue, mucociliary transport, IgA levels
Gastrointestinal tract	Decreased gastric acidity and motility
Genitourinary tract	Decreased estrogen effect on mucosa; decreased prostatic secretions, increased prostatic size
Immune system	Thymic involution; decreased antibody production, decreased T cell count, decreased mitogen stimulation, decreased fever response, decreased interleukin-2 levels, and increased levels of autoantibodies
Chronic illness	Diabetes, congestive heart failure, vascular insufficiency, chronic obstructive pulmonary disease, neurologic impairment, dementia
Nutritional impairment	Decreased cell-mediated immunity and wound healing
Functional impairment	Immobility, incontinence, impaired cognitive status, poor hygiene
Invasive devices	Indwelling urinary catheter, tracheostomy, feeding tube gastrostomy, central venous catheter
Institutionalization	Increased person-to-person contact

are used.[5] Some medications also have effects that may facilitate infection. For instance, antidepressants with anticholinergic side effects will dry oral secretions and increase the frequency of pharyngeal colonization. These drugs also inhibit bladder contraction and impede urine flow, which may predispose residents to urinary tract infections. Proton pump inhibitors and other medications that impair gastric acidity increase the risk of gastrointestinal infection.

In an institution, persons at increased risk of infection live on units in close proximity and have continuing interaction in dining areas, during recreational activities, and during physical or occupational therapy. This facilitates transmission of microorganisms from one resident to another, following introduction into the facility by residents, visitors, or staff. Transmission may occur through the air (eg, *Mycobacterium tuberculosis*), by way of the hands of staff or residents (eg, *Staphylococcus aureus* or uropathogens), or by way of contaminated items (eg, food and shared equipment).

Special Considerations for Infection Control Programs in LTCFs

There are differences between long-term care and acute care facilities that have a bearing on infection control programs.[6] Thus, some aspects of infection prevention and control may need to be approached differently in the LTCF. A fundamental difference is that the acute care facility is a high-technology environment with a goal of prompt patient discharge. The LTCF is a low-technology environment, and, for many residents, the facility is their permanent residence, with discharge not anticipated. The major goal of management is to maximize quality of life for the individual resident. Other specific differences are discussed in the rest of this section.

Resources

In LTCFs, the resources and expertise available for establishing and maintaining infection control programs are limited. Specific issues include the following:

1. In many facilities, individuals responsible for the infection control program are only employed part-time. These individuals have other responsibilities in the facility that limit the time they have available for infection control, which reduces its priority.
2. Some facilities have limited access to personnel with the necessary expertise in infectious diseases, microbiology, and epidemiology to assist in dealing with infections and infection outbreaks.
3. Some facilities have a minimal employee health program or none at all. These facilities may have limited resources to educate staff about prevention of infection, to conduct immunization programs, to screen for tuberculosis, or to manage healthcare workers who may be ill with an infection.
4. Access to equipment or special facilities for residents requiring isolation (eg, negative air-pressure rooms for residents with infectious tuberculosis) may be limited or not available at all on-site.
5. There are few clinical trials in the medical literature to provide evidence to support the effectiveness of specific practices to limit the transmission of infections in LTCFs.

Personnel

Compared with staff in acute care facilities, staff in LTCFs frequently have less training in infection control and other patient care practices. Furthermore, staff turnover in LTCFs is higher than in acute care facilities, making it difficult to maintain personnel adequately trained in infection control practices.

Patient-Related Issues

The clinical approach to the diagnosis and management of patients in LTCFs differs from that practiced in the community or in acute care facilities. The reasons for this include the following:

1. The ability to communicate may be diminished by residents' sensory impairment (vision and/or hearing) or cognitive impairment.
2. Clinical criteria to diagnose infections have been developed for younger populations with fewer comorbidities. Most LTCF residents have chronic symptoms that may make it difficult to identify and evaluate acute changes in clinical status that represent a new infection.
3. Infectious diseases in the older adult may have non-classical presentations. For instance, impaired elderly patients who have infections may present with confusion or a loss of appetite, rather than prominent localizing findings. Compared with younger patients, elderly patients have a relatively blunted body-temperature response, and a higher proportion of infected patients may be afebrile.
4. Access to radiologic and laboratory facilities may be limited, and specimens or patients may need to be transferred off-site for appropriate diagnostic testing.
5. Microbiologic findings may have limitations for diagnostic accuracy in this population. For example, gram-negative organisms frequently colonize the oropharynx of elderly LTCF residents.[2] Sputum specimens obtained from these patients may be contaminated by the colonizing organisms. In addition, 30%–50% of noncatheterized elderly residents of nursing homes have bacteriuria; for these patients, a positive urine culture has a low predictive value for diagnosis of symptomatic urinary infection.[7]
6. In general, the use of standard clinical diagnostic tests has not been assessed for patients in LTCFs. The optimal use of laboratory tests for management of infection in this population and the appropriate empirical approach to treatment are not well established.[8]

Developing an Infection Control Program

Infection control programs should be developed to serve the specific needs of a given institution and its resident population. The basic components of an infection control program for an LTCF are summarized in Table 26-4.

Table 26-4. The basic components of an infection control program

Infection control oversight (administration)
    Resources
    Responsibility and/or authority
    Reporting
Infection control practitioner
Surveillance of infections
    Process and outcome measures
Policies and procedures, development, and compliance monitoring
    Hand hygiene
    Outbreak control
    Isolation precautions
    Resident health
    Employee health
Antimicrobial stewardship
Emergency preparedness planning
Facility management
    Housekeeping, cleaning, linen
    Food handling
    Waste management
Disease reporting and other regulatory requirements

NOTE: Adapted from Smith et al.[6]

Administration

An effective reporting structure must be clearly defined. It is essential that the infection control program staff have a close working association with the medical director, administrator, or nursing supervisor for the LTCF. Responsibility and authority should be defined, and the structure should be developed to ensure an efficient and effective infection control program. Usually, continuing review and oversight is provided through an infection control committee that meets on a regular basis. There must be policies in place that allow prompt intervention, if necessary, both for management of infection in an individual patient and for outbreak control.

Personnel

The size and complexity of the institution will determine the number of individuals needed to support the infection control program. At least one individual must have this responsibility. If an individual is responsible for programs in the institution in addition to infection control, the specific time commitment to infection control must be defined clearly.

Personnel with responsibility for infection control must have a good understanding of the basic principles of microbiology, infectious diseases, epidemiology, program management, and patient and staff education. These individuals should have access to opportunities for education in infection control.

When not available on site, individuals with expertise in areas such as outbreak investigation, infectious diseases, antimicrobial use, and microbiology should be identified and consulted as needed.

Surveillance for Infections

Surveillance to identify both infected patients and outbreaks of infection is an essential component of the infection control program.[6] Standard definitions should be used to identify infections. Definitions appropriate for LTCFs have been published.[9]

Infection control personnel should use case-finding methods that are appropriate for the available resources and the characteristics of the institution. Methods may include walking rounds, nursing-generated reports, medical chart reviews, filing system reviews, and laboratory or medication record reviews. When an infection control program is initially being developed for a facility, prevalence surveys may be useful. Generally, incidence surveys are more useful for continuing programs.

Infection control personnel should analyze data, review findings, and report results to the administration and the infection control committee at least quarterly. Incidence rates should generally be reported as the number of infections per 1,000 resident-days. The surveillance program should enable infection control personnel to identify outbreaks of influenza, gastrointestinal illness, scabies, and other common problems in a timely manner so that appropriate control measures can be implemented promptly.

Policies and Practices

Policies addressing the identification and control of infections must be developed, reviewed, updated, and monitored to document adherence. Relevant guidelines are available through the Healthcare Infection Control Practices Advisory Committee (HICPAC) and other authoritative bodies, which may be incorporated into facility-specific policies and practices. Some specific issues to be addressed by institutional policies include the following.

- Hand hygiene
- Standard precautions and additional isolation practices
- Environmental cleaning, laundry, and waste disposal
- Food preparation, holding, and transport
- Preadmission screening of residents for infections (eg, tuberculosis)
- Vaccination policies (eg, influenza vaccine and pneumococcal vaccine)
- Management of patients with infections, especially those who may require special infection control precautions

- Use and care of invasive devices, such as urinary catheters, central lines, percutaneous feeding tubes, and tracheostomy tubes
- Identification and management of residents colonized or infected with antimicrobial-resistant organisms
- Outbreak identification, investigation, and control
- Review of antimicrobial use[10,11]
- Employee health issues, including tuberculosis screening, immunizations, and work restriction when employees have potentially transmissible infections

Education

Ongoing education of staff, residents, and visitors is an important component of the infection control program. All individuals working in the facility, including medical staff, should know risk factors for infection and methods for minimizing each resident's risk of infection. These programs should be developed with input from the groups of employees to whom they are targeted.

Regulations

The infection control program must be in compliance with all national, state or provincial, and local regulations. These will usually pertain to activities such as food handling, cleaning, and waste disposal, as well as specific disease reporting requirements and selected patient care practices, such as immunization.

Special Infection Problems

Endemic Infections

The most common endemic infections in LTCF residents are urinary tract infections, respiratory infections, and skin and soft-tissue infections.[2,3] Programs must be in place to minimize the frequency and impact of these infections.[6]

In addressing urinary tract infection, the high prevalence of asymptomatic bacteriuria in LTCF populations must be appreciated.[6,7] Urinary infection in the patient without an indwelling catheter should be diagnosed and treated with antimicrobials only if there are localizing symptoms referable to the genitourinary tract. Urinary infections in residents with long-term indwelling catheters may present as fever without localizing signs or symptoms. Infection may be prevented by discontinuing use of the catheter as soon as possible and by following appropriate catheter care practices to minimize trauma and identify obstruction early. For respiratory tract infections, programs are necessary to minimize aspiration through identification of patients with swallowing disorders, appropriate positioning of patients, and limiting use of sedative drugs.[12] Smoking should be discouraged, and all LTCF residents should receive influenza vaccine yearly and pneumococcal vaccine at least once. Skin infections may be prevented by maintaining skin integrity through optimal management of comorbid illnesses and prevention of trauma or other injury. In addition, a skin care program should include appropriate patient positioning and monitoring to prevent pressure ulcers.

Influenza

Influenza is the most frequent cause of outbreaks in LTCFs.[13] These outbreaks may be disruptive, because many patients and staff become ill within a short time. Outbreaks have been associated with substantial patient mortality. Each facility should develop a specific plan for managing influenza that includes the following elements.

- Yearly vaccination programs for patients and staff
- Clinical and epidemiologic definitions of influenza and influenza outbreaks
- Surveillance for possible outbreaks
- Criteria defining when and from whom diagnostic specimens should be collected
- Response to outbreaks, including when to place restrictions on residents or visitors
- Notification of local authorities
- Recommendations for prophylactic and therapeutic use of antiviral medications (eg, the agent, the dose, patient exclusions, triggers for initiating prophylaxis, and criteria for deciding which patients will receive prophylaxis)

Gastrointestinal Illness, Including *Clostridium difficile* Infection

Outbreaks of gastrointestinal illness are common in LTCFs.[14] Sporadic episodes of diarrhea may be caused by infectious bacteria or viruses or by noninfectious conditions, including medications or diet. A program to control diarrheal illness should include the following specific components.

- Clinical definitions to identify individual cases and outbreaks of diarrheal disease
- Criteria that define when specimens should be obtained and what laboratory tests should be performed
- A description of infection control practices, including isolation precautions required for patients with suspected infectious diarrheal illness, and when these can be discontinued
- A protocol for the use of oral rehydration therapy
- Protocols for antimicrobial therapy for infected patients and staff (eg, metronidazole therapy for moderate-to-severe *C. difficile* colitis)[15]
- Staff education programs

Food poisoning may present as clusters of patients with vomiting or diarrhea. It may be the result either of

contamination of food with bacteria that produce toxins (eg, *Bacillus cereus, S. aureus,* and *Clostridium perfringens*) or of infection following ingestion of foodborne bacteria, viruses, or parasites. Management of such clusters should include the following components:

- Epidemiologic investigation that collects data on the number of cases, the number of persons at risk, and the characterization of cases by time, place, and person
- Identification of the source of infection (eg, food item, person, or utensil)
- Elimination or treatment of the source
- Staff education to limit transmission and to prevent future outbreaks
- Appropriate management of infected residents and personnel

Scabies

To manage scabies[16] effectively, LTCF personnel must understand characteristics of the infestation and follow a systematic approach to identifying and treating affected patients and staff. Issues that must be addressed include clinical and microscopic diagnostic criteria; treatment for infected patients or staff (with 5% permethrin or 1% lindane); management of exposed patients, staff, and household contacts; handling of contaminated bedding and clothing; and follow-up of treated subjects.

Group A Streptococcal Skin Infections

LTCFs should define how they will identify individual cases and outbreaks of streptococcal disease.[17] Management practices should address when samples for culture will be obtained from infected or colonized patients and how a potential outbreak will be recognized. They also must identify infection control precautions to be instituted when streptococcal infections are identified.

For severe outbreaks, or when initial control measures are not effective in preventing further cases, institutional criteria for performing a culture survey to identify colonized patients and staff or to undertake mass treatment of residents must be developed.

Tuberculosis

The most important strategy for preventing the spread of tuberculosis is the early identification and treatment of possibly infectious persons.[18] LTCFs need to monitor for skin-test conversion among residents and staff, identify and promptly evaluate patients with pulmonary symptoms or radiologic findings consistent with tuberculosis, immediately isolate patients with potential or proven pulmonary tuberculosis, trace exposed patients and staff, and initiate prophylactic isoniazid therapy. If appropriate facilities for isolation of potentially infectious patients are not available on-site, an agreement should be in place to facilitate prompt transfer of such patients to alternate facilities.

Bloodborne Pathogens

Policies must be developed for the management of patients with potential bloodborne infections, including hepatitis B virus, hepatitis C virus, and human immuno-deficiency virus infection. These policies should describe the practice of standard precautions, include guidelines for the use of gloves and other barriers during patient care and should discuss methods to limit needlestick exposures and other percutaneous injuries. Appropriate postexposure prophylaxis must be accessible for employees.

Multidrug-Resistant Organisms

Some patients in nursing homes are at increased risk of colonization or infection with antimicrobial-resistant organisms.[2] Institutional programs should define multidrug-resistant organisms, have programs for surveillance to identify infections with these organisms, provide guidelines for the management of patients who are colonized or infected with drug-resistant organisms, and review and restrict antimicrobial use when appropriate.[6] Some patients may require isolation precautions; however, standard infection control precautions, such as appropriate hand hygiene, environmental cleaning, and wound care, will usually be sufficient.[6,19]

LTCFs should develop policies addressing the transfer of patients colonized or infected with resistant organisms to other institutions or for accepting such patients from other institutions. In general, detection of a drug-resistant organism in a patient should not preclude transferring or accepting that patient. However, the institution that transfers the patient should always notify the accepting institution before the patient is transferred, so that the staff of the latter facility can be prepared.[19]

References

1. Kemper P, Murtaugh C. Lifetime use of nursing home care. *N Engl J Med* 1991; 324:595–600.
2. Nicolle LE, Strausbaugh LJ, Garibaldi RA. Infections and antibiotic resistance in nursing home. *Clin Microbiol Rev* 1996; 9:1–17.
3. Stevenson K, Moore J, Colwell H, Sleeper B. Standardized infection surveillance in long term care: interfacility comparisons from a regional cohort of facilities. *Infect Control Hosp Epidemiol* 2005; 26:231–238.
4. High K, Bradley S, Loeb M, Palmer R, Quagliarello V, Yoshikowa T. A new paradigm for clinical investigation of infectious syndromes in older adults: assessment of functional status as a risk factor and outcome measure. *Clin Infect Dis* 2005; 40:114–122.
5. Mody L, Maheshwari S, Galecki A, Kauffman CA, Bradley SF. Indwelling device use and antibiotic resistance in nursing homes: identify a high-risk group. *J Am Geriatr Soc* 2007; 55:1921–1926.
6. Smith PW, Bennett G, Bradley S, et al. SHEA/APIC guideline: infection prevention and control in the long-term care facility. *Infect Control Hosp Epidemiol* 2008; 29:785–814.
7. Nicolle LE, SHEA Long Term Care Committee. Urinary tract infections in long term care facilities. *Infect Control Hosp Epidemiol* 2001; 22:167–175.
8. Bentley DW, Bradley S, High K, Schoenbaum S, Taler G, Yoshikawa TT. Practice guideline for evaluation of fever and infection in long term care facilities. *Clin Infect Dis* 2000; 31:640–653.
9. McGeer A, Campbell B, Emori TG, et al. Definitions of infection for surveillance in long-term care facilities. *Am J Infect Control* 1991; 19:1–7.
10. Nicolle LE, Bentley D, Garibaldi R, Neuhaus E, Smith P, SHEA Long Term Care Committee. Antimicrobial use in long-term care facilities. *Infect Control Hosp Epidemiol* 2000; 21:537–545.
11. Loeb M, Bentley DW, Bradley S, et al. Development of minimum criteria for the initiation of antibiotics in residents of long term care facilities: results of a consensus conference. *Infect Control Hosp Epidemiol* 2001; 22:120–124.
12. Marrie TJ. Pneumonia in the long term care facility. *Infect Control Hosp Epidemiol* 2002; 23:159–164.
13. Bradley SF, SHEA Long Term Care Committee. Prevention of influenza in long term care facilities. *Infect Control Hosp Epidemiol* 1999; 20:629–637.
14. Bennett RG. Diarrhea among residents of long-term care facilities. *Infect Control Hosp Epidemiol* 1993; 14:397–404.
15. Simor AE, Bradley SF, Strausbaugh LJ, Crossley K, Nicolle LE, SHEA Long Term Care Committee. *Clostridium difficile* in long term care facilities for the elderly. *Infect Control Hosp Epidemiol* 2002; 23:696–703.
16. Tjioe M, Vissers MH. Scabies outbreaks in nursing homes for the elderly: recognition, treatment options and control of reinfection. *Drugs Aging* 2008; 25:299–306.
17. Schwartz B, Ussery XT. Group A streptococcal outbreaks in nursing homes. *Infect Control Hosp Epidemiol* 1992; 13:742–747.
18. Thrupp L, Bradley S, Smith P, et al. Tuberculosis prevention and control in long-term-care facilities for older adults. *Infect Control Hosp Epidemiol* 2004; 25:1097–1108.
19. Strausbaugh LJ, Crossley KB, Nurse BA, Thrupp LD, SHEA Long Term Care Committee. Antimicrobial resistance in long-term care facilities. *Infect Control Hosp Epidemiol* 1996; 17:129–140.

Recommended Reading

1. Smith PW, ed. *Infection Control in Long-Term Care Facilities.* 2nd ed. Albany, NY: Delmar Publishers; 1994.
2. Strausbaugh LJ, Joseph C. Epidemiology and prevention of infections in residents of long-term care facilities. In: Mayhall G, ed. *Hospital Epidemiology and Infection Control.* 3rd ed. Philadelphia, PA: Lippincott, Williams & Wilkins; 2004:1855–1880.
3. Yoshikawa TT, Ouslander JG. *Infection Management for Geriatrics in Long-Term Care Facilities.* 2nd ed. New York: Marcel Dekker; 2006.

Chapter 27 Infection Prevention and Control in the Outpatient Setting

Deborah M. Nihill, RN, MS, CIC,
and Tammy Lundstrom, MD, JD

Healthcare delivery has undergone dramatic changes. Innovative medical technologies have allowed healthcare workers to perform many diagnostic and therapeutic procedures in the outpatient setting. Consequently, the proportion of medical care that is provided in the outpatient setting is increasing rapidly. For example, 80%–90% of all cancer care is delivered in the outpatient setting, and more than half of all hospital-based operations are done in ambulatory surgery centers.[1] From 1992 to 2001, the number of patient visits to hospital outpatient departments increased from 56.6 million to 83.7 million. This was a 48% increase, although the population in the United States increased by only 12% during the same time period. Office-based physician practices account for 80% of ambulatory health care delivered by non-Federal physicians: 880.5 million patient visits in 2001.[2,3]

Healthcare-associated infections are among the top 10 leading causes of death in this country, accounting for approximately 1.7 million infections and nearly 99,000 attributable deaths in 2002.[4] In a review of more than 1,000 outbreaks of healthcare-associated infection that occurred in the years 1966–2002, Gastmeier et al.[5] found that although the majority of reports (83%) were from hospitals, the remainder were from outpatient and nursing home sources. Unfortunately, the opportunity for healthcare-associated infections has similarly moved across the healthcare landscape. Although the federal Health and Human Services action plan to reduce the incidence of healthcare–associated infections is currently focused on inpatient care, the next phase will include outpatient arenas. The Centers for Medicare and Medicaid Services is also developing additional quality metrics, including "hospital outpatient healthcare-associated conditions," similar to those currently applied to the inpatient setting.

Given these statistics, hospital epidemiologists and infection preventionists must develop programs that address infection prevention and control issues across the entire spectrum of care, and not focus exclusively on the inpatient setting. In addition to the clinical and monetary incentives to address infection prevention and control issues in the outpatient setting, the Centers for Disease Control and Prevention (CDC) recommends, and the Occupational Safety and Health Administration (OSHA) mandates, that healthcare workers in the outpatient setting incorporate the Bloodborne Pathogen Standard[6] and the CDC tuberculosis guidelines into their practices.[7] Moreover, the Joint Commission requires that within a particular healthcare organization the infection control policies and procedures that are applied in the inpatient setting and in the outpatient setting are consistent in intent and application.[8]

Healthcare workers and infection prevention personnel traditionally have considered the risk of infection to be low in the ambulatory care setting. However, few investigators have evaluated systematically the rates of infection in outpatient populations. Investigators who have attempted to conduct surveillance in outpatient populations often have encountered substantial problems that have precluded instituting such programs in the routine practice of infection prevention. These include lack of resources, lack of education and training, lack of administrative support, and lack of time (because of multiple duties), and lack of computerized data sources.

As patients move in and out of various healthcare environments, often with unrecognized pathogen carriage, infection prevention and control in nonacute care settings is vitally important. Studies of the carriage of methicillin-resistant *Staphylococcus aureus* (MRSA) have demonstrated that there are a greater number of patients with MRSA carriage than patients with MRSA-positive clinical specimens.[9] Although the use of gowns and gloves has been shown to be an effective barrier for infection prevention in acute care environments,[10] what is unknown is the appropriate level of infection prevention needed to prevent ambulatory care–associated pathogen transmission.

To date, experts in infection prevention and control and agencies such as the CDC have not recommended specific surveillance systems for ambulatory care. However, essential elements of infection prevention and control programs in the outpatient setting have been developed and defined.[11-15] It has been noted that transmission of nosocomial infections in ambulatory care settings is more likely to occur in 1 of 2 manners: through congregation of patients in waiting rooms or other common areas, or in association with invasive procedures.[16] Despite the dearth of data and guidelines, infection prevention staff must develop programs that address the special needs of the ambulatory-care setting, because this is where most medical care will be given in the 21st century. Healthcare-associated infections are usually attributed to inpatient settings. However, they can occur in a healthcare setting where staff personnel fail to follow the principles of proper aseptic technique, hand washing, and environmental sanitation. As consumers of health care, the public are becoming increasingly more knowledgeable and concerned about infection transmission, especially the risk of acquiring MRSA.[17]

For the purposes of this chapter, we define outpatient or ambulatory care as any medical services provided to patients who are not admitted to inpatient hospital units. Infection prevention and control in the outpatient or ambulatory setting is a very broad topic, because the scope of medical services provided outside of the traditional inpatient setting and the types of facilities providing these services have expanded exponentially. Even conventional hospitals that specialize in hospital-based care have numerous areas where persons other than inpatients are waiting, visiting, or undergoing treatment. Although most hospitals have an infection prevention and control program, planning and provision of resources for expansion beyond inpatient units is often lacking. Numerous outpatient facilities have sprung up that are independent of hospitals, including ambulatory treatment centers, urgent-care centers, outpatient surgical centers, and radiology and imaging clinics. Third-party payers have encouraged this development and now

demand that a large proportion of medical care be given in the outpatient setting. Challenges for infection prevention and control in outpatient and ambulatory care settings include determining which infections to conduct surveillance, to whom the data will be reported, and who will be responsible for implementing the changes.[18]

This chapter discusses general principles of infection prevention and control in the outpatient setting, as well as specific considerations for some specific areas. However we do not discuss specific infection prevention and control issues for ambulatory surgery centers, home health care, or dental offices. Readers who need information on those topics are referred to other reviews.[19,20]

Common Problems Encountered in the Outpatient Setting

Persons who conduct infection control efforts in any outpatient setting will encounter problems that either do not occur in the inpatient setting or are exaggerated in comparison with the inpatient setting. We discuss some of these problems briefly.

The population of persons in the outpatient setting often is very difficult to define. For example, exposures to communicable diseases such as severe acute respiratory syndrome (SARS), measles, or tuberculosis can occur in clinic waiting areas, hospital registration areas, or in other large open areas where many people congregate. Infection prevention personnel may have difficulty identifying those patients, family members, and staff who were in the area at a particular time. Similarly, infection prevention personnel may have difficulty determining the incidence of device-related infections in outpatients, because patients who acquire these infections may be treated elsewhere and because the number of persons with devices (ie, the denominator) is very difficult to determine.

Most surveillance methodologies were developed for use in the inpatient setting and are of limited use in the outpatient setting. The surveillance methods that have been tested in the outpatient setting often are labor-intensive and lack sensitivity and specificity. Clinic schedules often are very busy and leave little time for formal educational programs. Moreover, unlike hospital-based programs, freestanding facilities might not have on-site continuing education programs.

Staff in some outpatient facilities change frequently, which makes it difficult to achieve and maintain an adequate level of infection prevention and control knowledge among staff members. Telephone calls and registration usually are handled by persons with the least medical knowledge. Thus, patients who have contagious diseases might not be identified and triaged appropriately.

Personnel in clinics often care for many patients and do many procedures during a day. Given current budget constraints, the number of staff and the amount of equipment often are limited. Thus, staff might cut corners to meet the demands of the schedule. For example, healthcare workers might not wash their hands when appropriate, or they might not clean, disinfect, or sterilize equipment properly.

Many outpatient medical facilities were not designed with infection prevention and control input, and thus are not adequately laid out with infection prevention principles in mind. For example, most outpatient facilities do not have negative air-pressure rooms, and the air often is recirculated without filtration. In addition, outpatient facilities might have only a single entrance and a single waiting area, making these clinics ideal places in which airborne pathogens (such as *Mycobacterium tuberculosis*, varicella-zoster virus, or SARS coronavirus) can spread. Moreover, many outpatient facilities have inadequate space to allow the staff to maintain separate clean and dirty areas, and storage is often a challenge.

Environmental Contamination With Pathogenic Organisms

The role the environment plays in transmission of organisms to patients has been examined in great detail in recent years. Several studies have determined that environmental contamination with organisms such as MRSA and vancomycin-resistant enterococci occurs regularly, and the environment may be an important source of pathogen acquisition by noncolonized patients.[21-27] Huang et al.[21] found MRSA to be viable on plastic charts, plastic-laminate tabletops, and polyester cloth curtains for periods greater than 1 week. Neely and Maley[22] studied the survivability of various organisms on common hospital materials and found enterococci and staphylococci to be particularly viable on hospital materials for periods ranging from at least 1 day to upwards of 90 days. Dietze et al.[28] found that MRSA can survive on sterile goods packing for more than 38 weeks.

Although most studies of environmental contamination and organism acquisition are conducted in the hospital setting, there is reason to believe outpatient settings are similarly affected by environmental contamination and suboptimal cleaning. Outpatient settings generally lack patients with an extended duration of contact with the environment; however, there are generally more patients per day visiting the same outpatient environment, in contrast to the situation with hospitalized patients. In the quest for faster appointment turnaround

times, effective cleaning may not consistently occur. Housekeeping personnel in outpatient settings are often contracted personnel; they may not be familiar with hospital environments and therefore may not be utilizing effective cleaning procedures. Observations of housekeeping practices have found inadequate cleaning of surfaces that patients frequently contact.[29] Cleaning interventions, such as education and observation of housekeeping personnel, have been shown to reduce the burden of pathogen contamination in the patient care environment.[30-32]

With the move toward use of more electronic medical records, computer keyboards have also become a source of contamination and potential transmission. Studies in intensive care units have demonstrated the widespread contamination of keyboards and faucet handles with many gram-negative and gram-positive organisms. Infection prevention recommendations include use of plastic keyboard covers, a policy of daily cleaning, and reinforcement of the need for proper hand hygiene after all environmental and patient contact.[33]

Pathogenic organism acquisition can occur because of environmental contamination from previous patients, suboptimal environmental cleaning, and suboptimal hand hygiene compliance among healthcare workers. Medical devices and equipment used on multiple patients are also important sources of transmission.[29,34]

Exposures to Bloodborne Pathogens

The OSHA Bloodborne Pathogen Standard specifies that all healthcare institutions, including outpatient facilities, must implement policies and procedures to protect healthcare workers from exposures to bloodborne pathogens.[6] Infection prevention personnel must ensure that the inpatient and outpatient facilities within their institution implement the same general policies and procedures (ie, a single standard of practice) to prevent exposures (for a discussion of bloodborne pathogen exposures, see Chapter 24 on occupational health). However, some of the specific policies might be different because staff provide different services in each type of setting.

Healthcare workers in the ambulatory care setting are at risk for exposure to blood and body fluids.[35-39] Staff in emergency departments may be at the highest risk, because they frequently provide acute care for persons who have traumatic injuries or are critically ill.[35,36] Despite their exposure to blood and body fluids, healthcare workers in the outpatient setting often do not comply with precautions designed to protect them from bloodborne pathogens.[38,39]

Devices with engineered safety features are required by OSHA and federal legislation in inpatient as well as outpatient settings.[40-41] Infection control professionals face challenges engaging frontline outpatient staff in the selection of particular devices. In addition, providing educational programs related to the safe use of these devices is problematic, since sites of care may be widely scattered, requiring a traveling "road show" for effective education.

In addition to healthcare workers, patients are also at risk for acquiring hepatitis B virus (HBV), hepatitis C virus (HCV) and HIV through medical care provided in the outpatient setting.[42-49] Contaminated multidose vials, equipment that is cleaned and disinfected improperly, and reuse of needles have been associated with transmission of these pathogens, both in isolated cases and in outbreaks.[50,51] In addition, lax infection control injection procedures have also been noted during employee vaccination programs.[52] These findings underscore the need for appropriate staff training and monitoring of infection control practices in these settings. A recent review of infection outbreaks in the past decade revealed multiple outbreaks of HBV or HCV infection in nonhospital healthcare settings, including outpatient clinics, hemodialysis centers, and long-term care facilities.[53] Lack of compliance with basic infection prevention techniques was cited as the primary means of patient-to-patient transmission in each instance.

Hand Hygiene

The CDC guidelines on hand hygiene are applicable to the inpatient as well as the outpatient setting.[54] Studies of hand hygiene compliance in the outpatient setting have found adherence rates ranging from 31% to 74%.[54-57] As in the inpatient setting, improvements are needed, and attention should be paid to installing waterless hand hygiene agents appropriately in outpatient settings (see Chapter 20, on hand hygiene).

Respiratory Infections

Several outbreaks have illustrated that bacterial and viral pathogens can be transmitted within outpatient facilities by airborne or droplet spread. Particular characteristics of the outpatient setting, such as those listed below, might enhance the likelihood that these pathogens will be transmitted.

- Many people congregate in waiting rooms.
- Many infectious patients come to outpatient facilities for evaluation and treatment, particularly during periods when viral infections are endemic or epidemic.
- Outpatient facilities frequently have inadequate triage systems.
- The number of air exchanges in the building often is low, and often the air is recirculated.

Viruses, such as influenza virus, have been demonstrated to survive on environmental surfaces for 24–48 hours[34] and that infection can occur by contact with contaminated fomites.[58] Increased attention has been focused on transmission of respiratory pathogens in the outpatient setting since the emergence of the SARS coronavirus and the avian influenza virus (influenza A virus subtype H5N1). The likelihood of transmission of respiratory pathogens in the outpatient setting can be reduced through the use of "respiratory etiquette" protocols.[59] These include the following interventions:

- Post visual alerts at the entrance to outpatient facilities instructing persons who are reporting for care to report respiratory symptoms.
- Cover the nose and mouth when coughing or sneezing.
- Use tissues to contain respiratory secretions and dispose of them in the nearest waste receptacle after use.
- Perform hand hygiene after contact with respiratory secretions and contaminated objects or materials.
- Provide tissues and no-touch receptacles for used tissue disposal.
- Provide conveniently located hand washing agents (either waterless agents or, if a sink is available, soap and towels).
- Offer masks to persons who are coughing.
- Triage coughing individuals out of the common waiting area as soon as possible.

During periods of increased respiratory infection activity in the community (eg, when there is increased absenteeism in schools and work settings and increased medical office visits by persons complaining of respiratory illness), offer masks to persons who are coughing. When space and the availability of chairs permit, encourage persons who are coughing to sit at least 1 m (3 feet) away from others in common waiting areas. Some facilities may find it logistically easier to institute this recommendation year-round.

Tuberculosis

Although the national trend in the incidence of tuberculosis is moving in a desirable direction, with rates at an all-time low, cases of multidrug-resistant tuberculosis and extensively drug-resistant tuberculosis still occur. In the time since drug-susceptibility reporting began in 1993, cases of extensively drug-resistant tuberculosis have been reported in the United States every year except 2003; 4 cases were reported in 2006 and 2 in 2007.[60]

Several outbreaks demonstrate that healthcare workers[61,66,67] and patients[62-65] can become infected when exposed to outpatients who have undiagnosed tuberculosis or to patients with known tuberculosis, if the isolation precautions are not optimal. Healthcare workers have transmitted M. tuberculosis to patients infrequently.

The most important element of a tuberculosis control program is early identification and isolation of patients with suspected or confirmed infectious tuberculosis, which includes use of respiratory etiquette protocols. Triage personnel and staff who perform the initial evaluation in ambulatory care facilities must be educated about how and when to institute respiratory etiquette protocols based on patient symptoms and risk factors.[68] Unfortunately, in many settings "tuberculosis is often unsuspected, and isolation measures are often not used."[69(p290)]

If staff in outpatient departments or facilities perform special procedures, such as sputum induction, bronchoscopy, aerosolized pentamidine treatments, or pulmonary function testing, the units must have adequate facilities (ie, booths or other enclosures meeting ventilation requirements for tuberculosis isolation) to prevent airborne spread of M. tuberculosis. However, a survey conducted by the CDC in 1992 revealed that patients with tuberculosis often were treated in emergency departments, but few emergency departments were equipped appropriately to care for patients with suspected tuberculosis.[70]

Infection control personnel should help staff in the ambulatory care areas of their healthcare center develop comprehensive protocols for identifying and treating patients with signs and symptoms of respiratory illness. As part of this process, infection control personnel should help create screening tools that staff can use to identify such patients.

Measles and Rubella

The incidence of measles and rubella in the United States has decreased substantially since appropriate vaccines were introduced. However, both continue to be transmitted within healthcare facilities and in the outpatient setting.[71-73]

Several infection control precautions can prevent measles and rubella transmission in medical settings: proper ventilation, infection control policies stipulating that all staff must be immune,[73] appropriate immunization protocols for patients, and prompt triage of patients who have febrile exanthems. However, proper ventilation may be difficult to achieve in many outpatient settings, because buildings are built to be energy efficient, which often means the buildings are airtight and the ventilation systems recirculate air. In addition, because the incidence of measles and rubella is low, few healthcare workers recognize these illnesses. Consequently, patients with measles may be misdiagnosed and may not be isolated appropriately.[71]

Appropriate triage begins with the telephone conversation during which clinic appointments are scheduled. The scheduler should ask the caller for the patient's chief complaint. If the patient has a fever and a rash, or respiratory symptoms, the appointment should be scheduled at the end of the day or during times of the day when few patients are present.[72] Patients who must come to the clinic when other patients are present should wear a mask, should enter through a separate entrance, and should go directly to an examination room. If a patient arrives unannounced at a clinic or emergency department, the staff at the registration desk should ask whether the patient has fever and a rash or respiratory symptoms. If the answer to this question is "Yes," the patient should put on a mask and immediately go to an examination room (preferably one with negative air-pressure).

Epidemic Keratoconjunctivitis

Adenovirus, particularly type 8, has caused numerous outbreaks of keratoconjunctivitis in ophthalmology clinics.[74-77] However, basic infection control precautions could prevent spread of this virus among patients and staff in those clinics. Infection control staff should ensure that their ophthalmology clinic and emergency departments have developed the appropriate policies and procedures listed below.[57,78]

- Healthcare workers should wear gloves for possible contact with the conjunctiva.
- Equipment, including tonometers, should be cleaned and disinfected according to recommendations by the CDC,[74] the American Academy of Ophthalmology,[78] or the manufacturers. (Note that the methods for cleaning and disinfecting tonometers vary with the type of tonometer. Moreover, tonometers must be rinsed appropriately and thoroughly to prevent chemical keratitis).[79]
- If a nosocomial outbreak is identified, all open containers of ophthalmic solution should be discarded, and the equipment and environment should be cleaned and disinfected thoroughly.
- During an outbreak, unit doses of ophthalmic solutions should be used.
- During outbreaks, patients with conjunctivitis should be examined in a separate room with designated equipment, supplies, and containers of ophthalmic solutions.
- During an outbreak, elective procedures, such as tonometry, should be postponed.
- Healthcare workers who work in any outpatient area and who have adenovirus conjunctivitis should not work until the inflammation has resolved, which may take 14 days or longer.

Other Transmissible Agents

Outbreaks of *Bordetella pertussis* infection in the community have spread to outpatient settings within hospitals.[80,81] Varicella-zoster virus, influenza virus, and parvovirus B19 all have been transmitted in the inpatient setting, and many exposures to varicella-zoster virus occur in clinics and physicians' offices. However, we have not found published descriptions of outbreaks in the ambulatory setting caused by varicella-zoster virus, influenza virus, or parvovirus B19. Outbreaks may in fact occur, but they might not be recognized, or the persons who investigate them might not report them. The general infection control precautions recommended to prevent the spread of respiratory pathogens will help prevent spread of these etiologic agents. Vaccines currently are available for all of these agents except for parvovirus B19.

Infection Control Issues Related to Devices

Bronchoscopes and Gastrointestinal Endoscopes

Eleven million endoscopic procedures are performed in the United States each year, many of which are done in the outpatient setting. Many inpatient healthcare facilities have bronchoscopy or gastrointestinal endoscopy suites that serve both inpatients and outpatients. Private physicians and physicians in freestanding clinics also do endoscopy in their offices. Moreover, the number and type of procedures performed and the number of patients undergoing these procedures continue to increase.

Given the nature of the procedures and the limitations of surveillance systems, infections associated with endoscopic procedures are very difficult to identify unless they occur in clusters. Despite these limitations, numerous investigators have reported outbreaks of infection related to endoscopic procedures.[65,82,85-87,91-97] Most outbreaks could have been prevented if the staff who performed endoscopy had followed basic infection control procedures.

Flexible fiberoptic endoscopes have allowed physicians to perform procedures that are not possible with rigid endoscopes. However, flexible endoscopes present problems not encountered with rigid endoscopes. The most important problem is that flexible endoscopes are very difficult to clean and disinfect because they have several long, narrow internal channels and because they cannot be steam-sterilized. Consequently, endoscopes must be reprocessed by persons who understand the internal structure of the devices and who take the time necessary to clean and disinfect these devices meticulously. Although professional societies have published guide-

lines for reprocessing endoscopes,[82-84] the protocols used for those procedures have not been standardized,[88,89] and different endoscopes must be reprocessed by different methods. Moreover, because endoscopes are very expensive, endoscopy staff may try to save money by using a small number of endoscopes to evaluate and treat a large number of patients. Staff may not be able to process the equipment properly in the time allowed between patients.[90]

This topic has been well reviewed in articles by Spach et al.,[85] Ayliffe,[86] Weber and Rutala,[87] and Bronowicki et al.[101] Factors that have allowed flexible endoscopes to transmit pathogenic organisms have included the following:

- Failure to clean and disinfect the blind channel of endoscopes that is used to examine the biliary tree[91]
- Failure to disinfect endoscopes and accessory equipment (eg, suction-collection bottles and rubber connection tubing) properly between procedures[92,93]
- Failure to sterilize biopsy forceps[91,92,102,103]
- Failure to dry endoscopes completely before storage
- Failure to wash endoscopes before disinfecting them[96]
- Use of contaminated water
- Use of tap water to rinse endoscopes[95]
- Use of unfiltered tap water in automatic endoscope washers[98]
- Improper maintenance of automatic endoscope washers, with subsequent contamination[97]
- Use of inadequate disinfectants[93]
- Use of contaminated suction valves
- Use of endoscopes with internal defects that prevent adequate cleaning and disinfection

Despite numerous reports of outbreaks, staff in endoscopy suites often do not reprocess endoscopes properly. Several studies conducted in different countries document that staff in many endoscopy suites continue to make the same errors that have caused the outbreaks listed above.[89,90,99,100,104]

The policies and procedures needed in an endoscopy suite for infection prevention and control, cleaning, and disinfection are not more complicated or substantially different from those needed in other healthcare settings. However, the procedures must be followed precisely, because endoscopes are such complex semicritical instruments that they are difficult to clean even when the staff reprocess these devices conscientiously.[105] Staff might be intimidated by the complexity of the cleaning and disinfection process. To overcome these barriers, infection control staff should read the published recommendations for cleaning and disinfecting endoscopes[82-84] and should understand the basic principles of disinfection and sterilization.[106] Infection prevention staff also should spend time in the endoscopy suite to learn from the staff and to identify potential infection control problems. Together, staff from both programs should review the policies and procedures to ensure that they are consistent with current recommendations and guidelines and that they are consistent in intent and application throughout the healthcare facility.

Infection prevention personnel should be aware that new sterilizers that use ion plasma, vaporized hydrogen peroxide, and 100% ethylene oxide do not reliably kill organisms in narrow lumens if serum and salt are present.[107,108] Thus, these methods may not be adequate for processing flexible endoscopes, especially if they have not been cleaned scrupulously. Infection prevention personnel should discuss the limitations of these technologies with staff in the endoscopy suite and, at present, probably should discourage their use for endoscope cleaning.

Endoscopy clinics that are not a part of a larger medical center may not have direct access to infection control personnel or to training in infection control. Regardless of the endoscopy suite's location, staff must be trained to clean and disinfect endoscopes meticulously. The endoscopy staff also should understand basic principles of infection prevention and control, including methods for preventing transmission of routine bacterial pathogens to patients and to staff, methods for preventing spread of bloodborne pathogens, and methods for cleaning and disinfection. Dedicated personnel should be responsible to ensure that staff are educated properly, that they always comply with the specified procedures, and that all equipment is handled and maintained properly.

Disposable Devices

With the increased pressure for cost containment, healthcare workers and administrators are questioning why many devices marketed for single use cannot be reprocessed and reused. This issue is not unique to the outpatient setting, but, in the outpatient setting, the pressure to reuse single-use items might be greater, or the infection control oversight might be substantially less than in the inpatient setting. Examples of single-use items that often are reused in the outpatient setting include syringes, plastic vaginal specula, mouthpieces for pulmonary function machines, cardiac catheters, and oral airways. An ambulatory care facility that chooses to reprocess single-use devices must develop a comprehensive quality-assurance program to ensure that the products are cleaned, disinfected, or sterilized adequately and that the items retain their integrity and function. Institutions that reprocess those devices, not the manufacturers, assume the liability.

Thus, infection control personnel would be wise to identify which single-use products are being reused and to ensure that the devices are reprocessed and tested for function and integrity appropriately. Readers interested in a more in-depth discussion of this topic should read the review by Green.[109]

Steam Sterilization of Surgical Equipment

Steam sterilization of patient-care items for immediate use, or "flash sterilization," was designed to process items that had become contaminated during a sterile procedure but were essential for that procedure. The time required for flash sterilization is very short. Thus, all of the sterilization parameters (eg, time and temperature) must be met precisely. The persons performing the flash sterilization must keep meticulous records to document that these parameters have been met. In addition, flash sterilization will not work if the device is contaminated with organic matter or if air is trapped in or around the device. The efficacy of flash sterilization also will be impaired if either the sterilizer or the flash pack are not working properly. Moreover, because the devices are used immediately (ie, before the results of biological indicators are known), personnel in ambulatory surgery centers that use flash sterilization must record which devices were used for specific patients, so that patients can be observed if a device was not processed properly.

Many institutions have expanded the use of flash sterilization well beyond its intended and approved uses. Some of the changes are reasonable, but some jeopardize the efficacy of sterilization. Infection preventionists whose outpatient facilities use flash sterilization should read 2 pertinent publications by the Association for the Advancement of Medical Instrumentation,[110,111] then assess whether their institution is using this process properly.

Other Infection Control Issues

Multidose Vials

Multidose vials are used frequently in the inpatient and outpatient setting, because they may be more convenient and cheaper than single-dose vials. However, studies in the literature indicate that up to 27% of used multidose vials are contaminated with bacterial pathogens.[112-113] Although the actual risk of infection associated with multidose vials remains unknown, the risk of infection appears to be low if multidose vials are used properly. However, staff in a busy outpatient clinic might not adhere to appropriate infection control precautions.

Antimicrobial Resistant Pathogens

Antimicrobial resistance is increasing among many gram-positive, gram-negative, viral, and fungal pathogens. Examples include vancomycin-resistant enterococci,[114-116] MRSA, glycopeptide-intermediate *S. aureus*, and vancomycin-resistant *S. aureus*.[116-121] Many of these organisms are spread on the hands of personnel and by way of contaminated patient-care equipment or environmental surfaces. Consequently, the organisms could be spread in both the inpatient and outpatient settings. Infection prevention personnel, therefore, must work across the spectrum of healthcare settings (eg, inpatient units, central surgical suites, ambulatory surgical centers, endoscopy suites, hospital-based and freestanding ambulatory clinics, long-term care facilities, and home care) if they want to control the spread of these organisms effectively. Efforts to control antimicrobial-resistant organisms in the outpatient setting will cross disciplines, departments, and organizations. Hence, these efforts may be very difficult to coordinate. General infection prevention and control measures, such as containment of patients' secretions and excretions, proper hand hygiene, and appropriate environmental cleaning, can impact cross-transmission of these important pathogens. Specific recommendations have been made for vancomycin-resistant enterococci,[114,115] MRSA,[115] and vancomycin-resistant *S. aureus*.[122]

Medical Waste in the Outpatient Setting

Like hospitals, healthcare facilities and agencies that operate in the ambulatory setting produce infectious waste. However, these institutions may not have an expert in waste management on their staff, and they may not have ready access to appropriate processing equipment (eg, autoclaves) and disposal facilities (eg, incinerators). In addition, because these facilities often operate for profit, staff might feel pressured to take shortcuts.

An in-depth discussion of the management of medical waste is beyond the scope of this chapter. Readers who must develop waste-management programs for outpatient facilities should read the several excellent reviews on this topic.[106,123,124] We review a few of the basic principles, most of which are discussed by Morrison.[123]

The control plan must meet city, county, state, and federal regulations. Healthcare organizations must comply with all regulatory requirements; penalties can be imposed for noncompliance.

Infection prevention personnel should help other staff to take the following steps for managing waste:

- Define which waste items are noninfectious and which are infectious.

- Develop protocols and procedures for separating infectious waste from noninfectious waste, for labeling the infectious waste properly, and for transporting, storing, and disposing of infectious waste safely.
- Develop contingency plans for managing waste spills and inadvertent exposures of visitors or healthcare workers.
- Develop programs to teach staff to handle infectious waste.

Infection prevention personnel also could help staff identify ways to minimize infectious waste. Examples include the following:

- Stop discarding noninfectious waste, such as wrappers and newspapers, in infectious waste containers.
- Substitute products that do not require special modes of disposal (eg, needleless intravenous systems) for those that must be discarded in the infectious waste (eg, needles).

Infection Control in Special Settings

Dialysis Centers

Dialysis centers can be associated with hospitals (on-site or at a distance), or they can be independent. In the former instance, the infection prevention staff should ensure that the policies and procedures of the dialysis unit are consistent with those in the hospital. Independent dialysis units resemble other freestanding medical facilities in that they usually are for-profit organizations, and they rarely have trained infection prevention staff.

Bloodborne Pathogens

The primary infection control issue in dialysis centers is transmission of bloodborne pathogens, such as HBV, HCV, and HIV. To date, transmission of HBV has been the most common problem. Use of good infection control precautions and the hepatitis B vaccine have decreased the risk of transmission substantially in this setting. Staff members always should use standard precautions when they handle blood and other specimens, such as peritoneal fluid, which can contain high levels of hepatitis B surface antigen (HBsAg) and HBV.[125] In this section, we review some basic infection control precautions for dialysis units by summarizing the main points from an excellent review by Favero et al.[125] Infection prevention personnel who need detailed information should read that chapter and several other articles.[125-128]

Infection prevention and control precautions designed to prevent transmission of hepatitis B in dialysis centers include the following[125]:

- Patients and staff members should be screened for HBsAg and antibody to HBsAg (anti-HBsAg) when they enter the unit.
- Dialysis centers should survey susceptible patients and healthcare workers routinely for HBsAg and anti-HBsAg to determine whether transmission of HBV has occurred in the unit.
- Patients who are HBsAg carriers should undergo dialysis in a separate room designated only for use by such patients. If this is impossible, patients who are HBsAg carriers should undergo dialysis on dedicated machines in an area that is separated from the area in which HBV-seronegative patients undergo dialysis.
- Medications in multidose vials should not be shared among patients.
- Staff members should not care for patients who carry HBsAg and care for seronegative patients during the same shift.
- Ideally, the same hemodialysis equipment should not be used for both HBsAg-seropositive and HBsAg-seronegative patients. However, if this is not possible, the machines should be disinfected using conventional protocols, and the external surfaces should be cleaned or disinfected using soap and water or a detergent germicide.
- HBsAg-positive patients should not participate in dialyzer reuse programs.
- Centers that perform peritoneal dialysis should separate HBsAg-positive patients from those who are HBsAg-negative patients and should observe precautions similar to those used for hemodialysis patients, because peritoneal fluid can transmit HBV to susceptible persons.

Hepatitis D virus has been transmitted in dialysis centers. Therefore, patients who are infected with hepatitis D virus should undergo dialysis on dedicated machines in an area that is separated from all other dialysis patients, especially those who are HBsAg-positive.[125] If there is evidence that hepatitis D virus has been transmitted within the unit, patients should be screened for delta antigen and antibody to delta antigen.

Hepatitis C virus has caused outbreaks in dialysis centers.[129] However, there is no consensus on the use of specific precautions for patients who have antibody to HCV or for those who are infected with HIV.[124,129,130] General procedures that are recommended include the following:

- Staff should use precautions that limit exposure to blood and body fluids.[129]
- Staff always should clean and disinfect all instruments and environmental surfaces that are touched routinely.[129]

- Items should not be shared among patients.[129]
- The center should measure liver enzyme levels for all patients to determine whether any patients have evidence of non-A, non-B hepatitis, including HCV infection.[125]
- If liver enzyme levels in several patients increase during a short time period, staff should identify the etiology of each patient's hepatitis, so that the staff can determine whether an outbreak exists.[125]
- Hemodialysis centers could conduct serological surveys of their patient and staff populations to determine the baseline prevalence of hepatitis C in their center and to determine whether the prevalence has changed over time.[125]

If transmission of bloodborne pathogens occurs in a dialysis unit, staff must determine how the organism was transmitted and what infection control procedures have been violated. To facilitate such investigations, Favero et al.[125] have recommended that dialysis centers maintain detailed records on the following events and identifying information:

- The lot number of all blood and blood products used
- All mishaps, such as blood leaks or spills and dialysis machine malfunctions
- The location, name, and/or number of the dialysis machine used for each dialysis session
- The names of staff members who connect and disconnect the patient to and from a machine
- Results of serologic tests for hepatitis
- All accidental needle punctures and similar accidents sustained by staff members and patients

S. aureus

Patients receiving dialysis have a high prevalence of *S. aureus* nasal carriage (up to 81% of patients). *S. aureus* also is a leading cause of infections in dialysis patients. Moreover, the risk of *S. aureus* infection is significantly higher for dialysis patients who carry this organism in their nares than for patients who do not carry it.[131]

Most dialysis patients are infected by the *S. aureus* strains that they carry in their nares.[132,133] Thus, the most effective preventive strategies involve decolonization of the nares. Several investigators have documented that prophylactic use of rifampin or mupirocin can prevent *S. aureus* infections in dialysis patients.[131-135] However, mupirocin resistance can develop rapidly.

Approximately 10%–20% of *S. aureus* infections in peritoneal dialysis patients occur in noncarriers or are caused by strains other than those carried by the patients. Thus, healthcare workers in dialysis units, family members, and home healthcare staff must practice basic infection prevention and control precautions to prevent transmission of *S. aureus* to dialysis patients.[136-137] Of note, a dialysis center patient was the source of the first vancomycin-resistant *S. aureus* infection in the world.[120]

Outbreaks in Dialysis Centers

Numerous outbreaks of infection have occurred in dialysis centers.[43,44,48,129,149-153] Most of the outbreaks were caused by major deficiencies in basic infection prevention and control practices. The primary errors were inadequate or improper disinfection processes, poor compliance with precautions to prevent spread of bloodborne pathogens, and allowing HBsAg-positive and HBsAg-negative patients to share multidose vials of medications. Bacterial pathogens other than *S. aureus* that have been reported to cause outbreaks in dialysis patients include coagulase-negative staphylococci,[138] gram-negative bacilli (*Enterobacter cloacae, Pseudomonas aeruginosa,* and *Escherichia coli* from contaminated dialysis machine valves),[139] and *Pasturella multocida*.[140] Hotchkiss et al.[156] found that decontaminating the patient's environment, particularly the chair, could markedly affect pathogen dissemination. Scheduling appointments later in the day for patients known to be infection risks was also found to be effective in reducing the transmission potential.

In 1999, the CDC initiated a voluntary national system to monitor and prevent infections in outpatient dialysis centers. Results identified a vascular access infection rate of 3.2 cases per 100 patient-months overall.[141] However, rates varied significantly by type of vascular access: rates were lowest for natural arteriovenous fistulas and highest for noncuffed catheters (0.56 vs 11.98 cases per 100 patient-months),[141] confirming a finding that had been previously reported.[142] Infection prevention professionals armed with this knowledge can impact infection rates through education of physicians and dialysis staff regarding access selection, where appropriate.

Infection Prevention in Physical Therapy

Physical therapy facilities can be located in hospitals, in clinics, or can be freestanding. Many of these facilities provide services to outpatients. Physical therapy facilities have not been documented to be the source of outbreaks except in the case of infections associated with hydrotherapy. However, the absence of data is no cause to be sanguine, because all the elements necessary for transmission of pathogens exist in physical therapy centers. In addition, few investigators have looked rigorously for evidence of transmission in this type of setting.

The Association for Professionals in Infection Control and Epidemiology has published the only recommendations for infection control in physical therapy.[143] However, these recommendations provide limited guidance, because little information is available regarding appropriate infection control precautions specific for this setting.[144] We outline a few basic infection control principles for physical therapy centers that are summarized from an excellent review written by Linnemann.[144]

Infection prevention professionals who advise physical therapy units should ensure that the staff implement the following recommendations:

- All mats, table tops, and equipment handles should be covered with impervious materials, so that these items can be cleaned frequently.
- Cleaning supplies should be stored where they are readily accessible, so that staff can clean the equipment whenever necessary, not just at the scheduled times.
- Sinks for hand washing or dispensers of alcohol-based hand rub should be located such that physical therapists can wash or disinfect their hands easily after caring for each patient, or between caring for patients if they are helping more than 1 patient at a time.
- Physical therapists should be educated regarding standard precautions and the mechanisms by which organisms are transmitted.
- Physical therapists should understand that they could transmit infectious agents as they move from patient to patient and should know what precautions are necessary to prevent spread of pathogens.
- Therapists should wear gloves and gowns when they could contaminate their hands or clothing.
- Patients who have active infections caused by transmissible organisms or who are infected or colonized with resistant organisms should not use the facility. If such patients are allowed to use the facility, the staff must use appropriate precautions to prevent spread of these organisms.

OSHA Inspections

A full discussion of OSHA inspections in the ambulatory care setting is beyond the scope of this chapter (see Chapter 31, on regulatory inspections). However, infection control personnel must understand that OSHA's regulations also apply to facilities that provide care to outpatients. Persons who must ensure that their outpatient facility complies with OSHA's regulations must understand the Bloodborne Pathogen Standard[6] and the recommendations of the CDC tuberculosis guideline.[7]

Outpatient facilities associated with hospitals might be at greater risk of an OSHA inspection than are physicians' offices. However, OSHA does inspect offices operated by doctors of medicine and osteopathy.[145-146] Most inspections result from complaints filed by employees. Common violations in physicians' offices "involve personal protective equipment such as gloves and gowns, employee information and training requirements, records of training sessions, storage of those records, development of an exposure control plan, and procedures for laundering of personal protective equipment."[146] It is important to remember that outpatient settings are not exempt from requirements for implementation of safety features for sharp devices.

Recommendations

By now, you probably are overwhelmed completely and are wondering where to begin. Obviously, one person or one program cannot assess all the areas we discuss in this chapter. Thus, we would recommend that infection prevention staff assess their own institutions to answer the questions below to identify the characteristics of the facility.

- What type of outpatient facilities are present in your medical center? (eg, only hospital-based clinics, units that provide services to both inpatients and outpatients, only freestanding clinics, or a mixture of on-site and off-site clinics owned by the medical center).
- What types of patients are seen in the outpatient facilities? (eg, young children, immunocompromised patients, or healthy preoperative patients).
- What types of procedures are performed in the outpatient facilities?
- Which procedures are performed most commonly?
- What types of infectious diseases are diagnosed and treated in the outpatient facilities?
- What resources are available for infection control?
- Do the administration and the clinicians support the infection control program?

Once you have identified the primary characteristics of your facility, you should review the policies and procedures in the major areas and conduct walking rounds to answer the following questions:

- Does the area have policies, procedures, and engineering controls to prevent transmission of bloodborne pathogens?
- Does the area have appropriate screening, triage, isolation protocols, and engineering controls to prevent the airborne spread of pathogens?
- Does the area have appropriate screening, triage, and isolation for patients who may be infected or colonized with other infectious agents?

- Does the area have appropriate exposure management plans for bloodborne pathogens, measles virus, *M. tuberculosis*, varicella-zoster virus, *B. pertussis*, lice, and scabies?
- Does the staff understand and practice principles of asepsis, including those required to use multidose vials safely?
- Does the staff understand and practice appropriate cleaning, disinfection, and sterilization?
- Are the policies and procedures in this area consistent in content and intent with those from other areas in the medical center?
- Does the area have educational programs to teach staff precautions for respiratory pathogens, bloodborne pathogens, and other infectious agents? Do they document these programs adequately?

Once you have answered these questions, you should set priorities. We would suggest that all outpatient facilities should focus first on the 4 preventive measures listed below.

1. Ensure that the staff complies with the recommendations of the CDC tuberculosis guideline[7] and the Bloodborne Pathogen Standard[6] published by OSHA.
2. Implement "respiratory etiquette" protocols to prevent unnecessary patient and employee exposures.
3. Ensure that the staff and patients have been vaccinated appropriately.
4. Ensure that the staff practices good aseptic technique when handling multidose vials and that protocols and practices for cleaning, disinfection, and sterilization are appropriate.

Once they have addressed these priorities, infection control personnel then can address other issues, including whether or not to develop surveillance for surgical site infections after ambulatory surgery. For example, infection control personnel might be able to work with home-healthcare agencies to identify infections that are associated with inpatient medical care or with treatments given in hospital-based clinics or ambulatory surgery centers.

Conclusion

Infection prevention and control personnel who begin to address infection prevention and control needs in the ambulatory setting have not only a huge task to accomplish but also an enormous opportunity. Infection prevention and control personnel who take on this challenge will help to improve safety for the patient and the healthcare professional, as well as the quality of care provided, and they will help their institutions survive in the current extraordinarily competitive environment. In addition, those intrepid persons will help shape the course of health care over the next decade.

References

1. Lamkin L. Outpatient oncology settings: a variety of services. *Semin Oncol Nurs* 1994;10:227, 229–235.

2. Hing E and Middleton K. National Hospital Ambulatory Medical Care Survey: 2006 Outpatient Department Summary. Advance data from vital and health statistics; no. 338. Hyattsville, Maryland: National Center for Health Statistics; 2006.

3. Cherry D, Burt CW, Woodwell DA. National Hospital Ambulatory Medical Care Survey: 2006 Summary. Advance data from vital and health statistics; no. 337. Hyattsville, Maryland: National Center for Health Statistics; 2006.

4. Klevens RM, Edwards J, Richards C, et al. Estimating health care–associated infections and deaths in US hospitals, 2002. *Public Health Rep* 2007;122: 160–166.

5. Gastmeier P, Stamm-Balderjahn S, Hansen S, et al. How outbreaks can contribute to prevention of nosocomial infection: analysis of 1,022 outbreaks. *Infect Control Hosp Epidemiol* 2005;26: 357–361.

6. Department of Labor, Occupational Safety and Health Administration. 29 CFR Part 1920.1030, Occupational exposure to bloodborne pathogens, final rule. *Fed Regist* 56:64,004–64,182.

7. Centers for Disease Control and Prevention. Guidelines for preventing the transmission of *Mycobacterium tuberculosis* in health care facilities, 1994. *MMWR Recomm Rep* 1994; 43:1–132.

8. Joint Commission. Leadership Standard. In: *Comprehensive Accreditation Manual for Hospitals: The Official Handbook.* Oakbrook Terrace, IL: Joint Commission; 2009.

9. Lucet JC, Grenet K, Armand-Lefevre L, et al. High prevalence of carriage of methicillin-resistant *Staphylococcus aureus* at hospital admission in elderly patients: implications for infection control strategies. *Infect Control Hosp Epidemiol* 2005;26:121–126.

10. Mundy LM. Contamination, acquisition, and transmission of pathogens: implications for research and practices of infection control. *Infect Control Hosp Epidemiol* 2008;29: 590–592.

11. Herwaldt LA, Smith SD, Carter CD. Infection control in the outpatient setting. *Infect Control Hosp Epidemiol* 1998; 19:41–74.

12. Friedman C, Barnette M, Buck AS, et al. Requirements for infrastructure and essential activities of infection control and epidemiology in out-of-hospital settings: a consensus panel report. Association for Professionals in Infection Control and Epidemiology and Society for Healthcare Epidemiology of America. *Infect Control Hosp Epidemiol* 1999;20:695–705.

13. Nguyen GT, Proctor SE, Sinkowitz-Cochran RL, Garrett DO, Jarvis WI. Status of infection surveillance and control programs in the United States, 1992–1996. Association for

14. Friedman C. Infection control outside the hospital: developing a continuum of care. *Qual Lett Healthc Lead* 2000;12: 12–13.

15. Health Canada. Nosocomial and Occupational Infections Section. Development of a resource model for infection prevention and control programs in acute, long term, and home care settings: conference proceedings of the infection prevention and control alliance. *Am J Infect Control* 2004;32:2–6.

16. Nafziger DA, Lundstrom T, Chandra S, Massanari RM. Infection control in ambulatory care. *Infect Dis Clin North Am* 1997; 11(2):279–296.

17. Gould DJ, Drey NS, Millar M, et al. Patients and the public: knowledge, sources of information and perceptions about healthcare-associated infection. *J Hosp Infect* 2009;72:1–8.

18. Jarvis WR. Infection control and changing health-care delivery systems. *Emerg Infect Dis* 2001;7(2):170–173.

19. Smith PW, Roccaforte JS. Epidemiology and prevention of infections in home health care. In: Mayhall CG, ed. *Hospital Epidemiology and Infection Control.* Philadelphia: Williams & Wilkins; 2004:1881–1888.

20. Kohn WG, Collins AS, Cleveland JL, Harte JA, Eklund KJ, Malvitz DM, Centers for Disease Control and Prevention. Guidelines for infection control in dental health-care settings 2003. *MMWR Recomm Rep* 2003;52(R-17):1–61.

21. Huang SS, Mehta S, Weed D, Price CS. Methicillin-resistant *Staphylococcus aureus* survival on hospital fomites. *Infect Control Hosp Epidemiol* 2006;27:1267–69.

22. Neely AN, Maley MP. Survival of enterococci and staphylococci on hospital fabrics and plastic. *J Clin Microbiol* 2000; 38:724–746.

23. Hardy KJ, Oppenheim BA, Gossain S, Gao F, Hawkey PM. A study of the relationship between environmental contamination with methicillin-resistant *Staphylococcus aureus* (MRSA) and patients' acquisition of MRSA. *Infect Control Hosp Epidemiol* 2006;27:127–132.

24. Drees M, Snydman D, Schmid CH, et al. Prior environmental contamination increases the risk of acquisition of vancomycin-resistant enterococci. *Clin Infect Dis* 2008;46:678–685.

25. Blythe D, Keenlyside D, Dawson SJ, Galloway A. Environmental contamination due to methicillin-resistant *Staphylococcus aureus* (MRSA). *J Hosp Infect* 1998;38:67–70.

26. Harris AD. How important is the environment in the emergence of nosocomial antimicrobial-resistant bacteria? *Clin Infect Dis* 2008;46:686–688.

27. Snyder GM, Thom KA, Furuno JP. Detection of methicillin-resistant *Staphylococcus aureus* and vancomycin-resistant enterococci on the gown and gloves of healthcare workers. *Infect Control Hosp Epidemiol* 2008;29:583–589.

28. Dietze B, Rath A, Wendt C, Martiny H. Survival of MRSA on sterile goods packaging. *J Hosp Infect* 2001;49:255–261.

29. Bhalla A, Pultz NJ, Gries DM, et al. Acquisition of nosocomial pathogens on hands after contract with environmental surfaces near hospitalized patients. *Infect Control Hosp Epidemiol* 2004;25:164–167.

30. Goodman ER, Platt R, Bass R, et al. Impact of an environmental cleaning intervention on the presence of methicillin-resistant *Staphylococcus aureus* and vancomycin-resistant enterococci on surfaces in intensive care unit rooms. *Infect Control Hosp Epidemiol* 2008;29:593–599.

31. Hayden MK, Bonten MJ, Blom DW, et al. Reduction in acquisition of vancomycin-resistant *Enterococcus* after enforcement of routine environmental cleaning measures. *Clin Infect Dis* 2006;42:1552–1560.

32. Rampling A, Wiseman S, Davis L, et al. Evidence that hospital hygiene is important in the control of methicillin-resistant *Staphylococcus aureus*. *J Hosp Infect* 2001;49:109–116.

33. Bures S, Fishbain JT, Uyehara CF, Parker JM, Berg BW. Computer keyboards and faucet handles as reservoirs of nosocomial pathogens in the intensive care unit. *Am J Infect Control* 2000;28:465–471.

34. Hota B. Contamination, disinfection, and cross-colonization: are hospital surfaces reservoirs for nosocomial infection? *Clin Infect Dis* 2004;39:1182–1189.

35. Kelen GD, Hansen KN, Green GB, Tang N, Ganguli C. Determinants of emergency department procedure- and condition-specific universal (barrier) precaution requirements for optimal provider protection. *Ann Emerg Med* 1995;25:743–750.

36. Kelen GD, Green GB, Purcell RH, et al. Hepatitis B and hepatitis C in emergency department patients. *N Engl J Med* 1992;326:1399–1404.

37. Sivapalasingam S, Malak SF, Sullivan JF, Lorch J, Sepkowitz KA. High prevalence of hepatitis C infection among patients receiving hemodialysis at an urban dialysis center. *Infect Control Hosp Epidemiol* 2002;23:319–324.

38. Kelen GD, Green GB, Hexter DA, et al. Substantial improvement in compliance with universal precautions in an emergency department following institution of policy. *Arch Intern Med* 1991;151:2051–2056.

39. Miller KE, Krol RA, Losh DP. Universal precautions in the family physician's office. *J Fam Pract* 1992;35:163–168.

40. OSHA Compliance Document. Occupational exposure to bloodborne pathogens; needlesticks and other sharp injuries; Final rule. 29 C.F.R. Part 1910 (January 18, 2001).

41. Federal Needlestick Safety and Prevention Act Pub. L. 106–430 (2000).

42. Canter J, Mackey K, Good LS, et al. An outbreak of hepatitis B associated with jet injections in a weight reduction clinic. *Arch Intern Med* 1990;150:1923–1927.

43. Danzig LE, Tormey MP, Sinha SD, et al. Common source transmission of hepatitis B virus infection in a hemodialysis unit [abstract]. *Infect Control Hosp Epidemiol* 1995;16(suppl):19.

44. Rosenberg J, Gilliss DL, Moyer L, Vugia D. A double outbreak of hepatitis B in a dialysis center [abstract]. *Infect Control Hosp Epidemiol* 1995;16(suppl):19.

45. Hlady WG, Hopkins RS, Ogilby TE, Allen ST. Patient-to-patient transmission of hepatitis B in a dermatology practice. *Am J Public Health* 1993;83:1689–1693.

46. Drescher J, Wagner D, Haverich A, et al. Nosocomial hepatitis B virus infections in cardiac transplant recipients transmitted during transvenous endomyocardial biopsy. *J Hosp Infect* 1994;26:81–92.

47. Ciesielski C, Marianos D, Ou CY, et al. Transmission of human immunodeficiency virus in a dental practice. *Ann Intern Med* 1992;116:798–805.

48. Velandia M, Fridkin SK, Cardenas V, et al. Transmission of HIV in dialysis centre. *Lancet* 1995;345:1417–1422.

49. Chant K, Lowe D, Rubin G, et al. Patient-to-patient transmission of HIV in private surgical consulting rooms. *Lancet* 1993;342:1548–1549.

50. Centers for Disease Control and Prevention. Transmission of hepatitis B and C viruses in outpatient settings—New York, Oklahoma, and Nebraska, 2000–2002. *MMWR Morb Mortal Wkly Rep* 2003; 52; 901–906.

51. Katzenstein TL, Jorgensen LB, Permin H, et al. Nosocomial HIV-transmission in an outpatient clinic detected by epidemiologic and phylogenetic analyses. *AIDS* 1999;13: 1779–1781.

52. Centers for Disease Control and Prevention. Improper infection-control practices during employee vaccination programs—District of Columbia and Pennsylvania, 1993. *MMWR Morb Mortal Wkly Rep* 1993;42:969–971.

53. Thompson ND, Perz JF, Moorman AC, Holmberg SD. Non-hospital health care—associated hepatitis B and C virus transmission: United States, 1998–2008. *Ann Intern Med* 2009;150:33–39.

54. Boyce JM, Pittet D; Healthcare Infection Control Practices Advisory Committee; HICPAC/SHEA/APIC/IDSA Hand Hygiene Task Force. Guideline for hand hygiene in health-care settings. *MMWR Recomm Rep* 2002; 51 (RR-16):1–45.

55. Cohen HA, Kitai E, Levy I, Ben-Amitai D. Handwashing patterns in two dermatology clinics. *Dermatology* 2002; 205(4):358–61.

56. Cohen HA, Matalon A, Amir J, Paret G, Barzilai A. Handwashing patterns in primary pediatric community clinics. *Infection* 1998;26:45–47.

57. Aizman A, Stein JD, Stenson SM. A survey of patterns of physician hygiene in ophthalmology clinic patients encounters. *Eye Contact Lens* 2003;29:221–222.

58. Bridges CB, Keuhnert MJ, Hall CB. Transmission of influenza: implications for control in health care settings. *Clin Infect Dis* 2003;37:1094–1101.

59. Centers for Disease Control and Prevention. Public health guidance document for community-level preparedness and response to severe acute respiratory syndrome (SARS). 2005. http://www.cdc.gov/ncidod/SARS/guidance/. Accessed October 7, 2009.

60. Centers for Disease Control and Prevention. Trends in tuberculosis—United States, 2008. *MMWR Morb Mortal Wkly Rep* 2009;58:249–253.

61. Calder RA, Duclos P, Wilder MH, Pryor VL, Scheel WJ. *Mycobacterium tuberculosis* transmission in a health clinic. *Bull Int Union Tuberc Lung Dis* 1991;66:103–106.

62. Fischl MA, Uttamchandani RB, Daikos GL, et al. An outbreak of tuberculosis caused by multiple-drug–resistant tubercle bacilli among patients with HIV infection. *Ann Intern Med* 1992;117:177–183.

63. Couldwell DL, Dore GJ, Harkness JL, et al. Nosocomial outbreak of tuberculosis in an outpatient HIV treatment room. *AIDS* 1996;10:521–525.

64. Moore M, the Investigative Team. Evaluation of transmission of tuberculosis in a pediatric setting—Pennsylvania. In: Program and abstracts of the 46th Annual Conference of the Epidemic Intelligence Service; April 14–18, 1997; Atlanta, GA:53.

65. Agerton TB, Valway S, Gore B, Poszik C, Onorato I. Transmission of multidrug-resistant tuberculosis via bronchoscopy. In: Program and abstracts of the 46th Annual Conference of the Epidemic Intelligence Service; April 14–18, 1997; Atlanta, GA:54.

66. Griffith DE, Hardeman JL, Zhang Y, Wallace RJ, Mazurek GH. Tuberculosis outbreak among HCW in a community hospital. *Am J Respir Crit Care Med* 1995;152:808–811.

67. Sokolove PE, Mackey D, Wiles J, Lewis RJ. Exposure of emergency department personnel to tuberculosis: PPD testing during an epidemic in the community. *Ann Emerg Med* 1994;24:418–421.

68. Sepkowitz KA. AIDS, tuberculosis, and the health care worker. *Clin Infect Dis* 1995;20:232–242.

69. Moran GJ, McCabe F, Morgan MT, Talan DA. Delayed recognition and infection control for tuberculosis patients in the emergency department. *Ann Emerg Med* 1995;26:290–295.

70. Moran GJ, Fuchs MA, Jarvis WR, Talan DA. Tuberculosis infection-control practices in United States emergency departments. *Ann Emerg Med* 1995;26:283–289.

71. Centers for Disease Control. Measles—Washington, 1990. *MMWR Morb Mortal Wkly Rep* 1990;39:473–476.

72. Miranda AC, Falcao JM, Dias JA. Measles transmission in health facilities during outbreaks. *Int J Epidemiol* 1994;23:843–848.

73. Krause PJ, Gross PA, Barrett TL, et al. Quality standard for assurance of measles immunity among health care workers. *Clin Infect Dis* 1994;18:431–436.

74. Centers for Disease Control and Prevention. Epidemic keratoconjunctivitis in an ophthalmology clinic—California. *MMWR Morb Mortal Wkly Rep* 1990;39:598–601.

75. Colon LE. Keratoconjunctivitis due to adenovirus type 8: report on a large outbreak. *Ann Ophthalmol* 1991;23:63–65.

76. Jernigan JA, Lowry BS, Hayden FG, et al. Adenovirus type 8 epidemic keratoconjunctivitis in an eye clinic: risk factors and control. *J Infect Dis* 1993;167:1307–1313.

77. Smith D, Gottsch J, Froggatt J, Dwyer D, Karanfil L, Groves C. Performance improvement process to control epidemic keratoconjunctivitis transmission [abstract]. *Infect Control Hosp Epidemiol* 1996;17(suppl):36.

78. American Academy of Ophthalmology (AAO). Updated recommendations for Ophthalmic Practice in Relation to the Human Immunodeficiency Virus and Other Infectious Agents. San Francisco: AAO; 1992.

79. Dailey JR, Parnes RE, Aminlari A. Glutaraldehyde keratopathy. *Am J Ophthal* 1993;115:256–258.

80. Christie CDC, Marx ML, Marchant CD, Reising SF. The 1993 epidemic of pertussis in Cincinnati. Resurgence of disease in a highly immunized population of children. *N Engl J Med* 1994;331:16–21.

81. Hardy IRB, Strebel PM, Wharton M, Orenstein WA. The 1993 pertussis epidemic in Cincinnati. *N Engl J Med* 1994;331:1455–1455.

82. Members of the American Society for Gastrointestinal Endoscopy Ad Hoc Committee on Disinfection. Position statement: reprocessing of flexible gastrointestinal endoscopes. *Gastrointest Endosc* 1996;43:540–546.

83. Rutala WA. APIC guideline for selection and use of disinfectants. *Am J Infect Control* 1996;24:313–342.

84. Nelson DB, Jarvis WR, Rutala WA, et al; Society for Healthcare Epidemiology of America. Multi-society guideline for reprocessing flexible gastrointestinal endoscopes. *Infect Control Hosp Epidemiol* 2003;24:532–537.

85. Spach DH, Silverstein FE, Stamm WE. Transmission of infection by gastrointestinal endoscopy and bronchoscopy. *Ann Intern Med* 1993;118:117–128.

86. Ayliffe GA. Nosocomial infections associated with endoscopy. In: Mayhall CG, ed. *Hospital Epidemiology and Infection Control*. Philadelphia: Williams & Wilkins; 1996:680–693.

87. Weber DJ, Rutala WA. Nosocomial infections associated with respiratory therapy. In: Mayhall CG, ed. *Hospital Epidemiology and Infection Control*. Philadelphia: Williams & Wilkins; 1996:748–758.

88. Muscarella LF. Advantages and limitations of automatic flexible endoscope reprocessors. *Am J Infect Control* 1996;24:304–309.

89. Reynolds CD, Rhinehart E, Dreyer P, Goldmann DA. Variability in reprocessing policies and procedures for flexible fiberoptic endoscopes in Massachusetts hospitals. *Am J Infect Control* 1992;20:283–290.

90. Foss D, Monagan D. A national survey of physicians' and nurses' attitudes toward endoscope cleaning and the potential for cross-infection. *Gastroenterol Nurs* 1992;15:59–65.

91. Struelens MJ, Rost F, Deplano A, et al. *Pseudomonas aeruginosa* and Enterobacteriacee bacteremia after biliary endoscopy: an outbreak investigation using DNA macrorestriction analysis. *Am J Med* 1993;95:489–498.

92. Langenberg W, Rauws EAJ, Oudbier JH, Tytgat GNJ. Patient-to-patient transmission of *Campylobacter pylori* infection by fiberoptic gastroduodenoscopy and biopsy. *J Infect Dis* 1990;161:507–511.

93. Akamatsu T, Tabata K, Hironga M, Kawakami H, Uyeda M. Transmission of *Helicobacter pylori* infection via flexible fiberoptic endoscopy. *Am J Infect Control* 1996;24:396–401.

94. Wang H-C, Liaw Y-S, Yang P-C, Kuo S-H, Luh K-T. A pseudoepidemic of *Mycobacterium chelonae* infection caused by contamination of a fiberoptic bronchoscope suction channel. *Eur Respir J* 1995;8:1259–1262.

95. Nye K, Shadha DK, Hodgkin P, Bradley C, Hancox J, Wise R. *Mycobacterium chelonei* isolation from broncho-alveolar lavage fluid and its practical implications. *J Hosp Infect* 1990;16:257–261.

96. Kolmos HJ, Lerche A, Kristoffersen K, Rosdahl VT. Pseudo-outbreak of *Pseudomonas aeruginosa* in HIV-infected patients undergoing fiberoptic bronchoscopy. *Scand J Infect Dis* 1994;26:653–657.

97. Umphrey J, Raad I, Tarrand J, Hill LA. Bronchoscopes as a contamination source of *Pseudomonas putida* [abstract]. *Infect Control Hosp Epidemiol* 1996;17(suppl):42.

98. Reeves DS, Brown NM. Mycobacterial contamination of fibreoptic bronchoscopes. *J Hosp Infect* 1995;30(suppl):531–536.

99. Kaczmarek RG, Moore RM, McCrohan J, et al. Multi-state investigation of the actual disinfection/sterilization of endoscopes in health care facilities. *Am J Med* 1992;92:257–100.

100. Rutala WA, Clontz EP, Weber DJ, Hoffmann KK. Disinfection practices for endoscopes and other semicritical items. *Infect Control Hosp Epidemiol* 1991;12:282–288.

101. Bronowicki JP, Venard V, Botte C, et al. Patient-to-patient transmission of hepatitis C virus during colonoscopy (published correction appears in *N Engl J Med* 2001;344:392). *N Engl J Med* 1997; 337:237–240.

102. Lessa F, Tak S, DeVader SR, et al. Risk of infections associated with improperly reprocessed transrectal ultrasound-guided prostate biopsy equipment. *Infect Control Hosp Epidemiol* 2008;29:289–293.

103. Gillespie JL, Arnold KE, Noble-Wang J, et al. Outbreak of *Pseudomonas aeruginosa* infections after transrectal ultrasound–guided prostate biopsy. *Urology* 2007;69:912–914.

104. Gorse GJ, Messner RL. Infection control practices in gastrointestinal endoscopy in the United States: a national survey. *Infect Control Hosp Epidemiol* 1991;12:289–296.

105. Favero MS. Strategies for disinfection and sterilization of endoscopes: the gap between basic principles and actual practice. *Infect Control Hosp Epidemiol* 1991;12:279–281.

106. Rutala WA. Disinfection, sterilization, and waste disposal. In: Wenzel RP, ed. *Prevention and Control of Nosocomial Infections*. 3rd ed. Philadelphia: Williams & Wilkins; 1997: 539–593.

107. Alfa MJ, DeGagne P, Olson N, Puchalski T. Comparison of ion plasma, vaporized hydrogen peroxide, and 100% ethylene oxide sterilizers to the 12/88 ethylene oxide gas sterilizer. *Infect Control Hosp Epidemiol* 1996;17:92–100.

108. Rutala WA, Weber DJ. Low-temperature sterilization technologies: do we need to redefine "sterilization"? *Infect Control Hosp Epidemiol* 1996;17:87–91.

109. Greene VW. Reuse of disposable devices. In: Mayhall CG, ed. *Hospital Epidemiology and Infection Control*. Philadelphia: Williams & Wilkins; 1996:946–954.

110. Association for the Advancement of Medical Instrumentation. AAMI—Good hospital practice: flash sterilization—steam sterilization of patient care items for immediate use (ST37). Arlington, VA: AAMI Steam Sterilization Hospital Practices Working Group, AAMI Sterilization Standards Committee; 1996.

111. Association for the Advancement of Medical Instrumentation (AAMI). Good Hospital Practice: Guidelines for the Selection and Use of Reusable Rigid Sterilization Container Systems (ST33). Arlington, VA: AAMI; 1996.

112. Melnyk PS, Shevchuk YM, Conly JM, Richardson CJ. Contamination study of multiple-dose vials. *Ann Pharmacother* 1993;27:274–277.

113. Mattner F, Gastmeier P. Bacterial contamination of multiple-dose vials: a prevalence study. *Am J Infect Control* 2004;32: 12–16.

114. Hospital Infection Control Practices Advisory Committee (HICPAC). Recommendations for preventing the spread of vancomycin resistance. *Infect Control Hosp Epidemiol* 1995; 16:105–113.

115. Muto CA, Jernigan JA, Ostrowsky BE, et al. SHEA guideline for preventing nosocomial transmission of multidrug-resistant strains of *Staphylococcus aureus* and *Enterococcus*. *Infect Control Hosp Epidemiol* 2003;24:362–386.

116. Frieden TR, Munsiff SS, Low DE, et al. Emergence of vancomycin-resistant enterococci in New York City. *Lancet* 1993:342:76.

117. Centers for Disease Control and Prevention. Update: *Staphylococcus aureus*: reduced susceptibility of *Staphylococcus aureus* to vancomycin—United States, 1997. *MMWR Morb Mortal Wkly Rep* 1997;46:813.

118. Sieradzki K, Roberts RB, Haber SW. The development of vancomycin resistance in a patient with methicillin-resistant *Staphylococcus aureus* infection. *N Engl J Med* 1999; 340:517.

119. Fridkin SK, Hageman J, McDougal LK, et al., Vancomycin-Intermediate *Staphylococcus aureus* Epidemiology Study Group. Epidemiological and microbiological characterization of infections caused by *Staphylococcus aureus* with reduced susceptibility to vancomycin, United States, 1997–2001. *Clin Infect Dis* 2003;36:429–39.

120. Chang S, Sievert D, Hageman J, et al. Infection with vancomycin-resistant *Staphylococcus aureus* containing the *vanA* resistance gene. *N Engl J Med* 2003;348:121.

121. Centers for Disease Control and Prevention. Public health dispatch: vancomycin-resistant *Staphylococcus aureus*—Pennsylvania, 2002. *MMWR Morb Mortal Wkly Rep* 2002; 51(40):902.

122. Centers for Disease Control and Prevention. Preventing the spread of vancomycin resistance-report from the Hospital Infection Control Practices Advisory Committee. *Fed Regist* May 17, 1994.

123. Morrison AJ Jr. Infection control in the outpatient setting. In: Wenzel RP, ed. *Prevention and Control of Nosocomial Infections*. 2nd ed. Philadelphia: Williams & Wilkins; 1993:89–92.

124. Reinhardt PA, Gordon JG, Alvarado CJ. Medical waste management. In: Mayhall CG, ed. *Hospital Epidemiology and Infection Control*. Philadelphia: Williams & Wilkins; 1996: 1099–1108.

125. Favero MS, Alter MJ, Bland LE. Nosocomial infections associated with hemodialysis. In: Mayhall CG, ed. *Hospital Epidemiology and Infection Control*. Philadelphia: Williams & Wilkins; 1996:693–714.

126. Alter MJ, Favero MS, Moyer LA, Bland LA. National surveillance of dialysis-associated diseases in the United States, 1989. *ASAIO Trans* 1991;37:97–109.

127. Band JD. Nosocomial infections associated with peritoneal dialysis. In: Mayhall CG, ed. *Hospital Epidemiology and Infection Control*. Philadelphia: Williams & Wilkins; 1996: 714–725.

128. Garcia-Houchins S. Dialysis. In: Olmsted RN, ed. *APIC Infection Control and Applied Epidemiology: Principles and Practice*. St Louis: Mosby–Year Book; 1996:89-1–89-15.

129. Niu MT, Alter JM, Kristensen C, Margolis HS. Outbreak of hemodialysis-associated non-A, non-B hepatitis and correlation with antibody to hepatitis C virus. *Am J Kidney Dis* 1992;19:345–352.

130. Delarocque-Astagneau E, Baffoy N, Thiers V, et al. Outbreak of hepatitis C virus infection in a hemodialysis unit: potential transmission by the hemodialysis machine? *Infect Control Hosp Epidemiol* 2002;23:328–334.

131. Luzar MA, Coles GA, Faller B, et al. *Staphylococcus aureus* nasal carriage and infection in patients on continuous ambulatory peritoneal dialysis. *N Engl J Med* 1990;322:505–509.

132. Ena J, Boelaert JR, Boyken L, Van Landuyt HW, Godard CA, Herwaldt LA. Epidemiology of *Staphylococcus aureus* infections in patients on hemodialysis. *Infect Control Hosp Epidemiol* 1994;15:78–81.

133. Pignatari A, Pfaller M, Hollis R, Sesso R, Leme I, Herwaldt L. *Staphylococcus aureus* colonization and infection in patients on continuous ambulatory peritoneal dialysis. *J Clin Microbiol* 1990;28:1898–1902.

134. Boelaert JR, Van Landuyt HW, Godard CA, et al. Nasal mupirocin ointment decreases the incidence of *Staphylococcus aureus* bacteremias in haemodialysis patients. *Nephrol Dial Transplant* 1993;8:235–239.

135. The Mupirocin Study Group. Nasal mupirocin prevents *Staphylococcus aureus* exit-site infection during peritoneal dialysis. *J Am Soc Nephrol* 1996;7:2403–2408.

136. Dryden MS, McCann M, Phillips I. Housewife peritonitis: conjugal transfer of a pathogen. *J Hosp Infect* 1991;17:69–70.

137. Herwaldt LA, Boyken LD, Coffman S. Epidemiology of *S. aureus* nasal carriage in patients on continuous ambulatory peritoneal dialysis who were in a multicenter trial of mupirocin. In: Program and abstracts of the 36th Interscience Conference on Antimicrobial Agents and Chemotherapy; September 15–18, 1996; New Orleans:233.

138. Spare MK, Tebbs SE, Lang S, et al. Genotypic and phenotypic properties of coagulase-negative staphylocooci causing dialysis catheter-related sepsis. *J Hosp Infect* 2003;54:272–278.

139. Wang SA, Levine RB, Carson LA, et al. An outbreak of gram-negative bacteremia in hemodialysis patients traced to hemodialysis machine waste drain ports. *Infect Control Hosp Epidemiol* 1999;20:746–751.

140. Kitching AR, Macdonald A, Hatfield PJ. *Pasturella multocida* infection in continuous ambulatory peritoneal dialysis. *N Z Med J* 1996;109(1016):59.

141. Tokars JI, Miller ER, Stein G. New national surveillance system for hemodialysis-associated infections: initial results. *Am J Infect Control* 2002;30:288–295.

142. Stevenson KB, Adcox MJ, Mallea MC, Narasimhan N, Wagnild JP. Standardized surveillance of hemodialysis vascular access infections: 18-month experience at an outpatient, multifacility hemodialysis center. *Infect Control Hosp Epidemiol* 2000;21:200–203.

143. Temple RS. Physical medicine and rehabilitation/occupational therapy/speech. In: Olmsted RN, ed. *APIC Infection Control and Applied Epidemiology: Principles and Practice*. St Louis: Mosby–Year Book; 1996:114-1–114-5.

144. Linnemann CC. Nosocomial infections associated with physical therapy, including hydrotherapy. In: Mayhall CG, ed. *Hospital Epidemiology and Infection Control*. Philadelphia: Williams & Wilkins: 1996:725–730.

145. Zuber TJ, Geddie JE. Occupational safety and health administration regulations for the physician's office. *J Fam Pract* 1993;36:540–550.

146. Favero MS, Sadovsky R. Office infection control, OSHA, and you. *Patient Care* 1993;27:117–134.

147. Wenger JD, Spika JS, Smithwick RW, et al. Outbreak of *Mycobacterium chelonae* infection associated with use of jet injectors. *JAMA* 1990;264:373–376.

148. Pegues DA, Carson LA, Anderson Rl, et al. Outbreak of *Pseudomonas cepacia* bacteremia in oncology patients. *Clin Infect Dis* 1993;16:407–411.

149. Gordon SM, Drachman J, Bland LA, Reid MH, Favero M, Jarvis WR. Epidemic hypotension in a dialysis center caused by sodium azide. *Kidney Int* 1990;37:110–115.

150. Longfield RN, Wortham WG, Fletcher LL, Nauscheutz WF. Clustered bacteremias in a hemodialysis unit: cross-contamination of blood tubing from ultrafiltrate waste. *Infect Control Hosp Epidemiol* 1992;13:160–164.

151. Flaherty JP, Garcia-Houchins S, Chudy R, Arnow PM. An outbreak of gram-negative bacteremia traced to contaminated O-rings and reprocessed dialyzers. *Ann Intern Med* 1993;119:1072–1078.

152. Beck-Sague CM, Jarvis WR, Bland LA, Arduino MJ, Aguero SM, Verosic G. Outbreak of gram-negative bacteremia and pyrogenic reactions in a hemodialysis center. *Am J Nephrol* 1990;10:397–403.

153. Fridkin SK, Kremer FB, Bland LA, Padhye A, McNeil MM, Jarvis W. *Acremonium kiliense* endophthalmitis that occurred after cataract extraction in an ambulatory surgical center and was traced to an environmental source. *Clin Infect Dis* 1996;22:222–227.

154. Hopkins DP, Cicirello H, Dievendorf G, Kondracki S, Morse D. An outbreak of culture-negative peritonitis in dialysis patients—New York. In: Program and abstracts of the 46th Annual Conference of the Epidemic Intelligence Service; April 14–18, 1997; Atlanta, GA.

155. Hammons T, Piland N, Small S, Hatlie M, Burstein H. Ambulatory patient safety—what we know and need to know. *J Ambul Care Manage* 2003; 26(1):63–82.

156. Hotchkiss JR, Holley P, Crooke PS. Analyzing pathogen transmission in the dialysis unit: time for a (schedule) change? *Clin J Am Soc Nephrol* 2007;2:1176–1185.

Additional Resources

American Academy of Ophthalmology (AAO). Updated recommendations for ophthalmic practice in relation to the human immunodeficiency virus and other infectious agents. (May be obtained from AAO at 655 Beach St, San Francisco, CA 94109. http://www.eyenet.org.)

Association for Professionals in Infection Control and Epidemiology (APIC). Guide to the elimination of methicillin-resistant *Staphylococcus aureus* (MRSA) transmission in hospital settings. Washington, DC: APIC; 2007.

Association for Professionals in Infection Control and Epidemiology (APIC). Guide to the elimination of *Clostridium difficile* in hospital settings. Washington, DC: APIC; 2008. http://www.apic.org/AM/Template.cfm?Section=APIC_Elimination_Guides. Accessed October 7, 2009.

American Society of Gastrointestinal Endoscopy (ASGE), Society for Healthcare Epidemiology of America. Multi-society guideline for reprocessing of flexible gastrointestinal endoscopes. *Gastrointest Endosc* 2003;58:1–8. http://www.asge.org/PublicationsProductsIndex.aspx?id=352. Accessed October 7, 2009.

Canadian Healthcare Association (CHA). *The Reuse of Single-Use Medical Devices: Guidelines for Healthcare Facilities*. Ottawa: CHA Press, 1996.

Centers for Disease Control and Prevention (CDC). *Core Curriculum on Tuberculosis: What the Clinician Should Know*. 3rd ed. Atlanta, GA: CDC; 1994. (This print version is in revision; the most up-to-date information is available as an interactive course, at http://www.cdc.gov/tb/webcourses/corecurr/index.htm. Accessed October 2, 2009.)

Centers for Disease Control and Prevention (CDC). *Pseudomonas aeruginosa* infections associated with transrectal ultrasound-guided prostate biopsies—Georgia, 2005. *MMWR Morb Mortal Wkly Rep* 2006;55:776–777.

Department of Labor, Occupational Health and Safety Administration. Occupational exposure to tuberculosis. *Fed Regist* 1997;62(201).

Diosegy AJ, Lord MC. What physicians need to know about OSHA: how to avoid tough new penalties. *N Carol Med J* 1993;54:251–254.

Emergency Care Research Institute (ECRI). Medical Device Special Report: FDA issues statement on reuse of single-use devices [Normal Priority Medical Device Alert]. 2006. https://www.ecri.org/Documents/Device_Reuse/single_use_hda_alert_10_06.pdf. Accessed October 7, 2009.

FDA public health notification: reprocessing of reusable ultrasound transducer assemblies used for biopsy procedures. June 19, 2006. http://www.fda.gov/cdrh/safety/061906-ultrasound-transducers.html.

Harris AD, McGregor JC, Furuno JP. What infection control interventions should be undertaken to control multidrug-resistant gram-negative bacteria? *Clin Infect Dis* 2006; 43 (suppl 2): S57–S61.

Heroux DL. Ambulatory care. In: Olmsted RN, ed. *APIC Infection Control and Applied Epidemiology: Principles and Practice.* St Louis: Mosby–Year Book; 1996:83-1–83-15.

Heroux D, Garris J, Nahan J, Vivolo P. *Ambulatory Care Infection Control Manual.* Seattle: Group Health Cooperative of Puget Sound; 1993.

Hidron AI, Edwards JR, Patel J, et al. Antimicrobial-resistant pathogens associated with healthcare-associated infections: annual summary of data reported to the national healthcare safety network at the centers for disease control and prevention, 2006–2007. *Infect Control Hosp Epidemiol* 2008;29: 996–1011.

Jensen PA, Lambert LA, Iademarco MF, Ridzon R, Centers for Disease Control and Prevention (CDC). Guidelines for preventing the transmission of *Mycobacterium tuberculosis* in health-care settings, 2005. *MMWR Recomm Rep* 2005;54(RR-17):1-141.

Morris J, Duckworth GJ, Ridgway GL. Gastrointestinal endoscopy decontamination failure and the risk of transmission of blood-borne viruses: a review. *J Hosp Infect* 2006;63:1–13.

Muto C, Jernigan J, Ostrowsky B, et al. SHEA guideline for preventing nosocomial transmission of multidrug-resistant strains of *Staphylococcus aureus* and *Enterococcus. Infect Control Hosp Epidemiol* 2003;24:362–386.

Rothschild P. Preventing infection in MRI–best practices: infection control in and around MRI suites. 2008. http://www.patient-comfortsystems.com/. Accessed October 7, 2009.

Rutala WA, Weber DJ. How to assess risk of disease transmission to patients when there is a failure to follow recommended disinfection and sterilization guidelines. *Infect Control Hosp Epidemiol* 2007;28:146–155.

Siegel J, Rhinehart E, Jackson, M. Chiarello L, Healthcare Infection Control Practices Advisory Committee. Management of multidrug -resistant organisms in healthcare settings, 2006. http://www.cdc.gov/ncidod/dhqp/guidelines.html. Accessed: October 7, 2009.

Siegel JD, Rhinehart E, Jackson M, Chiarello L, Healthcare Infection Control Practices Advisory Committee 2007. Guideline for isolation precautions: preventing transmission of infectious agents in healthcare settings. *Am J Infect Control* 2007;35(10 suppl 2):S65–S164.

Stone PW, Braccia D, Larson E. Systematic review of economic analysis of health care-associated infections. *Am J Infect Control* 2005;33:501–509.

Vanhems P, Gayet-Ageron A, Ponchon T, et al. Follow-up and management of patients exposed to a flawed automated endoscope washer disinfector in a digestive diseases unit. *Infect Control Hosp Epidemiol* 2006;27:89–92.

Wong ES, Rupp ME, Mermel L, et al. Public disclosure of healthcare-associated infections: the role of the Society for Healthcare Epidemiology of America. *Infect Control Hosp Epidemiol* 2005;26:210–212.

Chapter 28 Infection Prevention in Resource-Limited Settings

Anucha Apisarnthanarak, MD, M. Cristina Ajenjo, MD, and Virginia R. Roth, MD

Healthcare-associated infections (HAIs) affect hundreds of millions of people worldwide and are a major patient-safety issue.[1-3] The risk of acquiring HAIs exists in every healthcare facility and system throughout the world, regardless of the resources available. Although many institutions, even those in developed countries,[4] can be considered resource-limited settings depending on the definitions applied, resource-limited settings in this chapter are defined as countries with low- and middle-income economies on the basis of their gross national product per capita in 2006, as reported by the World Bank.[5] With a few exceptions, this categorization includes most countries in the following regions: Africa, Asia (excluding Japan, Saudi Arabia, South Korea, and Taiwan), some Eastern European countries, Latin America, the Caribbean, and Oceania (excluding Australia and New Zealand) (Figure 28-1). Resource-limited countries, including the regions mentioned above, are often called "developing countries." That term is not ideal for discussion of infection prevention issues, because infection prevention programs are impacted not only by economic resources and education but also by sociocultural differences. However, most of the concepts and recommendations discussed in this chapter are presented in a generic manner so that they can be applied in different resource-limited settings by trained infection preventionists.

In resource-limited settings, the risk of HAI has been estimated to be 2–20 times higher than that in developed countries.[6-12] Prevalence studies conducted in some developing countries have generally reported HAI prevalences of greater than 15% (range, 6%–27%).[7-13] Very few studies have evaluated the mortality associated with HAIs in resource-limited settings, but reported figures in international studies indicate excess mortality rates of 14.3%–27.5% for catheter-associated bloodstream infection (CA-BSI) and 12%–28% for ventilator-associated pneumonia (VAP).[14,15] According to some investigators, case-fatality rates for HAI in developing countries may exceed 50% among neonates.[16] These figures are particularly sobering in comparison with data from developed countries.[17-19]

Substantial progress has been made in recent years in improving infection prevention programs in resource-limited settings. National infection prevention initiatives have gained considerable importance, particularly in Asia and Latin America.[20-24] Numerous factors have helped focus attention on the importance of infection prevention, including the emergence of highly drug-resistant microorganisms, the increasing perception of occupational hazards among healthcare workers, and public demands for improved quality and cost-effectiveness of health care.[25] Initiatives from the Centers for Disease Control and Prevention, the World Health Organization (WHO), the Society for Healthcare Epidemiology of America (SHEA), the Association for Professionals in Infection Control and Epidemiology (APIC), and the International Federation of Infection Control have been successful in increasing public and professional awareness of the need for organized infection prevention programs in resource-limited settings, establishing training courses for infection preventionists, and promulgating guidelines. However, well-organized, effective infection prevention programs are largely confined to select healthcare facilities, usually academic hospitals. The number of

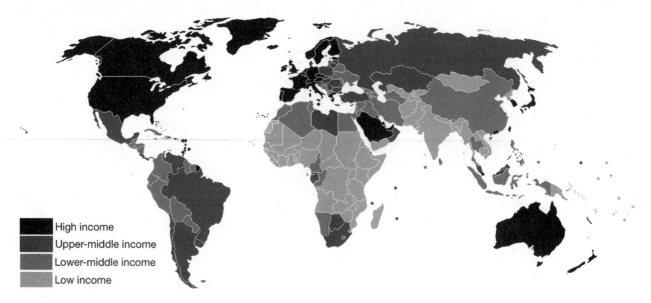

Figure 28-1. Income groupings of the countries of the world, based on World Bank income groupings for 2006 (calculated by gross national income per capita, Atlas method). Reproduced under GNU free documentation license from http://en.wikipedia.org/wiki/File:World_Bank_income_groups.svg (accessed March 2, 2009).

trained infection prevention physicians and nurses is currently insufficient in some countries, and training is limited or nonexistent in some regions, especially sub-Saharan Africa.[26] A growing awareness of the lack of infection prevention programs in resource-limited settings prompted the WHO to create the World Alliance for Patient Safety.[27] Prevention of HAIs is the target of the First Global Patient Safety Challenge from the Alliance, "Clean Care is Safer Care," which was launched in October 2005. The challenge consists of specific actions in 5 major areas to promote patient safety in healthcare settings: blood safety, clinical procedure safety, injection safety, sanitation and waste management safety, and promotion of safe hand hygiene practices during patient care. A primary objective of this challenge is to launch a practical approach to improve hand hygiene in health care throughout the world.[27]

Epidemiology of Nosocomial Infections

Limited epidemiologic data on HAIs are available from developing countries. Existing HAI surveillance and outbreak data from resource-limited settings are difficult to interpret. Some of the limitations of the existing data include the use of different surveillance definitions, the diversity of the patient populations included, variability in the laboratory support available, the range of available expertise in data collection and analysis, and the paucity of peer-reviewed publications in this field. Language barriers, lack of protected time to conduct

research, and other obstacles may discourage infection preventionists from publishing their hospital's experiences. Despite these limitations, some important efforts in recent years have resulted in publications that expand our knowledge of the incidence and prevalence of HAIs in resource-limited settings. The International Nosocomial Infection Control Consortium (INNIC)[14,24,28] is a multinational, multicenter, and collaborative HAI control program that uses a surveillance system based on that of the US National Healthcare Safety Network (NHSN; formerly called the National Nosocomial Infection Surveillance System).[18,25,29] The methodology used by this group is quite rigorous, and the data provided come from 98 intensive care units (ICUs) in 18 different developing countries. INNIC data are probably the best available representative sample of HAI data from resource-limited settings, although they do not necessarily represent the extent of the impact of HAIs in any entire single country.

Results from other important infection prevention surveillance initiatives have been published by other investigators from around the world.[30-36] These groups have reported data according to specific sites of infection and specific populations, using the Centers for Disease Control and Prevention's NHSN surveillance system definitions. Use of "benchmarking" and comparison with NHSN system data puts significant pressure on developing countries to address specific local problems in order to achieve infection rates comparable with those in the United States and Western Europe. Table 28-1 and Table 28-2 summarize the data reported by these authors and

Table 28-1. Incidence rates of ventilator-associated pneumonia (VAP), catheter-associated urinary tract infection (CAUTI), and catheter-associated bloodstream infection (CA-BSI) in intensive care units (ICUs) in resource-limited settings (RLSs), compared with National Healthcare Safety Network (NHSN) data

Type of infection, reference (year[s]), type of hospital unit	No. of patients	No. of device-days	Incidence rate,[a] by percentile			Pooled mean incidence rate (IQR)[a]	
			10th	50th (median)	90th	In the RLS	NHSN
VAP							
INNIC[28] (2002–2007)							
Medical ICU	2	3,117	6.7	25.5	44.4	40.7 (6.7–44.4)	3.1 (0.9–4.6)
Surgical ICU	4	5,214	5.9	15.1	24.4	18.0 (8.5–21.7)	5.2 (1.8–6.4)
Medical-surgical ICU	60	90,905	0.0	16.5	51.4	19.8 (9.6–24.1)	3.6 (1.3–5.1)
Pediatric ICU	9	7,898	1.3	6.1	15.5	7.8 (3.0–14.2)	2.5 (0.0–2.8)
Danchaivijitr et al.[12] (2003–2004)							
Medical ICU	37	28,174	0.0	11.8	26.4	13.1 (5.0–18.4)	3.1 (0.9–4.6)
Surgical ICU	27	20,295	0.0	11.7	25.5	13.1 (6.0–20.4)	5.2 (1.8–6.4)
Pediatric ICU	23	13,113	0.0	8.7	26.3	11.2 (0.0–18.9)	2.5 (0.0–2.8)
Starling[30] (1996[b])							
Medical-surgical ICU	4	NA	NA	NA	NA	22.7 (NA)	3.6 (1.3–5.1)
Pediatric ICU	1	NA	NA	NA	NA	9.7 (NA)	2.5 (0.0–2.8)
CAUTI							
INNIC[28] (2002–2007)							
Medical ICU	2	6,646	0.0	5.3	10.5	9.6 (0.0–10.5)	4.4 (1.8–5.6)
Surgical ICU	4	8,808	0.3	12.0	27.8	4.2 (3.1–22.9)	4.0 (1.2–6.1)
Medical-surgical ICU	60	155,722	0.0	5.2	22.8	6.61 (2.5–8.3)	3.4 (1.9–4.5)
Pediatric ICU	9	4,777	0.0	0.8	8.0	3.98 (0.0–3.3)	5.2 (0.0–6.0)
Danchaivijitr et al.[12] (2003–2004)							
Medical ICU	37	25,826	0.0	3.7	14.1	5.7 (0.0–8.5)	4.4 (1.8–5.6)
Surgical ICU	27	24,205	0.0	0.0	13.1	3.0 (0.0–5.3)	4.0 (1.2–6.1)
Starling[30] (1996[b])							
Medical-surgical ICU	4	NA	NA	NA	NA	8.55 (NA)	3.4 (1.9–4.5)
Pediatric ICU	1	NA	NA	NA	NA	0.0 (NA)	4.0 (1.2–6.1)
CA-BSI							
INNIC[28] (2002–2007)							
Medical ICU	2	2,364	2.1	7.4	12.7	10.5 (2.1–12.7)	2.9 (0.8–4.2)
Surgical ICU	4	7,526	1.3	18.2	41.6	17.1 (1.3–41.6)	2.7 (0.9–4.4)
Medical-surgical ICU	60	132,061	0.0	9.7	34.3	8.9 (3.7–16.5)	2.4 (0.6–3.1)
Pediatric ICU	9	16,012	0.0	9.5	24.4	6.8 (7.9–19.2)	5.3 (1.1–6.5)
Danchaivijitr et al.[12] (2003–2004)							
Medical ICU	14	5,567	0.0	0.0	11.7	2.7 (0.0–4.9)	2.9 (0.8–4.2)
Surgical ICU	15	6,763	0.0	0.0	11.5	3.3 (0.0–2.7)	2.7 (0.9–4.4)
Pediatric ICU	14	4,851	0.0	0.0	16.9	5.2 (0.0–6.4)	5.3 (1.1–6.5)
Starling[30] (1996[b])							
Medical-surgical ICU	4	NA	NA	NA	NA	2.13 (NA)	2.4 (0.6–3.1)
Pediatric ICU	1	NA	NA	NA	NA	4.1[c]	5.3 (1.1–6.5)

NOTE: INNIC, International Nosocomial Infection Control Consortium; IQR, interquartile range (ie, 25th and 75th percentile); NA, data not available.

a Incidence rates are expressed as follows: for VAP, number of cases per 1,000 ventilator-days; for CAUTI, number of cases per 1,000 catheter-days; for CA-BSI, number of cases per 1,000 central line–days.

b Only data from 1996 are shown.

c Data from 1994.

Table 28-2. Procedure-specific National Nosocomial Infections Surveillance (NNIS) System risk index–adjusted rates of surgical site infection (SSI) in resource-limited settings (RLSs), compared with NNIS benchmarks

Type of procedure, reference (year[s]), risk index category	RLS data		NNIS data[a] (1992–2004)	
	No. of procedures	No. of SSIs per 100 procedures	No. of procedures	No. of SSIs per 100 procedures
Craniotomy				
Danchaivijitr et al.[12] (2003–2004)				
Category 0	435	0.69	4,717	0.91
Category 1	800	1.88	14,864	1.72
Category 2	184	3.80	4,666	2.40[c]
Starling[30] (1994–1996[b])				
Category 0–1	541	2.8	14,864	1.72[d]
Category 2–3	541	2.8	4,666	2.40
Cardiac surgery				
Danchaivijitr et al.[12] (2003–2004)				
Category 0	205	0.98	2,147	0.70
Category 1	175	1.71	49,135	1.50
Category 2	80	2.5	15,215	2.21[c]
Febré et al.[33] (1998–1999[e])				
Category 0	14	0	2,147	0.70
Category 1	161	6.83	49,135	1.50
Category 2	108	9.26	15,215	2.21[c]
Category 3	40	7.5	15,215	2.21[c]
Starling[30] (1994–1996[b])				
Category 0–1	214	5.6	49,135	1.50
Category 2–3	193	5.7	15,215	2.21
Mastectomy				
Danchaivijitr et al.[12] (2003–2004)				
Category 0	212	0.47	16,287	1.74
Category 1	140	1.47	10,700	2.2
Starling[30] (1994–1996[b])				
Category 0	369	0.5	16,287	1.74
Category 1	230	2.2	10,700	2.2
Herniorraphy				
Danchaivijitr et al.[12] (2003–2004)				
Category 0	1,296	0.23	12,659	0.81
Category 1	416	0.24	8,397	2.14
Starling[30] (1994–1996[b])				
Category 0	698	0.1	12,659	0.81
Category 1	348	0.9	8,397	2.14
Laparotomy				
Hernández et al.[34] (1998[f])				
Category 0	56	3.6	6,414	1.71
Category 1	277	22	8,802	3.08

(continued)

Table 28-2. (continued)

Type of procedure, reference (year[s]), risk index category	RLS data		NNIS data[a] (1992–2004)	
	No. of procedures	No. of SSIs per 100 procedures	No. of procedures	No. of SSIs per 100 procedures
Laparotomy				
Starling[30] (1994–1996[b])				
Category 0	56	1.8	6,414	1.71
Category 1	135	5.2	8,802	3.08
Abdominal hysterectomy				
Danchaivijitr et al.[12] (2003–2004)				
Category 0	112	1.16	49,024	1.36
Category 1	112	1.16	24,064	2.32
Starling[30] (1994–1996[b])				
Category 0	119	0.0	49,024	1.36
Category 1	329	0.9	24,064	2.32

a NNIS data given in Danchaivijitr et al.[12]
b Only data from Hospital A are shown.[25]
c Risk index category 2–3.
d Risk index category 1.
e Data on cardiac surgery and thoracic surgery are included.
f Data on different types of abdominal surgeries and postdischarge surveillance are included.

the INNIC for the HAIs occurring most commonly in developing countries.[13] Table 28-1 shows that VAP incidence rates from developing countries are 3–13 times higher than those reported by the NHSN in 2006 for ICUs. Rates of catheter-associated urinary tract infection (CAUTI) reported by the INNIC and by Starling et al.[30] are 2–13 times higher than those reported by the NHSN in 2006 for medical-surgical ICUs but are lower than rates in pediatric ICUs. In a Thai study, CAUTI rates were similar to the rates reported by the NHSN.[35] Incidence rates of CA-BSI reported by the INNIC are 2–6 times higher than those reported by the NHSN in 2006 for medical-surgical ICUs. These data provide a baseline and demonstrate the great potential to design and implement interventions to decrease the incidence of HAIs in resource-limited settings.[37-39] Table 28-2 shows that surgical site infection (SSI) following cardiac surgery and SSI following abdominal surgery are also important problems to address in some Latin American countries, with particular attention given to modifiable SSI risk factors.[40]

Risk factors for HAIs have been infrequently studied in resource-limited settings. In a prospective study involving a critically ill pediatric population in Brazil, investigators found an HAI rate of 13% (incidence rate, 31.7 HAIs per 1,000 patient-days).[41] Independent risk factors for development of HAI that they identified included the device utilization ratio (relative risk, 1.6), receipt of parenteral

nutrition (relative risk, 2.5), and greater length of stay in the healthcare facility (relative risk, 1.7). The authors concluded that the main primary preventive measures to be adopted in that hospital unit should focus on reduction of the use of invasive devices, a more restrictive parenteral nutrition policy, and reduction in the length of stay. Length of stay has been reported to be longer in resource-limited settings, compared with in US hospitals, for multiple reasons.[42]

Bacterial infections are the most frequent HAIs in resource-limited settings, especially bacterial infections associated with invasive devices or procedures.[2,14,28] In a prevalence survey in Thailand, gram-negative bacteria were responsible for 70.2% and gram-positive bacteria for only 19.9% of a total of 699 HAIs identified.[12] Aggregated data from all INICC ICUs showed that 80.8% of all *Staphylococcus aureus* isolates were methicillin resistant.[2,3] Infections caused by *Enterococcus* species are less common in developing countries than in developed countries; however, vancomycin-resistant enterococci have recently emerged as important pathogens in Latin America.[43-45] The most commonly reported gram-negative bacterial HAIs are due to *Pseudomonas aeruginosa*, *Enterobacter* species, *Klebsiella pneumoniae*, *Escherichia coli*, or *Acinetobacter baumannii*.[14,28,46] High levels of antimicrobial resistance and high mortality rates have been reported for *P. aeruginosa* and *A. baumannii* in Latin America and Asia.[14,46,47]

Nosocomial transmission of common communicable infections in resource-limited settings is particularly problematic for pediatric patients.[48-58] There are multiple reports of communicable respiratory viral infections[49,52,53] and bacterial infections,[54] as well as gastrointestinal infections[55-57] and systemic viral infections.[51,58,59] In adults, nosocomial dissemination of multidrug-resistant *Mycobacterium tuberculosis* is also a serious threat.[60-63]

Outbreaks are also significant in resource-limited settings, but they may not be recognized in the absence of effective surveillance systems. Investigators in Mexico described 12 outbreaks of nosocomial infection over a 14-year period,[64] with an outbreak incidence almost 3 times higher than that reported in US hospitals.[65] The overall mortality rate was 25.8%; half of deaths were due to pneumonia. The incidence rate was 3 outbreaks per 10,000 hospital discharges, and outbreak-related infections accounted for 1.56% of all HAIs. The investigators reported only 2 outbreaks in the years 1985–1991 but 10 outbreaks in the years 1992–1996; the increase was probably the result of improved surveillance.

Nosocomial transmission of bloodborne pathogens remains a major but often underappreciated problem in resource-limited settings.[66-69] Despite the WHO Safety Injection Global Network, which has been promoted since 1999,[70] unsafe injection practices—such as reusing disposable needles and syringes, using multidose vials of medication, recapping needles, and discarding needles and syringes into the general waste system—remain common in some developing countries.[70-73] It is estimated that up to 160,000 human immunodeficiency virus (HIV) infections, 4.7 million hepatitis C virus infections, and 16 million hepatitis B virus infections each year are attributable to unsafe injection practices.[68] A comprehensive review of all identifiable studies related to injection practices in resource-limited settings reported that, for 14 of these 19 countries, at least 50% of injections given were considered unsafe.[66] These findings reinforce the importance of using standard precautions and educating healthcare workers about injection safety in resource-limited settings.[74]

Implementing Infection Control Programs

Interventions to Reduce the Incidence of Nosocomial Infections

Reducing the incidence of major nosocomial infections—VAP, CAUTI, CA-BSI, and SSI—is achievable by implementing simple, affordable, nondevice interventions that are feasible and cost-saving in various resource-limited settings.[75-81]

Reducing the incidence of VAP can be achieved by means of educational interventions, use of continuous quality improvement models to create a multidisciplinary nosocomial pneumonia team, and implementing a VAP prevention "bundle" (ie, a combined group of prevention measures).[82-85] Interestingly, although these interventions have been shown to be effective in preventing VAP,[82-85] they are not widely implemented. Factors associated with successful implementation include active participation by respiratory therapists, physicians, and nurses and other key leaders; the use of evidence-based educational programs with the VAP prevention bundle; and continuous monitoring of nursing care practices to prevent VAP.[82-85] Together, these findings emphasize the importance of improving the management and care of patients who undergo ventilation, rather than eliminating a particular nosocomial reservoir of infection.

Several nondevice interventions have been shown to be effective in reducing the incidence of CAUTI.[86-91] These simple approaches include providing education, performance feedback to physicians and nurses about catheter care, written reminders for physicians about catheter indications, antibiotic guidelines tailored to specific units, and reminders to physicians to remove unnecessary catheters.[86-91] A recent meta-analysis revealed that these nondevice interventions had a significant impact on the duration of catheter use: there were 1.69 fewer catheter-days in the case group than in the control group.[92] Additionally, the relative risk of CAUTI was 0.68 (95% confidence interval, 0.45–1.02; $P=.06$), suggesting a trend toward fewer cases of CAUTI after these interventions.[92] As these reminder systems appear to significantly reduce the duration of urinary catheter use and possibly even the incidence of CAUTI, without evidence of harm, they should be adopted for implementation in resource-limited settings.

Education interventions to reduce the incidence of CA-BSIs in single institutions and within a single city have been examined in several studies.[93-99] These interventions have used didactic training sessions[93,97,99] or a combination of both didactic and hands-on training,[95,96] and have targeted various groups of healthcare workers, including resident physicians and medical students,[96] physicians-in-training and nursing staff,[93,95,97] intensivists and nurses,[98] and nurses alone.[99] Six of these studies reported a 28%–72% decrease in the incidence of CA-BSI after the intervention.[93-98] A multicenter study in the United States demonstrated that an education-based intervention based on evidence-based practices can be successfully implemented in a diverse group of medical and surgical units and can reduce CA-BSI rates.[100] Alternatively, one ICU in the United States significantly reduced the incidence of CA-BSI by implementing a CA-BSI prevention bundle consisting of staff education, use of a mobile catheter-insertion cart, daily inquiries to care providers about whether catheters could be removed, use of a checklist to ensure adherence to evidence-based guidelines for preventing CA-BSIs, and empowerment of nurses to stop the catheter insertion procedure if they observe a violation of the guidelines.[101] A CA-BSI prevention bundle implemented in one multicenter study[102] included clinician education about practices to control infection and harm resulting from CA-BSIs, use of a mobile catheter care cart with necessary supplies, use of a checklist to ensure adherence to infection control practices, stopping catheter insertion (in nonemergency situations) if recommended practices were not being followed, discussion of catheter removal at daily rounds, and providing feedback regarding the number of cases of CA-BSI at monthly team meetings and regarding the rate of CA-BSI at quarterly team meetings. Implementation of this bundle of measures resulted in a large and sustained reduction in the incidence of CA-BSI (up to 66%) that was maintained throughout the 18-month study period. Although most of these studies were performed in developed countries,[93-98,100-102] application to resource-limited settings seems feasible.

Strategies that have been successful in reducing the incidence of SSI, regardless of resources, have a common theme: process improvement.[103] Despite several national initiatives and wide dissemination of evidence and guidelines, rates of compliance with recommended prevention measures remain low.[103] However, many unnecessary practices, such as formaldehyde fogging or installation of UV lights in the operating room, have been found to be common in resource-limited settings.[104] Studies have shown that implementing standardized protocols can help increase the reliability of the processes (eg, reduction in the expenditure for antibiotics used during surgery and reduction in the rate of SSI) in developing countries.[105-107] Thus, evidence-based protocols may serve as a useful tool to help standardize processes for SSI prevention and achieve higher performance. For example, procedure-specific protocols for preoperative antibiotic administration that are implemented by nursing and/or pharmacy staff eliminate the need for surgeons to remember to order the antibiotic and reduce variation in orders written in different ways at different times. Clinical exceptions can be designed into the protocol, providing guidance to staff as to when an alternate treatment path should be followed, such as for a patient allergic to β-lactam agents, or giving instructions to contact the physician. Options to document contraindications can be incorporated into process tools (eg, preprinted orders), serving as a reminder and making it easy for physicians to note exceptions. Implementing such protocols, together with strict adherence to basic infection control measures (eg, proper hand hygiene

and scrub time), can result in SSI-prevention process improvements in resource-limited settings.

Interventions to Reduce the Incidence of Infection With Multidrug-Resistant Microorganisms

Inappropriate use of antibiotics contributes to a very high prevalence of antibiotic-resistant pathogens in resource-limited settings.[108] However, this is poorly documented, because most resource-limited settings lack reliable surveillance systems. It is well demonstrated that appropriate use of antimicrobials ("antimicrobial stewardship") and infection control programs are essential components in the effort to reduce the incidence of infection with drug-resistant microorganisms, regardless of resource availability.[109] Although guidelines on preventing the transmission of drug-resistant microorganisms, particularly gram-positive microorganisms, are available,[110,111] application of these guidelines in resource-limited settings might not be practical. Issues that complicate guideline implementation in resource-limited settings include the lack of antimicrobial stewardship initiatives, lack of resources to meet the cost of implementing some processes (eg, use of active surveillance cultures, or providing isolation gowns for patient cohorts), and the lack of evidence-based recommendations to reduce the prevalence of drug-resistant gram-negative microorganisms, which is far higher in developing countries than in developed countries.[112,113]

Despite these limitations, several reports from resource-limited settings describe success with a multifaceted intervention program that featured education and use of antibiotic order forms, with or without audits and prescriber feedback, to promote appropriate antibiotic use.[114] It is also important to emphasize that antimicrobial stewardship educational programs should rely on 3 treatment principles. First, the choice of empirical therapy should be based on the prevalence, patterns, and risks of infection with drug-resistant microorganisms in the particular setting. Second, procurement of specimens for culture before the start of treatment is essential for implementing and evaluating strategies for de-escalation of antimicrobial use. Third, subsequent de-escalation should focus on appropriate duration of treatment and monitoring for adverse events. Use of these 3 principles appears to be essential in minimizing the emergence of drug-resistant microorganisms in resource-limited settings.[115-117] Together, these relatively simple interventions have been shown to reduce the incidence of infection with resistant microorganisms in various resource-limited settings.[112-117]

Several multidrug-resistant gram-negative microorganisms, such as *A. baumannii* and *P. aeruginosa*, have a propensity to cause outbreaks of nosocomial infection.[109,118] Outbreaks of infection with these microorganisms can occur in institutions where an effective antimicrobial stewardship program has been established,[109,114] demonstrating the importance of additional infection control interventions for outbreak control. Although molecular epidemiologic analysis may be unavailable, practitioners in resource-limited settings need to adopt and adapt other infection control components to help control outbreaks of infection with these microorganisms, such as use of selective environmental cultures to determine if a common environmental source is present, use of enhanced contact isolation, use of enhanced environmental cleaning, and use of modified active surveillance to identify and isolate colonized patients among high-risk groups.[112-116,118] Monitoring adherence to these infection control interventions is also important. These infection control measures, together with antimicrobial stewardship programs, have been shown to reduce the rate of transmission of multidrug-resistant gram-negative microorganisms in many resource-limited settings.[112-116,118]

Developing a Cost-Effective Infection Prevention Program

Multiple studies from throughout the developed world have shown that HAIs increase the cost of medical care because of excess lengths of stay and increased morbidity and mortality.[119-132] It is well established that integrated infection prevention programs that include HAI surveillance and multifaceted interventions, including education, can significantly reduce the incidence of HAIs.[12,83,98,102,117,133-136] Thus, it is reasonable to assume that infection prevention programs can lead to decreased healthcare costs. The cost benefit of infection prevention programs has been demonstrated in developed countries.[137-141] The cost-effectiveness of infection prevention interventions is more complex to study, but it too has been demonstrated in some articles.[142-146]

New infection prevention programs in resource-limited settings usually have to prove at least their ability to be cost-neutral in order to receive enough resources to become established. Few studies on this topic from resource-limited settings have been published, and most of them are related to the cost benefit of infection prevention programs. A case-control study in Turkey indicated that HAIs increased the average hospital stay by approximately 4 days.[147] Using an incremental cost estimate, the hospital sector had to spend an additional US$48 million for medical management of HAIs; the cost-to-benefit ratio for an infection prevention program was found to be approximately 4.6.[147] Clearly, a program for preventing HAIs will not only pay for itself but also generate other direct and indirect benefits, both for patients and for society as a whole.

Another recent case-control study from India demonstrated that case patients with hospital-acquired bacteremia had a significantly longer total stay (mean duration, 22.9 days), a significantly longer ICU stay (mean duration, 11.3 days), a significantly higher mortality rate (mean rate, 54%), and significantly greater costs (mean cost, US$14,818) than did uninfected control subjects.[148] The authors estimated that hospital-acquired bacteremia increased costs by US$980,000 in their cardiothoracic unit and illustrated that, although the cost of health care is much lower in India than in Western countries, the costs of HAIs were similar in the 2 regions.[149] Considering the socioeconomic situation of their country, their study shows that HAIs place an even greater burden on resources in India than in Western countries.

Given this background, we recommend several steps to develop a cost-effective infection prevention program in resource-limited settings.

1. Perform a risk assessment for your particular setting. Initially, it is impossible to do everything. One must first establish priorities and perform a risk assessment in order to develop a cost-effective infection prevention program. It is necessary to have a baseline assessment of the healthcare system, including the number of beds, the number of procedures performed, the patient population served, and the rate of device utilization, as well as to understand the history of previous efforts to establish infection prevention programs. In the absence of regular surveillance, HAI point-prevalence surveys will be a useful starting point to clarify the extent of the problem and to identify priority areas to target for initial interventions.[12,150]

2. Engage hospital leaders and government authorities. The ultimate responsibility for infection prevention rests within individual facilities, but external political, economic, and social forces may have a significant impact on the development of these initiatives. Healthcare managers must be convinced of the potential to reduce costs and save resources by implementing infection prevention programs, as well as the safety and quality benefits.[151]

3. Establish an infection prevention and control program and committee. The characteristic features of highly effective infection prevention programs have been recognized since 1980.[134,135] In 1998, a consensus panel provided updated and specific recommendations in the following areas: managing critical data and information, including surveillance data; developing, implementing, and monitoring policies and procedures; following guidelines and meeting

accreditation requirements; collaborating with the employee health program; applying interventions to prevent transmission of infectious diseases; educating and training healthcare workers; and dedicating resources to infection prevention programs.[152]

4. Adjust the surveillance plan to match the characteristics of your particular setting. Surveillance should focus on high-risk hospital populations, for instance patients in ICUs and patients with device-associated HAIs.[18,153] Computer and statistical support is desirable, as is a plan to communicate results and implement interventions according to the highest priorities.

5. Implement evidence-based measures first. Hand hygiene programs, standard precautions, and isolation precautions are the basic fundamentals of any infection prevention program to reduce the cross-transmission and spread of infections in healthcare settings.[27,75,154] In addition, unnecessary practices should be eliminated, especially if they are expensive or do not "add value"—that is, do not aid in preventing infection.[104,155]

6. Provide education and training in infection prevention for healthcare workers. Education is a priority, and infection prevention education should first target local infection prevention leaders, then target doctors, nurses, and students.[156,157] Academic centers should develop and implement infection prevention education for the curriculum in medical and nursing schools.[158]

7. Be prepared for obstacles and problems. Multiple barriers have been reported by infection prevention teams during the development of new infection prevention programs. A study from Egypt described numerous constraints, such as noncooperation of medical staff, underreporting of SSIs and BSIs, lack of communication between the infection prevention team and laboratory staff, and lack of advocacy and political support.[150] In a study from Thailand, the main obstacles for infection prevention programs were a lack of incentives (reported by 66.7% of respondents) and lack of support from administrators (reported by 30.2%).[159] Many of these same barriers exist in developed countries as well. There are numerous examples of studies on how to build a business case for infection prevention and on establishing priorities in the infection prevention literature from developed countries.[151,160-162] Existing models and tools from developed countries, the Centers for Disease Control and Prevention, the WHO, and resource-limited settings with successful infection prevention programs should be used to help address these barriers. New infection prevention programs should focus on establishing a track record of reporting infection rates, providing education, and developing interventions to reduce infections, to demonstrate their achievements. Initial successes and cost savings can then be used to provide data to expand the infection prevention program.

Establishing Occupational Health Programs

Healthcare epidemiologists in resource-limited settings should advocate for strong occupational health programs, as healthcare workers are a scarce resource and are costly to train and replace. Furthermore, communicable diseases are readily transmitted between healthcare workers and patients, and the incidence of occupationally acquired infection reflects the success of a hospital's infection prevention program. Although healthcare workers may be exposed to a number of occupational hazards, this section focuses on selected infectious hazards encountered in resource-limited settings.

Hospitals in resource-limited settings may be unable to implement use of costly protective measures, such as airborne-infection isolation rooms or safety-engineered needleless devices. Nevertheless, many effective interventions are relatively inexpensive. These include correct use of hand hygiene and standard precautions, vaccination of healthcare workers, performance of risk assessments, and reducing "presenteeism" (ie, healthcare workers working when they are ill with a communicable disease). All healthcare workers, including physicians, should receive training on hand hygiene and standard precautions. The nosocomial transmission of vaccine-preventable illnesses is well documented; thus, WHO-recommended vaccines should be provided free of charge to all healthcare workers.[163] Risk assessment—that is, the process of evaluating the risk of potential exposure to an infectious disease before each patient contact—requires that healthcare workers be educated on the infectious syndromes and patient-care activities that may warrant use of additional protective measures.[74] Presenteeism may be discouraged by hospital policies, peers, or a personal sense of duty, and healthcare epidemiologists should help policy makers and healthcare workers understand the risks associated with this practice.

Prevention of M. tuberculosis Infection

M. tuberculosis infection poses a serious occupational risk in many resource-limited settings. *M. tuberculosis* infection rates are higher among healthcare workers with longer length of employment or more exposure to patients infected with *M. tuberculosis*.[63] However, *M. tuberculosis* infection may go unrecognized as an occupational infection if there is a high prevalence of

M. tuberculosis infection in the community or a lack of nosocomial surveillance.

Surveillance for nosocomial *M. tuberculosis* infection is useful to evaluate the effectiveness of *M. tuberculosis* control measures, the ongoing risk of acquisition, and the need for additional interventions. Since most healthcare workers who acquire *M. tuberculosis* will have latent infection, screening for *M. tuberculosis* disease (ie, active *M. tuberculosis* infection) will greatly underestimate the occupational risk. Surveillance for latent infection can be conducted by use of the tuberculin skin test or interferon-γ release assays. A positive result of an interferon-γ release assay is more specific for *M. tuberculosis* infection than is a positive tuberculin skin test result, but the assay is costly, time-sensitive, and unavailable in many resource-limited settings. Tuberculin skin test results may be affected by previous receipt of BCG vaccine, infection with nontuberculous mycobacteria, or HIV coinfection. Nevertheless, several studies have demonstrated that the tuberculin skin test is a useful tool to quantify the risk of nosocomial *M. tuberculosis* transmission, even among healthcare workers who have received BCG vaccine,[164-166] and to identify the areas of highest risk to be targeted for control measures.[167]

Although implementing *M. tuberculosis* control measures may be challenging in resource-limited settings, studies from Brazil, Malawi, and Thailand suggest such measures can reduce occupational risk for healthcare workers.[167-170] The WHO recommends that the highest priority be given to administrative control measures; in particular, prompt triage, diagnosis, and treatment of patients with suspected *M. tuberculosis* infection.[171] A study from Brazil found that the time from patient presentation to receipt of the first dose of medication active against *M. tuberculosis* can be reduced from days to hours and that implementing administrative control measures, without the use of more costly engineering controls, can result in a significant reduction in the number of healthcare worker tuberculin skin test conversions.[168] The use of personal respirators varies widely between and within resource-limited settings. Because of cost considerations, most hospitals require that healthcare workers reuse their respirators. If use of respirators is poorly accepted by healthcare workers, they may wear them improperly, not wear them when use is indicated, or lose them.[164,172] Thus, hospitals that provide respirators should ensure that healthcare workers are properly trained in their use and should replace lost, torn, or dirty respirators. Many engineering controls are expensive and technically complex. However, some simple steps can include maximizing natural ventilation (eg, by opening windows, where the climate permits), maximizing exposure to natural UV light (ie, sunlight), and installing simple exhaust systems.[171]

Prevention of Transmission of Bloodborne Pathogens and Postexposure Management

Exposure to bloodborne pathogens is a serious occupational hazard in countries with a high prevalence of infection with HIV, hepatitis B virus, and/or hepatitis C virus. The WHO estimates that approximately 3 million healthcare workers are occupationally exposed to bloodborne pathogens each year, resulting in 15,000 hepatitis C virus infections, 70,000 hepatitis B virus infections, and 1,000 HIV infections; more than 90% of these occur in resource-limited settings.[173] Needlestick injuries are the most common source of exposure to bloodborne pathogens, and most of these injuries are preventable.[174] It has been estimated that healthcare workers in Africa, the Eastern Mediterranean, and Asia sustain, on average, 4 needlestick injuries per healthcare worker per year.[174]

Prevention strategies for reducing occupational exposure to bloodborne pathogens should focus on eliminating unnecessary injections, observing standard precautions, and training healthcare workers in the safe use and disposal of needles and sharp devices ("sharps").[173-175] Special effort should be made to train healthcare workers who are less experienced, as they are at higher risk of sustaining a needlestick injury.[176] Training resources are available from the WHO.[177]

Healthcare workers should receive hepatitis B vaccine, free of charge, as early in their career as possible.[173] Healthcare workers occupationally exposed to a bloodborne pathogen should have immediate access to a medical assessment, counseling, confidential testing, and follow-up. Hospitals should ensure that medication for postexposure prophylaxis is readily available for workers with exposures deemed high risk.[174,178]

Management of Respiratory Virus Infection

Respiratory viruses cause frequent outbreaks of infection both in the community and in healthcare settings. In addition to seasonal and endemic respiratory viruses, hospitals must be prepared to deal with new, emerging viruses that can cause severe illness or a future pandemic. Hospitals may be the vanguard of both the recognition and the control of an emerging virus.

The outbreak of severe acute respiratory syndrome (SARS) demonstrated that lack of adherence to standard infection prevention practices in both developing and developed countries facilitated nosocomial spread of the disease to patients and healthcare workers.[179] To prevent nosocomial transmission, healthcare workers should be trained to screen patients for febrile respiratory illness and adhere to standard precautions, which include compliance with hand hygiene, the appropriate use of facial protection, and containment of respiratory secretions (eg, by teaching the patient cough etiquette or by masking the patient). Patients with febrile respiratory illness should be spatially separated from patients without febrile respiratory illness as much as possible.[180] Furthermore, healthcare workers should be trained to identify unusual clusters of patients with severe febrile respiratory illness, particularly if there is a history of contact with other severely ill persons (eg, a family cluster). These clusters of infection should be reported to local public health authorities. Additional measures recommended by the WHO include immediately placing the affected patients in a well-ventilated private room and the use of personal respirators by healthcare workers.[180] In resource-limited settings, healthcare worker preparedness training should address the modes of transmission and should specify how to implement appropriate infection prevention strategies to prevent and control the spread of avian influenza. Pandemic influenza preparedness plans must include health care administrative support, mechanisms to rapidly create temporary isolation facilities, systems to restrict access to exposed healthcare workers, and plans to involve specialists to screen and identify case patients early, to provide for continuous monitoring to ensure adherence to optimal infection prevention practices, and to provide regular feedback to healthcare workers.[181] Hospitals should be aware of current recommendations on antiviral prophylaxis for exposed healthcare workers, as these will depend on the nature of the virus.[182-184]

Annual vaccination against seasonal influenza is recommended for all healthcare workers.[185] A study from Thailand estimates that the costs incurred in a nosocomial influenza outbreak are 10-fold higher than the costs of a universal annual healthcare worker vaccination program.[186] Thus, vaccination would be cost-effective, particularly in tropical countries where influenza occurs year-round.

Disinfection and Sterilization

The principals of disinfection and sterilization are discussed in Chapter 7. This section focuses on challenges that may confront resource-limited settings more frequently: the impact of water quality on disinfection and sterilization and the reuse of single-use medical devices.

Hospitals in resource-limited settings may face difficulty in ensuring a reliable supply of clean water for cleaning and reprocessing medical devices. The risk of bacterial contamination is lower if deionized or distilled water is used instead of tap water. However, even these sources of water may have high levels of endotoxin. Outbreaks of pyrogenic reactions have been linked to the use of water with high endotoxin levels to reprocess cardiac catheters.[187,188] Monitoring the quality of the water

used for reprocessing by measuring both bacterial contamination and endotoxin levels is essential for preventing adverse outcomes.

Hospitals may feel compelled to reprocess and reuse medical devices intended "for single use only" as procedures involving increasingly complex and costly medical devices become more widely available in resource-limited settings. This practice raises ethical and patient safety concerns. Safety data supporting the reuse of many such devices are lacking, but not reusing such a device may deny a patient a potentially life-saving procedure, where resources are limited. Inadequate local practices may further increase the risk of adverse patient outcomes.[189] At a minimum, hospitals that reprocess and reuse single-use devices should train the personnel who do the reprocessing; implement a standardized, written reprocessing protocol; ensure adequate cleaning, rinsing, drying, sterilizing, and packaging methods; develop criteria for discarding reused devices so they are not used repeatedly until they malfunction or break; and conduct surveillance for adverse patient events.

Finally, it is important to emphasize that syringes and needles must never be reprocessed or reused. The WHO estimates that as many as 40% of injections worldwide are given with reused, unsterilized syringes and needles and that these practices cause an estimated 1.3 million early deaths, 26 million years of life lost, and US$535 million in direct medical costs each year.[190]

Implementing Hand Hygiene Programs

Hospitals in resource-limited settings may face formidable barriers to hand hygiene, including a lack of access to clean (or any) water, lack of administrative support, overworked staff, and crowded wards. A successful hand hygiene program will require a step-by-step approach, with successful completion of each step before moving on to the next (Table 28-3). The first step is to garner the support of hospital administrators. Since the WHO launched the First Global Patient Safety Challenge in 2005, the Ministries of Health in over 100 countries have pledged to address hand hygiene in their hospitals.[191] Hospitals in these countries can use this national-level commitment to help convince local administrators of the importance of hand hygiene. Tools and resources from the WHO Global Patient Safety Challenge[192] and the strong association between improved hand hygiene compliance and fewer healthcare-acquired infections may also be useful to rally administrative support (see Chapter 21 on hand hygiene).

Once hospital leadership is supportive, access to appropriate hand hygiene agents must be provided. In most cases, use of plain soap and water is the cheapest

Table 28-3. Steps to implementing a successful hand hygiene program in a resource-limited setting

Step 1. Seek administrative support ("buy-in")

Promote the World Health Organization campaign (adopt and/or adapt it)

Present an evidence-based case for hand hygiene

Step 2. Provide access to hand hygiene products

Assess the availability of clean water and soap

Implement use of alcohol-based hand rub where it is feasible

Place hand hygiene supplies at the point of care

Step 3. Address barriers to good hand hygiene compliance

Lack of understanding: educate healthcare workers about the chain of transmission, the effectiveness of hand hygiene in breaking the chain, and the appropriate moments for hand hygiene

Cultural norms: identify and groom advocates ("champions") and address religious or cultural barriers to use of alcohol-based hand rub

Forgetfulness: use reminders (eg, posters and promotional materials)

Step 4. Conduct audits and provide feedback on compliance rates

method of hand hygiene. However, the feasibility of providing alcohol-based hand rub should be considered, as alcohol-based hand rub has many advantages over soap and water. Alcohol-based hand rub may provide a solution for hospitals without a reliable source of clean, running water, as there is no need for plumbing or concern about water quality. Effective alcohol-based rubs can be made in-house at a fraction of the cost of commercial product.[193] Along with choosing and providing the hand hygiene agent(s) to be made available, careful consideration must be given to their placement. It is essential that healthcare workers have convenient access to hand hygiene agents at the point where they are providing care to the patient.[193]

Once appropriate hand hygiene agents have been provided, it is necessary to identify and address other potential barriers to compliance. These may include a lack of understanding of the role of hand hygiene in preventing healthcare-acquired infections, lack of awareness of when hand hygiene should be performed, poor hand hygiene practices as a cultural norm, or cultural or religious concerns with the use of alcohol-based hand rub.[194] Healthcare workers should be educated on the importance of hand hygiene in preventing the transmission of infections to patients, as well as protecting themselves from occupational infections. They should be aware of the appropriate moments during patient care activities when hand hygiene should be performed.[195] Cultural norms within a hospital can sometimes be

successfully changed by identifying hand hygiene advocates or "champions." Champions should be well-respected individuals within the hospital whose hand hygiene practices provide a model for others to follow and who can motivate others to change their attitudes and behavior. Other approaches to changing the cultural norms of a hospital may include poster campaigns, visible support of hospital leadership, and informing patients that hand hygiene is a hospital priority so that patient expectations help drive behavior. In hospitals that promote the use of alcohol-based hand rub, any cultural or religious concerns about the use of this type of agent should be addressed.[194]

Once the barriers to hand hygiene have been identified and addressed, observing hand hygiene practices and providing feedback to healthcare workers are an effective method of modifying behavior and further improving compliance. This should be done in a positive and not punitive manner. Performance data may be better received when provided in aggregate fashion (eg, by unit or occupational group). Competition may drive further improvement (eg, comparing the performance of one unit or healthcare worker group against another). Hand hygiene compliance rates should become a performance indicator for hospitals committed to improving patient safety and quality of care.

International Infection Control Networks

The first national, multidisciplinary infection prevention societies were formed in the early 1970s and are responsible for much national and global progress; professionals are more likely to be effectively focused on the problem and on education, research, and solutions than are government agencies, and they are less likely to be distracted by political concerns. Just as new infection preventionists need support, education, and a network of more-experienced individuals, the new infection prevention societies soon found that they also needed some of these same resources. Infection prevention nurses from the United States, the United Kingdom, Sweden, Canada, and Denmark requested WHO support for an international infection prevention meeting. The WHO sponsored such a meeting in 1978, which was attended by 75 professionals from 25 countries. The development of the International Federation of Infection Control[196] soon followed; the members are the national infection prevention societies of more than 55 countries.[197]

As regional and international societies are an integral parts of the International Federation of Infection Control, so several regional and international infection control networks exist in many regions around the world. These are nonprofit, open, international, professional organizations established to help distribute education on infection control to practitioners in each region, and some offer evidence-based guidelines on infection prevention or offer the opportunity for research networks around the world. Practitioners in resource-limited settings can participate in the activities developed by these societies. Organizations from around the world that sponsor infection prevention networks are listed in Table 28-4.

Conclusions

Healthcare-associated infections in resource-limited settings, particularly in developing countries, represent a huge, unrecognized threat to patient safety. Successful research in resource-limited settings, combined with intensive ongoing efforts to more consistently implement simple and inexpensive measures for prevention, will lead to wider acceptance of infection prevention practices in resource-limited settings. Existing evidence suggests that infection prevention measures are feasible and effective in reducing the incidence of HAIs and improving healthcare outcomes. However, several unanswered questions remain, and additional studies are required to help understand and improve infection prevention programs in resource-limited settings. Because many interventions do not require expensive technology, resource-limited settings should not delay the implementation of basic infection control interventions while awaiting additional data. Resource-limited countries should develop national infection prevention guidelines to reduce the incidences of nosocomial infections and infections with drug-resistant microorganisms. Guidelines to improve the quality of disinfection and sterilization processes, as well as guidelines for preexposure and postexposure diagnostic work-ups for occupational exposures are needed. Resource-limited settings should implement practical, evidence-based, low-cost, and simple preventive strategies first. Additional studies to explore the long-term effect and cost benefit of specific infection prevention interventions in resource-limited settings are needed. Studies of antimicrobial stewardship programs and other interventions to reduce the development and transmission of drug-resistant microorganisms are also warranted. Such studies should include analyses of economic, behavioral, communication, and organizational strategies to optimize the implementation of and adherence to the best practices.

Table 28-4. Infection prevention networks sponsored by professional societies worldwide

Network	URL
Society for Healthcare Epidemiology of America	http://www.shea-online.org
Association for Professionals in Infection Control and Epidemiology	http://www.apic.org
The International Nosocomial Infection Control Program	http://www.inicc.org
Asia Pacific Society of Infection Control	http://www.apsic2009.org
New Zealand National Division of Infection Control Nurses	http://www.infectioncontrol.co.nz
Eastern Mediterranean Regional Network for Infection Control	http://www.emro.who.int/emrnic/about.htm
Southeastern Europe Infection Control	None
Baltic Network for Infection Control and Containment of Antimicrobial Resistance	http://balticcare.wordpress.com/
Australian Infection Control Association	http://www.aica.org.au/
Infection Control Association (Singapore)	http://www.icas.org.sg/
The Nosocomial Infection Control Group of Thailand	http://www.idthai.org/
The Hospital Infection Society	http://www.his.org.uk/
Infection Control Nurses Association	http://www.ips.uk.net/
Middle Eastern Society of Infection Control	None

References

1. Burke JP. Infection control—a problem for patient safety. *N Engl J Med* 2003;348:651–656.
2. Christenson M, Hitt JA, Abbott G, Septimus EJ, Iversen N. Improving patient safety: resource availability and application for reducing the incidence of healthcare-associated infection. *Infect Control Hosp Epidemiol* 2006;27:245–251.
3. Yokoe DS, Classen D. Improving patient safety through infection control: a new healthcare imperative. *Infect Control Hosp Epidemiol* 2008;29(Suppl 1):S3–S11.
4. Anderson DJ, Kirkland KB, Kaye KS, et al. Underresourced hospital infection control and prevention programs: penny wise, pound foolish? *Infect Control Hosp Epidemiol* 2007;28:767–773.
5. World Bank. *World Bank Development Indicators 2006.* Washington, DC: World Bank; 2006.
6. Mayon-White RT, Ducel G, Kereselidze T, Tikomirov E. An international survey of the prevalence of hospital-acquired infection. *J Hosp Infect* 1988;11(Suppl A):43–48.

7. Faria S, Sodano L, Gjata A, et al. The first prevalence survey of nosocomial infections in the University Hospital Centre 'Mother Teresa' of Tirana, Albania. *J Hosp Infect* 2007; 65:244–250.

8. Farhat CK. Nosocomial infection [in Portugese]. *J Pediatr (Rio J)* 2000;76:259–260.

9. Rezende EM, Couto BR, Starling CE, Modena CM. Prevalence of nosocomial infections in general hospitals in Belo Horizonte. *Infect Control Hosp Epidemiol* 1998; 19:872–876.

10. Kallel H, Bahoul M, Ksibi H, et al. Prevalence of hospital-acquired infection in a Tunisian hospital. *J Hosp Infect* 2005;59:343–347.

11. Gosling R, Mbatia R, Savage A, Mulligan JA, Reyburn H. Prevalence of hospital-acquired infections in a tertiary referral hospital in northern Tanzania. *Ann Trop Med Parasitol* 2003;97:69–73.

12. Danchaivijitr S, Judaeng T, Sripalakij S, Naksawas K, Plipat T. Prevalence of nosocomial infection in Thailand 2006. *J Med Assoc Thai* 2007;90:1524–1529.

13. Ponce de Leon-Rosales SP, Molinar-Ramos F, Dominguez-Cherit G, Rangel-Frausto MS, Vazquez-Ramos VG. Prevalence of infections in intensive care units in Mexico: a multi-center study. *Crit Care Med* 2000;28:1316–1321.

14. Rosenthal VD, Maki DG, Mehta A, et al. International Nosocomial Infection Control Consortium report, data summary for 2002–2007, issued January 2008. *Am J Infect Control* 2008;36:627–637.

15. Aygun C, Sobreyra Oropeza M, Rosenthal VD, Villamil Gomez W, Rodriguez Calderon ME. Extra mortality of nosocomial infections in neonatal ICUs at eight hospitals of Argentina, Colombia, Mexico, Peru, and Turkey: findings of the International Nosocomial Infection Control Consortium (INICC). *Am J Infect Control* 2006;34:E135.

16. Zaidi AK, Huskins WC, Thaver D, Bhutta ZA, Abbas Z, Goldmann DA. Hospital-acquired neonatal infections in developing countries. *Lancet* 2005;365:1175–1188.

17. Horan TC, Gaynes RP. Surveillance of nosocomial infections. In: Mayall CG, ed. *Hospital Epidemiology and Infection Control*. 3rd ed. Philadelphia, PA: Lippincot Williams & Wilkins; 2004:1659–1702.

18. Emori TG, Culver DH, Horan TC, et al. National nosocomial infections surveillance system (NNIS): description of surveillance methods. *Am J Infect Control* 1991;19:19–35.

19. Klevens RM, Edwards JR, Richards CL Jr, et al. Estimating health care-associated infections and deaths in U.S. hospitals, 2002. *Public Health Rep* 2007;122:160–166.

20. Danchaivijitr S. Hospital infection control in Thailand. *Can J Infect Control* 1991;6:97–99.

21. Danchaivijitr S, Tangtrakool T, Waitayapiches S, Chokloikaew S. Efficacy of hospital infection control in Thailand 1988–1992. *J Hosp Infect* 1996;32:147–153.

22. Ajenjo MC. Infecciones intrahospitalarias: conceptos actuales de prevencion y control. *Rev Chil Urol* 2006;72: 95–101.

23. Marcel JP, Alfa M, Baquero F, et al. Healthcare-associated infections: think globally, act locally. *Clin Microbiol Infect* 2008;14:895–907.

24. Rosenthal VD, Maki DG, Graves N. The International Nosocomial Infection Control Consortium (INICC): goals and objectives, description of surveillance methods, and operational activities. *Am J Infect Control* 2008;36: e1–e12.

25. Starling C. Infection control in developing countries. *Curr Opin Infect Dis* 2001;14:461–466.

26. Newman MJ. Infection control in Africa south of the Sahara. *Infect Control Hosp Epidemiol* 2001;22:68–69.

27. Pittet D, Allegranzi B, Storr J, et al. Infection control as a major World Health Organization priority for developing countries. *J Hosp Infect* 2008;68:285–292.

28. Rosenthal VD. Device-associated nosocomial infections in limited-resources countries: findings of the International Nosocomial Infection Control Consortium (INICC). *Am J Infect Control* 2008;36:S171.e7–e12.

29. National Healthcare Safety Network (NHSN) Web site. http://www.cdc.gov/nhsn/. Accessed September 30, 2009.

30. Starling CE, Couto BR, Pinheiro SM. Applying the Centers for Disease Control and Prevention and National Nosocomial Surveillance system methods in Brazilian hospitals. *Am J Infect Control* 1997;25:303–311.

31. Lopes JM, Tonelli E, Lamounier JA, et al. Prospective surveillance applying the National Nosocomial Infection Surveillance methods in a Brazilian pediatric public hospital. *Am J Infect Control* 2002;30:1–7.

32. Ercole FF, Starling CE, Chianca TC, Carneiro M. Applicability of the National Nosocomial Infections Surveillance system risk index for the prediction of surgical site infections: a review. *Braz J Infect Dis* 2007;11:134–141.

33. Febré N, de Medeiros ES, Wey SB, Larrondo M, Silva V. Is the epidemiological surveillance system of nosocomial infections recommended by the American CDC applicable in a Chilean hospital? [in Spanish]. *Rev Med Chil* 2001;129: 1379–1386.

34. Hernández K, Ramos E, Seas C, Henostroza G, Gotuzzo E. Incidence of and risk factors for surgical-site infections in a Peruvian hospital. *Infect Control Hosp Epidemiol* 2005; 26:473–477.

35. Danchaivijitr S, Rongrungruang Y, Pakaworawuth S, Jintanothaitavorn D, Naksawas K. Development of quality indicators of nosocomial infection control. *J Med Assoc Thai* 2005;88(Suppl 10):S75–S82.

36. Kasatpibal N, Jamulitrat S, Chongsuvivatwong V. Standardized incidence rates of surgical site infection: a multicenter study in Thailand. *Am J Infect Control* 2005;33:587–594.

37. Coffin SE, Klompas M, Classen D, et al. Strategies to prevent ventilator-associated pneumonia in acute care hospitals. *Infect Control Hosp Epidemiol* 2008;29(Suppl 1):S31–S40.

38. Lo E, Nicolle L, Classen D, et al. Strategies to prevent catheter-associated urinary tract infections in acute care hospitals. *Infect Control Hosp Epidemiol* 2008;29(Suppl 1): S41–S50.

39. Marschall J, Mermel LA, Classen D, et al. Strategies to prevent central line-associated bloodstream infections in acute care hospitals. *Infect Control Hosp Epidemiol* 2008; 29(Suppl 1):S22–S30.

40. Anderson DJ, Kaye KS, Classen D, et al. Strategies to prevent surgical site infections in acute care hospitals. *Infect Control Hosp Epidemiol* 2008;29(Suppl 1):S51–S61.

41. Gilio AE, Stape A, Pereira CR, Cardoso MF, Silva CV, Troster EJ. Risk factors for nosocomial infections in a critically ill pediatric population: a 25-month prospective cohort study. *Infect Control Hosp Epidemiol* 2000;21:340–342.

42. Barnum H. *Public Hospitals in Developing Countries: Resource Use, Cost, Financing.* Baltimore, MD: Johns Hopkins University Press; 1993.

43. Fica A, Jemenao MI, Bilbao P, et al. Emergency of vancomycin-resistant *Enterococcus* infections in a teaching hospital in Chile [in Spanish]. *Rev Chilena Infectol* 2007;24:462–471.

44. Zárate MS, Gales A, Jordá-Vargas L, et al. Environmental contamination during a vancomycin-resistant enterococci outbreak at a hospital in Argentina [in Spanish]. *Enferm Infecc Microbiol Clin* 2007;25:508–512.

45. Corso AC, Gagetti PS, Rodriguez MM, et al. Molecular epidemiology of vancomycin-resistant *Enterococcus faecium* in Argentina. *Int J Infect Dis* 2007;11:69–75.

46. Lizaso D, Aguilera CK, Correa M, et al. Nosocomial bloodstream infections caused by gram-negative bacilli: epidemiology and risk factors for mortality [in Spanish]. *Rev Chil Infect* 2008;25:368–373.

47. Bello H, González G, Dominguez M, Zemelman R, Garcia A, Mella S. Activity of selected beta-lactams, ciprofloxacin, and amikacin against different *Acinetobacter baumannii* biotypes from Chilean hospitals. Diagn Microbiol Infect Dis 1997; 28:183–186.

48. Dutta P, Mitra U, Rasaily R, et al. Prospective study of nosocomial enteric infections in a pediatric hospital, Calcutta. *Indian Pediatr* 1993;30:187–194.

49. Cotton MF, Berkowitz FE, Berkowitz Z, Becker PJ, Heney C. Nosocomial infections in black South African children. *Pediatr Infect Dis J* 1989;8:676–683.

50. Shekhawat PS, Singh RN, Shekhawat R, Joshi KR. Nosocomial infections in pediatric and neonatal ward of Umaid Hospital, Jodhpur: a cross-sectional study. *Indian Pediatr* 1992;29:384–385.

51. Aaby P, Bukh J, Lisse IM, Smits AJ. Introduction of measles into a highly immunised West African community: the role of health care institutions. *J Epidemiol Community Health* 1985;39:113–116.

52. Madhi SA, Ismail K, O'Reilly C, Cutland C. Importance of nosocomial respiratory syncytial virus infections in an African setting. *Trop Med Int Health* 2004;9:491–498.

53. Avendaño LF, Larrañaga C, Palomino MA, et al. Community- and hospital-acquired respiratory syncytial virus infections in Chile. *Pediatr Infect Dis J* 1991;10:564–568.

54. Reichler MR, Rakovsky J, Slacikova M, et al. Spread of multidrug-resistant *Streptococcus pneumoniae* among hospitalized children in Slovakia. *J Infect Dis* 1996;173:374–379.

55. Rodrigues A, de Carvalho M, Monteiro S, et al. Hospital surveillance of rotavirus infection and nosocomial transmission of rotavirus disease among children in Guinea-Bissau. *Pediatr Infect Dis J* 2007;26:233–237.

56. Abiodun PO, Omoigberale A. Prevalence of nosocomial rotavirus infection in hospitalized children in Benin City, Nigeria. *Ann Trop Paediatr* 1994;14:85–88.

57. Gusmao RH, Mascarenhas JD, Gabbay YB, et al. Rotaviruses as a cause of nosocomial, infantile diarrhoea in northern Brazil: pilot study. *Mem Inst Oswaldo Cruz* 1995; 90:743–749.

58. Biellik RJ, Clements CJ. Strategies for minimizing nosocomial transmission of measles [in Spanish]. *Rev Panam Salud Publica* 1998;4:350–357.

59. Navarrete-Navarro S, Avila-Figueroa C, Ruiz-Gutiérrez E, Ramírez-Galván L, Santos JI. Nosocomial measles: a proposal for its control in hospitals [in Spanish]. *Bol Med Hosp Infant Mex* 1990;47:495–499.

60. Ritacco V, Di Lonardo M, Reniero A, et al. Nosocomial spread of human immunodeficiency virus-related multidrug-resistant tuberculosis in Buenos Aires. *J Infect Dis* 1997;176:637–642.

61. Harries AD, Kamenya A, Namarika D, et al. Delays in diagnosis and treatment of smear-positive tuberculosis and the incidence of tuberculosis in hospital nurses in Blantyre, Malawi. *Trans R Soc Trop Med Hyg* 1997;91:15–17.

62. Wilkinson D, Crump J, Pillay M, Sturm AW. Nosocomial transmission of tuberculosis in Africa documented by restriction fragment length polymorphism. *Trans R Soc Trop Med Hyg* 1997;91:318.

63. Joshi R, Reingold AL, Menzies D, Pai M. Tuberculosis among health-care workers in low- and middle-income countries: a systematic review. *PLoS Med* 2006;3:e494.

64. Ostrosky-Zeichner L, Baez-Martinez R, Rangel-Frausto MS, Ponce-de-Leon S. Epidemiology of nosocomial outbreaks: 14-year experience at a tertiary-care center. *Infect Control Hosp Epidemiol* 2000;21:527–529.

65. Wenzel RP, Thompson RL, Landry SM, et al. Hospital-acquired infections in intensive care unit patients: an overview with emphasis on epidemics. *Infect Control* 1983; 4:371–375.

66. Simonsen L, Kane A, Lloyd J, Zaffran M, Kane M. Unsafe injections in the developing world and transmission of blood-borne pathogens: a review. *Bull World Health Organ* 1999;77:789–800.

67. Kane A, Lloyd J, Zaffran M, Simonsen L, Kane M. Transmission of hepatitis B, hepatitis C and human immunodeficiency viruses through unsafe injections in the developing world: model-based regional estimates. *Bull World Health Organ* 1999;77:801–807.

68. Kermode M. Unsafe injections in low-income country health settings: need for injection safety promotion to prevent the spread of blood-borne viruses. *Health Promot Int* 2004; 19:95–103.

69. Kermode M. Healthcare worker safety is a pre-requisite for injection safety in developing countries. *Int J Infect Dis* 2004;8:325–327.

70. Hutin YJ, Chen RT. Injection safety: a global challenge. *Bull World Health Organ* 1999;77:787–788.

71. Lakshman M, Nichter M. Contamination of medicine injection paraphernalia used by registered medical practitioners in south India: an ethnographic study. *Soc Sci Med* 2000; 51:11–28.

72. Khan AJ, Luby SP, Fikree F, et al. Unsafe injections and the transmission of hepatitis B and C in a periurban community in Pakistan. *Bull World Health Organ* 2000;78: 956–963.

73. Dicko M, Oni AQ, Ganivet S, Kone S, Pierre L, Jacquet B. Safety of immunization injections in Africa: not simply a problem of logistics. *Bull World Health Organ* 2000; 78:163–169.

74. Siegel JD, Rhinehart E, Jackson M, Chiarello L. 2007 Guideline for isolation precautions: preventing transmission of infectious agents in health care settings. *Am J Infect Control* 2007;35:S65–S164.

75. Mayhall CG. In pursuit of ventilator-associated pneumonia prevention: the right path. *Clin Infect Dis* 2007;45:712–714.

76. Apisarnthanarak A, Thongphubeth K, Sirinvaravong S, et al. Effectiveness of multifaceted hospitalwide quality improvement programs featuring an intervention to remove unnecessary urinary catheters at a tertiary care center in Thailand. *Infect Control Hosp Epidemiol* 2007;28:791–798.

77. Apisarnthanarak A, Suwannakin A, Maungboon P, Warren DK, Fraser VJ. Long-term outcome of an intervention to remove unnecessary urinary catheters, with and without a quality improvement team, in a Thai tertiary care center. *Infect Control Hosp Epidemiol* 2008;29:1094–1095.

78. Gastmeier P, Sohr D, Schwab F, et al. Ten years of KISS: the most important requirements for success. *J Hosp Infect* 2008;70(Suppl 1):11–16.

79. Gastmeier P, Geffers C, Brandt C, et al. Effectiveness of a nationwide nosocomial infection surveillance system for reducing nosocomial infections. *J Hosp Infect* 2006;64:16–22.

80. McKee C, Berkowitz I, Cosgrove SE, et al. Reduction of catheter-associated bloodstream infections in pediatric patients: experimentation and reality. *Pediatr Crit Care Med* 2008;9:40–46.

81. Lobo RD, Levin AS, Gomes LM, et al. Impact of an educational program and policy changes on decreasing catheter-associated bloodstream infections in a medical intensive care unit in Brazil. *Am J Infect Control* 2005;33:83–87.

82. Zack JE, Garrison T, Trovillion E, et al. Effect of an education program aimed at reducing the occurrence of ventilator-associated pneumonia. *Crit Care Med* 2002;30:2407–2412.

83. Babcock HM, Zack JE, Garrison T, et al. An educational intervention to reduce ventilator-associated pneumonia in an integrated health system: a comparison of effects. *Chest* 2004;125:2224–2231.

84. Resar R, Pronovost P, Haraden C, Simmonds T, Rainey T, Nolan T. Using a bundle approach to improve ventilator care processes and reduce ventilator-associated pneumonia. *Jt Comm J Qual Patient Saf* 2005;31:243–248.

85. Institute for Healthcare Improvement (IHI). Implement the Ventilator Bundle. http://www.ihi.org/IHI/Topics/CriticalCare/IntensiveCare/Changes/ImplementtheVentilatorBundle.htm. Accessed September 29, 2009.

86. Huang WC, Wann SR, Lin SL, et al. Catheter-associated urinary tract infections in intensive care units can be reduced by prompting physicians to remove unnecessary catheters. *Infect Control Hosp Epidemiol* 2004;25:974–978.

87. Rosenthal VD, Guzman S, Safdar N. Effect of education and performance feedback on rates of catheter-associated urinary tract infection in intensive care units in Argentina. *Infect Control Hosp Epidemiol* 2004;25:47–50.

88. Goetz AM, Kedzuf S, Wagener M, Muder RR. Feedback to nursing staff as an intervention to reduce catheter-associated urinary tract infections. *Am J Infect Control* 1999;27:402–404.

89. Stephan F, Sax H, Wachsmuth M, Hoffmeyer P, Clergue F, Pittet D. Reduction of urinary tract infection and antibiotic use after surgery: a controlled, prospective, before-after intervention study. *Clin Infect Dis* 2006;42:1544–1551.

90. Saint S, Kaufman SR, Thompson M, Rogers MA, Chenoweth CE. A reminder reduces urinary catheterization in hospitalized patients. *Jt Comm J Qual Patient Saf* 2005;31:455–462.

91. Garibaldi RA, Mooney BR, Epstein BJ, et al. An evaluation of daily bacteriologic monitoring to identify preventable episodes of catheter-associated urinary tract infection. *Infect Control* 1982;3:466–470.

92. Meddings J, Macy M, Rogers MA. Reminder systems to reduce urinary catheter use and catheter-associated urinary tract infection in hospitalized patients: a systematic review and meta-analysis. In: Program and abstracts of the 19th Annual Scientific Meeting of the Society for Healthcare Epidemiology of America; March 19–22, 2009; San Diego, CA. Abstract 141.

93. Parras F, Ena J, Bouza E, et al. Impact of an educational program for the prevention of colonization of intravascular catheters. *Infect Control Hosp Epidemiol* 1994;15:239–242.

94. Maas A, Flament P, Pardou A, et al. Central venous catheter-related bacteraemia in critically ill neonates: risk factors and impact of a prevention programme. *J Hosp Infect* 1998;40:211–224.

95. Eggimann P, Harbarth S, Constantin MN, et al. Impact of a prevention strategy targeted at vascular-access care on incidence of infections acquired in intensive care. *Lancet* 2000;355:1864–1868.

96. Sherertz RJ, Ely EW, Westbrook DM, et al. Education of physicians-in-training can decrease the risk for vascular catheter infection. *Ann Intern Med* 2000;132:641–648.

97. Coopersmith CM, Rebmann TL, Zack JE, et al. Effect of an education program on decreasing catheter-related bloodstream infections in the surgical intensive care unit. *Crit Care Med* 2002;30:59–64.

98. Warren DK, Zack JE, Cox MJ, et al. An educational intervention to prevent catheter-associated bloodstream infections in a non-teaching, community medical center. *Crit Care Med* 2003;31:1959–1963.

99. Rosenthal VD, Guzman S, Pezzotto SM, et al. Effect of an infection control program using education and performance feedback on rates of intravascular device–associated bloodstream infections in intensive care units in Argentina. *Am J Infect Control* 2003;31:405–409.

100. Warren DK, Cosgrove SE, Diekema DJ, et al. A multicenter intervention to prevent catheter-associated bloodstream infections. *Infect Control Hosp Epidemiol* 2006;27:662–669.

101. Berenholtz SM, Pronovost PJ, Lipsett PA, et al. Eliminating catheter-related bloodstream infections in the intensive care unit. *Crit Care Med* 2004;32:2014–2020.

102. Pronovost P, Needham D, Berenholtz S, et al. An intervention to decrease catheter-related bloodstream infections in the ICU. *N Engl J Med* 2006;355:2725–2732.

103. Griffin FA. 5 Million Lives Campaign: reducing methicillin-resistant *Staphylococcus aureus* (MRSA) infections. *Jt Comm J Qual Patient Saf* 2007;33:726–731.

104. Kunaratanapruk S, Silpapojakul K. Unnecessary hospital infection control practices in Thailand: a survey. *J Hosp Infect* 1998;40:55–59.

105. Gomez MI, Acosta-Gnass SI, Mosqueda-Barboza L, Basualdo JA. Reduction in surgical antibiotic prophylaxis expenditure and the rate of surgical site infection by means of a protocol that controls the use of prophylaxis. *Infect Control Hosp Epidemiol* 2006;27:1358–1365.

106. Hermsen ED, Smith Shull S, Puumala SE, Rupp ME. Improvement in prescribing habits and economic outcomes associated with the introduction of a standardized approach

for surgical antimicrobial prophylaxis. *Infect Control Hosp Epidemiol* 2008;29:457–461.

107. Apisarnthanarak A, Jirajariyavej S, Thongphubeth K, Yuekyen C, Warren DK, Fraser VJ. Outbreak of postoperative endophthalmitis in a Thai tertiary care center. *Infect Control Hosp Epidemiol* 2008;29:564–566.

108. Apisarnthanarak A, Danchaivijitr S, Bailey TC, Fraser VJ. Inappropriate antibiotic use in a tertiary care center in Thailand: an incidence study and review of experience in Thailand. *Infect Control Hosp Epidemiol* 2006;27:416–420.

109. Paterson DL. The role of antimicrobial management programs in optimizing antibiotic prescribing within hospitals. *Clin Infect Dis* 2006;42(Suppl 2):S90–S95.

110. Calfee DP, Salgado CD, Classen D, et al. Strategies to prevent transmission of methicillin-resistant *Staphylococcus aureus* in acute care hospitals. *Infect Control Hosp Epidemiol* 2008;29(Suppl 1):S62–S80.

111. Cohen AL, Calfee D, Fridkin SK, et al. Recommendations for metrics for multidrug-resistant organisms in healthcare settings: SHEA/HICPAC position paper. *Infect Control Hosp Epidemiol* 2008;29:901–913.

112. Apisarnthanarak A, Pinitchai U, Thongphubeth K, Yuekyen C, Warren DK, Fraser VJ. A multifaceted intervention to reduce pandrug-resistant *Acinetobacter baumannii* colonization and infection in 3 intensive care units in a Thai tertiary care center: a 3-year study. *Clin Infect Dis* 2008;47: 760–767.

113. Apisarnthanarak A, Warren DK, Fraser VJ. Creating a cohort area to limit transmission of pandrug-resistant *Acinetobacter baumannii* in a Thai tertiary care center. *Clin Infect Dis* 2009;48:1487–1488.

114. Apisarnthanarak A, Danchaivijitr S, Khawcharoenporn T, et al. Effectiveness of education and an antibiotic-control program in a tertiary care hospital in Thailand. *Clin Infect Dis* 2006;42:768–775.

115. Apisarnthanarak A, Mundy LM. Inappropriate use of carbapenems in Thailand: a need for better education on de-escalation therapy. *Clin Infect Dis* 2008;47:858–859.

116. Apisarnthanarak A, Fraser VJ. Feasibility and efficacy of infection-control interventions to reduce the number of nosocomial infections and drug-resistant microorganisms in developing countries: what else do we need? *Clin Infect Dis* 2009;48:22–24.

117. Apisarnthanarak A, Pinitchai U, Thongphubeth K, et al. Effectiveness of an educational program to reduce ventilator-associated pneumonia in a tertiary care center in Thailand: a 4-year study. *Clin Infect Dis* 2007;45:704–711.

118. Peleg AY, Seifert H, Paterson DL. *Acinetobacter baumannii*: emergence of a successful pathogen. *Clin Microbiol Rev* 2008;21:538–582.

119. Rose R, Hunting KJ, Townsend TR, Wenzel RP. Morbidity/mortality and economics of hospital-acquired blood stream infections: a controlled study. *South Med J* 1977;70:1267–1269.

120. Spengler RF, Greenough WB 3rd. Hospital costs and mortality attributed to nosocomial bacteremias. *JAMA* 1978;240:2455–2458.

121. Kappstein I, Schulgen G, Beyer U, Geiger K, Schumacher M, Daschner FD. Prolongation of hospital stay and extra costs due to ventilator-associated pneumonia in an intensive care unit. *Eur J Clin Microbiol Infect Dis* 1992;11:504–508.

122. Kappstein I, Schulgen G, Fraedrich G, Schlosser V, Schumacher M, Daschner FD. Added hospital stay due to wound infections following cardiac surgery. *Thorac Cardiovasc Surg* 1992;40:148–151.

123. Poulsen KB, Bremmelgaard A, Sorensen AI, Raahave D, Petersen JV. Estimated costs of postoperative wound infections: a case-control study of marginal hospital and social security costs. *Epidemiol Infect* 1994;113:283–295.

124. Pittet D, Tarara D, Wenzel RP. Nosocomial bloodstream infection in critically ill patients: excess length of stay, extra costs, and attributable mortality. *JAMA* 1994;271:1598–1601.

125. Coello R, Glenister H, Fereres J, et al. The cost of infection in surgical patients: a case-control study. *J Hosp Infect* 1993;25:239–250.

126. Girard R, Fabry J, Meynet R, Lambert DC, Sepetjan M. Costs of nosocomial infection in a neonatal unit. *J Hosp Infect* 1983;4:361–366.

127. Medina M, Martínez-Gallego G, Sillero-Arenas M, Delgado-Rodríguez M. Risk factors and length of stay attributable to hospital infections of the urinary tract in general surgery patients [in Spanish]. *Enferm Infecc Microbiol Clin* 1997;15:310–314.

128. Digiovine B, Chenoweth C, Watts C, Higgins M. The attributable mortality and costs of primary nosocomial bloodstream infections in the intensive care unit. *Am J Respir Crit Care Med* 1999;160:976–981.

129. Plowman R, Graves N, Griffin MA, et al. The rate and cost of hospital-acquired infections occurring in patients admitted to selected specialties of a district general hospital in England and the national burden imposed. *J Hosp Infect* 2001; 47:198–209.

130. Pirson M, Dramaix M, Struelens M, Riley TV, Leclercq P. Costs associated with hospital-acquired bacteraemia in a Belgian hospital. *J Hosp Infect* 2005;59:33–40.

131. Sheng WH, Wang JT, Lu DC, Chie WC, Chen YC, Chang SC. Comparative impact of hospital-acquired infections on medical costs, length of hospital stay and outcome between community hospitals and medical centres. *J Hosp Infect* 2005;59:205–214.

132. Warren DK, Quadir WW, Hollenbeak CS, Elward AM, Cox MJ, Fraser VJ. Attributable cost of catheter-associated bloodstream infections among intensive care patients in a nonteaching hospital. *Crit Care Med* 2006;34:2084–2089.

133. Eickhoff TC. General comments on the study on the efficacy of nosocomial infection control (SENIC Project). *Am J Epidemiol* 1980;111:465–469.

134. Haley RW, Quade D, Freeman HE, Bennett JV. Study on the Efficacy of Nosocomial Infection Control (SENIC Project): summary of study design. *Am J Epidemiol* 1980;111: 472–485.

135. Haley RW, Culver DH, White JW, et al. The efficacy of infection surveillance and control programs in preventing nosocomial infections in US hospitals. *Am J Epidemiol* 1985; 121:182–205.

136. Harbarth S, Sax H, Gastmeier P. The preventable proportion of nosocomial infections: an overview of published reports. *J Hosp Infect* 2003;54:258–266.

137. Wakefield DS, Helms CM, Massanari RM, Mori M, Pfaller M. Cost of nosocomial infection: relative contributions of laboratory, antibiotic, and per diem costs in serious

Staphylococcus aureus infections. *Am J Infect Control* 1988;16:185–189.

138. Shulkin DJ, Kinosian B, Glick H, Glen-Puschett C, Daly J, Eisenberg JM. The economic impact of infections: an analysis of hospital costs and charges in surgical patients with cancer. *Arch Surg* 1993;128:449–452.

139. Wenzel RP. The Lowbury Lecture: the economics of nosocomial infections. *J Hosp Infect* 1995;31:79–87.

140. Pittet D, Sax H, Hugonnet S, Harbarth S. Cost implications of successful hand hygiene promotion. *Infect Control Hosp Epidemiol* 2004;25:264–266.

141. Obasanjo O, Perl TM. Cost-benefit and effectiveness of nosocomial surveillance methods. *Curr Clin Top Infect Dis* 2001;21:391–406.

142. VandenBergh MF, Kluytmans JA, van Hout BA, et al. Cost-effectiveness of perioperative mupirocin nasal ointment in cardiothoracic surgery. *Infect Control Hosp Epidemiol* 1996;17:786–792.

143. Daschner FD. How cost-effective is the present use of antiseptics? *J Hosp Infect* 1988;11(Suppl A):227–235.

144. Graves N. Economics and preventing hospital-acquired infection. *Emerg Infect Dis* 2004;10:561–566.

145. Graves N, Halton K, Lairson D. Economics and preventing hospital-acquired infection: broadening the perspective. *Infect Control Hosp Epidemiol* 2007;28:178–184.

146. Nyamogoba H, Obala AA. Nosocomial infections in developing countries: cost effective control and prevention. *East Afr Med J* 2002;79:435–441.

147. Khan MM, Celik Y. Cost of nosocomial infection in Turkey: an estimate based on the university hospital data. *Health Serv Manage Res* 2001;14:49–54.

148. Kothari A, Sagar V, Ahluwalia V, Pillai BS, Madan M. Costs associated with hospital-acquired bacteraemia in an Indian hospital: a case-control study. *J Hosp Infect* 2009;71: 143–148.

149. Puskas JD, Williams WH, Mahoney EM, et al. Off-pump vs conventional coronary artery bypass grafting: early and 1-year graft patency, cost, and quality-of-life outcomes: a randomized trial. *JAMA* 2004;291:1841–1849.

150. Talaat M, Kandeel A, Rasslan O, et al. Evolution of infection control in Egypt: achievements and challenges. *Am J Infect Control* 2006;34:193–200.

151. Murphy DM, Alvarado CJ, Fawal H. The business of infection control and epidemiology. *Am J Infect Control* 2002; 30:75–76.

152. Scheckler WE, Brimhall D, Buck AS, et al. Requirements for infrastructure and essential activities of infection control and epidemiology in hospitals: a consensus panel report. Society for Healthcare Epidemiology of America. *Infect Control Hosp Epidemiol* 1998;19:114–124.

153. National Healthcare Safety Network (NHSN) Web site. Updated May 11, 2005. http://www.cdc.gov/nhsn/. Accessed August 13, 2009.

154. Boyce JM, Pittet D. Guideline for hand hygiene in health-care settings: recommendations of the Healthcare Infection Control Practices Advisory Committee and the HICPAC/SHEA/APIC/IDSA Hand Hygiene Task Force. *Infect Control Hosp Epidemiol* 2002;23:S3–S40.

155. Issack MI. Unnecessary hospital infection control practices in developing countries. *J Hosp Infect* 1999;42:339–341.

156. Picheansatian W, Moongtui W, Soparatana P, Chittreecheur J, Apisarnthanarak A, Danchaivijitr S. Evaluation of a training course in infection control for nurses. *J Med Assoc Thai* 2005;88(Suppl 10):S171–S176.

157. Pethyoung W, Picheansathian W, Boonchuang P, Apisarnthanarak A, Danchaivijitr S. Effectiveness of education and quality control work group focusing on nursing practices for prevention of ventilator-associated pneumonia. *J Med Assoc Thai* 2005;88(Suppl 10):S110–S114.

158. Danchaivijitr S, Chakpaiwong S, Jaturatramrong U, Wachiraporntip A, Cherdrungsi R, Sripalakij S. Program on nosocomial infection in the curricula of medicine, dentistry, nursing and medical technology in Thailand. *J Med Assoc Thai* 2005;88(Suppl 10):S150–S154.

159. Danchaivijitr S, Assanasen S, Trakuldis M, Waitayapiches S, Santiprasitkul S. Problems and obstacles in implementation of nosocomial infection control in Thailand. *J Med Assoc Thai* 2005;88(Suppl 10):S70–S74.

160. Perencevich EN, Stone PW, Wright SB, Carmeli Y, Fisman DN, Cosgrove SE. Raising standards while watching the bottom line: making a business case for infection control. *Infect Control Hosp Epidemiol* 2007;28:1121–1133.

161. Furuno JP, Schweizer ML, McGregor JC, Perencevich EN. Economics of infection control surveillance technology: cost-effective or just cost? *Am J Infect Control* 2008;36:S12–S17.

162. Dunagan WC, Murphy DM, Hollenbeak CS, Miller SB. Making the business case for infection control: pitfalls and opportunities. *Am J Infect Control* 2002;30:86–92.

163. World Health Organization (WHO). WHO Recommendations for Routine Immunizations—Summary Tables. Updated April 21, 2009. http://www.who.int/immunization/policy/immunization_tables/en/index.html. Accessed August 13, 2009.

164. Joshi R, Reingold AL, Menzies D, Pai M. Tuberculosis among health-care workers in low- and middle-income countries: a systematic review. *PLoS Med* 2006;3:e494.

165. Apisarnthanarak A, Thongphubeth K, Yuekyen C, Mundy LM. Postexposure detection of *Mycobacterium tuberculosis* infection in health care workers in resource-limited settings. *Clin Infect Dis* 2008;47:982–984.

166. Khawcharoenporn T, Apisarnthanarak A, Thongphubeth K, Yuekyen C, Mundy LM. Tuberculin skin tests among medical students with prior bacille Calmette Guérin vaccination in a setting with a high prevalence of tuberculosis. *Infect Control Hosp Epidemiol* 2009;30:705–709.

167. Roth VR, Garrett DO, Laserson KF, et al. A multicenter evaluation of tuberculin skin test positivity and conversion among health care workers in Brazilian hospitals. *Int J Tuberc Lung Dis* 2005;9:1335–1442.

168. da Costa PA, Trajman A, Mello FC, et al. Administrative measures for preventing *Mycobacterium tuberculosis* infection among HCWs in a teaching hospital in Rio de Janeiro, Brazil. *J Hosp Infect* 2009;72:57–64.

169. Harries AD, Hargreaves NJ, Gausi F, Kwanjana JH, Salaniponi FM. Preventing tuberculosis among health workers in Malawi. *Bull World Health Organ* 2002;80:526–531.

170. Yanai H, Limpakarnjanarat K, Uthaivoravit W, Mastro TD, Mori T, Tappero JW. Risk of *Mycobacterium tuberculosis* infection and disease among health care workers, Chiang Rai, Thailand. *Int J Tuberc Lung Dis* 2003;7:36–41.

171. World Health Organization (WHO). Guidelines for the Prevention of Tuberculosis in Health Care Facilities in Resource-Limited Settings. 1999. WHO/CDS/TB/99.269. http://www.who.int/tb/publications/who_tb_99_269/en/index.html. Accessed August 13, 2009.

172. Biscotto CR, Pedroso ER, Starling CE, Roth VR. Evaluation of N95 respirator use as a tuberculosis control measure in a resource-limited setting. *Int J Tuberc Lung Dis* 2005;9:545–549.

173. World Health Organization (WHO). AIDE-MEMOIRE for a Strategy to Protect Health Workers From Infection With Bloodborne Viruses. December 2003. WHO/EHT/03.11. http://www.who.int/immunization_safety/publications/safe_injections/en/. Accessed August 13, 2009.

174. Wilburn SQ, Eijkemans G. Preventing needlestick injuries among healthcare workers: a WHO-ICN collaboration. *Int J Occup Environ Health* 2004;10:451–456.

175. Apisarnthanarak A, Babcock HM, Fraser VJ. The effect of nondevice interventions to reduce needlestick injuries among health care workers in a Thai tertiary care center. *Am J Infect Control* 2008;36:74–75.

176. Clarke SP. Hospital work environments, nurse characteristics, and sharps injuries. *Am J Infect Control* 2007;35:302–309.

177. World Health Organization. Protecting Healthcare Workers: Preventing Needlestick Injuries Toolkit. 2005. http://www.who.int/occupational_health/activities/pnitoolkit/en/index.html. Accessed August 13, 2009.

178. Panlilio AL, Cardo DM, Grohskopf LA, Heneine W, Ross CS. Updated US Public Health Service guidelines for the management of occupational exposures to HIV and recommendations for postexposure prophylaxis. *MMWR Recomm Rep* 2005;54(RR-9):1–17.

179. Chan-Yeung M. Severe acute respiratory syndrome (SARS) and healthcare workers. *Int J Occup Environ Health* 2004;10:421–427.

180. World Health Organization (WHO). Avian influenza, including influenza A (H5N1), in humans: WHO interim infection control guidelines for health-care facilities. Updated May 10, 2007. http://www.who.int/csr/disease/avian_influenza/guidelines/infectioncontrol1/en/. Accessed August 13, 2009.

181. Apisarnthanarak A, Warren DK, Fraser VJ. Issues relevant to the adoption and modification of hospital infection-control recommendations for avian influenza (H5N1 infection) in developing countries. *Clin Infect Dis* 2007;45:1338–1342.

182. Harper SA, Bradley JS, Englund JA, et al. Seasonal influenza in adults and children—diagnosis, treatment, chemoprophylaxis, and institutional outbreak management: clinical practice guidelines of the Infectious Diseases Society of America. *Clin Infect Dis* 2009;48:1003–1032.

183. Novel Swine-Origin Influenza A (H1N1) Virus Investigation Team. Emergence of a novel swine-origin influenza A (H1N1) virus in humans. *N Engl J Med* 2009;360:2605–2615.

184. Shinde V, Bridges CB, Uyeki TM, et al. Triple-reassortant swine influenza A (H1) in humans in the United States, 2005–2009. *N Engl J Med* 2009;360:2616–2625.

185. World Health Organization. Influenza vaccines. *Wkly Epidemiol Rec* 2005;80:279–287.

186. Apisarnthanarak A, Puthavathana P, Kitphati R, Auewarakul P, Mundy LM. Outbreaks of influenza A among nonvaccinated healthcare workers: implications for resource-limited settings. *Infect Control Hosp Epidemiol* 2008;29:777–780.

187. Duffy RE, Couto B, Pessoa JM, et al. Improving water quality can reduce pyrogenic reactions associated with reuse of cardiac catheters. *Infect Control Hosp Epidemiol* 2003;24:955–960.

188. Archibald LK, Khoi NN, Jarvis WR, et al. Pyrogenic reactions in hemodialysis patients, Hanoi, Vietnam. *Infect Control Hosp Epidemiol* 2006;27:424–426.

189. Amarante JM, Toscano CM, Pearson ML, Roth V, Jarvis WR, Levin AS. Reprocessing and reuse of single-use medical devices used during hemodynamic procedures in Brazil: a widespread and largely overlooked problem. *Infect Control Hosp Epidemiol* 2008;29:854–858.

190. World Health Organization. Injection Safety. Revised October 2006. http://www.who.int/mediacentre/factsheets/fs231/en/. Accessed August 13, 2009.

191. World Health Organization. Clean Care Is Safer Care: Support From Countries and Territories Worldwide. May 2009. http://www.who.int/gpsc/statements/en/index.html. Accessed August 13, 2009.

192. World Health Organization. Clean Care is Safer Care: Tools and Resources. 2009. http://www.who.int/gpsc/5may/tools/en/index.html. Accessed August 13, 2009.

193. World Health Organization (WHO). World Health Organization guidelines on hand hygiene in health care. Geneva: WHO; 2009. http://www.who.int/gpsc/5may/en/. Accessed September 29, 2009.

194. Allegranzi B, Memish ZA, Donaldson L, et al. Religion and culture: potential undercurrents influencing hand hygiene promotion in health care. *Am J Infect Control* 2009;37:28–34.

195. Sax H, Allegranzi B, Uçkay I, Larson E, Boyce J, Pittet D. "My five moments for hand hygiene": a user-centred design approach to understand, train, monitor and report hand hygiene. *J Hosp Infect* 2007;67:9–21.

196. International Federation of Infection Control (IFIC). http://www.theIFIC.org. Accessed September 28, 2009.

197. Hambraeus A. Establishing an infection control structure. *J Hosp Infect* 1995;30(Suppl):232–240.

Chapter 29 Infection Control and Patient Safety

Darren R. Linkin, MD, MSCE, and P. J. Brennan, MD

There is currently a strong, growing interest in patient safety in the United States. Fortunately, for healthcare epidemiologists, the practice of infection control is already a patient safety effort: surveillance for adverse events and interventions to prevent harm to patients in the future. In this chapter, we discuss the history and importance of patient safety, terms and techniques unique to the field, and the role of infection control.

Space Shuttle Disasters: The Importance of System Errors

The crash of the *Challenger* space shuttle in 1986 resulted in an investigation that found substantial problems with how the National Aeronautics and Space Administration (NASA) managed significant safety threats in the space shuttle program. In particular, it was found that problems with the O-rings that led to the crash had been known but ignored by NASA for years preceding the disaster.[1]

On February 1, 2003, the *Columbia* became the second space shuttle to be destroyed in flight, killing all on board. This is one of a number of prominent disasters that highlights the complex nature of the problems that affect safety. An investigation found that a falling foam chunk from another part of the shuttle damaged the heat-resistant surface of the left wing, allowing superheated air into the structure of the damaged wing, which eventually led to the destruction of the shuttle. Foam debris was known to have fallen in a similar manner from the space shuttle on multiple earlier flights, but the prob-

lem was never adequately addressed. During the flight, after the foam debris was spotted on video of the shuttle's take-off, engineers repeatedly asked for photographs of the shuttle wing to be obtained so they could assess for damage. Because of lack of communication and a culture of downplaying potential risks, NASA management did not allow the photographs to be taken.

Errors in management and politics also predated the last flight of the *Columbia*. The NASA safety program personnel reported to managers who ran the program that was being assessed, which created a lack of independence in those evaluating and reporting on safety issues. There was a lack of strategic planning by NASA and the executive branch of the federal government. The space shuttle was originally designed as part of a broader (American) space station plan that was rejected; a huge amount of money and effort was funneled into a space shuttle program that had little reason to exist. The budget approved by Congress, however, was insufficient to allow for a robust safety program. In summary, the *Columbia* disaster was primarily the result of multiple preexisting errors in the system for how the space shuttle program was run by NASA and the US government.[2]

Many poor infection control outcomes apparently "caused" by individual healthcare workers can similarly be traced back to latent system errors. For instance, inadequate sterile technique during placement of a central venous catheter (an individual error) that leads to a catheter-associated bloodstream infection (an adverse event) may actually be attributed to factors such as inadequate training, missing or incorrect equipment parts or additions to central-line insertion kits, and understaffing (such that there is no assistant and/or a fatigued staff member performing the line insertion), which are all potential latent system errors.

Introduction to Patient Safety

While the field of safety has been an active source of investigation and planning in nonmedical fields since the 1960s, it was only in the past decade that the issue gained national attention in the medical industry. In 1999, the Institute of Medicine released a report that estimated that 44,000–98,000 inpatients die each year from medical errors.[3] Even the lower estimate would make medical errors the eighth leading cause of death in the United States, ahead of motor vehicle accidents and breast cancer. The cost of adverse events due to medical errors is estimated to be between $17 billion and $29 billion per year; the lower estimate is 2% of US annual healthcare expenditures.[3] Clearly, medical errors lead to substantial mortality and cost. The well-publicized Institute of Medicine report[3] brought a renewed focus on patient safety, as well as an outpouring of studies and reviews addressing the topic. However, the response of the healthcare industry to introduce patient safety practices has been slow, leading to external pressures on healthcare organizations to examine and improve the safety of patients. Regulation has come in the form of patient safety requirements for healthcare organizations from the Joint Commission (formerly the Joint Commission on Accreditation of Healthcare Organizations, or JCAHO).[4] In addition, many states now mandate public reporting of healthcare-associated infections, and federal legislation is under consideration.[5] Businesses have also joined in the fray: the Leapfrog Group, a consortium of large businesses that purchase health care for their employees, recommends that its patients be cared for in hospitals that have taken particular patient safety measures, such as using computerized physician-order entry.[5]

Compared with the healthcare industry, other industries have a far better safety record. For instance, for air travel, the fatality rate is 0.43 deaths per million opportunities, and the baggage mishandling rate is approximately 1 instance per 100 opportunities. The accuracy of inpatient medication delivery and the adequate use of post–myocardial infarction medication are compromised in more than 1 opportunity in 10. These differences were noted in the Institute of Medicine report.[3] The report suggested that significant improvements in patient safety were needed and would require large changes in the paradigm of how safety is addressed in healthcare organizations.

Patient safety is more than the use of computer physician-order entry or other interventions to prevent errors.

It is a fundamental change in the way errors and adverse events are viewed. Previously, individuals have been singled out as the cause of an adverse event, leading to an organizational culture that emphasized blame and resulted in silence by healthcare workers when errors or adverse events were made. In the newer paradigm for patient safety, it is assumed that most workers are trying to do good work within the constraints of their job and that problems with the system are often the cause of adverse events. Optimally, a nonpunitive culture should emphasize reporting of problems and improving the design of systems so that adverse events in patients will be prevented.[6]

Patient Safety Terminology

To further understand errors and their consequences, a basic understanding of terms used in the safety field is necessary. Examples and definitions are given in Table 29-1.

Although errors may lead to adverse events, each can occur without the other. The failure to give appropriate preoperative antibiotics is an error that does not invariably lead to a postoperative surgical site infection (an adverse event). This error is then a near miss. Conversely, surgical site infection can occur even in the absence of errors. But, if an infection is due to an error, it is a preventable adverse event.

Postsurgical Wound Infection Outbreak: A Medical Adverse Event

An example of an adverse event relating to infection control was reported by the Agency for Healthcare Research and Quality in its online morbidity and mortality forum.[7] The case vignette describes an increase in postoperative sternal wound infections. An investigation by the infection control team determined that the outbreak was occurring in patients operated on by one surgeon and his team. When observed during surgery, this team used "sloppy" technique, including having loose hair and jewelry while performing surgery. Although contamination of surgical wounds by the operating room personnel was not proven in this case, studies have suggested that bacteria shed from surgical personnel—including from skin and hair[8,9]— is common and has led to outbreaks of postoperative infection.[10,11] Although the operating room staff committed active errors (ie, not properly covering their hair and not removing jewelry), these were likely related to errors in the system. For instance, the culture of the operating room (ie, the entire surgical team) and leadership of the operating room (ie, the individual surgeon and surgical chief) permitted sloppy personal surgical attire. Other

Table 29-1. Patient-safety terminology

Term	Definition	Example
Error	The failure of a planned action to be competed as intended (ie, error of execution) or the use of a wrong plan to achieve an aim (ie, error of planning)	Error of execution: right medication administered to the wrong patient by a nurse Error of planning: wrong medication ordered by a physician
Active error	An error by an individual at the "front line" of a complex process	Nurse administers a toxic dose of an aminoglycoside (that was incorrectly ordered and dispensed)
Slip	An error of implementation (ie, failure to perform a semiautomatic, low-level behavior)	Failure to order contact isolation for an inpatient with a new culture growing vancomycin-resistant *Enterococcus*
Mistake	An error of higher functioning during a nonstereotypic behavior	Physician made the wrong diagnosis although the right diagnosis was evident
Latent error	A system error that leads to adverse events if combined with another factor or factors	Chronic understaffing of nursing, which may increase the risk of subsequent active errors by overworked nurses
Adverse event	An unexpected negative outcome of a process; in the case of health care, the processes are medical interventions	Postoperative pneumonia in an otherwise healthy patient
Preventable adverse event	An adverse event caused by an error	Postoperative wound infection after failure to administer appropriate preoperative antibiotic prophylaxis
Sentinel event	An adverse event that is serious and unexpected	Death from postoperative pneumonia in an otherwise healthy patient
Near miss	An error that does not result in a preventable adverse event but could if the error were repeated in the future	Use of inadequately sterilized surgical instruments that does not lead to a subsequent infection in the patient

NOTE: Information is from Kohn et al.[3] and Reason.[26]

potential system errors may have included lack of sleep and rest by surgical team members, a rushed schedule that compromised safe practices in order to facilitate rapid patient turnover, lack of access to properly fitting head coverings, and inadequate training of the operating room staff in infection control procedures.

Latent System Errors

In both examples given above, one individual did not act in error alone. Instead, there were multiple active errors because there were several latent errors in the system. For instance, in the *Columbia* disaster, there were many faulty decisions made before and during the flight. However, these decisions were made in a system of poor communication (between engineers and managers) and a culture that downplayed risks instead of actively exploring potential problems. The investigation concluded that the disaster was primarily the result of NASA's "culture," not the act of any one individual.[2] In the instance of the outbreak of sternal wound infections, bacteria shed from the hair of multiple individuals likely caused the wound contamination. These errors could only occur because of latent errors in training, leadership, group culture, and/or patient scheduling.

Adverse events are usually preceded by an active error by an individual. However, these errors are typically the result of latent errors that both increased the risk of the active error (eg, understaffing) and allowed the error(s) to progress to adverse events (eg, lack of engineering or procedural safety checks). Thus, system errors are thought to account for most adverse events.[6]

Multiple errors usually need to occur together for major adverse events to occur. James Reason's "Swiss cheese" model analogizes the multiple errors that precede an adverse event to holes in slices of Swiss cheese lined up like dominoes.[12] The layers of cheese represent multiple barriers and safeguards to prevent adverse events from taking place. The holes in the cheese are active and latent errors. A hole in any one slice does not lead to an adverse event. Only when the holes in multiple layers momentarily line up (ie, when multiple active and latent errors occur together) do adverse events happen.

Techniques for Detecting and Investigating Adverse Events

Since sentinel events represent severe adverse events, they are worthy of investigation, to prevent further injuries or deaths. Two techniques used to investigate past and potential future adverse events are root cause analysis (RCA) and failure modes and effects analysis (FMEA). Infection control programs have traditionally used surveillance to compare infection rates against past rates ("benchmarks") at a single institution or from other medical centers, such as those available through the National Healthcare Safety Network (NHSN) program at the Centers for Disease Control and Prevention (CDC).[13,14] After a discussion of RCAs and FMEAs, we contrast these methods with "benchmarking," and describe how the 3 approaches are complementary. While full instruction on how to perform an RCA and a FMEA is beyond the scope of this chapter, we provide a conceptual overview of the techniques, illustrated with theoretical examples that may be encountered in infection control.

RCA

RCA is used to investigate a sentinel event in order to determine and correct its causes, and thus to prevent the event or decrease the likelihood that the event will recur. An RCA starts with creation of a flow diagram of events that led to the adverse event. Next, a separate cause and effect diagram is made. Starting with the adverse event, the RCA team traces the causes of the event backward sequentially, elaborating the "roots of the tree" to determine underlying root causes of the event. Causal statements are then constructed to describe how a root cause(s) led to the adverse event(s), with emphasis on latent system errors. Finally, recommendations are made for how to correct the root causes to prevent the adverse event from recurring. It is important that the investigators both review the medical record and interview people involved in the process or event under study.[15]

An investigation of an unexpected death from a post-surgical wound infection could be performed using an RCA. In fact, the Joint Commission mandates that an RCA be performed for all such sentinel events (whether the event is infection related or not). The flow of events is mapped, starting with the need for surgery, through the details of the infection control practices and preoperative antibiotic prophylaxis used, and concluding with the postoperative diagnosis and the management of the infection up through the time of death. A causal diagram would elaborate the root cause of the infectious death; for example: death from infection, *caused by* (1) wound contamination, *caused by* lack of hospital guidelines mandating that surgical attire cover all head and face hair; and (another branch) *caused by* (2) preoperative administration of antibiotics that was started after the skin incision, *caused by* a schedule that did not allow time for preoperative administration of antibiotics. Causal statements and recommendations may include

the following: "Death from wound infection was caused by lack of adequate procedures for protection from wound contamination and lack of proper timing of preoperative antibiotics. Recommendation will be for use of a surgical checklist prior to skin incision that includes checking for proper surgical attire by entire surgical team and completion of preoperative antibiotic administration prior to skin incision."

FMEA

In general terms, FMEA is a systematic method of identifying and preventing product, equipment, and process problems before they occur. Each way a system can fail is called a "failure mode." Each failure mode has a potential "effect" (adverse event). As with an RCA, the first step is creating a flow chart of the system. Next, the FMEA team brainstorms to think of failure modes for each step and their potential effect(s). As with an RCA, people involved with the process under study should be interviewed (and/or included on the team). The "severity" of the effect, risk of "occurrence," and ease of "detection" are then assessed on a scale (eg, a scale of 1 to 10, where 10 is the worst). Each effect is then assigned a risk priority number (RPN) which is the product of the 3 scores (the severity score multiplied by the occurrence score multiplied by the detection score). In our example, the RPN (or "criticality") of the effect will be on a scale from 1 to 1,000. The potential effects of a system are then ranked from highest to lowest RPN score. The effects with the highest RPN—and all effects with an absolutely high RPN—are targeted for corrective action. After the intervention(s), the RPNs of the effects are recalculated; these are referred to as the "resulting RPNs." Corrective actions should continue until the RPN or resulting RPN for all potential effects is at an acceptable level. An acceptable RPN level is not a number set in stone. The team performing the FMEA has to decide what an acceptable level of risk is in the system they are evaluating.

An example of an FMEA is the evaluation of a hospital's system for sterilizing surgical instruments. There are many steps in the process, including cleaning, sterilization, and the evaluation of sterilization using biological tests (ie, determining if the sterilizer properly killed a standard test sample of bacteria). Failure to list all sterilized instruments in a log may be a common occurrence (occurrence score, 7), but it is easily detected (detection score, 2) and does not lead to a severe effect (severity score, 2), giving a low RPN of 28 (ie, 7 x 2 x 2). Failure to exchange the ethylene oxide canister in the ethylene oxide sterilizer during a sterilizer run may be an uncommon occurrence (occurrence score, 3), but it may be difficult to detect without an automatic alarm system (detection score, 8), and operating with nonsterile instruments is likely to have a severe effect (severity score, 8), leading to a relatively higher RPN of 192 (ie, 3 x 8 x 8). An FMEA of this system would first target the latter step for corrective action.

Contrasting RCA, FMEA, and Benchmarking

The advantages and disadvantages of using these 3 investigative techniques are listed in Table 29-2. Surveillance with benchmarking detects trends in infection rates even without an unexpected death or disability (ie, without a sentinel event). An RCA can uncover system errors that may not be explored in an outbreak case-control study that focuses on patient-level risk factors. Since it can be triggered by a single sentinel event, an RCA investigation can also be initiated sooner, before a benchmarking-based system would have detected the new problem. Finally, an FMEA investigates the potential for adverse events that have not yet occurred, preventing patient harm before it happens. Thus, the 3 techniques are complementary.

Healthcare-associated infections have traditionally been tracked using surveillance with benchmarking, then investigated with a retrospective cohort study or a case-control study. An outcome is (at least in part) attributed to or "caused by" a risk factor if the probability of finding the observed association between risk factor by chance alone is less than 0.05 (ie, if the P value is less than .05).[16] This quantitative approach can be contrasted with the qualitative techniques of RCA and FMEA, which base decisions on the consensus of an investigative team. However, the determination of the actual "causality" is done on the basis of the sum of the available evidence using Hill's classic criteria,[17] and it is not typically estab-

lished by a single study. Furthermore, the results of any investigation need to be interpreted in the context of the study methods. While epidemiologists may be more comfortable with the results of a quantitative study, both qualitative and quantitative techniques have a useful role in investigating healthcare-related infections.

Other Approaches to Improving Safety

In addition to the use of FMEAs and RCAs, there are multiple other quality improvement systems that have begun to be adapted for use in the medical setting. For example, "Six Sigma" is an approach focused on reducing the error rate to less than 3.4 errors per 1 million events (ie, 6 standard deviations [sigma] from the mean) by "design, measure, analyze and improving" processes (for ongoing processes); in one study, Six Sigma methodology improved hospital hand hygiene compliance[18] Another example is Toyota Production Systems (TPS), which seeks to reduce "overburden" and "inconsistency" in order to decrease waste and thus improve the efficiency of a process. TPS and real-time error reporting across a large healthcare system was used by the Pittsburgh Regional Healthcare Initiatives to decrease infection rates.[19] Another example is Positive Deviance, which identifies individuals in a group with an uncommon approach that leads to a better solution or outcome without using more resources. Along with TPS, Positive Deviance is being utilized by selected Veterans Affairs hospital sites as part of a national Veterans Affairs collaborative to decrease the incidence of methicillin-resistant *Staphylococcus aureus* transmission and infections (personal observation, D.R.L.). These approaches represent potential systems of

Table 29-2. Comparison of techniques for detecting and investigating adverse events

Method	Advantages	Disadvantages
Surveillance and benchmarking (and outbreak investigation)	• Detects trends in adverse events that are not sentinel events • Detects patient-level risk factors • Yields quantitative results	• Requires multiple events to trigger an investigation, with harm to those patients • May not detect system-level errors • Time-consuming and costly
Root cause analysis	• Can initiate after one sentinel event • Detects system errors	• Requires wait until sentinel event occurs, such that at least 1 patient is harmed • Qualitative results may be susceptible to hindsight bias • May not detect patient-level risk factors • Time-consuming and costly
Failure modes and effects analysis	• Can be initiated prior to an adverse or sentinel event • Detects system errors	• Risk of potential events may not be anticipated until errors occur • Qualitative results may be susceptible to hindsight bias • May not detect patient-level risk factors • Time-consuming and costly

improvement or cultural change that can be learned and potentially adapted to improve hospital patient safety with respect to infections.

Infection Control and Patient Safety

The field of infection control is already working toward improving patient safety by focusing on the prevention of healthcare-associated infections. The Joint Commission's National Patient Safety Goals include infection control goals.[4] Reviewers have emphasized the role of infection control as a critical component of patient safety.[20,21] The following list gives examples of infection control activities that promote patient safety by decreasing the risk of healthcare-associated infections.

1. Surveillance for healthcare-associated infections with feedback of data on infection rates to clinicians
2. Investigating and controlling outbreaks
3. Ensuring proper sterilization or disinfection of equipment for procedures and surgeries
4. Vaccination of vulnerable patients against preventable infectious diseases
5. Evaluating and improving infection control practices that protect patients
   A. Fostering adherence to hand hygiene recommendations
   B. Ensuring proper placement and care of invasive devices (eg, central venous catheters)
   C. Using contact isolation for patients with infectious diseases that are spread by healthcare workers
   D. Ensuring proper adherence to environmental infection control procedures by operating room staff (eg, complete coverage of head and face hair)
   E. Administering preoperative antibiotic prophylaxis when indicated
   F. Ensuring judicious use of antimicrobials
   G. Vaccination of staff against influenza

Joint Commission Regulations Addressing Infection Control

The Joint Commission is the major source of accreditation and stands at the intersection of patient safety and infection control. It first published 6 National Patient Safety Goals that became effective on January 1, 2003. For 2010, there are now 16 goals, of which 2 directly address issues in infection prevention.[4] Goal 7 is *"Reduce the risk of healthcare-associated infections."* There are 5 requirements to this goal 2 of which we review here in detail. Requirement 07.01.01 is *Meeting hand hygiene guidelines.* Compliance with either CDC or World Health Organization hand hygiene guidelines by the hospital will be assessed by interviews and observations of hospital staff. If there is more than a "sporadic" miss in compliance by staff, the hospital will be scored as noncompliant. The hand hygiene recommendations of the most recent World Health Organization guidelines (published in 2009) are discussed and summarized in Chapter 21 of this book.

The second requirement (07.02.01) is *"Manage as sentinel events all identified cases of unanticipated death or major permanent loss of function associated with a healthcare-acquired infection."* In their regulations, the Joint Commission specifies that all sentinel events must be evaluated with an RCA. The Joint Commission emphasizes that this requirement is not new; it simply clarifies that an unanticipated death or loss of function should be reported even if it is due to a healthcare-associated infection. Whether a death is "unanticipated" depends on the patient's condition on admission. The death of an otherwise healthy adult admitted for an elective procedure would be unanticipated. If the patient was not likely to survive the hospitalization because of their medical conditions at baseline and at admission, then their death would not be a sentinel event. Importantly, the Joint Commission emphasizes that this requirement should not increase the surveillance already being performed (by infection preventionists and/or other hospital personnel). Thus, a hospital's current sentinel event reporting system should clearly include events due to healthcare-associated infections. Another potential source of surveillance for these types of sentinel events is for infection preventionists to report whether infections discovered through surveillance activities are related to subsequent patient deaths.

The second National Patient Safety Goal that directly addresses issues in infection prevention is Goal 10, which is *"Reduce the risk of influenza and pneumococcal disease in institutionalized older adults."* This goal has 2 requirements: the first (10.01.01) is to develop protocols to identify whether to administer vaccine and to vaccinate patients at "high risk" for influenza and for pneumococcal disease, and the second (10.02.01) is to develop protocols to identify new cases of influenza and manage an outbreak.

Information Technology and Sentinel Event Reporting

The role of technology in improving the quality and safety of patient care in general has been emphasized in recent reviews and reports publications.[6,22] One example is electronic reporting of adverse events by healthcare workers to the hospitals. The University of Pennsylvania Health System uses a computer-based system to collect spontaneous reports of adverse events. This system provides several advantages over alternative mediums (eg, paper-based reports). The hospital's internal computer network can be readily accessed at computer workstations, and the information is quickly transmitted to administrators (including hospital epidemiologists) who can respond to the events. The reports can also be transmitted directly into a database. Thus, single sentinel events as well as trends in adverse events can be communicated efficiently, allowing for rapid response by the hospital administration and infection control team.

Getting Started: Patient Safety and Infection Control at Your Healthcare Institution

1. By performing surveillance for healthcare-associated infections, investigating outbreaks, and promoting good infection control practices, you are already contributing to patient safety.
2. Ensure that a mechanism for reporting and investigating healthcare-associated infection sentinel events is in place at your hospital. Various mechanisms are possible, including passive surveillance by healthcare workers (through a telephone hotline, paper form, or computer-based form) or active surveillance by the infection control team or other hospital-based personnel. Other administrators may primarily run sentinel-event reporting, but input from infection control will be vital. The infection control team should also become familiar with RCA and FMEA techniques and may be called upon by the hospital to lead or at least participate on teams performing these investigations.
3. Establish a nonpunitive culture with regard to medical errors. Healthcare workers do not report to work with the intention of making mistakes. However, they routinely perform complex tasks under less-than-ideal conditions (eg, with inadequate training or as part of an understaffed department). Active errors by healthcare workers that led to adverse events were likely caused by (or allowed to happen by) system errors in how care is delivered. Although active errors should be evaluated and corrected with improvements in systems, including retraining of staff, they should not result in punitive action against the healthcare worker. Only in such a nonpunitive culture will errors and adverse events be reported, allowing for investigation and system corrections to protect future patients.[6]

Conversely, there are blameworthy behaviors that merit action by healthcare organizations (personal communication, James Bagian). The behaviors include errors by healthcare workers that occur because the worker is impaired by alcohol or other drugs, because the worker

is working outside the scope of their responsibility, because of reckless behavior, or because of intentionally harmful behavior. A nonpunitive environment does not mean that people are not responsible for their actions but that staff who commit errors will be encouraged to report mistakes and slips rather than hide them for fear of reprisal and that they will be valued for their reporting.

Other Resources for FMEA and RCA

The review article by Spath,[23] "Using failure mode and effects analysis to improve patient safety," summarizes the history and methodology of FMEA using simple language and concepts.

The Veterans Affairs Patient Safety Web site[24] has explanations and instructions for performing RCAs and FMEAs, as well as links to other resources. The site describes a trademarked FMEA methodology referred to as a Healthcare FMEA.

The chapter "Making Health Care Safer: A Critical Analysis of Patient Safety Practices," in the report on patient safety from the Agency for Healthcare Research and Quality,[25] describes the RCA process, then critically evaluates the evidence supporting its use as a tool to investigate medical sentinel events and to improve patient safety.

References

1. GPOAccess. Selected congressional hearings and report from the Challenger space shuttle accident. 2003. http://www.gpoaccess.gov/challenger/index.html. Accessed October 12, 2009.
2. Columbia accident investigation board. Report. 2003. http://caib.nasa.gov/. Accessed October 12, 2009.
3. Kohn LT, Corrigan JM, Donaldson MS, eds. Committee on Quality Health Care in America, Institute of Medicine. To err is human: building a safer health system. Report 0309068371. Washington, DC: National Academy Press; 1999.
4. The Joint Commission. National Patient Safety Goals (NPSG): hospital. 2010. http://www.jointcommission.org/PatientSafety/NationalPatientSafetyGoals/. Accessed October 12, 2009.
5. Eikel C, Delbanco S. The Leapfrog Group for patient safety: rewarding higher standards. *Jt Comm J Qual Saf* 2003;29(12):634–639.
6. Committee on Quality Health Care in America, Institute of Medicine. Crossing the quality chasm: a new health system for the 21st century. Washington, DC: National Academy Press; 2001. http://www.iom.edu/Reports.aspx. Accessed October 12, 2009.
7. Agency for Healthcare Research and Quality (AHRQ). AHRQ Web M&M: Morbidity and Mortality Rounds on the Web. http://www.webmm.ahrq.gov/. Accessed October 12, 2009.

8. Bitkover CY, Marcusson E, Ransjo U. Spread of coagulase-negative staphylococci during cardiac operations in a modern operating room. *Ann Thorac Surg* 2000;69(4): 1110–1115.

9. Hubble MJ, Weale AE, Perez JV, Bowker KE, MacGowan AP, Bannister GC. Clothing in laminar-flow operating theatres. *J Hosp Infect* 1996;32(1):1–7.

10. Mastro TD, Farley TA, Elliott JA, et al. An outbreak of surgical-wound infections due to group A *Streptococcus* carried on the scalp. *N Engl J Med* 1990;323(14):968–972.

11. Dineen P, Drusin L. Epidemics of postoperative wound infections associated with hair carriers. *Lancet* 1973;2(7839): 1157–1159.

12. Reason J. Human error: models and management. *BMJ* 2000;320(7237):768–770.

13. National Nosocomial Infections Surveillance (NNIS) System Report, data summary from January 1992 to June 2002, issued August 2002. *Am J Infect Control* 2002;30(8): 458–475.

14. Hidron AI, Edwards JR, Patel J, et al.; National Healthcare Safety Network Team; Participating National Healthcare Safety Network Facilities. Antimicrobial-resistant pathogens associated with healthcare-associated infections: annual summary of data reported to the National Healthcare Safety Network at the Centers for Disease Control and Prevention, 2006–2007 [published correction appears in *Infect Control Hosp Epidemiol* 2009;30(1):107]. *Infect Control Hosp Epidemiol* 2008;29(11):996–1011.

15. Vincent C. Understanding and responding to adverse events. *N Engl J Med* 2003;348(11):1051–1056.

16. Rosner B. *Fundamentals of Biostatistics*. 6th ed. Pacific Grove, CA: Duxbury Press; 2005.

17. Hill AB. The environment and disease: association or causation? *Proc R Soc Med* 1965; 58:295–300.

18. Eldridge NE, Woods SS, Bonello RS, et al. Using the Six Sigma process to implement the Centers for Disease Control and Prevention guideline for hand hygiene in 4 intensive care units. *J Gen Intern Med* 2006;21(suppl 2):S35–S42.

19. Sirio CA, Segel KT, Keyser DJ, et al. Pittsburgh Regional Healthcare Initiative: a systems approach for achieving perfect patient care. *Health Aff (Millwood)* 2003;22(5):157–165.

20. Burke JP. Patient safety: infection control—a problem for patient safety. *N Engl J Med* 2003;348(7):651–656.

21. Gerberding JL. Hospital-onset infections: a patient safety issue. *Ann Intern Med* 2002;137(8):665–670.

22. Bates DW, Gawande AA. Improving safety with information technology. *N Engl J Med* 2003;348(25):2526–2534.

23. Spath PL. Using failure mode and effects analysis to improve patient safety. *Aorn J* 2003;78(1):16–37.

24. Patient Safety Web site. Veterans Affairs National Center for Patient Safety. http://www.patientsafety.gov. Accessed October 9, 2009.

25. Shojania KG, Duncan BW, McDonald KM, Wachter RM, Markowitz AJ. Agency for Healthcare Research and Quality. Making health care safer: a critical analysis of patient safety practices. *Evid Rep Technol Assess (Summ)* 2001;43:i–x, 1–668. http://www.ahrq.gov/CLINIC/PTSAFETY/. Accessed October 9, 2009.

26. Reason J. *Human Error*. Cambridge, UK: Cambridge University Press; 1990.

Administrative Issues

Chapter 30 Administering an Infection Prevention Program

Virginia R. Roth, MD, FRCPC, and Mark Loeb, MD, MSc, FRCPC

The infection control committee plays an important role in ensuring patient safety through the prevention and control of infections in a healthcare facility. Members of the committee should hold a leadership position in the hospital in order to serve as opinion leaders and to effect change when necessary. This committee is a mechanism for the infection control program to report activities, including statistics, outbreaks, and proactive control measures. The infection control committee also drafts and disseminates policies and procedures on infection surveillance, prevention, control, and education. Because its responsibilities reach virtually all hospital departments, the infection control committee can serve as a liaison between departments responsible for patient care and supporting departments, such as pharmacy, housekeeping, and maintenance. The committee should report to the facility's medical board or medical advisory committee and/or senior management.

Infection control committees vary, in that some serve single healthcare facilities (a hospital or long-term care facility), and others are regional committees. As a general rule, the former are smaller committees (eg, 8–12 members), while the latter are much larger (eg, 15–25 members) because representation from various hospitals, public health, and/or long-term care facilities is required. Regardless of size, most infection control committees function in a reporting capacity. That is, a typical meeting will not involve a working group brainstorming over policy. Rather, the findings and results obtained by subcommittees formed to address specific issues (eg, policies on staff vaccination) will be presented and discussed. The outcome of such deliberations may be the approval of the policy or a request for the subcommittee to revise a policy for discussion at the next meeting.

Membership

The infection control committee should be multidisciplinary: at a minimum, it should have representation from senior facility management, the physician group, and nursing. In addition, consideration should be given to representation from the departments of critical care, surgery, and medicine; the microbiology laboratory; the pharmacy; the departments of occupational health and central processing; housekeeping; and the local public health department. When establishing an infection control committee in your healthcare facility, it is important to take an inventory of patient care activity, particularly where infection control challenges are likely to occur. For example, if your hospital has a large hematology/oncology department, then inviting a representative of this department might be wise. Representatives from hospital departments may be nominated by the departments and preferably should be in a position to make decisions.

The committee may be most effective if the chair is a physician leader and if both the chair and members are appointed or approved by the medical board. The chair may be the hospital epidemiologist. It is important to note that all members of the committee need not be present at all meetings, particularly if the committee is very large. Members from certain groups (eg, emergency services) may have ad hoc status; if their presence is required, they are notified before the meeting.

Function of the Committee

The functions of the infection control committee should include the following:

- Formulating and recommending policy on all matters pertaining to infection control
- Development of policy and adherence to legislation about pubic reporting
- Serving in an advisory capacity to the medical and senior administration of the facility
- Reviewing infection control surveillance data and developing an appropriate action plan
- Reviewing outbreak situations and developing outbreak control measures
- Approving infection control–related policy and procedures
- Approving the annual goals and objectives of the infection control program

The infection control committee is responsible for making final recommendations to the medical board with respect to policy. The actual deliberations about the specific details of policy do not necessarily occur during the committee meetings. Often, it is more efficient to have smaller working groups iron out a draft of the policy. This is circulated to committee members well in advance of the next meeting, where committee members voice their approval or concern about the policy. Policies may range from isolation precautions for multidrug-resistant organisms to administering skin-tests for tuberculosis to hospital employees.

Agenda items to be discussed at the infection control committee meeting may include the need to change policy in response to information gathered during the day-to-day work of infection preventionists (eg, lack of adherence to guidelines for maximal sterile precautions at the insertion of central venous catheters) or in response to concerns of administrators (eg, lack of standardization of active surveillance for multidrug-resistant organisms among different hospital sites in a health system). The committee recommendations on policy are then sent to the medical board for approval. The committee, therefore, serves an advisory role for senior clinicians and administrators. Generally speaking, the recommendations of a good infection control committee are adhered to and usually are not challenged by hospital administration. Senior hospital administrators place a high value on the expertise of the infection control committee. Surveillance data collected by the infection control program are regularly reviewed at committee meetings. Quarterly surveillance reports may be attached as an appendix to the agenda for an upcoming meeting or may be distributed at the meeting. It is important that

the data are thoroughly discussed and interpreted. For example, an increase in the incidence of methicillin-resistant *Staphylococcus aureus* colonization and infection in July may be due to lack of adherence to infection control precautions among new house staff and may signal the need for a more intensive educational effort targeted at this group.

Reviewing the handling of outbreaks is another important function of the infection control committee (see Chapter 12 on outbreak investigations). During an outbreak, the outbreak management team is responsible for the investigation, including the implementation of appropriate infection control measures. This almost certainly will involve some but not all members of the infection control committee. The committee itself may not be charged with making immediate "real time" decisions in the midst of an outbreak (although provisions should be made for calling an emergency meeting if needed) but definitely should review how the outbreak was handled as part of a quality improvement process. The committee is charged with providing a strategy for managing outbreaks, such as notification procedures and possibly the establishment of a formal outbreak team. This is particularly important when planning for serious threats, such as bioterrorism and pandemic influenza.

In addition to approving infection control procedures and policies, it is important that the committee review the annual goals and objectives of the infection control program. Regular meetings should be held (eg, monthly). The infection control committee must also ensure that all appropriate facilities are available to the hospital personnel to maintain good infection control practices. The committee might, for example, have to advocate on behalf of the infection control team for more resources, such as hiring additional staff or updating computer software. The committee can coordinate the development of infection control rules and practices. It is important that the committee maintain a strong liaison with occupational health staff, because often there is substantial overlap in content between these areas. Having the occupational health physician on the infection control committee is a good way to ensure effective communication. Whether establishing a new infection control committee or taking over as chair of an existing committee, it is important to either establish or review the mission statement of the committee. The mission statement should summarize the overall goals of the committee.

Responsibilities of the Chair

Ideally, the chair should have a strong background in infection control and/or infectious diseases in order to provide leadership and direction to members. The chair's responsibilities should be specified in writing and should include the following:

- Ensuring the plans and actions decided upon by the committee are implemented
- Reviewing the membership list annually to ensure adequate representation from appropriate departments
- Replacing members who have poor attendance or are unable to fulfill their mandate
- Replacing members who have poor attendance or are unable to fulfill their mandate
- Representing the committee as spokesperson
- Appointing special subcommittees or task forces, as required

It is important that the chair have respect from committee members and the hospital administration generally. This will help ensure that the policies and recommendations of the committee are acted upon promptly. An ideal chair should be someone who is a good facilitator and who is able to convince others of the importance of new policies or practices, without being overly dogmatic or rigid. Usually, there are many items on the agenda during an infection control committee meeting. The chair must have the ability to facilitate important deliberation but at the same time move the agenda forward. The chair needs to have good delegation ability and must be capable of assigning tasks to subcommittees. The chair should also reserve the right to call an emergency infection control committee meeting at his or her discretion. The chair of the committee is usually the official spokesperson for the committee. It is usually the chair who will sit on the facility's medical board and may be called upon to defend or explain the committee's recommendations on certain issues. This may also include communications with the media, particularly likely if the chair is the healthcare facility's hospital epidemiologist.

The chair should review the makeup of the committee on an annual basis. New committee members may need to be added to the committee. For example, if the hospital undergoes extensive renovation, it may be reasonable to have a representative from the facility planning group on the committee. Individuals who do not contribute (by not attending meetings or by changing roles in the facility) should be replaced.

It is important that the chair of the committee have good administrative support. An administrative assistant needs to organize the meetings, distribute the minutes, and compile reports for distribution. Because these are legal documents, complete and accurate minutes of committee meetings must be maintained and distributed to committee members, the medical board, and department heads. The chair is also responsible for setting the agenda for each infection control committee meeting and making sure this is distributed to committee members well in advance of the meeting.

Responsibilities of the Members

The responsibilities and office terms of the members should be specified in writing. Responsibilities should include the following:

- Participating in regular meetings
- Working with other members on current infection control issues
- Identifying a delegate to attend in his or her absence
- Reporting on committee activities and decisions to those whom they represent and initiating any necessary action
- Bringing forward concerns from the population they represent

It is important that members of the committee be carefully selected. For example, when it is possible to choose between members to represent a particular department, it is important to consider the time commitments of the potential members. It might not always be advantageous to select the chair of a department because he or she likely has substantial administrative or clinical commitments. Often, such an individual will have limited time for attending meetings, let alone participating on subcommittees.

It is important to select people who have a track record of working collaboratively with others. An individual who is dogmatic and inflexible or who tends to dominate group discussions may be a liability, because open discussion and a free exchange of information are essential. Members should delegate others to attend in their place when they cannot attend the meeting. It is important that members have good communication skills, as they will be responsible for communicating committee policies to other members of their department. Members of the committee should also be well respected by their peers and be approachable, so that concerns can be brought through the member to the committee.

Compliance With Benchmarks

The infection control committee should play a key role in reviewing compliance with national and local set performance benchmarks. This may include reviewing semiannual or annual rates of adherence to hand hygiene, annual staff uptake of influenza vaccination, or appropriate timing of preoperative antimicrobial prophylaxis. The committee can help address barriers to achieving the goals, and this can provide insight and help to the infection control team implementing the interventions.

Compliance With Regulatory Requirements

Laws, local ordinances, regulations, and standards of municipal, county, state, and federal authorities influence activities in healthcare institutions, so the infection control committee may need to develop guidelines to help clinicians comply with extramural regulations. Such examples include institutional guidelines on the reporting of communicable diseases, consent requirements for human immunodeficiency virus antibody testing, policies governing professional activities of health care workers with bloodborne diseases, reporting of communicable diseases to emergency services personnel under the Ryan White Act, and procurement of cadaveric organs for transplantation.

Meeting Accreditation Requirements

The Joint Commission currently requires that hospitals have written infection control policies and documentation of procedures needed to conduct the organization's mission effectively. The infection control officer should ensure the existence of these documents. Important examples of the required documentation include a defined program for nosocomial infection surveillance, prevention, and control; departmental infection control guidelines; and the institutional statement of the infection control officer's authority.

Reporting

Minutes should be kept of each committee meeting. Timely reporting of any actions and decisions to the medical board, senior management, and other relevant committees (eg, quality improvement committee) should be ensured. Reporting to outside agencies, such as the local public health department, also may be necessary to comply with reportable disease requirements.

Suggested Reading

Boyce JM. Hospital epidemiology in smaller hospitals. *Infect Control Hosp Epidemiol* 1995; 16:600–606.

Friedman C, Barnette M, Buck AS, et al. Requirements for infrastructure and essential activities of infection control and epidemiology in out-of-hospital settings: a consensus panel report. Association for Professionals in Infection Control and Epidemiology and Society for Healthcare Epidemiology of America. *Infect Control Hosp Epidemiol* 1999; 20:695–705.

Girouard S, Levine G, Goodrich K, et al. Infection control programs at children's hospitals: a description of structures and processes. *Am J Infect Control* 2001; 29:145–151.

Haley RW. The "hospital epidemiologist" in U.S. hospitals, 1976–1977: a description of the head of the infection surveillance and control program. Report from the SENIC project. *Infect Control* 1980; 1:21–32.

Scheckler WE, Brimhall D, Buck AS, et al. Requirements for infrastructure and essential activities of infection control and epidemiology in hospitals: a consensus panel report. Society for Healthcare Epidemiology of America. *Infect Control Hosp Epidemiol* 1998; 19:114–124.

Society for Healthcare Epidemiology of America. Guidelines for Public Reporting. In: Joint Commission on Accreditation of Healthcare Organizations (JCAHO). *Surveillance, Prevention, and Control of Infections: Comprehensive Accreditation Manual for Hospitals: The Official Handbook*. Oakbrook Terrace, IL: JCAHO; 1996:IC1–IC25.

Chapter 31 An Overview of Important Regulatory and Accrediting Agencies

Judene Bartley, MS, MPH, CIC, Tammy Lundstrom, MD, JD, Mary Nettleman, MD, MS, MACP, and Gina Pugliese, RN, MS

The healthcare epidemiologist must interact with an enormous number of government agencies and other accrediting groups on an ongoing basis. In this chapter we provide an overview of the US federal and state agencies that have a major influence on infection prevention and control programs and offer practical advice on how to develop programs that meet regulatory and legislative requirements. The financial consequences from lack of compliance or deficiencies are also discussed, including reduced reimbursement or fines. We begin with a broad overview of key agencies before focusing on 3 that have the most relevance on infection prevention and control programs: the Centers for Medicare and Medicaid Services (CMS), the Joint Commission (TJC), and the Occupational Safety and Health Administration (OSHA). This chapter provides specific details on each agency's scope and on applicable legislative, regulatory, and accreditation requirements.

Government and Accrediting Agencies of Interest to Infection Prevention and Control Programs

Governmental Agencies

The Department of Health and Human Services (HHS) is the principal agency for protecting the health of all Americans. The HHS provides a range of services, including research, public health, food and drug safety, grants and funding, and health insurance. It includes several agencies, offices, and centers, many of which impact infection prevention and control programs. These include the Agency for Healthcare Research and Quality, the Food and Drug Administration, the CMS, the National Institutes of Health, and the Health Resource Service Administration. Perhaps the best known HHS agency is the Centers for Disease Control and Prevention, which also has oversight of the National Institute of Occupational Safety and Heath.

Other governmental agencies distinct from the HHS include OSHA, within the Department of Labor; the Environmental Protection Agency; the US Department of Agriculture; the Department of Transportation; and the recently established Department of Homeland Security.

Nongovernmental Accrediting Agencies

In addition to CMS, there are several nongovernmental agencies that also survey and accredit hospitals, such as TJC. To avoid duplication of efforts, CMS has given "deeming authority" to such groups, so that successful surveys and accreditations are "deemed" to be equivalent to a survey by CMS and hence qualify an institution for reimbursement.

The Food and Drug Administration

This agency develops, implements, monitors, and enforces standards for the safety, effectiveness, and labeling of all drugs and biologics, including food, blood and blood products, medical and radiological devices, antimicrobial products, and chemical germicides used in conjunction with medical devices. Blood safety and medical devices are areas that frequently involve infection prevention and control programs.

Blood safety. The Food and Drug Administration is responsible for the safety of the nation's blood supply and has specific standards for collection, testing, and distribution of blood as well as for disposal of contaminated or untested blood. These standards apply to all facilities that have blood-banking operations.

The Medical Device Act (1974) and Safe Medical Device Act of 1990. The Medical Device Act initially required the classification of medical devices according to their potential to cause harm. The Safe Medical Device Act expanded the Food and Drug Administration's authority in this area by improving incident reporting, removing defective or dangerous devices, and preventing such devices from entering the marketplace.

Vaccine Adverse Event Reporting System. The Food and Drug Administration also monitors vaccine safety. The Vaccine Adverse Event Reporting System requires all providers to report adverse events suspected to be related to the administration of vaccines.

The Environmental Protection Agency

This governmental agency is responsible for regulation and registration of chemical germicides formulated as sterilants and disinfectants used on devices or environmental surfaces as part of the Federal Insecticide, Fungicide, and Rodenticide Act. However, the Environmental Protection Agency may also involve infection prevention and control programs because of requirements for medical and pharmaceutical waste disposal and for medical and pathological incinerators. Through the Resource Conservation and Recovery Act of 1976, the Environmental Protection Agency has authority to develop regulation for solid-waste management, including regulated medical waste. As of 2009, there is no federal policy on medical waste. Instead, regulations are established by each state and interact primarily with the Department of Transportation. New attention to pharmaceutical waste disposal as part of general concerns regarding the environment may involve infection prevention and control programs.

Table 31-1 provides a selected list of federal, state, and national accreditation agencies that impact infection prevention and control programs along with details about their regulatory, accreditation, or guideline-setting activities, as well as their Internet addresses.

CMS and Conditions of Participation

Beyond other related responsibilities for quality and measurement, CMS is responsible for oversight and reimbursement monitoring of Medicare and Medicaid

programs through its 10 regional offices. The regional offices maintain close working relationships with state health department surveyors, who act as agents for state enforcement activity.

CMS develops "Conditions of Participation" and "Conditions for Coverage," which are the minimum health and safety standards that healthcare organizations must meet in order to be CMS-certified and receive reimbursement.[2] CMS also maintains standards for infection prevention and control in hospitals, ambulatory surgical centers, long-term care facilities, and home care facilities and/or agencies and enforces compliance with these as conditions for payment. CMS considers the "Infection Control Condition of Participation" (below) and related standards to be critical. An Infection Control Condition citation can risk a hospital's CMS standing and/or closure to new patient admissions.

§482.42 Condition of Participation: Infection Control: The hospital must provide a sanitary environment to avoid sources and transmission of infections and communicable diseases. There must be an active program for the prevention, control, and investigation of infections and communicable diseases.[3]

Interpretive Guidelines

Although the actual infection control standards have not been updated for decades, CMS publishes a set of "Interpretive Guidelines" to guide surveyors during inspections. CMS's Survey and Certification Group published a revision in November 2007, stating that though there was no change in the Conditions of Participation regulation for infection control, changing infectious disease threats, as well as new mechanisms to confront these threats, had emerged in recent years. As a result, CMS felt it was necessary to update these guidelines to better reflect current conditions within hospitals as well as contemporary infection prevention and control standards of practice. CMS also noted that the update included requirements for the infection control condition and related standards, as well as survey procedures to be used to determine compliance. The interpretive guidelines also include discussion and examples of practices that hospitals are encouraged to adopt but that are not necessarily required by the regulation (see Table 1, CMS Infection Control Interpretive Guidelines).

Deemed status and state exemptions. As noted earlier, if a national accrediting organization such as TJC has and enforces standards that meet the federal Conditions of Participation, CMS may grant the accrediting organization "deeming" authority and "deems" each accredited healthcare organization as meeting CMS

Table 31-1. Selected regulatory and accrediting agencies influencing infection prevention and control programs

Agency	Program(s) and/or jurisdiction	URL and telephone number	Regulatory or voluntary oversight	Key areas of focus	Comments
HHS					
Food and Drug Admin. (FDA)[a]	Safe Medical Device Act (SMDA); Safe Blood Supply (both FDA and CDC); food safety for all but meat, poultry, and eggs; drugs; biologics	http://www.fda.gov	Regulatory: enforces SMDA and regulates germicides as antiseptics, medical devices, medical device systems, and blood products; investigates food contamination, vaccine-related adverse events	Interacts with safety enforcement and recalls, including infection issues; needle-device safety per health department, OSHA	Major focus on needles, surgical implants, latex issues, sterile gloves, and medical devices, including reuse of single-use devices
Health Resources and Services Admin. (HRSA)	Health Delivery Services	http://www.hrsa.gov	Regulatory: National Practitioner Data Bank; Organ procurement transplant/funding aspects	Licensure; privileges; conditions for Ryan White funding	Reports on adverse actions; may include infection-related events
Centers for Medicare and Medicaid Services (CMS)[a]	Oversight for Medicare/Medicaid Conditions of Participation for reimbursement; includes Hospital Acquired Conditions and Healthcare Associated Infections Web pages; Infection Control Interpretative Guidelines Web page; CLIA standards (See the State level, below, for inspection and enforcement)	http://www.cms.gov (access to Medicare/Medicaid databases for all sites) http://www.cms.hhs.gov/ HospitalAcqCond/01_ Overview.asp http://www.cms.hhs.gov/ SurveyCertificationGenInfo/ PMSR/list.asp#TopOfPage http://www.cdc.gov/clia/regs/ toc.aspx	Regulatory: hospital certificate and licensure; quality screens; infection control and quality assurance standards; hospital-acquired conditions and healthcare-associated infections; Infection Control Interpretative Guidelines, CLIA standards	Enforces "Safe and Sanitary" facility in Infection Control standards; CLIA standards for laboratories and offices, hospital-acquired conditions and health-care-associated infections for inpatient and future outpatient settings	CMS rules; Conditions of Participation for all applicable standards; infection control standards and Interpretive Guidelines for acute care; ambulatory surgical centers; long-term care
Environmental Protection Agency[a]	Regulated medical waste; disinfectants for hard surfaces; antimicrobial pesticides	http://www.epa.gov http://www.epa.gov/osw/ nonhaz/industrial/medical/ mwfaqs.htm http://www.epa.gov/ oppad001/dis_tss_docs/ dis-01.htm http://www.epa.gov/ oppad001/regpolicy.htm Antimicrobial hotline: 1-703-308-0127	Regulatory: environmental chemicals; disinfectants and cleaners. Regulates disinfectants and antimicrobials as pesticides under the Federal Insecticide, Fungicide, and Rodenticide Act	Regulation of medical waste and incinerators, chemical cleaners and disinfectants; has a hotline for antimicrobial pesticides	Major impact on programs for waste disposal; claims for chemical cleaners and disinfectant; and label checks. Approves disinfectants as "hospital use"

(continued)

Table 31-1. (continued)

Agency	Program(s) and/or jurisdiction	URL and telephone number	Regulatory or voluntary oversight	Key areas of focus	Comments
Dept. of Labor					
Occupational Health and Safety Admin. (OSHA)[a]	OSHA standards: general safety and health standards	http://www.osha.gov http://www.osha.gov/SLTC/ bloodbornepathogens/ index.html	Regulatory: bloodborne pathogens, also environmental chemical germicides for blood spills	Enforces standard precautions; hepatitis B vaccination and vaccination offered to all workers; Exposure follow-up	Enforced by federal OSHA or by states with state-plan OSHA agencies
[same]	OSHA standards: respiratory protection: Standard 1910.134, General Industry Respiratory Protection Standard (GIRPS)	www.osha.gov/SLTC/ tuberculosis/standards.html	Regulatory: enforces under the General Duty Clause[9]; the GIRPS includes medical evaluation, annual fit-testing of respirators, training, and record keeping	Enforces 2005 CDC tuberculosis guideline with a compliance survey document based on the General Duty Clause	Enforced in states through state plan OSHA agencies; 24 states have their own plan
Dept. of Transportation	Research and Special Programs	http://www.dot.gov	Regulatory: regulates hazardous and medical waste crossing state lines	Regulates infectious substances; impacts state medical waste regulations.	Enforces regulated medical waste crossing state lines. Issue for facilities is the training of workers and availability of the waste manifest
US Dept. of Agriculture	Food Safety Inspection Service	http://www.fsis.usda.gov/	Regulatory: food inspections for commercial food providers	Recall of food products; enforcement	Healthcare facility cafeterias inspected if open to public; enforced by local health departments.
Dept. of Homeland Security	Emergency preparedness and response	http://www.dhs.gov/ index.shtm	Regulatory	Interacts with Federal Emergency Management Admin. and bioterrorism planning; pandemic influenza plan	Created by Public Law 110-53, Aug 3, 2007
Voluntary organizations **Nongovernmental accreditation**					
The Joint Commission (TJC)	TJC Committee of leading medical professional societies	http://www.jointcommission. org/	Voluntary: major organization providing deemed status for Medicare/ Medicaid in lieu of CMS agent	Accredits healthcare organizations for CMS Medicare and Medicaid reimbursement through "deemed status"; core measures: enforces government regulations and the organizations' policies	TJC core indicators include National Patient Safety Goals: device-related infection, infection due to multidrug-resistant organisms, and surgical site infection

Organization	Function	URL	Type	Reimbursement/Standards	Notes
American Osteopathic Association	Healthcare Facilities Accreditation Program	http://www.osteopathic.org/	Same as above	Same as above	...
Det Norske Veritas Healthcare	National Integrated Accreditation for Healthcare Organizations	http://www.dnv.us	Same as above	Same as above	...
HMO					
National Committee on Quality Assurance	Accredits HMO and outpatient settings; as of 1998, collaborates with TJC and the American Medical Accreditation Program for quality and measurement	http://www.ncqa.org	Voluntary: provides for reimbursement agreements for private and possibly government funding	Accredits organization for CMS reimbursement; HEDIS provides measurement set used for HMO reimbursement	Indicators such as immunization rates; infection control input critical
College of American Pathologists	Certifies laboratories; recognized as having "deemed status"	http://www.cap.org	Voluntary: sets standards for laboratory tests	Third-party payer reimbursement requires TJC, CAP, or COLA; must meet CLIA standards	Requirement for meeting basic of laboratory safety and infection control; asepsis; medical waste
Commission on Office Laboratory Accreditation	Certifies laboratories; recognized as having "deemed status"	http://www.cola.org	Voluntary: sets standards for laboratory tests	Third-party payer reimbursement requires TJC, CAP, or COLA; must meet CLIA standards	Requirement for meeting basic of laboratory safety and infection control; asepsis; medical waste
State					
Dept. of public or community health; disease control; laboratory services	Infectious disease control services; eg, Communicable Disease Control	URL is state-specific	Regulatory: CDC reporting AIDS/HIV or HIV infection; MRSA, VRE infection	Tuberculosis testing; DNA typing; HIV reporting	Communicable disease reporting; immunization issues; state guidelines
State agencies charged with health facility enforcement[a]	State licensing bureau: construction codes; Office of Fire Safety	URL is state-specific	Regulatory: enforce ventilation codes; authority for clinical and physical plant surveys	Enforce CMS Conditions of Participation for Medicare/Medicaid CLIA standards, and state codes and standards. Follow up with validation surveys after accreditation surveys as part of validating "deemed status"	CMS and Certificate of Need; facility construction review; state codes based on American Institute of Architects guidelines[1]; enforces CLIA standards

(continued)

Table 31-1. (continued)

Agency	Program(s) and/or jurisdiction	URL and telephone number	Regulatory or voluntary oversight	Key areas of focus	Comments
Agency charged to enforce medical waste program and/or incinerators[a]	State plan: environmental quality agency, such as Department of Natural Resources	URL is state-specific	Regulatory: medical waste, hazardous materials and chemical waste, and incinerators	Incinerators and medical waste are controversial issue	Major impact on medical waste program; safety programs
State occupational health dept.; radiation health	State plan: 24 states have state plan enforcement; remaining states have enforcement by federal OSHA	URL is state-specific	Regulatory: Bloodborne Pathogen Standard enforcement; tuberculosis enforcement	Major enforcer of CDC guidelines; Bloodborne Pathogen Standard and tuberculosis; focus is on healthcare personnel	Enforces negative air pressure for tuberculosis control; regulated medical waste occupational injury for healthcare personnel
State Labor dept.: general safety program[a]	State plan: general safety program is within the Labor division, but refer to Occupational Health as needed	URL is state-specific	Regulatory: safety programs, workers compensation; levies fines	Labor inspections, wall-to-wall surveys every 3–5 years, involves infection control; frequently calls in Occupational Health Division	Includes barrier-free hallways, trip hazards etc.; impacts on isolation; traffic
Local					
Public health dept.; jurisdiction may be separate from state health depts.[a]	Local public health dept. communicable disease agency	URL is state-specific and may have local links	Regulatory: communicable disease reporting and follow-up	Laboratory: infection prevention and control interaction	Reporting of communicable diseases according to local rules and regulations
City or municipality dept. of public health[a] or local jurisdiction for safety	Local public health codes and inspection for food, water, waste regulations	URL is state-specific and may have local links	Regulatory: inspections for food safety, water department for effluent, etc.	Infection control and interaction with dietary, facility services, environmental services, etc.	Concerned with spills of hazardous materials and/or chemicals (eg, mercury or formaldehyde) into waste water

NOTE: CDC, Centers for Disease Control and Prevention; CLIA, Clinical Laboratory Improvement Amendments; HEDIS, Healthcare Effectiveness Data and Information Set; HHS, US Department of Health and Human Services; HIV, human immunodeficiency virus; HMO, health maintenance organization; MRSA, methicillin-resistant *Staphylococcus aureus*; URL, uniform resource locator; VRE, vancomycin-resistant *Enterococcus*. HHS designations based on the HHS organizational chart, available at http://www.hhs.gov/about/orgchart/ (accessed 5/10/2009).
a Survey agency.

certification requirements. The healthcare organization is then not subject to the routine Medicare survey and certification process.

Accrediting agencies. Most hospitals subscribe to TJC, the American Osteopathic Accreditation's Healthcare Facilities Accreditation Program, or Det Norske Veritas Healthcare, an agency approved by CMS in 2008. The accreditation program of Det Norske Veritas Healthcare, called the National Integrated Accreditation for Healthcare Organizations, is the first accrediting agency to integrate the International Organization for Standards' ISO 9001 quality management system standards with CMS conditions of participation.

CMS State Inspections: When, Why, and Preparation

CMS confers "deeming authority" but also conducts inspections in hospitals for a variety of reasons. For example, state health department surveyors act as agents of CMS to enforce Conditions of Participation in facilities that have foregone voluntary accreditation by one of the organizations noted earlier. CMS conducts inspections for the following reasons:

1. *Validation surveys.* Random validation surveys are carried out in 5% of organizations after an accrediting agency survey. These are full, comprehensive surveys that may involve many surveyors and can last a full week, depending on the size of the facility.
2. *Full survey.* State surveyors will conduct a full survey if a hospital does not participate in a survey of an accrediting agency, such as TJC or Det Norske Veritas Healthcare.
3. *Complaints.* State surveyors must conduct a partial or full survey to follow up on complaint investigations.
4. *Licensure.* If a hospital has lost its license as the result of findings from a survey of any type for any reason, CMS must conduct a full survey as an initial step to restoring licensure and/or certification.

Preparation for Surveys

The key to preparation is being familiar with the Infection Control Conditions and Standards and ensuring that all aspects of the Interpretative Guidelines (usually called "IGs") are accounted for in the written program of the institution's infection prevention and control program. CMS puts a great deal of focus on documentation, particularly on any action taken for an identified problem. One example is the requirement to maintain an "infection log" required by Standard §482.42(a)(2), which states "the infection control officer or officers must maintain a log of incidents related to infections and communicable diseases."[3] The interpretive guideline provides a broader interpretation to what has been deemed an outmoded and onerous standard. CMS requires the linking the Infection Control Standards to the Quality Assurance and Performance Improvements standards, holding the institutional leadership accountable for follow-up on all infection prevention and control program recommendations.

The Joint Commission

When preparing for accreditation or reaccreditation with TJC, the best advice is to start early and work continuously. Although TJC offers accreditation for a wide array of healthcare settings, we will focus on hospital accreditation. To maintain accreditation, hospitals undergo on-site reviews known as surveys. Surveys are unannounced and occur approximately every 18–39 months. The survey team usually includes a physician, a nurse, and a life safety code specialist or hospital administrator who has senior management experience. Between surveys, hospitals are expected to send TJC interim reports based on the results of the institution's quality improvement program and the previous TJC survey.

Prior to the on-site visit, institutions send performance measurement data on "core" measures to TJC for review through its ORYX initiative. TJC uses these data in conjunction with other measures to focus priorities for the visit. Such measures include results of previous surveys; information on the facility's programs, as contained in its application; and reported sentinel events. Surveyors tour patient care areas to observe compliance with the facility's policies and procedures, TJC standards, and National Patient Safety Goals. Surveyors use "patient tracers" who follow the patient through all of the facility's care processes, to assess whether staff understand and follow infection prevention and control standards and comply with policies such as hand hygiene and isolation requirements. Surveyors will also review medical records and interview patients.

During the "infection control tracer," which occurs near the end of the survey, surveyors will interview the infection prevention and control staff about issues they have uncovered regarding employee knowledge or policy and procedure breaches. At the end of the visit, surveyors meet with the hospital leadership team to summarize their findings and provide a preliminary report. The TJC has a central committee that issues formal notices of noncompliance and determines the accreditation status of the hospital.

Infection prevention and control. TJC sets specific standards for infection prevention and control. The standards are revised and published annually in *The Comprehensive Accreditation Manual for Hospitals*, which can be purchased in paper or electronic format.[4] Each standard includes the rationale for why the goal was chosen and details the elements of performance that are considered necessary to meet the standard. The infection prevention and control team must be familiar with the standards and collaborate with the quality team to adjust policies and procedures in an ongoing manner. In order to remain continually ready, it is helpful to assess whether the infection prevention and control program meets the intent of the standard, whether the program complies with each element of performance, and how compliance is documented. Standards impacting the infection prevention and control program are not limited to the section labeled "Infection Prevention and Control." Sections on patient rights, information management, environment, emergency management, leadership, and others often involve infection prevention and control program activities.

TJC resources. In addition to the published standards, TJC provides newsletters and conferences through its knowledge arm, Joint Commission Resources. Conferences are helpful, and the speakers often have insight into future changes. Such events are a valuable introduction to the standards and to the survey process. Many hospitals avail themselves of consulting services prior to their survey and conduct "mock surveys." Changes in proposed standards, National Patient Safety Goals information, and frequently asked questions are available on TJC's Internet site.[4] Facility reports are available to the public via TJC's "Quality Check" Internet site as part of TJC's transparency initiative.[5]

Scoring System

Overview. In general, surveyors score performance on a standard as either "compliant" or "not compliant." Standards are made up of individual elements of performance, each of which is also scored. Elements that are not applicable to an institution are not scored. Each element of performance is scored as an "A" standard (Yes or No for compliance) or as a "C" standard for which multiple observations are required for a score of noncompliance. Some standards require a "measurement of success." Two components are scored for each element of performance: compliance with the standard and the track record of compliance. The track record of compliance refers to the underlying processes and program design that support performance. Elements of performance may be scored as satisfactory, partial, or insufficient compliance. Partial or insufficient compliance will result in a "Requirement for Improvement" (usually referred to as an "RFI").

Weighting. Standards are weighted on the basis of how critical the issue is to the safety of patients. There are 2 levels of "criticality": immediate impact requirements and less-immediate impact requirements. Each Requirement for Improvement must be addressed by submission of "Evidence of Standards Compliance." Immediate impact requirements must be addressed within 45 days, whereas less-immediate impact requirements must be addressed within 60 days. Failure to submit an acceptable Evidence of Standards Compliance within the specified time frame will lead to an unfavorable accreditation decision. A second site visit may be required for hospitals that have a large number of citations or that have citations in highly critical areas.

Decision on status. Ultimately, TJC will use the information it obtains from the site visit and the organization's response and corrective actions to determine which accreditation category the institution should receive: Accreditation, Preliminary Accreditation, Provisional Accreditation, Conditional Accreditation, Preliminary Denial of Accreditation, or Denial of Accreditation. These categories are no longer established by simply counting the number or type of instances of noncompliance. Rather, the information from the site visit serves as an initial "screen" to guide the decision process. The decision process takes into account the size and complexity of the organization and the institution's response to deficiencies identified by TJC. If the institution receives a Denial of Accreditation, this will ultimately lead to a separate CMS review (see the section CMS and Conditions of Participation, below).

TJC Expectations for Infection Prevention and Control

Although standards are modified annually, there are some key points that remain constant. For example, TJC considers infection prevention and control to be a hospital-wide program. As such, there should be evidence that the hospital leadership is involved and fully engaged. This includes assigning adequate resources to the program. The infection prevention and control team is expected to demonstrate that it involves key hospital personnel in decision making. Examples include individuals from nursing, the laboratory, occupational health, and other departments.

The minutes of Infection Control Committee meetings should document discussion, conclusions, actions, and the results of actions. There should be evidence that projects have improved the quality of care and impacted patient outcomes. Falsification of data is considered grounds to deny accreditation. Under no circumstances should minutes be altered after the committee has approved them. The flow of reports through the hospitals' review and approval structure should ultimately lead to a regular report to the organization's Board of Trustees.

Key Issues

Written plan. A key document required for infection prevention and control is a written plan that is updated at least annually. TJC has specific requirements for the content of the plan but intends that its execution should closely fit the overall mission of the hospital.

Risk assessment. TJC will look at whether a risk assessment has been carried out and how priorities are set, as well as how strategies are implemented to achieve specific goals. Goals and priorities are expected to be based on the risk of transmission and acquisition of infectious agents. The analysis of risk should result in a formal document that is reviewed at least annually.

Evaluation. The program must be evaluated for effectiveness on a regular basis, and goals must be realigned accordingly.

Key activities. TJC standards for infection prevention and control span the entire spectrum of hospital activity. Examples include the appropriate storage and disinfection and/or sterilization of medical equipment, use of personal protective equipment, vaccination of staff, and management of infected employees and patients who are potentially contagious. A more recent standard is the requirement for the infection prevention and control team to be involved in biological disaster preparedness, specifically addressing preparations for the influx of potentially infectious patients.[6] TJC standards emphasize the need to coordinate with the community and to establish a communications plan (see Chapter 22).

National Patient Safety Goals

TJC established the National Patient Safety Goals in 2002 to stimulate organizational improvement activities for the most pressing patient safety issues. The National Patient Safety Goals are updated annually by TJC and include specific patient safety indicators and requirements that are typically later incorporated into Infection Control Standards. Many of the goals have impact on the infection prevention and control program, such as those pertaining to hand hygiene compliance and the prevention of device-associated infections, infections with multidrug-resistant organisms, and surgical site infections.

Common Pitfalls in Infection Prevention and Control Programs

Infection prevention and control is based on performing surveillance, analyzing data, and taking action. The TJC will review surveillance data and will expect to see inter-ventions that have prevented and reduced healthcare-associated infections. Interventions should address both control of outbreaks and prevention of endemic infections. Merely having an infection control committee and surveillance data will not suffice; it is essential to document success in improving care.

Hand hygiene standards are frequently difficult to meet. In addition to reviewing policies and data, surveyors (patient tracers) will tour patient care areas to identify episodes of noncompliance. If personnel are observed to move from patient to patient without practicing appropriate hand hygiene, then the surveyor will score the hospital as noncompliant with that element. Although this seems like a simple issue, this is one of the more common citations. The best approach is to have a program that evaluates and promotes appropriate hand hygiene on a continuous basis. A continuous monitoring process should include some "stealth" observers who are unlikely to be recognized by employees.[7]

Another common problem is for the infection prevention and control team to limit their interactions to passive responses to questions asked by the TJC surveyor. This is a missed opportunity. Do not allow the surveyor to overlook your best projects. If a problem-prone project is highlighted, the infection prevention and control team should be prepared to answer questions about why the problem was selected, who was involved in designing the intervention, and the current status of the project. Infection prevention and control issues that are high volume, high risk, or problem prone are considered high priority by TJC. Poor documentation is another pitfall; surveyors will expect a successful intervention to be mentioned in minutes and other documents.

TJC will require the hospital to follow its own policies. This is a cautionary tale for large organizations that might make policies that are impractical to implement. The infection control committee should have significant input into all policies that deal with healthcare-associated infections. Surveyors will find nurses, physicians, or trainees and may ask him or her how to clean up a blood spill or dispose of infectious waste. They might ask healthcare providers when they were last tested for tuberculosis or ask a physician what he or she would do if needle disposal boxes were too full to use. Surveyors may also talk with patients to ensure that patient educational goals have been met.

Other common problems include using nonstandard definitions of healthcare-associated infections, failing to analyze trends in data to identify outbreaks, failing to use appropriate strategies to address infection prevention, or allotting insufficient time or resources to infection prevention and control activities.[8] A common request from surveyors is to explain how hospital staff members are involved in infection prevention and control. The response should highlight the fact that staff involvement is a continuous process that begins at the time of hiring and continues with in-service training and vaccination assessments. Staff should be involved at all levels of the program, including identifying infections, providing input on control of infections, and participating in interventions.

Tips for the On-Site Survey

Surveyors bring their experiences and personalities with them. They spend a lot of time away from home and frequently endure stressful, even hostile, situations. Courtesy and hospitality will create a good working environment. Most surveyors appreciate time at the end of the day to summarize their work and begin written reports. Meetings and tours should begin on time. Committee minutes and policies should be organized and easy to access. Presentations should be concise and given by knowledgeable individuals.

Results of TJC surveys are available to the public. Detailed descriptions of compliance with core measures are available on the Internet[5] or by request. At the current time, TJC does not publicize nosocomial infection rates; however, the public is invited to comment on the performance of individual institutions.

OSHA

OSHA is the federal agency authorized to conduct workplace inspections in order to determine whether employers are complying with the agency's safety and health standards. The General Duty Clause of the Occupational Safety and Health Act of 1970 requires that employers provide every worker with a safe and healthful workplace.[9] OSHA may adopt a specific standard or regulation, such as the Bloodborne Pathogens Standard, on which it bases all its inspections and enforcement actions. When a specific standard does not exist, such as protection against occupational exposure to tuberculosis, OSHA must rely on the General Duty Clause for the authority to inspect the workplace and assess compliance with recognized "guidelines or standards of care," such as the Centers for Disease Control and Prevention guidelines.

OSHA can issue citations under the General Duty Clause, if its compliance safety and health officers (CSHOs) can demonstrate that the employer failed to keep the workplace free of a recognized hazard that was causing, or was likely to cause, death or serious physical harm and that a feasible and useful method of abatement existed.[9] All work sites, including hospitals, must be in compliance with all standards that may apply to the work site, such as its standards for respira-

tory protection (29 CFR 1910.134; 29 CFR)[10]; hazard notification (29 CFR 1910.145); record keeping (29 CFR 1910.20), which requires that facilities allow CSHO access to employee exposure and medical records; and 29 CFR 1904, which requires a log of occupational injuries and illnesses, called the "OSHA 300 log." All standards are easily accessed from the OSHA Internet site.[11]

OSHA has the authority to inspect work sites for occupational risks of tuberculosis under the General Duty Clause[9] and continues to assess compliance with the Centers for Disease Control and Prevention 2005 tuberculosis guidelines.[12] Although there is no formal tuberculosis standard, OSHA's revised Respiratory Protection Standard (29 CFR 1910.134) has been in effect since April 8, 1998.[13] This standard currently requires healthcare workers likely to be exposed to tuberculosis to undergo annual fit-testing in addition to the initial fit-test for N-95 respirators, despite the fact that this is not an explicit recommendation in the 2005 the Centers for Disease Control and Prevention guidelines.[12]

Understanding OSHA Through the Bloodborne Pathogens Standard

The Bloodborne Pathogens Standard[14,15] represented the first time that OSHA was involved in regulating a biological hazard. Because it is the most comprehensive standard that is directly applicable to the healthcare setting, it serves as a model to illustrate OSHA's processes and requirements.

OSHA first published its Occupational Exposure to Bloodborne Pathogens Standard in the December 6, 1991, *Federal Register*. When OSHA publishes a standard, it also publishes a compliance directive to assist CSHO staff in interpreting the intent of the standard and to establish uniform inspection procedures. The compliance directive also assists employers in determining the requirements for compliance and thus is valuable for developing specific procedures. The compliance directive for the Bloodborne Pathogens Standard was first issued in February 1992 and has undergone numerous revisions; the most recent revision was issued in November 2001 and reflects the requirements for the use of sharps safety devices outlined in the revised Bloodborne Pathogens Standard.[14,15] These changes were mandated by the Needlestick Safety and Prevention Act signed into law on November 2000.[16] This law authorized OSHA to revise the Bloodborne Pathogens Standard to require the use of devices with engineered injury protection and maintain a log with details of injuries related to sharp devices.

A basic program to meet the requirements of the Bloodborne Pathogens Standard includes the following elements:

- A written exposure-control plan
- Protocols that mandate healthcare workers practice universal precautions (now included in standard precautions)
- A program to provide personal protective equipment
- A hepatitis B vaccination program
- A postexposure evaluation and follow-up program
- A comprehensive hazard-communication program
- A record-keeping system that is well-maintained and accessible

State-Level OSHA Plans

Twenty-four states have state-approved OSHA plans. These state-level plans must incorporate regulations that are "at least as effective" (at least as strict) as those set forth by OSHA at the federal level. States with state-approved OSHA plans include: Alaska, Arizona, Arkansas, California, Connecticut, Hawaii, Indiana, Iowa, Kentucky, Maryland, Michigan, Minnesota, Nevada, New Mexico, New York, North Carolina, Oregon, South Carolina, Tennessee, Utah, Vermont, Virginia, Washington, and Wyoming. In Connecticut and New York, state programs cover public employees, and OSHA covers private employees. It is critical to be knowledgeable of state and federal differences since many states enact regulations that go beyond federal rules.

State Laws

A number of states passed laws that mandated the use of sharps-injury prevention devices. California was the first state to enact a law that mandated that the state OSHA program revise its a to require employers to implement sharps-injury prevention technology, including needleless systems when applicable for intravenous access and needles with engineered sharps-injury protection.

By December 2001, a total of 21 states had passed needle-safety legislation; the bulk of the bills are patterned after California's. Many of these state laws have requirements similar to those in the revised OSHA standard as mandated by the new Federal Needlestick Safety and Prevention Act.[16] If a state needle-safety law has requirements that are more stringent than what the federal law requires, then the additional state requirements must be followed.

Bloodborne Pathogens Standard: Implementation Strategies

OSHA is currently citing healthcare organizations for failure to utilize safety devices. OSHA may be flexible in issuing citations in some situations if there is evidence that safety devices are already being used in some clinical applications and that the facility has a written plan with a realistic time line that outlines the process for completion of the selection, evaluation, and adoption of safety devices in all areas where sharp devices are used. The bloodborne pathogens exposure control plan should be revised to reflect the process that will be used to achieve these goals.

Involvement of Frontline Workers

OSHA wants to ensure that management does not select devices without input from nonmanagerial workers—those responsible for direct patient care or potentially exposed to injuries from contaminated sharp devices. Input may be obtained from these frontline workers in any manner appropriate to the circumstances of the workplace. This input will be needed for identifying devices to consider, performing some type of assessment or evaluation of the devices, and selecting devices whose use will be implemented. Such input may be formal or informal; OSHA has explained that it does not prescribe any specific procedures for obtaining worker input. Frontline worker involvement in the evaluation and selection of safety devices can help promote acceptance of these devices when their use is implemented. Although it may not be feasible to involve every worker, a representative sample of workers must be included.

Device Evaluation

The device evaluation process can be formal or informal (see Chapter 33). A formal evaluation might include a pilot study on a particular unit, with written evaluation forms completed by each worker. An informal evaluation might include bringing sample devices to the department or setting for a representative sample of frontline workers to evaluate them and provide informal feedback. There are no exact formulas for the number of workers needed to evaluate a device, the number of devices to be evaluated, or the length of time an evaluation should be conducted. What *is* important is having a mechanism in place to solicit input from workers on an ongoing basis regarding their needs and preferences for safety devices. This input will be combined with data from exposure incidents and the sharps injury log and employee feedback and will guide future decisions on selection and implementation of safety devices. In some cases, it may be necessary to replace the device that was originally selected with a more suitable device. This determination can be made only by the individual facility or work site, on the basis of its own data and experiences. A key step is documenting how the selection was made.

Preferences may vary for a single device, depending on the department and workers evaluating the device.

The preferences are influenced by a number of factors: for example, prior experience with safety devices, type of clinical procedures being performed, noise or lighting in the clinical setting, or even the size of the workers' hands. Other factors that might be considered include the characteristics listed in Table 31-2.

Sharps Injury Log
OSHA intends the sharps injury log to be used as a tool for identifying high-risk areas and providing information that may be helpful in evaluating devices. Each entry in the confidential sharps injury log must include, at a minimum, the following information:

1. The type and brand of the device that caused the injury (if known)
2. The department or work areas where the incident occurred
3. A description of the events surrounding the injury—including, for example:
   a. The procedure being performed
   b. The body part affected
   c. The objects or substances involved in the exposure

OSHA permits the log to be kept in any format, such as electronic or paper. Employers may also use existing mechanisms for data collection, such as incident reports, provided that the necessary data are collected. The data from the sharps injury log are only one source of information for assessing the effectiveness of engineering controls. Employee interviews and informal feedback are other examples. Trends in the data may be helpful in making a general assessment of the effectiveness of the sharps injury prevention program; however, calculation of rates of injury by device or brand is often inaccurate and misleading for a number of reasons:

Table 31-2. Safety characteristics of devices

Suitability for a range of uses across patient populations and procedures
Single-handed or 2-handed use
Extent of change in technique required
Undefeatable safety feature
Permanent coverage of the sharp
Patient safety
Breadth of product line
Active versus passive
Positioning of hands behind sharp
Indication of activation
Packaging
Interference with procedure
Right-handed or left-handed use
Studies in the literature on efficacy

---

injuries are significantly underreported (up to 70% are not reported, according to some studies), and individual facilities usually do not have enough data to calculate rates that are statistically significant and could not have occurred by chance, so in many cases, it is impossible to compare the relative ability of devices to reduce needle-stick injuries.[17]

Conventional Needles and Exceptions
Conventional needles may still be needed. Safety devices are only required for situations in which the needle may become contaminated. It is important to document those situations where conventional needles and devices may still be appropriate.

OSHA allows exceptions to the use of safer sharps technology under the following conditions: (1) there is no safety device commercially available to perform the specific procedure, (2) employee safety is compromised by use of the safety device during a specific procedure, or (3) patient safety is compromised by use of the safety device during a specific procedure. Exceptions allowed in a facility should be carefully researched and documented. For example, use of blunt-needle sutures or securement devices must have input from clinicians, such as reports of interference with a clinical procedure, since these are considered safety devices. In these situations, it is essential that there be an annual review of new devices that come out that may solve the problem. The review process and decision must be documented, not only for clinicians' safety, but also to meet the requirement of the OSHA rule.

Overview of OSHA Inspections

OSHA lists its inspection priorities as follows: imminent danger situations, catastrophes and fatal accidents, employee complaints, programmed inspections, and follow-up inspections. Rather than being scheduled inspections, nearly all healthcare inspections related to the Bloodborne Pathogens Standard have been triggered by employee complaints. If an employee believes he or she is in imminent danger from a hazard or believes there is a violation of an OSHA standard that may result in physical harm to workers, the employee may ask for an inspection. OSHA will withhold the employee's name from the employer if the employee so requests. OSHA will inform the employee of any actions taken and also will hold an informal review with an employee of any decision not to inspect.

Inspection process. An inspection consists of 3 parts: an opening conference with the employer, a tour of the facility, and a closing conference. At the opening conference, the CSHO first will explain the reasons for the inspection, the scope of the inspection, and any applicable OSHA standards. Although a complaint regarding the Bloodborne Pathogens Standard may have triggered the visit, OSHA almost certainly will conduct a complete survey, including hazardous chemicals, radiation safety, hazard communication, and so on. In some cases, the inspection may be terminated at this point, if the CSHO finds that an exemption is appropriate. If an employee's complaint triggered the inspection, the CSHO will give a copy of the complaint to the employer. The CSHO will ask the employer to designate an employee representative. An employee representative (selected by the bargaining unit [union] employee members of the safety committee, or by the employees) is entitled to attend the opening conference and to accompany the CSHO during the inspection.

Facility tour. Healthcare epidemiologists and infection preventionists should accompany the CSHO during the tour to answer technical or clinical questions that may arise. They should be aware that in their interactions with the CSHO, it is likely they will be viewed by the CSHO as a representative of management, not as independent professionals. Infection prevention experts may help to avert a citation simply by clarifying the hospital's protocols.

During the inspection, the CSHO will review policies, procedures, and training records; survey engineering controls; and observe employee practices. The CSHO will interview employees privately about their safety and work practices and likely solicit their input into selecting safer devices to prevent needlestick injuries. The CSHO will review records of work-related injuries, illnesses, and fatalities and will check the OSHA 300 log. This log records incidents such as needlestick injuries, tuberculin skin test conversions, and cases of tuberculosis among employees. During the tour, the CSHO will point out any unsafe working conditions to the employer and may take photographs or measurements to document the problem. The CSHO may specify corrective actions and may allow the hospital to correct the problems at this point. Still, OSHA may cite and fine the hospital for these deficiencies.

Closing conference. At the closing conference, the CSHO meets with the representatives of the employer and employees to discuss problems and needs. The CSHO provides an OSHA document that explains the employer's rights and responsibilities following the inspection. The CSHO will discuss all apparent violations but will not indicate the proposed penalties at this time; the OSHA area director later reviews the report and determines citations and proposes penalties. A facility spokesperson should explain to the CSHO how the employer has attempted to comply with the standards and should provide any information that can help OSHA to determine how much time may be needed to abate an apparent violation.

Citations and penalties. The area director will send citations and notices of proposed penalties to the hospital by certified mail. The hospital must post these citations on or near the areas where the alleged violations occurred. Penalties vary according to the seriousness of the violation. The area director may propose substantial fines, up to $70,000.00, for willful or repeated violations of a standard. Both the hospital and the employees have the right to appeal and should do so, since this is the expected process OSHA uses for resolution. Employees may request an informal review if OSHA decides not to issue a citation and also may contest the time frame allowed for the hospital to correct the hazardous conditions.

Appeals. If the facility decides to contest a citation, an abatement period, or a proposed penalty, it must submit a written "Notice of Contest" to the area director within 15 working days from the time of the citation. The area director will forward this notice to the Occupational Safety and Health Review Commission (which operates independently of OSHA); this commission will assign the case to an administrative law judge. If the hospital fails to file a "Notice of Contest" within 15 days, the citation and proposed penalty will become a final order that cannot be appealed.

Organizations should not be afraid to contest citations. This is simply the process used by OSHA, and the local area offices expect further communication. Challenging of citations judged to be unfair or incorrect can lead to substantial reductions in fines. Sometimes it is possible to discuss issues informally with the area director or to meet face-to-face with additional documentation to support the reason for disagreement.

Tips for Preparing for OSHA

Accountability. The healthcare epidemiologist and the administration must understand that the employer is responsible for compliance; employees have no responsibility whatsoever. It is the employer who controls the workplace, and the employer must exercise control to ensure that employees comply. Therefore, the hospital administration must understand, promulgate, and enforce the regulations. The infection prevention and control department, the safety committee, and the occupational/employee health unit will bear much of the responsibility for developing and implementing the exposure control plan and the tuberculosis and respiratory protection program. If there are employees who repeatedly fail to comply with regulations, OSHA will listen to efforts taken to educate and retrain such individuals, but documentation of training and communication records are essential to avoid citations.

Multidisciplinary teams. A multidisciplinary group supports a proactive approach to an otherwise purely regulatory function. Many facilities have a subcommittee that includes representatives from infection prevention and control, safety, occupational health, risk management, and administration to improve interdepartmental communications, and implement new policies and procedures. During an actual inspection, members of the group communicate with facility representatives accompanying the compliance officer and provide additional information when needed. After the inspection, members implement proposed corrective measures and help to prepare responses to any citations.

Mock inspections. The healthcare epidemiologist must understand the key elements of the standards and may want to conduct periodic surveys or mock OSHA inspections to ensure compliance, check documentation, and identify problem areas. A comprehensive checklist based on the Bloodborne Pathogens Standard may help evaluate each of the areas subject to inspection (see the list of Additional Reading, at the end of the chapter). Records of these mock inspections should be maintained.

Exposure management. The management of employee exposures to blood, other potentially infectious material, and tuberculosis will vary according to each institution's plan, but infection control, employee health, and occupational medicine personnel should work together to establish a postexposure prophylaxis protocol. Infection prevention and control personnel should develop policies for handling prehospital exposures of nonemployed emergency medical personnel, visitors, and students, and nonemployee physicians and their personnel. The administration and the risk management and legal departments should review these policies.

Education and training. In larger institutions, the infection prevention and control department will not be able to assume the entire burden of initial employee training and the required annual updates. OSHA coordinators are designated in each department and are responsible for implementing regulations and teaching personnel about the standards that apply to their departments. If individuals at the departmental level are involved, their staff is more likely to understand the regulations and to comply.

Videos made by the infection prevention and control department or online web training may be helpful. Use of such resources, including computerized self-training modules, must include site-specific information, and a knowledgeable individual must be available to answer questions at the time the employee is participating in the training session. Many resources are available for all types of training (see the list of Additional Reading, at the end of the chapter). Training sessions must be conducted during working hours, on the employer's time, and must cover topics specific to the employer's workplace, including details of the protection plan, the names of persons whom the employee must contact after an exposure, and the method of medical follow-up.

Healthcare epidemiologists should seek to develop a cordial working relationship with their regional or state OSHA. By understanding the occupational health paradigm, finding common ground, and promoting dialogue, healthcare facilities can change a "regulatory burden" into a proactive safety program that affects the overall safety culture of an organization. OSHA provides multiple resources to understand and comply with its inspection process.[18] Additional resources include OSHA's "Safety and Health Information Bulletins," which address specific issues that need clarification. For example, OSHA jointly published a safety and health information bulletin with the Centers for Disease Control and Prevention and the National Institute for Occupational Safety and Health on the use of blunt-tip suture needles to decrease injury rates.[19]

References

1. American Institute of Architects. Guidelines for design and construction of healthcare facilities. Washington, DC: American Institute of Architects Press, 2006.
2. Centers for Medicaid and Medicare Services (CMS). Conditions for Coverage (CfCs) & Conditions of Participations (CoPs). http://www.cms.hhs.gov/CFCsAndCOPs/. Accessed September 8, 2009.
3. Centers for Medicaid and Medicare Services (CMS). Conditions for Coverage (CfCs) & Conditions of Participations (CoPs): Hospitals. http://www.cms.hhs.gov/CFCsAndCoPs/06_Hospitals.asp#TopOfPage. Accessed September 8, 2009.
4. The Joint Commission. 2009 Hospital Accreditation Standards. Joint Commission Resources. Chicago: TJC; 2008. http://www.jointcommission.org. Accessed May 10, 2009.
5. The Joint Commission. Quality Check Web page. http://www.qualitycheck.org. Accessed September 8, 2009.
6. Braun BI, Wineman NV, Finn NL, Barbera JA, Schmaltz SP, Loeb JM. Integrating hospitals into community emergency preparedness planning. *Ann Intern Med* 2006;144:799–811.
7. Kohli E, Ptak J, Smith R, Taylor E, Talbot EA, Kirkland KB. Variability in the Hawthorne effect with regard to hand hygiene performance in high- and low-performing inpatient care units. *Infect Control Hosp Epidemiol* 2009;30:222–225.
8. Yokoe DS, Mermel LA, Anderson DJ, et al. A compendium of strategies to prevent healthcare-associated infections in acute care hospitals. *Infect Control Hosp Epidemiol* 2008;29(suppl 1):S12–S21. http://www.journals.uchicago.edu/toc/iche/2008/29/s1. Accessed May 10, 2009.
9. Occupational Health and Safety Admnistration (OSHA). OSHA Act of 1970—General Duty Clause. http://www.osha.gov/pls/oshaweb/owadisp.show_document?p_table=OSHACT&p_id=3359. Accessed May 10, 2009.
10. Occupational Health and Safety Administration (OSHA). General Industry Respiratory Protection Standard. http://www.osha.gov/pls/oshaweb/owadisp.show_document?p_table=STANDARDS&p_id=12716. Accessed May 10, 2009.

11. Occupational Health and Safety Administration (OSHA). OSHA standards and regulations. http://www.osha.gov/html/a-z-index.html. Accessed May 10, 2009.

12. Centers for Disease Control and Prevention (CDC). Guidelines for preventing the transmission of *Mycobacterium tuberculosis* in health-care settings, 2005. *MMWR Recomm Rep* 2005;54(RR-17);1–141. http://www.cdc.gov/tb/publications/guidelines/list_date.htm. Accessed September 8, 2009.

13. Occupational Health and Safety Administration (OSHA). Tuberculosis and Respiratory Protection Enforcement. OSHA Standard Interpretation (2008, March 24). Resumes full enforcement of the entire Respiratory Protection standard, including 1910.134(f)(2). http://www.osha.gov/SLTC/tuberculosis/index.html. Accessed May 10, 2009.

14. Enforcement Policy and Procedures for Occupational Exposure to Bloodborne Pathogens. OSHA Directive CPL 2–2.69. Washington, DC: Occupational Safety and Health Administration; November 27, 2001. http://www.osha.gov/pls/oshaweb/owadisp.show_document?p_table=DIRECTIVES&p_id=2570. Accessed May 10, 2009.

15. US Department of Labor, Occupational Health and Safety Administration. Occupational exposure to bloodborne pathogens: needlesticks and other sharps injuries; final rule 29 CFR Part 1910. *Fed Regist* 2001;66:5318–5325.

16. Needlestick Safety and Prevention Act. Law 106 430, Nov 6, 2000. 106th Congress of the United States of America. http://thomas.loc.gov/. Accessed May 10, 2009.

17. Pugliese G, Germanson TP, Bartley J, et al. Evaluating sharps safety devices: meeting OSHA's intent. *Infect Control Hosp Epidemiol* 2001;22(7):456–459.

18. Occupational Health and Safety Administration (OSHA). Bloodborne Pathogens and Needlestick Prevention. http://www.osha.gov/SLTC/bloodbornepathogens/index.html. Accessed May 10, 2009.

19. Safety and Health Information Bulletin. DHHS (NIOSH) Publication No. 2008–101. Use of Blunt-Tip Suture Needles to Decrease Percutaneous Injuries to Surgical Personnel. http://www.cdc.gov/niosh/docs/2008-101/pdfs/2008-101.pdf. Accessed May 10, 2009.

Additional Reading

Occupational Health and Safety Administration (OSHA). Bloodborne Pathogen Program Assessment Tool. Premier Safety Institute. http://www.premierinc.com/quality-safety/tools-services/safety/topics/needlestick/osha_compliance.jsp. Accessed May 10, 2009.

Centers for Disease Control and Prevention (CDC). Sharps Injury Prevention Program. http://www.premierinc.com/quality-safety/tools-services/safety/topics/needlestick/cdc-sharps-injury-prevention.jsp. Accessed May 10, 2009.

Premier Safety Institute. Needlestick Prevention [brochure]. http://www.premierinc.com/quality-safety/tools-services/safety/topics/needlestick/non-acute-care.jsp. Accessed May 10, 2009.

Chapter 32 Government Mandates and Infection Control

Stephen G. Weber, MD, MSc

Despite comprehensive and evidence-based control efforts, the incidence of healthcare-associated infections (HAIs) in the United States has remained unacceptably high during the past 3 decades.[1] Perhaps of greater concern is the increasing frequency of HAIs attributable to multidrug-resistant organisms such as methicillin-resistant *Staphylococcus aureus* (MRSA) and vancomycin-resistant enterococci (VRE).[2] Further complicating and hindering efforts at control is the constantly evolving epidemiologic characteristics of these potentially lethal pathogens.[3]

These disturbing trends have focused attention on the need for even greater innovation and improved effectiveness in the fields of infection prevention and healthcare epidemiology. Increasing concern for the ongoing HAI epidemic has been widely embraced in the context of patient safety and clinical quality. The clinical quality movement, the goals of which were powerfully articulated in a 2000 Institute of Medicine report,[4] focuses on gaps in efficiency, expertise, and effectiveness across the healthcare system. Recognition of the chasm between ideal and actual practice has been accompanied by a sense of urgency in uniformly promoting high standards of clinical excellence across the healthcare system.

In many jurisdictions across the United States and internationally, the continued proliferation of HAIs, the perception that existing control strategies are not sufficient, and the effort to enhance clinical quality through reduced practice variation have coalesced to prompt governmental intervention. These initiatives have taken many forms, ranging from efforts to promote transparency and accountability to more controversial laws that mandate specific practices for infection control and pre-

vention (Table 32-1). The goal of this chapter is to provide the clinical, political, and social context for government infection control mandates; an overview of the types of measures that have been enacted; and a practical guide for those in the field to both respond to and help shape mandates.

Factors Influencing Government Infection Control Mandates

Before examining the specific nature and provisions of government infection control mandates enacted in the United States, it is worthwhile to consider factors that helped shape the evolution of this strategy during the past 2 decades. For some observers, infection prevention mandates from the government are viewed as the byproduct of years of insufficient accomplishment and relative ineffectiveness in preventing infection in US hospitals. Commentators and financial stakeholders point to the relatively unchecked incidence of healthcare associated infections in hospitals during this period, especially those caused by multidrug-resistant organisms.[1,2]

Making matters worse, there is evidence that specific strategies proven to be effective in preventing HAIs have not been universally deployed and adopted across the US healthcare system. For example, hand hygiene has been established to be the single most effective strategy for preventing HAI and the dissemination of multidrug-resistant organisms in hospitals. Nonetheless, adherence to widely accepted hand hygiene standards remains far below optimal (or even acceptable) levels at most

Table 32-1. Notable elements of state legislative measures to control healthcare-associated infection (HAI)

Illinois

Mandatory establishment of an MRSA control program at every hospital

Inclusion of MRSA bloodstream infection and aymptomatic colonization as reportable diseases

Pennsylvania

Requirement that all hospitals collect and submit data regarding 14 different categories of HAI, including surgical site infection and catheter-associated urinary tract infection

California

Mandatory screening of all hospitalized patients at high risk for MRSA infection and colonization

Phased in quarterly reporting of specific HAIs, using the National Healthcare Safety Network of the Centers for Disease Control and Prevention

NOTE: MRSA, methicillin-resistant *Staphylococcus aureus*.

institutions. Although some hospitals do self-report excellent adherence rates, levels of adherence determined by rigorous trials reported in the peer-reviewed literature remain unacceptably poor.[5] Unfortunately, the experience with hand hygiene is the rule and not the exception when it comes to assessing the adoption of many basic infection prevention techniques.

Given the prevalent perception that prevention strategies have not stemmed the spread of infections in hospitals and the undeniable failure of many healthcare workers to adhere to even the most basic practices, it is no surprise that the HAI epidemic has attracted attention from concerned parties outside of clinical medicine and hospital administration. Many point to a series of articles published in the *Chicago Tribune* in 2002 as the watershed event in raising public awareness about HAI.[6] These articles, interspersed with compelling stories from affected patients, highlighted the proliferation of HAI and the shortcomings of the prevention strategies used by hospitals and healthcare workers. Since this initial prompting, the seriousness of the HAI epidemic has continued to draw the attention of both the media and the general public. Innumerable subsequent news stories have dramatized the human toll of HAI and attempted to cast serious doubt on not only the effectiveness but, in some instances, even the competence of those working in infection control and healthcare epidemiology.

With time, the public outcry over inadequate HAI control has given rise to a robust advocacy movement in many parts of the United States. Fueled by personal ex-

perience with the devastating consequences of HAI and by frustration at perceived inaction by the healthcare establishment, these passionate individuals and groups have profoundly impacted public policy. In Illinois, a small but vocal patient advocacy group nearly single-handedly prompted state legislators to draft a bill mandating active surveillance for MRSA, lobbied to see the bill passed by a majority of state representatives, and even successfully engaged the state hospital association to endorse the measure.

Unfortunately, these well-intentioned efforts have on occasion become imbued with a more negative tone. For example, the well-respected Consumer's Union has championed efforts to draw attention to the problem of HAI and multidrug-resistant organisms through their Stop Hospital Infections campaign. Remarkably, the program's Web site promotes the slogan "End hospital secrecy and save lives."[7] This type of approach, together with the failure of many epidemiologists and infection preventionists to seriously engage with advocacy groups, has fostered an adversarial relationship between many patient advocates and clinical experts.

Simultaneous to the increasing public demand for better control of HAI, reform of the healthcare system has once again emerged as a central issue on the US political landscape. Increasingly, lawmakers have sought to identify and embrace health policy issues that not only promote improved efficiency and cost savings but also quality care and patient safety. As an aside, serendipitous factors have also helped draw the attention of lawmakers to the regulation of infection control in hospitals and healthcare facilities. In the Illinois State House, an individual who, early on, recognized the potential clinical and economic benefit of infection prevention legislation was a then little-known state senator named Barack Obama. The Illinois Hospital Report Card Act, a measure introduced by the junior legislator when he was the chair of the state's Health and Human Services Committee, mandates public reporting of rates of certain HAIs as well as other quality metrics. His subsequent ascendancy in the US political system has done little to dissuade other up-and-coming politicians from aligning themselves with the movement toward government intervention in the prevention of HAIs.

Unfortunately, there is also the potential for other, more worrisome influences on the proliferation of government infection prevention mandates. One of the less frequently discussed forces that could potentially shape government intervention in hospital infection control mandates is that of industry groups. Through their promotion of new pharmaceutical products to treat or prevent HAI or advanced technology for more-rapid or more-sensitive diagnostic tests, innumerable commercial entities have the potential to gain or profit, according to whether specific mandates related to their products are adopted by local, state, or federal lawmakers. In this complex but lucrative interplay, best clinical practices and patient safety cannot be compromised in favor of commercial gain and profit.

Types of Legislation

Public Reporting of Infection Prevention Performance and Outcomes

At the time of this writing, more than 30 US states have already enacted legislative measures that compel the public reporting of data concerning HAIs in one form or another.[8] The subsequent proliferation of these laws suggests that this approach has been widely accepted as a potentially useful tool in promoting improved infection prevention practices.

Central to the healthcare quality and safety movement is the importance of transparency in methods, practices, and outcomes as a powerful motivator for promoting performance improvement. The rationale for openness with patients, the public, and key financial and regulatory stakeholders is to promote better performance by unleashing market forces to drive practice improvement. The specific influence of these forces is best understood by considering the potential impact of public reporting of infection rates and adherence to best practices on 2 key audiences.

First, reporting of hospital infection data may directly influence the selection of healthcare professionals and institutions on the part of individual consumers (ie, patients and potential patients). In a completely unrestricted market, individuals would seek out reliable data regarding hospital infection rates and adherence to prevention practices, integrate this information with other evidence, and rationally select a healthcare professional, such as a primary care doctor or transplant surgeon, or a healthcare institution, such as a dialysis center or hospital. In doing so, the consumer would presumably seek to maximize safety and quality by selecting institutions and professionals associated with a lower risk for infection (and other poor outcomes). The reality of course is that health care in the United States is not provided or purchased in a completely free market. Patients are typically restricted by other factors, most notably the clinicians and facilities with whom their insurance provider has contracted for services. As a result, even the most informed consumer of health care might be led to choices that would not be viewed as rational from a purely economic or clinical standpoint.

Perhaps even more influential is the potential impact of the availability of infection control data to third-party insurance payers. Insurance providers are motivated by their own market pressures to maximize financial returns and improve performance. In this context, the rationale for public reporting of infection rates or performance data is based on the premise that payers will seek to contract with healthcare professionals and institutions who demonstrate excellence in infection prevention. The potential impact of this approach can be beneficial to the payer in 2 respects. First, and most obviously, insurers can avoid the costs associated with complications of care such as HAI. Second, by leveraging contractual relationships on the basis of promoting quality, insurers can market themselves more favorably to consumers. During the past 2 decades, this strategy of value-based purchasing has been embraced across the insurance industry, prompting further calls for transparency.

From the perspective of an activist lawmaker, support for legislation that mandates public reporting of infection prevention practices or HAI outcomes has few disadvantages. Even without establishing complicated rules or infrastructure for enforcement or for matching rewards or penalties with various levels of performance, policy makers can reason that market forces will provide sufficient incentive to influence practice and improve outcomes.

Public reporting may have appeared at first to offer a noncontroversial approach to promoting efforts for the prevention of HAI. However, with the proliferation of such measures, questions have emerged as to whether consumers and payers are given adequate and even valid data on which to make such critical market decisions to drive improvement. A number of potential pitfalls related to each of the various approaches to government infection control mandates are summarized in Table 32-2.

One of the greatest potential pitfalls to any public reporting mandate is the validity and reliability of the data being collected and reported. Standardization of surveillance methods, specification of populations covered, and development of tools to adjust for patient complexity or severity of illness represent elusive challenges that have not yet been definitively addressed. The collection of reliable and accurate data regarding infection rates remains difficult at many healthcare centers. For example, in the case of surgical site infections, case finding often depends on self-reporting by surgical staff.

Even where the capacity exists for adequate data collection, more-nuanced issues remain regarding the reporting of accurate infection control data. For example, in some instances it is challenging to determine whether a particular infection was actually acquired in the hospital as opposed to originally arising in the community. Inasmuch as government mandates for public reporting are designed to promote best practices in infection prevention in the hospital, the inappropriate

Table 32-2. Potential challenges to various government mandates for infection control

Public reporting

Nonstandardized surveillance methods

Inadequate case definitions

Insufficient risk adjustment tools

Distinction of community versus hospital cases

Challenges to the use of administrative data

Selection of appropriate performance indicators

Mandates for specific practices

Concerns about suitability of mandated practice for all populations

Potential for unintended clinical and economic consequences

Compromised autonomy for institutional risk assessment

Government-sponsored pay-for-performance programs

Largely based on imprecise administrative data

"Gaming of the system" to achieve high performance without achieving improvements in infection prevention

---

inclusion of infectious cases that originated outside of the hospital is doubly problematic. First, this will artificially increase the number of infections ascribed to a facility, which unnecessarily penalizes clinicians and administrators at that hospital. Perhaps more troubling is the prospect that without strict and appropriate definitions to reliably identify infections that are acquired in the community, the possibility exists that hospitals and healthcare professionals will seek to avoid patients with predisposing conditions associated with the risk of infection. As a result, patients may be cherry-picked to avoid those who might eventually be counted against a facility's HAI case total.

In meeting these challenges, 2 sources have been heralded as potentially valuable in the accounting of infection rates within hospital, with one more promising than the other. Because of the difficulties associated with surveillance, some have advocated for the use of administrative or billing data to track infection rates. In most US healthcare facilities, repositories of data are available regarding patient diagnoses, hospital charges, and other clinical information. These data can be mined to identify patients with likely HAI. However, this method is problematic because serious inaccuracies have been reported when comparisons are made between infection rates determined using clinical surveillance data and rates determined using administrative data.[9]

A more promising source for public reporting of infection prevention process or outcome measures may be to make use of existing clinical data collection and surveillance programs. The most often cited example is the National Healthcare Safety Network (NHSN), estab-

lished and maintained by the Centers for Disease Control and Prevention (CDC). The NHSN offers Internet functionality that allows hospitals to enter data about a number of infectious conditions, including central line–related bloodstream infections and ventilator-associated pneumonia, and even the frequency of infections caused by multidrug-resistant organisms. The reporting system uses case definitions that rigorously conform to CDC standards. The integrated nature of the tool allows for the possibility of benchmarking and comparisons between institutions and within hospitals over time.

Unfortunately, even the NHSN has met with criticism. A 2008 report from the Government Accountability Office articulated concerns about the adoption of the NHSN as the primary tool to satisfy state or federal reporting requirements.[10] The Government Accountability Office suggested that, because the NHSN was originally designed as a surveillance system for only specific types of infection in a limited number of clinical care areas, it is not an appropriate tool for the purpose of reporting and comparing global performance in preventing infections. The Government Accountability Office and a number of outside critics have also pointed out that not all hospitals may be relied on to provide valid data, because of the financial disincentives associated with disclosure of high infection rates.

Where data regarding specific infectious outcomes have been difficult to precisely collect and report, the use of performance metrics has been advocated as an alternative and potentially more reliable means by which to disclose and compare the risk of infections in hospitals. In essence, these performance standards, which quantify adherence to evidence-based best practices that have been associated with reduced rates of infection, offer a discrete metric that can be more easily measured, even at underresourced healthcare facilities. Examples include determining the frequency with which proper aseptic techniques are used when inserting central venous lines and the timely administration of appropriate antibiotics to prevent perioperative infection for patients undergoing specific surgical procedures. Performance measures have been adopted as part of the Centers for Medicare and Medicaid Services (CMS) Core Measures to assess quality at US hospitals and have also been widely promoted by the Institute for Healthcare Improvement and the Leapfrog Group, among other organizations[11,12] (see Chapter 31, on regulatory and accrediting agencies).

Mandates for Specific Infection Prevention Strategies and Practices

Arguably the most controversial approach to governmental intervention in infection prevention in hospitals is the mandating of specific infection prevention practices.

At the time of this writing, at least 5 states have enacted public measures requiring adherence to specific infection prevention standards and practices.[13] In many cases, these laws mandate the implementation of active surveillance programs to check the spread of MRSA. However, broader measures have been proposed and enacted in some jurisdictions.

The motivation for this approach, as articulated by proponents of these measures, is to compel institutions and individual healthcare professionals to meet specific practice standards and performance objectives that are proven to reduce the risk of HAI. The penalty for nonadherence to the specific practice is linked to the loss of licensure or other disincentives related to reimbursement or accreditation. Proponents cite the failure of many healthcare institutions and professionals to meet even minimal standards of adherence to best practices in infection prevention, as was discussed earlier in this section. They argue that in the absence of government mandates, professionals and institutions might otherwise elect to not adopt the particular standard or practice and, because of this, would fail to provide the highest level of protection for patients.

In principle, government mandates compelling healthcare institutions and professionals to adopt specific prevention practices could offer a powerful tool that is closely aligned with the objectives of individuals who work in institutional infection control and healthcare epidemiology. Rather than repeatedly campaigning with hospital leadership to provide resources for a new or enhanced prevention standard, members of the institution's infection control team might instead leverage the compulsory nature of the governmental mandate to secure the necessary support (see Chapter 2). Moreover, when managed optimally, resources allocated to support the mandatory practice could also be used to strengthen and enhance other closely aligned infection prevention standards. For example, a nurse coordinator hired to manage a hospital's mandatory MRSA active surveillance program could be cross-trained to contribute to other infection control activities, including surveillance, performance improvement, education, and outbreak investigation.

Unfortunately, the experience to date with governmental standards mandating specific infection prevention practices has been somewhat more complicated and quite a bit more controversial than this introduction would suggest. Many of the concerns were articulated in the 2006 position statement published jointly by the Society for Healthcare Epidemiology of America (SHEA) and the Association for Professionals in Infection Control and Epidemiology (APIC) regarding legislative mandates for active surveillance for MRSA and VRE.[14] The issue, and indeed the position paper itself, stirred controversy, as supporters and opponents of active surveillance

squared off over the suitability of such legislation. Table 32-2 summarizes some of the challenges related to government-mandated infection control interventions.

Regardless of one's personal outlook regarding active surveillance as an infection prevention tool, the debate surrounding this focus does serve to highlight some of the potential downsides of these directives in general. More specifically, although it has already been acknowledged that government mandates can be quite useful in promoting the dissemination of unequivocally effective practices, what happens when the mandated practice does not meet this high standard for adoption? This was the position embraced by many individuals who originally opposed mandatory active surveillance legislation. Specifically, they argued that although active surveillance appears to be an effective tool for multidrug-resistant organisms control in specific patient populations and care settings, the relative efficacy of this approach had not yet been adequately proven (or even tested) in all populations that would be covered by some of the laws.[15,16] Moreover, the relative contribution of active surveillance to the prevention of MRSA infection had not been sufficiently well studied relative to the benefit attributable to other more basic practices such as hand hygiene and isolation precautions. Critics of the proposed legislation asked whether it is sensible to mandate a resource-intensive strategy when the very practices on which the effectiveness of that strategy is based are themselves not adequately adhered to or enforced.

Additional concerns surrounding other practical aspects of mandatory active surveillance shed light on some more-pervasive concerns about government mandates for specific infection prevention practices. Even well-intentioned legislation mandating a specific infection prevention practice may lead to unintended consequences that could not be reasonably anticipated by even the most savvy or well-briefed legislator. In the case of mandatory MRSA screening, concerns have been raised about the flow of patients through facilities adhering to a strict and compulsory surveillance standard. Will the detection of colonization, which sometimes presents a barrier in assigning patients to rooms, transferring patients between care areas, and even discharging patients from the hospital, create a ripple effect in a busy hospital and result in unexpected backups in key clinical areas (eg, performance of procedures or treatment in the emergency department) or unnecessarily prolong length of stay?

Some of the unintended consequences of legislation mandating specific practices may be more subtle. For example, concerns have been raised that widespread mandatory MRSA screening could prompt concerned clinicians to overuse the nasal decolonizing agent mupirocin, the administration of which may be linked to subsequent increases in the incidence of antibiotic-resistant strains of staphylococcal species.[17] Of course, even such seemingly esoteric consequences can generally be navigated with thoughtful follow-up and appropriate modifications to the newly adopted practice. Unfortunately, legislative mandates do not typically embrace or even permit the kind of constant reevaluation and improvement that is so crucial in applying rigorous science to the most effective bedside practice.

Regrettably, the fundamental issue regarding the suitability of government regulations mandating any particular infection prevention practice was never completely addressed during the sometimes rancorous debate about active surveillance mandates. The work of the most effective institutional infection prevention programs is specifically tailored to the needs of the institution as determined by the results of periodic and comprehensive risk and performance assessment. Put another way, infection control is not a one-size-fits-all approach, and local autonomy in needs assessment and resource allocation is vital. Legislative mandates for specific practices can essentially short circuit this approach. As was stated in the APIC/SHEA position paper, "legislation mandating use of one particular infection control strategy is no different than legislation insisting on use of one specific operative approach by cardiovascular surgeons, one pain regimen by palliative care specialists, or one particular chemotherapy agent by oncologists."[14(p255)]

Government mandates of specific infection prevention practices will likely remain controversial. For the foreseeable future, two powerful forces will continue to collide: the autonomy of individual healthcare professionals and institutions to develop and deploy customized infection prevention programs, and the urgency of ensuring that a high standard of safety and protection from infection is offered to all patients. For now, the jurisdictions in which such measures have already been deployed should be examined very intensely and very carefully. In addition to quantification of the effectiveness and costs of meeting the stated aims of the legislation, understanding the unintended consequences of state mandates will be most valuable.

Government-Sponsored Pay-for-Performance Programs

Strictly speaking, value-based purchasing programs adopted by the government cannot be considered a specific mandate for infection prevention and healthcare epidemiology. Nonetheless, because these programs have so significantly impacted the practice of infection prevention at many US hospitals, it appears appropriate to address the implications of government "pay for performance" initiatives.

The Deficit Reduction Act of 2005 included provisions that require the Secretary of Health and Human Services to identify healthcare conditions that (1) are high-cost, high-volume, or both; (2) result in the assignment of a case to a Medicare or Medicaid reimbursement that has a higher payment when present as a secondary diagnosis; and (3) could reasonably have been prevented through the application of evidence-based guidelines. For hospital discharges occurring on or after October 1, 2008, hospitals would no longer receive additional payment for cases in which one of the selected conditions was not already present at the time of hospital admission. In other words, the hospital would be reimbursed as though the secondary diagnosis were not present. In an example provided by the CMS, a patient admitted to the hospital for an acute stroke who also had a stage 3 pressure ulcer would be reimbursed more than $8,000 if the ulcer was present on admission but only $5,300 if it developed during the hospitalization.[18]

After some deliberation, a number of conditions were identified as part of the initial rollout of the CMS value-based purchasing system. Several selected conditions are related to infection prevention and healthcare epidemiology. These are detailed in Table 32-3 along with a list of other infectious conditions that have been proposed for addition to the list of preventable "hospital-acquired conditions" in the future.

The rationale for the provisions of the Deficit Reduction Act are closely aligned with the themes related to value-based purchasing introduced earlier in this section. In essence, the provisions of this law specifically hold hospitals and inpatient-care professionals financially accountable for the harm that befalls patients when under their care. With this measure, the era in which hospitals could actually recoup extra financial return as a result of complications (including infection) during hospitalization came to an abrupt end.

Table 32-3. Centers for Medicare and Medicaid Services list of preventable hospital-acquired conditions

---

Surgical site infection following certain orthopedic procedures or bariatric surgery for obesity[a]

Mediastinitis after coronary artery bypass graft surgery[a]

Vascular catheter–associated bloodstream infection[a]

Catheter-associated urinary tract infection[a]

Staphylococcus aureus septicemia (proposed)

Clostridium difficile infection (proposed)

Methicillin-resistant *S. aureus* colonization and infection (proposed)

Ventilator-associated pneumonia (proposed)

---

a Eligible for reduced payment beginning in October 2008.

In the experience of many infection preventionists and epidemiologists, some favorable impact of the Deficit Reduction Act has already been felt just a relatively short time after enactment. Many experts attribute this rapid impact to the fact that the provisions of the law establish a direct link between excellence in preventing infection and enhanced financial reimbursement. As anyone who has ever needed to make a business case for infection control to a skeptical hospital administrator can attest, it is considerably easier to frame this appeal when the impact is not on cost-avoidance but when it actually relates to adding value through increased revenue generation.[19] The theme of how this phenomenon can be leveraged by the savvy infection preventionist or epidemiologist is discussed again in the last section of this chapter.

This is not to say that the first tentative moves toward value-based purchasing by the US federal government have proceeded seamlessly. Shortcomings have been identified, including the measures selected, the manner in which data are collected and tracked, and the unintended consequences in terms of some hospital responses (Table 32-2). First, all of the information considered for the CMS program is based on hospital administrative (ie, billing) data. These data have been shown to be relatively insensitive and nonspecific for the detection of HAI. As a result, increases or decreases in infection rates as documented with these metrics are difficult to interpret and not necessarily reliable. This phenomenon invites hospital interventions that may not be targeted to improved actual performance but in fact may be directed at "gaming the system" to maximize scores through manipulation of billing and documentation. Needless to say, this approach defeats the purpose of value-based purchasing.

There is an additional, more specific concern related to the provisions of the Deficit Reduction Act that not only could limit the intended effect of the regulation but could create new pitfalls for patients and hospitals. Because the program is specifically aimed to detect avoidable complications of care that occur during hospitalization, such diagnoses are excluded from nonreimbursement when they are documented to be already present at the time of hospital admission. In this context, hospitals could theoretically attempt to maximize reimbursement by either preventing the occurrence of these complications during hospitalization (as is intended) or by modifying the approach to detecting and documenting the presence of preexisting conditions at the time of admission to the hospital.

This phenomenon is illustrated by the response at many hospitals to the inclusion of catheter-associated urinary tract infections in the bundle of "do not pay" conditions. At these institutions, clinical and administrative leadership, working from a misconceived notion of the regulation, actively promote the policy that all patients be screened using urine microscopy at the time of hospital admission, to establish that bacteriuria (and presumably urinary tract infection) was already present at admission. Remarkably, at many of these facilities, the true evidence-based measures that can actually reduce the risk of this common HAI are not promoted or allocated resources.

This approach represents not just a misguided allocation of resources and effort but a potential risk to patients. Specifically, asymptomatic bacteriuria, while common, is clinically unimportant for many patients. If patients are screened for urinary contamination in the absence of symptoms at the time of hospital admission, clinicians could feel compelled to offer antimicrobial treatment for what they presume to be an infection. As a result, patients may be exposed to the undue harm of drug toxicities and interactions that come with any needless pharmacologic exposure. More concerning may be the risk of subsequent antimicrobial resistance that endangers the patients receiving the antibiotics and could pose a threat to others with whom these individuals have contact.

These concerns do not undermine the overall potential for benefit associated with value-based purchasing. Rather, they serve as a stark reminder that even well-intentioned regulations and mandates need to be conceived and executed with extraordinary caution. Unintended consequences, if not anticipated and managed, could ultimately pose greater risk than the benefit of the measures as originally intended.

Activism and the Healthcare Epidemiologist

Given the rationale, scope, and potential consequences of government mandates for infection prevention, what are the practical steps that healthcare epidemiologists and infection preventionists must take to ensure that hospitals and other healthcare facilities are prepared to respond? More importantly, how can experts in the field actively engage in the policy process to help shape future mandates to ensure that policies are rational, practical, and equitable? In the sections that follow, specific steps for individual infection preventionists and epidemiologists to meet this challenge are described briefly (Table 32-4).

Keep Abreast of Local Legislative Issues

Many practicing epidemiologists and infection preventionists do not feel that they have the time or expertise to maintain awareness about pending legislative or regulatory mandates. Of course, given the plethora of

Table 32-4. Activities for remaining aware of government mandates on infection control

Keep abreast of local legislative issues

Inform key stakeholders at the institution

Identify opportunities to align institutional priorities with legislative mandates

Get involved in the regulatory process

responsibilities for which these individuals have oversight, this may come as no surprise. However, considering the degree to which government mandates can suddenly and dramatically affect the practice of infection prevention, routine surveillance for "outbreaks" of government intervention should be a priority.

One practical and sustainable way for hospital-based experts to remain vigilant for pending legislative action is to integrate consideration of government mandates into the periodic risk assessment performed for the hospital's infection prevention program. Institutional risk assessment is a crucial step in determining the most appropriate allocation of program resources and personnel.

Information about pending legislative measures can be obtained from a number of sources. Local chapters of professional societies, such as SHEA and APIC, typically provide updates for members and can also provide a forum for discussions about pending measures and response strategies. Obviously, SHEA, APIC, and other professional societies remain engaged at the national level (as discussed below) and provide a number of resources about legislative mandates. While surfing the

Internet, epidemiologists and infection preventionists might also examine the Web sites of various consumer advocacy groups who sponsor and promote positions regarding such legislation. A number of useful Internet resources are listed in Table 32-5.

Information is also available directly from the local and federal government. After accessing the Web site of state legislatures or the US Congress, a keyword search of the online database of pending legislation, using terms such as "infection" or "hospital" or even more specific terms such as "MRSA," can point even to bills just going before first committee review. An institutional government affairs officer not only can direct infection preventionists and epidemiologists to these resources but may be an invaluable source of more direct information from government officials.

Inform Key Stakeholders in the Institution

For infection preventionists and healthcare epidemiologists, knowledge about current or pending infection control legislation is not fully actionable until it is shared across the institution. In much the same manner that the practice of infection prevention at the bedside relies on the participation of expert clinicians and staff from other disciplines, so too does the effective response to government mandates depend on the complete engagement of other institutional experts and key stakeholders.

To best disseminate such information, it is appropriate to work in close collaboration with the institution's government affairs office. Establishing mechanisms to communicate to leadership not only the basic content but also the clinical and financial implications of

Table 32-5. Internet resources relating to government mandates on infection control

Source	URL
Society for Healthcare Epidemiology of America Public Policy and Government Affairs Division	http://www.shea-online.org/news/publicpolicy.cfm
Association for Professionals in Infection Control and Epidemiology Public Policy Program	http://www.apic.org/AM/Template.cfm?Section=Government_Advocacy
Infectious Diseases Society of America Policy and Advocacy Division	http://www.idsociety.org/pa/toc.htm
Consumer's Union	http://www.stophospitalinfections.org/learn.html
US House of Representatives	http://www.house.gov/
US Senate	http://www.senate.gov/
Centers for Medicare and Medicaid Services Hospital-Acquired Conditions Overview	http://www.cms.hhs.gov/hospitalacqcond/
MRSA Survivor's Network (advocacy)	http://www.mrsasurvivors.org/

NOTE: MRSA, methicillin-resistant *Staphylococcus aureus*; URL, uniform resource locator.

proposed legislative measures is crucial. At some institutions in jurisdictions where legislative activity is especially active, updates about pending legislative measures are integrated into the standing agenda of the hospital's infection control committee. In this fashion, the potential critical impact of such legislation is routinely incorporated into both the operational and strategic planning for infection prevention at the hospital.

When possible, efforts should be undertaken to ensure that leaders at the highest levels of the organization are made aware of the actual or potential implications of current or pending infection control legislative mandates. Ultimately, this represents another of the countless reasons why experts in the hospital's infection prevention program need to have direct access not only to clinical leaders and directors but to senior officers of the management team, as well as to members of the board of governance. This direct contact with institutional leaders can have an additional practical and favorable effect. The same individuals who serve in these capacities at hospitals often have access, through professional and personal contacts, to the legislative leaders and other influential persons who help play a role in shaping legislation.

Identify Opportunities to Align Institutional Priorities With Legislative Mandates

Without question, government intervention in the practice of infection prevention, whether through public reporting, mandate of specific practices or pay for performance rules, provides powerful incentive for hospitals to improve performance. Regardless of whether an individual hospital epidemiologist or infection preventionist completely supports a particular infection control mandate, the authority behind these external forces offers a unique opportunity to highlight the value to the institution of all infection prevention activities. At hospitals in jurisdictions in which government mandates have been enacted, successful infection preventionists and epidemiologists have been able to leverage focus about these measures, to strengthen the overall commitment to infection prevention at the institution.

The section on types of legislation, above, mentions an example in which personnel hired specifically to manage the mandates of a specific new infection control law could be concurrently used for additional infection prevention activities. Other opportunities to leverage government mandates to promote HAI prevention exist. For example, hand hygiene promotion should be highlighted as an evidence-based strategy not only to reduce the incidence of nearly all infections that might be publicly reported but also to prevent the types of HAIs that might no longer be reimbursed according to the provisions of

the Deficit Reduction Act. The challenge to the hospital epidemiologist and infection preventionist is to help senior budget leaders understand the interconnectedness of these activities, to ensure that the hospital's response is not unnecessarily narrow or entirely dedicated to "gaming the system."

Get Involved in the Regulatory Process

Increasingly, successful infection preventionists and healthcare epidemiologists will need to work outside the institution to maximize their effectiveness in controlling and eliminating HAI. By engaging in the policy process itself, these content experts can help ensure that any government mandates are well reasoned and appropriately deployed.

A critical first step in this process is to gain familiarity with the people and process by which laws are developed and implemented in the local jurisdiction. At minimum, institutional experts in infection control and prevention should be able to identify local and federal legislators with whom to make contact when issues arise regarding the practice of infection control. It is especially valuable that contact is established not only with the lawmakers representing the district in which the facility is located but also with policy makers who represent the home district where the infection preventionists and hospital epidemiologists live (and vote). Legislators generally provide increased attention to individuals who represent key voting constituencies in their home district.

It is also useful to establish contact (if not build a relationship) with key legislators and their aides long before the first mandate related to infection control becomes active. Familiarity with local experts may encourage the legislator to call on such individuals for advice and guidance when related laws come before them for consideration.

In addition to reaching out to local representatives, it is important that contact be established with the legislative leaders who may play the greatest role in influencing the likelihood that infection prevention legislation is passed in the local jurisdiction. Such individuals might include the chairpersons of healthcare-related legislative committees or representatives with large healthcare facilities in their home district. These individuals and their staff members may be particularly receptive to input from experts in the field.

In all cases, hospital epidemiologists and infection preventionists should exercise their right to speak out to local legislators with an informed outlook on the proposed measure(s). Before doing so, it may be useful to discuss the matter with the hospital's legal affairs or government relations officer so that the expert can have assurance as to whether they are free to speak as a representative of the institution or whether they must advocate only from the perspective of a concerned and knowledgeable private citizen.

For readers for whom the personal touch with local lawmakers is either unappealing or not practical, it is reasonable to reach out in other manners to try to influence proposed legislative measures. In nearly all cases, every public policy item is subject to public review and feedback prior to full consideration or adoption by a particular legislative body, whether a local town council or the US Congress. Examples include open forums and fact-finding sessions (which are typically a part of local politics and lawmaking), listening sessions (such as those run by the Department of Health and Human Services before adoption of new elements of the Deficit Reduction Act for healthcare-associated conditions), and public comment (such as is done for federal laws and regulations through the *Federal Register*).

Arguably the most effective means by which to shape and inform government mandates regarding infection control relates to engagement through the resources of professional societies. Both SHEA and APIC are very active in the work of shaping legislation. For both organizations, there is a well-staffed and well-supported committee responsible for public policy and government affairs. On occasion, these societies have collaborated to shape policy, often in conjunction with other professional organizations (such as the Infectious Diseases Society of America).

Volunteering to serve on or to collaborate with the public policy committees of these groups is an important opportunity that can have enormous implications regarding the effectiveness and productivity of the societies in influencing legislation. Even when not named to these bodies, society members can be certain to keep leaders of the public policy groups up to date on the situation in their local area. Moreover, active members serve as a key check on the policies and positions adopted by the societies and their committees. Give feedback and engage not only the public policy representatives but also the society leadership as a whole. Make certain that the interests of healthcare professionals at the bedside are well represented.

Conclusion

In the future, government mandates for infection prevention are likely to become both more common and potentially more prescriptive. Whether the outlook of an individual infection preventionist or healthcare epidemiologist regarding such intervention is based on overall principles (such as a concern for compromised autonomy) or the perceived merit of an individual govern-

ment mandate, the ability to effectively respond and adapt to such measures must be developed and honed. However, simply to react to government mandates will not be sufficient to ensure that future government mandates are applied both rationally and equitably.

The past 2 decades provide numerous examples in which the inaction, real or perceived, of individuals charged with preventing HAIs has resulted in progressively increasing intrusion from external forces. To persons in the field, the charge is clear and the consequences of inaction are apparent. Through rigorous scientific investigation, evidence-based best practices must be established and disseminated. Historical barriers—financial, clinical, and cultural—must be set aside to ensure widespread adoption of the most effective strategies. Nonadherence by individuals or institutions must not be tolerated. Finally, leaders and frontline experts in infection prevention and healthcare epidemiology must earn and demand a seat at the table to ensure that, when mandates do come, they reflect the best scientific findings and clinical practice recommendations. Anything less than this degree of engagement can only lead to misplaced priorities, unproductive intrusions, and, most alarmingly, inadequate care of the patients who entrust us with their health and safety.

References

1. Edwards JR, Peterson KD, Andrus ML, Dudeck MA, Pollock DA, Horan TC. National Healthcare Safety Network (NHSN) Report, data summary for 2006 through 2007, issued November 2008. *Am J Infect Control* 2008;36:609–626.

2. Hidron AI, Edwards JR, Patel J, et al. Antimicrobial-resistant pathogens associated with healthcare-associated infections: annual summary of data reported to the National Healthcare Safety Network at the Centers for Disease Control and Prevention, 2006–2007. *Infect Control Hosp Epidemiol* 2008;29:996–1011.

3. Pépin J, Valiquette L, Alary ME, et al. *Clostridium difficile*-associated diarrhea in a region of Quebec from 1991 to 2003: a changing pattern of disease severity. *CMAJ* 2004;171:466–472.

4. Kohn LT, Corrigan JM, Donaldson MS. *To Err Is Human: Building a Safer Health System*. Washington, DC: National Academy Press; 2000.

5. Rupp ME, Fitzgerald T, Puumala S, et al. Prospective, controlled, cross-over trial of alcohol-based hand gel in critical care units. *Infect Control Hosp Epidemiol* 2008;29:8–15.

6. Berens MJ. Infection epidemic carves deadly path. *Chicago Tribune* July 20, 2002.

7. Consumers Union. Ten Years Later, and We're Still Dying. http://www.stophospitalinfections.org. Accessed May 23, 2009.

8. Association for Professionals in Infection Control and Epidemiology. HAI Reporting Laws and Regulations.

http://www.apic.org/am/images/maps/mandrpt_map.gif.
Accessed May 22, 2009.

9. Fraser TG, Fatica C, Gordon SM. Necessary but not sufficient: a comparison of surveillance definitions of *Clostridium difficile*–associated diarrhea. *Infect Control Hosp Epidemiol* 2009;30:377–379.

10. US Government Accountability Office. *Health-care–Associated Infections in Hospitals: Leadership Needed From HHS to Prioritize Prevention Practices and Improve Data on These Infections.* Washington, DC: Government Printing Office; 2008.

11. The Leapfrog Group Web site. http://www.leapfroggroup.org/. Accessed May 23, 2009.

12. The Institute for Healthcare Improvement. http://www.ihi.org/ihi. Accessed May 23, 2009.

13. Association for Professionals in Infection Control and Epidemiology. MRSA Laws & Pending Legislation—2009. http://www.apic.org/am/images/maps/mrsa_map.gif. Accessed May 22, 2009.

14. Weber SG, Huang SS, Oriola S, et al. Legislative mandates for use of active surveillance cultures to screen for methicillin-resistant *Staphylococcus aureus* and vancomycin-resistant enterococci: position statement from the Joint SHEA and APIC Task Force. *Infect Control Hosp Epidemiol* 2007;28: 249–260.

15. Robicsek A, Beaumont JL, Paule SM, et al. Universal surveillance for methicillin-resistant *Staphylococcus aureus* in 3 affiliated hospitals. *Ann Intern Med* 2008;148:409–418.

16. Harbarth S, Fankhauser C, Schrenzel J, et al. Universal screening for methicillin-resistant *Staphylococcus aureus* at hospital admission and nosocomial infection in surgical patients. *JAMA* 2008;299:1149–1157.

17. Ammerlaan HS, Kluytmans JA, Wertheim HF, Nouwen JL, Bonten MJ. Eradication of methicillin-resistant *Staphylococcus aureus* carriage: a systematic review. *Clin Infect Dis* 2009; 48:922–930.

18. Centers for Medicare and Medicaid Services. Hospital-Acquired Conditions (Present on Admission Indicator): Overview. http://www.cms.hhs.gov/HospitalAcqCond/. Accessed May 22, 2009.

19. Perencevich EN, Stone PW, Wright SB, et al. Raising standards while watching the bottom line: making a business case for infection control. *Infect Control Hosp Epidemiol* 2007;28:1121–1133.

Chapter 33 Product Evaluation

William M. Valenti, MD

Healthcare facilities remain under increasing pressure to manage the costs of care as new reimbursement models are implemented. In addition to managed care and capitated reimbursement strategies, the 2008 Medicare rule to cease reimbursement to hospitals for the treatment of certain preventable conditions, including some hospital-acquired infections,[1] underscores the need for clinicians and administrators to spend healthcare dollars wisely, without sacrificing the quality of care or safety. Healthcare-associated infections increase the cost of health care substantially, and the cost of these infections is not reimbursed under capitated programs. Thus, hospitals now have additional incentives to prevent nosocomial infections rather than treat them after they occur.

In addition, healthcare facilities and infection control programs are under increasing pressure to address new and emerging infectious diseases or so-called epidemiologically important organisms.[2] These agents include multidrug-resistant pathogens, agents of bioterrorism, avian influenza virus, and the hemorrhagic fever viruses.

In theory, one way to reduce the incidence of healthcare-associated infections is to use devices that have a lower risk of infection than other products. These products usually cost more than the standard products. Administrators and staff in healthcare facilities must evaluate such devices and other products that may affect infection rates to determine whether they are efficacious and, thus, worth the added cost.

While new and improved equipment and devices can make a contribution to reducing the incidence of healthcare-associated infections, the fundamentals of infection control practice cannot be overemphasized. These time-honored practices include adherence to hand hygiene practices, education of healthcare providers, adherence to proper disinfection and sterilization procedures, and ongoing maintenance of equipment and the inanimate environment, to name a few.

Transmission of Healthcare-Associated Pathogens

When evaluating a product, infection preventionists should consider the various modes by which pathogens are transmitted in healthcare settings (eg, by means of a common vehicle, droplets, or airborne particles) to determine whether a new device or product might increase or decrease transmission of important pathogens. Most pathogens acquired in healthcare facilities are transmitted by people and medical devices. The inanimate environment (eg, walls, floors, counter tops, food, and water) plays a minor role in transmission of pathogens. The inanimate environment should be cleaned and maintained routinely as part of the healthcare facility's overall maintenance plan, as much as to control infection. However, infection control activities designed to decrease the incidence of infections should focus actively on the role of people, invasive procedures, and medical devices, such as needles, catheters, endoscopes, and surgical equipment. In addition, the infection control literature documents a variety of outbreaks and pseudo-outbreaks attributed to problems with disinfection and sterilization of equipment and invasive devices. Some outbreaks have occurred because the equipment was designed poorly or because of inadequate cleaning and disinfection.[3]

Some steps to consider in evaluating the performance of invasive medical devices and equipment are the following:

- An assessment of design flaws that could make the device hard to clean and disinfect
- An assessment to determine whether instructions for cleaning and disinfection meet current standards
- An assessment of the ability to clean and disinfect the device adequately (ie, the facility has the equipment needed to reprocess the device; the staff have been, or will be, trained to reprocess the device; and staff have adequate time to reprocess the device properly)
- A literature review to see whether similar devices have caused outbreaks of infection at other institutions and what factors (eg, a design flaw, improper use, or improper disinfection) were implicated in the outbreaks.

Some equipment-related outbreaks have occurred because of user error (eg, staff did not know how to clean and disinfect the equipment, took short cuts to save time or money, or did not maintain the device properly). In this era of cost containment, staff may be tempted to save time and money by changing disinfection and sterilization protocols and procedures. The outbreaks that occur emphasize that adhering to the basic, time-honored principles of infection control practice, such as proper reprocessing of reusable patient-care devices, regular maintenance of equipment, and effective staff education, are as important as, or perhaps more important than, purchasing more-expensive devices purported to lower infection rates.

A Standardized Method for Analyzing Products and Interventions

The evaluation process should start with a review of the evidence that supports the product's performance from an infection control perspective. While an evidence-based approach is not always possible, common sense and experience can play important roles, in addition to an assessment of the product's priority for implementation.

Healthcare providers may have difficulty evaluating whether products and equipment will be cost-effective in their facilities. The task is especially difficult if there are no peer-reviewed publications to support the manufacturer's claims. In general, advertisements, promotional materials, and anecdotal information (eg, testimonials) are not objective enough for the evaluation process. Whenever possible, the healthcare facility should form

a multidisciplinary evaluation team that will do the following:

- Define priorities based on the facility's rates of certain categories of infections, rate of antibiotic-resistant infections, rate of injuries from sharp devices, or other pertinent data
- Develop criteria regarding product design and performance
- Assess published data regarding the product or similar products
- Gather information regarding the experiences of similar facilities
- Review Health Device Alerts distributed by the US Food and Drug Administration and other independent organizations, such as the ECRI[4] (formerly the Emergency Care Research Institute), a nonprofit health services research agency
- Consider a trial period for use of the product in the appropriate clinical settings to determine whether the product functions well and is accepted by the primary user
- Assess the results of the trial period
- Analyze the cost benefit or cost-effectiveness of the product

The multidisciplinary team should include members who are either affected by the purchase or need to be involved to ensure seamless implementation (eg, the purchasing and stores departments, the central sterile supply department, the infection control program, and the staff who will use the product). In some situations, representatives from the facility's administration and from the finance department also should participate. Infection preventionists play a pivotal role on this team by assessing the product's impact on important parameters, such as the rates of infections and injuries from sharp devices. The person who represents the primary users can provide valuable input on how likely the product will meet their needs and on its ultimate acceptance.

After the multidisciplinary team has identified products or equipment that meet the predefined criteria, the team members should decide whether these items must be tested before purchase. Chiarello[5] notes that healthcare workers are likely to reject new devices, despite infection control or safety advantages, if they do not like to use them. The evaluation team should assess the results of the trial period, including cost, staff satisfaction, adverse reactions, the infection rate, the injury rate, the frequency of malfunction, the ease of cleaning and disinfection, and other appropriate information, before deciding whether or not to recommend the product.

In this era of declining reimbursement, the cost and the cost-effectiveness of a product have become increasingly important. The cost-effectiveness analysis is easy if the same product can be obtained cheaper from a different supplier or if an equivalent product made by another company is cheaper. In these situations, the multidisciplinary team could plan for the substitution, inform and educate the primary users of the minor change, and work to ensure a seamless transition.

The multidisciplinary team may need to analyze some products or equipment more thoroughly, particularly those that are more expensive than products currently in use. A more formal cost-effectiveness analysis methodology is often used for this purpose.[6] This type of economic analysis determines a ratio of the incremental costs for devices on the basis of the expected change in outcomes. Furthermore, cost-effectiveness analysis can help the multidisciplinary team demonstrate to the administration that a more expensive product should be purchased, either because the expected outcome would be improved significantly or because the complication rate would be decreased significantly, and thus the overall cost to the institution would be decreased.

Laufer and Chiarello[6] used a fairly simple formula to estimate the cost of preventing a needlestick injury with each of 3 devices:

$$\frac{\text{incremental cost of new product or intervention}}{(\text{no. of events without product}) - (\text{no. of events with product})}.$$

In this case, the cost to prevent a needlestick injury varied substantially between the 3 different devices. The authors estimated the yearly cost of preventing one needlestick injury to be $984 if protective injection equipment was used, $1,574 if an intravenous system with recessed needles was used, or $1,877 if a needleless intravenous system was used.

Using a version of this formula (Figure 33-1), it can be calculated that, if the incremental cost of a needleless system is $250,000, a total of 130 needlestick injuries (ie, the number of injuries without the product minus the number of injuries with the product) would need to be prevented to reach the $1,877 threshold.

Additional data on the costs of occupational needlestick injuries can be factored into the analysis. The costs of medical office visits, preventive treatment, diagnostic testing for the injured worker and source patient, time lost from work, and other costs, when subtracted from initial costs per injury prevented, reduce the overall cost and help support the case for incurring the increased costs of using the device rather than maintaining the status quo.

In any event, beyond tracking costs, prospective monitoring of the impact of various interventions (in this example, needlestick injuries) is warranted once the product is in use.

Protecting Patients, Healthcare Workers, and Visitors From Pathogens in the Environment

Three examples of infection control products or interventions are discussed below: (1) disinfectants and the healthcare environment, (2) preventing transmission of *Mycobacterium tuberculosis*, and (3) advances in technology to manage methicillin-resistant *Staphylococcus aureus* (MRSA). These examples of products or

---

Assumptions

1. The incremental cost of the system is $250,000.
2. The number of needlestick injuries reported and evaluated each year is 260.
3. The needleless intravenous system will reduce needlestick injuries by 50%.

Calculation

Using this formula:

$$\frac{\text{incremental costs of a needleless intravenous system}}{(\text{no. of injuries without product}) - (\text{no. of injuries with product})},$$

calculate the financial impact with the assumptions:

$$\frac{\$250,000}{260 - 130} = \$1,877 \text{ per injury prevented.}$$

Figure 33-1. Example of an evaluation of the financial impact of a needleless intravenous system to prevent needlestick injury, using the formula of Laufer and Chiarello.[6] The yearly cost of preventing 1 needlestick injury was estimated to be $1,877 for the needleless system used. A simple calculation with the assumptions and the formula shown indicated that a total of 130 needlestick injuries would need to be prevented to reach that $1,877 threshold cost.

interventions range from fairly low-technology solutions, in the case of disinfectants and tuberculosis control, to newer, high-technology solutions for control of MRSA. However, the caveat is that the literature in this field changes rapidly. Therefore, infection preventionists who evaluate products or interventions should not rely solely on this assessment but also should review current literature to ensure up-to-date decision making.

Disinfectants and the Healthcare Environment

Manufacturers often claim that their products will reduce the risk of infection from organisms found in the environment. Occasionally, some of these claims find their way into credible publications, only to confuse individuals who are assessing the value of these products for their facilities. A report on the bactericidal effects of copper-based paints illustrates this point.[7] The author showed that copper-based paints kill bacteria and then concluded that "such paints could be used to render surfaces self-disinfecting in strategic locations where environmental causation of nosocomial infections is suspected."[7] This conclusion conflicts with the current infection control consensus on the role of the environment in infection transmission. In an accompanying editorial, Rutala and Weber[8] wisely point out that walls always harbor bacteria, but they never have been linked by scientific data to nosocomial infection.

Companies that produce disinfectants may try to capitalize on healthcare workers' anxiety over human immunodeficiency virus (HIV) infection by claiming that their products kill HIV on contact. However, most agents that are used to clean floors, counter tops, and other surfaces readily kill HIV. In addition, HIV is inactivated rapidly by several products that are inexpensive and readily available, including 10% chlorine bleach, 50% ethanol, 35% isopropyl alcohol, hydrogen peroxide, and soap and water.[9] Thus, an expensive new product is not needed to clean areas contaminated by blood, because existing products are cheaper and do the job adequately. Rather than buying an expensive new product, personnel from the infection control program and from the housekeeping department should intensify their educational efforts regarding the infection control practices relevant to HIV. Furthermore, policies and procedures should be in place that clearly specify when and how to clean the environment and how to manage blood spills and other emergencies.

Preventing Transmission of *M. tuberculosis*

In most instances, the multidisciplinary team will need to evaluate products in context with other infection control measures when evaluating the infection-associated risks and benefits of particular products or interventions. For example, when developing a tuberculosis control program, healthcare facilities must consider the nature of the individual facility, the frequency with which tuberculosis patients are seen in the facility, and tuberculin (purified protein derivative) test conversion rates of employees. Guidelines from regulatory or advisory agencies, such as the occupational health and safety administration and the Centers for Disease Control and Prevention, are also helpful (see Chapter 25, on control of tuberculosis).

Data in the literature indicate that UV radiation kills *M. tuberculosis*. However, most experts do not think that use of UV light alone is adequate,[10] and the guidelines from the Centers for Disease Control and Prevention state that use of UV lights cannot substitute for use of proper air-handling units, but it could be an adjunctive measure.[11] In other words, despite recent data on the effectiveness of UV irradiation in preventing transmission of *M. tuberculosis* in animal models,[12] proper attention to the environment is still required. In this case, UV irradiation is not a substitute for the infection control fundamental of using adequate air-handling systems.

MRSA and Advances in Technology

The contemporary infection control problem of MRSA and the evaluation process for a new intervention offers a final example. This example underscores the importance of staying current with the available infection control literature as newer technologies are developed to enhance the traditional "low-tech" approach to this problem. As for other multidrug-resistant bacteria, the traditional approach for MRSA control has been administrative support, staff education, judicious use of antimicrobial agents, and surveillance.[2] In addition, most experts agree that some kind of active screening program for potential MRSA carriers is preferable to managing clusters of infection after MRSA transmission.[2,13-15] More recently, the adoption of the practice of screening selected patients for MRSA, including the use of rapid screening technologies, has been used as a component of MRSA control, particularly in intensive care units. Although patient screening increases the direct costs of care, the emerging body of evidence suggests that this intervention is cost-effective.[13]

Regarding the cost of MRSA infections, Chaix et al.[13] performed a case-control study in a French university hospital with a 4% rate of MRSA colonization at admission among intensive care unit (ICU) patients. They compared patients who had ICU-acquired MRSA infection with control patients hospitalized at the same time who did not have MRSA infection. The mean increased cost attributable to MRSA infection in ICU patients (in 1999 dollars) was US $9,275.00. Using a time-and-motion study, the investigators determined the excess cost of contact isolation precautions were up to $705 per patient assigned to isolation for an average of 20 days in the ICU, including the costs of screening and contact-isolation materials. The obvious conclusion is that use of screening and contact isolation has the potential to avoid some of the costs of treating hospital-acquired MRSA infection. In this instance, isolation was cost-effective if the number of MRSA infections was reduced 5-fold.

Moving beyond isolation assumptions and costs, a study by Clancy et al.[14] reported the results of a MRSA screening program in ICUs in the United States. During a 15-month period, all patients admitted to adult medical and surgical ICUs were screened for MRSA nasal carriage on admission and weekly thereafter. The overall rates of all MRSA infection and of nosocomial MRSA infection in the 2 adult ICUs and the general wards were compared with the corresponding rates during the 15-month period prior to the start of routine screening. The percentage of patients colonized or infected with MRSA on admission and the cost avoidance of the surveillance program were also assessed. The overall incidence of MRSA infection for all 3 areas combined decreased significantly, from 6.1 infections per 1,000 census-days in the preintervention period to 4.1 infections per 1,000 census-days in the postintervention period ($P < .01$). Also, MRSA would not have been detected in 91% of the patients if screening had not been performed.[13] At a cost for the program of $3,475 per month, they concluded that they averted a mean of 2.5 MRSA infections per month for the ICUs combined, avoiding $19,714 per month in excess cost in the ICUs, on the basis of the cost data from Chaix et al.[13]

Rapid MRSA screening technologies have been developed, ideally to provide faster screening results and reduce the need for unnecessary isolation of suspected carriers. Harbath et al.[16] performed a study that used the end point of change in the time from ICU admission to notification of test results among patients screened with a rapid molecular test for MRSA. Screening on admission identified the prevalence of MRSA to be 6.7% (detected in 71 of 1,053 patients) in 2 ICUs.[16] Without admission screening, 55 patients with previously unknown MRSA carriage would have been missed. The median time from ICU admission to notification of test results decreased from 87 to 21 hours in the surgical ICU ($P < .001$) and from 106 to 23 hours in the medical ICU ($P < .001$). In the surgical ICU, use of the rapid MRSA test saved 1,227 preemptive isolation days for 245 patients not colonized with MRSA.

After the findings were adjusted to account for colonization pressure, the systematic on-admission screening and preemptive isolation policy was associated with a reduction in the number of MRSA infections acquired in the medical ICU but had no effect on the number in the surgical ICU. Harbath et al.[16] concluded that the rapid MRSA test decreased median time to notification from 4 days to 1 day and helped to identify previously undetected MRSA carriers rapidly. A strategy linking the rapid screening test results with preemptive isolation and cohorting of MRSA patients substantially reduced cross-transmission of MRSA infection in the medical ICU but not in the surgical ICU. Although no cost data were given in this study, the clear advantage of rapid testing was shown to be the more appropriate isolation of patients, by virtue of the test's faster turnaround time.

Making the case for an intervention such as screening with a rapid test requires a hierarchical approach to the problem and provides a good example of the importance of the multidisciplinary team approach to state the case. For this evaluation, key team members are infection control staff; ICU staff; and administration, infectious diseases, laboratory, and hospital administration and finance staff.

The central question to be answered is how much more aggressive the facility's administration wants to be to manage the problem. The next level of control would be some kind of surveillance of selected patients. Since we know that patient surveillance has been shown to be cost-effective,[13,14] a cost-effectiveness study in the institution is not likely to add much new information to support the case. Instead, the critical decision point for the patient surveillance case is whether to use the traditional culture-based method or to adopt the rapid-test method. If the goal of reducing unnecessary isolation time to improve patient care is important enough administratively and clinically, a simple calculation, using some of the assumptions mentioned previously, will help make the case for rapid MRSA testing. As a minimum, a trial period for rapid MRSA testing might also support its value, assuming that the laboratory is suitably equipped to perform the test and is willing to participate.

Using the formula proposed by Laufler and Chiarello[6] and assumptions specific to the facility may be all that is necessary to support the case. Once use of the rapid test is in place, ongoing surveillance of the intervention's impact will help sustain it as an ongoing component of MRSA control. However, if the decision is made to continue with no patient surveillance or to continue with surveillance based on results of culture rather than the rapid test, the infection control team should be poised to bring the issue up at a later time as more facilities begin to adopt the rapid-screening approach.[17]

Summary

In many cases, purchasers cannot choose a product solely on the basis of the manufacturer's claims or the cost. Instead, healthcare facilities should consider a multidisciplinary team approach to evaluate products carefully. The team would review the literature, review benchmark experience from other healthcare settings, and review the manufacturer's materials to determine whether the product could perform as required, has an equal or lower risk of infection compared with the current product, and would be cost-effective. Products that meet these criteria could be tested by the staff who will use them, to ensure that the products will be accepted and used properly after they are introduced into general use.

Infection preventionists may find the process of evaluating new products or interventions to be complex and time consuming. However, this activity will assume greater importance as healthcare organizations come under increased pressure to address key healthcare-associated infections or face the consequences of reduced reimbursement. As new emerging infectious diseases (eg, avian influenza and MRSA infection) and diagnostic technologies are identified, or anticipated (eg, smallpox or agents of bioterrorism), infection control personnel must ensure that the risk of infections does not increase as healthcare facilities drastically reduce the cost of care.

Many infection control programs are already active internal consultants to their facilities in the evaluation of products, procedures, or interventions with infection control impact. Infection preventionists who help their institutions assess products carefully will enable these healthcare facilities to survive the current and future financial crises without sacrificing the quality of care or employee safety.

References

1. Centers for Medicare and Medicaid Services. Medicare program: changes to the hospital inpatient prospective payment systems and fiscal year 2008 rates. 42 *CFR* Parts 411, 412, 413, 489. http://www.cms.hhs.gov/center/hospital.asp or http://www.cms.hhs.gov/AcuteInpatientPPS/downloads/CMS-1533-FC.pdf. Accessed March 30, 2009.

2. Siegel JD, Rhinehart E, Jackson M, Chiarello L, Healthcare Infection Control Practices Advisory Committee. 2007 Guideline for isolation precautions: preventing transmission of infectious agents in healthcare settings, June 2007. http://www.cdc.gov/ncidod/dhqp/pdf/guidelines/Isolation2007.pdf. Accessed March 30, 2009.

3. Srinivasan A, Wolfenden LL, Song X, et al. An outbreak of *Pseudomonas aeruginosa* infections associated with flexible bronchoscopes. *N Engl J Med* 2003;348:221–227.

4. ECRI (formerly the Emergency Care Research Institute) Web page. http://www.ecri.org. Accessed March 30, 2009.

5. Chiarello LA. Selection of needle stick prevention devices: a conceptual framework for approaching product evaluation. *Am J Infect Control* 1995;23:386–395.

6. Laufer FN, Chiarello LA. Application of cost-effectiveness methodology to the consideration of needle stick prevention technology. *Am J Infect Control* 1994;22:75–82.

7. Cooney TE. Bactericidal activity of copper and noncopper paints. *Infect Control Hosp Epidemiol* 1995;16:444–446.

8. Rutala WA, Weber DJ. Environmental interventions to control nosocomial infections. *Infect Control Hosp Epidemiol* 1995;16:442–443.

9. Centers for Disease Control and Prevention. Guideline for environmental infection control in health-care facilities, 2003. http://www.cdc.gov/ncidod/dhqp/gl_environinfection.html. Accessed March 30, 2009.

10. Sepkowitz K. Tuberculosis control in the 21st century. *Emerging Infect Dis* 2001;7:259–262. http://www.cdc.gov/ncidod/eid/vol7no2/sepkowitz.htm. Accessed March 30, 2009.

11. Centers for Disease Control and Prevention. Guidelines for preventing the transmission of *Mycobacterium tuberculosis* in health-care settings, 2005. *MMWR Recomm Rep* 2005; 54(RR-17):1–141. http://www.cdc.gov/mmwr/preview/mmwrhtml/rr5417a1.htm?s_cid=rr5417a1_e. Accessed March 30, 2009.

12. Escobe AR, Moore DAJ, Gilman RH, et al. Upper-room ultraviolet light and negative air ionization to prevent tuberculosis transmission. *PLoS Med* 2009;6(3): e1000043.http://www.plosmedicine.org/article/info:doi/10.1371/journal.pmed.1000043. Accessed March 30, 2009.

13. Chaix C, Durand-Zaleski I, Alberti C, et al. Control of endemic methicillin-resistant *Staphylococcus aureus*: a cost-benefit analysis in an intensive care unit. *JAMA* 1999;282:1745–1751.

14. Clancy M, Graepler A, Wilson M, et al. Active screening in high-risk units is an effective and cost-avoidant method to reduce the rate of methicillin-resistant *Staphylococcus aureus* infection in the hospital. *Infect Control Hosp Epidemiol* 2006;27:1009–1017.

15. Muto CA, Jernigan JA, Ostrowsky BE, et al. SHEA guideline for preventing nosocomial transmission of multidrug-resistant strains of *Staphylococcus aureus* and *Enterococcus. Infect Control Hosp Epidemiol* 2003;24:362–386.

16. Harbarth S, Masuet-Aumatell C, Schrenze J, et al. Evaluation of rapid screening and pre-emptive contact isolation for detecting and controlling methicillin-resistant *Staphylococcus aureus* in critical care: an interventional cohort study. *Critical Care* 2006;10:R25. http://ccforum.com/content/10/1/R25. Accessed March 30, 2009.

17. Paule SM, Mehta M, Hacek DM, Gonzales TM, Robicsek A, Peterson LR. Chromogenic media vs real-time PCR for nasal surveillance of methicillin-resistant *Staphylococcus aureus*: impact on detection of MRSA-positive persons. *Am J Clin Pathol* 2009;131:532–539.

Chapter 34 Infection Prevention in Construction and Renovation

Loie Ruhl, RN, BS, CIC, and Loreen A. Herwaldt, MD

Construction, renovation, and maintenance in healthcare facilities challenge infection prevention personnel. These activities can increase the risk of healthcare-associated infections. Obviously these activities increase the risk to patients, but the construction activity itself is not the only consideration. Construction and renovation projects in healthcare facilities must meet guidelines and regulations established by the local (ie, city and county) governments and state governments, as well as those established by the federal government and by regulatory and accreditation agencies. The infection preventionist must collaborate with engineers, architects, administration, nurse managers, physicians, construction personnel, and maintenance staff before, during, and after the project.

During, maintenance, renovation, and construction, bacterial or fungal microorganisms in the dust and dirt can contaminate air handling or water systems, which can transmit these organisms to susceptible persons. Seemingly benign activities or changes in the healthcare environment can increase the risk of infection for susceptible patients. While the big construction and renovation projects usually receive the most attention, infection prevention staff should forget that simple daily activities that may be considered general maintenance can also put patients at risk for healthcare-associated infections. For example, moving a ceiling tile to replace a telephone line or pulling up an old carpet can release *Aspergillus* species into the air and ventilation system. Activities such as cutting into walls may disturb mold growing in areas where the plumbing or the windows leaked. Capping off a plumbing line or shutting down the water system for repairs can create dead spaces in the system, leading to the growth of *Legionella* species.

When such systems are nonfunctional, routine prevention measures, such as handwashing, may be difficult to maintain. Restarting these systems after maintenance or renovations also may increase the risk of infections such as legionnaires disease. Furthermore, routine clinical practice and traffic patterns may need to be substantially modified during construction projects to ensure that basic infection prevention precautions are maintained. Current patient populations in hospitals, clinics, and care centers are sicker than those in the past. The numbers of elderly patients, immunocompromised patients, and patients with significant underlying illnesses have increased, and these patients are at high risk of acquiring infections that are associated with maintenance, renovation, and construction. Thus, the infection prevention staff, especially the infection preventionist, have a tremendous opportunity and responsibility to protect patients, visitors, and staff members during such projects. This chapter identifies the potential risks involved in maintenance, renovation, and construction activities and provides practical solutions to decrease these risks. Although not specifically discussed in this chapter, excavation and demolition projects near patient-care areas create similar infection prevention issues. Infection prevention personnel who must deal with such projects should read the article on demolition issues by Streifel et al.[1]

Role of the Infection Prevention Team

The primary goal of the infection prevention team during maintenance, renovation, or construction in healthcare facilities is to protect susceptible patients, visitors, and healthcare workers from acquiring infections. The 1999 state-of-the-art report from the Association for Professionals in Infection Control and Epidemiology,[2] an excellent resource, suggests that the role of the infection prevention team is to provide infection prevention expertise throughout a project (ie, from the design phase until the area worked on is ready to use). Thus, infection prevention personnel should participate in construction projects from the inception, so that they can identify potential infection prevention problems created by the project and can design solutions prospectively. In addition, infection prevention personnel should understand the purpose of the project, so they can assess whether or not the design will facilitate good infection prevention practice.

The infection prevention team must collaborate with the architects, engineers, and maintenance staff to develop comprehensive maintenance, renovation, and construction policies that define the procedures necessary to maintain a safe environment. An infection control risk assessment (ICRA) is an essential part of these policies.[2,3] The ICRA helps the infection prevention personnel and other members of a multidisciplinary planning team determine the infectious risks associated with each project. By forcing the team to identify the patient populations at risk and the magnitude of the project, the ICRA helps the team identify important preventive strategies such as what type of barriers are necessary, whether workers need to wear protective attire and use special entrances and exits, and whether obtaining particle counts is necessary.

A caveat is necessary. During construction projects, infection prevention personnel will be asked to judge the value of many designs or products. Often, they will be asked to determine how much space is necessary for a certain function, which products should be used (eg, vinyl floor covering or carpet), and what air handling requirements must be met, and any of a number of other questions. The persons asking the questions may genuinely want to know the answers, or they may have hidden agendas that they hope the infection prevention program will endorse. In addition, the infection prevention staff members may find themselves mediating between opposing sides, such as department directors who want vast amounts of space and the latest innovations and administrators who want to limit costs. To avoid costly mistakes and political land mines, infection prevention personnel must ask many questions to determine what the real issues are; how the product, equipment, room, or clinic will be used; what possible solutions are available; what the budgetary limitations are; and what infection prevention principles or external regulations apply. In addition, infection prevention personnel may need to review the medical literature, governmental codes, guidelines from architectural and engineering societies and accrediting agencies, and product descriptions to determine which of the products or designs is within the project's budget and also balances the infection prevention requirements with patient and employee safety and satisfaction.

As healthcare budgets shrink, the expertise of infection prevention personnel will become more important during construction and renovation projects. Simultaneously, infection prevention personnel will feel increasing pressure to choose the least expensive products or design. Despite the pressures, they must remember their primary goals and recommend the products or design that will achieve these goals most effectively. The appropriate products or designs may be more expensive initially, but, in the long run, they probably will be less costly, as they may prevent outbreaks or may last longer and require less maintenance.

Infection prevention personnel often are the only clinical personnel who work on all construction and renovation projects. Thus, they may have to be the watchdogs for the entire project to make sure that the design and the construction meets the appropriate standards.

Many of the comments above are based on common sense. However, our experience and the medical literature testify that common sense answers often are not chosen during construction projects.[4-34] Table 34-1 lists design and construction errors that the authors of this chapter have encountered in the practice of infection prevention.

Table 34-1. Design and construction errors that have been encountered in the practice of infection prevention

No airborne infection isolation rooms in a medical intensive care unit

Entrance to dirty utility room is through clean utility room

Air intakes placed too close to exhausts or other mistakes in the placement or air intakes

Incorrect number of air exchanges

Air handling system functions only during the week or on particular days of the week

Air vents not reopened after construction completed

No negative air-pressure rooms in a large new inpatient building

Carpet placed where vinyl should be used

Wet-vacuum system in the operating suite pulls water up one floor into a holding tank, rather than down one floor

Aerators on faucets

Sinks located in inaccessible places

Patient rooms or treatment rooms do not have sinks in which healthcare workers can wash their hands

Room or elevator doors too narrow to allow beds and equipment to be moved in and out of rooms or elevators

Waterless alcohol hand rub locations not included in room layout

Risks Associated With Maintenance, Renovation, and Construction

Persons at Risk of Acquiring Construction-Related Infections

In a healthcare setting, special precautions are needed to protect susceptible or immunocompromised patients from acquiring infections related to maintenance, renovation, or construction. The persons who are most susceptible to these infections have immunologic disorders (infection with human immunodeficiency virus or congenital immune deficiency syndromes) or are receiving immunosuppressive therapy (radiation, chemotherapy, steroids, anti–organ rejection drugs, anti–tumor necrosis factor antibodies). Patients with severe neutropenia (defined as an absolute neutrophil count of 500 cells/mL or less), such as patients who have undergone allogeneic or autologous hematopoietic stem cell transplant or patients with leukemia who are receiving intensive chemotherapy, are at highest risk of these infections.[35] However, patients with underlying diseases such as chronic obstructive pulmonary disease, cancer, cardiac failure, or diabetes are also at increased risk, compared with healthy persons.[6]

Organisms That Cause Construction-Related Infections

The 2 microorganisms that most commonly cause outbreaks of healthcare-associated infection during construction-type activities are Legionella species[4-7] and Aspergillus species.[6-34] Legionella species are ubiquitous aquatic microorganisms that can be isolated from 20%–40% of freshwater environments and from soil and dust. There are 42 species of Legionella and 54 serogroups. Legionella pneumophila serogroup 1 causes 90% of the 10,000–20,000 cases of legionnaires disease (legionellosis) that occur annually.[3] Cooling towers, potable water systems, and heating and air-conditioning systems can be contaminated with Legionella species. These organisms can be transmitted by aerosols, which are inhaled, or by potable water, which is ingested. During construction, Legionella organisms can be introduced directly into the water when pipes are disrupted and become contaminated with soil. If a water system is already contaminated, organisms in the biofilm can be released into the water by changes in water pressure (eg, when a plumbing system is repressurized). Legionella can multiply rapidly in stagnant water. Therefore, pipes that have not been used for a considerable period of time should be flushed for more than 5 minutes before the water is used.[37] In general, outbreaks of legionellosis have been related to contaminated water. However, one outbreak was associated with the installation of a lawn sprinkler system;

investigators postulated that L. pneumophila was aerosolized during excavation and inhaled by susceptible people, who developed legionnaires disease.[6]

Fungi are ubiquitous in both indoor and outdoor environments. There are approximately 900 fungal species that cause mycosis,[38] but Aspergillus fumigatus and Aspergillus flavus are the species that cause invasive disease most frequently. Aspergillus species can be found anywhere in a hospital; however, during construction activities, dust, dirt, and debris that harbor these organisms can be released into the air in quantities that can be harmful to susceptible patients. In general, healthy persons are not susceptible to Aspergillus infection, but immunocompromised patients may become severely ill and may die from it.

Additional Construction-Related Health Risks

There are other problems with fungi and mold that can occur during maintenance and renovation of healthcare facilities. Fungi may be growing behind walls, above false ceilings, or in any area that may have had water leaks or high humidity. Mold grows quickly and can contaminate water-soaked building materials within 48 hours. Organisms such as Penicillium,[39] Fusarium, Trichoderma, and Memnomiella species and Stachybotrys chartarum can produce potent mycotoxins that are harmful to persons who inhale them or touch them with bare skin. Mold-related illness can range from mild allergic rhinitis symptoms—with symptoms such as runny

Table 34-2. The organizations that provide the most important resources for infection prevention personnel who are helping with maintenance, renovation, and construction projects

Organization	URL
American Institute of Architects	http://www.aia.org/index.htm
Association of Professionals in Infection Control and Epidemiology	http://www.apic.org/
American Society for Healthcare Engineering of the American Hospital Association	http://www.ashe.org/
American Society of Heating, Refrigeration, and Air-Conditioning Engineers	http://www.ashrae.org/
Centers for Disease Control and Prevention	http://www.cdc.gov/
The Joint Commission	http://www.jointcommission.org/
Occupational Safety and Health Agency	http://www.osha.gov/

nose, sneezing, and itchy eyes—to hypersensitivity pneumonitis, an allergic reaction to mold that becomes worse with repeated exposures and can cause permanent lung damage. Toxins produced by molds can cause a severe illness called organic dust toxic syndrome, which can start after exposure to a single heavy dose of allergen. The signs and symptoms are abrupt onset of fever, influenza-like symptoms, and respiratory difficulty within hours after exposure.

If employees discover discoloration or a musty odor in an area that is undergoing maintenance or is being renovated, the area needs to be assessed and mold remediation must be done before the project is finished. The workers should tape a tight barrier of plastic around the affected area and report it immediately to the project manager. Only persons trained in mold remediation should clean the area. If the area is small, trained maintenance or housekeeping staff who are wearing goggles without venting holes, N-95 respirators, and gloves can clean the area with a mild detergent or a 10% solution of bleach (sodium hypochlorite). Large areas of mold may

need to be addressed by a professional mold remediation contractor who uses protective attire and engineering controls, such as barriers and high efficiency particulate air (HEPA)–filtered, negative-airflow machines. Infection prevention policies should define when these special precautions are needed.[40]

Overview of Guidelines, Standards, and Regulations

A number of agencies have produced important resources for infection prevention personnel who are helping with maintenance, renovation, and construction projects. The most important documents are provided by the organizations listed in Table 34-2.

American Institute of Architects Guidelines

The American Institute of Architects (AIA), with assistance from the US Department of Health and Human

---

Table 34-3. Guidelines, recommendations, and standards for maintenance, renovation, and construction projects in healthcare facilities

Source and section; key points	Summary information and relevant subsection
AIA Guidelines[3]	
ICRA	AIA Guidelines (Sect. 1.5–2.1.1)
Toilet rooms	Patient access without entering a hallway (Sect. 2.1–2.2.1); staff toilets (Sect. 2.1–2.4.2)
Handwashing stations	Required in all patient bathrooms (Sect 2.1–2.1.2); in nursing locations, handwashing stations shall be conveniently accessible to the nurse station, medication station, and nourishment area (Sect 2.1–3.1.5.5)
Emergency Service	At least one AII room (Sect. 2.1–5.1.2.6)
Finishes	Floor materials shall be readily cleanable and appropriately wear-resistant for the location (Sect. 2.1–7.2.3.2); wall finishes shall be washable and, in the proximity of plumbing fixtures, shall be smooth and moisture-resistant (Sect. 2.1–7.2.3.3)
Ventilation requirements	Table 2.1–5 lists all the special requirements needed for room ventilation
Clean and soiled workrooms	Such rooms shall be separate from and have no direct connection with clean work rooms or clean supply rooms (Sect. 2.1–2.3.8)
Housekeeping rooms	Housekeeping rooms shall be directly accessible from the unit or floor they serve and may serve more than one nursing unit on a floor (Sect 2.1–2.3.10.1)
AII room(s) and protective environment rooms	The ICRA shall address number, location, and type of AII and protective environment rooms (Sect. 1.5–2.2.1.1)
Protective environment rooms	Each protective environment room shall have an area for handwashing, gowning, and storage of clean and soiled materials located directly outside or immediately inside the entry door to the room (Sect. 2.1–3.2.3.50)

(continued)

Table 34-3. (continued)

Source and section; key points	Summary information and relevant subsection
Clean linen storage	Location of the designated area within the clean workroom, a separate closet, or an approved distribution system on each floor shall be permitted; if a closed cart system is used, storage of clean linen carts in an alcove shall be permitted; this cart storage must be out of the path of normal traffic and under staff control (Sect. 2.1–2.3.9.1)
CDC/HICPAC Guideline[41]	
Recommendations—Air: Section II, Construction, Renovation, Remediation, Repair, and Demolition	
ICRA	Convene a multidisciplinary team including infection prevention to coordinate the project
Education	Educate the construction team about dispersal of fungal spores
Mandatory adherence agreements	Written into the contract or contractor safety policy
Infection prevention surveillance	Review microbiologic data and other means of surveillance for fungal infections
Control measures	Define scope of activity; determine barrier and infection prevention requirements; relocate patients and/or staff; conduct measures to prevent contamination through HVAC systems; create negative air-pressure in work zones; and monitor barriers
Monitor the construction environment	Infection prevention professional should make rounds
Conduct epidemiologic investigations in cases of healthcare acquired *Aspergillus* infection or other fungal disease	Use airborne-particle sampling to evaluate barrier integrity; conduct an environmental assessment as indicated; perform conductive measures to eliminate fungal contamination
Recommendations—Water: Section VII, Cooling Towers and Evaporative Condensers	
Planning construction of new healthcare facilities	Locate cooling towers so that drift is directed away from air-intake system; design to minimize the volume of aerosol drift
Recommendations—Environmental Services	
Construction activities	Develop strategies for pest control
The Joint Commission Standard[42]	
EC.7.10: The hospital plans for managing utilities	
Promote a safe, controlled environment of care	Develop a process for designing, installing and maintaining appropriate utility systems: for example, domestic water, cooling towers and ventilation systems, including pressure relationships, air exchanges, and air filtration efficiencies
Reduce the potential for hospital-acquired illness	Control of elements used in health care: biological agents, gases, fumes, and dust
EC.8.30: The organization manages the design and building of the environment when it is renovated, altered, or newly created	
This Standard refers to the AIA Guidelines,[3] state and county regulations, and codes or standards that provide equivalent design standards	Follow AIA Guidelines[3] and local rules and regulations
Identify hazards that could compromise patient care	Development of an ICRA to address the effect of construction activities on air quality, infection prevention and control, utility requirements, noise, vibration and emergency procedures

NOTE: AIA, American Institute of Architects; AII, airborne infection isolation; CDC, Centers for Disease Control and Prevention; HVAC, heating, ventilating, and air-conditioning; ICRA, infection control risk assessment.

Services, has developed guidelines on the design and construction of health care facilities.[3] Some states have adopted these guidelines as their codes, while other states have specific codes and use the AIA guidelines in conjunction with them. Infection prevention teams should obtain a copy of the AIA guidelines and their state's regulations and make sure that their facility complies with both.

The AIA guidelines[3] indicate that healthcare facilities must incorporate infection prevention considerations into every project. For example, they recommend appropriate locations and requirements for clean and soiled workrooms, for storage facilities for clean linen, for handwashing stations, for housekeeping rooms, for patient toilet rooms, for airborne infection isolation (AII) rooms, and for protective environment rooms (see Table 34-3). Chapter 5 of the AIA guidelines[3] mandates that a multidisciplinary team complete an ICRA for each project.

CDC/HICPAC Guideline

The Centers for Disease Control and Prevention (CDC) and the Healthcare Infection Control Practices Advisory Committee (HICPAC) published a guideline on environmental infection control in healthcare facilities in 2003, an extensive document that describes environmental infection prevention strategies and engineering controls to help prevent transmission of infectious agents.[41] Discussions of construction issues are interspersed throughout the document, but most of the recommendations regarding construction are in the section "Recommendations—Air: Section II, Construction, Renovation, Remediation, Repair, and Demolition" (see Table 34-3).

The Joint Commission Standard

The Joint Commission (formerly the Joint Commission on Accreditation for Healthcare Organizations [JCAHO]) evaluates and accredits nearly 17,000 healthcare organizations and programs in the United States. The Joint Commission cites the AIA guidelines in their standard on management of the environment of care[42] and recommends that healthcare organizations follow the guidelines when planning to renovate existing space or construct new facilities. The Joint Commission's surveyors assess whether healthcare facilities comply with the Environment of Care Standard,[42] the AIA guidelines,[3] and the CDC's recommendations[41] for protecting patients, visitors, and healthcare workers during maintenance, renovation, and construction projects (see Table 34-3).

Infection Control Risk Assessment

Team Development

Whether the project involves remodeling an existing area or building a new one, the staff members must complete an ICRA before the project begins. A multidisciplinary team that includes persons with expertise in infection prevention, risk management, facility design, construction, heating, ventilating, and air-conditioning (HVAC), and safety (Figure 34-1) should complete the ICRA to ensure that the project meets all the standards and codes.[3] Infection prevention personnel should help complete the ICRA and should participate in projects from their inception so that they can identify infection prevention issues early and they can make suggestions prospectively, rather than after the design is complete.

Notification of Team Members

Infection prevention personnel should be notified of all major and/or high-risk projects so that they can determine which precautions are needed. However, infection prevention personnel must be available to consult on any project.[2] Maintenance personnel can do minor maintenance and renovation projects that have low risk for patients without direct input from the infection prevention team, if infection prevention personnel previously developed clear policies and procedures describing how to manage the

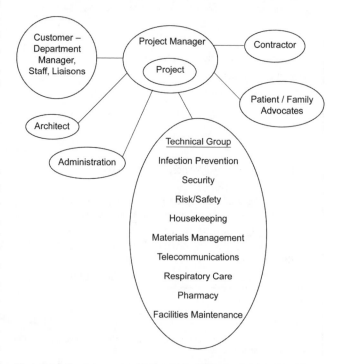

Figure 34-1. Schematic of the multidisciplinary team that should complete the infection control risk assessment to ensure that the project meets all the standards and codes.

infection prevention risks created by these projects and if they educated the maintenance personnel who will do these minor projects. The facilities we practice (Barnes-Jewish Hospital, St. Louis, Missouri, and the University of Iowa Hospitals and Clinics, Iowa City, Iowa) require that an ICRA be completed for all projects, regardless of size. Additionally, the facilities' maintenance departments have been educated about ICRAs and the infection prevention requirements needed for each project.

Each facility needs a mechanism whereby infection prevention personnel are notified that the planning phase of a project will begin soon. In small hospitals, the person responsible for renovation and construction could simply call the infection preventionist and invite him or her to the first meeting or could notify the infection preventionist and other persons who should be on the team at a meeting of another committee such as the Environment of Care, Risk or Safety. Large hospitals that have many projects under way at the same time should develop a more formal process for notifying infection prevention personnel and other ICRA team members to ensure that they can participate. To facilitate planning, some healthcare facilities hold regularly scheduled meetings at which new projects are discussed and updates are provided on current projects. Project managers should be educated as to the importance of having infection prevention input early in the process. After all, design changes are much cheaper before construction begins.

The multidisciplinary team must address the following key points when completing the ICRA:[2,9]

1. The team must assess the type of patients treated in the affected area (ie, their risk factors for infection) and the services that are provided there. In particular, the team should ask whether the area or facility cares for patients who are highly susceptible to fungal or *Legionella* infections.
2. The team should assess whether essential services, such as power, medical gas, water, sewer, and fire protection, might be disrupted, and it should develop a contingency plan to provide these services if one or more of them are affected by the project.
3. The team should evaluate areas that are above, below, and adjacent to the affected area and determine whether any phase of the project will affect these areas adversely. The team must develop a plan to minimize problems and infection risks on the other units.
4. The team must determine whether the patients on the affected or nearby units need to be relocated during the project to protect them from possible infectious risks or to provide an environment conducive to healing (eg, to limit noise).
5. The team must decide what types of barriers are necessary to decrease the risk of infection and should assign responsibility for inspecting the barriers and the cleanliness of both the work area and the area immediately adjacent. The barriers and detailed descriptions of other infection control measures should be included in the project specifications so that the costs of these measures can be included in the cost of the project.
6. The team must discuss how the ventilation system will be affected by the project and must determine what measures should be taken to protect the ventilation system and to maintain good quality air in the surrounding areas. For example, the team should determine whether the supply and return ducts need to be sealed. They should also ensure that the ventilation system around the affected area is balanced and that appropriate pressure relationships are maintained. On the basis of the types of patients in adjacent areas, the team should decide whether to obtain particle counts before and during the project.

These questions are formalized in the ICRA matrix. This is a tool that can help infection prevention personnel and other members of the multidisciplinary team systematically identify the infection precautions needed for the project. In one style of ICRA matrix (Figure 34-2), the first task is to identify the type of construction project activity. There are 4 types (A–D) that range from noninvasive procedures or simple inspection to major demolition and construction projects. Next (step 2) the team must identify the areas affected by the project and thereby determine how susceptible the patients in these areas are to construction-related infections (ie, determine the patients' level of risk). Subsequently, (step 3), the team uses the information about the project type and the patients' risk group to determine the class of precautions necessary for the project. The matrix specifies the precautions necessary for each classification. The matrix then requires the team to answer 11 questions about the project that will help them prospectively identify most important infection prevention issues.

Another type of ICRA matrix (Figure 34-3) is tailored to the institution. Because smaller projects do not have a team to investigate the potential risks, tailoring the ICRA to the institution allows nonclinical personnel to assess the potential risks for the specified task.

Infection prevention personnel need to understand that the ICRA matrix is only a guide. The team must assess each project individually and must apply the principles in the matrix flexibly. Not all projects fit exactly into the parameters listed in the matrix. Thus, infection prevention personnel and the team must use their best judgment when their project falls between classifications.

Infection prevention personnel should educate unit staff, architects, engineers, maintenance personnel, and construction workers about infection risks associated with construction and about appropriate methods for minimizing these risks. Education is a continual process, because different hospital staff members will be involved in each project and because many people involved with maintenance, renovation, and construction are contract workers. Infection prevention personnel could develop a brochure that discusses basic infection prevention issues in construction projects also and develop a checklist that itemizes infection prevention essentials, to answer particular questions and prevent problems that occur frequently (eg, the number and location of sinks, the type of ceiling tiles to use, and the types of floor and wall coverings to use).

Healthcare facilities should include the infection prevention requirements in the written contract for the project, so the contractors know what they are expected to do. If the construction team consistently ignores infection prevention policies, the hospital should levy a fine or the contractors should not be allowed to do other projects in the hospital.

Step One: Using the following table, *identify* the <u>Type</u> of construction project activity (Type A-D)

Type A	**Inspection and noninvasive activities** Includes, but is not limited to: ▪ Removing ceiling tiles for visual inspection limited to 1 tile per 50 square feet ▪ Painting (but not sanding) ▪ Working on wallcoverings, electrical trim work, minor plumbing, and activities that do not generate dust or require cutting into walls or access to ceilings other than for visual inspection
Type B	**Small-scale, short duration activities that create minimal dust** Includes, but is not limited to: ▪ Installing telephone and computer cabling ▪ Accessing chase spaces ▪ Cutting into walls or ceilings where dust migration can be controlled
Type C	**Work that generates a moderate to high level of dust or requires demolition or removal of any fixed building components or assemblies** Includes, but is not limited to: ▪ Sanding walls for painting or wall covering ▪ Removing floorcoverings, ceiling tiles, and casework ▪ Constructing new walls ▪ Working on ducts or electrical wiring above ceilings (minor) ▪ Moving or placing cables (major) ▪ Any activity that cannot be completed within a single workshift
Type D	**Major demolition and construction projects** Includes, but is not limited to: ▪ Activities that require consecutive work shifts ▪ Activities that require heavy demolition or removal of a complete cabling system ▪ New construction

Step 1 _____

Figure 34-2. Infection control risk assessment (ICRA) matrix that can be used for systematic identification of the infection precautions needed for a construction and/or renovation project. Steps 1–3 are adapted with permission from V. Kennedy and B. Barnard, St. Luke Episcopal Hospital, Houston Texas; and C. Fine, California. Steps 4–14 are adapted with permission from Fairview University Medical Center, Minneapolis, Minnesota. Forms modified and provided courtesy of Judene Bartley (ECSI; Beverly Hills, Michigan), 2002. (*Continued on the next 3 pages.*)

Step Two: Using the following table, *identify* the <u>Patient Risk</u> Groups that will be affected. If more than one risk group will be affected, select the higher risk group.

Low Risk	Medium Risk	High Risk	Highest Risk
▪ Office areas	▪ Cardiology ▪ Echocardiography ▪ Endoscopy ▪ Nuclear Medicine ▪ Physical Therapy ▪ Radiology/MRI ▪ Respiratory Therapy ▪ Other patient care areas not identified in **High Risk** or **Highest Risk** categories	▪ Cardiac Care Unit ▪ Emergency Room ▪ Labor and Delivery ▪ Laboratories (specimen) ▪ Newborn Nursery ▪ Outpatient Surgery ▪ Pediatrics ▪ Pharmacy ▪ Post Anesthesia Care Unit ▪ Surgical Units	▪ Any area caring for immunocompromised patients ▪ Burn Unit ▪ Cardiac Cath Lab ▪ Central Sterile Supply ▪ Intensive Care Units ▪ Medical Unit ▪ Negative air-pressure isolation rooms ▪ Oncology ▪ Operating rooms, including C-section rooms

Step 2 _____

Step 3: <u>Match</u> the

   Patient risk group (low, medium, high, highest) with the planned …
   Construction project type (A, B, C, D) on the following matrix, to find the …
   Class of precautions (I, II, III, IV) or level of infection control activities required.

Class I-IV precautions are delineated on the following page.

IC Matrix – Class of Precautions: Construction Project by Patient Risk

Patient risk group	Construction project type			
	Type A	Type B	Type C	Type D
LOW risk group	II	II	II	III/IV
MEDIUM risk group	I	II	III	IV
HIGH risk group	I	II	III/IV	IV
HIGHEST risk group	II	III/IV	III/IV	IV

Note. Infection Control approval will be required when the construction activity and risk level indicate that Class III or Class IV control procedures are necessary.

Step 3 _____

Figure 34-2. *(Continued)*

Description of required infection control precautions by <u>Class</u>

	During construction project	Upon completion of project
Class I	1. Use methods that minimize dust. 2. Immediately replace a ceiling tile displaced for visual inspection	1. Clean work area when task is completed.
Class II	1. Prevent dust from dispersing into air. 2. Use mist (water) on work surfaces to control dust while cutting. 3. Seal unused doors with duct tape. 4. Block off and seal air vents. 5. Place dust mat at entrance and exit of work area 6. Remove or isolate HVAC system in work areas. 7. Contain construction waste in tightly covered containers before transport.	1. Wipe work surfaces with disinfectant. 2. Wet mop and/or vacuum area with HEPA filtered vacuum before leaving. 3. Reintegrate HVAC system.
Class III	1. Isolate HVAC system in area where work is being done to prevent contamination. 2. Complete all critical barriers (ie, sheetrock, plywood, plastic) to seal work area from non-work area or use control cube method before construction begins. 3. Maintain negative air pressure within work site; use HEPA equipped air filtration units. 4. Contain construction waste in tightly covered containers before transport. Cover transport receptacles or carts.	1. Do not remove barriers from work area until completed project is inspected by the owner's Safety Department and Infection Control Department and thoroughly cleaned by the owner's Environmental Services Department. 2. Remove barrier materials carefully to minimize spreading dirt and debris created by construction. 3. Vacuum work area with HEPA filtered vacuums. 4. Wet mop area with disinfectant. Do not sweep. 5. Reintegrate HVAC system.
Class IV	1. Isolate HVAC system in area where work is being done to prevent contamination of duct system. 2. Complete all critical barriers ie, sheetrock, plywood, plastic, to seal area from non-work area or implement control cube method before construction begins. 3. Maintain negative air pressure within work site; use HEPA equipped air filtration units. 4. Seal holes, pipes, conduits, and punctures. 5. Construct anteroom. All personnel must use anteroom so they can be vacuumed using a HEPA vacuum cleaner before leaving work site or they can wear cloth or paper coveralls that are removed each time they leave the work site. 6. All personnel entering work site are required to wear shoe covers and change them each time they exit the work area. 7. Contain construction waste before transport in tightly covered containers. Cover transport receptacles or cart.	1. Do not remove barriers from work area until completed project is inspected by the owner's Safety Department and Infection Control Department and thoroughly cleaned by the owner's Environmental Services Department. 2. Remove barrier material carefully to minimize spreading dirt and debris created by construction. 3. Vacuum work area with HEPA filtered vacuums. 4. Wet mop area with disinfectant. Do not sweep. 5. Reintegrate HVAC system.

Figure 34-2. *(Continued)*

Step 4. Identify the areas surrounding the project area, assessing potential impact

Unit below	Unit above	Lateral	Lateral	Behind	Front
Risk group	Risk group	Risk group	Risk group	Risk group	Risk group

Step 5. Identify specific site of activity: eg, patient rooms, medication room, etc.

_____

Step 6. Identify issues related to: ventilation, plumbing, electrical systems (ie, are outages likely to occur during the project?).

_____

Step 7. Identify necessary containment measures based on the classification of the project. What types of barriers (eg, solid wall barriers) are needed? Is HEPA filtration required?

_____

(Note. Renovation/construction area shall be isolated from the occupied areas during construction and shall be at negative pressure with respect to surrounding areas)

Step 8. Consider potential risk of water damage. Is there a risk due to compromising structural integrity (eg, wall, ceiling, roof)?

Step 9. Work hours: can or will the work be done during non-patient care hours?

Step 10. Do plans allow for adequate number of isolation and/or negative-airflow rooms?

Step 11. Do the plans allow for the required number and type of handwashing sinks?

Step 12. Does the infection control staff agree with the minimum number of sinks for this project? (Verify against AIA Guidelines for types and area)

Step 13. Does the infection control staff agree with the plans relative to clean and soiled utility rooms?

Step 14. Plan to discuss the following containment issues with the project team: traffic flow, housekeeping, debris removal (how and when), etc.

_____

_____

_____

Appendix. Identify and communicate the responsibility for project monitoring that includes infection control concerns and risks. The ICRA may be modified throughout the project. Revisions must be communicated to the Project Manager.

Figure 34-2. _(Continued)_

449

STEP 1 - Choose work area

1 - Work Area					
1	2	3	4	5	6
Office Areas	Kitchens, cafeterias 2300	Wound Center, CDL	Endo / DDCC	Outside near air intakes	Outside away from air intakes
		Admitting, ED	CSPD		
	6300, 6400	Pulmonary Rehab	Pharmacy Main AND all satellites		
	7300, 7400, 7500	Resp Care, Lab	Dialysis		
	10100, 10200	PACU, Shukar, SDSA	ALL ORs		
	11100, 11200	All Radiology and CT	Perfusion, Card Cath		
	11400, 11500	Physical Therapy	All ICUs and CCUs		
	12100, 12200	5200, 6200, 6500	Chemo Center		
	All nurseries	7100, 71OU			
	13100, 17400	Nuclear Medicine			
	14400, 14500	9100, 9200			
	15300, 15400, 15500	10100, 10200, 10500			
	17400	4900			
	Radiation Oncology	Ultrasound			
	Gamma Knife				
	3200, 5900				
	6900, 7900				
	8100, 9400				
	89ICU & PICRU				
	Bronch, MPC				

NOTE: DAMAGED ASBESTOS FOUND WILL BE ABATED OR ENCAPSULATED PER ENVIRONMENTAL HEALTH AND SAFETY POLICY

STEP 2 - Choose type of work

2 - Type of Work	
A	Non-invasive, inspection, wallcovering, electrical trim, minor plumbing. NO sanding, wall cutting or dust generation at all
B	Small scale, short duration, minimal dust generation. Including but not limited to: cables or wires above ceilings, accessing chase spaces, wall cutting, ceiling penetrations where dust can be controlled
C	Work that generates moderate to high levels of dust, removal or demolition of fixed building components or assemblies including but not limited to; sanding walls, removing floorcovering, ceiling tiles and casework, new wall construction, minor duct or electrical work above ceilings, major cabling or wiring activities, or anything that cannot be completed within a single work shift.
D	Major demolition or construction. Including but not limited to; activities which require consecutive work shifts, heavy demolition or removal of complete cabling or wiring systems, new construction.

STEP 3 - Find Precaution Level:

STEP 4 - Color level of project:

3 - Infection Control Precautions

GREEN

1A, 2A, 3A, 5A, 6A, 6B, 6C, 6D

* Ceiling tile sprayed with bleach prior to displacing, replaced immediately after after visual inspection.
* Visitor and patient traffic routes should avoid work area.
* Clean supplies transported and stored away from contaminated materials.
* Water interruption scheduled during times of low activity.

YELLOW

1B, 1C, 2B, 5B

In addition to GREEN precautions;
* Mist work surfaces when cutting
* Air vents or returns blocked and sealed – If air handler supplies construction area only, it should be shut down (monitor filters during construction and change if necessary)
* 1 room areas to have walls from floor to ceiling, door closed, frame and door duct taped
* Debris removed in covered, sealed or taped containers, use service elevators or non-patient elevator.
* Walk-off mats at entrance
* Penetrations in walls not open for >4 hours. Cover with plastic and tape if more.
* Ceiling tiles replaced ASAP, if open >4 hours, must be covered.

STEP 5

3 - Infection Control Precautions

ORANGE

4A, 3B, 2C, 5C

In addition to GREEN and YELLOW precautions;
*Infection Control consulted
* Dust minimized - partitions erected to deck above.
* Optimal: Chute for debris removal, with HEPA filter or see YELLOW elevator guidelines.
* Barriers - dampers closed, assure adjacent air filters functioning, airtight plastic or drywall barriers from floor to ceiling, seams sealed with duct tape or ECUs – NOTE barriers must be removed carefully to minimize dust generation. Barriers to be disposed of with other debris.
*Negative pressure in const. area with HEPA filters
*Increase air filter changing.
* Vents cleaned prior to occupancy
* Water lines at site and adjacent areas flushed prior to occupancy
*Outside demolition >75 feet from air intakes.

RED

4B, 3C, 4C, 1D, 2D, 3D, 4D, 5D

In addition to GREEN, YELLOW, and ORANGE precautions;
*Infection Control consulted
*Patients relocated to remote area away from construction.

STEP 6

Please send to Infection Control

STEP 5 - Sign and authorize risk assessment below

STEP 6 - Return to Infection Control - addresses below

Date:

Project Name:
Project Location:

Date
Contractor
I am aware of and will enforce Infection Control precautions required for this project.

Figure 34-3. Infection control risk assessment (ICRA) matrix tailored to the specific institution that can be used to identify the infection precautions needed for a construction and renovation project.

Additionally, contractors need to be aware that if proper infection prevention requirements are not followed, the infection preventionist can halt work on the project.

Infection Prevention Construction Permit

The infection preventionist or another ICRA team member should fill out the Infection Prevention Construction Permit (Figure 34-4) during the design phase of the project. The permit identifies the project's location, start and end date, project manager, the type of construction activity, the patient risk group, the class selected for the particular project, and the necessary precautions. Either the infection prevention team or the project manager should keep this permit on file, and the contractor should post a copy at the job site.

Construction Site Monitoring Tool

Someone (often an infection preventionist) should inspect the work site to make sure that the construction workers are following the guidelines. A Construction Site Monitoring Tool (Figure 34-5) is useful for documenting the construction workers' compliance with the infection prevention policies and for identifying the contractors whose workers are most compliant with infection prevention practice. This information can be used when choosing contractors to do new projects.

Each facility should determine how often the site should be inspected and who will do the inspections. The hospital's size, the type of project, and the nature of surrounding patient-care areas will affect this decision. In a small hospital, daily monitoring may be feasible. In a larger hospital, weekly monitoring may be more feasible. However, if the project is done in a highly sensitive area, such as a bone marrow transplant unit or an operating suite, daily inspections may be necessary. If infection prevention staff members teach the staff on the surrounding units what to look for, they can monitor the site almost continuously and save the infection prevention team considerable effort. If the unit staff members note breaches in infection prevention practice, they can contact the infection prevention personnel, who can address the issue.

Unit/Area Opening Worksheet

A checklist such as the Unit/Area Opening Worksheet (Figure 34-6) can help infection prevention personnel determine whether the unit or area is ready to open after the project is complete. For example, the tool directs the user to determine whether the unit has been cleaned thoroughly, the water system has been flushed, the ventilation system has been cleaned and balanced, and whether any deficits need to be corrected before the unit or area can open.

Major Infection Control Issues to Consider During Planning of Projects

Air Handling Systems

Generally, air handling systems do not transmit nosocomial pathogens. However, at times these systems can transmit pathogens such as *Mycobacterium tuberculosis,* *Aspergillus* species, *Legionella pneumophila*, and varicella zoster virus. Air handling systems can increase the risk of infection in other ways. For example, if the humidity level is high and the number of air exchanges is inadequate, walls, ceilings, and vents may drip water onto sterile supplies or clean surfaces. Thus, infection prevention personnel should make sure that the air handling systems planned for new or renovated buildings will meet basic infection prevention requirements.

During the planning phase, infection prevention and engineering personnel should ensure that the air handling systems will be adequate to provide the ventilation required for that area.[3] Patient rooms should have 6 air changes per hour (ACH), of which 2 ACH must be outside air. Operating rooms and cesarean section delivery rooms require 15 ACH with 3 ACH of outside air. The relative humidity in operating rooms should be 30%–60%, and the airflow should move from the operating room to adjacent areas. AII rooms require 12 ACH with 2 ACH of outside air and negative airflow with respect to adjacent areas (ie, air flows to the adjacent areas). Air from these rooms should be exhausted to the outside away from air intakes or be recirculated after passing through HEPA filters. In addition, rooms in which high-risk procedures (such as bronchoscopy and aerosolized pentamidine treatments) are performed should have negative air-pressure with respect to adjacent areas or should have a flexible ventilation system that allows the pressure to be changed from neutral to negative. Bronchoscopy suites are required to have 12 ACH with outside air comprising 2 ACH. If a room that has special ventilation is renovated, the engineers should measure the number of air changes per hour and the pressure relationships before and after the renovation to ensure that the renovation did not disrupt the air handling system within the room.[43]

When reviewing designs for a project, infection prevention personnel should ensure that exterior air intakes are placed at least 8 m upwind of the exhaust outlets (according to the prevailing wind). The bottom of an intake should be at least 2 m above the ground or 1 m above the roof level. Intakes should be located away from cooling towers, trash compactors, loading docks, heliports, exhaust from biological safety hoods,[44] ethylene oxide sterilizers, aerators, and incinerators. Personnel from infection prevention and engineering should evaluate the

INFECTION CONTROL CONSTRUCTION PERMIT

				Permit No.:	
Location of Construction:				Project Start Date:	
Project Coordinator:				Estimated Duration:	
Contractor Performing Work:				Permit Expiration Date:	
Supervisor:				Telephone:	

YES	NO	CONSTRUCTION ACTIVITY	YES	NO	INFECTION CONTROL RISK GROUP
		TYPE A: Inspection, non-invasive activity			GROUP 1: Low Risk
		TYPE B: Small scale, short duration, moderate to high levels			GROUP 2: Medium Risk
		TYPE C: Activity generates moderate to high levels of dust, requires greater 1 work shift for completion			GROUP 3: Medium/High Risk
		TYPE D: Extended duration and major construction activities requiring consecutive work shifts			GROUP 4: Highest Risk

	During Construction	Upon Completion of Construction
CLASS I	1. Use methods that minimize dust. 2. Immediately replace any ceiling tile displaced for visual inspection.	1. Clean work area when task is completed.
CLASS II	1. Prevent dust from dispersing into air. 2. Use mist (water) on work surfaces to control dust while cutting. 3. Seal unused doors with duct tape. 4. Block off and seal air vents. 5. Place dust mat at entrance and exit of work area. 6. Remove or isolate HVAC system in work areas. 7. Contain construction waste in tightly covered containers before transport.	1. Wipe work surfaces with disinfectant. 2. Wet mop and/or vacuum with HEPA filtered vacuum before leaving work area. 3. Re-integrate HVAC system.
CLASS III **Date** **Initial**	1. Isolate HVAC system in area where work is being done to prevent contamination of the duct system. 2. Complete all critical barriers or implement control cube method before construction begins. 3. Maintain negative air pressure within work site; use HEPA equipped air filtration units. 4. Contain construction waste in tightly covered containers before transport. Cover transport receptacles or carts.	1. Don't remove barriers until project is inspected and thoroughly cleaned. 2. Remove barrier materials carefully to minimize spreading dirt and debris created by construction. 3. Vacuum work area with HEPA filtered vacuums. 4. Wet mop with disinfectant. Do not sweep. 5. Re-integrate HVAC system.
CLASS IV **Date** **Initial**	1. Isolate HVAC system in area where work is being done to prevent contamination of duct system. 2. Complete all critical barriers or implement control cube method before construction begins. 3. Maintain negative air pressure within work site; use HEPA equipped air filtration units. 4. Seal holes, pipes, conduits, and punctures appropriately. 5. Construct anteroom. All personnel must go through anteroom and be vacuumed with a HEPA vacuum cleaner before leaving work site or they can wear cloth or paper coveralls that are removed each time they leave the work site. 6. All personnel entering work site are required to wear shoe covers and change them each time they exit the work site. 7. Contain construction waste before transport in tightly covered containers. Cover transport receptacles or carts.	1. Do not remove barriers from work area until completed project is inspected and thoroughly cleaned. 2. Remove barrier materials carefully to minimize spreading dirt and debris created by construction. 3. Vacuum work area with HEPA filtered vacuums. 4. Wet mop with disinfectant. 5. Re-integrate HVAC system.

Additional Requirements:

Date Initials	_____ Exceptions/Additions to this permit Date Initials are noted by attached memoranda
Permit Request By:	Permit Authorized By:
Date:	Date:

Figure 34-4. Example of an infection prevention construction permit to document details of the construction project and the required infection prevention precautions.

Date:_____   **Time:**_____   **Time:**_____

Barriers

Construction signs posted for the area	☐ Yes ☐ No	☐ Yes ☐ No
Doors properly closed and sealed	☐ Yes ☐ No	☐ Yes ☐ No
Floor area clean, no dust tracked	☐ Yes ☐ No	☐ Yes ☐ No

Air handling

All windows closed behind barrier	☐ Yes ☐ No	☐ Yes ☐ No
Negative air-pressure at barrier entrance	☐ Yes ☐ No	☐ Yes ☐ No
Negative-airflow machine running	☐ Yes ☐ No	☐ Yes ☐ No

Project area

Debris removed in covered container daily	☐ Yes ☐ No	☐ Yes ☐ No
Designated route used for debris removal	☐ Yes ☐ No	☐ Yes ☐ No
Trash in appropriate container	☐ Yes ☐ No	☐ Yes ☐ No
Routine cleaning done on job site	☐ Yes ☐ No	☐ Yes ☐ No

Traffic control

Restricted to construction workers and necessary staff only	☐ Yes ☐ No	☐ Yes ☐ No
All doors and exits free of debris	☐ Yes ☐ No	☐ Yes ☐ No

Dress code

Appropriate for the area (OR, CSS, OB, BMTU)	☐ Yes ☐ No	☐ Yes ☐ No
Required to enter	☐ Yes ☐ No	☐ Yes ☐ No
Required to leave	☐ Yes ☐ No	☐ Yes ☐ No

Note. OR, operating room; CSS, central sterile supply; OB, obstetrics; BMTU, bone marrow transplant unit.

Comments:

Surveyor:_____

Figure 34-5. Example of a construction site survey tool to document construction workers' compliance with the infection prevention policies and to identify the contractors whose workers are most compliant with those policies.

INFECTION CONTROL UNIT / AREA OPENING WORKSHEET

Area Surveyed_____   Date_____

Surveyors are to check Yes, No, or NA for each criterion. A satisfactory review is required prior to reopening any unit or department.

Criteria	Yes	No	NA	Comments
I. Contractor final cleanup				
a. Horizontal surfaces free of residual construction dust				
b. Installed equipment and cabinets properly cleaned				
c. Barriers cleaned and removed				
II. HVAC system				
a. HVAC system cleaned if not isolated				
b. New filters in place and operational				
c. HVAC system balanced as specified				
III. Plumbing system				
a. No visible leaks				
b. Plumbing system flushed within 24 hours prior to occupancy				
c. Sinks functional				
IV. Equipment				
a. Soap / towel dispensers / hand sanitizers installed and filled				
b. Refrigerators – checklist for temperature control				
c. Ice machine cleaned and flushed				
V. Final cleaning				
a. Housekeeping final cleaning completed				
VI. Environmental Rounds				
a. Completion of Environmental Rounds				

Surveyors – additional comments:

Date_____
□   Satisfactory review
□   Unsatisfactory review

Unit Administrator_____
Infection Control_____
Facilities Management_____
Housekeeping_____

Figure 34-6.   Example of an infection control unit/area opening worksheet. This can help infection prevention personnel determine whether the unit or area of the construction project is cleaned and ready to open and/or identify any deficits that need to be corrected first. Used with permission from J. Pottinger and L. Goergen.

design and operation of ventilation systems carefully to ensure that potentially contaminated air is discharged safely, to prevent airborne disease transmission.[45] Infection prevention personnel may need to tour the air intake and exhaust sites to ensure that they are placed properly.

Isolation Rooms

Given the resurgence of tuberculosis and the emergence of multidrug-resistant bacterial pathogens and new viral pathogens, infection prevention personnel must ensure that the number, type, and placement of isolation rooms is adequate. The advent of new syndromes such as severe acute respiratory syndrome suggests that healthcare facilities may need more AII rooms than they did previously. In the past, some facilities have transferred patients with tuberculosis rather than create AII rooms. However, it may not be possible (or safe) to transport patients with severe acute respiratory syndrome or smallpox from one facility to another, and each facility may need AII rooms to accommodate such patients.

Infection prevention personnel should assess the patient population served by the facility to determine how many single rooms and how many AII rooms are necessary. Typically, a hospital should have 1 isolation bed for every 30 acute care beds.[3] Pediatric areas require more single rooms (which can be used for isolation) relative to the total number of beds than do other areas of the hospital,[44] because respiratory or enteric infections that require isolation are more frequent among children than among adults.[46] The number of patients needing isolation on pediatric units will vary with patient age and the season.[46] In general, isolation rooms for patients with nonrespiratory diseases do not require special features. If the isolation room has an anteroom, both the room and the anteroom should have handwashing sinks. Built-in nurse-server cabinets or isolation carts can be used to store necessary supplies, such as gowns, masks, and gloves.

Patients who have infectious diseases that are spread by respiratory droplets or droplet nuclei often are seen first in the emergency department or an outpatient clinic. Thus, appropriate isolation rooms will be beneficial in these areas. In fact, the AIA Guideline[3] states that emergency departments need at least 1 AII isolation room.

The requirements for air exchanges in AII rooms are discussed in the subsection on air handling systems (above). Notably, some hospitals use flexible ventilation systems that allow some patient-care rooms to have either neutral or negative air-pressure. For such rooms, care must be taken to ensure the airflow is appropriate for the patient in the room. Each AII room must have a handwashing station and storage area for clean and soiled material located directly outside or immediately inside the entry door. Each room must have a separate toilet room with a tub or shower and a handwashing station. The door must have a self-closing device, and penetrations in walls, ceilings, and floors must be tightly sealed so that room is maintained at negative air-pressure with respect to the surrounding environment.[45,47] A permanent monitor must assess the air pressure status of the room continuously, and a staff member must check the monitor daily. Anterooms are not required for AII rooms, except for those intended for immunosuppressed patients who need AII. Infection prevention personnel must evaluate proposed AII isolation rooms during the design phase, because retrofitting regular patient rooms to meet the requirements of AII isolation rooms can be very costly.

Protective Environments

Protective environments include operating rooms and bone marrow transplant rooms. These areas must remain as free of airborne infectious agents as possible. The airflow within these areas must flow from clean to less-clean.[43]

Handwashing Facilities

Each patient-care room, examination room, procedure room, and toilet room needs at least 1 handwashing sink, which should be located as close to the room's exit as possible. Sinks should be large enough to prevent splashing. All sinks must have an associated soap dispenser (built-in stainless steel soap dispensers should not be used) and a paper-towel holder that are located at a level that is comfortable for the user. A trash receptacle should be placed near the sink, so paper towels can be discarded properly. Alcohol-based hand rub dispensers should be located away from electrical outlets and switches[48] in places where healthcare workers are likely to use them.

A variety of mechanisms exist to control water flow. Conventional hand controls are the least expensive but may not be appropriate for all areas. Foot, knee, or electric-eye controls allow staff members to wash their hands or scrub without touching the sink ("no-touch" methods). Such sinks would be appropriate in operating suites, isolation rooms, and critical-care units. The electric-eye devices break frequently, and they are more expensive than the foot or knee controls. In addition, unless these devices can be bypassed, these faucets are essentially impossible to flush. Infection prevention personnel should help unit staff and architects select the best equipment for the location and purpose.

Water Supply and Plumbing

Occasionally, the hospital's water supply will be disrupted intentionally or accidentally during construction projects. Hospitals should have emergency plans that are activated if the water supply to the hospital is disrupted or contaminated. Infection prevention personnel should help develop this plan, because water is crucial to many infection prevention practices and because contaminated water can spread pathogenic organisms. During the summer of 1993, the University of Iowa Hospitals and Clinics, in Iowa City, was faced with the possibility of losing its water supply because the Iowa River was flooding. The hospital's contingency plan included water conservation measures, such as shutting down drinking fountains and ice machines, replacing showers and full-tub or bed baths with partial baths, using alcohol-based hand cleaners rather than soap and water, and serving meals on disposable dishes. The alternative water system was a well, which was tested for coliform organisms, nitrates, and iron. The hospital's plant operations department was prepared to adjust the plumbing system in order to use the well water. Hospitals that do not have wells must design alternative plans. If the water supply will be disrupted only for a short time, staff can fill large plastic containers with water to be used while the water system is turned off. If the water supply will be off for a longer period of time, the hospital may need to have a company deliver bottled water. During emergencies, agencies such as the US National Guard may be able to provide water. If, during a construction project, the water supply will be turned off for more than 4 hours, the contractor should do this work during times of non-peak water use, such as evenings, nights, or weekends.

Space for Personal Protective Equipment

All patient-care areas should store personal protective equipment, such as gloves, in areas where they are readily accessible. A container for disposal of sharp devices (sharps) must be accessible to workers who use, maintain, or dispose of sharps. The number of containers must be adequate, the size must be large enough for the sharps that will be disposed in that area, and healthcare workers must be able to safely access the opening when they need to discard a sharp. Sharps containers should be 132–142 cm (52-56 inches) above the floor for use by healthcare workers who are standing, or 96–107 cm (38-42 inches) above the floor for use by healthcare workers who are sitting.[49]

Waste

Infection prevention personnel should help clinical staff plan how urine and feces will be discarded in patient-care areas, clinics, and laboratories. A variety of bedpan flushing devices are available, such as spray hoses and spray arms. Some of these options create splash hazards, clean poorly, or allow water to pool in hoses or nozzles. Alternatives such as disposable pulp bedpan liners can be quite expensive. Soiled utility rooms should contain a clinical sink or a flushing-rim fixture and also a separate handwashing sink.[3] Additionally, containers must be available for biohazardous and nonbiohazardous waste. There may be local regulatory agency requirements relating to biohazardous waste. A nonbiohazardous waste container must be large enough to prevent overfilling and must be located so it is easily accessible to both staff who generate the waste and staff who remove the waste.

Finishes

General Considerations

During the design and development phase, infection prevention personnel should help the clinicians and architects choose the finishes: for example, flooring, wall coverings, and ceiling tiles. Ideal finishes are those that are washable and easy to clean.[3] Porous or textured materials can be difficult to clean and thus may allow bacteria and fungi to grow. The finishes should be durable and able to withstand repeated cleaning. In addition, counter tops, backsplashes, and floors should have as few joints as possible, so they are easy to clean.

Ceilings

Ceiling tiles should be appropriate for the areas in which they are being placed. Acoustical tiles may be used in hallways, waiting rooms, and standard patient rooms. Ceilings in semirestricted areas, such as central sterile supply, radiology procedure rooms, minor surgical procedure rooms, and clean corridors in operating suites, should not be perforated or have crevices where mold and bacteria could grow. The ceiling must be made of smooth, nonabsorptive material that can be washed and is capable of withstanding cleaning with chemicals. Perforated, serrated, cut, or highly textured ceilings are not permitted. In restricted areas, such as operating rooms, all ceilings should be monolithic, washable, and capable of withstanding cleaning chemicals.[3] Cracks and perforations are not allowed. Ceilings in protective isolation rooms should be monolithic.

Floors

Floors should be easy to clean and should resist wear. Floors where food is prepared should be water-resistant, and floors that are walked on while wet should have a nonslip surface. The housekeeping department should use their cleaning procedures on samples of flooring to determine whether the materials can withstand cleaning and disinfection with germicidal cleaning solutions. Floors in operating rooms and delivery rooms used for cesarean section delivery should be monolithic and should not have joints. Floors in kitchens, soiled workrooms and other areas, which are frequently washed with water, should have tightly sealed joints.

Carpets decrease noise and have become popular in healthcare facilities. There is no conclusive evidence that links carpet to illness, but carpets can harbor microorganisms. Additionally, the vacuuming of the carpet can create microbursts of dust and potentially put patients at risk. Carpets should not be used in isolation rooms, protective environments, operating rooms, critical care units, kitchens, laboratories, autopsy rooms, or dialysis units. The CDC recommends that carpet should not be used in any high traffic areas or in areas where people might spill liquids.[41,56]

Walls

Walls should be washable and the finish should be smooth. Wall finishes in areas where blood or body fluids could splatter (eg, operating rooms and cardiac catheterization laboratories) should be fluid resistant (ie, vinyl) and easy to clean. Wall finishes around plumbing fixtures should be smooth and water resistant.[3] Wall bases and floors, especially around small pipes, should not have joints or should have joints that are sealed tightly.[3] In food preparation areas, walls should be free of spaces that harbor insects and rodents. In operating rooms, cesarean section delivery rooms, and sterile processing rooms, walls should not have fissures, open joints, or crevices that permit dirt particles to enter the room.

Counter Tops

Counter tops should typically be composed of a nonporous solid material, such as thermoset polymer (Corian; DuPont), stainless steel, or laminate, with a protective sealant.

Minimizing the Risk of Infection During Projects

Air Handling Systems

Infection prevention personnel should collaborate with the facility's HVAC specialist to decide whether the HVAC system needs to be isolated during construction. During renovation or construction projects, selected air intakes (particularly those near excavation sites) and air ducts in the construction area need to be protected from dust by 1 of 3 methods: shutting them down, equipping them with additional filters, or covering them with plastic. Engineering or maintenance personnel also should check air filters frequently and change them when necessary. Air handling units in areas that care for immunocompromised patients should contain HEPA filters, to decrease the amount of particulate matter and the number of microbes in the air.

For projects in Class III or Class IV, according to the ICRA matrix (Figure 34-2), the airflow should move from outside the construction site into the site, because air should flow from a clean area into a dirty area. Negative air-pressure, or airflow into the construction site, can be achieved by placing a HEPA-filtered negative-airflow machine within the work zone. Ideally, the air from this machine would be exhausted directly to the outside. If this is not possible, the air can be exhausted into the air ducts and recirculated. A pressure monitor should be placed within the work zone to ensure that this area is maintained at negative air-pressure with respect to the adjacent areas. The contractor is responsible to monitor the airflow and to make sure the HEPA filters are clean and working properly.

A HEPA-filtered negative-airflow machine may be required for other projects. For example, such machines would be necessary during work on the ceilings within patient-care areas in Group 3 (ie, medium high risk) or Group 4 (highest risk) of the ICRA matrix for Class I or Class II construction activities (see Figure 34-2. Each contractor should own one or more of these machines, and healthcare maintenance departments should have at least one machine that they can use during maintenance activities that generate dust.

The air quality must be maintained and monitored carefully in areas where immunocompromised patients are cared for, including patients receiving treatment for malignancies, patients with bone marrow or solid organ transplants, and premature neonates. We recommend that infection prevention personnel work with staff in the appropriate departments to develop policies that describe in detail what must be done when any modifications, renovations, demolition, or construction are done in their areas of the facility. Activities as seemingly minor as installing computer cables or conduits in the ceiling space could stir up *Aspergillus*-laden dust that would be hazardous for immunocompromised patients. When work is being done in areas that house immunocompromised patients, some precautions are needed in addition to those used in all patient care areas (see the subsection on barriers, below). For example, existing air ducts and the space above the ceiling tiles must be cleaned with a HEPA-filtered vacuum cleaner before undertaking any project that involves opening these areas. The area inside the barrier must be cleaned and vacuumed (with a HEPA-filtered vacuum cleaner) before the barrier is removed and again after the barrier is removed. In addition to these precautions, portable HEPA filters could be placed in patients' rooms to ensure that the air is as clean as possible. Facilities that cannot implement appropriate precautions must move units to other areas of the hospital for the duration of the project.

Controlling Dust and Dirt

Construction and renovation projects create tremendous amounts of dust or debris that may carry microorganisms, such as *Aspergillus* spores.[55] Infection prevention personnel must collaborate with other staff to devise ways to prevent the dust and dirt from contaminating clean or sterile patient-care surfaces, supplies, and equipment. Some general measures to limit dust and dirt and to minimize the risk of fungal infections in healthcare facilities during maintenance, renovation, or construction include the following:

- Wet mop the area just outside the door to the construction site daily, or more often if necessary.
- Use a HEPA-filtered vacuum to clean adjacent carpeted areas daily, or more often if necessary.
- Shampoo carpets when the construction project is completed.
- Transport debris in containers with tight-fitting lids, or cover debris with a wet sheet.
- Remove debris as it is created; do not let it accumulate.
- Do not haul debris through patient-care areas, if possible.
- Remove debris through a window when construction occurs above the first (ground) floor, if possible.
- Remove debris after normal work hours, if possible, through an exit restricted to the construction crew.
- Designate an entrance, an elevator, and a hallway for use only by construction workers.
- If workers must traverse patient-care areas, they must remove dust from their bodies and clothes and then put on gowns, shoe covers, and head covers before walking through the unit. In particular areas of the hospital (eg, the operating suite), workers may need to wear protective clothing while working in the construction site.
- For small projects, the construction-tool carts should be cleaned before entering the unit and left at the exit through the barrier (see the subsection on barriers, below). For larger projects, the carts and equipment should go into the area and stay behind the barrier until the project is done. Before removing carts and equipment from inside the barrier, the construction crew should clean the items and cover them with

moist sheets. They should be moved off the unit by the designated route.

Barriers

Barriers are needed during maintenance, renovation, and construction projects to minimize the dispersion of dust. Commercially available, portable drop-down cubicle barriers can be used for small, quick jobs, such as removing ceiling tiles to install computer cables. These units are equipped with either small HEPA-filtered negative-air-flow machines or connections for HEPA vacuums so that the space inside each cubicle is at negative air-pressure with respect to the surrounding area. A closed door that is sealed with tape is an adequate barrier for enclosed short-term projects that generate minimal dust. Plastic sheeting that is 3-8 mil (120 μm) thick or canvas barriers made specifically for dust control can be used for short-term projects that have minimal traffic and do not require fire rating. If plastic is used, contractors must inspect the integrity of the barrier several times during a work shift and they must repair holes, tears, or any defects in the plastic or canvass barrier immediately. A door can be created in the plastic barrier with a zipper or with an overlap of the plastic of 61 cm (2 feet). Anterooms, made of plastic, allow workers to don or remove protective attire or clean dust and debris off of their clothes and their carts before they leave the construction zone and enter a patient-care area. Solid drywall barriers that are taped and finished are required for longer, more extensive projects. If the area is a high-risk area (eg, a bone marrow transplant unit or an operating suite), plastic barriers should be erected and the drywall barriers should be built behind the plastic barriers. Plastic barriers should be wiped down with a moist cloth before they are removed.

Facilities that are undergoing construction or renovation in particularly sensitive areas (eg, bone marrow transplant units if patients are on the unit) may want to document that the barriers are adequate. Particle counters that determine the number of particles suspended in the air can be used for this purpose. The infection prevention personnel on the multidisciplinary team should determine during the ICRA whether it is necessary to obtain particle counts. Particle counts should be obtained outdoors and compared with the indoor counts to ensure that the 90% filters are functioning properly. Thereafter, particle counts should be obtained in the areas adjacent to the work site before construction begins, during several days at the start of the project, and weekly until completion. Cultures of air samples are not as useful for this task and should be reserved for special circumstances (eg, before opening a bone marrow transplant unit after construction) to document that the environment is not contaminated by fungi.

The work site should be kept as clean as possible to enhance the effectiveness of the barriers. HEPA-filtered vacuums should be used for cleaning the construction site and the workers' clothing before they leave the work site. If HEPA-filtered vacuums are not available, the site can be wet mopped, but it should never be swept, because sweeping disperses dust into the air. In addition, sticky "walk-off" mats should be placed just inside the entrance to the work site to clean shoes and the wheels of equipment. A new mat should be used every day. Moist "walk-off" mats do not adequately remove dust from the wheels of carts as they exit a work site.

Traffic Patterns

To reduce the amount of dust and dirt in the hospital and the risk of exposure to infectious agents, patients, visitors, and staff may need to traverse the hospital by alternate routes. Infection prevention personnel should help identify the appropriate detours before construction begins. Staff should design these routes in a logical manner, so that they do not inadvertently increase the risk of nosocomial infection or of noninfectious hazards, such as falls. They also should consider whether housekeeping personnel can maintain the new route, whether the new route interferes with the work done in the area, and whether the route meets minimum aesthetic requirements. If construction is necessary in or near operating suites, surgical personnel must be able to move from place to place without contaminating their surgical attire (scrubs).

The routes by which inanimate items are transported throughout the hospital may need to be altered during construction. In general, all materials, including food, linens, medical supplies and equipment, and janitorial supplies and equipment, must be handled in a manner that minimizes the risk of contamination.[52] Before the construction project begins, infection prevention personnel should help the staff from the affected units plan the routes by which various supplies and equipment will be transported. Clean or sterile supplies and equipment must be transported to storage areas by a route that minimizes contamination from the construction site and prevents contact with soiled or contaminated trash and linens. To prevent unnecessary contamination with dirt and dust, used supplies and equipment should be moved in enclosed containers from the point of use to the point at which they will be processed.

Traffic patterns in critical areas, such as the operating suite, labor and delivery rooms, nurseries, laboratories, and pharmacies, may not be easy to alter to meet these infection prevention requirements. In such circumstances, the construction crew may need to work during off hours and on weekends. If infection prevention requirements still cannot be met, some areas may need to relocated or closed temporarily.

Storage Areas

During construction, basic principles of infection prevention still apply. Thus, clinical areas must maintain appropriate storage areas, which may be difficult, because the allotted space may be small or may lack essential features. Before construction begins, infection prevention personnel should help the staff identify the locations in which they will store equipment and supplies. Temporary storage areas should allow staff to do the following:

- Easily monitor the supplies (look at expiration dates)
- Store sterile supplies and equipment away from soiled items (separate clean and dirty areas must be maintained)
- Store clean or sterile supplies at an appropriate distance from sinks to prevent the supplies from becoming wet
- Store contaminated wastes in a designated dirty area outside of direct patient-care areas
- Move items without placing them on the floor (have adequate work space)

In addition, the temporary storage space should be clean, have adequate temperature and humidity control, and should be free of insects and rodents.

An outbreak of 4 surgical and burn wound infections that occurred when a large tertiary-care hospital renovated its central inventory control area illustrates the importance of storing supplies properly during construction.[22] The investigators identified several *Aspergillus* species on the outside of packages of materials from the main floor of Inventory Control: on bags of intravenous preparations, the outsides of sterile paper wrappers, and storage bins in the pharmacy, which was adjacent to the area under construction; and on the outsides of packages containing burn dressings, elastic adhesive, Elastoplast (Beiersdorf AG), gloves, and disposable scissors that were stored on the burn unit and in the intensive care unit. The investigators postulated that the supply boxes were contaminated during construction. The outside of the packages became contaminated when the boxes were opened, and the fungus was inoculated directly into the patients' wounds when the packages were torn open during dressing changes.[53]

One of the authors of this chapter recently consulted with a hospital that was renovating several nursing stations during a time when the units were empty. The barriers were thin plastic that was not sealed and the areas being renovated were at positive air-pressure with respect to the surrounding areas. When the units closed, the staff left the supply carts in place and did not cover them. Consequently, the packages of clean and sterile supplies were covered with dust and grime. Supplies that could not be wiped with a disinfectant (ie, those that had paper wrappers) had to be discarded, costing the hospital thousands of dollars.

Final Check

After the project is completed, infection prevention personnel should inspect the area to ensure that all requirements have been met. Infection prevention personnel should verify the following steps have been taken:

1. Check the location of soap, alcohol-based hand hygiene products, towel dispensers, the sharps disposal container, and the wastebasket.
2. Check all areas to ensure that the appropriate flooring, ceiling tiles, and wall finishes have been installed.
3. Check all procedure rooms, kitchens, and utility rooms to ensure that they have the appropriate washable flooring and splash guards on sinks.
4. Inspect water faucets to ensure that they do not have aerators.
5. Check pressure and drainage in the water system.
6. Have personnel from maintenance or housekeeping run all faucets the day before patients occupy the unit to decrease the risk of infection from *Legionella* species.
7. Evaluate the direction of airflow in negative air-pressure rooms and ensure that the air pressure monitors are placed and functioning properly.
8. Review the HVAC balance reports to ensure that system meets the specification.

In sensitive areas, such as a bone marrow transplant unit or operating rooms, air sampling can be performed with an air sampling device, such as the SAS compact air sampler (PBI International), to check for contaminated air. Alternatively, sampling can be done by placing settle plates in various areas throughout the room for 30 minutes to 1 hour while the ventilation system is running and the room is vacant. The door should be closed and taped shut, so that persons do not enter the room while the settle plates are in place. If the ventilation system is running properly, the settle plates should be negative for pathogenic microbial growth.[44] A certified engineer can evaluate the effectiveness of a laminar airflow system.

Cost of Construction-Related Infection Control Measures

There is no rule of thumb to determine the cost of the ICRA and the infection prevention measures. Douglas Erickson, a fellow of the American Society for Health-care Engineering, estimated that when the ICRA was first introduced, costs per contract increased by 25%, but by last year this figure was down to 5%. He noted recently that some contractors were reporting that these measures actual reduced costs because they increased the pace of the project and prevented delays[54] (D. Erickson, personal communication).

The most accurate way to determine the cost of infection prevention measures is to add the cost of all the components (ie, the barriers, the HEPA-filtered negative-airflow machines, vacuums, "walk-off" mats, modifications to the HVAC system, cleaning of the work zone and surrounding areas, monitoring of air pressure, and protective attire). Other costs to consider are special methods needed to minimize noise, vibration, and dust. For example, chipping masonry by hand rather than with a rotary hammer will take more time but will protect nearby patients from excessive noise and vibration. Generally, infection prevention precautions for large projects are best priced on the basis of a fixed set-up cost (ie, the cost of barriers and duct work) plus the cost per day (ie, the daily cost of renting negative-airflow machines, maintaining barriers, cleaning work zones, and replacing "walk-off" mats). The longer the project takes, the greater the cost.

Infection prevention precautions can make otherwise simple projects, such as carpet replacement and installing electrical wiring, more complicated and costly. For example, a project to replace 35,000 square feet (3,252 m²) of flooring at the authors' hospital was bid at $147,000 without infection prevention measures and $180,000 once infection prevention precautions were included. The infection prevention measures added $33,000 to the cost of the project, which was about $1 extra per square foot, or an increase of 22% over the cost of conventional carpet replacement. At Barnes-Jewish Hospital (St. Louis, Missouri), a carpet replacement project on the bone marrow transplant unit was bid without taking infection prevention precautions into account. Fortunately the project was discussed with the infection preventionist, and, after looking at all of the costs and options, staff decided to install new carpet over the old carpet. When the unit is renovated in the future, the patients will be moved and both layers of carpet can be removed safely.

Conclusion

Construction and renovation projects pose special challenges for infection prevention personnel. In many hospitals, they are the only clinical staff members who assist in all construction and renovation projects. Therefore, they may find themselves having to ensure that both infection prevention guidelines and general building codes are met. We would encourage infection prevention personnel to be involved in all phases of these projects to avert outbreaks of infection and to ensure that newly constructed or renovated areas allow staff to follow good infection prevention practices. We would also encourage infection prevention personnel to maintain good relationships with the architects, contractors, facility maintenance personnel, and others involved in construction and renovation of healthcare facilities so that together they can ensure that the area is safe and well designed.

We think the role of infection prevention personnel in these projects will increase as the clinical complexity of hospitalized patients and the proportion with immunosuppression increase at the same time that hospitals are required to decrease their budgets drastically and regulatory and accrediting agencies are increasing the number of infection prevention guidelines. Infection prevention aspects of construction and renovation projects require large amounts of time and hard work. We would argue that the time and energy invested before and during the project will save hours of time, huge sums of money, and the lives of patients and healthcare workers after the project is finished.

Acknowledgments

We recognize Sherry A. David, RN, BS, CIC, and Jose A. Fernandez, RA, for their contributions to this chapter in earlier editions of this textbook.

References

1. Streifel AJ, Lauer JL, Vesley B, Juni B, Rhame FS. *Aspergillus fumigatus* and other thermotolerant fungi generated by hospital building demolition. *Appl Environ Microbiol* 1983;46: 375–378.

2. Bartley JM. APIC state-of-the-art report: the role of infection control during construction in health care facilities. *Am J Infect Control* 2000;28(2):156–169.

3. American Institute of Architects (AIA). Guidelines for design and construction of hospital and health care facilities, 2006 edition. Washington, DC: AIA; 2006. http://www.fgiguidelines.org/guidelines.html. Accessed September 16, 2009.

4. Mermel LA, Josephson S, Giorgio C, et al. Association of legionnaires' disease with construction: contamination of potable water? *Infect Control Hosp Epidemiol* 1995;16:76–91.

5. Parry MF, Stampleman L, Hutchinson JH, et al. Waterborne *Legionella bozemanii* and nosocomial pneumonia in immunosuppressed patients. *Ann Intern Med* 1985;103:205–210.

6. Thacker SB, Bennett JV, Tsai TF, et al. An outbreak in 1965 of severe respiratory illness caused by legionnaires' disease bacterium. *J Infect Dis* 1978;138:512–519.

7. Haley CE, Cohen ML, Halter J, et al. Nosocomial legionnaire's disease: a continuing common-source epidemic at Wadsworth Medical Center. *Ann Intern Med* 1979;90:583–586.

8. Kistemann T, Huneburg H, Exner M, Vacata V, Engelhart S. Role of increased environmental aspergillus exposure for patients with chronic obstructive pulmonary disease (COPD) treated with corticosteroids in an intensive care unit. *Int J Hyg Environ Health* 2002;204:347–351.

9. Oren I, Haddad N, Finkelstein R, Rowe JM. Invasive pulmonary aspergillosis in neutropenic patients during hospital construction: before and after chemoprophylaxis and institutions of HEPA filters. *Am J Hematol* 2001;66(4):257–262.

10. Bryce EA, Walker M, Scharf S, et al. An outbreak of cutaneous aspergillosis in a tertiary-care hospital. *Infect Control Hosp Epidemiol* 1996;17:170–172.

11. Lueg EA, Ballagh RH, Forte V. Analysis of the recent cluster of invasive fungal sinusitis at the Toronto Hospital for Sick Children. *J Otolaryngol* 1996;25:366–370.

12. Loo VG, Bertrand C, Dixon C, et al. Control of construction-associated nosocomial aspergillosis in an antiquated hematology unit. *Infect Control Hosp Epidemiol* 1996;17:360–364.

13. Sessa A, Meroni M, Battini G, et al. Nosocomial outbreak of *Aspergillus fumigatus* infection among patients in a renal unit? *Nephrol Dial Transplant* 1996;11:1322–1324.

14. Alvarez M, Lopez Ponga B, Raon C, et al. Nosocomial outbreak caused by *Scedosporium prolificans (inflatum)*: four fatal cases of leukemic patients. *J Clin Microbiol* 1995;33: 3290–3295.

15. Berg R. Nosocomial aspergillosis during hospital remodel. In: Soule BM, Larson EL, Preston GA, eds. *Infections and Nursing Practice: Prevention and Control*. St. Louis: Mosby; 1995: 271–274.

16. American Health Consultants. Aspergillosis: a deadly dust may be in the wind during renovations. *Hosp Infect Control* 1995;22:125–126.

17. American Health Consultants. Construction breaches tied to bone marrow infections. *Hosp Infect Control* 1995;22:130–131.

18. Iwen PC, Davis JC, Reed EC, et al. Airborne fungal spore monitoring in a protective environment during construction and correlation with an outbreak of invasive aspergillosis. *Infect Control Hosp Epidemiol* 1994;15:303–306.

19. Gerson SL, Parker P, Jacobs MR et al. Aspergillosis due to carpet contamination. *Infect Control Hosp Epidemiol* 1994; 15:221–223.

20. Flynn PM, Williams BG, Hethrington SV, Williams BF, Giannini MA, Pearson TA. *Aspergillus terreus* during hospital renovation [letter]. *Infect Control Hosp Epidemiol* 1993;14: 363–365.

21. Dewhurst AG, Cooper MJ, Khan SM, et al. Invasive aspergillosis in immunosuppressed patients: potential hazard of hospital building work. *BMJ* 1990;301:802–804.

22. Jackson L, Klotz SA, Normand RE. A pseudoepidemic of *Sporothrix cyanescens* pneumonia occurring during renovation of a bronchoscopy suite. *J Med Vet Mycol* 1990;28: 455–459.

23. Hospital Infection Control. APIC coverage: dust from construction site carries pathogen into unit. *Hosp Infect Control* 1990;17:73.

24. Humphreys H, Johnson EM, Warnock DW, Willats SM, Winter RJ, Speller DC. An outbreak of aspergillosis in a general intensive therapy unit. *J Hosp Infect* 1991;18:167–168.

25. Barnes RA, Rogers TR. Control of an outbreak of nosocomial aspergillosis by laminar air-flow isolation. *J Hosp Infect* 1989; 14:89–94.

26. Weems JJ, David, BJ, Tablan OC, et al. Construction activity: an independent risk factor for invasive aspergillosis and zygomycosis in patients with hematologic malignancy. *Infection Control* 1987;8:71–75.

27. Perraud M, Piens MA, Nicoloyannis N, Girard P, Sepetjan M, Garin JP. Invasive nosocomial pulmonary aspergillosis: risk factors and hospital building works. *Epidemiol Infect* 1987;99:407–412.

28. Opal SM, Asp AA, Cannady PB Jr, Morse PL, Burton LJ, Hammer PG II. Efficacy of infection control measures during a nosocomial outbreak of disseminated aspergillosis associated with hospital construction. *J Infect Dis* 1986;153:634–637.

29. Krasinski K, Holzman RS, Hanna B, Greco MA, Graff M, Bhogal M. Nosocomial fungal infection during hospital renovation. *Infect Control* 1985;6(7):278-282.

30. Grossman ME, Fithian EC, Behrens C, et al. Primary cutaneous aspergillosis in six leukemic children. *J Am Acad Dermatol* 1985;12(2 part 1):313–318.

31. Sarubbi FA, Kopf HB, Wilson MB, et al. Increased recovery of *Aspergillus flavus* from respiratory specimens during hospital construction. *Am Rev Respir Dis* 1982;125:31–38.

32. Lentino JR, Rosenkranz MA, Michaels JA, Kurup VP, Rose HD, Rytel MW. Nosocomial aspergillosis: a retrospective review of air-borne disease secondary to road construction and contaminated air conditioners. *Am J Epidemiol* 1982; 116:430–437.

33. Arnow PM, Sadigh M, Costas C, et al. Endemic and epidemic aspergillosis associated with in-hospital replication of aspergillus organisms. *J Infect Dis* 1991;164:998–1002.

34. Aisner J, Schimpff S, Bennett J, et al. *Aspergillus* infections in cancer patients. *JAMA* 1976;235:411–412.

35. Centers for Disease Control and Prevention, Infectious Disease Society of America, American Society of Blood and Marrow Transplantation. Guidelines for preventing opportunistic infections among hematopoietic stem cell transplant recipients. *MMWR Recomm Rep* 2000;49(RR-10):1–125.

36. Miscellaneous gram-negative bacilli. In: Murray PR, Rosenthal KS, Kobayashi GS, Pfaller MA. *Medical Microbiology*. 4th ed. St. Louis: Mosby; 2002:325–333.

37. American Society of Heating, Refrigerating and Air-Conditioning Engineers (ASHRAE). *Minimizing the Risk of Legionellosis Associated with Building Water Systems*. ASHRAE Guideline 12–2000. Atlanta, GA: ASHRAE; 2000. http://www.ashrae.org/publications/page/1285. Accessed September 16, 2009.

38. Opportunistic mycoses. In: Murray PR, Rosenthal KS, Kobayashi GS, Pfaller MA. *Medical Microbiology*. 4th ed. St. Louis: Mosby; 2002:664–672.

39. Fox BC, Chamberlin L, Kulich P, et al. Heavy contamination of operating room air by *Penicillium* species: identification of the source and attempts at decontamination. *Am J Infect Control* 1990;18:300–306.

40. D'Andrea C, New York City Department of Health and Mental Hygiene Bureau of Environmental and Occupational Disease and Epidemiology. *Guidelines on Assessment and Remediation of Fungi in Indoor Environments*. New York: 2000:1–17.

41. Centers for Disease Control and Prevention (CDC), Healthcare Infection Control Practices Advisory Committee. Guidelines for environmental infection control in healthcare facilities: recommendations of CDC and the Healthcare Infection Control Practices Advisory Committee (HICPAC). *MMWR Recomm Rep* 2003;52(RR-10):1–44.

42. The Joint Commission. *Comprehensive Accreditation Manual for Hospitals: The Official Handbook*. Chicago: The Joint Commission; 2003.

43. Streifel A. Health-care IAQ guidance for infection control. *HPAC Engineering* 2000;72(Apr-May):28–36.

44. Soule BM, ed. *The APIC Curriculum for Infection Control Practice, Vol II*. Dubuque, IA: Kendall/Hunt Publishing; 1983.

45. Neill HM. Isolation-room ventilation critical to control disease. *Health Facil Manage* 1992;5:30–38.

46. Langley JM, Hanakowski M, Bortolussi R. Demand for isolation beds in a pediatric hospital. *Am J Infect Control* 1994;22: 207–211.

47. Centers for Disease Control and Prevention. Guidelines for preventing the transmission of *Mycobacterium tuberculosis* in health care facilities, 1994. *MMWR Recomm Rep* 1994; 43(RR-13):1–132.

48. Boyce JM, Pittet D, Healthcare Infection Control Practices Advisory Committee, HICPAC/SHEA/APIC/IDSA Hand Hygiene Task Force. Guideline for hand hygiene in health-care settings. *MMWR Recomm Rep* 2002;51(RR-16):1–45.

49. US Department of Health and Human Services (DHHS), National Institute for National Safety and Health (NIOSH). Selecting, evaluating and using sharps disposal containers. DHHS (NIOSH) publication 97–111. Cincinnati, OH: NIOSH; 1998.

50. Madden CS. Environmental considerations in critical care interiors. *Crit Care Nurs Q* 1991;14(1):43–49.

51. Opal SM, Asp AA, Cannady PB Jr, Morse PL, Burton LJ, Hammer PG II. Efficacy of infection control measures during a nosocomial outbreak of disseminated aspergillosis associated with hospital construction. *J Infect Dis* 1986; 153:634–637.

52. Fitch H. Hospital and industry can benefit by sharing contamination control knowledge. *Clean Rooms* 1993;7:8–9.

53. Bryce EA, Walker M, Scharf S, et al. An outbreak of cutaneous aspergillosis in a tertiary-care hospital. *Infect Control Hosp Epidemiol* 1996;17:170–172.

54. Downs P. Infection control expert hosted by local architecture, engineering firms. *CNR News* 2003. http://www.stlconstruction.com.

55. Haiduven D. Nosocomial aspergillosis and building construction. *Med Mycol* 2009;47(suppl 1):S1–S6.

56. Noskin GA, Peterson LR. Engineering infection control through facility design. *Emerg Infect Dis* 2001;7:354–357.

Index